HORTICULTURAL PLANT BREEDING

HORTICULTURAL PLANT BREEDING

THOMAS J. ORTON, PH.D.
Professor and Extension Specialist
Department of Plant Biology
School of Environmental and Biological Sciences
Rutgers, The State University of New Jersey
New Brunswick, NJ, United States

ELSEVIER

ACADEMIC PRESS
An imprint of Elsevier

Academic Press is an imprint of Elsevier
125 London Wall, London EC2Y 5AS, United Kingdom
525 B Street, Suite 1650, San Diego, CA 92101, United States
50 Hampshire Street, 5th Floor, Cambridge, MA 02139, United States
The Boulevard, Langford Lane, Kidlington, Oxford OX5 1GB, United Kingdom

Library of Congress Cataloging-in-Publication Data
A catalog record for this book is available from the Library of Congress

British Library Cataloguing-in-Publication Data
A catalogue record for this book is available from the British Library

ISBN 978-0-12-815396-3

For information on all Academic Press publications
visit our website at https://www.elsevier.com/books-and-journals

Publisher: Charlotte Cockle
Acquisition Editor: Nancy Maragioglio
Editorial Project Manager: Susan Ikeda
Production Project Manager: Divya KrishnaKumar
Cover Designer: Miles Hitchen

Typeset by SPi Global, India

Working together
to grow libraries in
developing countries

www.elsevier.com • www.bookaid.org

Contents

Preface and Acknowledgments

Everyone takes a unique pathway through life that is a consequence of circumstances, opportunities, conscious decisions, relationships, and plain luck. My decision to become a plant breeder was in response to a job opportunity at the University of California, Davis in 1978. I had not been formally trained as a plant breeder, nor was I attracted to the discipline as an undergraduate or graduate student. My academic training at Michigan State University was primarily in plant biology with an emphasis on genetics, although I had enrolled in a few plant breeding courses. After taking the position at UCD as an Assistant Professor/Geneticist/Plant Breeder, my transition from esoteric science into the world of plant breeding started with on-the-job learning experiences starting a program in cool-season vegetables and teaching the UCD plant breeding class.

My pathway diverted in 1982 when I made a conscious decision to leave UCD for a position in the rapidly expanding agricultural biotechnology industry. Ultimately, I remained in the agricultural biotech industry for 12 years, starting with forging applications of cellular and molecular biology in crop improvement. I gravitated progressively to the product development and commercialization facet of the business since there was no shortage of scientists, but there was a paucity of broadly trained professionals to bridge the gap from research to products and services. Along the way, I experienced the entire biotechnology spectrum from basic science to marketing and sales. I learned to respect the challenges and rigors of all links in the product development chain from basic sciences to finished products. None of them are trivial or easy.

My return to academia in 1995 came as an administrator at the School of Environmental and Biological Sciences (SEBS), Rutgers University, a position I held until 2002. Following two years as an acting county extension agent, I returned to the research/extension/teaching faculty at Rutgers after an absence of 22 years as a practicing academic. I assumed the responsibility to co-teach Plant Breeding, and embarked on applied plant breeding efforts in processing and fresh market tomatoes. Over time, I also initiated targeted breeding efforts at Rutgers in *Capsicum* sp. and seedless table grape.

My Plant Breeding co-instructor and I experienced difficulty in selecting a textbook that could serve as an effective resource to students; to reinforce the topics we covered in class and fill in the details we did not. The plant agricultural industry in New Jersey is driven heavily by horticultural species, but most plant breeding textbooks focused only on agronomic species. Further, available textbooks were woefully outdated with regard to the integration of cell and molecular biology applications.

During the transition year from extension agent to research/extension/teaching (2005), I decided to write this book. Time was available while the research program was planned and launched, but I knew it would soon disappear as the program expanded. I had a vision for how the book would be crafted and a lot of ideas spiraling in my head, so I simply drafted a table of contents and started writing. By late 2005, a very rough draft of the book had been completed. It consisted of a download of virtually everything that was in my memory banks, rife with personal experiences. The draft was not replete, however, with documentation or visual examples. It needed a lot of work to become a finished product.

I was correct that available discretionary time would quickly vanish. My attempts to keep the book project on the front, then the back burners, after 2005 were unsuccessful. Research, extension, and teaching obligations occupied progressively more of my calendars and the book project was correspondingly shelved. My department chair urged me to rekindle my interest in the project and, in 2017, I took a 6-month sabbatical leave to finish the book. I had already expended a lot of time and effort on the draft, and did not want the investment to be squandered.

After a couple of chapters were completed to my satisfaction, I sought a publishing partner. Academic Press/Elsevier responded quickly with interest. I'd published before with Elsevier, and the experience had been a good one. We came to an agreement by August, 2017.

The original project was more ambitious than what ended up between these covers. A third section with specific crop species examples was planned and partially written. I underestimated the time requirements particularly of documentation of facts presented in the book. Many chapters consumed over a month to re-write, document, and illustrate. The sabbatical ended in December 2017, and the book was far from completed. The third section was suspended in the interests of getting the book finished and published in a reasonable time frame. Plant breeding is accelerating, and new discoveries are changing the discipline almost daily. I couldn't afford to take months to add more content at the risk of allowing the rest of it to become obsolete. Eventually, I would like to complete and publish this compendium.

The Introduction to Section 1 explains the impetus for the book. There are many excellent plant breeding textbooks in the marketplace. The overwhelming majority of them are focused on agronomic crop species. Breeding of horticultural crop species is an adaptation of strategies developed in agronomic crops to account for high individual plant value, emphasis on quality over quantity, and broad range of domesticates. The book is comprehensive enough that it may be used, I believe, as a general plant breeding reference. Since the foundations of plant breeding were built on maize and small grains, the book includes many examples from breeding of agronomic crop species. One book reviewer requested that these examples be replaced with analogous examples in the world of horticulture. In many instances, such examples are either weak or nonexistent.

The book is almost entirely the original product of the author, including many of the graphic illustrations. I borrowed or adapted some material and examples, and have efforts to ensure that these sources are appropriately cited or that permissions were obtained to use copyrighted entities. The strength of a single-author book is that it is cohesive and consistent in style, unlike an edited multi-authored book that benefits from a multitude of experiences and specialties. This book is far from perfect. Each time I re-read passages, my mind conjures new and potentially better ways to state ideas and concepts. Perhaps there will be a future edition that will improve on this version, but I doubt it.

Every Preface I have read includes a section devoted to thanking all the people who contributed in a meaningful way to the project. I am likewise indebted to everyone that took the time to help me along the way, including family, friends, and colleagues. Many people supported me directly with this book project, and I will take the time and space to recognize their efforts. I have learned valuable lessons from virtually every other plant breeder and geneticist I have worked with.

Specific thanks are due to Dr. Mark Robson for encouraging me to revisit the book and to the SEBS leadership at Rutgers for allowing me to take the 6-month sabbatical leave to work on the project: Drs. Robert Goodman, Bradley Hillman, and Donald Kobayashi. My Plant Breeding co-instructors at Rutgers were particularly important since many of my ideas originated from the planning and presentation of the course over many years: Drs. Stacy Bonos and Thomas Molnar. They have been very supportive of the project, even during the years on hiatus and doubt. Much of the content of Chapter 12 originated from lecture material from Dr. William Meyer of Rutgers University. I appreciate the willingness of Drs. Molnar, Robert Pyne, James Simon, and C. Andrew Wyenandt for permission to use their research results as examples of disease resistance breeding in Chapter 19.

An advanced draft of the book was reviewed by Dr. Derek Barchenger (currently at the World Vegetable Center), Jennifer Paul (of Rutgers), and Eileen Boyle (of Mt. Cuba Center; my wife). Their remarks were absolutely invaluable for elevating the book to its present state. Finally, my editors at AP/Elsevier have been exceedingly enthusiastic, patient, and supportive: Nancy Maragiolio and Susan Ikeda. My experiences with AP/Elsevier with this project have been superb.

A web site will be established for the book that will include links to slide sets used in the Rutgers University Plant Breeding course. The web site will also serve as a forum for any comments on the book and posting of clarifications and errata. Readers are also encouraged to contact the author directly with any comments or suggestions; these will be posted on the web site.

For the students who read this book while considering a career in plant breeding, I urge you to give the discipline strong consideration. Few careers present the practitioner with the broad range of possibilities for basic and applied research, invention, and personal gratification accorded by plant breeding. The profession is exciting and dynamic, with new discoveries and applications for solving problems and building new genomes appearing constantly. Plant Breeders thrive when they are independent and are generally accorded abundant freedom to operate. Very skilled and lucky plant breeders may invent one or more cultivars that soar in the marketplace and provide personal and professional monetary rewards. Money is not the driving force, however, behind plant breeding. Rather, it is the primal urge to create and to take the process of genetic enhancement further to serve the needs of humans and domesticated animals for food, fiber, pharmaceuticals, and aesthetic pleasure. The conscious choice I made to become a plant breeder in 1978 led to a long and gratifying career that I have never regretted.

Thomas J. Orton

SECTION I

ELEMENTS AND UNDERPINNINGS OF PLANT BREEDING

Introduction to Section I

PLANTS AND PEOPLE

Plant breeding is defined as the heritable change of plant populations from the activities of humans. The history of plant breeding will be covered in a cursory manner in Chapter 1, but it began through the activities of farmers at the dawn of agriculture over 20,000 years ago (Gallant, 1990; Malthus, 1993). Early plant breeding was nothing more than keeping seeds of the most desirable plants and discarding seeds of everything else. Since the traits that appealed to farmers were, in part, controlled by genes, it is not surprising that cultivated populations of plants changed over time to something more desirable. Progress was slow, however, and often not apparent during the lifetime of the farmer.

Seeds and other plant propagules became the legacy of cultures as the populations of plants became progressively more differentiated from wild progenitors. Agricultural cultures were tied to the land and people collected seeds of the wild plant species that were endemic to the settlement region (Stearn, 1965). We know from the discoveries of Nikolai Vavilov, and other botanists studying plant ranges, that species are not distributed equitably on Earth (Harlan, 1976). Therefore, each culture developed its own mix of useful plants domesticated from the wild, some for food and others for fiber, structural products, and medicine.

As human cultures evolved and adopted technologies, long-distance travel began to link cultures in trade networks. This social change broadened awareness of different plant forms and introduced plant species to new cultures and non-native regions of Earth. The scientific enlightenment swept over Europe and Asia starting in the 15th century CE and continued to the present. By the early 19th century CE, science began to eclipse religion as the foundation of the human experience and quality of life (Jacob, 2000). The field of biology began to flourish, and important discoveries during the 19th century CE established most of the enduring underpinnings regarding the nature of life. The systematic classification of plants by Carolus Linnaeus (aka Carl von Linné) during the mid-18th century CE and the articulation of the Theory of Evolution by Charles Darwin in the mid-19th century CE stand out as major advances that portended the scientific discipline of plant breeding. Gregor Mendel also conducted his seminal experiments on the laws of inheritance during the 19th century CE, but the results were ignored until they were rediscovered at the dawn of the early 20th century CE (Serafini, 2001; Chapter 3).

BRIEF HISTORY OF PLANT BREEDING

Scientific discoveries ignited more inquiry, and advances during the 20th century CE accelerated. Plant breeding became an occupation, not just a life skill, as Mendel's Laws were applied vigorously during the early- and mid-20th century CE, resulting in most of the framework within which breeding is still practiced. The discovery of DNA as the heritable information storage macromolecule by Watson and Crick in 1949 opened a massive door to a new era of biological discoveries that have been effectively co-opted and applied by plant breeders. The rate of discovery is surging ever faster, rendering any static summary of the body of knowledge to be quickly obsolete.

As applications of biological technologies in plant breeding compounded, the rate of heritable population improvement increased. Using unselected wild populations as a baseline, mass selection practiced by early human agriculturalists produced slow and steady progress (Fig. I1.1). Following the early scientific advancements during the 18th and 19th centuries CE, new methods like pedigree and ear-to-row contributed to a faster rate of plant population improvement. During the 20th century CE, the discoveries of Mendelian inheritance followed by early phases of cellular and molecular biology further accelerated the rate of advancements in plant breeding. With the advent of genome sequencing and editing in the early 21st century CE the rate will increase again dramatically.

The Impact of Technologies on Domesticated Plant Performance

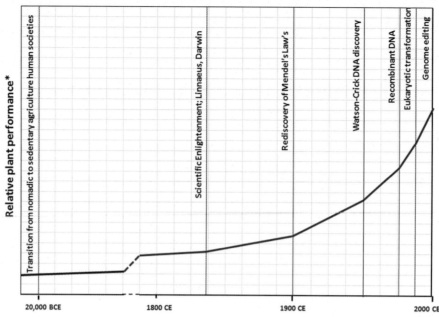

FIG. I1.1   The relative impacts of compounding biological discoveries and applied technologies on desirable heritable population improvements during the history of human agriculture on Earth.

As technology advances and is applied to plant breeding, the roles fulfilled by the plant breeder in the development of new cultivars are changing. Most high-powered plant breeding programs in the private sector are now multi-faceted and multi-tiered endeavors that include a team of diverse specialists including agronomists/horticulturists, plant pathologists, entomologists, plant physiologists, and molecular biologists. The "traditional" plant breeder who excels at interpreting genotype based on whole-plant phenotype is often a part of this team, but doesn't necessarily lead it. 21st century crop agriculture, however, continues to be based mostly on plants growing in variable natural soils and uncontrolled environments. The generalized plant breeder has grown to become a professional who appreciates and utilizes (through other specialists) the benefits of advanced technologies, but who also understands, appreciates, and appropriates the vagaries of agriculture.

FUTURE CHALLENGES TO PLANT BREEDERS

As biological technologies have advanced and helped humans to live longer, more healthful lives, many other technologies have been spawned that have changed the world for humans. The industrial revolution is largely credited with the most dramatic changes in human lifestyles. The internal combustion engine increased the area of land that humans could cultivate and bridged human cultures through amazing improvements in transportation. While food shortages persist in the present day, they are not nearly as prevalent or impactful as they were historically. Technology applications in medical sciences also prolonged life, culminating in markedly reduced infant mortality rates and longer life spans. The most striking manifestation of all of these trends is the growth of the human population on Earth (Fig. I1.2). By 1000 CE, Earth's human population was less than 500 million (Ehrlich and Ehrlich, 1990). By 2000 CE, the total had skyrocketed to nearly 7 billion (Gerland et al., 2014). Much of the impetus for plant breeding during the 20th century CE was propelled by questions of how to feed, clothe, and otherwise nurture this rapidly growing human population. The forces driving plant genetic improvement in the 21st century CE and beyond will be progressively much more diverse.

Human lifestyle and socio-economic trends that are technology-driven have been concurrent with this stunning increase in population. If agricultural output is a factor limiting the growth of human populations, why are there progressively fewer farmers? In 1900 CE, over 50% of all U.S. residents were employed directly in the agriculture industry. By 2000 CE, the proportion of U.S. residents directly employed in agriculture had dropped to less than 2% (Orton, 2017). During this same period in the U.S., the proportion of total disposable income spent on food dropped from over 24% to less than 10%, considering both market and food service sources (Orton, 2017).

Human Population on Earth 10,000 BCE to 2,000 CE

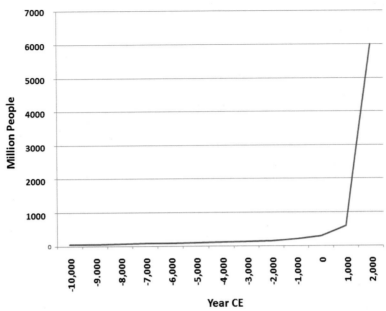

FIG. I1.2 Estimated total number of humans on planet Earth from 10,000 BCE to 2000+ CE.

Where has all this disposable income gone that is no longer being spent on food? The answer lies in lifestyle that has changed tremendously in a relatively short time period. People are no longer merely residents or citizens; they are economic units, consumers. One of the new classes of products that humans now voraciously consume that was virtually absent in 1900 is nursery and ornamental plants and landscaping services. The availability of leisure imparted by new machines and growing appreciation for aesthetic entities in the 21st century CE has spawned a huge industry that barely existed a century ago.

AGRONOMIC VS. HORTICULTURAL PLANT BREEDING

Agriculture changed very slowly from 20,000 BCE to approximately 1800 CE, but has evolved dramatically since then, including the rise of plant genetics and breeding in the early 20th century CE (Spiertz, 2014). Most of the theoretical framework for plant breeding was established in the early- to mid-20th century CE by scientists experimenting with agronomic crops such as maize and wheat. Most of the textbooks on the subject continue to adapt this framework to advancing technologies with a focus on applications in large-acreage agronomic crops. A research study has demonstrated that, among all technological advances, plant breeding has contributed nearly 90% of the added value of cereals and oilseeds over the past 50 years (Mackay et al., 2011). The same study concluded that plant breeding and agronomy contributed approximately equally to the expanded values and profitability of maize and sugar beet. Economic growth in agriculture, however, and especially in developed countries such as the U.S., has occurred primarily in horticultural and niche crops over the past 50 years (Dimitri et al., 2005). This is the reason "Horticultural Plant Breeding" was produced, to better serve the specific needs of this expanding industry sector.

Fortunately, much of the theoretical plant breeding framework established for agronomic crops also applies to horticultural crop species. What are the distinctions between the worlds of agronomy and horticulture? Horticulture is the science and art of growing plants (fruits, vegetables, flowers, and any other cultivar). Horticulture may be practiced on a relatively large (thousands of hectares) scale or by the square centimeter in the home garden. Horticultural crops are used to diversify human diets and to enhance our living environment. The field of horticulture also includes plant conservation, landscape restoration, soil management, landscape and garden design, construction, and maintenance, and arboriculture (Preece and Read, 2005; Shry and Reiley, 2016). To state the case more succinctly, horticulture is the art and science of growing plants that are intensively, as compared to extensively, grown (Janick, 2005). In contrast, agronomy is defined as the application of plant and soil science to crop production. Agronomy is generally applied on a larger scale and is focused on crops that generate commodity grains, plant-based oils, sugar, and fiber (Scheaffer and Moncada, 2012).

Jules Janick articulated further differences between breeding horticultural and agronomic crops (Janick, 2005): "Horticultural crops are those that serve to fit the special food and esthetic needs of humans. They are crops that not only make life possible but make life worth living. In horticultural crops, quality is supreme."

"Differences between horticultural and agronomic crops are reflected in genetic improvement objectives. Agronomic crops become commodities in which the product is interchangeable. Breeding objectives are based on increasing yield, often determined by resistance to biotic and non-biotic stress. For example improvement in hybrid maize yields are based on increasing yield stability under high populations. In horticultural crops, breeding objectives must be consumer directed because consumers make individual decisions about consumption and make choices between different cultivars and alternate crop species. There are many examples of large breeding efforts that have had little grower acceptance because they have not been able to compete in the marketplace based on quality. For example, consumers have no interest in disease resistance of food crops but buy on the basis of their eyes, and continued purchase based on their palate. Thus, unique quality rather than yield per se must be the overriding breeding objective. It also must be stressed that grower-directed traits can often be solved by non-genetic means, while consumer-directed traits, especially quality, are often not amenable to alternate solutions.

"A further profound influence on plant breeding is the effect of individual breeder's imagination in proposing startling new innovations…(R)emarkable achievements are based on the inspiration and research of individuals. These successes emphasize the impact of imagination and skill combined with the plant breeder's art and science on the future direction of horticulture."

Horticulturists apply their knowledge, skills, and technologies used to grow intensively produced plants for human food and non-food uses and personal or social needs. The horticulture profession involves plant propagation and cultivation with the aim of improving plant growth, yields, quality, nutritional value, and resistance to insects, diseases, and environmental stresses. Horticultural professionals work as gardeners, growers, therapists, designers, and technical advisors in the food and non-food sectors of horticulture. The term "horticulture" may be broadly inferred as the practice by home owners of growing of plants in landscapes or gardens.

The range of biodiversity represented by the plant taxa included within horticulture is much larger than that of the field of agronomy. The industries served by horticulture are inherently diverse, including perishable produce, fruit and vegetable juices and wines, dehydrated fruits and vegetables, nuts, spices and condiments, specialty oils and essences, phytopharmaceuticals, and nursery and ornamentals. The spectrum of taxonomic phyla includes mosses, ferns, gymnosperms, and angiosperms. The total number of plant species that have been domesticated has been estimated at 35,000 or about 10% of the estimated total of 353,000 angiosperm and gymnosperm species on Earth (Khoshbakh and Hammer, 2008; Christenhusz and Byng, 2016). Of all of the plant species domesticated, the overwhelming majority (about 96%) would be considered "horticultural" (Table I1.1). Agronomic crops supply the bulk of the food calories consumed by humans and domesticated animals and also most of the natural fibers used in industry (Prescott-Allen and Prescott-Allen, 1990), but the horticultural crops are largely responsible for what would be considered nutrition. Horticulture is also aimed at a large fraction of what we would define as "quality of life".

The chapters in Section 1 first cover the history, context, and biological underpinnings of plant breeding. The elements of the basic plant breeding algorithm are presented next and detailed accounts of the roles of germplasm,

TABLE I1.1 Breakdown of Domesticated Plant Species According to Human Uses and Use Classifications

Domesticated use	Number of domesticated forms
Grains, food (including cereals and pseudocereals)	21
Fruit; food	395
Nut; food	60
Vegetable; food	208
Spice; food	185
Medicinal	161
Ornamental (nursery, land-scape, turf, Indoor)	2324
Industrial; food (e.g. sweeteners, brewing, oils, fragrances, chemicals, rubber, fiber, structural, other ingredients or uses)	248
Agronomic	145
Horticultural	3457
Total	3602

selection, and mating in the algorithm. Section 1 concludes with a description of the endpoint of the algorithm, the testing and release of the new cultivar and the protection of intellectual property and new inventions. Examples are presented, in most cases, from the realm of horticulture, but since much of the underlying theories and principles of plant breeding were developed using agronomic crop species, there will necessarily be some use of these examples as well. The textbook's point of view will be skewed toward applications to breeding horticultural crops, but presents information of broader pertinence to all cultivated plant species.

References

Christenhusz, M.J.M., Byng, J.W., 2016. The number of known plants species in the world and its annual increase. Phytotaxa 261 (3), 201–217.

Dimitri, C., Effland, A.B.W., Conklin, N.C., 2005. The 20th century transformation of US agriculture and farm policy. In: Economic Information Bulletin No. 3. Economic Research Service, U.S. Department of Agriculture, Washington, DC. 14 pp.

Ehrlich, P., Ehrlich, A., 1990. The Population Explosion. Simon and Schuster, New York, NY. 320 pp.

Gallant, R., 1990. The Peopling of the Planet Earth. Macmillan Publishing Company, New York, NY. 163 pp.

Gerland, P., Raftery, A.E., Evikova, H., Li, N., Gu, D., Spoorenberg, T., Alkema, L., Fosdick, B.K., Chunn, J., Lalic, N., Bay, G., Buettner, T., Heilig, G.K., Wilmoth, J., 2014. World population stabilization unlikely this century. Science 346 (6206), 234–237.

Harlan, J.R., 1976. The plants and animals that nourish man. Sci. Am. 235, 88–97.

Jacob, M.C., 2000. The Enlightenment: A Brief History With Documents. MacMillan Learning, New York, NY. 253 pp.

Janick, J., 2005. Horticultural plant breeding: past accomplishments, future directions. Acta Horticulturae. 294, 61–65.

Khoshbakh, K., Hammer, K., 2008. How many plant species are cultivated? Genet. Res. Crop Evol. 55 (7), 925–928.

Mackay, I., Horwell, A., Garner, J., White, J., McKee, J., Philpott, H., 2011. Reanalyses of the historical series of UK variety trials to quantify the contributions of genetic and environmental factors to trends and variability in yield over time. Theor. Appl. Genet. 122 (1), 225–238.

Malthus, T.R., 1993. In: Gilbert, G. (Ed.), Essay on the Principle of Population. Oxford University Press, Oxford. 208 pp.

Orton, T.J., 2017. Pathways to collaboration in agricultural research and extension. In: Fowler, J., Holowinsky, R., Channell, A., Crocomo, O., Kreier, J., Sharp, W. (Eds.), Pathways to Collaboration. vol. 2. ScienTech Publishers, Columbia, SC, pp. 323–368.

Preece, J.E., Read, P.E., 2005. The Biology of Horticulture: An Introductory Textbook, second ed. John Wiley & Sons, New York, NY. 528 pp.

Prescott-Allen, R., Prescott-Allen, C., 1990. How many plants feed the world? Conserv. Biol. 4, 365–374.

Scheaffer, C.C., Moncada, K.M., 2012. Introduction to Agronomy: Food, Crops, and Environment, second ed. Delmar Cengage Learning, Clifton Park, NY. 720 pp.

Serafini, A., 2001. The Epic History of Biology. Basic Books, New York, NY. 408 pp.

Shry, C.L., Reiley, H.E., 2016. Introductory Horticulture, ninth ed. Delmar Cengage Learning, Clifton Park, NY. 780 pp.

Spiertz, H., 2014. Agricultural sciences in transition from 1800 to 2020: exploring knowledge and creating impact. Eur. J. Agron. 59, 96–106.

Stearn, W.T., 1965. The origin and later development of cultivated plants. J. R. Hortic. Soc. 90. 279–291, 322–341.

C H A P T E R

1

Introduction

THE BEGINNING

During the course of evolutionary history, organisms, including vascular plants, developed genetic systems that gained them a place in earth's dynamic natural habitats. Mechanisms emerged and were perpetuated that allowed terrestrial plants to survive, to thrive, and to reproduce in a myriad of environments. Light, dark, hot, cold, dry, wet, saline, solute-less, acidic, alkaline, rich soil, gravel, calm air, wind, etc. As populations flourished, challenges of competition for resources, parasitism, and herbivory ultimately arose. Organisms reproduced themselves with mechanisms that preserved and perpetuated "successful" genes or gene combinations and allowed for new genetic variability to constantly appear. Thus, only the fittest, most competitive individuals contributed offspring to the next generation, but populations also came to be adequately buffered for the inevitable and perpetual flux of both the abiotic and biotic factors with which they coexist. Our understanding of this process is the legacy of the groundbreaking theories advanced by Charles Darwin in the mid 1800s.

Terrestrial vascular plants had been undergoing this process for well over 100 million years before the recent ancestors of *Homo sapiens* gained the ability to reason. Primitive humans foraged among local offerings and developed preferences among plant and animal species for uses as food, health maintenance, fiber, and shelter. Eventually, nomadic social behavior gave way to sedentary cultures, and agriculture was born. Paleobiologists speculate that widespread primary human dependence on agriculture began about 6000–10,000 years ago (Vasey, 1992). The independent rise of all major human cultures has been inextricably linked to agriculture, the steady flow of food from which allowed societies to become segmented and occupations to appear and flourish (Heiser, 1990). Agriculture begot the process of domestication, and the practice of *plant breeding* as we know it began (Fussell, 1966). For the purposes of this textbook, plant breeding is defined as *"any method of plant population reproduction that results in a permanent, heritable change."*

Long before Linnaeus, Darwin, and Mendel, humans understood the basic principles of inheritance (Sauer, 1969). Offspring tend to resemble their parents. Hybrids tend to be intermediate between two parents. Certain notable traits tend to "run in families", such as stature, hair, eye, and skin coloration, and hemophilia. While the academics of the day argued the tenets that governed these phenomena, agricultural practitioners worldwide continued to select and mate individuals, affecting population genetic changes that ultimately gave us all of our major food and fiber crops. Ornamental crops are somewhat more recent, humans becoming adequately fed, clothed, and sheltered (in many parts of the world) to afford things more aesthetic.

While modern cotton, wheat, corn (synonymous with maize), soybeans, rice, millet, sorghum, potatoes, and tomatoes retain resemblances to plants found in the wild, the cultivated populations are mostly dependent on humans for their sustenance (Vasey, 1992). The needs of humans tend to run counter to characters that make plants competitive in nature. Compare the wild bison with the dairy cow, or the toy terrier with the timber wolf and the differences are illustrative. Our ability to feed people has become dramatically more efficient, requiring progressively less land and labor to attain equivalent calories. It is argued that much of this efficiency is owed to the rampant overuse of fossil energy reserves, but a substantial body of work demonstrates that improved varieties have also played a significant role (Fussell, 1966; Poincelot et al., 2001). The progressive reduction in the percentage of food producers and the proportion of disposable income going to raw food demonstrates how successful agriculture, including plant breeding, has become (Tables 1.1 and 1.2). Consequently, the reduction in the proportion of the population engaged in agriculture in developed countries since 1900 has been staggering.

With time, the connection between cultivated crop species and the wild progenitors from whence they came has become obscured or lost. While it is relatively simple to trace the ancestry of some crops, such as rice, tomatoes, and soybeans, the wild populations that were the sources of crops such as corn and wheat no longer exist. Progressively, the importance of maintaining a living historical crop archive, heirloom varieties, has come to light.

New tools and technologies have greatly impacted the rate of genetic changes in our domesticated plant populations. Machinery, transportation, communication have all accelerated the process (Fussell, 1966). Mendel's laws of inheritance have led to countless iterations in breeding methods that improved efficiencies dramatically, for example, hybrid varieties. Within the past 30 years, remarkable progress has been realized in the understanding of genes and their molecular behavior, and in the ability to manipulate them at the level of nucleotide bases. We are ever more skilled at finding specific genes, changing them in test tubes, and putting them back into plants and animals. Certain traits have already been successfully engineered into crop species and commercialized, such as herbicide and insect resistance.

None of these new molecular technologies or tools, however, will supplant the fundamental skill set required for the successful development of new, commercially viable plant populations, at least not in the foreseeable future. Moreover, these new tools alone will not give us the ability to develop any new crop species from those that currently exist in the wild. Following nearly 10 years of intensive cultivation of Bt corn and cotton and Roundup-Ready® soybeans the consequences of strong selection pressures on pests are starting to appear in the form of mutant herbicide-resistant weeds and insects that have overcome host GMO resistance genes.

TABLE 1.1   Proportion of Disposable Income Spent on Food in The U.S.; 1930–2014

Year	$Billion disp income	% in house	% out of house	% total
1930	74.9	21.09	3.07	24.17
1935	59.8	20.23	3.01	23.24
1940	77.7	17.37	3.09	20.46
1945	156.3	15.08	3.65	18.73
1950	215.0	16.59	3.53	20.12
1955	291.7	14.70	3.36	18.06
1960	376.5	13.67	3.34	17.01
1965	513.2	11.38	3.30	14.69
1970	761.5	9.92	3.47	13.39
1975	1219.3	9.63	3.77	13.39
1980	2018.0	8.96	4.22	13.18
1985	3098.5	7.55	4.15	11.70
1990	4311.8	7.29	4.06	11.35
1995	5532.6	6.50	4.14	10.63
2000	7400.5	5.83	3.94	9.77
2005	9400.8	5.68	4.06	9.74
2010	11,237.9	5.52	4.02	9.54
2014	12,913.9	5.48	4.26	9.74

Source: Calculated by the Economic Research Service, USDA. USDA, from various data sets from the U.S. Census Bureau and the Bureau of Labor Statistics.

I. ELEMENTS AND UNDERPINNINGS OF PLANT BREEDING

TABLE 1.2   Disposable Income and Proportion of Income Spent on Food Among Selected Countries (2014)

Country	% food	% EtOH Tob[a]	Disp Income	$ on food
United States	6.42	2.00	37,253.4	2391.8
United Kingdom	8.22	3.93	26,975.6	2217.2
Switzerland	8.74	3.61	43,061.3	3762.6
Canada	9.12	3.32	23,763.9	2168.4
Australia	9.81	3.43	29,709.3	2914.8
Austria	9.90	3.37	23,443.3	2322.0
Qatar	11.73	0.25	12,917.9	1515.9
Norway	12.27	4.08	28,836.1	3538.2
Belgium	12.87	4.23	19,675.2	2532.1
Bahrain	12.95	0.37	11,409.1	1477.3
South Korea	13.03	2.15	12,584.4	1639.9
France	13.22	3.51	20,022.5	2646.7
United Arab Emirates	13.77	0.20	22,944.6	3158.6
Japan	14.15	2.55	18,588.6	2630.9
Brazil	15.51	2.20	5256.5	815.3
Poland	16.50	7.37	7195.9	1187.6
Colombia	17.43	2.93	3731.0	650.4
Uruguay	18.21	1.20	11,163.1	2033.1
Bulgaria	18.37	6.89	4425.2	812.9
Venezuela	19.59	3.71	842.8	165.1
Costa Rica	19.94	0.94	7542.7	1503.7
Malaysia	20.63	1.79	5436.5	1121.5
Turkey	21.54	5.10	6688.2	1440.4
Tunisia	22.33	3.22	2877.8	642.5
Mexico	23.13	2.56	6143.5	1420.7
Iran	24.22	0.41	2232.5	540.7
China	24.96	3.47	3005.1	750.0
Saudi Arabia	24.97	0.62	7826.9	1954.1
Russia	28.02	9.20	4707.8	1319.0
India	30.47	2.32	1033.3	314.8
Indonesia	32.89	5.22	1912.8	629.1
Ukraine	38.06	8.09	1347.1	512.7
Pakistan	40.91	0.97	1118.6	457.6
Philippines	41.89	1.17	2134.1	894.0
Cameroon	45.55	2.10	934.3	425.6
Nigeria	56.41	1.33	2006.3	1131.8

[a] % EtOH Tob = % alcoholic and tobacco products.

Sources: Euromonitor.com and USDA/EMS.

Plant breeding as a professional pursuit has graduated largely from being a purely academic exercise during the early 20th century to a mostly commercial endeavor. The seed and planting stock industry has sprung up to develop and market new genetic products to farmers, leaving public sector entities such as universities to focus on economic species of marginal value, the development of "germplasm" for private companies, and new technologies. This trend is fostered by liberal extensions of laws for the protection of intellectual property to DNA nucleotide sequences. As universities continually remold themselves to be ever relevant, ever on the "cutting edge", more mature disciplines such as plant breeding tend to be left behind.

Will industry train practitioners in the face of fading participation in the field of Plant Breeding by academia? Major seed companies that historically recruited academically trained plant breeders now hire "New Trait Developers" in lieu of plant breeders. These are Ph.D. level scientists charged primarily with the pursuit of heritable characteristics that can be patented and incorporated across a broad spectrum of crops. The ability to simultaneously address thousands of interacting genes that condition the whole organism and the population remains tantamount to success, and that is presently the sole providence of plant breeding. Microarrays and other technologies will someday supplant the eye and the yardstick, but a profound technology gap exists that cannot be easily overcome.

As will be duly demonstrated in this textbook, much of plant breeding is driven by probability and chance. Success can be greatly affected by quantities, such as time, space, and especially labor. Plant breeding has adopted the model developed for the manufacturing of hard goods wherein resource-, commodity-, labor- consumptive activities are relegated to developing countries. Reductions in the level of intellectual inputs, it is reasoned, are more than made up for by cheap land and labor inputs. If it is presumed that economic plants will no longer be produced in developed countries, this strategy will work for a short period of time, until developing countries mature into developed ones. When the last outpost on earth is thusly exploited, where will we go next?

The human brain is the most powerful integrating tool on earth, and can be very effectively trained to incorporate seemingly infinite amounts of new information. Most practicing plant breeders surmise that what they do is a balance of science and art, art being that which defies succinct technical explanation. Art is likened to the ability (mostly acquired) to simply look at an individual within a given population and fully assess its genetic potential as a contributor to an idealized, abstract future "perfect" population. Further, art is the ability to visualize how best to utilize individual plants in subsequent population development. A good analogy is a card game where the player has the ability to undertake the improvement of a hand dealt, such as poker or bridge. The science of statistics and probability is exceedingly helpful, but most excellent players attribute success to well-developed intuition (Duvick, 2002).

How can a book such as this address the "art" aspects of plant breeding? Intuition building is an intensely personal process, but seasoned plant breeders suggest that there are some ways that it may be accelerated. First and foremost, the plant breeder must get to know the biological entity he/she is working with as intimately as possible. Preferred habitats, range of characteristics, how to grow the best possible plant, biotic and abiotic stresses, ecological niches, and commensal and symbiotic relationships to name a few aspects. Most importantly, the plant breeder must be thoroughly familiar with the reproductive system, both within the organism and at the level of the population. Genetically, a grasp of the body of knowledge, cytogenetics, marker genes, genome mapping, inheritance mode of economic traits is essential. Being well-versed in genetic principles does not in itself guarantee success, however. Recall that 90% or more of all heritable crop improvements were made without any formal training in modern genetics. The author once advised a graduate student who was a highly accomplished, successful plant breeder in a large private seed company. He returned to my university to earn a graduate degree, but could not pass entry-level genetics courses. His story will be described further in Chapter 3.

With that brief perspective in mind, let us begin.

HISTORICAL PERSPECTIVE

The scales of time serve to warn us of the hasty use of new-found technologies to make irreversible changes in our genetic crop heritage. The earth is 4.5 billion years old, and is still progressing through a myriad of changes that defy human prediction (Vartanyan, 2006). Angiosperms first appeared 125 million years ago, and direct ancestors of *Homo sapiens* about 100,000 years ago. As was stated earlier, plant and animal domestication began as recently as 6000 years ago, and the principles of genetics were only discerned 135 years ago. Since 1890, plant breeding has progressed from a process of germplasm evaluation and mass population selection into various advanced iterations of controlled mating and selection schemes (MacKey, 1963; Jensen, 1988). Directed recombinant DNA was first demonstrated in 1974, and plant transformation in 1984. The first GMO releases into agriculture were made in the late 1980s, and they were permitted into the food chain in 1994. If not for a curious political sideshow wherein technology antagonists gained the upper hand, GMO crops would surely have predominated planet Earth by now. However, history will no doubt repeat itself, and GMO crops will eventually come to dominate the crop variety landscape.

Agriculture and the domestication of plants and animals are the cornerstones of human societies and culture. Put into perspective, however, plant breeding has only been practiced for 0.01% of all the time that angiosperms have existed, and the principles of genetics have only been applied for about 1% of that time. If one were driving across the U.S. from New York to Los Angeles, the relative distances would be respectively less than a one-half mile and 50 ft. Over 100% yield increases have typically been realized as a consequence of plant breeding efforts over the past 100 years, greatly accelerated by a better understanding of the principles of inheritance. With increasing knowledge of genes and their functions, improvements in the coming 100 years should be enormous (Borojevic, 1990).

New technologies, such as microarrays, will ultimately take their place as important tools in the plant breeding arsenal, and progress will be incrementally quickened. What will our crops look like 100 years from now? Will we still be growing corn, wheat, and soybeans in agricultural systems, or will our food be produced in entirely new ways, perhaps without need of conventional agricultural at all?

Plant breeding is an interaction of humans with the plants that are used in agriculture. As such, it is fitting to address history by discussing chronologically the people who made notable impacts. Hundreds, perhaps thousands contributed to the chain of domestication, and we have "stood on the shoulders of giants" ever since. Each generation inherits the populations of ancestors and changes it slightly to fit the whims of the present. Since latent genetic variability is not purposefully retained in our commercial populations, it is hoped that these incremental changes will benefit the ages and not only the present. No written history exists, of course, of these countless individuals who made incremental improvements in time, some quite remarkable, such as the development of modern monoecious corn from hermaphroditic ancestral species.

Any author puts themselves at risk for criticism when developing a list, for the criteria for inclusion and exclusion are always arbitrary. In the case of plant breeding (circa 1700 to present), many individuals were undoubtedly involved, and competition from several independent groups often contributed greatly to the discoveries. Singling out any one person is emblematic of the collective breakthrough. The list is culturally biased, since languages and politics have greatly affected our appreciation of individual accomplishments exclusive of Europe and North America. Suffice it to say, then, that here is a chronological list of important people in the field of plant breeding, and a brief summary of what they did, and what impacts (positive and negative) the work had:

Carolus Linnaeus (1707–78). Linnaeus was a Swedish botanist and physician who laid the foundations for the modern naming and classification scheme for life. He is also considered one of the fathers of modern ecology. Linnaeus was born in southern Sweden and was groomed as a youth for the clergy, but he showed little enthusiasm for it. His interest in botany impressed a physician from his town and he was sent to study at Lund University and transferring to Uppsala University after 1 year. During this time Linnaeus became convinced that in the stamens and pistils of flowers lay the basis for the classification of plants, and he wrote a short work on the subject that earned him the position of adjunct professor. In 1732 Linnaeus explored Lapland, then virtually unknown. The result of this was the Flora Laponica published in 1737. Thereafter Linnaeus moved to the Netherlands where he met **Jan Frederik Gronovius** and showed him a draft of his work on taxonomy, the Systema Naturae that introduced the now familiar Latinized genus-species names. Higher taxa were constructed and arranged in a simple and orderly manner. He continued to work on his classifications, extending them to the kingdom of animals and the kingdom of minerals. Linnaeus' research had begun to take science on a path that diverged from what had been taught by religious authorities, and the local Lutheran archbishop had accused him of "impiety." Nonetheless, the Swedish king, Adolf Fredrik, ennobled Linnaeus in 1757, and after the privy council had confirmed the ennoblement Linnaeus took the surname von Linné, later often signing just **Carl Linné**. The lasting impact of Linnaeus in the realm of plant breeding was in paving the way for the species taxon to become the cornerstone in the concept of evolution, and the intermediate steps that result in population genetic changes.

Chevalier de Lamarck (1744–1829). He was a major 19th-century French naturalist, who was one of the first to use the term "biology" in its modern sense. Lamarck was born in Picardy, France and died in Paris. He is remembered today mainly in connection with a discredited theory of heredity, the "inheritance of acquired traits", but Charles Darwin and others acknowledged him as an early proponent of ideas about evolution. Lamarck's theory of evolution was in fact based on the idea that individuals adapt during their lifetimes and transmit traits they acquire to their offspring. Offspring then adapt from where the parents left off, enabling evolution to advance. As a mechanism for adaptation, Lamarck proposed that individuals increased specific capabilities by exercising them, while losing others through disuse. While this conception of evolution did not originate wholly with Lamarck, he has come to personify pre-Darwinian ideas about biological evolution, now called Lamarckism. Ironically, with the discovery of molecular phenomena such as snRNAs, certain of the precepts of Lamarckism appear to hold validity.

Alphonse de Candolle (1806–93). In 1855 the French botanist de Candolle published Géographie botanique raisonnée (Reasoned Geographical Botany). This was a ground-breaking book that for the first time brought together the large mass of data being collected by worldwide scientific expeditions. The natural sciences had become highly specialized during the mid-19th century, but de Candolle's book explained living organisms within their environment and why plants were distributed geologically the way they were. The book had a significant impact on the teachings of eminent Harvard College botanist **Asa Gray**. De Candolle's findings and speculations also stimulated Russian geneticist **Nikolai N. Vavilov** to investigate the genetic bases of the distribution of organisms, leading to his Theory on Centers of Origin and Diversity.

Charles Darwin (1809–82). Darwin was a British naturalist who achieved lasting fame as the originator of the theory of evolution through natural selection. He developed his interest in natural history while studying first medicine, then theology. Darwin's 5-year voyage on the HMS Beagle brought him eminence as a geologist and fame as a popular author. His biological observations led him to study the transmutation of species and develop his theory of natural selection in 1838. Fully aware of the likely reaction, he confided only in close friends and continued his research to meet anticipated objections, but in 1858 the information that a rival scientist, **Alfred Russel Wallace**, now had a similar theory forced early joint publication of Darwin's theory. His 1859 book *The Origin of Species by Means of Natural Selection*, or *The Preservation of Favored Races in the Struggle for Life* (usually abbreviated to *The Origin of Species*) established evolution by common descent as the dominant scientific theory of diversification in nature. He was made a Fellow of the Royal Society, continued his research, and wrote a series of books on plants and animals, including humankind, notably *The Descent of Man*, and *Selection in Relation to Sex* and *The Expression of the Emotions in Man and Animals*. Darwin also performed many experiments that lent a base of knowledge to plant breeding, such as some of the earliest studies of the phenomenon of heterosis. In recognition of Darwin's pre-eminence, he was buried in Westminster Abbey, close to William Herschel and Isaac Newton.

Louis de Vilmorin (1816–60). The practical plant breeder Louis de Vilmorin lived and worked in France. Using new methods he developed to estimate the sugar content of expressed plant sap, Vilmorin and other plant breeders were able to make rapid progress in improving sugar beet through continuous mass selection. From the 1830s, beet sugar production increased dramatically in France and Germany. Consequently, Europe was able to become independent of the vagaries of sugar production from sugar cane in distant tropical colonies. Vilmorin was also engaged in the breeding of other forms of beet (table, fodder) and also many vegetable and flower crops. He is credited with the first demonstration of the use of progeny tests as a basis for selection, and the "ear to row" method in corn was based on his findings. The Vilmorin Seed Company still is in business to this day.

Gregor Mendel (1822–84). He was an Austrian Augustinian monk who is widely called the "father of genetics" for his study of the inheritance of traits in pea plants. Mendel was inspired by both his professors at university and his colleagues at the monastery to study variation in plants. He commenced his study in his monastery's experimental garden, located in what is now the Czech Republic. Between 1856 and 1863 Mendel cultivated and tested some 28,000 pea plants. His experiments brought forth two generalizations (Segregation and Dominance) which later became known as Mendel's Laws of Inheritance. The significance of Mendel's work was not recognized until the turn of the 20th century. Its rediscovery by **William Bateson** and **Hugo DeVries** prompted the foundation of the scientific discipline of genetics.

Luther Burbank (1849–1926). Considered the first American plant breeder, Burbank was born in Lancaster, Mass and traveled all over North America during his professional life. He experimented with thousands of plant selections and developed many new cultivars of prunes, plums, raspberries, blackberries, apples, peaches, and nectarines. Besides the 'Burbank' potato (and 'Russet Burbank' selection thereof that is still widely grown to this day), he produced new tomato, corn, squash, pea, and asparagus cultivars; a spineless cactus useful in cattle feeding; and many new flowers including lilies and the famous 'Shasta' daisy. His methods and results are described in his books—*How Plants Are Trained to Work for Man* (8 vol., 1921) and, with Wilbur Hall, *Harvest of the Years* (1927) and *Partner of Nature* (1939)—and in his descriptive catalogs, *New Creations* (Fig. 1.1).

Wilhelm L. Johannsen (1857–1927). He was a Danish biologist who provided the first sound scientific basis for selection in self-pollinated plant species when he defined "pure lines" in 1903 and described the genetic mechanism by which they are established. In a notable series of experiments on garden bean (*Phaseolus vulgaris*, a highly inbreeding species) Johannsen studied the effects of selection for seed weight. He discovered that the progenies of heavy-seeded individuals also tended to also have heavy seeds, and the same held true for other seed weights. Individuals that bred true for seed weight were termed pure lines. Further eloquent studies demonstrated that some bean cultivars are actually mixtures of pure lines. Later, Johannsen introduced and defined essential terms that pervade plant breeding: *Phenotype* and *genotype*. Subsequently, he went on to develop compelling hypotheses about the nature of the gene, many of which ultimately contributed to our contemporary understanding of structure and function.

FIG. 1.1   (A) Carolus Linnaeus; (B) Chevalier de Lamarck; (C) Alphonse de Candolle; (D) Charles Darwin; (E) Louis de Vilmorin; (F) Gregor Mendel; (G) Luther Burbank. *(A) From https://en.wikipedia.org/wiki/Carl_Linnaeus#/media/File:Carl_von_Linn%C3%A9,_1707-1778,_botanist,_professor_(Alexander_Roslin)_-_Nationalmuseum_-_15723.tif. (B) From https://www.antwiki.org/wiki/File:Lamarck.jpg. (C) From https://picryl.com/media/candolle-augustin-pyrame-de-9dca6c. (D) From Portrait of Charles Darwin, Late 1830s, Origins, Richard Leakey and Roger Lewin, https://commons.wikimedia.org/wiki/File:Charles_Darwin_by_G._Richmond.jpg. (E) From https://en.wikipedia.org/wiki/Louis_de_Vilmorin#/media/File:Louis_de_Vilmorin00.jpg. (F) From Bateson, W., 1909. Mendel's Principles of Heredity. (G) From https://en.wikiquote.org/wiki/Luther_Burbank#/media/File:Burbank-_Luther_btwn_420_and_421.jpg.*

William Bateson (1861–1926). Bateson was born in England and entered Cambridge University in 1878. Even as a student, he was interested in species variation and heredity. He traveled to the Central Asian steppe and collected data on how environmental conditions relate to variation. In 1894, he published a book *Materials for the Study of Variation* based on his observations. In this book, he outlined the experimental approach that should be used to study inheritance. He did not yet know that he was actually recapitulating experiments that Mendel had already done years earlier. When he read Mendel's papers and conferred with **Hugo DeVries** (the scientist who rediscovered Mendel's work independently with **Correns** and **von Tschermak** in 1900), Bateson recognized the importance of "Mendelian Laws" and by 1902, he had translated Mendel's works into English and was a strong proponent of the Mendelian laws of inheritance. Bateson is credited with coining the terms "genetics," "allelomorph" (later shortened to allele), "zygote," "heterozygote" and "homozygote." In 1908, as a Professor of Biology at Cambridge, Bateson helped establish the first School of Genetics. Bateson left Cambridge in 1910 to accept the Directorship of the John Innes Horticultural Institute. He continued to have ties to Cambridge and co-founded the *Journal of Genetics* in 1910 with **Reginald Punnett**. Bateson was reluctant to believe in the chromosomal theory of inheritance. He was vocally antagonistic to the theory, and it was not until 1922 after a visit to **Thomas Hunt Morgan's** laboratory in Cold Spring Harbor, NY that he publicly accepted chromosomes and their role in heredity.

Herman Nilsson Ehle (1873–1949). A Swedish scientist, he is known mostly for an experiment he conducted that was the first demonstration of the compatibility of Mendel's Laws with continuous phenotypic variability. The trait studied by Nilsson Ehle was kernel color in wheat. The results clearly were most readily explained if the character was determined by more than one gene, thus paving the way to contemporary quantitative genetics. Nilsson Ehle conducted his research at a plant breeding station in Svalof, Sweden. The Svalof station had been renowned for its 'scientific' breeding methods, which basically consisted of an elaborate system of record-keeping through which the offspring of individual plants were traced over generations while being meticulously described. Inspired by a translation of Mendel's paper on peas, Nilsson-Ehle began further experiments in 1900 and published a first, major synthesis of his findings on the results of organized hybridization and selection in 1908. This system was ultimately refined into what is now known as the *Pedigree Method*.

George H. Shull (1874–1954). Shull was born in Ohio and attended Antioch College and the University of Chicago. His first actual job beyond working on the family farm was as Botanical Assistant at the U.S. National Herbarium. He also worked at the U.S. Bureau of Plant Industry as a Botanical Expert examining flora and fauna of the Chesapeake Bay and Currituck Sound. During this period, Shull had become interested in the statistical analysis of variation in plants, the subject of his later Ph.D. research. In 1904, Shull was appointed to oversee plant work at the Station of Experimental Evolution. He published his uncompleted Ph.D. thesis as a paper then proceeded to a position at the Cold Spring Harbor Laboratory on Long Island, NY. While there, Shull studied and bred a large variety of plants, for example, the evening primrose, shepherd's purse, corn, and peas. He published many papers on his observations of plant traits and inheritance. In 1905, he began work in parallel with **E.M. East** of the Connecticut Agricultural Experiment Station on corn, maize, with the intent of examining the quantitative inheritance of corn traits. Following Mendel's example, Shull obtained pure-bred lines of corn through self-pollination. The pure-bred lines were less vigorous and productive, but when he crossed the pure-bred lines, the hybrid yields were better than any of the parents or those pollinated in the open fields. He immediately recognized the potential for using this strategy to improve crop yields, and vigorously promoted the strategy of hybrid development starting in 1909. By the 1930s and 40s, most U.S. and western Europe corn farms were growing hybrid stocks, and improved yields contributed greatly to the allies during World War II and the rehabilitation of post-war Europe. In 1915, Shull accepted a professorship at Princeton University and began the publication of a new journal, *Genetics*, still one of the top international science journals.

Reginald Punnett (1875–1967). Born in England, as a child Punnett was fascinated with biology and naturalism. He attended Cambridge University as a medical student but ultimately graduated with a degree in zoology in 1898. He remained at Cambridge after graduation to conduct research on marine worms and discovered two species that bear his name. He later became interested in the experimental work being done by **William Bateson** and began a collaboration that resulted in the establishment of the first program of genetics. Punnett and Bateson published the first account of gene linkage in sweet peas and developed the "Punnett Square" to depict the number and variety of genetic combinations. Punnett was asked at a lecture in 1908 to explain why recessive phenotypes still persist, and he turned to his friend, mathematician **G.H. Hardy** for an explanation that was crafted into the Hardy-Weinberg Law.

Edward M. East (1879–1938). East and **Donald F. Jones** worked together in 1919 on monographs in experimental biology including a book on inbreeding and outbreeding and their genetic significance. This work covered the mechanism of heredity, mathematical considerations of inbreeding, inbreeding experiments with animals and plants, hybrid vigor or heterosis, possible causes of hybrid vigor, sterility and its relations to inbreeding and crossbreeding, the role of inbreeding and outbreeding in evolution, and the value of inbreeding and crossbreeding in plant and animal improvement. East concluded that the increased homozygosity caused by inbreeding should generally be accompanied by detrimental effects. His theory also explained that unless deleterious recessive alleles were present in the genotype, inbreeding caused no ill effects. Outcrossing had the effect of increasing heterozygosity and was often accompanied by heterosis, or hybrid vigor. Ultimately, East played an important role in the development of hybrid corn. His main interest was in the theoretical interpretations from inbreeding leading to reductions of and hybridization leading to increases of vigor (Fig. 1.2).

Harry V. Harlan (1882–1944). Harry Harlan was born in rural Kansas and attended Kansas State University for both B.S. and M.S. degrees, earning a Ph.D. from the University of Minnesota in 1914. From 1910 to 1944 Harlan Sr. was the leader of barley investigations for the U.S. Department of Agriculture, Washington, D.C., as well as a plant explorer. He collected seeds in South America, Asia, Europe, and Africa and amassed the largest collection of barley germplasm, or any crop species for that matter, in the world. Harlan's barley cultivars occupied the overwhelming majority of U.S. acreage during the mid-20th century. He was a renowned expert on the origin of grain crops in general, and became an exceedingly popular author in the 1920s and 1930s, penning many articles for National Geographic and other popular science periodicals. Harlan published the seminal paper with **M.N. Pope** in 1922 on the backcross method of plant breeding.

Nikolai Ivanovich Vavilov (Russian **Николай Иванович Вавилов**) (1887–1943). He was a prominent Russian botanist and geneticist. After his education in Moscow and postgraduate work for the Russian Ministry of Science, Vavilov traveled in 1913–14 to England to study plant diseases in collaboration with **William Bateson**. Vavilov organized a series of botanical-agronomic expeditions all over the world in pursuit of the development of his theory about Centers of Origin of Cultivated Plants and created the largest collection of plant seeds (which was tragically lost during the Siege of Leningrad during World War II) in the world. He was a prominent member of the USSR Central Executive Committee, President of All-Union Geographical Society, a recipient of the Lenin Prize, and President of the Institute of Botany during the 1920s and 1930s. Vavilov was prosecuted in 1940 as a defender of "bourgeois pseudoscience" genetics in a political struggle with **Trofim Lysenko** (see below) and died in 1943 as a political prisoner in Siberia.

FIG. 1.2 (A) Wilhelm L. Johannsen; (B) William Bateson; (C) Herman Nilsson Ehle; (D) George H. Shull; (E) Reginald Punnett; (F) Edward M. East. *(A) From https://commons.wikimedia.org/wiki/File:Wilhelm_Johannsen_1857-1927.jpg. (B) From John Innes Archives, Courtesy of the John Innes Foundation. (C) From https://commons.wikimedia.org/wiki/File:Herman_Nilsson-Ehle.jpg. (D) From Harry, S. Truman Library and Museum. (E) From https://en.wikipedia.org/wiki/Reginald_Punnett#/media/File:Reginald_Crundall_Punnett,_1875%E2%80%931967.jpg. (F) "Edward Murray East", 2012-06-11, History of the Marine Biological Laboratory, https://hdl.handle.net/1912/17098.*

Henry A. Wallace (1888–1965). Born on a farm in Iowa, Wallace graduated from Iowa State College at Ames in 1910. In 1915 he devised the first corn-hog ratio charts indicating the probable course of markets. He worked on the editorial staff of Wallace's Farmer (a publication started by his father, a former U.S. Secretary of Agriculture) and served as Editor-in-Chief from 1924 to 1929. Wallace experimented with breeding high-yielding hybrid cultivars of corn and authored many publications on agriculture. The company he founded during this time, now known as Pioneer Hi-Bred (a division of Dow/DuPont), was among the most profitable agriculture corporations in the world. Subsequently, Wallace was appointed U.S. Secretary of Agriculture in 1933, and Vice President with Franklin D. Roosevelt from 1940 to 1945. Later, he assumed the position of Secretary of Commerce in Truman's cabinet but was ultimately dismissed. After that, Wallace resumed his farming interests in upstate New York. Wallace forged a number of advances during his later years in the field of agricultural science. His many accomplishments included a strain of chicken that at one point accounted for the overwhelming majority of all egg-laying poultry worldwide and the introduction of the honeydew melon to China, where it is still referred to as the "Wallace" melon.

Sewell Wright (1889–1988). He was an eminent American geneticist and one of the founders of the field of modern theoretical population genetics. While a faculty member at the University of Wisconsin Wright researched the effects of inbreeding and crossbreeding with guinea pigs and later on the effects of gene action on inherited characteristics. He adopted statistical techniques to develop evolutionary theories. He also adapted theories to practical uses, such as the prediction of performance of synthetic populations based on the combining ability of constituent inbreds. Wright is best known for his concept of genetic drift, also known as the Sewell Wright effect. The **Sewall Wright Award** was established in 1991 is given annually by the American Society of Naturalists in honor of his contributions to the conceptual unification of the biological sciences.

Henry A. Jones (1889–1981). The accidental discovery of a male-sterile onion plant by Jones in 1925 marked the beginning of hybrid onion breeding, and of many hybrid breeding programs in crop species other than corn. He used the backcross method to transfer the genetic determinants for cytoplasmic male sterility to parental lines and forged the first demonstration of the use of the use of male sterility for the production of hybrid seeds. Jones started the USDA/ARS onion breeding program in 1936 in Beltsville, MD where he is credited with being the first person to hybridize the flat Bermuda and the top-shaped grano into a new onion Jones called the "Granex". The Granex onion variety was a deeper-shaped onion, with sweetness and a longer shelf-life in markets than previous cultivars. When Jones retired from the USDA he became Research Director of the Desert Seed Company in the California Imperial Valley and continued to develop new onion cultivars. He developed the F_1 Hybrid 'Yellow Granex' that is more popularly known as the "Vidalia" after the Georgia town where it is grown. Jones' model for the use of cytoplasmic male sterility for large-scale hybrid seed production was ultimately adapted to many other important crop species.

Sir Ronald Aylmer Fisher (1890–1962). Fisher was a British eugenicist, evolutionary biologist, geneticist, and statistician. He has been described by historians as "The greatest of Darwin's successors" and "… a genius who almost single-handedly created the foundations for modern statistical science." Because of his poor eyesight, he was tutored in mathematics without the aid of paper and pen, which developed his ability to visualize problems in geometrical terms as opposed to using algebraic manipulations. Fisher developed a strong interest in biology and especially evolution. He graduated from Cambridge University in 1913 with a degree in mathematics but could not find a suitable position until after World War I. Starting in 1919 he published an astonishing number of articles on biometry, including the ground-breaking *The Correlation to be Expected Between Relatives on the Supposition of Mendelian Inheritance*. This paper showed very convincingly that the inheritance of continuous variables was consistent with Mendelian principles and laid the foundation for what came to be known as biometrical genetics. Later, Fisher wrote a series of reports under the general title *Studies in Crop Variation* in which he pioneered the principles of the design of experiments and elaborated his studies of analysis of variance and partitioning of phenotypic variance. He invented the techniques of maximum likelihood and analysis of variance and originated the concepts of sufficiency, ancillarity, Fisher's linear discriminator, and Fisher information. He developed ingenious computational methods that were as practical as they were theoretically novel. In 1925, this work culminated in the publication of his first book, *Statistical Methods for Research Workers*. This volume went into many editions and translations in later years and became a standard reference work for scientists in many disciplines including plant breeding and genetics. In 1935, this was followed by *The Design of Experiments* that also became an academic standard. His work on the theory of population genetics made him one of the great figures of that field as well (Fig. 1.3).

Donald F. Jones (1890–1963). Jones was a United States corn geneticist and practical breeder at the Connecticut Agricultural Experiment Station, New Haven. He made high-yielding hybrid corn practical by his invention in 1914 of the double-cross hybrid. In this method four inbred parents were used in an (AxB)x(CxD) configuration. The seed from two initial crosses were used to produce partially heterotic parental hybrids for the second hybridization that resulted in seeds distributed to farmers. In this manner, seed production fields could yield seed in sufficient quantity to make the scheme practical. Prior to the double-cross method the seed yield of the parent lines (the inbreds) was insufficient to allow practical production of hybrid corn seed. Within a few years following the introduction of the double-cross method corn-breeding programs were initiated by USDA/ARS and many of the state experiment stations. By 1933 hybrid corn was in commercial production on a substantial scale. By 1949 78% of the total U.S. corn acreage was planted in hybrid cultivars. More than 95% of corn acreage was hybrid by 1959 and the average yield of corn in the U.S. was double that of 1929 (Mangelsdorf 1975). Jones was the president of the Genetics Society of America in 1935 and was elected to the U.S. National Academy of Sciences.

Fred N. Briggs (1896–1965). Briggs was a plant breeder and geneticist working on small grains who worked for his entire professional career at the University of California, Berkeley (Davis Farm). Coincidentally with **Harry Harlan** and **M.N. Pope's** 1922 paper on the backcross method, Briggs began what was to become one of the longest and most famous experiments in the field of plant breeding and population genetics. He set out to develop and new variety of wheat that was resistant to bunt, a debilitating fungal disease of small grains. Several populations of bunt resistant wheat had been successfully developed by 1930 through application of the new backcross breeding method, thus verifying the theoretical basis and practical significance Harlan and Pope's speculations. Briggs went on to use the backcross method to develop several important disease resistant wheat varieties that predominated in the western U.S. during the 1930s through the 1950s. He and **Robert Allard**, who later authored one of the most widely used plant breeding texts, published a landmark paper in 1953 that spelled out the most critical factors for achieving success with the backcross method. Briggs also co-published a popular textbook with **Paul F. Knowles** in 1967 entitled *Introduction to Plant Breeding*.

Trofim Denisovich Lysenko (Russian: Трофи́м Дени́сович Лысе́нко) (1898–1976): Lysenko was a Russian biologist who rose to great power and influence within the former Union of Soviet Socialist Republics. At the age of 29 while working at an experiment station in Azerbaijan in 1927 Lysenko was credited by the Soviet newspaper Pravda with having discovered a method to energize crops without using fertilizers or minerals (considered a "miracle") and that a winter crop of peas could be grown in Azerbaijan despite previous crop failures. Lysenko led a campaign

FIG. 1.3   (A) Harry V. Harlan; (B) Nikolai Ivanovich Vavilov; (C) Henry A. Wallace; (D) Sewell Wright; (E) Henry A. Jones; (F) Sir Ronald Aylmer Fisher. *(A) Image 0007063, Courtesy of the University of Illinois Archives, Jack R. & Harry V. Harlan Papers, RS 8/6/25. (B) From Russiapedia, https://russiapedia.rt.com/prominent-russians/science-and-technology/nikolay-vavilov/. (C) From The Wallace Centers of Iowa, https://wallace.org/who-are-the-wallaces/ henry-a-wallace/. (D) From https://en.wikipedia.org/wiki/Sewall_Wright#/media/File:Sewall_Wright.jpg. (E) From Whitaker, T.W., 1983. Dedication: Henry A. Jones (1889–1981) Plant Breeder Extraordinaire. In: Janick, J. (Ed.), Plant Breeding Reviews, vol. 1. (F) From https://en.wikipedia.org/wiki/Ronald_Fisher#/ media/File:Youngronaldfisher2.JPG.*

during the 1930s of agricultural science, now known as "Lysenkoism", that was contrary to contemporary agricultural genetics. Lysenkoism was an amalgam of Lamarckism and Darwinism with political overtones. Though scientifically unsound on a number of levels, Soviet journalists and agricultural officials were interested in and supported Lysenko's theories since they sped up laboratory work and cheapened it considerably. Lysenko was subsequently appointed Director of the Soviet Academy of Agricultural Sciences and charged with ending the propagation of "harmful" ideas among Soviet scientists. Lysenko served this purpose faithfully, causing the expulsion, imprisonment, and death of hundreds of scientists and the demise of genetics (a previously flourishing field) throughout the Soviet Union. In this role he was directly responsibility for the death of the greatest Soviet biological scientist of that time, **Nikolai I. Vavilov**. After Stalin's death in 1953 Lysenko retained his position and enjoying a relative degree of trust from new Soviet Premier Khrushchev, but he died with his theories discredited. Lysenko's legacy was to greatly retard the growth of plant genetics and breeding worldwide and cause great damage to humanity as a consequence.

Sir Otto H. Frankel (1900–98). Frankel was a geneticist by training, a plant breeder by occupation, a cytologist by inclination and a genetic conservationist by acclaim. From 1910 to 1918 Frankel attended the Piaristen Staatsgymnasiums Wien (Vienna) VIII and was admitted to Munich University (1919–20) to study chemistry, botany and physics. After three semesters he had lost his enthusiasm for chemistry and wanted to do something more practical like agriculture. Frankel began his studies in genetics at the Agricultural University of Berlin in 1922. He started graduate studies in 1923 working on the genetics of snapdragon, the results of which turned out to be strikingly similar to those of **B. McClintock** on maize. Frankel then worked as a plant breeder of sugar beets in 1925–27 but also practiced wheat and barley breeding, leading to his later appointment to the New Zealand Science and Industry Research Institute in 1928 where he was to work until 1951. Frankel began a wheat breeding program at DSIRO in 1929 that led to the release of the widely grown cultivar 'Cross 7' in 1934. He also began a groundbreaking analysis of the yield components in wheat. In 1951, he was named Chief of the Australian CSIRO Division of Plant Industry, where he remained until 1962. Among his many honors was knighthood by the British government. Beyond all of his scientific accomplishments, Frankel was extremely active in the preservation of Jewish rights worldwide.

Harold H. Flor (1900–91). Born in St. Paul, MN, he received his M.S. Degree from the University of Minnesota in 1922 and his Ph.D. from the same institution in 1929. Flor began his career with the USDA at Washington State University studying bunt of wheat and transferred to the North Dakota Agricultural College in 1931 to study flax diseases. His research on flax rust showed resistance in flax was dominant to susceptibility and the genes conditioning reaction occurred as multiple alleles at five loci. Flor was the first to study simultaneously the genetics of the host and parasite, which allowed him to deduce what is popularly known as the gene-for-gene hypothesis. His interpretation of host-parasite genetic interaction has proven to be a critically important paradigm in plant pathology and of extraordinary utility in the breeding of disease resistant cultivars. It has been used extensively to explain genetic relationships in different rusts and other diseases, as well as in diverse symbiotic relationships such as plants and herbivorous insects. **J.E. Vanderplank** expounded in 1963 on the differences between simply inherited modes of disease resistance and multigenic forms, coining the terms "horizontal" and "vertical" resistance. In recognition of his valuable contributions, Flor was named Fellow of the American Phytopathological Society in 1965 and was elected President of APS in 1968.

Barbara McClintock (1902–92). She was a pioneering U.S. scientist and one of the world's most distinguished cytogeneticists. McClintock studied chromosomes and how they change during reproduction in maize during the late 1920s. Her work was groundbreaking: she developed an effective technique to visualize maize chromosomes and used microscopic analysis to demonstrate many fundamental genetic concepts, including genetic recombination by crossing-over during meiosis. She produced the first genetic map for maize, linking regions of the chromosome with physical traits, and she demonstrated the role of the telomere and centromere. During the 1940s and 1950s, McClintock discovered the phenomenon of genetic transposition and using this system showed how genes are responsible for turning on or off physical characteristics. She developed theories to explain the repression or expression of genetic information from one generation of maize plants to the next. These theories met with skepticism until proven unequivocally in the 1960s and 1970s. McClintock was awarded the Nobel Prize in Physiology or Medicine in 1983 for the discovery of genetic transposition.

George F. Sprague (1902–98). Sprague received a B.S. degree from the University of Nebraska then a doctorate at Cornell University as a student with **R.A. Emerson**. He worked as maize geneticist for the U.S. Department of Agriculture, the University of Illinois, and Iowa State University. Starting with research on basic inheritance, including chromosomal inversions, Sprague went on to pioneer much of the knowledge base that underpins modern corn breeding methodology such as recurrent selection for combining ability. In recognition of his productive career and tremendous impacts on corn productivity, George Sprague was elected to the National Academy of Sciences and was a recipient of the Wolf Foundation Award from Israel (Fig. 1.4).

Cyril D. Darlington (1903–81). He was a British biologist and geneticist who discovered the mechanics of chromosomal crossover and its importance in evolution. Darlington rose from humble beginnings to Chair the Department of Cytology and later serve as Director at the John Innes Institute in England. He co-founded the journal Heredity with colleague **R.A. Fisher**. In 1931 he began writing the book that would establish his worldwide reputation as an eminent scientist, *Recent Advances in Cytology*. It was published in 1932 and created a firestorm of controversy at first but was then nearly universally accepted as a work of great importance. Darlington showed that the mechanisms of evolution that acted at the level of the chromosome created possibilities far more substantial than the simple mutations and deletions that affect single genes. His writings served as the underpinnings for cytogenetics and evolution for over 50 years. Later, Darlington was an outspoken international critic of Lysenkoism and an authority on genetics and evolution. Darlington was awarded the Darwin Medal in honor of his important discoveries.

George W. Beadle (1903–89): Born in Nebraska, Beadle was the son of a farmer and was educated at the University of Nebraska. In 1926 he earned a B.Sc. degree and subsequently worked for a year with **F.D. Keim**, who was studying hybrid wheat. Beadle earned his M.Sc. degree in 1927 and pursued further graduate studies at Cornell University where he worked until finishing his doctorate in 1931 with **R.A. Emerson** and **L.W. Sharp** on maize genetics. Beadle went on to the California Institute of Technology in 1931 where he continued his work on Indian corn and began, in collaboration with **T. Dobzhansky, S. Emerson,** and **A.H. Sturtevant** work on crossing-over in the fruit fly, *Drosophila melanogaster*. In 1935 Beadle visited Paris for six months to work with **Boris Ephrussi** on the development of eye pigment in *Drosophila*. This collaboration led to the work on the biochemistry of the genetics of the fungus *Neurospora* for which Beadle and **Edward Tatum** were together awarded the 1958 Nobel Prize for Physiology or Medicine. In later years, after stints at Harvard, Stanford, and Cal Tech, Beadle was appointed as President of the University of Chicago. In retirement, he engaged in a high-profile debate with colleague **Paul Mangelsdorf** on the origin of domesticated maize. Beadle's hypothesis that domesticated corn descended from a *Z. mays* teosinte ancestor is supported by more recent genomic evidence.

Glenn Burton (1910–2005). He grew up on a cattle ranch in Nebraska perhaps leading Burton to work on behalf of the livestock industry for parts of his entire career. Following his formative and learning years Burton spent over

FIG. 1.4 (A) Donald F. Jones; (B) Fred N. Briggs; (C) Trofim Denisovich Lysenko; (D) Sir Otto H. Frankel; (E) Harold H. Flor; (F) Barbara McClintock; (G) George F. Sprague. *(B) From UC Davis Plant Breeding, https://plantbreeding.ucdavis.edu/major-contributors. (C) From Wiki Pseudosciencia, http://es.pseudociencia.wikia.com/wiki/Trofim_Lysenko. (D) Used with permission from Australian Academy of Science. (E) From https://en.wikipedia.org/wiki/ Harold_Henry_Flor#/media/File:Harold_Henry_Flor.jpg. (F) Courtesy of the Barbara McClintock Papers, American Philosophical Society. (G) From https:// web.archive.org/web/20160412082557/http://www.ars.usda.gov/careers/hof/browse.htm.*

60 years of his career as a USDA/ARS Plant Geneticist at the University of Georgia Coastal Plain Experiment Station in Tifton, Georgia. 'Coastal' Bermuda grass was released in 1943, primarily as an effort by Burton to improve cattle performance in Georgia. Currently 'Coastal' is grown on over 10 million acres across the southeast, and is still the standard that most new forage grass cultivars are compared with for yield, persistence and quality. Burton was a leader in utilizing techniques to improve forage quality as well as production quantity. He has bred and released a great number of warm-season perennial and annual grasses that have increased yield, disease resistance and improved quality, ultimately serving as standards for entire industries. His plant breeding work has had broad industry and public benefits including livestock, homeowners, nursery and landscape professionals, and golf enthusiasts.

Norman Borlaug (1914–2009). He is widely regarded as the father of the "Green Revolution". Borlaug received his Ph.D. in plant pathology and genetics from the University of Minnesota in 1942 and subsequently accepted an agricultural research position as a plant breeder at CYMMYT in Mexico with the Rockefeller Foundation. While there, he developed semi-dwarf high-yield, disease-resistant wheat varieties and led the introduction of his grain breeding techniques and modern agricultural production methods to Mexico, Pakistan, and India. Mexico became a net exporter of wheat by 1963 as a consequence of Borlaug's research results. Wheat yields nearly doubled in Pakistan and India between 1965 and 1970, greatly improving food security in those nations. These collective increases in yield and self-sufficiency have been labeled the Green Revolution, and Borlaug's advancements at CYMMYT are often credited with saving over a billion people from starvation. He was awarded the Nobel Peace Prize in 1970 in recognition of his contributions to increases in world food supply. More recently, he has helped apply methods to increase food production to Asia and Africa.

Charles M. Rick (1915–2002). Born in Reading, Pennsylvania, Rick grew up working in local fruit orchards. He earned a B.S. degree at Pennsylvania State University and a Ph.D. at Harvard in 1940, concentrating on botany and plant genetics. He subsequently joined the faculty of the Vegetable Crops Department at the University of California, Davis, where he remained for his career of more than 60 years. Much of Rick's most fascinating work came from a firsthand knowledge of the plants' roles in local environments and their evolving reproductive strategies using tomato and wild relatives as examples. He traveled tirelessly to South America and trekked within the nooks and hollows of the northern Andes Mountains and lowland deserts and rain forests. Over time, Rick's seminal work on tomato genetics established this plant as an important model organism in the era of genomics. His fundamental contribution was to characterize and utilize the incredible array of genetic variability that exists in wild progenitor populations of domesticated species. Rick received many honors during his career including membership in the U.S. National Academy of Sciences and recognition from dozens of universities and learned societies. He received the Alexander von Humboldt Award and was also the first recipient of the Filipo Maseri Florio World Prize in Agriculture in 1997.

Jack Harlan (1917–98): **Harry V. Harlan's** youngest son was born in Washington, D.C. He earned a B.S. degree (with distinction) from George Washington University, Washington, D.C., in 1938 and his Ph.D. in genetics from the University of California, Berkeley in 1942. He was the first graduate student to complete a Ph.D. under the guidance of eminent evolutionary geneticist **G. Ledyard Stebbins**. He was greatly influenced in his choice of career by the professional activities of his father. During the summer months, **Harlan Sr.** brought young Jack to the barley stations in Aberdeen, Idaho, and Sacaton, Arizona and he met the great Russian botanist, **N.I. Vavilov** in 1932. Harlan Jr. planned to study in St. Petersburg but the imprisonment and subsequent death of Vavilov in 1943 ended these plans. His professional career began with the U.S. Department of Agriculture (USDA) at Woodward, Oklahoma, and in 1951, while still with the USDA, Harlan Jr. transferred to Oklahoma State University. It was during this period and while teaching that he developed his philosophy concerning the evolution of crop plants and civilization. In 1966 Harlan Jr. moved to the University of Illinois where he remained until retirement as professor of plant genetics in the Department of Agronomy. With Professor **J.M. deWet**, a colleague from Oklahoma State and then at the University of Illinois, he founded the internationally renowned Crop Evolution Laboratory. Harlan Jr. explored for and introduced plants from Africa, Asia, Central America, South America, and Australia into the United States. In 1948 he led a USDA-sponsored plant exploration trip to Turkey, Syria, Lebanon, and Iraq and in 1960 to Iran, Afghanistan, Pakistan, India, and Ethiopia. He was a fellow of several professional societies, including the American Association for the Advancement of Science, American Society of Agronomy, Crop Science Society of America, and the American Academy of Arts and Sciences. Jack Harlan was elected to membership in the National Academy of Sciences in 1972. He served as president of the Crop Science Society of America in 1965–66. In addition he received a medal for service to the U.N. Food and Agriculture Organization and the International Board for Plant Genetic Resources and a medal at the N.I. Vavilov Centennial Celebration.

Oliver E. Nelson (1920–2001). Nelson graduated from Colgate University and subsequently received his M.S. and Ph.D. degrees from Yale University under the direction of the renowned plant breeder **Donald F. Jones**. He joined the faculty at Purdue University in 1947 and then the University of Wisconsin-Madison in 1969 from which he retired in 1991. Nelson's research focused on the *waxy1* locus to address questions of fundamental importance to biologists of the time, as well as to test predictions about transposable elements (see **Barbara McClintock**). His construction of the first fine structure map of *wx1* provided the most detailed glimpse of the structure of a higher organism gene before the advent of DNA sequencing. Nelson and his colleagues conducted a systematic investigation of the deficiencies of a number of corn endosperm kernel mutations over the course of three decades and identified the biochemical defects associated with eight starch mutants: *wx1*, *sh1*, *sh2*, *sh4*, *bt1*, *bt2*, *du1*, and *su1*. His identification of the *wx1* lesion with a starch-bound ADP-glucose glucosyl transferase represented the first association of mutation with biochemical function in plants. Nelson was also the first reports that transposable elements inserted into coding genes led to the production of a structurally altered protein. In 1984, Nelson isolated for the first time a plant gene by the novel procedure of transposon tagging in collaboration with **Nina Fedoroff**. Nelson's research showed that levels of the essential amino acids lysine and tryptophan could be enhanced by mutation, opening the door for plant breeders to address food nutritional quality traits. In recognition of his extraordinary contributions to plant genetics and breeding, Nelson was awarded the Herbert Newby McCoy Award and the John Scott Medal in 1967, elected to the U.S. National Academy of Sciences in 1972, awarded the Thomas Hunt Morgan Medal from the Genetics Society of America in 1997, and awarded the Stephen Hales Prize from the American Society of Plant Physiologists in 1998 (Fig. 1.5).

Stanley J. Peloquin (1921–2008). He was a highly productive potato geneticist and breeder whose career spanned over 30 years at the University of Wisconsin. Peloquin discovered the phenomenon of parallel spindles during meiotic cell divisions leading to the formation of true haploid gametes from tetraploid cultivated potato, controlled by a single gene. Subsequently, he demonstrated the advantages of performing selection and mating at the diploid level, then reconstituting the autotetraploid after the most desirable genotype had been attained. He also discovered male sterility genes in potato, and championed the conversion of the species from a purely asexual crop to one that utilized the benefits of sexual reproduction. A significant applied aspect of Peloquin's work is his development of methods

FIG. 1.5 (A) Cyril D. Darlington; (B) George W. Beadle; (C) Glenn Burton; (D) Norman Borlaug; (E) Charles M. Rick; (F) Jack Harlan; (G) Oliver E. Nelson. *(A) From John Innes Archives. Courtesy of the John Innes Foundation. (B) Courtesy of American Philosophical Society. (C) From http://tifton. caes.uga.edu/about/campus-overview/history/global-impact/glenn-burton.html, University of Georgia College of Agricultural and Environmental Sciences. All rights reserved. (D) From https://en.wikipedia.org/wiki/Norman_Borlaug#/media/File:Norman_Borlaug,_2004_(cropped).jpg. (E) Qualset, C., McGuire, P., Warburton, M., 1995. California agrobiodiversity key to agricultural productivity. Calif. Agric. 49 (6), 45–49. https://doi.org/10.3733/ca.v049n06p45. Copyright © 1995 Regents of the University of California. Used by permission. (F) Portrait photograph of Jack Harlan, courtesy of the University of Illinois at Urbana-Champaign Archives. (G) From University of Wisconsin https://genetics.wisc.edu/history/.*

to reliably propagate potatoes from true seed rather than asexually. His work has made the production of virus-free potatoes affordable for poor farmers in tropical countries where the production of potatoes in the past had only been possible using prohibitively expensive imported tubers. His scientific and applied contributions yielded significant recognition, most notably his election to the U.S. National Academy of Sciences in 1984.

Norman W. Simmonds (1922–2017): A British botanist, Simmonds became an expert on the subject of plant biodiversity and it's utilization for the long term sustenance of human cultures. He also was an influential and respected botanist and plant breeder better known in the banana community as the author of *Bananas*, the standard monograph on the crop, and *The Evolution of the Bananas*. In 1976 he published a popular book *Evolution of Crop Plants*. Simmonds was a world expert in the biodiversity of tropical crop species, and authored a very popular and influential plant breeding textbook in 1978 *Principles of Crop Improvement* that was replete with a fresh approach to what had become dull subjects mired in mathematical theory. He was also a highly accomplished potato and sugar cane breeder. In 1970 Simmonds was elected a Fellow of the Royal Society of Edinburgh in recognition of his illustrious and productive career.

Cyril Reed Funk (1928–2012). His turfgrass cultivars can be found from urban lawns to the White House and from Arlington National Cemetery to the Rose Bowl. Born and raised in Utah, Funk's career spanned nearly 50 years. He was the world-wide leading authority in turfgrass breeding. While on the faculty of Rutgers University, Funk developed and released more than 75 turfgrass cultivars and was awarded eight U.S. Plant Patents for Kentucky Bluegrass cultivars, nearly 60 Plant Protection certificates (USDA patent-like protection for sexually reproduced plants), and numerous plant registrations. His development of intraspecific hybrids of Kentucky Bluegrass led to the release of cultivars with resistance to devastating Bluegrass diseases such as striped smut, leaf rust, crown rot and powdery mildew. Other research led to the breeding of fescues and rye grasses which possess greater resistance to insect, disease and drought, better plant vigor, and wider adaptability to growing conditions. For these accomplishments, Funk was the recipient of the U.S. Secretary of Agriculture's Distinguished Service Award for Scientific Research, the nation's most prestigious award in agricultural research in 1990. In 1995, Funk initiated a program in perennial nut tree breeding and genetics and later founded the non-profit foundation "Improving Perennial Plants for Food and Bioenergy".

Stanley N. Cohen (1935–). He obtained a B.S. degree at Rutgers University then a medical degree at the University of Pennsylvania in 1960. After his residency, Cohen accepted a position at Stanford University's medical school in 1968 and began experimenting with bacterial plasmids. Cohen surmised during the course of his research that if DNA could be first introduced into plasmids and then transformed into bacteria then large quantities of introduced DNA could be produced by natural mechanisms of replication. Cohen sat in on a talk by **Herbert Boyer** in 1972 about how a restriction enzyme, *Eco*RI, generated "sticky ends" that could be re-annealed to possibly produce recombinant DNA molecules. A group including Cohen and Boyer met later that night and discussed various ways they could collaborate. Recombinant DNA technology was born on a deli napkin. Cohen and Boyer eventually patented their technique—one of the first biotech patents granted, and one of the most lucrative of recent history. Among the many awards Cohen has received in recognition of his enormous accomplishments in biology are the Albert Lasker Award for Basic Medical Research, the Wolf Prize in Medicine, the U.S. National Medal of Science for Biological Sciences (with Herbert Boyer), and the National Medal of Technology and Innovation.

Herbert Boyer (1936–). Boyer was born in Pennsylvania and attended the University of Pennsylvania and Yale. In 1966 Boyer accepted a faculty position at the University of California, San Francisco where he conducted pioneering research on restriction enzymes in the human enteric bacteria *Escherichia coli*. Boyer's discovery of the uneven cutting of DNA by *Eco*RI was a pivotal step towards achieving chimeric nucleic acids. The chance meeting with **Cohen** was described earlier along with the birth of the concept of recombinant DNA. Boyer and a group of investors formed Genentech, Inc. in 1975, a company that continues to be one of the largest biotechnology organizations in the world. Among many lifetime achievements, Boyer was awarded the U.S. National Medal of Science in 1996 (with Cohen).

Ananda Mohan Chakrabarty (1938–). Chakrobarty was a microbiologist working for General Electric during the 1970s when he developed a bacterium capable of breaking down crude oil that he proposed to use to treat accidental oil spills. He requested a patent for the new strain of the bacterium in the U.S. but was turned down by a patent examiner who believed that living organisms were not patentable. The Patent Office Board of Appeals agreed with the original decision but the Court of Customs and Patent Appeals overturned the case in Chakrabarty's favor, writing that *"the fact that micro-organisms are alive is without legal significance for purposes of the patent law."* Sidney A. Diamond, Commissioner of Patents and Trademarks, appealed this decision to the Supreme Court and the case was decided on June 16, 1980 that Chakrobarty's invention was indeed patentable. In writing for the majority Chief Justice Warren Burger cited the congressional report accompanying the 1952 Patent Act that Congress intended statutory subject matter to include "anything under the sun that is made by man". After *Diamond v. Chakrobarty*, the biotech industry grew phenomenally, not a coincidence. With the help of the Supreme Court's *Diamond v. Chakrobarty* decision and the "Bayh-Dole Act", the biotech industry sky-rocketed. Today, there are more than 1300 biotechnology companies in the U.S. alone. This legal decision paved the way for the extension of patent law for the protection of plant and animal genotypes that imparted human utility. Chakrabarty was later appointed as a distinguished Professor at the University of Chicago and went on to garner many prestigious career awards, and continues to serve the U.N., industry and nations as an international expert on the application and proprietary protection of biological inventions (Fig. 1.6).

Kary Mullis (1944–). While Mullis was working for the Cetus Corporation as a chemist he had the idea to use a pair of primers to bracket the desired DNA sequence and to copy it using DNA polymerase; a technique which would allow rapid amplification of a small strand of DNA and become a standard procedure in molecular biology labs. He succeeded in demonstrating this invention, called "Polymerase Chain Reaction" (PCR) in December 1983 and he garnered broad patent protection for applications of PCR in biology. For this discovery, Mullis was awarded the Nobel Prize for Chemistry in 1993. The invention of PCR paved the way for all the genomics and marker-assisted plant breeding applications that would follow in the 1990s to the present.

Kenneth J. Kasha (1945–). He attended the University of Alberta for B.S. and M.S. degrees then earned a Ph.D. at the University of Minnesota in plant genetics. Subsequently, he accepted a faculty position at the University of Guelph where he remained for the rest of his career. Kasha developed the technique of haploid development by interspecific hybridization, and later furthered the frontiers of plant microspore culture. He received a number of prestigious awards for his pioneering research in establishing haploid techniques to reduce the time required for barley breeding time by up to 50%. By 1996, at least 55 new doubled-haploid cultivars had been released around the world using this technique developed by Kasha and colleague **Kuo-Nan Kao**. Kasha was the recipient of the Ernest Manning award for outstanding innovation by a Canadian (1983), was elected to the Royal Society of Canada (1990), was awarded the Genetics Society of Canada award of excellence (1994) and became an Officer of the Order of Canada (1994).

FIG. 1.6 (A) Stanley J. Peloquin; (B) Norman W. Simmonds; (C) Cyril Reed Funk; (D) Stanley M. Cohen; (E) Herbert Boyer; (F) Ananda Mohan Chakrabarty. *(A) From Palta, J.P., 1994. Am. Potato J. 71, 485. https://doi.org/10.1007/BF02849102; https://rd.springer.com/article/10.1007/ BF02849102; Used with permission from the Board of Regents of the University of Wisconsin. (B) From William Spoor Frederick England, Dedication: Normal Willison Simmonds Plant Breeder, Teacher, Administrator. In: Janick, J. (Ed.), Plant Breeding Reviews, vol. 20. John Wiley and Sons, 2000. (C) From Rutgers University, https://sebsnjaesnews.rutgers.edu/2012/10/in-memoriam-c-reed-funk-1928-2012/. (D) From The Lasker Foundation, http:// www.laskerfoundation.org/awards/show/cloning-genes-by-recombinant-dna-technology/. (E) From https://en.wikipedia.org/wiki/Herbert_Boyer#/me-dia/File:Herbert_Boyer_HD2005_Winthrop_Sears_Medal.JPG. (F) From University of Illinois Champaign-Urbana, http://microbiology.uic.edu/cms/ One1fc1-2.html?portalId=506244&pageId=30177682.*

Robert T. Fraley (1953–). He grew up on a 350-acre soybean and corn farm in Illinois. Fraley was at UCSF during a 2-year postdoctoral appointment while **Boyer** and others were reporting the first gene-splicing successes in bacteria. While Boyer and most others in the tiny field at the time viewed biotechnology as a novel way to make human medicines, Fraley was curious about its applications to agriculture. In 1980, he was hired as a Research Scientist at Monsanto Corporation. Fraley and colleague **Robert Horsch** launched Monsanto's biotechnology program in the early 1980s. At the time, Monsanto was still primarily a maker of industrial chemicals and many, even some inside the company, saw agricultural research as folly. The first transformed eukaryotic organism, a petunia, was successfully developed by Fraley and Horsch in 1984. Both are still senior executives at Monsanto Corporation. U.S. President Bill Clinton awarded both Fraley and Horsch with the National Medal of Technology in recognition of this feat in 1999.

Stephen D. Tanksley (1954–). A plant geneticist, Tanksley received his Ph.D. in 1979 under the tutelage of **Charles M. Rick** at the University of California, Davis. Following a short stint at New Mexico State University, Tanksley moved to Cornell University where he remained until 2011. Early in his career, Tanksley added many new isozyme markers to the tomato linkage map, then later pioneered the applications of RFLPs, RAPDs, and other molecular marker systems in marker-assisted selection schemes to improve the efficiency and effectiveness of plant breeding schemes such as backcross. Tanksley was among the first scientists to verify the validity and utility of molecular markers for the study and breeding of traits conditioned by quantitative trait loci (QTL). He also developed the use of molecular markers as predictive benchmarks to ascertain the effects of plant breeding operations, e.g., selection and controlled mating at the level of genomic organization. Further, Tanksley was among the first to use genomics as a tool to study evolution and domestication of plant species. He was appointed to the U.S. National Academy of Sciences in 1995 and has garnered many prestigious awards including the Wolf Prize in Agriculture, the Martin Gibbs Medal, and the Japan Prize (Fig. 1.7).

FIG. 1.7 (A) Kary Mullis; (B) Kenneth J. Kasha; (C) Robert T. Fraley; (D) Stephen D. Tanksley. *(A) From Gairdner Foundation https://gairdner. org/award_winners/kary-b-mullis/. (B) Used with permission from the University of Guelph. (C) Fraley, R.T., 2003. Improving the nutritional quality of plants. In: Vasil I.K. (Eds.), Plant Biotechnology 2002 and Beyond. Springer, Dordrecht. (D) From The Japan Prize Foundation, http://www.japanprize.jp/en/ prize_prof_2016_tanksley.html.*

References

Borojevic, S., 1990. Principles and Methods of Plant Breeding. Elsevier, Amsterdam (NETH), pp. 13–17.

Duvick, D.N., 2002. Theory, empiricism, and intuition in professional plant breeding. In: Cleveland, D.A., Soleri, D. (Eds.), Farmers, Scientists, and Plant Breeding. CABI Publ, Cambridge, MA, pp. 189–211.

Fussell, G.E., 1966. Farming Techniques From Prehistoric To Modern Times. Pergamon Press, New York, NY, p. 181.

Heiser Jr., C.B., 1990. Seed To Civilization: The Story of Food. Harvard Univ. Press, Cambridge, MA, pp. 1–13.

Jensen, N.F., 1988. Plant Breeding Methodology. John Wiley & Sons, New York, NY, 676 pp.

MacKey, J., 1963. Autogamous plant breeding based on already highbred material. In: Akerberg, E., Hagberg, A. (Eds.), Recent Plant Breeding Research, Svalof, 1946–1961. John Wiley, New York, NY.

Mangelsdorf, P.C., 1975. Donald Forsha Jones. Biographical Memoirs. National Academy of Science of the United States of America, Washington, DC, pp. 135–156.

Poincelot, R.P., Horne, J., Mcdermott, M., 2001. The Next Green Revolution: Essential Steps to a Healthy, Sustainable Agriculture, 1st ed. CRC Press, Boca Raton, FL, 312 pp.

Sauer, C.O., 1969. Spades, Hearths, and Seeds: The Domestication of Animals and Foodstuffs. MIT Press, Cambridge, MA, 74–83.

Vartanyan, G., 2006. Contemporary concepts of the geological environment: basic features, structure, and system of links. In: Zekster, I.S. (Ed.), Geology and Ecosystems. Springer, Berlin (GER), pp. 1–8.

Vasey, D.E., 1992. An Ecological History of Agriculture, 10,000 BC–10,000 AD. Iowa State University Press, Ames, IA, pp. 23–43.

2

The Context of Plant Breeding

INTRODUCTION

The successful plant breeder must integrate information from many sources and many levels. Fundamentally, plant breeding is a human endeavor (it is not clear whether any other species on Earth are so engaged) wherein naturally occurring genetic variability is used to craft a population that is useful. Utility is gauged by acres of production, volume of seeds or propagules sold, the value of the product in trade, and the quality of life in terms of health improvement and aesthetic enjoyment. For now, the plant breeder must rely mainly on the phenotype to make selections on desirable genotypes. The plant breeder must be aware of the phylogeny of the targeted species and the degree and location of extant genetic variability. Thorough knowledge of how phenotype relates to genotype through reproduction and the developmental process is essential. An appreciation of how the individual relates to populations and ecosystems is extremely important. Because plant breeding is a human endeavor and produces something of economic value, overarching social and political factors must be considered in all decisions about the objective, process, and how the product will be disseminated to users. This chapter will address the issues that the plant breeder must contend with conceptually to be successful.

EVOLUTION AND SPECIATION

Plant breeding is an extension of the evolutionary processes articulated by Darwin (1859). The now traditional view is that life originated in earth's primordial soup, and has been perpetually changing in a directional manner ever since. This textbook will be presented within a framework of facts that are substantiated purely by science, including evolution. This in no way obviates the possibility, or even probability, that phenomena are influenced by forces that we cannot currently perceive, or can never perceive, with humanoid scientific methods and innate limitations of the human senses. Factors that emanate beyond the realm of science will not be addressed in this book. Readers and students who cling to beliefs that include those not supported by science are urged to suspend disbelief in evolution and embrace the notion that humans can transform plants and animals into useful inventions.

Generally speaking, organisms, including plants, have evolved in the direction of more complexity, although conspicuous exceptions have recently been discovered, for example prions and parasitic RNAs (Weber, 1998). As evolution proceeds, the legacy of genetic lineage sometimes persists, the so-called "living fossils". Seed ferns, horseshoe crabs, and coelacanth fishes are but a few of the hundreds of examples. Such species must have encountered selective pressures that perpetuated themselves over evolutionary time in a form that is similar to that captured in the fossil record. In most cases, the functional intermediates that led to flora and fauna of a given day do not persist, and can only be surmised from the fossil record. Humans are exemplary of this observation; our direct links to other primates were lost as the species evolved. The debate continues on our origins as new archeological finds unearth skeletal remains of ever more human-like ape species. No evidence can be found that humans such as we existed more than one million years ago, a veritable snap in evolutionary time (Briggs and Walters, 1997).

A fundamental tenet is that life perpetuates itself from generation to generation, but not with absolute fidelity. Changes in the DNA code creep in all the time: mutations, transposon insertions, chromosome rearrangements, recombinants, segregants. In the short term, the differences between organisms may be barely perceivable, but over time geographical isolation of subpopulations and differences in environment results in permanent genetic changes and reproductive isolation. The extent to which natural selection as opposed to genetic drift are involved in this process is not known with any certainty (Salthe, 1998).

Unlike modern agriculture, plants in the wild live in a highly heterogeneous environment and in competition with many other species for available resources (Coates and Byrne, 2005). Anyone who has done field experiments has been frustrated with encroachment by weeds. In our anthropomorphic view weeds are the interlopers and must be eradicated for our success to be realized. The same field left to fallow is, over many years, colonized by an array of species characteristic of that global coordinate, soil type, and adjacent vegetation. This new ecosystem will usually be quite stable from year to year, and may not include the aforementioned weed species, that are specialists in colonizing disturbed or unstable ecosystems (Briggs and Walters, 1997).

From single celled organisms, multicellular forms arose, phylogeny a reflection of evolutionary history: bryophytes to mosses to ferns to seed ferns to gymnosperms to angiosperms (de Queiros, 1999). Terrestrial plants evolved from aquatic habitats and progenitor species. In primitive eukaryotes, either the gametophyte or sporophyte or both may be conspicuous. Within Gymnospermae and Angiospermae, the conspicuous gametophyte is reduced to almost nothing while the sporophyte becomes the predominant life cycle phase (Weber, 1998). Within the angiosperms, fossil evidence supports the progressive reduction in floral units over evolutionary time, and subsequent fusion into plural structures (Willis and McElwain, 2002). It is presumed that these macro trends were driven to a large extent by the forces of natural selection.

Many or even most of the 300,000 or more described plant species on earth has been scrutinized at one time or another for human utility (Hammond, 1995; Cracraft, 2002; Henry, 2005). From that enormous range of diversity, we now consume over 90% of all calories from only seven plant species (Simmonds, 1979). Those species happen to be extremely adaptable, and are produced successfully in a very broad range of the planet's climates and solar exposures. Consumer demand for more diversity of food alternatives has not lead to more domesticated crop species but instead an endless palette of processed admixtures of ingredients from existing domesticates, milled and seasoned to the desired point of differentiation. It is much easier to manufacture diversity that way than to develop and introduce entirely new crops. The scrap heap is littered with the carcasses of new crop species touted as the next coming at your local supermarket. There are a few success stories as well. Consumers are wary of new crops, and generally don't trust scientists (or plant breeders) in white lab coats.

Ancillary technologies, such as transportation and communication, have played a major role in the globalization of food. Most crop species are now produced in lands distant from whence they originated. Soybeans from China are grown in South and North America; Tomatoes, peppers, and potatoes from Peru are produced in Europe and Asia; Rice from India cultivated in China and Africa, Coffee from Africa reared in South and Central America. What would Italian food be without the tomato, but the first seeds only arrived there in the 17th century (Frankel et al., 1995). Instability and uncertainty in the costs of energy for transportation coupled with dwindling fossil fuels, however, are spawning renewed interest in locally sustainable food systems.

Governments and policies also play a role in the diet. In the U.S., the executive branch Department of Health and Human Services develops and practices enforcement on dietary guidelines. The U.S. Food and Drug Administration oversees the safety of the food supply, banning entire crop species due to health concerns, for example canola (low erucic acid rapeseed) oil. Before 1985 the sale of canola was not permitted in the U.S. because the oil contained what was considered to be unhealthy levels of erucic acid, a C-22 fatty acid associated with cardiac dysfunction in rodents. Plant breeders successfully reduced the content of erucic acid, and canola oil was added to the list of GRAS (generally recognized as safe) foods. Canola is now the second largest oilseed crop in the USA behind soybean and, and currently the fastest growing in terms of per capita consumption and land area (Anon, 2015).

SYSTEMATICS AND NOMENCLATURE

Ideally, the classification of organisms is an accurate reflection of their evolutionary interrelationships (Semple and Steel, 2003). The field of taxonomy has matured into systematics, wherein many factors other than floral and foliar morphology are considered in deciding how to classify organisms. Examples of convergent evolution abound. The field of systematics is dynamic and has been enriched dramatically by recent breakthroughs in molecular genetics (Schuh, 2000). Since organisms became different by virtue of the accumulation of genetic differences, the direct comparison of DNA sequences has become a primary tool in determining how they are related. The dendrogram and parsimony statistical analyses have lent a quantitative aspect to a heretofore subjective field (Simpson, 2006).

The fundamental system of nomenclature has been remarkably durable, however, since it was first introduced by Carolus Linnaeus in 1762. The classification of higher plants within this framework, however, has changed dramatically over the past 100 years. The species is the primary unit, consisting of all individuals that share a common gene pool (Campbell, 1993). This has also remained a basic tenet of plant breeding, the main source of genetic variability, or germplasm, is all organisms of the same species (Fig. 2.1). Going up the ladder from species to kingdom, the units of sameness or difference become relatively subjective and variable. What distinguishes certain families within one class might not distinguish families in another. Of concern to the plant breeder are aspects of similarity, "is it genetically accessible?", and difference, "is the trait not available at the level of species?"

The plant breeder often contemplates the capture of desirable attributes from different taxa, usually an esoteric exercise. "How could I move the salt tolerance of the mangrove into my wheat program?" or "What about nitrogen fixation in corn?". Such attributes are inevitably determined by a large number of genes that interact in specific and subtle ways, and in a strictly defined context. Simply lifting the trait up from one species and plopping it down into another and expecting the same result is unrealistic.

The notion of "wide crosses" has probably existed ever since humans realized that they could control the outcome of seed parentage; since the early phases of agriculture. Over time, it was discovered what worked and what didn't, and the curious outcomes that fell somewhere in between. The reasons for this will be discussed in the "Plant Reproduction" section below, but it is instructive to turn to animals to illustrate the phenomena. The *Animalia* Order Perissodactyla includes horse and camel-like beasts. It is well known that horses and donkeys can be hybridized, resulting in a mule. The two distinct species have different chromosome numbers (see Chapter 3 for coverage of cytogenetics) and the hybrid is completely sterile, producing no functional gametes. Zebras can also be hybridized with donkeys, and the result is intermediate in appearance (the so-called "zedonk"), and is at least partially fertile.

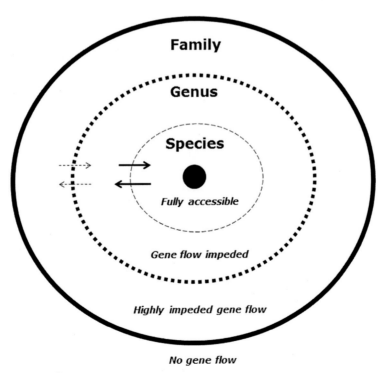

FIG. 2.1 Phylogenetic barriers to gene flow.

Perhaps it would be possible to find desirable genes in wild zebras that could be used for breeding better horses? But it is not possible to interbreed horses with camels or llamas. Conceptually, horse-like and camel-like species exist outside the concentric circles in Fig. 2.1.

Zebras, horses, and donkeys all belong to the genus *Equus*. Thus, while the species all have distinct gene pools, it is possible to find biological bridges between them, and to move, or "introgress", genes from one pool to another, via interspecific hybridization. Plants are considerably more tolerant of genetic imbalances such as chromosome number changes and mutations than are animals, and plant breeders have taken advantage of this abundantly, particularly for economic traits governed by single genes, such as resistance to disease pathogens. Periodically, reports of successful intergeneric hybridizations appear, but the veracity of the classification comes into question, or the viability of the offspring for subsequent breeding efforts is diminished. The utilization of a normal sexual reproduction system to achieve successful crosses beyond the level of the genus (family or higher) is virtually impossible.

The pathway from the species level to the individual plant is likewise replete with subjectivity (de Queiros, 1999). Systematic "lumpers" and "splitters" have clashed continuously on the existence of subspecies. Pathologists have often resorted to classification systems based on virulence groups, such as "formae specialis" (e.g., forms of a given species that attack only certain hosts) or races (forms of a given species that attack only certain host genotypes). Plant systematists often distinguish between geographical forms of a species that are separated spatially (e.g., eastern vs. western, northern vs. southern).

At the level of genetic distinctions between populations within species, a free-for-all breaks out where such populations often become economic entities. Groups of academics or quasi-governmental officials have interceded to bring order to species that also fall into commodity groups, such as corn, soybeans, wheat, and cotton. Cute anthropomorphic names are attached to these populations to distinguish them in the marketplace such as 'Supernova', 'Hyperspace', and 'Galaxy' (listed with apologies to bona fide trademark owners). If a breeder seeks to sell seed of a major crop commodity, he/she inevitably will butt heads with a seed certification organization that polices the genetic and overall purity of seed on behalf of farmers. Plant species that command a small piece of the pie are associated with progressively less oversight, usually none at all, except for the players that are in direct competition, or by the farmers themselves (Chase, 2005).

PLANT GROWTH AND DEVELOPMENT

While striving to understand what makes a particular crop species tick, plant breeders must relate to the design and phenology of organisms on many levels. The subject of partitioning total phenotypic variation will be addressed in a later chapter. On the most simple of terms, organisms develop both intrinsically and extrinsically. Intrinsically according to the genes and control DNA sequences inherited from parents, and extrinsically according to interaction with a multitude of factors abstracted as the "environment" (Cronk, 2002). Plant physiologists are fascinated with the responses plants exhibit to fluctuating environmental cues, such as light, temperature, salt, growth regulators. Thus, it is not surprising that we can make genetically identical plants look and behave quite differently by virtue of the environmental factors they are exposed to. Grow a flat of vegetable transplants and put half in the field, leaving the other half in the confined cells and differences will very quickly become evident.

At the subcellular level, plant breeders must constantly appreciate the fact that certain organelles, plastids and mitochondria, contain expressed genomes that are inherited maternally. While the vast majority of genes are localized in the eukaryotic nucleus many of critical importance have been shown to be maternally inherited and some have even been mapped in organelle genomes. Traits such as anther and pollen development (abortion, leading to male sterility), herbicide resistance, energy transduction, and photosynthetic efficiency are among prominent examples. Organelle genomes generally exhibit mechanisms of replication and transcription/translation that are identical to those of prokaryotes, consistent with the bacterial capture theory of organelle origin (Lang and Burger, 2012). Nuclear and organelle gene products can interact in the cytoplasm to condition phenotypes, for example nuclear restoration of fertility in cytoplasmic male sterility (Havey, 2004). Further, over long periods of evolutionary time, DNA sequences have been discovered to translocate among the nucleus and organelles, the mechanism of which is still unknown (Cullis et al., 2009).

Cells organize into tissues, and tissues into organs. The concert of developing organs and tissues determines the overall phenotype of a given individual plant. Plants enlarge through cell division and expansion, and individual cells communicate and exchange with each other via plasmodesmata (Epel, 1994). Cell division is generally presumed to be mitotic, or clonal. Instances of specific genetic changes during somatic development, however, are well documented. Certain animal organs contain polyploidy cells, for example mammalian liver and arthropod salivary gland. In some tissues, and under the influence of stress factors, transposable DNA elements are activated that precipitate somatic mutations (Pritham, 2009). When plant tissues are removed from the developmental context and cultured in vitro, all bets are off.

Plant cells are generally impervious due to the thick, rigid cell walls that encase them, except for the plasmodesmata that connect them. This feature imparts turgor to plants and permits them to assume structure on land, in deference to gravity, wind, negative water potential. Chemical signals are induced, synthesized, and transported within plants that mitigate growth, transpiration, and developmental processes (Srivastava, 2002). In some cases, the plant breeder/geneticist wishes to introduce genetic material directly into a plant cell, and the cell wall presents a serious barrier to this objective. Not to be deterred, scientists strip away the cell walls with microbial cellulases, hemicellulases and pectinases, the genetic material is blasted through the cell wall with ballistics, or the genes are injected using a biological syringe, better known as *Agrobacterium* sp.

The plant is a complex, highly efficient energy transducer, changing solar into chemical energy then channeling it into the packaging of genes for future generations (i.e., reproduction). The prospective utility of plants for humans generally stems from a subversion of the energy flow into a desired pool, such as grain endosperm, seed oil or protein, enlarged taproot, fruit, flower, foliage, or cellulose fibers. The plant breeder must be ever cognizant of the realities of energy balance sheet and devised selection strategies. Selecting for the higher yield of one component alone usually comes at the expense of another; it is uncoupling and rebalancing of energy flow that leads to true breakthroughs.

We generally see only the aerial portions of the plant, and make judgments about the total potential based on those observations alone. The root system is nearly impossible to visualize in a complete, intact, state but it is every bit as important to overall plant performance as are the stems, leaves, flowers. Breeders of woody perennial plant species have known this for decades (Beckman and Lang, 2003). Soil type, fertility, water and micronutrient availability, and pH are all factors that plants react to and have range preferences for. Further, the ancient art of grafting has been elevated to commercial status, and most commercial tree fruit plantings in the U.S. are grafted onto rootstocks that dramatically alter, and improve the performance of the aerial parts (scion). Recently, grafting has been successfully adapted to herbaceous vegetables (e.g., tomato, pepper, eggplant, squash, cucumber, watermelon, muskmelon, and other species) as a viable tool for combining synergistic rootstocks and scions (Kubota et al., 2008). Rootstock resistance to soil-borne disease agents is a prominent example.

Cultivated plants were all derived from wild ancestors in native habitats. The metabolic "machinery" of domesticates is conditioned to work best under a specific range of environmental parameters. As was stated earlier, the plant species that humans chose for domestication were generally forgiving and broadly adaptable. The plant phenotype is conditioned by the interaction of genotype and environment. The plants utilized by the plant breeder for genetic improvement are selected based on their phenotype, although technology has made it possible to select directly on genotype or genomic DNA sequence. Thus, the plant breeder must always strive for one of two ideal situations: Elimination of environmental variation or selection purely in the environment in which the crop will be produced. Another way to state the first alternative is *broad adaptation*; the plant behaves similarly across a diverse array of environments. Another way to state the second is *narrow or specific adaptation*: it will have the desired phenotype in a specific place and time of the year, not if produced elsewhere.

The plant breeder is not patient with life cycles. Any given program requires a number of generations to reach the endpoint. For research-adapted species such as *Arabidopsis thaliana*, that has a life cycle of fewer than 30 days, this is not very problematic. But if you are charged with breeding woody perennial species wherein a 10-year life cycle is typical, progress is measured in entire careers, and not years. With an understanding of the factors that accelerate and retard development, it is possible to speed things up. Many economic plant species have retained the mechanisms that couple environmental cues, such as temperature and day length, with reproduction. Thus, it is possible to impose artificial environments to accelerate flowering, and shorten the life cycle (e.g., "seed-to-seed"). Many of the exemplary breeding programs described later take advantage of this, and also use alternating breeding sites in the northern and southern hemispheres or tropics to advance generations. The plant breeder must take steps to ensure that any measures employed strictly to accelerate a program do not inadvertently interpose selection pressures that result in undesirable changes to the population genetic structure.

PLANT REPRODUCTION

All angiosperm and gymnosperm plants that constitute our major economic plant species feature an alternation of haploid (gametophyte) and diploid (sporophyte) generations. With evolutionary time, the sporophyte has remained as the most conspicuous entity in both gymnosperms and angiosperms. Gametophytes are reduced to the point that they are vestigial, and are virtually ignored except for the mating process. While scientific evidence has demonstrated that genetic selection does occasionally exert changes at the level of the microgametophyte (or pollen) and megagametophyte (or egg), the plant breeder is focused exclusively on the sporophyte. Perhaps most importantly, the plant breeder frequently benefits from the ability to store pollen for extended periods of time, using it to make crosses when the optimal timing presents itself.

Organisms are necessarily selfish entities, but are programmed to generate offspring that carry forward their genes in time, a seemingly selfless notion. Sporophytes are also programmed with very limited life spans. This allows evolution to progress in an orderly fashion. Such is the nature of life on Earth. I know some folks who want their children to be just like them. Not similar, but exactly the same, like a clone. While children do resemble their parents, however, they are also different in striking ways, both physically and mentally. Periodically, a truly exceptional individual is born, brilliant, insightful, kind, well balanced, productive. Why wouldn't some natural process kick in to fix that person's genotype, to ensure that more like he or she populated the earth from that point henceforth? But, alas, the children of genius' while usually bright, are often as dull and flawed as the rest of us. The truly gifted individuals are doomed to perish as well.

Allele fixation and perpetuation of fixed genotypes do not generally occur in humans but they do in many plant species during discrete stretches of evolutionary time. These species have jettisoned the mechanisms that force genes to be recombined, usually through floral mechanisms that result in self-pollination. As will be seen in forthcoming chapters, this leads to genetic fixation and, usually, a reduction in the degree of genetic variability (total number of alleles) in a population. In other cases, sexual reproduction is completely abandoned and somatic cells alone are used to produce propagules, also known as asexual propagation or cloning. Evolutionary forces have not been kind, however, to species that forego the long-term benefits of sex for the immediate gratification of habitat exploitation.

Planet Earth is not static; it undergoes constant, profound changes. Continents separate and drift apart; mountain ranges spring up; volcanoes belch out magma and new gases into the atmosphere; fertile plains become ocean depths; rainforests are transformed into deserts; jungles are overrun with ice sheets; and asteroids hurdle at us constantly from deep space, occasionally hitting and wreaking unimaginable biological havoc. With each abiotic shift comes corresponding changes in biotic factors, such as competitors in ecosystems, parasites, and herbivores. The clone or genetically fixed population sits with "clay feet" while the environmental and ecological changes in their midst gradually render them obsolete. Those populations that have dutifully maintained balanced reproductive strategies that continually generate a modicum of new variability can cope with the changes, change themselves, and live on.

The maintenance of genetic variability comes at a high cost. The biological processes that generate and perpetuate genetic variability are less efficient and require more energy inputs than self-pollination and clonal propagation. Species that are predominantly cross pollinated often feature a delicate chemical system to prevent self-pollination at an additional energy cost. Usually, a physical vector is required to physically move gametes about in the population, for example an insect or wind. For every superior individual that is born, another inferior one emerges. Alleles that impart benefits in certain contexts are deleterious in others. And the truly exceptional individual is extremely rare. That is where the plant breeder comes in.

It is probable that domestication has converted predominantly outcrossing into predominantly inbreeding species, especially in seed crops such as grains. The constant selection for yield and broad adaptation culminates in populations with high seed number and individual mass. As was stated, self-pollination is more energy-efficient than is cross pollination, and does not require a pollen vector. This phenomenon points out the high level of genetic variability present in wild plant populations for reproductive traits, the mutability of the phenotypes, and the tolerance for multiple reproductive forms. In buckwheat, a single gene (the "S" locus) conditions flower development into "pin" (gynoecium superior) or "thrum" (androecium superior) configurations, both of which are physical attributes that promote outcrossing (Li et al., 2011).

Following the union of gametes, fertilization, and formation of the zygote, mitotic cell divisions commence and the embryo begins to develop. It is at this juncture that plants have drawn the imaginary line between generations, or "seed to seed". This is more a convenience than a biological paradigm, since the next generation has already begun. In most cases, however, the embryo will develop to a certain degree, then dehydrate and enter into quiescence in the seed. The seed is a package to be opened later, when seasons or environmental conditions are ripe for the completion of the sporophyte to the next sexual cycle and seed. It is also a method used by species to disperse themselves, by wind on the wings of giant appendages, or by animals, in the feces from digested fruit tissues that surround them (Desai, 2004).

Plant breeders commonly use natural mechanisms of reproduction to increase genetically pure populations for commercial purposes, often termed a "seed increase". The same mechanisms can also introduce unwanted genetic variation, such as with cross-pollinated crop species. For example, new varieties of sweet corn with different mutations in starch biosynthesis must be segregated physically during reproduction because endosperm genes will complement to enable starch to be synthesized (Tracy, 1997). This obviates the intended phenotype, sweetness, returning the seed to the original state: storage of starch as chemical energy for sustenance of the germinating seedling.

Not surprisingly, plant breeders also use seeds as convenient means of genetic storage, since they are resistant to the ravages of time. Seeds are also relatively cheap and easy to store for long periods of time. Thousands of seeds can usually be stored for protracted periods in small envelopes. If ideal storage conditions can be achieved and maintained, the life span of seeds can be prolonged. While such conditions vary with species, cool temperatures (0–10 °C) and low relative humidity (15–30%) is usually conducive to longer viability spans (Roos, 1989). When maintained under such conditions, cotton seeds have been demonstrated to retain viability and ability to germinate for over 20 years (Janick et al., 1974). Seeds are also convenient ways to synchronize plant populations, and for the delivery of treatments such as mutagens. "Earliness" is a trait of some economic importance, since time is money, and is usually measured as the period from seed germination to the point that the targeted product is harvested.

As was stated earlier, not all plants use sexual reproduction exclusively for the generation and management of genetic variability. Many species have co-opted floral biology and gametes or diverted somatic tissues and organs to minimize new genetic variability, in anthropomorphic terms, for an ultra-short termed evolutionary strategy and resorted to cloning. A broad spectrum of strategies was developed by plant species over evolutionary time that were aimed at achieving asexual reproduction. Cloning can be highly desirable to the plant breeder, who creates a superior genotype that is not sexually sustainable.

The ability to propagate ultimately becomes an important consideration in any plant breeding program. This is where much of the debate on breeding objectives occurs. Just because a certain genotype can be made doesn't mean that it is economically feasible or socially advisable to do so. One example would be carrot (*Daucus carota*), where micropropagation, or asexual, clonal propagation on a small (cell or tissue) scale, is possible. Why not develop the perfect genotype, then clone it? It would be very expensive to do that, however, and value of carrots at this juncture does not support the requisite investment.

Hybrid tomato (*Solanum lycopersicum*) lies at the other end of the feasibility spectrum. The only effective way to make hybrid tomato seed is by hand crossing, a very labor-intensive operation. By training and contracting workers in China and India, where hand labor costs are relatively low, it is possible to produce cost-effective seeds of hybrid tomato varieties. The economics are favorable in tomato since each individual hand-pollination can produce hundred seeds. Also, the value of the crop is sufficient to support the additional costs.

POPULATION BIOLOGY AND ECOLOGY

It is useful and mostly correct for plant breeders to regard varieties or cultivars as populations (Chapter 4). It is more correct to refer to them as artificial populations, for most of the present day varieties and cultivars are far from being natural populations. A population is a group of individual organisms, members of a given species that is relatively stable over time. The behavior of a population is determined by the genotypes of constituent individuals in concert with the environment. Individuals interact, however, in unpredictable ways and population genetics is the study of the elements that govern the additive and non-additive heritable and non-heritable aspects of population dynamics.

All economically-valuable plant populations have been subjected to artificial selection and mating schemes that are not found, as such, under natural conditions. Consequently, these artificial populations all have genetic structures that are unstable and unsustainable unless human energy is invested into their maintenance. At the most extreme end of the spectrum, such populations are a genetic monoculture, all constituent individuals of the same genotype. Examples of genetic monocultures include F_1 hybrids and pure lines. Such a population structure carries the theoretical requirement of absolute uniformity of environmental conditions to maximize performance, a condition that necessitates extensive human intervention.

Mankind has learned first-hand of the dire consequences of profound or continuous genetic monocultures. The Southern Corn Leaf Blight Epidemic of 1970 is featured prominently in every textbook as an example of genetic vulnerability (Simmonds, 1979; Oldfield, 1989). Although many different corn hybrids were grown in 1970 and preceding years, most were produced using T-CMS, and the genes responsible for susceptibility to the pathogen *Cochliobolus heterostroplus* are mitochondrial. It is not the only example of the potentially disastrous consequences of genetic uniformity—others include the Irish potato famine in the mid-1800s, the 1917 wheat stem rust epidemic, and the 1943 Bengal Indian rice brown spot famine (Borojevic, 1990).

The endless game of one-upmanship known as vertical disease resistance is another example. The noble plant breeder develops a new disease-resistant crop variety based on a simply-inherited resistance gene. A mutation occurs in the disease pathogen that results in a new virulent form (see Chapter 19 for an explanation of the gene-for-gene theory). The plant breeder then finds a new resistance gene, and the pathogen overcomes that one as well. And on and on it goes. The resistance monoculture presents the pathogen with a very strong selective pressure for new mutations conferring virulence that overcomes the original resistance gene.

Competing strategies have been developed to strike a compromise between the requirement for phenotypic uniformity and genetic variability, such as mixtures, synthetics, multilines, pyramiding resistance genes, etc. The extremes of population phenotypic variability are illustrated in Figs. 2.2 and 2.3, comparisons of genetically variable (landrace) vs. genetically monolithic populations. Few examples have been successful in persuading agriculture to move away from genetically monolithic populations. Most contemporary plant breeding strategies are, in fact, predicated on the development of genetically monolithic populations. These strategies will be covered in this book with the caveat that monocultures may ultimately be found to be ecologically and environmentally unsustainable.

The term "monoculture" is used more commonly in an ecological context. Environments or habitats on earth are occupied by a myriad of organisms of varying complexities, from viroids and bacteria, to saprophytic fungi, algae, lichens, mosses, vascular plants, worms, arthropods, birds, reptiles, mammals, etc. The naïve view of agriculture is that it consists of the crop plant produced on an environmental substrate, consisting of soil matrix, water, and nutrients, and driven by energy from the sun. Huge industries have been built to provide farmers with weapons to do battle with the organisms who defy the monoculture. They are the weeds, the pests, the pathogens, the animal interlopers. Every weed scientist, entomologist, plant pathologist I have ever met is pleased that the demand for their services will never diminish as long as we derive our food, fiber, and aesthetic products from monolithic agriculture. The crop monoculture in an open field ecosystem is like a piece of cake in a room of unattended children; what fool could blame them for breaking off a piece?

In contrast, natural ecosystems are fascinating studies of subtle interactions of biotic and abiotic factors, so much so that they are often viewed in terms of the flow of energy or fundamental elements such as carbon or nitrogen. Some constituent organisms of the ecosystem fix atmospheric carbon and release oxygen while others fix atmospheric nitrogen. These organisms are consumed and degraded over and over. No particular organism comes to predominate because all components of the ecosystem are dependent on each other for survival. Ecosystems in rarified environments, such as the arctic, tend to consist of fewer species than do those in the tropics. Species within desert ecosystems have sometimes been discovered to emit compounds from their roots into the soil that inhibit other species from growing within a prescribed distance, called allelopathy (Chou, 2006).

The organic food movement in the U.S. touts engineered biodiversity as a defensive strategy against attack from unwanted intruders. Intercropping two or more species is an example of this strategy that has been developed also in cultures where arable land is scarce. The theory is that diversity somehow foils the pest or pathogen, impeding its reproduction, feeding frenzy, much like a natural ecosystem would. In reality, any artificial habitat is not sustainable, and will require energy inputs to be channeled in the desired direction.

FIG. 2.2　A comparison of two populations of wheat at grain maturity. Panel (A) depicts a modern genetic monolith cultivar and (B) is a genetically-variable landrace from central Asia. *(A) From https://en.wikipedia.org/wiki/Common_wheat#/media/File:A_field_of_wheat.JPG.*

(A)   (B)

FIG. 2.3   A comparison of two populations of corn at tasseling. Panel (A) depicts a modern genetic F$_1$ hybrid monolith cultivar and (B) is a genetically-variable open-pollinated landrace from southern Mexico. *(A) From https://commons.wikimedia.org/wiki/File:Agriculture_-_Corn_Field_ (45691292921).jpg.*

What sort of population will the plant breeder of the future be targeting? The paradigms of the 20th century are gradually changing as the development of new agrichemicals to manage weeds, arthropods, and disease agents becomes cost-prohibitive. Horticultural crop species will be the first to undergo the paradigm shift since volumes and acreages are smaller than those of agronomic plant species.

SOCIAL AND POLITICAL SCIENCES

As human cultures have become more sociologically complex and imbued with technology so have the corresponding methods to breed and produce crop plants. When plant breeders were the farmers, plants and seeds thereof were regarded as global resources, free for all the planet's inhabitants and passed along from generation to generation. Farmers retained a portion of each year's harvest to glean seeds for the next year's crop, usually selecting progeny of the best individuals in the process. This cycle repeated itself for thousands of years (Cotton, 1996) and ultimately led to crop domestication.

The first documented seed company, Vilmorin, was established in France in 1743 to produce and sell seeds to urban residents of Paris and to be the chief suppliers of seeds to King Louis XV. The proprietor Louis de Vilmorin was also instrumental in early demonstrations of plant breeding methodologies, employing mass selection to successfully increase sugar content of sugar beets during the mid-19th century (see Chapter 1). The concept of selling seeds gained momentum in the late 1800s as the industrial revolution made it possible for farmers to work more land, and buying seed saved time for more important pursuits (Pottier, 1999). The discovery of the principles of genetics led to the development and successful commercialization of hybrid corn in the 1930s, the first plant populations that farmers could not reproduce on their own. From that point on, seeds have ceased to be regarded as a planetary resource, and became instead a product in a marketplace and germplasm became subject to property laws.

Breeders fought for and were granted laws that protected the populations they developed from piracy, the same as any other invention. These laws preserved the historical rights of farmers to produce their own seeds, but forbade them from selling seed to other farmers. Not satisfied, seed companies, engulfed progressively by mammoth multinational corporations and made more accountable to near-term profits, turned to contract law. Buying the seed was an implied contract by the farmer to refrain from using it to produce more seed for planting. All of this was supposed to stimulate more investment into new product development, and most believe that it has worked. As the plant populations become more technologically advanced, and expensive, however, the trend toward farm consolidation was fostered. The varieties that are currently produced in mainstream agriculture are not sold in the small mom and pop stores, or the big home improvement chain stores either. They are sold in large quantities by large dealers from large companies directly to large farmers (Dodds, 2003).

The progression of economic plant populations from family heirloom to market commodity has had a coattail effect on the free flow of germplasm. Plants do not situate themselves according to political boundaries, and many of the "Centers of Origin" (Chapter 7) of our important crop species are governed by those who have historically had little appreciation for their native wild vegetation. Developed countries such as the U.S. took advantage of this indifference with the establishment of plant exploration in 1898, originally with the Department of Naval Forces, that later became the Department of Agriculture National Plant Germplasm System. Plant biologists were enlightened and intrigued in the 1920s by the theories of Nikolai Vavilov (Chapters 1 and 6), and traveled about the world collecting seeds of the wild progenitors and relatives of crop species with little impediments and few questions from the endemic governments (White et al., 1989).

By 1980, developing countries had grown suspicious of incessant requests for exploration on their native soil, and imputed values for the genes being extracted for commercial uses. They aligned with the World Bank and formed what came to be known as the "Consultative Agreement" that mandated exchange agreements as a condition for the removal of germplasm from points of origin. Since then, the situation has become much more complicated, with governments and so-called "non-government organizations" requiring a specific and strict paper trail to avert risk and limit future potential liability. The free flow of germplasm within and among nations has been crimped enormously, probably permanently so. With persistence and patience, however, germplasm is still acquired and exchanged but the pace of plant breeding is negatively impacted (Pistorius and Van Wijk, 1999).

As certain human cultures progressed, individuals were inevitably distressed with the horrible reality that the earth produces a bounty of food, more than enough for everyone, but that millions still starve to death. Simply giving food away hadn't worked, and was mostly impractical due to enormous distribution costs and cultural barriers. After World War II, the Rockefeller Foundation took on the world hunger challenge, and embraced the concept that people needed the tools to feed themselves, tools that had been developed and were used successfully in the west. One of these tools was modern plant breeding. A consortium of centers was established in developing countries around the globe, each charged with the deployment of technologies to regional food producers that would increase production and close the "hunger gap" (Borlaug, 2007).

"Dream teams" of plant breeders were assembled, and they went to work applying the formula that had proven successful in Europe and U.S.: establish genetically attainable breeding objectives, gather and evaluate germplasm under local conditions (for they knew instinctively that western varieties would not work), apply the principles of genetics to the development of new adapted populations, and produce and disseminate seed of improved varieties to farmers. By 1960, the newly developed varieties were beginning to make an impact, most notably short, stiff-strawed grain crops, such as wheat and rice that made it possible to channel nitrogen and other nutrients directly into seed yield, not lost vegetative growth (Borlaug, 2007). The effort came to be known as the "green revolution", and forever left a positive impression on the field of plant breeding. In 1970, Norman E. Borlaug of CYMMYT (one of the original world centers) was awarded the 1970 Nobel Prize for World Peace in recognition of these landmark achievements (see Chapter 1).

The "World Centers" are still in existence (Chapter 6), and are considered to be major players in certain crop species, mainly grains and tropicals. But did the green revolution have a lasting impact on world hunger? Plant breeding is but a tool, and its effectiveness depends on the practitioner that wields it. Populations increased concomitantly with the expanded food supply, and government support waned as the news media lost interest. Hunger and starvation are still prevalent on our planet, and we persist in our efforts toward effective, sustainable alleviation. In reality, the unequal distribution of calories is mostly a byproduct of human greed, and not the inaccessibility of local farmers to technology (Horne and McDermott, 2001). But more productive plant varieties can certainly have a positive impact, and the need for plant breeding will continue.

Following the breathtaking advances in biological sciences during the mid-20th century, especially the discovery of DNA, recombinant DNA, eukaryotic transformation technologies, and polymerase chain reaction plant breeding appeared to be poised to undergo a major change to accommodate and capitalize on advances. The first hint that trouble might be bubbling under the surface of our social context occurred with backlashes against tests of recombinant organisms in open environments during the late 1980s and early 1990s.

Recombinant plants were approved by the US FDA for human food consumption during this period, but a series of gaffes and missteps while introducing transgenic crops in Western Europe and North America galvanized political opposition to transgenic organisms in foods, drugs, and the environment (Davison, 2010). Opposition groups successfully stymied attempts to integrate new technologies into the fabric of plant and animal genetic improvement programs. Since food costs and availability and quality are not major social issues in developed countries, no overriding need for these technologies has yet been demonstrated, and the negative public perception of transgenic organisms persists (Hilbeck and Andow, 2004). Transgenic plants have found certain niches in agriculture, e.g., herbicide-resistant soybeans, pest-resistant cotton, and other specific uses, but generally outside the arena of mainstream human food products (soy oil contains no transgenic DNA).

The mantra of the future, however, will be long-term sustainability. Plant breeding is a tool that can change plants and animals to suit the wants and desires of humans but must be practiced within the contexts and ranges of social acceptance described above. One major need for the future of humans is the preservation of ancestral genes and genotypes. We should not risk losing what has taken nature 4.5 billion years to produce even if we see no immediate utility. Plant breeders will continue to play a vital role in the preservation of genes, both in collections and in nascent economic plant populations.

References

Anon, 2015. The 2015–2016 Outlook For Fats and Oils. Business Trend Analysts, Inc, North Adams, MA (USA). 545 pp.

Beckman, T.G., Lang, G.A., 2003. Rootstock breeding for stone fruits. Acta Hortic. 622, 531–551.

Borlaug, N.E., 2007. Sixty-two years of fighting hunger: personal recollections. Euphytica 157 (3), 287–297.

Borojevic, S., 1990. Principles and methods of plant breeding. Elsevier, Amsterdam (Netherlands), pp. 18–31.

Briggs, D., Walters, S.M., 1997. Plant Variation and Evolution. Cambridge Univ. Press, Cambridge (UK). 512 pp.

Campbell, N.A., 1993. Biology. Benjamin-Cummings, Redwood City, CA (USA), pp. 456–473.

Chase, M., 2005. Relationships between the families of flowering plants. In: Henry, R.J. (Ed.), Plant Diversity and Evolution. CABI Publ., Oxfordshire (UK), pp. 7–23.

Chou, C.H., 2006. Introduction to allelopathy. In: Roger, M.J., Reigosa, M.J., Pedrol, N., González, L. (Eds.), Allelopathy: A Physiological Process with Ecological Implications. Springer, Dordrecht (Netherlands), pp. 1–10.

Coates, D.J., Byrne, M., 2005. Genetic variation in plant populations: assessing cause and pattern. In: Henry, R.J. (Ed.), Plant Diversity and Evolution. CABI Publ., Oxfordshire (UK), pp. 139–164.

Cotton, C.M., 1996. Ethnobotany: Principles and Applications. John Wiley & Sons, Inc., West Sussex (UK). 424 pp.

Cracraft, C., 2002. The seven great questions of systematic biology: an essential foundation for conservation and sustainable use of biodiversity. Ann. Missouri Bot. Gard. 89, 127–144.

Cronk, Q.O.B., 2002. Perspectives and paradigms in plant evo-devo. In: Cronk, Q.O.B., Bateman, R.M., Hawkins, J.A. (Eds.), Developmental Genetics and Plant Evolution. Taylor & Francis, London (UK), pp. 1–14.

Cullis, C.A., Vorster, B.J., Van Der Vyver, C., Kunert, K.J., 2009. Transfer of genetic material between the chloroplast and nucleus: how is it related to stress in plants? Ann. Bot. 103 (4), 625–633.

Darwin, C.W., 1859. The Origin of Species (by Means of Natural Selection or the Preservation of Favoured Races in the Struggle for Life). Republished in 1958 by Penguin Books, New York, NY (USA). 479 pp.

Davison, J., 2010. GM plants: science, politics, and EC regulations. Plant Sci. 178 (2), 94–98.

Desai, B., 2004. Seeds Handbook—Biology, Production, Processing, and Storage, second ed. Marcel Dekker, Inc., New York, NY (USA), pp. 7–110.

Dodds, J., 2003. Intellectual property rights: biotechnology and the green revolution. In: Serageldin, I., Persley, G.J. (Eds.), Biotechnology and Sustainable Development: Voices of the North and South. CABI Publ., Cambridge, MA (USA), pp. 149–160.

Epel, B.L., 1994. Plasmodesmata: composition, structure, and trafficking. Plant Mol. Biol. 26 (5), 1343–1356.

Frankel, O.H., Brown, A.H.D., Bardon, J.J., 1995. The Conservation of Plant Biodiversity. Cambridge Univ. Press, Cambridge (UK). 299 pp.

Hammond, P., 1995. The current magnitude of bio(diversity). In: Heywood, V.H., Wootson, R.T. (Eds.), Global Biodiversity Assessment. Cambridge Univ. Press, Cambridge (UK), pp. 113–138.

Havey, M.J., 2004. The use of cytoplasmic male sterility for hybrid seed production. In: Daniell, H., Chase, C. (Eds.), Molecular Biology and Biotechnology of Plant Organelles: chloroplasts and Mitochondria. Springer, Norwell, MA (USA), pp. 623–634.

Henry, R.J., 2005. Importance of plant diversity. In: Henry, R.J. (Ed.), Plant Diversity and Evolution. CABI Publ., Oxfordshire (UK), pp. 1–5.

Hilbeck, A., Andow, D.A., 2004. Environmental Risk Assessment of Genetically Modified Organisms. Vol. 1. CABI Publ., New York, NY (USA). 281 pp.

Horne, J.E., McDermott, M., 2001. The Next Green Revolution: Essential Steps To A Healthy, Sustainable Agriculture. Food Products Press, New York, NY (USA), pp. 33–65.

Janick, J., Schery, R.W., Woods, F.W., Ruttan, V.W., 1974. Plant Science. Freeman Publ., San Francisco, CA (USA), p. 173.

Kubota, C., McClure, M.A., Kokalis-Burelle, N., Bausher, M.G., Rosskopf, E.N., 2008. Vegetable grafting: history, use, and current technology status in North America. HortSci. 43 (6), 1664–1669.

Lang, B.F., Burger, G., 2012. Mitochondrial and eukaryotic origins: a critical review. Adv. Bot. Res. 63, 1–20.

Li, J., Webster, M.A., Smith, M.C., Gilmartin, P.M., 2011. Floral heteromorphy in Primula vulgaris: progress towards isolation and characterization of the S locus. Ann. Bot. 108 (4), 715–726.

Oldfield, M., 1989. The Value of Conserving Genetic Resources. Sinauer Association, Inc., Sunderland, MA (USA), pp. 12–52.

Pistorius, R., Van Wijk, J., 1999. The Exploration of Plant Genetic Information: Political Strategies in Crop Improvement. CABI Publ., Cambridge, MA (USA). 231 pp.

Pottier, J., 1999. Anthropology of Food: The Social Dynamics of Food Security. Polity Press, Cambridge (UK). 230 pp.

Pritham, E.J., 2009. Transposable elements and factors influencing their success in eukaryotes. J. Hered. 100 (5), 648–655.

de Queiros, K., 1999. The gener linear concept of species and defining properties of the species category. In: Wilson, R.A. (Ed.), Species. MIT Press, Cambridge, MA (USA), pp. 49–89.

Roos, E.E., 1989. Long-term seed storage. Plant Breed Rev. 7, 129–158.

Salthe, S.N., 1998. The role of natural selection theory in understanding evolutionary systems. In: Van de Vijver, G., Salthe, S.N., Delpus, M. (Eds.), Evolutionary Systems. Kluwer Academic Publ, Dordsecht (GER), pp. 13–20.

Schuh, R.T., 2000. Biological Systematics: Principles and Applications. Cornell Univ. Press, Ithaca, NY (USA). 236 pp.

Semple, C., Steel, M., 2003. Phylogenetics. Oxford Univ. Press, Oxford (UK). 239 pp.

Simmonds, N.W., 1979. Epidemics, populations, and the genetic base. In: Simmonds, N.W. (Ed.), Principles of Crop Improvement. Longman, Harlow (UK), pp. 262–269.

Simpson, M.G., 2006. Plant Systematics. Elsevier Academic Press, Amsterdam (NETH). 590 pp.

Srivastava, L.M., 2002. Plant Growth and Development. Academic Press, New York, NY (USA). 772 pp.

Tracy, W.F., 1997. History, genetics, and breeding of supersweet (shrunken2) sweet corn (su). In: Plant Breeding Reviews. 1997, pp. 189–236.

Weber, B.H., 1998. Emergence of life and biological selection from the perspective of complex systems dynamics. In: Van de Vijver, G., Salthe, S.N., Delpus, M. (Eds.), Evolutionary Systems. Kluwer Academic Publ, Dordsecht (Germany), pp. 59–66.

White, G.A., Shands, H.L., Lovell, G.R., 1989. History and operation of the national plant germplasm system. Plant Breed Rev. 7, 5–56.

Willis, K.J., McElwain, J.C., 2002. The Evolution of Plants. Oxford Univ Press, Oxford (UK). 378 pp.

3

Review of Genetics (From The Perspective of A Plant Breeder)

Plant breeding is a discipline and a profession that is supported by a broad base of knowledge: biological, physical, chemical, artistic/aesthetic, and social. A solid understanding of Mendelian genetics is considered to be among the most important of prerequisites since advances in plant breeding occurred interdependently with genetics. My earlier true story about Bob, the successful private sector plant breeder who could not pass genetics 101, illustrates an important counterpoint to this tenet. Plant breeding is as much about becoming one with the genetic potential of an individual plant and developing an intuition about transforming potential into reality as it is about an understanding of how genes are transformed into phenotypes. Many plant breeders do not fully appreciate the power of the ability to translate plant phenotype into genotype. They will probably not be successful without the undefinable spark (or intuition) that Bob possessed. Perhaps a thorough knowledge of genetics imparts a sense of mastery that cripples the aspiring plant breeder; a false security that lulls the plant breeder into believing that phenotypic inheritance is already known or finished, that every observation is easily explained by permutations of Mendel's laws? The new frontier lies in the world of genomics, transcriptomics, and proteomics, and the days of the field-trained plant breeder are over? It is not necessarily so.

MENDELIAN INHERITANCE

Mendel's Laws of Inheritance are as follows:

- Genes exist in two copies
- One copy from each parent
- Some genes mask others, or are dominant over others

The laws seem so intuitive now, 150 years later, but so do many other monumental discoveries like fire and the wheel. So simple, in fact, that it is hard to believe that humans engaged in the saving and replanting seeds for thousands of generations didn't have the same hunches as Mendel. All the glory goes to the one who conducts the experiments and publishes the results and analysis in a scientific paper, in this case Gregor Mendel. In his day, however, Mendel did not attract much attention for his new theory. It was only after Bateson, DeVries, and Correns rediscovered Mendel's lost manuscript on the inheritance of pea plant traits in 1902, 18 years after Mendel's death, that the science of genetics became known as "Mendelian" (Fairbanks and Rytting, 2001).

Mendel was lucky in his choice of plant species, cultivars used in the experiments, and the traits chose to study. Pea (*Pisum sativum* L.) is 100% self-pollinated (see Chapter 10), so the cultivars that differed in the traits he measured were genetically fixed, or homozygous. If he had experimented with corn, peppers, or cabbage instead, the outcome would have been quite different, and not nearly as definitive or noteworthy. The traits he reported on were all simply inherited. Perhaps Mendel tried to study trait inheritance other species but was stymied by results that did not conform to his newly formed laws, so ignored them. Pea is also diploid meaning that genes are present in two copies. Polyploidy, a condition that will be addressed later in this chapter and also in Chapter 8, is common in plants and results in multiple gene copies and more complicated inheritance patterns.

Much like a mathematical theorem, most of the rest of genetics is an extrapolation, deduction, corollary, or extension of these original laws. What about the joint inheritance of two traits? They behave as if determined by two independent genes unless they are physically linked. Linkage was initially a theoretical concept until experimental results unambiguously unified linkage with cytogenetics. The number of linkage groups was found to be identical to the number of chromosomes in a broad spectrum of diploid eukaryotes. This led to the conclusion that genes reside on chromosomes, thus unifying the centuries-old science of cytology with the new science of genetics (Darlington, 1937).

It is presumed that the reader/student is familiar with basic biochemistry, particularly DNA replication, transcription of genomic DNA sequence into mRNA, and translation of the mRNA sequence into polypeptides, or proteins. Many important physiological functions and morphological traits are manifested as a consequence of a biochemical pathway, or series of steps, each catalyzed by a specific enzyme. It is instructive to think about intra- and inter-genic "interactions" within the context of gene products and gene-encoded enzymes that collectively mitigate the end product, or by-products, of biochemical pathways. Quotations have been placed around *interactions* since what is observed at the level of the phenotype may not be an interaction at all in the strict sense, but merely what happens when two different alleles or genes exist within the same organism. This is also helpful for sorting out terminology that is sometimes overlapping and can cause confusion. A *locus* is a physical location where a specific gene, DNA sequence, exists within the genome of a given species of organism. Thus, when referring to alleles at a gene or alleles at a locus, it is the same thing.

Intragenic interactions are also known as *inter-allelic* interactions or different *alleles* at a given locus within the same genome. Alleles are analogous DNA sequences that are similar but not identical. They may differ with regard to a single base pair or many base pairs. Alleles are sometimes silent. Since most crop species have two copies of each gene (some species and genes are present or have more than two copies), it is possible that they correspond to either identical or different alleles. The "gene product" of the two alleles may be identical if the nucleotide base occurs in a locus within a codon that does not result in an amino acid change. Moreover, two analogous polypeptides may have equivalent function despite having been encoded by allelic DNA sequences since certain changes in amino acids in a polypeptide impart the same properties. A given locus may have many different alleles, as the DNA sequence is composed of hundreds or thousands of nucleotide bases, each base locus mutable.

We can identify alleles by sequencing genomic DNA or mRNA or, alternatively, when they are associated with different phenotypes. Mendel noticed morphological differences among his pea populations, some exhibited either short or tall stature, some had round or wrinkled seeds, etc. The importance of his discovery was that "short" vs. "tall" and "round" vs. "wrinkled" were derived from differences in the DNA sequences that gave rise to the gene products that, in turn, affected stature and seed shape. When he crossed tall with short plants, the progeny (first filial generation, or F_1) were all tall. What became of the heritable determinants of shortness in F_1 plants? Were they forever lost, obliterated by the determinants of tallness?

These F_1 plants were self-pollinated and the resulting progeny constituted the second filial, or F_2, generation. Short plants reappeared but in only 25% of the F_2 plants. So the tall F_1 plants had somehow stored shortness, or remembered that one of the parents had been short, passing shortness to some, but not all F_2 progeny. Mendel's fundamental genius was to surmise that this could happen if one hypothesized that his peas always had two separate genes that controlled plant stature. The plant could have two copies of the tall gene and be tall, or two copies of the short gene and be short, but if the plant had one copy of each, the tall form was dominant over the short form, and the resulting plant was tall. Thus, both parents were *homozygous*, or possessed two identical alleles (TT or tt), while the F_1 was *heterozygous*, or possessed two different alleles (Tt). Complete dominance is defined as the two-copy plants looking identical to the one-copy plants, such as the case with Mendel's example (i.e., TT = Tt). He went further to hypothesize that each parent of the original cross had contributed one copy of the stature gene, that we will call "T": *Tall* TT x *Short* tt → F$_1$ all *Tall* Tt self-pollinate → F$_2$ ¾ *Tall* (1/4 TT + 2/4 Tt) + ¼ *Short* (tt).

Since Mendel's initial studies, millions of experiments have been performed, and thousands of resulting scientific papers have appeared, that collectively demonstrate the generality of his laws on our planet. As with most biological phenomena, there are necessary extensions and important exceptions, but the ever-expanding body of genetic knowledge remains highly robust and almost entirely consistent with Mendel's Laws of Inheritance.

Many studies have been conducted on the inheritance of stature since the studies of Mendel. Most experiments have found that stature is not manifested in the same neat ratios as in Mendel's experiments with peas. When tall and short humans hybridize the offspring are not necessarily all tall. More often than not the offspring are intermediate but also may be as tall or as short as the parents. Sometimes they may even be taller or shorter. There are almost always differences in stature among siblings although the conditions under which they are raised are nearly identical. How can this difference in phenotype inheritance patterns be explained in ways that do not violate Mendel's Laws?

Alleles do not always behave in a strictly dominant way. Rather, they interact in a manner that can be viewed as a continuum (Fig. 3.1). Complete dominance occurs when the heterozygote has the same value as one of the homozygotes. Alleles are considered to be *additive* if the heterozygote is exactly halfway between the two homozygotes. If the heterozygote falls somewhere in between halfway and one of the parents, the interaction of alleles is referred to as *incomplete dominance*. If the heterozygote falls outside of the range defined by the parental homozygotes, the interaction is referred to as *overdominance* (Fig. 3.2).

Studies on humans have shown that stature is not explained entirely by inter-allelic, intra-genic interactions (Visscher, 2008). But what if more than one gene were somehow involved in the determination of a given trait? What if 2, 3, 10, 100 or more genes were involved? Let us assume further that each of these genes acts independently in that the presence of a given allele at one gene does not bias which allele is present at another. So if the short parent had additive alleles at ten genes for shortness, and the tall parent had additive alleles at the same ten genes for tallness, the progeny should all be intermediate between the two parents. Never mind that sexual dimorphism in humans also affects stature.

Stature in plants is heavily influenced by the environment, but genes that influence stature and components of stature (e.g., internode length) have been characterized (Busov et al., 2008). For example, stature in rose (*Rosa* sp.) exhibits continuous phenotypic variability such as was illustrated in Fig. 3.1 (Kawamura et al., 2015). For the sake of

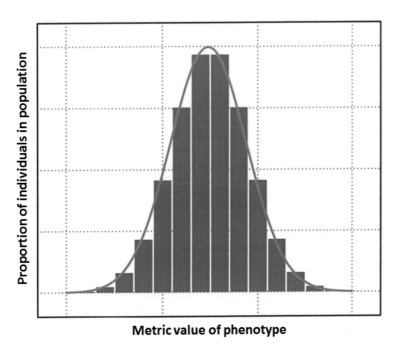

FIG. 3.1 Normal distribution of a random sample of individuals from a mixed population for yield, a quantitatively inherited trait. When the observations are grouped into small frequency classes they form a typical bell-shaped curve. *Adapted from https://math.stackexchange.com/questions/2123873/is-the-maximum-of-a-probability-distribution-function-of-a-binomial-distribution.*

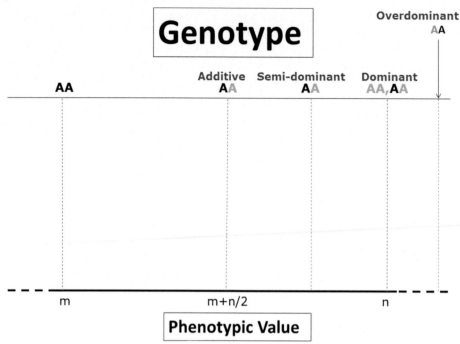

FIG. 3.2   A metric representation of allelic interactions.

example, let us postulate that stature in *Rosa* sp. is controlled by ten loci that contribute more-or-less equally to the phenotype. If we hybridized a short with a tall stature rose, what would we expect to see among the F_2 progeny with respect to stature? Each of the ten genes, or loci, for stature, are heterozygous in the F_1, then would segregate 1 homozygous tall:2 heterozygous:1 homozygous short in the F_2. The progeny would have a variable number of tall and short alleles, ranging from 100% tall in a continuum to 100% short. The distribution would be centered on the average between the two extremes (parents), and would actually constitute a perfect binomial distribution (Fig. 3.1). As the number of loci that condition a given trait increases, the number of genotypes increases and the distinction between them decreases, until the distribution is a smooth curve, a normal or Gaussian distribution. Also, we must factor in the axiom that all traits are determined by a combination of genotype and environment (see Chapter 2). In the case of our rose stature example, if the F_2 progeny were reared under different conditions, received different cultural inputs, and faced different challenges over time (e.g., herbivory or disease), the influence of environment on determining individual stature could be quite significant. Some progeny would grow taller than their genes would predict, some shorter. This has the effect of further smoothing the distribution curve and erasing the discreet lines that distinguish genotypic and corresponding phenotypic classes.

We have already discussed that alleles can have different interactions, for example, dominance, incomplete dominance, and additive. It should come as no surprise that different genes, or loci, interact with each other in different ways as well. Some would seem to have little to do with one another, like Mendel's stature and seed shape in peas. Just because a given plant is tall, one would not expect there to be any secondary or tertiary effects on seed shape or vice versa. In many instances, however, and some that are surprising and even unexplainable, seemingly unrelated genes interact to modulate phenotypes. This is easy to visualize where gene products are enzymatic steps in a single biochemical pathway. Gene A encodes enzyme P that catalyzes one molecular transformation, and gene B encodes enzyme Q that catalyzes another, in the same pathway further down the road. The existence of the end product depends on the functioning of both genes. The loss of either will eliminate the end product, unless the plant has some sort of redundant pathway.

There are many ways in which different genes, or loci, have been observed to interact. One example is carotenoid pigmentation in tomato (also many other plant species). Carotenoids are synthesized by plants from carbon skeletons via the isoprenoid pathway (also known as the mevalonate or HMG-CoA reductase pathway; Goldstein and Brown, 1990). Isoprene derivatives have many functions in light capture and intermediate metabolism and many absorb light of specific wave spectra and tend to reflect light of unabsorbed wave lengths. In other words many carotenoid compounds are pigments, usually in the yellow-red portion of the visible spectrum. Carotenoids are what

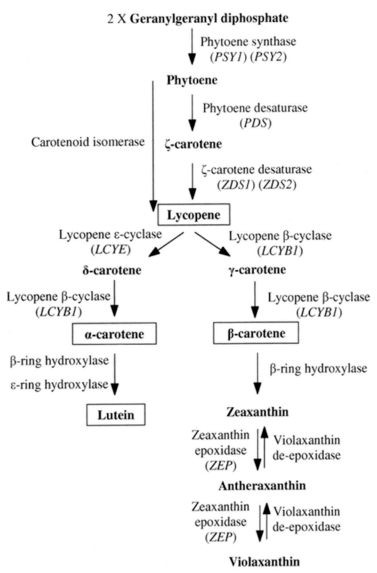

FIG. 3.3 The fundamental carotenoid biosynthetic pathway in plants. *Modified from Diretto, G., et al., 2006. Metabolic engineering of potato tuber carotenoids through tuber-specific silencing of lycopene epsilon cyclase. BMC Plant Biol. 6, Article number: 13.*

make tomato fruits orange-red and pumpkins orange. The isoprenoid pathway starts with acetyl-Coenzyme A, and the carbon skeleton is lengthened and modified in a stepwise enzymatic fashion (Fig. 3.3).

Biosynthetic intermediates of this pathway are used in the synthesis of many different end products. The main products, however, are known as carotenoid pigments. Examples include β-carotene and lycopene, both well known for their human health benefits, but also α-carotene, β-ionone, zeaxanthin, lutein, etc. Alleles have been discovered in tomato in the loci that encode the enzymes that catalyze many of these specific transformations. One (Og^c) is known as "Crimson" because it is characterized by a more intense, pinkish red internal fruit color as compared to "conventional", or wild-type tomatoes. Biochemical studies have shown that the relative concentrations of carotenoids in fruit tissues are shifted from β-carotene, a yellow-orange compound, to red lycopene. It would be desirable to have high concentrations of both compounds since they impart distinct benefits, but the number of isoprene skeletons is limited and an enzymatic modulator to enable this phenotype to be possible has not yet been discovered. The enzymatic step that is affected by the Og^c mutation is chromoplast-specific lycopene βcyclase, wherein lycopene accumulates because the enzyme to convert it to β-carotene is missing or nonfunctional. Since the phenotypes controlled by the genes in this pathway are dependent on the interaction of gene products (enzymes) in a sequential manner, alleles of the genes that encode these enzymes will exhibit joint inheritance patterns that embody epistasis (see "Two Genes" below).

To summarize, Mendelian concepts are exemplified by the following:

Single gene

a. Dominance:

$$Red\ RR \times White\ rr \rightarrow (F_1)\ Red\ Rr\ self-pollinate \rightarrow (F_2)\ \tfrac{3}{4}\ Red\ (1/4RR + 1/2Rr) + \tfrac{1}{4}\ White\ (rr)$$

b. Co-dominance:

$$Red\ RR \times White\ rr \rightarrow (F_1)\ Pink\ Rr\ self-pollinate \rightarrow (F_2)\ \tfrac{1}{4}\ Red\ (RR) + \tfrac{1}{2}\ Pink\ (Rr) + \tfrac{1}{4}\ White\ (rr)$$

Two genes

a. Unlinked with dominance:

$$Yellow/wrinkled\ YYss \times Green/smooth\ yySS \rightarrow (F_1)\ Yellow/smooth\ YySs\ self-pollinate \rightarrow$$

$$(F_2)\ 9/16\ Yellow/smooth\ (1/16\ YYSS + 2/16\ YySS + 2/16\ YYSs + 4/16\ YySs)$$

$$+ 3/16\ Yellow/wrinkled\ (1/16\ YYss + 2/16\ Yyss) + 3/16\ Green/smooth\ (1/16\ yySS + 2/16\ yySs)$$

$$+ 1/16\ Green/wrinkled\ (yyss)$$

b. Linked (absolutely; no recombinants) with dominance:

$$Yellow/wrinkled\ YYss \times Green/smooth\ yySS \rightarrow (F_1)\ Yellow/smooth\ YySs\ self-pollinate \rightarrow$$

$$(F_2)\ \tfrac{1}{4}\ Yellow/wrinkled\ (YYss) + \tfrac{1}{2}\ Yellow/smooth\ (YySs) + \tfrac{1}{4}\ Green/smooth\ (yySS)$$

c. Unlinked with dominance and epistasis:

$$Red\ RRbb \times Blue\ rrBB \rightarrow (F_1)\ Purple\ RrBb\ self-pollinate \rightarrow$$

$$(F_2)\ 9/16\ Purple\ (1/16\ RRBB + 2/16\ RrBB + 2/16\ RRBb + 4/16\ RrBb)$$

$$+ 3/16\ Red\ (1/16\ RRbb + 2/16\ Rrbb) + 3/16\ Blue\ (1/16\ rrBB + 2/16\ rrBb) + 1/16\ white\ (rrbb)$$

Punnett Square or Binomial equation

For the mathematically uninclined, the numerical aspects of genetics can be daunting, or at least confusing. We are fortunate that sex on planet earth always culminates in the fusion of two gametes to form the next generation, never more. Therefore, all possible outcomes can be reduced to a binomial equation, as follows:

$$Aa \rightarrow 50\%\ A\ gametes + 50\%\ a\ gametes$$

$$Aa\ self\ pollinated:\ (0.5\ A + 0.5\ a)(0.5A + 0.5\ a) = 0.25AA + 0.50\ Aa + 0.25\ aa$$

$$AaBb \rightarrow 25\%\ AB\ gametes + 25\%\ Ab\ gametes + 25\%\ aB\ gametes + 25\%\ ab\ gametes$$

$$AaBb\ self\ pollinated:\ (0.25\ AB + 0.25\ Ab + 0.25\ aB + 0.25\ ab)^2 = 0.0625\ AABB + 0.125\ AABb + 0.0625\ AAbb$$

$$+ 0.125\ AaBB + 0.250\ AaBb + 0.125\ Aabb + 0.0625\ aaBB + 0.125\ aaBb + 0.0625\ aabb$$

Punnett (1922) was among the first quantitative geneticists to point out the binomial nature of gametic arrays and progenies. A conceptual device known as a *Punnett square* has been devised to help the non-mathematically inclined to calculate the frequencies of phenotypic classes among progeny when the proportions of gametes are known (Fig. 3.4). Even the most experienced plant breeders and geneticists are occasionally stymied by predicted segregation ratios and resort to the Punnett square. The student should use whatever tool you are most comfortable with to visualize the behavior of genes and alleles in specific examples.

When studying a trait of unknown genetic determination, a common question is: "How many genes are responsible for the observed phenotypic differences? Mendel's pea seed and stature traits were governed by single genes, but most phenotypes are more complicated. If it is assumed that the underlying genes and their alleles interact additively (a huge and generally violated assumption), it is possible to use the F_2 population distribution relative to the parents and F_1 to answer this question. If a given phenotype is determined by N loci, with two alleles per locus, the number

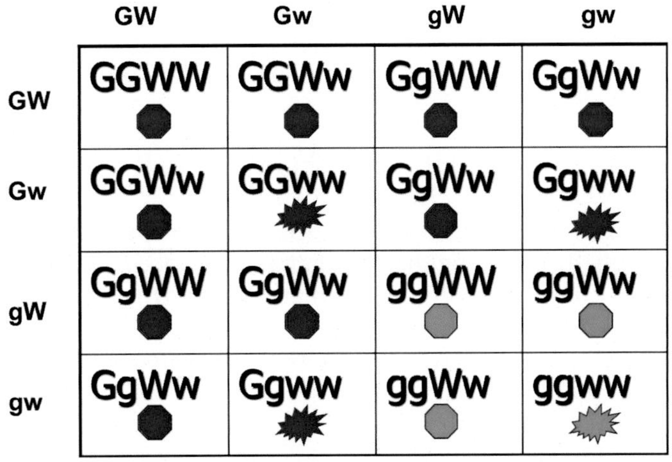

	GW	Gw	gW	gw
GW	GGWW	GGWw	GgWW	GgWw
Gw	GGWw	GGww	GgWw	Ggww
gW	GgWW	GgWw	ggWW	ggWw
gw	GgWw	Ggww	ggWw	ggww

FIG. 3.4 Example of a Punnett square, illustrating the outcome of the cross GgWw x GgWw where GG or Gg results in red seeds and gg results in green seeds; WW or Ww conditions smooth seeds and ww conditions wrinkled seeds.

of discrete phenotypes in the F_2 progresses as the following series: 2^N ($N = 4, 16, 64, 256$, etc.). This can be extrapolated to the following equation to estimate N:

$$N = \left(X_{P1} - X_{P2}\right)^2 \Big/ \left\{ 8\left[\left(\sigma_{F2}\right)^2 - \left(\sigma_{F1}\right)^2 \right] \right\}$$

where N = number of genes, X_{P1} and X_{P2} are the means of inbred parental lines, and σ_{F2} and σ_{F1} are the standard deviations of F_2 and F_1 populations.

It should be noted that this relationship is only valid if the F_1 and F_2 population distributions relative to the phenotype under consideration are normal (i.e., a bell curve). Linkage disequilibrium, dominance, and epistasis can distort phenotype distributions beyond the range of statistically valid normalcy. A test to ascertain the validity of the assumption of a normal distribution is recommended, and can be found in applied agricultural statistics texts (Hoshmand, 1994; Petersen, 1994; Quinn and Keough, 2002).

LINKAGE

We have thus far only considered the case of absolute linkage. Linkage is defined only as a departure of the classes of gametes from complete independent assortment. In the case of gametes from a cross of AAbb x aaBB depicted above, the F_1 is AaBb, and the gametic classes are expected to be 25% each if *A* and *B* are not linked. If locus *A* and locus *B* are linked, the proportion of *parental* gametes (in this case, Ab and aB) will be higher than the proportion of non-parental, or *recombinant* gametes (AB and ab). The greater the degree of departure, the closer the linkage, until it is absolute, as in the earlier example, and no recombinant gametes are present.

The gametic genotype is referred to as a *haplotype* in a broad spectrum of literature on eukaryotes. A haplotype is defined as a set of alleles that is inherited from a single parent (Li et al., 2003). Hence if the parents AAbb and aaBB are hybridized, the haplotype is the association of alleles without the presumption of linkage since *A* and *B* may or may not be linked. The two haplotypes, in this case, are Ab and aB.

In reality linkage between two or more loci is rarely absolute. Even when loci are on the same chromosome, recombination is extremely prevalent within the chromosome, or linkage group. Single chromosomes may experience recombination events many times over their respective lengths during the process of meiosis (see below). If loci are far enough apart, the propensity for crossing over completely obscures the linkage, and the loci behave as if they are not linked at all (i.e., the frequency of parental and recombinant gametes and progeny are equal).

In general, the plant breeder prefers situations where few genes (loci) control the targeted trait. Alleles should, ideally, interact in an additive fashion, and the outcome of combinations of loci should be predictable from their separate phenotypes (also referred to as additive). There should ideally be no linkage because this impedes the random assortment of desirable alleles. All phenotypes should be determined solely by genotype with no effects imparted by the environment. Unfortunately, the ideal plant breeding scenario is only rarely encountered. On the bright side, however, advances in genomics, transcriptomics, proteomics, and phenomics are progressively unraveling the intertwined and complicating effects of allelic, intergenic, and environmental interactions on phenogensis.

The term *linkage disequilibrium* is used to describe a situation where a given population exhibits non-random associations of alleles at different loci. This situation is encountered if recombination is suppressed following hybridization of two haplotypes or if the hybridization event was very recent.

POPULATION GENETICS

As will be discussed in Chapter 4 plant breeding is a process that results in genetically unique populations. This has been accomplished historically by enriching the population for a specific allele or alleles at loci that control the genotype, or set of genotypes, as compared with the population from whence they were derived. Another way to view this process is through the lens of allelic frequencies. Mendelian genetics carries the tacit assumption that if a locus has two alleles within an individual they will be represented equally, or 50% each. In populations of individuals, however, alleles are rarely present in equal proportions.

If locus *A* has two alleles A and a, the relative frequencies of each in a given population are said to be "*p*" and "*q*". Since $p + q = 1.0$ one only has to know p to be able to calculate q. By convention p denotes the frequency of the predominant allele. Rather than considering the outcome of individual matings we are now concerned with the collective outcome of a multiplicity of matings within a population. Since all matings are singularly pairwise the population will be described by the term $(p + q)^2$. If $p = 0.6$ and $q = 0.4$, and matings within the population are random with no selective advantages to either allele, it will be possible to predict the genotypic proportions and, hence, the phenotypic proportions from one generation to the next:

$$(0.6A + 0.4a)^2 = (0.6A + 0.4a) \times (6A + 4a)$$
$$= 0.36\,AA + 0.48\,Aa + 0.16\,aa$$

Such a population will continue to contribute .6A and .4a gametes, so the genotypic and phenotypic frequencies of the corresponding sporophytes within the population will be stable from generation to generation, a phenomenon first predicted independently by Hardy and Weinberg, and now known as *Hardy-Weinberg equilibrium* (Hardy, 1908; Weinberg, 1908).

In reality, these assumptions are invalid. Mating is almost never completely random and in plants is often skewed by floral mechanisms that favor self-pollination over outcrossing. Different phenotypes inevitably impart different relative fitness to individuals who are challenged with specific environmental factors, for example spring flowering date in woody perennials. One could suppose that plants are constantly seeking a longer period of time during which to carry out seed and fruit development following sexual fertilization, favoring earlier flowering dates. On occasion, however, late spring frosts will eliminate progeny from any bud that flowers too early. Selection would favor plants that have genotypes that condition intermediate flowering dates, not too early or too late.

In nature, one can imagine, theoretically, that alleles impart relative fitness to the individuals that carry them. In the case of locus *A*, let us suppose that AA and Aa have a relative fitness of 1.0 and that aa has a relative fitness of 0.8. If we start with the above population:

$$36\,AA + 48\,Aa + 16\,aa$$

Only 0.8×16 out of 100 aa individuals will contribute gametes to the next generation. Therefore, the allelic frequencies are immediately changed from $p = 0.6$ and $q = 0.4$ to $p = 0.62$ and $q = 0.38$. The next generation would be changed to

$$384\,AA + 471\,Aa + 144\,aa$$

Over time, if the selection continues to have the same consequences, $p \to 1.0$ and $q \to 0.0$, but this will take many generations to occur. In nature, selection is usually not consistent, varying from generation to generation. Under these conditions, most alleles fluctuate in relative frequency, waxing and waning. If the environment changes in a directional way over time, however, such as hotter or colder, one could expect allelic frequencies at a multitude of loci affected by temperature to change over time. If $p = 1.0$, however, and the allele is deleterious to the individual under the new environmental conditions, extinction is inevitable unless new alleles are somehow introduced.

Plant breeding proceeds by assigning tremendous differences in the relative fitness of genotypes within populations. It is not unusual for one allele to have a relative fitness of 1.0 and another 0.0. Thus, the genetic composition of populations subjected to plant breeding change very rapidly as compared to those in nature. As it happens, selection is practiced directly on individual phenotypes, and thus only indirectly on genotypes. If the phenotype is not an accurate portrayal of the underlying genotype, the selection may not be very effective at all in changing gene frequencies. The plant breeder would prefer situations of high phenotype-genotype fidelity (i.e., high h²). The importance of correspondence of phenotype to genotype will be addressed in the next section.

QUANTITATIVE GENETICS

Most phenotypic variation does not conform to simple models of inheritance. When two parents that differ with respect to a given trait, the patterns of segregation observed in F_2 and backcross (F_1 x parent) populations are not explained by discrete 3:1 or 1:1 phenotypic ratios. Rather, patterns more similar to continuums or better explained by other discrete ratios are common. Generally, this is because most phenotypes are conditioned by a large number of genes that interact with each other and the environment.

It is important for the plant breeder to understand as much as possible about the inheritance of the targeted trait prior to formulating a plan aimed at genetic population improvement. How many genes are involved in the trait? What are their relative contributions to the phenotype? How do alleles and genes interact? What proportion of the phenotype is conditioned by genetic as opposed to environmental factors? The breeding plan will be much different if few genes with high effects than if many genes with minor effects are responsible.

The field of quantitative genetics has progressed historically much like that of physics, wherein seemingly continuous phenomena are broken down into smaller discernible units. In physics, the unit would be atoms or subatomic particles or energy packets such as photons. In genetics, the unit would be the gene. In physics, we observe the behavior of matter and energy, and it manifests itself to us as electrical and gravitational fields and motion of mass over time. Correspondingly, in genetics, phenotypes are generally expressed in a continuum, not discretely. Quantum mechanics unifies particles and packets with the behavior of matter and energy, whereas phenotypic continuums are rectified with Mendelian theories by the tenets of quantitative genetics.

In the most simplistic cases, one assumes that a given phenotypic trait is conditioned by N genes, each with identical additive effects. For example, if $N = 4$ (loci A, B, C, and D), then a diploid organism can have up to 8 doses of the alleles that contribute to the phenotypic range. Let us apply this to the genetic determination of stature, a trait controlled in reality by hundreds, perhaps even thousands, of genes.

The relative effect of genes on a given phenotype is expressed by the variance terms attributable to these effects. This is because variances can be partitioned into constituent sources that sum to the whole. The variance of a population is imputed from a random sample of n constituent individuals as follows:

$$V_T = 1/n \sum (x - \mu)^2$$

where x = value of the individual and μ = the mean of the population (estimated by sample).

A more thorough treatment of the partitioning of phenotypic variances will be presented later. In simplified form, because variances are additive we may represent the total (T) or phenotypic (P) variance as being the sum of variances due to genotype (G) and environment (E). Beginning with $V_T = V_P = V_G + V_E$ we first assume that $V_E = 0$. If AABBCCDD results in a stature of 6.5 ft and aabbccdd results in 4.5 ft then we can suppose that the following genotypes will give rise to the corresponding statures:

$$AaBBCCDD = AABbCCDD = AABBCcDD = AABBCCDd \rightarrow 6.25\,ft$$

$$aaBBCCDD = AaBbCCDD = AAbbCCDD = AABbCcDD\,(etc.) \rightarrow 6.00\,ft$$

$$aaBbCCDD = AabbCCDD = AABBCcdd = AABbCcDd\,(etc.) \rightarrow 5.75\,ft$$

$$aabbCCDD = AabbCcDD = AaBBCcdd = AaBbCcDd\,(etc.) \rightarrow 5.50\,ft$$

$$aabbCcDD = AabbccDD = AaBbccDd = AaBbCcdd\,(etc.) \rightarrow 5.25\,ft$$

$$aabbccDD = AabbccDd = AabbCcdd = AaBbccdd\,(etc.) \rightarrow 5.00\,ft$$

$$aabbccDd = Aabbccdd = aaBbccdd = aabbCcdd\,(etc.) \rightarrow 4.75\,ft$$

If $V_E > 0$ then each discrete class will be subject to phenotypic dispersion and as V_E increases progressively a continuum will be approximated. If all phenotypic variation is environmental (i.e., $V_P = V_E$) the population distribution will vary only according to different environments. In our example of human stature this could result if the phenotype was conditioned purely by daily caloric intake and nutritive value. In reality, we know that, beyond additive genes, human stature is also conditioned by both non-additive genetic and environmental factors. From our example progeny will always be intermediate to the parents, but we all know of instances where adult children are shorter or taller than their respective parents.

The stature phenotype can thus, in our simplified example, be predicted from genotype, and the outcome of matings duly explained. For example the cross of an individual 4.5 ft in height (aabbccdd) with an individual 6.5 ft in height (AABBCCDD) will give rise to progeny that are intermediate (5.5 ft) (AaBbCcDd). If one is able to obtain a true F_2 population, a binomial distribution would be observed, with nine phenotypic classes in a 1:4:28:64:92:64:28:4:1 ratio. This distribution of stature would resemble a normal curve with a range of 4.5–6.5 ft and a mean of 5.5 ft.

As we mentioned above, V_E for human stature is greater than zero, being much affected by factors such as diet, health maintenance, and physical activity. As C_{VE} (the coefficient of environmental variation) approaches 0.5, the distribution becomes indistinguishable from Gaussian, or normal. The actual distribution of human stature does approximate a normal curve. A normal curve is also approximated using the same assumptions about gene action and as $N \rightarrow \infty$. As $N \rightarrow 0$, $V_E > 0$ tends to smooth the curve and to create a phenotypic continuum (Fig. 3.1).

While most phenotypes exist in a continuum rather than in discrete classes, the assumption of complete additivity is rarely if ever realized, according to actual research results on quantitative trait loci (QTL, Chapter 9). If ten loci are involved in the expression of a given phenotype, perhaps only three loci will account for >90% of the range, while the remaining seven loci might have relatively minor effects. Within each locus, alleles may not interact in a strictly additive fashion, and epistasis may be observed with regard to intergenic interactions. To complicate things further, while environment might affect a phenotype independent of genotype, certain genotypes might interact in different ways within a given environment. For example, if a population of humans is administered a diet consisting of 500 cal per day the mean individual weight of the population will decrease because humans consume, on average, approximately 1200 cal per day for resting metabolism. Basal metabolism rate (BMR) varies and is under genetic control. The 500 cal diet will affect individuals with a high BMR more than it affects individuals with a low BMR. Therefore, the effects of environment on individual weight are dependent on genotype.

The field of quantitative genetics began in large part with the pioneering research of Sir Ronald A. Fisher (see Chapter 1) in the early 20th century. Fisher theorized that a population of individuals expressed a given phenotype that could be characterized by a mean and variance (Fisher, 1930). The value of any individual can be expressed as:

$$P = \mu + G + E + GE$$

where P is the individual's phenotypic value, μ is the population mean, G is the effect of genotype, E is the effect of environment, and GE the effect of genotype x environment interaction. While P and μ are usually positive numbers, the rest of the terms may be positive or negative. The plant breeder is concerned with G and to some extent with GE since these are the only terms of the equation that are heritable. The "breeding value" of an individual, therefore, traces to that proportion of the genotype that contributes predictably to the phenotype of interest. The agronomist, horticulturist, and plant physiologist are more interested in E and GE, since they seek to understand and apply environmental factors that maximize performance. GE may be observed and measured by conducting performance trials of populations with a range of fixed genotypes in multiple locations and time frames.

For single loci, the G term can be further partitioned into additive and dominance components. If the alleles at the locus do not interact, they are said to be additive, and the heterozygote is metrically intermediate between the homozygotes. If the heterozygote is greater or less than the mathematical mean, some degree of dominance is present. The equation, therefore, expands to:

$$P = \mu + A + D + GE + E$$

where A is the additive portion of G and D is the dominance portion of G. As we shall see later, the plant breeder is usually most concerned with the additive term, since dominance tends to obscure the ability to determine genotype based on an individual's phenotype.

The following section will delve into the mathematical models that support the theoretical bases of quantitative genetics. The treatment is scaled back to emphasize the most critical factors. A basic understanding of these basics is extremely helpful to the plant breeder who seeks to unravel the complexities of multiple phenotypes, multiple genes, and multiple environments. For a single locus, the theoretical phenotypic values are presumed to be:

Genotype	A_1A_1	A_1A_2	A_2A_2
Metric Value	a	d	−a

If d = 0, the condition is additive; if 0 < d < a, the condition is incomplete dominance; if d = a, the condition is complete dominance, if d > a, the condition is overdominance (alternatively, d may tend to –a). Now, if the population frequencies of A_1 and A_2 are p and q respectively and are in Wardy-Weinberg equilibrium (a common assumption, but rarely proven), the corresponding population genotypic frequencies will be p^2, $2pq$, and q^2. The mean of such a population will be $(ap^2 + 2dpq - aq^2)/(p + q)^2 = a(p-q) + 2dpq$.

If a phenotype is conditioned by n genes that do not interact with each other, the mean (M) may be expressed as:

$$M = \sum a_i (p_i - q_i) + 2 \sum d_i p_i q_i$$

If the value of all individual a_i, d_i, p_i, and q_i terms are known, any departure of the theoretical and actual mean is attributed to epistasis, or intergenic interactions. Thus, when two or more genes are involved in the determination of phenotypic value, the theoretical equation expands to:

$$P = \mu + A + D + I + GE + E$$

where I is the value of intergenic or epistatic interactions. The equation can be expanded further into constituent components, and even further than the following:

$$P = \mu + A + D + I + AE + DE + IE + E$$

With each additional term, a correspondingly larger experimental design is necessary for isolation and estimation and only the most compelling and economically significant phenotypes are usually accorded such attention due to resource limitations. While it is difficult or impossible to measure these components within individuals, it is much easier to measure their effects on dispersion within populations. Thus, most empirical studies of quantitatively inherited traits are conducted via the estimation of phenotypic variances. By comparing the variances of populations of hypothetical genotypic constitution, it is possible to deduce the relative effects of genotype, environment, and inter/intragenic interactions. The plant breeder may then use this information in the formulation of breeding plans and strategies.

Let's attempt to put this all into mathematical terms. The dispersion of a phenotype can best be described by variance V. The total phenotypic variance within a population V_T or $V_P = \sum(x - mean)^2/(n-1)$. Variance is a statistical parameter that can be partitioned into constituent components provided that certain assumptions are satisfied about the nature of the phenotypic distribution, that it closely emulates a normal curve. For our purposes, we will simply assume that all of our examples are consistent with this assumption.

The following tenet has been raised pertaining to the determination of the phenotype of an individual: Phenotype = genotype + environment. The concept is extended to the partitioning of phenotypic variance of a population: $V_P = V_G + V_E$. As we discussed above, there is another term, usually small, that is attributable to the interaction of genotype with the environment, or V_{GxE}. As we also discussed, total genotypic variance, V_G, has an additive, a non-additive intragenic component (referred to as dominance), and a non-additive intergenic component (referred to as epistasis), V_A, V_D, and V_I, such that $V_G = V_A + V_D + V_I$. So the equation can be expanded to:

$$V_P = V_A + V_D + V_I + V_E + V_{AxE} + V_{DxE} + V_{IxE}$$

Quantitative geneticists can slice and dice variances even further, but that is as far as this treatment will venture in this volume. As the researcher seeks to estimate variance terms of higher order of interaction, correspondingly more complex and large experiments are required. At some point, such variances are usually attributed to residual, or unaccounted for.

The components of phenotypic variance may be estimated directly if all of the above parameters are known:

$$V_G = p^2 a^2 + 2pq d^2 + q^2 a^2 - M^2 = 2pq a^2 + (2pq d)^2$$

$$V_A = p^2 [2qa]^2 + 2pq [(q-p)a]^2 + q^2 (2pa)^2 = 2pq a^2$$

and

$$V_D = p^2 (2q2d)^2 + 2pq(2pqd)^2 + q^2 (2p2d)^2 = (2pqd)^2$$

note that $V_G = V_A + V_D$, by definition and as expected.

If V_E is known as the average dispersion of fixed genotypes, then V_I may be estimated as the unaccounted residual variance:

$$V_I = V_P - (V_G + V_E)$$

Each component of total phenotypic variance has specific meaning to the plant breeder. V_G is the portion of V_P that can be used for improvement. V_A is the most predictable component of V_G, and most readily addressed using existing breeding methods that depend on observations of phenotype to infer genotype. V_D and V_I are generally considered to be less useful for plant improvement than V_A. Both sources of heritable variation result from interactions, alleles, and distinct genes respectively. If the relative contribution to V_P is high, progeny do not necessarily behave as predicted from the phenotypes of parents. When breeding hybrid cultivars or long-lived perennial species, however, V_D and V_I are considered to be as important as V_A.

HERITABILITY

A useful parameter known as *heritability* has been developed to measure and compare the relative amenability of phenotypes to improvement by breeding methods. The mathematical term h^2 is used to denote heritability, somewhat confusing because there is no such thing as h. h^2 is based on a ratio of variances or mean square terms, hence the terminology. In some textbooks, heritability is abbreviated differently to alleviate this confusion. In simple terms, heritability is the proportion of a phenotype that is determined by genotype. In mathematical terms, it is V_G/V_P, varying from 0 to 1.0. Specifically, this is referred to as *broad sense heritability* or broad sense h^2.

The reader has likely guessed that there must also be a *narrow sense heritability* (narrow sense h^2). Narrow sense h^2 = V_A/V_P, or the proportion of total phenotypic variance attributable to additive genetic effects. This also varies from 0 to 1.0, and narrow sense $h^2 \leq$ broad sense h^2. If narrow sense $h^2 \leq 0.2$, it is usually said to be indicative of low heritability. Conversely, if narrow sense $h^2 \geq 0.7$, it is usually said to be indicative of high heritability.

To illustrate heritability conceptually, let us evoke a set of hypothetical decks of playing cards. The standard deck consists of 4 suits that each contain 13 cards of varying value from 2 to ace, 52 in total. The card in the standard deck also features an image on the opposite side of the one that denotes card identity, or value. The image is the same on all 52 cards but varies from one standard deck to another. Let us now turn our attention to a marked deck of cards. The card values are the same as in the unmarked, standard deck. The opposite side of the card, however, conveys information that tells the skilled player exactly what the value of the card is.

The typical marked deck features 52 cryptic images on the cards' backs that correspond exactly to the value of the card on the other side. Let us imagine further that decks of marked cards exist that are not so precise with regard to predicting actual card value. In deck "C", the cryptic image on the back of a given card indicates that the value is 8D, but upon turning it over, you discover a 7D instead. The one marked 7D is actually a 9D in value. In each case the value indicated on the back of the card is close to the actual value and of the same suit but is not the precise value. In another deck "D" the marking system is even less precise. The cryptic back indicates 8D but is actually a 6H. The cryptic marking on another in the same deck predicts a 4C but is actually a 7S.

One can imagine a range of fidelities among a set of marked decks from absolute, as with the typical marked deck, to deck C where the fidelity was close, but not exact, and to deck D where the fidelity was even less precise. At the other extreme lies the standard deck, where it is impossible to surmise the value of any card based on the image on the opposite side. To complete the analogy, let us suppose that the deck of cards is a population of individuals. The actual card value is the genotype and the image on the opposite side is the phenotype of the individual. The value, or genotype, determines the cryptic marking image, or phenotype but depends on the skill of the cardmaker, the environment in this analogy.

A player (plant breeder) spreads a deck of cards out on a table face down. Depending on the game, certain hands are more desirable than are others, and the player's challenge is to choose the best possible hand of cards based on the image on the back of the card. If the deck is a standard one and the cards are well mixed on the table top, the selected hand of cards will be governed by probability alone. The chances of any given card are 1/52, 1/51, 1/50, etc. If the deck is marked the player can easily choose the desired hand based on the cryptic backside images. For decks C and D, however, selected hands of cards based on the backs will be closer to the desired composition than if by chance alone but not as precise as with the typical marked deck.

The game is five-card draw poker. The player chooses five cards from a standard deck and the hand is 3H, 7D, 10C, JC, AS. Next, the player chooses a hand from a typical marked deck and the result is 10D, JD, QD, KD, AD, a royal flush. Using deck C the player again attempts to pick out the cards to fashion a royal flush but falls short: 8D, 9D, QD, KD, AD; a flush but not a "straight" (unbroken sequence), thus diminishing the hand value. The same hand is attempted from deck D with the result: 9D, 10S, QS, KD, AS. This time the resulting hand is closer to a royal flush than if by chance but still virtually worthless in poker.

Completing the analogy, h^2 of the standard deck is 0.0 and of the typical marked deck is 1.0. Decks C and D fall somewhere in between, but h^2 for deck C is higher than it is for deck D, perhaps 0.8 and 0.6 respectively. The plant breeder is constantly faced with a similar challenge, to select genotypes from populations based on corresponding phenotypes. If the celestial cardmaker is not very skilled (e.g., low h^2), the group selected may be better than the average of the original population, but will probably not be of significantly greater value. The playing card analogy may be taken even further. The cards being selected will depend on the game being played since hands of cards have different values in different games. So it is with plant breeding as well. A selected population may be well suited for one purpose but ill-suited for another.

The plant breeder would obviously much prefer to work with phenotypes that have high vs. low heritability. Conceptually, this is because the higher the value of narrow sense h^2, or broad sense h^2 to a lesser extent, the more likely that the phenotype of an individual is purely a consequence of underlying genotype. Therefore, the progeny of that individual will have derived genotypes, are more likely to exhibit corresponding phenotypes that are similar to that of the selected individual. If the breeder selects the shortest plants in a population, then, if $h^2 > 0$ for stature, their progeny should be relatively and collectively shorter than the original population. If h^2 had been low, the differences among individuals were mostly determined by non-heritable factors that will not be passed along to progeny.

The V_{GxE} terms of the phenotypic variance equation also have significance to the plant breeder. Relatively high V_{GxE}/V_P values indicate that the population has high *specific adaptation*. In other words, the population performs better in certain environments than in others. The breeder may actually covet such a situation, for example a population that performs well only in a certain set of environments or geographical locations. If V_{GxE}/V_P is relatively low performance is not dependent on environment and it should be possible to develop populations that have broad adaptation or can be successfully grown in many different environments. While narrowly adapted populations may be outstanding in certain environments they generally exhibit average or even poor performance in others. Thus, the breeder struggles with the demand to justify such a variety that will inevitably command a small market size. Conversely, the broadly adapted population usually does not excel in any specific environment and the breeder has problems demonstrating superiority to potential customers.

Very generally, agronomic crops tend to be genetically programmed for higher heritability across a broad range of traits as compared to horticultural crop species. Agronomic species were domesticated and bred more intensively over the past 100 years for resistance to environmental fluxes and adaptation over larger ranges than horticultural species. Another way to portray this is that changes in the environment will have a proportionately larger effect on phenotype in horticultural than in agronomic species. The proportionate contribution of V_E and V_{GxE} to V_P tends to be larger in horticultural species as compared to agronomics, leading to reduced h^2. While there are scores of exceptions to this rule, the generalization is instructive for the application of strategies to horticultural vs. agronomic species.

How does one go about measuring all of these parameters? The set of genotypes and environments under study must first be clearly defined since all parameters are strictly dependent on the datasets used. V_P of maturity date of peaches grown in Georgia is less than that of all *Prunus persica* grown in North America. Once these are defined, it is possible to vary one set while holding the other constant: one genotype grown in a multitude of environments that collectively encompasses the whole range; many genotypes that approximate the whole of the gene pool under consideration grown in a specific environment. The former will yield an estimate of V_E, while the latter will estimate V_G. If either V_E or V_G is known it is possible to use the equation $V_P = V_G + V_E$ to estimate the other.

It is much more challenging to arrive at credible estimates of V_A, V_D, V_I, and the V_{GxE} interaction terms. Quantitative geneticists have devised elaborate statistical designs that allow each term to be mathematically isolated. The plant breeder, however, generally uses indirect or imprecise estimation methods. If one assumes that V_A is the best genetic predictor of phenotype, then proportionally higher V_A should lead to a higher correlation of parent and progeny phenotypes. Therefore, the slope of the parent-progeny regression:

$$Y = aX + b$$

where Y = progeny phenotype, X = midparent, a = slope, b = Y intercept is often used as an estimate of narrow sense h^2. V_A can then be inferred by multiplying by V_P.

Another indirect method is by comparing the means of a source population and of selected progeny plants. If narrow sense $h^2 = 1.0$ then the phenotypic mean of progeny of selected individuals should be equal to the mean of the selected plants. If the mean of progeny is less than the mean of the selected population then narrow sense h^2 must be <1.0. This is termed *realized heritability*, an estimate of narrow sense h^2, and is extremely useful to plant breeders due to simplicity and accuracy.

Calculation of realized heritability:

$$h^2 \text{ (realized)} = \text{(response to selection)} \div \text{(selection differential)}$$

Or

$$h^2 \text{ (realized)} = \left[\text{(avg 1st generation)} - \text{(avg 2nd generation)} \right] \div \left[\text{(avg 1st generation)} - \text{(avg selected parents)} \right]$$

The accepted equation for estimating narrow sense h^2 based on response to selection is as follows:

$$h^2 = \text{cov}(X,Y) / V_{PX}$$

Cov(X,Y) is the covariance of the parent and progeny populations and V_{PX} is the phenotypic variance of the parent population (Holland et al., 2003).

It must be fully appreciated that heritability estimates are truly empirical. The estimates are relevant only within the range of genotypes and environments under which the data were generated. Thus, heritability for a given trait in a given crop species may vary tremendously depending on the range of varieties tested and where and when the testing was done. Heritability can even change during the course of a single breeding program, as the range of variability is progressively attenuated. The parameter will also vary according to the method used to derive the estimate.

IMPLICATIONS OF QUANTITATIVE GENETICS TO POPULATION GENE FREQUENCIES

Natural populations of cross-pollinated plants tend to carry (store, preserve, and generate) genetic variability. Recessive alleles are masked by dominant counterparts, thus not imparting reduced or enhanced fitness. The cumulative result of all genetic variability carried by a population always results in an overall net fitness that is less than optimum. This is because certain individuals within the population at any given point in time will be homozygous for the more deleterious allele at a locus. The mean fitness of a population relative to the theoretical optimum is termed "load". The parameter is not quantifiable in a strict sense and is always relative to a defined set of environmental assumptions. Rather, plant breeders and geneticists refer to load in a general sense as being relatively large or small. Load is the price a population or species must pay for buffering against environmental shifts and long-term fitness. Inbreeding depression (see Chapter 16) is a direct manifestation of genetic load at the level of the fitness of individuals within populations.

The plant breeder applies the harshest selection pressures of all. If narrow sense $h^2 = 1.0$, then the relative fitness of undesirable genotypes at loci under artificial selection is 0.0. This scenario is highly illustrative of why V_A is preferable to V_D to the plant breeder. Any dominance will mask an undesirable allele, perpetuating it to the next generation, such as with allele a above. For the plant breeder relative selective fitness would be 1.0 for AA and Aa and 0.0 for aa. Starting from the same point, the phenotypic frequencies in the following generation would be

$$510\, AA + 0.408\, Aa + 0.082\, aa$$

If instead A and a were additive and the relative fitness is 1.0 for AA and 0.0 for both Aa and aa, the next generation would immediately become 1.00 (or 100%) AA. Thus, a large advantage is imparted by additivity as compared to complete dominance.

Generally, the plant breeder will apply selection to a population in attempts to move the mean of populations of progeny in a defined direction. For example, if the mean nut weight of a pecan population is 10.0 g, the breeder may wish to incrementally increase mean nut mass in an effort to increase overall orchard yield. Naturally, the individuals exhibiting the highest nut mass, exceeding 10.0 g would be selected and their progeny used to constitute the next generation. Depending on heritability, the mean would increase with each ensuing generation until a limit is reached, an asymptote. No further gains are usually realized, evidence that all V_G has been exploited, and none remains, only leaving $V_P = V_E$. Woodworth et al. (1952) demonstrated, however, that response to selection may resume following a long period of futility. Such a "second spurt" of response to selection is usually due to a rare recombination event that aligns pre-existing desirable alleles in trans to cis. A second but less likely explanation is the appearance of a new mutation that positively affects the character under selection.

In certain instances, the plant breeder may apply a *truncating* selection to a population that is centered around an already desirable mean. This situation would apply for example in fruit tree crops where considerable investment has been made in orchard management and harvesting equipment. These machines have been manufactured with a certain plant ideotype (see Chapter 6) in mind, and the grower only wants individuals that fall within the specifications that are consistent with this ideotype. In this case, the plant breeder reduces V_P by decreasing V_G, selecting for the desired mean $\pm X(\sigma)$, where X usually falls between 0 and 1 and σ is the standard deviation of the sample.

The plant breeder also uses the concepts of population genetics when striving to preserve germplasm and for the reconstitution of populations. Specifically, this pertains to the assumption of random mating and the maintenance of allelic frequencies over generations. Most of our food crops are propagated sexually by seed, and seed is used by growers to establish the production populations that ultimately will be harvested for commercial purposes. The grower carefully considers that attributes of all possible varieties, commercial breeding populations that he/she could produce. Each presents a special set of opportunities and risks and it is important that the attributes that are espoused be the same as those exhibited.

Two special cases come into play that are both related to reduced population size: *drift* and *founders effect*. If a population exists wherein 100 loci have two alleles with frequencies p^N and q^N are 0.9 and 0.1, then the genotypic class frequencies at each locus are 0.81 AA + 0.18 Aa + 0.01 aa. The probability of any one individual having at least one of all 100 of the scarcer alleles is exceedingly low $[(0.19)^{100} = 7.505 \times 10^{-73}]$. The corresponding population size necessary to ensure that the allelic frequencies will be faithfully maintained in the next generation, the one planted by the grower, is exceedingly large, much greater than even the reciprocal of this number ($>10^{72}$). It is recommended that this number be increased by another 1000% (i.e., multiply by ten) to ensure that all alleles have been represented within a reasonable doubt. Greenbaum et al. (2014) published an example of the theoretical statistical implications of both genetic drift and founders effect.

Utilizing population sizes smaller than this threshold will likely culminate in genetic drift or a generation change in allelic frequencies due to inadequate sample sizes. The plant breeder and seed production coordinator (see Chapter 11) are faced with this issue when undertaking increases of any population that is comprised of genetic variability (e.g., an "open-pollinated" population). Estimates of the frequencies of critical alleles must be considered in deciding how many individuals should be used to generate the increased population. The plant explorer collecting germplasm in the wild is also faced with this dilemma: how to sample the wild population to efficiently and effectively capture the maximum degree of available genetic variability? In general, minimum population size necessary to reconstitute the gene pool increases exponentially as the total number of loci deemed important (see above) and the number of alleles per locus increases, and with progressively lower allelic frequencies (e.g., $q < 0.1$).

Unfortunately, plant breeding is not practiced effectively or efficiently in the world of the theoretical. Time, space, and work all require resources, or money, that will always be limiting. Compromises must be made nearly every minute of every day in a plant breeding project. Where compromises to conserve resources are necessary it is desirable to make them with the full knowledge of what is being sacrificed, or placed at risk, rather than simply guessing.

CYTOGENETICS

Long before Mendel experimented with his pea plants in the mid-19th century, the light microscope had been invented and the wonders of cells and their components had been intensively studied and characterized. The significance of the curious dance of the rod-like colored bodies (*chromosomes*) in eukaryotic somatic cells that could be readily stained with many popular cytological dyes had been documented (by Walther Flemming in 1882; Elliot, 1958) but the significance was not yet known. It was concluded that these rod-shaped bodies played a role in cell division and proliferation since they split into two units apiece that migrated to the corresponding daughter cells. A similar, but not identical, set of observations on the behavior of dyed rod-like bodies was made on certain cells within stamens and pistils of immature flowers.

The genetic significance became evident when, as alluded to above, the congruencies with the behavior of genes and alleles with that of chromosomes were illuminated. Following the articulation of the Sutton-Boveri Chromosomal Theory in 1902 (Crow and Crow, 2002) and conclusion by Morgan et al. (1915) that genes were physically located on chromosomes, interest in studying their behavior increased tremendously. During the period 1910–60, an enormous body of work pertaining to eukaryotic cytogenetics was generated, the net results of which mostly confirm what may be studied nowadays with more precise or direct molecular methods. Since 1960, the number of new papers pertaining purely to cytogenetics has dwindled and there are very few students currently being trained in classical cytogenetics, especially in plants.

The field of cytogenetics was considered to be so important that many popular textbooks were authored. One that focused specifically on the plant breeding implications of the genomic structure was entitled *Plant Breeding and*

Cytogenetics (Elliot, 1958). Despite the decreased interest in cytogenetics this field continues to be extremely useful in plant breeding, primarily as a conceptual way to view plants and the genomes that underlie their genetic behavior both in the individual and in the population. Although recent advances in cell and molecular biology have largely eclipsed or superceded the field of cytogenetics, it is still historically and conceptually important to understand the behavior of chromosomes within genomes. Many plant geneticists and breeders continue to appropriate classical cytogenetic concepts and terminology to illustrate evolution, speciation, and barriers to gene flow. The mitotic and meiotic cell cycles are studied in elementary science classes as early as primary school grades.

The fundamental unit of cytogenetics is the chromosome, now known to be an organelle in its own right. The chromosomes of vascular plants are nearly identical to those of other eukaryotes in terms of general structure, composition, and function. The chromosomal "backbone" consists of a linear strand of DNA that is characterized by specific functional sequences and domains (Fig. 3.5). The chromosome may be either single stranded, as during anaphase, telophase, and interphase I of mitosis or double-stranded as during interphase II, prophase, and metaphase. While the chromosome is visible as a double-stranded or bi-modal structure, each strand is referred to as a chromatid. This terminology is particularly useful in following the individual products of meiosis. Each strand is the product of DNA

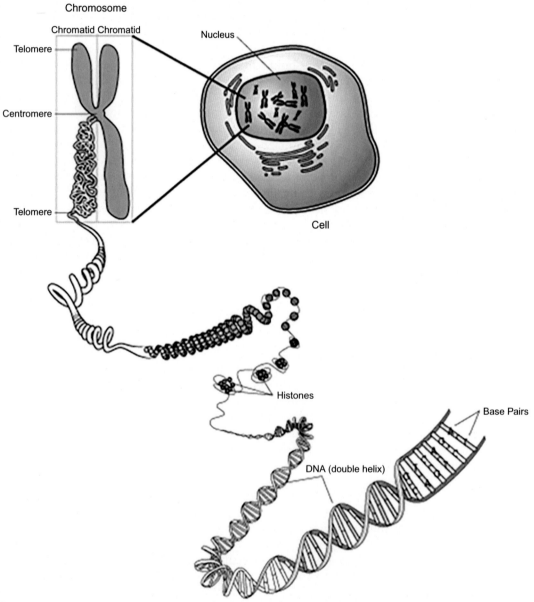

FIG. 3.5 The primary, secondary, tertiary, and quaternary structure of the eukaryotic chromosome. *Source: National Institutes of Health, National Human Genome Research Institute, Division of Intramural Research, https://commons.wikimedia.org/wiki/File:Chromosome.gif.*

replication and is considered to be an exact copy. Therefore, when chromatids separate at anaphase the daughter cells receive identical genetic information.

The DNA strand is associated with, and chemically bonded to, proteins and RNA, and more infrequently, carbohydrates. The types of proteins and RNAs vary according to location and may recur in patterns. The entire associated substance, DNA, protein, and RNA, is called *chromatin*. The complex of molecules plays a role in the storage, replication, and transcription of genes, and in other specific functions such as homologous pairing and cell division (Nicodemi and Pombo, 2014).

The ends of the chromosomes, or chromatids, are called *telomeres* (Alberts et al., 2014). They have been found to be associated with specific DNA sequences, usually simple sequence repeats (SSR), and with specific classes of ribosomal polypeptides. A suborganelle known as a *nucleolus* has been observed to be associated cytologically at or near the telomeres of specific chromosomes in somatic tissues. This has been termed the *nucleolus organizer* (NOR) chromosomal region and has subsequently been shown to be a predominant site at which the rRNA genes are located. Nearly all chromosomes are characterized by a distinct constriction, and sometimes two or more. The constriction is usually the locus at which spindle microtubules attach to chromatids at metaphase, and the effective "hook" used to pull them to the opposite poles that will constitute the new daughter cells. The chromosomal constriction used for microtubule attachment is called a *centromere*. This region has been studied extensively, and, like the telomeres, also consists of specific DNA sequences and polypeptides. The centromere bisects the chromosome into two arms, each comprising the segment from the centromere to one of the telomeres. The point in the chromosome wherein the centromere is located is variable, from central (known as metacentric) to closer to a telomere (known as acrocentric) to indistinguishable from the telomere (known as telocentric). The two arms are usually unequal in length and are referred to simply as long and short (corresponding to the P and Q arms).

Other chromosomal constrictions are usually closer to the telomeres and have no apparent function (except that the NOR region is often a constriction; Alberts et al., 2014). These secondary constrictions are quite useful, however, for identifying specific chromosomes. Any chromosome that contains two functioning centromeres is inherently unstable during anaphase, since microtubules originating from different poles can and will attach to them and pull the chromosome apart, forming a rupturing bridge-like structure at anaphase. Since the break is not necessarily mediated by DNA sequence this can result in the unequal distribution of genes to daughter cells.

The genes, or genetic loci, have been shown to inhabit the chromosomal DNA strand that runs from one telomere through the centromere and to the other telomere (Alberts et al., 2014). Not all DNA is transcriptionally active, or even qualitative similar within the chromosome. Some chromatin is silent and never transcribed. Such DNA domains are often associated with chromatin that is condensed, tightly wound, associated with proteins known as histones, and can be viewed physically as staining more intensely than the surrounding chromatin domains. These chromosomal regions, or domains, are known as *heterochromatin*. Some heterochromatin is constitutive, never changing, while other heterochromatic regions are changeable, or facultative. Recent findings suggest that the structure and function of heterochromatin, including centromeres and telomeres, is mediated by small interface (si) RNAs that bind and interact with it (Kanno and Habu, 2011). In contrast, chromatin that is not condensed, that does not bind intensely with cytological stains, is known as *euchromatin*. As the name suggests, it is presumed to consist of DNA sequences that are functional or transcribed.

The chromosome during mitotic or meiotic metaphase is, however, not in a state where the genes are being actively transcribed. Metaphase chromatin is wound into a discrete package for distribution at anaphase. Immediately following the cytological drama of anaphase, chromosomes begin to lose their discreet appearance and become elongated and more thread-like. Cytological stains no longer are useful in distinguishing the rod-like bodies in telophase and interphase that are visible during metaphase. The nucleus assumes a diffused look during interphase. Functionally, it is assumed that genes go back to a state of activity, including transcription of mRNAs. DNA, and chromosome, replication and repair also occur during interphase.

Ultrastructurally, the metaphase chromosome is not a linear DNA backbone at all but a twisted structure that looks more like a gathered hem (Fig. 3.5). Despite this structural configuration, the linear sequence of genes in the chromosome is mostly retained at metaphase, as has been demonstrated by in situ hybridization (Jiang and Gill, 1994; see Gene Mapping below). Patterns of constitutive heterochromatin domains on specific are highly reproducible and have been used extensively to assist in the identification of specific chromosomes and subchromosomal regions. Several different stains that interact distinctly with chromatin have been used to visualize an array of patterns, combinations of which impart more acuity for the microscopic identification of individual chromosomes and study of the substructure (Alberts et al., 2014).

Each species has a characteristic number of chromosomes. The somatic cell number in vascular plants varies from four to several hundred and does not appear to be correlated with any particular morphological or physiologically features. The number of chromosomes in the gametophyte (or gamete) is referred to as *n* or the haploid number. As a direct consequence the number of chromosomes in the sporophyte is 2n or diploid. Hence, each cell of the sporophyte contains two nearly identical copies of each distinct chromosome. If the gametophyte contains seven chromosomes the sporophyte will contain 14 corresponding to seven pairs. The two nearly identical chromosomes are called *homologues*, the corresponding adjective being *homologous*.

The relative physical size of chromosomes is also highly variable among vascular plant species. Chromosomes in some species can be as much as ten orders of magnitude larger or smaller than another. In general, chromosome number and size are negatively associated (the higher the number, the smaller the size) but exceptions to this rule abound. The total number of transcribed, functioning genes has not been found to vary to nearly the extent of chromosome number and size. Plant species with higher DNA content (C value), manifested as higher chromosome numbers, larger chromosomes, or both, usually have larger quantities of as repetitive or non-transcribed DNA, such as in heterochromatin (Alberts et al., 2014).

While homologous chromosomes have been shown to interact during mitotic cell cycles, the net result is usually genetically neutral. In other words homologues behave independently for the most part. One homologue condenses into a metaphase chromosome and sister chromatids separate at anaphase then are replicated again during interphase. The other homologue is doing the same thing. Homologues have been shown to exchange DNA sequences during mitotic cell division cycles (known as *somatic recombination*; Pucha and Hohn, 2012), but the prevalence of the phenomenon is not known and may be variable during development or under different environmental conditions.

Each different chromosome has a unique DNA sequence backbone but has a similar appearance to other chromosomes under the microscope. They can be distinguished from one another cytologically by comparing other attributes, such as relative physical total and arm length and secondary constrictions. The repeatable patterns of staining intensity within each chromosome, known as *banding*, is also used especially in human amniocentesis where the ability to identify specific chromosomes in the developing fetus is essential to the accurate diagnoses of genetic disorders.

Humans characteristically have 46 chromosomes and females have 23 pairs but males have 22 pairs plus two chromosomes that that look very different from each other. These oddball chromosomes play a role in the determination of sex in humans, females being XX (two copies of the X chromosome) and males being XY (one copy of the X chromosome, one copy of the Y chromosome). The names "X" and "Y" are arbitrary. In other animals the female is the one that has different chromosomes. This is referred to as sexual chromosome dimorphism. The other 22 pairs of chromosomes do not play a direct role in sex determination in humans and are called autosomes. Each chromosome pair consists of two homologous chromosomes. While sexual chromosome dimorphism has been observed rarely in angiosperms (Charlesworth and Charlesworth, 2012) the rule is that plants do not have sex chromosomes or autosomes (just "chromosomes").

Following human embryonic amniocentesis prospective parents are presented with a *karyotype* of the developing fetus. A karyotype is a pictorial or graphic ordered representation of the genome of an individual, population, or species that depicts their number, size, and secondary features of individual chromosomes at mitotic metaphase. Human cytogeneticists have become so skilled at identifying specific chromosomes that each is given a number. When a cytogeneticist in the U.S. discusses human chromosome #13 in a patient with another cytogeneticist in Asia, they both understand the distinguishing aspects of human chromosome #13. They will even know whether extremely tiny segments of each chromosome that carry names according to the arm and distance from other distinguishing features are present or missing (Alberts et al., 2014). Plant cytogeneticists have not, in general, reached the same advanced level of knowledge and sophistication as have human cytogeneticists.

Meiosis is a specialized kind of cell division that occurs only in sexual organs and during the formation of the gametophyte from the sporophyte. Nearly all cytological studies of meiosis in plants have involved *microsporogenesis* or male gamete formation. Logistically, it is much easier to visualize microsporogenesis than is *megasporogenesis* (female gamete formation) because the former is associated with much higher numbers both within individual anthers and among androecia within a given plant. In contrast, megasporogenesis usually occurs in a single cell in each ovule/carpel. Finding many megaspore derivative cells at the desired stages of meiosis to permit the observation of meiotic metaphase is not an easy task.

Both microsporogenesis and megasporogenesis proceed according to two phases, known as meiosis I and meiosis II. The following discussion pertains specifically to microsporogenesis; exceptions pertaining to megasporogenesis will be addressed later, below. A layer of cells in the tapetum of the anther, located on the inner wall next to the lumen, becomes dissociated cytologically, and individual cells that will undergo meiosis are referred to as *microspore derivative cells*. During meiosis I (first, or reduction division), the homologous chromosomes undergo pairing that is driven by their nearly identical DNA sequence. This pairing is mitigated by polypeptides that are encoded by the DNA. Even if the DNA sequence is slightly different, such as minor additions or deletions, the chromosomes will not normally pair. The pairing process itself has been found to be under genetic control. The functioning or non-functioning of a single gene (*Ph*) in wheat has drastic effects on the pairing of homologous chromosomes (Sears, 1976).

The paired structure of homologous chromosomes is plain to see cytologically as the chromosomes enter meiotic prophase. As meiosis I continues, it is often possible to see where crossing over occurs physically, visible as arched structures called *chiasmata* (Fig. 3.6). Further condensation of chromatin occurs, concomitantly with the attachment of spindle microtubules to the paired chromosome centromeres and polar bodies at the opposite poles of the cell. The "crossed-over" chromatin strands are shifted toward the telomeres (known as "terminalized chiasmata") that play a

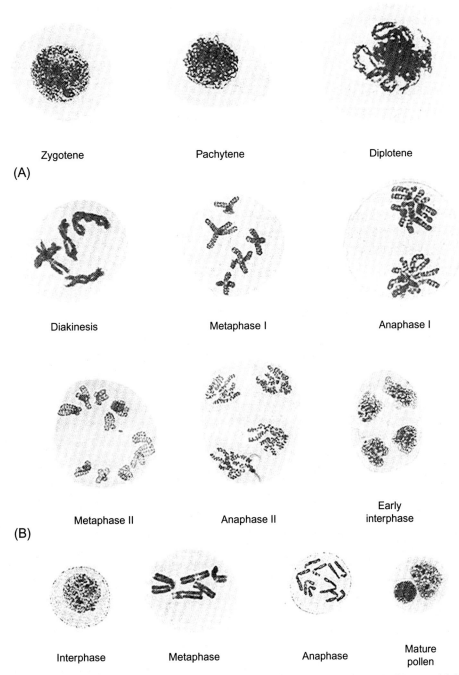

FIG. 3.6 Meiosis as observed during microsporogenesis in Trillium (*n* = 5). (A) Meiosis I; (B) Meiosis II. *Courtesy of A. H. Sparrow. Redrawn from Sparrow, A.H., Michigan State University, https://s10.lite.msu.edu/res/msu/botonl/b_online/e09/meioset.htm.*

role in binding the paired structures (known as *bivalents*) together. Bivalents can take on a ring-like or rod-like shape depending on whether chiasmata are present on both chromosome arms or only one arm (Fig. 3.6). Ring bivalents have terminalized chiasmata at both telomeres, while rod bivalents have them at only one end.

Anaphase I of meiosis is distinct from mitotic metaphase because chromosomes consisting of two distinct chromatids can be observed to migrate to the poles. At this point in development the sporophyte has ended and the gametophyte begun. The diploid has spawned two haploid cells each with half the number of chromosomes and genes. Genetically, the resemblance of the daughter cells to the parent is also obscured since thousands of alleles have been effectively reshuffled and assorted, the combinations of which vary with each incipient gamete.

Telophase I and interphase I are extremely brief, since little or no cellular business will be accomplished during this period, and no new gene transcription or translation takes place (Alberts et al., 2014). Meiosis II usually proceeds immediately upon the completion of meiosis I, and often within minutes, but rarely longer than hours (Elliot, 1958). The 2-chromatid chromosomes in each of the two newly haploid daughter cells again become condensed and undergo a process that looks nearly identical to mitotic cell division. Condensed metaphase chromosomes line up at the metaphase plate clearly suspended in a network of microtubules and connected to each centromere and the polar bodies at opposite poles. The metaphase plate in each of the two daughter cells of meiosis I is usually perpendicular and the polar bodies also perpendicular, and on a displaced plane (Fig. 3.6 metaphase II). This orientation is important because the divisions proceed so quickly that the new cell walls are not yet in place while the nuclear divisions are occurring. Certain mutant types have been found in which the metaphase II plates are not perpendicular, but parallel, leading to the fusion of two haploid nuclei back into diploids known as *restitution gametes* and sometimes employed by plant breeders to obtain polyploids (see Chapter 8).

At metaphase II, the chromatids then separate from each other and migrate to the poles, of which there are now 4 (Elliot, 1958). Cell walls are formed around the nuclei, usually forming a tetrahedral structure, called a *tetrad* or *pollen quartet*. The "tetrad" terminology is analogous to the "ordered tetrad" terminology that has been applied to genetic studies of meiosis in ascomycetes, although gametes in angiosperms are not ordered per se. From this point, the individual gametophytes begin to develop separately, each into an individual pollen grain, also considered the mature angiosperm male gametophyte. Many aspects of the phenotype of the mature male gametophyte are, however, determined by the genes of the sporophyte from which it was derived. One example is self-incompatibility the phenotype of pollen of gametophytic origin in some plant species and of sporophytic origin in others (see Chapter 10).

Each incipient pollen cell begins as a single haploid nucleus, single-stranded chromosomes entering prophase following meiosis. Following a round of DNA replication, one mitotic division, known as *pollen mitosis*, takes place, but no additional cell walls are deposited to physically separate the resulting nuclei. The mature pollen grain, male gametophyte, consists of two haploid nuclei that are usually distinct. One nucleus assumes a diffused appearance (generative nucleus), while the other (sperm nucleus) becomes more compact. In some angiosperm species, the generative nucleus divides again mitotically to result in three nuclei. During gametophyte maturation, the cell wall of the pollen grain is impregnated with molecules that play a role in the pollen-stigma recognition process. As was stated earlier, these may originate from gene sequences encoded in the gametophyte itself or may have come from the sporophyte. Hence, the phenotype and genotype may differ.

Megasporogenesis is quite different from microsporogenesis except for the fundamental process of meiosis (Elliot, 1958). The gynoecium prior to egg formation features a single *megaspore derivative cell* near the incipient micropyle. Meiosis proceeds in a similar fashion as described above for microsporogenesis to the 4-nucleate stage that remains devoid of cell walls. Evidence has appeared that meiosis may not be equivalent during microsporogenesis and megasporogenesis (Gohil and Ashraf, 1984). The physical location of the individual nuclei is extremely crucial, however, as only one of the four nuclei will go on to form an egg cell. Evidence has shown that which nucleus becomes the egg is not random, but preferentially includes nuclei that are genetically, and cytogenetically functional (Golubovskaya et al., 1992; Barrell and Grossniklaus, 2005). If any cytological anomalies occur during meiosis, such as chromatin bridges and unequal distribution of chromosomes, the resulting nuclei are excluded from subsequent egg development (Huang and Sheridan, 1996). Thus, microsporogenesis is more tolerant of genetic changes than is megasporogenesis.

Each of the four haploid nuclei from meiosis undergoes a mitotic cell division cycle to yield eight *coenocytic* haploid nuclei (with no cell walls or membranes separating the nuclei). Three nuclei migrate to one end of the ovule then degenerate, three to the other end, the closest to the micropyle, and 2 to the middle. One of the three nuclei at the other pole becomes surrounded by a cell membrane and becomes the egg, or female gametophyte. The other two nuclei are termed *polar nuclei* and will play an important role in seed development (i.e., fusion with a male haploid nucleus to form the triploid endosperm). The polar nuclei remain in a coenocytic state awaiting fertilization. The two nuclei in the middle are called the *antipodals* and become transcriptionally active for a period following meiosis then eventually degenerate.

To summarize, the mature male gametophyte, usually the pollen grain, contains two or three haploid nuclei, each chromosome of which was derived from a single DNA strand in one of the recombined chromatids at metaphase I of meiotic microsporogenesis. The sperm nucleus will undergo one more mitotic division during pollen tube growth (see below), and one of those daughter nuclei will eventually fuse with the egg to form the zygote or new sporophyte. The mature female gametophyte, or egg cell, contains a single haploid nucleus that was derived from a single DNA strand in one of the recombined chromatids at metaphase I of meiotic megasporogenesis, similar to the male. While genetic and cytogenetic anomalies can result in pollen inviability no mechanisms exists that preferentially selects for any one of the four haploid nuclei that are produced by meiosis in microsporogenesis. In contrast, mechanisms have been found that do select for the most fit genotype among the four haploid nuclei in megasporogenesis meiosis (Noher de Halac and Harte, 1985).

STRUCTURE AND FUNCTION OF CHROMOSOMES

Chromosomes are packages of functioning genes and replication/repair control and structural elements that have arisen and been perpetuated during evolutionary history. Thus, it is probable that the packaging of certain genes and elements on different vs. the same chromosome has some effect on the fitness of the organism that possesses it. The packaging of specific genes together and the linear sequence of the genes tend to be preserved among closely related species, even when they have different chromosome numbers. Specific chromosome segments are highly preserved but may be packaged differently among the chromosomes. A picture emerges of speciation by the geographic isolation of an originally allopatric population. With time, the genomes of reproductively separated subpopulations undergo independent new chromosomal configurations: fusions, fissions, translocations, duplications, inversions, deletions, mutations, etc. As different environmental factors contribute to different selection pressures, the subpopulations begin to diverge, one having a cytogenetic configuration (number and structure of chromosomes) different from the other. At a point, they are no longer able to intermate, the gene pools become independent, and new species are established.

Large chromosome segments of closely related but reproductively isolated species typically have remained intact. The DNA sequence of these chromosome segments is highly conserved or even syntenic. For example, the family Solanaceae includes the genera *Nicotiana*, *Solanum*, *Capsicum*, and *Petunia*, among others (these include the most important economic species on which research has been conducted: tobacco, nightshade, potato, eggplant, pepper, tomato). They have been classified together since the earliest iterations of systematics, mostly back to the original work of Linnaeus that was based mostly on floral structure. The haploid chromosome numbers in these four genera range from 7 to 32 (Huskins and La-Cour, 1930; Goodspeed, 1933; Chapman, 1958; Maizonnier, 1984). While chromosome numbers vary among species within the Solanaceae, certain subchromosomal segments have been found to be almost entirely conserved (Wu and Tanksley, 2010). Thus, the cytological changes that are associated with speciation within this family involved the movement of large chromosomal segments, not the wholesale reshuffling of genes. Chromosome number and structure, therefore, are good criteria for distinguishing taxa that are closely, but not distantly related.

Plants and animals differ in their relative tolerance of cytological changes. The most intensely studied animal, *Homo sapiens*, is characterized by an extremely narrow range of cytogenetic configurations. The overwhelming majority of humans, regardless of continental or cultural origins, have 46 chromosomes; 44 autosomes and two sex chromosomes. A comparison of the karyotypes of two humans of the same sex reveals that they are almost always identical. Approximately one in 700–800 humans born alive, or 0.125–0.143% has 47 chromosomes, one extra #21, the second-smallest human autosome. Such individuals do not develop "normally" and present the physical and mental characteristics of Downs Syndrome. Less frequently, individuals with extra copies of sex chromosomes, usually X, are born and survive to adulthood. They are often fully functional although reproductive anomalies and secondary sexual characters are affected. The Y chromosome is only marginally functional, and only one X chromosome is active in adult females, so this is perhaps not surprising. Cytogenetic of miscarried fetuses have revealed a range of more profound changes such as monosomics (one chromosome missing) and trisomics (one extra copy). About half of all human miscarriages up to week twelve of gestation have been determined to be attributable to cytogenetic/genomic structural aberrations (van Den Berg et al., 2012).

While the morphology and fertility of plants are affected by changes in chromosome number and structure, the occurrence of individuals that carry such changes is not particularly rare. One cytogenetic feature that has been utilized beneficially is *polyploidy* or multiple genome sets. It is not unusual to find plants, even within an interbreeding diploid population, that are triploid (three haplotypes) or tetraploid (four haplotypes). More infrequently other ploidy levels may be found such as octaploid (eight sets) and pentaploid (five sets). Haploid (one set) plants may also be found but this important and interesting case will not be addressed at this juncture.

In some cases, polyploidy appears to have supplanted diploidy as the predominant genomic status within a population, or more often, species. An examination of chromosome numbers of species that are obviously related reveals that they differ by orders of two. One of many such examples is tomato (*Solanum lycopersicum*) and potato (*Solanum tuberosum*). While the tomato has 24 chromosomes, potato has 48. Upon closer scrutiny the tomato clearly has two copies each of 12 morphologically different chromosomes while the potato has four copies each of 12 different chromosomes that are correspondingly similar to those of tomato (Figs. 3.7 and 3.8). Thus, the tomato is considered to be diploid and the potato to be tetraploid.

A dilemma is immediately apparent. If the potato is tetraploid what is the ploidy of the corresponding sporophyte and gametophyte? How does the potato behave that is different from tomato with regard to meiosis and the subsequent fusion of gametes? Sporophytes must behave as diploids or else all of our assumptions about reproduction, mating, genetics must be changed to account for more than two copies of chromosomes. It turns out that polyploids generally do behave like diploids during meiosis, with some important exceptions; and some of these exceptions have been very instructive and useful for studying species interrelationships. Even though the potato has four copies of each chromosome cytological observations of meiosis reveal that they generally behave more like diploids. Instead of seeing 12 *quadrivalents* (four homologous chromosomes bound by terminalized chiasmata at meiotic metaphase I) 24 bivalents are usually observed at meiotic metaphase I. The probable reasons for this will be discussed later.

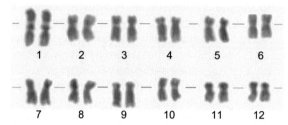

FIG. 3.7 Karyotype of tomato (*Solanum lycopersicum*) showing 12 pairs of chromosomes (2n = 2× = 24). *From Karsburg Isane, V., Roberto, C.C., Ronildo, C.W., 2009. Identification of chromosomal deficiency by flow cytometry and cytogenetics in mutant tomato (Solanum lycopersicum, Solanaceae) plants. Aust. J. Bot. 57, 444–449. https://doi.org/10.1071/BT08223. Reproduced with permission from CSIRO Publishing.*

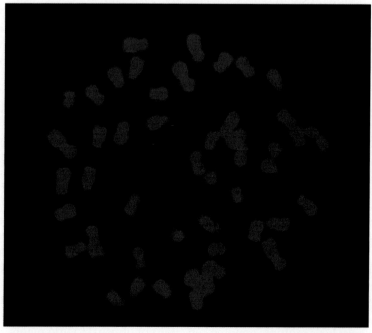

FIG. 3.8 Metaphase chromosomes of white potato (*Solanum tuberosum*) showing 12 groups of four identical chromosomes (2n = 4× = 48). *Source: https://commons.wikimedia.org/wiki/File:Karyotype_of_Potato_(Solanum_tuberosum).png.*

The potato is tetraploid but behaves reproductively more like a diploid. A convention has been adopted for the comparison of such related species. The letter "n" is used to denote the gametophytic chromosome number, while the letter "x" is used to denote the number of chromosomes in a theoretically basic set of chromosomes that consists of one copy of each unique DNA sequence. Thus, the tomato is described, therefore, as $2n = 2x = 24$ and the potato as $2n = 4x = 48$.

Recent advances in molecular biology have made it possible to study the structure of genomes at the level of DNA sequence. Among the fascinating discoveries that has emerged from the molecular study of genomes is rampant pseudo-homology. If a DNA sequence of given length is compared with the rest of the genome other similar sequences are often found. While not identical in nucleotide base sequence statistical models prove unconditionally that such similar sequences must share a common evolutionary ancestor sequence. The science of parsimony has been adapted to infer the genetic and phylogenetic relatedness of entities by comparing base sequences of homeologous DNA (Cracraft and Helm-Bychowski, 1991).

In certain cases ancestrally related sequences are associated with functional DNA that is disparate, for example genes that encode enzymes that catalyze different biochemical reactions. In other cases one sequence copy may have a known function while the other related sequence is either silent or has no known function. Hence, the concept of "basic genome or set of chromosomes" is blurred.

Polyploidy intuitively seems to make adaptive sense in biology. If two gene copies are good, why wouldn't three, four, or more be as good or even better? The natural order of life on Earth has answered this question. All organisms, including plants, are constantly experimenting with new genetic variability and genomic configurations during the course of evolution. It is likely that the mechanisms that seem to result in genetic "mistakes" are also perpetuated to ensure that life is not static. Polyploidy is one of many aspects of the total array of genetic variability.

Polyploids can often be found within predominantly diploid plant populations. Yet polyploidy has not come to predominate on Earth; rather, diploidy appears to predominate. Apparently multiple copies of genes does not necessarily impart evolutionary fitness to the individual. Naturally-occurring polyploids are, apparently, intermediates in long-term evolutionary processes. Mutations are more easily tolerated in polyploids because dysfunctional genes are buffered by multiple copies of functional counterparts. Maybe polyploids arise constantly from diploids, new mutations arise, and diploid derivatives evolve from these polyploids. These derivatives progressively behave more and more like diploids and shed or adapt the surplus genetic information over evolutionary time. These are speculations but not inconsistent with observations.

Another answer to the question "…why wouldn't three, four, or more (copies) be as good or even better than 2?" is provided by the accumulated observations of the comparative performance of diploids and *induced polyploids*. Induced polyploids? This is one of the "magic bullets" available to the plant breeder. While many compounds (e.g., oryzalin; Ascough et al., 2008) and some physical treatments (e.g., temperature shock) have been shown to affect mitotic anaphase, none is as effective or as inexpensive as *colchicine*. Colchicine (Fig. 3.9) is a secondary metabolite in the alkaloid biochemical family that has been isolated and purified from extracts of corms of Autumn Crocus (*Colchicum autumnale*). *Crocus* extracts have been used in herbal medicine for the treatment of gout (uric acid accumulation in joints) and colchicine was found to be the active ingredient in these extracts.

Colchicine has a number of primary and secondary modes of biological activity, one of which is the inactivation of mitotic spindle microtubules, the proteinaceous muscle-like strands that mediate the movement of chromatids to opposite poles of a dividing cell. Since tubulin structure is common across a broad array of eukaryotes colchicine can likewise inactivate mitotic spindles in an astonishing array of organisms. When colchicine is applied in an aqueous solution to active growing plant tissue, polyploidy cells inevitably appear. Cytological observations of cells treated

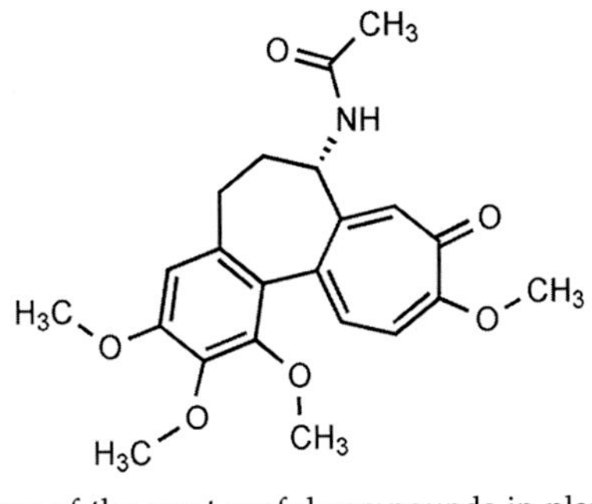

FIG. 3.9 The chemical structure of colchicine, one of the most useful compounds in plant breeding. *From Perfuratriz.ml http://perfuratriz.ml/gexa/o-colchicine-huw.php.*

with colchicine show that mitosis proceeds normally to metaphase. Chromatids separate but fail to migrate to the opposite poles, congregating instead at the cellular equator. Consequently, a tetraploid cell is derived from the original diploid.

If colchicine remains in contact with the tissue for prolonged periods of time additional mitotic cycles may be similarly affected resulting in higher ploidy levels (8×, 16×, etc.). The compound is usually applied as a pulse to minimize the probability of *mixoploidy* (tissues consisting of cells with different ploidy levels). As with any *mutagenic agent* (a substance or treatment that causes mutations) colchicine treatments are usually performed on germinating seeds. That way, meristematic cells that are the progenitors of all apical plant tissues and organs will be targeted for conversion to tetraploidy. The plant breeder usually wants the new ploidy condition to be carried to the reproductive organs, and to the gametes, thus allowing genetic studies and breeding to proceed.

A comparison of diploid and induced *autotetraploid* (4 nearly identical chromosome sets or genomes) plants shows many interesting differences. The most obvious question is: are the 4× plants larger? The answer is: sometimes, not always, and never as much as twice as large or larger (either in stature, volume, or mass). Upon examining tissues of diploids and autotetraploids, however, the cells do appear to differ by order of two with respect to volume. The number of cells that make up a given organ within an autotetraploid plant, however, is reduced as compared to the diploid, thus compensating for the differences in cell volume. Developmentally, the physical size of plants and constituent organs seems to be programmed into the genome and cell division stops when that size is reached (Stebbins, 1949).

Experimental studies of autotetraploids have revealed a range of findings on plant performance as compared to diploid counterparts. Are polyploids, for example, more tolerant of pests, diseases, or environmental stresses? Reports of enhanced tolerances to environmental stresses have appeared (Xiong et al., 2006; Sattler et al., 2016), but other evidence suggests that polyploidy per se is not the determining factor (Hilu, 1993). Many polyploids have, in fact, been found to be less tolerant of biotic and abiotic insults than are comparable diploids. While most biological systems operate comparably in polyploids as compared to diploids, self-incompatibility (see earlier discussion) is a possible exception. Polyploids often lose self-incompatibility, even though the underlying alleles are still present (Chen, 2007). Miller and Venable (2000) have proposed that polyploidy acts as a bridge to sexual dimorphism by disrupting self-incompatibility, allowing other mutations affecting floral development to enforce outcrossing.

Eukaryotic organisms strive over time to behave like diploids, even if they are actually polyploids. When a diploid is induced to become an autotetraploid, observations of meiosis in the early generations reveals a preponderance of multivalent chromosome pairing, with trivalents and quadrivalents observed frequently. The pairing process seems to be somewhat haphazard, as microspore mother cells from the same individual may be observed to have a variable number of such associations. For example, for an autotetraploid individual 2n = 4x = 4, metaphase I composites may be observed that consist of 1 quadrivalent, one trivalent plus one univalent, or two bivalents (Fig. 3.10).

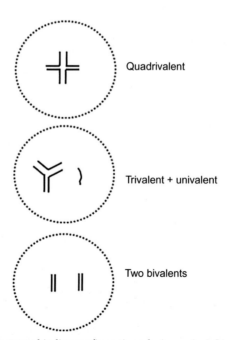

Quadrivalent

Trivalent + univalent

Two bivalents

FIG. 3.10   Possible chromosome binding configurations during meiosis I in an individual 2n = 4x = 4.

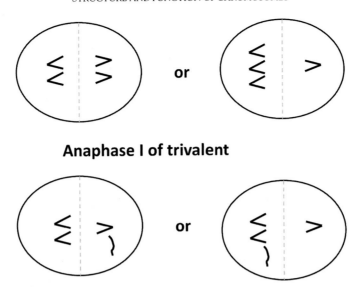

Anaphase I of trivalent

FIG. 3.11 Alternative disjunction patterns of bound chromosomes in quadrivalent and trivalent configurations during anaphase I in an autotetraploid.

The quadrivalents, trivalents, and univalents all present challenges to the equal distribution of genes to daughter cells (Fig. 3.11). The potential problems with trivalents and univalents are obvious, but even quadrivalents can become tangled in the mitotic spindle, and chromosome bridges and lagging chromosomes can result during anaphase I. Not surprisingly, reproductive fecundity is often reduced in autotetraploids as compared to diploids, a direct consequence of such cytogenetic phenomena.

As autotetraploids (and *allopolyploids*, polyploids that combine whole genomes from different species; see below) are reproduced recurrently and selected for sexual fertility, the incidences of multivalent chromosome pairing are reduced. This process is referred to as *diploidization*, and it is probably common in nature (Hilu, 1993). Selection for sexual fertility appears to select indirectly for diploid cytogenetic behavior, and there are likely subtle gene interactions that are responsible for the enforcement of bivalent vs. multivalent pairing associations. In some cases, major genes have been found that control and mitigate the process of homologous chromosome pairing (Dvorak et al., 2006; Bozza and Pawlowski, 2008; Feldman and Levy, 2012; see below).

Autotetraploidy complicates inheritance because Mendel's Laws are violated. There are no longer two copies of each gene, but four. If meiosis proceeds "normally" (as it would in a true diploid), and genes are assorted to gametes in an orderly, predictable fashion, it is possible, with certain assumptions, to generalize a model for inheritance in autotetraploids. Such models will differ for other levels of autopolyploidy. If two alleles exist at a given gene (A in this example), an individual may have five different genotypes: AAAA, AAAa, AAaa, Aaaa, and aaaa. These genotypes have been given names, though not in broad usage: *quadriplex, triplex, duplex, simplex,* and *nulliplex*, in obvious reference to the number of dominant alleles. A determination of the relative proportions of gametes is not necessarily intuitive but can be easily calculated. For example in a simplex individual (Aaaa) three different gametes are possible: AA, Aa, and aa. Assuming random assortment the proportions should be, respectively, $0.25 \times 0.25 = 0.0625$, $2(0.25 \times 0.75) = 0.375$, $0.75 \times 0.75 = 0.5625$. If a simplex were then selfed or crossed with another simplex, the expected relative frequencies of genotypes would be 0.0625^2 AAAA + $2(0.375 \times 0.0625)$ AAAa + 0.375^2 AAaa + $2(0.375 \times 0.5625)$ Aaaa + 0.5625^2 aaaa. By knowing the relationships between genotype and phenotype, it is possible to predict the frequency of phenotypes as well. If A were completely dominant to a in this example, the progeny of Aaaa selfed would be phenotypically 0.684 A- and 0.316 aa.

Rather than dwelling on the many permutations of the same concept, we will end the discussion at this point. The plant breeder must understand the ploidy level of the species he/she is working with, and the resulting impact ploidy has on the inheritance behavior of targeted genes. As was stated at the beginning of the discussion on autotetraploids, inheritance patterns are distorted from those of diploids but still conform to Mendel's Laws if certain assumptions are made. These assumptions are often violated, and the plant breeder is challenged with explaining results that diverge from expectations.

With regard to crop species that depend on reproduction for their value, such as the major grain and fruit crops, autopolyploids are decidedly undesirable because sexual function is impaired. The reader asks: "Are there any possible reasons at all that one would want to induce polyploidy in a seeded/fruited plant species?" The most prominent example of the practical use of autopolyploidy is in parthenocarpic seedless fruit. Fruits were not meant to be void of seeds, for that is why they exist in the first place, as a means to attract agents of dispersal. Seeds often get in the way of pleasurable culinary experience. Fruit and seed development are physiologically and genetically coupled to the extent that fruits will usually not develop unless sexual reproduction has been consummated, resulting in the next sporophytic generation, housed in the seeds.

In certain species, however, fruit development and seed set have been uncoupled, the condition is known as *parthenocarpy*. Fruit growth is generally initiated in concert with seed development, so how can this be accomplished? Triploids (three copies of each chromosome) are highly sterile because the condition violates the even number rule: odd numbers of chromosomes cannot pair and distribute equally at meiosis I. So it would be possible to obtain seedless fruits if the plant bearing the fruit were a triploid, and could not yield functional gametes.

How is it possible to obtain a triploid? By simple mathematics: 2 + 1. Theoretically, an autotetraploid will give rise to a 1n = 2x gamete that, when fertilized with a 1n = 1x gamete from a diploid, will produce a triploid. The plant breeder uses colchicine to obtain a 4x plant from a 2x, then crosses the 4x with a 2x to get a 3x that is sterile and does not produce seeds, but does produce parthenocarpic fruit. This has been accomplished commercially in bananas, navel oranges, and watermelons (Maynard and Elmstrom, 1992; Aleza et al., 2012). Seedless grapes exemplify an alternative genetic strategy: mutations that result in premature seed abortion (e.g., *stenospermocarpy*) prior to the deposition of the seed coat (Loomis and Weinberger, 1979). Triploid seedless grapes have been successfully developed but the fruit is extremely small due to the lack of embryonic gibberellic acid in the developing fruit (Ledbetter and Ramming, 1989). Plant breeders also use haploidy and dihaploidy as tools to facilitate plant breeding. These methods will be covered in a later chapter (Chapter 14).

Other uses for induced polyploidy exist within the realm of plant breeding methodology, namely the combination and bridging of distinct gene pools. Not all naturally occurring polyploids are autopolyploids. Allopolyploids (see definition above) are combinations of genomes from different, but usually closely related, species. The most thoroughly studied example is wheat (*Triticum aestivum*), an *allohexaploid* (2n = 6x = 42). Early cytogenetic observations showed that wheat did not behave like autopolyploids that had been characterized, and that minor structural differences existed in the constitutive sets of chromosomes. Thus, while the numbers added up, with x = 7 in the family Gramineae, it was concluded that the three diploid sets of chromosomes that contributed to the hexaploid were similar, but different from each other (Fig. 3.12).

Classical cytogenetic experiments were conducted wherein wheat was crossed with different diploid species within the genus *Triticum* (Sears, 1941). Hybridization between species, also referred to as *interspecific hybridization or wide crosses*, has been demonstrated to be easily accomplished in the family Gramineae, although it is sometimes necessary to rescue the embryo from a seed suffering from an endosperm incompatibility.

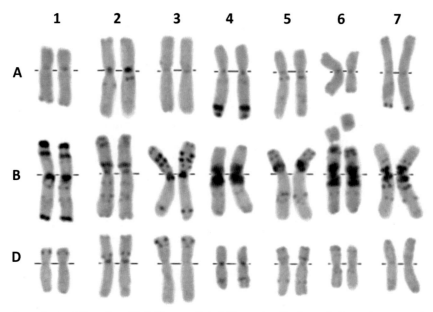

FIG. 3.12   Banded karyotype of wheat (2n = 6x = 42; *Triticum aestivum*) illustrating the three distinct subgenomes (A, B, and D). *From Badaeva, E.D., Dedkova, O.S., Koenig, J., et al., 2008. Analysis of introgression of Aegilops ventricosa Tausch. genetic material in a common wheat background using C-banding. Theor. Appl. Genet. 117, 803. https://doi.org/10.1007/s00122-008-0821-4.*

Subsequently, chromosome pairing behavior was observed in the wheat interspecific hybrids during meiosis I. Certain species would pair with one of the wheat genomes, but not the others, supporting the conclusion that wheat was an allohexaploid consisting of three distinct genomes, designated A, B, and D (Fig. 3.12). Since the actual parental populations that contributed to cultivated wheat and it's immediate progenitors were not available, cytogenetic, and later, molecular, studies were used to determine that the original diploid species components were, likely, *Triticum urartu* (AA) and another possibly extinct *Triticum* sp.(BB, from the *Sitopsis* section of *Triticum*) to produce the AABB allotetraploid. This intermediate was then hybridized with *Triticum tauschii* (DD) to culminate in AABBDD (Talbert et al., 1998; Matsuoka, 2011). Each constituent genome has seven pairs of homologous chromosomes, and these homologues are structurally and sequentially similar to homologous pairs in the other two genomes, referred to as *homeologous* chromosomes, or *homeologues*.

Cytogeneticists and plant breeders have found it convenient to represent composite genomes with letters, such as the A, B, and D genomes of wheat. Thus, the genome formula for wheat would be AABBDD. Barley (*Hordeum vulgare*) is another species in the family Gramineae that is related to wheat. *H. jubatum* is a weedy relative of cultivated barley commonly found on North American roadsides. It is a tetraploid (2n = 4x = 28) that has been demonstrated to be a *segmental allotetraploid*. The two constitutive genomes are very closely related, but not identical, and the genome formula is JJJ'J' Rajathy and Morrison, 1959). When *H. vulgare* and *H. jubatum* are interspecifically hybridized, the genome formula of the resulting interspecific hybrid is VJJ' (Orton and Steidl, 1980).

Classical cytogenetic experiments on wheat also led to important observations pertaining to the genetic control of chromosome pairing during meiosis. When interspecific hybrids were self pollinated or crossed with the diploid parent, chromosomes often paired and separated in such a manner that breakage occurred and chromosome segments ended up in some progeny but not others. When the chromosome segment from the long arm of the 5th chromosome of the D genome was absent high levels of multivalent pairing was observed at meiosis I (Riley et al., 1961). When the 5DL segment was present, pairing tended to be of a lower order, e.g., bivalents and univalents. It was speculated that the 5DL chromosome segment contained a gene, named *Ph* (homologous pairing), that promoted pairing between homologous chromosomes but discouraged pairing between homeologous chromosomes. When the *Ph* gene was absent homeologues were permitted to pair (Sears, 1976). Genes such as *Ph* are crucial for sexual fertility in allopolyploids since multivalent chromosome pairing is associated with the unequal distribution of genes during meiosis, and therefore greatly reduces gamete viability. With the advent of the tools of molecular biology, it has been possible to isolate the wheat PH genes and to study function, find orthologs (*Ph1*, *Ph2*, etc.), and map PH loci to other genomic locations (Hao et al., 2011).

Aside from the control of homologous pairing during meiosis, other mechanisms appear to active in allopolyploids to create genome sequence changes and rearrangements. He et al. (2017) showed that genome structural rearrangements occur frequently in allopolyploid crops. DNA sequence exchanges were found to occur most frequently where homoeologous chromosome segments are collinear to telomeres and in plants derived from doubled haploids (see Chapter 14).

Ideally, the plant breeder would like to expand the pool of genes accessible for plant improvement to include all species, not limited to the range of reproductive compatibility. After all, if a species targeted for improvement lacks an attribute that another species possesses, it is logical to assume that the genes responsible for the phenotype could be transported from one species to the other and still function in a desirable way. Speciation and reproductive isolation were addressed earlier (Chapter 2), and serve as intractable barriers to the unrestricted transfer of genes in the plant kingdom (and other kingdoms as well). Not to be discouraged plant breeders incessantly try to make wider and wider crosses regardless of these limitations and sometimes succeed in spite of the textbooks. When sexual hybridization failed, the cell walls were removed to produce protoplasts that were fused to produce the "pomato" (potato + tomato). The somatic hybrid appears to be useful mainly as a prospective intermediate in the transfer of useful genes between the two species (Garriga-Caldere et al., 1999).

If an interspecific hybrid can be successfully synthesized and grown to sexual maturity, it will most likely be sterile (*amphiploid*), unable to produce functioning gametes. If the reasons for infertility rest with chromosome pairing abnormalities during meiosis it may be possible, and has been demonstrated to restore fertility, by simply doubling the chromosome number of the interspecific hybrid (*amphidiploid*). Thus, each distinct chromosome will now have a homologue with which to pair and meiosis I can, theoretically, proceed to the formation of viable gametophytes. Interspecific hybrids have not usually proven to be of much direct value (like the pomato), but are useful as conduits for gene flow.

An exception to this generalization is the case of Triticale, a new allopolyploid (2n = 6x = 42; AABBRR) small grain crop species that was developed from the interspecific hybrid of durum wheat (*T. durum* 2n = 4x = 28) and rye (*Secale cereale* 2n = 2x = 14). The original goal in the development of triticale was to combine the endosperm characteristics of durum wheat with the cold-hardiness of rye. The resulting entity (triticale) has met with limited success, following the commercial introduction of triticale in the 1950s amid sensational fanfare, and the crop is still grown on small acreage in northern Canada (Oettler, 2005). The *T. aestivum* x *S. cereale* hybrid has also been accomplished, resulting in the AABBDDRR allohexaploid. A small sustained market has also emerged for triticale flour. It has a certain unique flavor, texture, and baking attributes that have attracted a culinary following.

Most plant breeders are understandably daunted by the prospects and challenges that attend the development of an entirely new crop species, such as triticale. Interspecific hybridization, however, has also proven to be an effective intermediate for the transfer of genes between distinct gene pools. Such a program generally begins with a cross between a crop species and a wild relative that possesses a desired attribute, not present in the immediate gene pool. The diploidized interspecific hybrid (amphidiploid; chromosomes doubled with colchicine) may be crossed ("*backcrossed*", a very useful term that is the subject of Chapter 18) to the economic plant species and the resulting population often regains a level of sexual fertility. Cytogenetic observations demonstrate that a higher level of bivalent pairing is evident, and a modicum of viable gametes can be produced. With each successive backcross to the economic parent, more fertility and characteristics of the domesticated parent are recovered. If the program is successful, the domesticated parent is reconstituted with the simple addition of the desired attribute from the wild relative. It is not surprising that simply inherited traits tend to be the most readily adaptable to this approach, since the meiotic transfer of small homeologous chromosome segments from the wild species to chromosomes of the domesticated species must occur.

Such incremental improvements may also involve the addition, or substitution, or entire homeologous chromosomes. This evokes the concepts of *euploidy* and *aneuploidy*. Euploidy is defined as changes in chromosome number in increments of whole sets or genomes. Auto- and allopolyploids are examples of euploid changes. Any change in chromosome number that is not an exact multiple of the basic chromosome number is considered to be aneuploid. Aneuploid cells and whole plants are common during the course of a program to transfer traits from one gene pool to another via interspecific hybridization. As the backcrosses to the economic parent proceed, the homeologues that fail to bind, or bind rarely or marginally with the chromosomes of the economic parent are progressively lost (Fig. 3.13). During the backcross process, intermediates are aneuploids, containing the whole genome of the economic parent with varying numbers of chromosomes of the wild parent. When one entire homologous pair in the economic parent is supplanted by the homologous pair of homeologues from the wild parent, the entity is called a *substitution line* (Fig. 3.14). When pairs of chromosomes are found along with a whole genomic set from the economic parent, the entity is called an *addition line*. Both entities are aneuploids. While addition and substitution lines are interesting and can yield information on the genetic organization they rarely embody anything of practical significance. Sort of like switching or adding body parts with or from another person; the nerves would not necessarily connect the same way. Substitution lines will have duplicated and missing genetic information, while addition lines will have extra copies of some genes, but not most others.

FIG. 3.13   Cell division during early embryo development in *Hordeum vulgare* x *H. jubatum* showing losses of whole chromosomes from aberrant spindle activity.

Addition and Substitution Lines

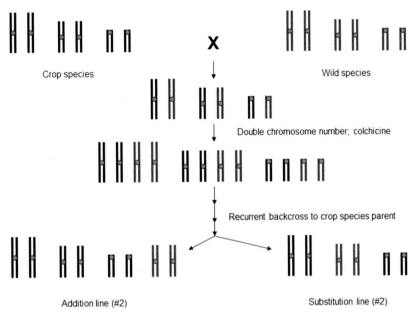

FIG. 3.14   The general process for the development of cytogenetic addition and substitution lines of a theoretical plant species of the genomic structure 2n = 2x = 6.

SUPERNUMERARY OR B CHROMOSOMES

Most of the nuclear genome of eukaryotes, including angiosperms, is contained within the confines of chromosomes that behave as described above. With varying frequency, chromatin may also be found associated with a distinctly different structure known as a *supernumerary* or *"B" chromosome*. These are usually physically much smaller than the main complement of chromosomes and lack the organizational features such as centromeres and telomeres. Despite the lack of centromeres or any obvious mechanisms to perpetuate themselves, supernumerary chromosomes have been found to persist during somatic and germ line development in angiosperms. In lily, it has been reported that a supernumerary chromosome was inherited by over 73% of progeny (Xie et al., 2014). This same paper reported that this B chromosome carried homologues or orthologs to 5S rDNA loci. It has been hypothesized that supernumerary chromosomes arose from an amalgam of genomic sources and evolved to become self-perpetuating; a genetic parasite (Martis et al., 2012). Perhaps supernumerary chromosomes may lead to a mechanism to introduce and express desired genes, but this is speculative at present.

MATERNAL INHERITANCE

The classical test for nuclear vs. maternal inheritance is to compare the progeny of *reciprocals* ($P_1♀ \times P_2♂$ vs. $P_1♂ \times P_2♀$) of a given pair of parents in a cross. If the phenotypes of the progeny are different between the two reciprocal crosses then it is suspected that the phenotype observed is *maternally inherited*. A related term *cytoplasmically inherited* is perhaps more modern in describing the phenomenon (see below) since not all cytoplasmic inheritance is maternal. When eukaryotic gametes fuse to produce a zygote, the relative contribution of near- and long-term heritable factors are not equal between the pollen and egg cells. Much more cytoplasm and, therefore, metabolic machinery is contributed by the egg cell, and this influence of the female sporophyte can affect the phenotype of the developing embryo for many cell division cycles until the genotype of the embryo exerts itself (Foolad and Jones, 1992).

Certain cytoplasmic organelles, namely plastids and mitochondria, possess small genomes. The relative sizes and functional complexity of cytoplasmic genomes as compared to the nuclear genome are small and very low. The typical plastid and plant mitochondrial genomes are in the range of 10^5 and 2×10^5 bp respectively (Gray et al., 1999;

Shaw et al., 2007), whereas the typical nuclear genome is 10^8–10^{11} bp (Greilhuber et al., 2005). The vast majority of heritable information, therefore, is contained within the nuclear genome as compared with those of cytoplasmic organelles.

Until the advent of sophisticated DNA sequence recognition capabilities, it was presumed that most cytoplasmic traits in eukaryotes, including plants, were maternally inherited. Recent findings have demonstrated that this is a gross oversimplification (Mogensen, 1996). Mitochondrial inheritance is mostly maternal in eukaryotes, with rare exceptions (Breton and Stewart, 2015). The inheritance of plastid genes and genome, however, is much more complex. Progeny can inherit predominantly paternal or maternal cpDNA depending on how a given species has evolved (Nagata, 2010). Recombinants of maternal and paternal cpDNA were found in some species at a high frequency.

Not surprisingly, there are not very many plant phenotypes of economic interest that are inherited cytoplasmically. The main example is cytoplasmic male sterility, or gynoecy, found to be controlled by several mtDNA energy transductions and unknown (*orf*) genes in concert with certain nuclear genes (Kubo et al., 2011; Touzet, 2012; see Chapter 10). Other examples of maternally-inherited economic phenotypes are disease resistance in sunflower (Deglene et al., 1999), sugar levels in sugar beets (Jassem et al., 2000) and levels of organic acids in *Pyrus* (Liu et al., 2016). Since cytoplasmic inheritance violates Mendel's Laws of Inheritance, substantial alterations in breeding methods must be invoked when the plant breeder encounters such traits. Probabilities are extremely low that maternally/cytoplasmically inheritance will be encountered except in cases involving cytoplasmic male sterility that are common in seed production of hybrid varieties (Chapters 10, 11, and 16).

GENOME MAPPING

Genetics is an amalgam of biological structure and function. Ever since it was known that genes were located on chromosomes, the burning questions were: Which genes on which chromosomes? How are they ordered? Is there any overarching significance to the way that nature has selected organisms with specific gene organizations? Plant breeders were also fascinated with gene organization. The meaning of distorted segregation ratios with a preponderance of parental types was immediately apparent after the appearance of landmark papers by Thomas Hunt Morgan in the early 20th century, and the concept of linkage was spawned. Situations were visualized wherein linkage could be used in plant improvement strategies.

Historically, the first genetic maps in eukaryotes were physical; the *polytene* (multiple homologous DNA strands) salivary chromosomes of certain insect *Diptera* species, including the common fruit fly (*Drosophila melanogaster*). The juxtaposition of euchromatic and heterochromatic regions in the parallel *polytene* chromatin strands gave rise to cytologically visible bands, the pattern of which varied with developmental stage and physiological condition. While no similar state of massive polyteny has been discovered in plants that could be used to replicate this kind of work, karyotypes were taken to higher levels, and subchromosomal structures described for important plant crop species such as corn and wheat.

Subsequently, enormous quantities of joint inheritance data were parlayed into linkage maps in maize (Emerson et al., 1935) and *Drosophila melanogaster* (Bridges and Brehme, 1941). Many other crop species followed suit until virtually all have at least some sort of linkage map. Highly studied species have all linkage groups saturated with *markers* (the loci of specific genes or sequences) while lesser studied species may not yet have all linkage groups mapped. Once a map is developed it is necessary to maintain the plant populations that contain the markers so that future studies to extend the accuracy of the map are possible. More and more markers may be added to the map imbuing it with ever more detail and informational value. Theoretically, mapping is an ongoing endeavor that does not end until the entire DNA sequence and its function are known.

Prior to 1970, all the markers used in plant genetic mapping studies were functional genes that conditioned gross morphological variation, such as height, habit, pigmentation, foliage types, floral types, disease resistance, etc. Advances in molecular biology provided the possibility to visualize DNA sequences directly, initially through proteins (*isozymes*: forms of the same enzyme that migrate at different velocities in gel media). Technical advances allowed researchers to distinguish *polymorphisms* (different alleles) at the level of DNA and RNA. Arrays of different marker systems were developed that employed alternative methods to visualize nucleotide base differences in DNA (e.g., RFLPs, RAPDs, AFLPs, SSRs, etc.). Marker systems are described in more detail in Chapter 9. These new molecular marker loci have been added to the original linkage maps many found to be closely linked to functional genes. The utility of such linkages will also be covered in Chapter 9. Unlike functional genes, DNA polymorphisms sequences provide many magnitudes of more loci for mapping studies. Further, polymorphism in functional genes often is manifested in reduced viability rendering linkage studies difficult to execute.

Scientists have developed a technique known as in situ hybridization, wherein DNA sequences can be brought into contact with immobilized mitotic metaphase chromosomes. The chemical attraction of nucleotide bases (adenine to thymine, cytosine to guanine), in a sequence that is complementary to a genomic DNA sequence, causes the sequences to bind to chromosomes at the location where they reside. If the hybridizing DNA sequence is linked with a visual detector, such as radioactivity or a special stain, it is possible to view the physical location of genes on chromosomes. Interestingly, genes tend to hybridize to chromosomes in locations consistent with predictions from linkage maps.

Molecular biology has inevitably made it possible to bypass sexual reproduction entirely for the study of the organization of genes. By cutting the genome into progressively smaller pieces, and annealing the pieces at homologous "tails", it is possible to reconstruct the genome one linkage group at a time. When comparing all forms of maps, cytogenetic with in situ, linkage, and physical, they all tend to be consistent with regard to the linear arrangement of functional genes, but tend to be distorted with regard to relative distances (Fig. 3.15). This distortion is likely because recombination frequencies are higher in some regions of the genome than others.

Within the foreseeable future, the entire genomic sequences of all crop species will be known. As the sequence is developed, the matrix of functioning genes, control sequences, replicons, transposons, will be overlaid onto these sequences. Ultimately, we will understand what "makes a genome tick" and this will impart clues about how to improve one, or even to engineer genomes to meet desired specifications (Barabaschi et al., 2016).

We are well on our way to discovering how genomes are organized and how to manipulate them, but do we understand if "…there is any overarching significance to the way that nature has selected organisms with specific gene organizations?" Classical mapping studies of closely related species have shown that genes involved in similar pathways or developmental/physiological processes tend to cluster at the same relative locations. Developmental studies have determined that genes expressed at the same time or place are not necessarily clustered physically, but they often are. Rather, they are controlled by cis-acting DNA sequences located at the 5′ and 3′ ends of the genes interacting with transcription factors and other components of the cell. The mechanisms that generate new genetic variability, the manifested differences in chromosome number and structure, the arrangements of genes, the DNA sequence appear to operate randomly (Heslop-Harrison, 2000). The environment then provides a standard for phenotypes to aspire to, and genotypes are dragged along for the ride, but it is a moving target.

The nascent organization of genetic variation we now see is but a snapshot in evolutionary time, an ever-changing and ever-moving front of extant organisms under selection by ever-changing forces on Earth. While origins may have been in the primordial soup, the end is the end to life on Earth itself. In the natural sense, the genotype has no state of absolute perfection or detriment. What does not work today, may work 100,000 years from now, or in 100,000,000 years. In the sense of humans, however, the genotype may well have an attainable state of perfection, at least in the near-term, and we will pursue it with all of the energy and ingenuity we can muster.

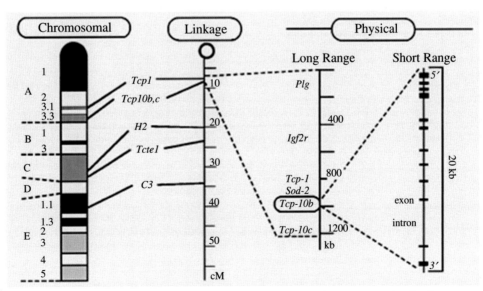

FIG. 3.15 Colinearity of genomic structure: Chromosomes (cytogenetic ideograms), linkage maps, and molecular constructs. *From Silver, L.M., 1995. Mouse Genetics, Concepts and Applications. Oxford University Press.*

I. ELEMENTS AND UNDERPINNINGS OF PLANT BREEDING

Returning to the dilemma of Bob, the highly successful but genetically challenged plant breeder, the reader might be tempted to write off this chapter as an unnecessary esoteric exercise. It has been estimated, however, that while plant breeding has been practiced for over 10,000 years, over 50% of the genetic progress in major crop species has been realized in the last 100 years alone, and that accomplished largely with the benefit of genetics. Bob may have been successful in his day, but he still needed a context to be successful in.

References

Alberts, B., Bray, D., Hopkin, K., Johnson, A., Lewis, J., Raff, M., Roberts, K., Walter, P., 2014. Essential Cell Biology, fourth ed W. W. Norton & Company, New York, NY (USA). 864 pp.

Aleza, P., Juarez, J., Cuenca, J., Hernandez, M., Ollitrault, P., 2012. Extensive citrus triploid breeding program by 2Xx4X sexual hybridizations. Acta Hortic. 961, 67–72.

Ascough, G.D., van Staden, J., Erwin, J.E., 2008. Effectiveness of colchicine and oryzalin at Inducing polyploidy in Watsonia lepida N.E. Brown. HortScience 43 (7), 2248–2251.

Barabaschi, D., Tondelli, A., Desiderio, F., Volante, A., Vaccino, P., 2016. Next generation breeding. Plant Sci. 242, 3–13.

Barrell, P.J., Grossniklaus, U., 2005. Confocal microscopy of whole ovules for analysis of reproductive development: the elongate1 mutant affects meiosis II. Plant J. 43 (2), 309–320.

Bozza, C.G., Pawlowski, W.P., 2008. The cytogenetics of homologous chromosome pairing in meiosis in plants. Cytogenet. Genome Res. 120 (3–4), 313–319.

Breton, S., Stewart, D.T., 2015. A typical mitochondrial inheritance patterns in eukaryotes. Genome 58 (10), 423–431.

Bridges, C.B., Brehme, K.S., 1941. The Mutants of Drosophila melanogaster. Carnegie Inst. Washington Pub, Washington, DC, p. 552.

Busov, V.B., Brunner, A.M., Strauss, S.H., 2008. Genes for control of plant stature and form. New Phytol. 177 (3), 589–607.

Chapman, G.P., 1958. Chromosome Studies in the Genera Lycopersicon and Solanum. Ph.D. Thesis, Univ. of Birmingham. Archived at http://etheses.bham.ac.uk/701/.

Charlesworth, B., Charlesworth, D., 2012. Elements of Evolutionary Genetics. Roberts & Co. Publ, Greenwood Village, CO (USA). 734 pp.

Chen, Z.J., 2007. Genetic and epigenetic mechanisms for gene expression and phenotypic variation in plant polyploids. Annu. Rev. Plant Biol. 2007, 377–406.

Cracraft, J., Helm-Bychowski, K., 1991. Parsimony and phylogenetic inferences using DNA sequences: some methodological strategies. In: Miyamoto, M., Cracraft, J. (Eds.), Phylogenetic Analysis of DNA Sequences. Oxford Univ. Press, New York, NY, pp. 184–220.

Crow, E.W., Crow, J.F., 2002. 100 years ago: Walter Sutton and the chromosome theory of heredity. Genetics 160, 1–4.

Darlington, C.D., 1937. Recent Advances in Cytology. Blakiston's Sons and Co., Philadelphia, PA. 671 pp.

Deglene, L., Alibert, G., Lesigne, P., Tourvieille de Labrouhe, D., Sarrafi, A., 1999. Inheritance of resistance to stem canker (Phomopsis helianthi) in sunflower. Plant Pathol. 48 (4), 559–563.

van Den Berg, M.M.J., van Maarle, M.C., van Wely, M., Goddijn, M., 2012. Genetics of early miscarriage. Biochim. Biophys. Acta 1822 (12), 1951–1959.

Dvorak, J., Deal, K.R., Luo, M.C., 2006. Discovery and mapping of wheat Ph1 suppressors. Genetics 174 (1), 17–27.

Elliot, F.C., 1958. Cytogenetics and Plant Breeding. McGraw-Hill, New York, NY (USA). 395 pp.

Emerson, R.A., Beadle, G.W., Fraser, A.C., 1935. A Summary of Linkage Studies in Maize. Cornell University Agricultural Experiment Station, Ithaca, NY. Mem. No. 180.

Fairbanks, D.J., Rytting, B., 2001. Mendelian controversies: a botanical and historical review. Am. J. Bot. 88 (5), 737–752.

Feldman, M., Levy, A.A., 2012. Genome evolution due to allopolyploidization in wheat. Genetics 192 (3), 763–774.

Fisher, R.A., 1930. The Genetical Theory of Natural Selection. Clarendon Press, Oxford, UK. 272 pp.

Foolad, M.R., Jones, R.A., 1992. Models to estimate maternally controlled genetic variation in quantitative seed characters. Theor. Appl. Genet. 83 (3), 360–366.

Garriga-Caldere, F., Huigen, D.J., Jacobsen, E., Ramanna, M.S., 1999. Prospects for introgressing tomato chromosomes into the potato genome: an assessment through GISH analysis. Genome 42 (2), 282–288.

Gohil, R.N., Ashraf, M., 1984. Chromosome behaviour during micro- and megasporogenesis and the development of embryo sac in Vicia faba L. Cytologia 49 (4), 697.

Goldstein, J.L., Brown, S.B., 1990. Regulation of the mevalonate pathway. Nature 343, 425–430.

Golubovskaya, I., Avalkina, N.A., Sheridan, W.F., 1992. Effects of several meiotic mutations on female meiosis in maize. Dev. Genet. 13 (6), 411–424.

Goodspeed, T.H., 1933. Chromosome number and morphology in Nicotiana VI. Chromosome numbers of forty species. Genetics 19, 649–653.

Gray, M.W., Berger, G., Lang, B.F., 1999. Mitochondrial evolution. Science 283 (5407), 1476–1481.

Greenbaum, G., Templeton, A.R., Zarmi, Y., Bar-David, S., 2014. Allelic Richness Following Population Founding Events—A Stochastic Modeling Framework Incorporating Gene Flow and Genetic Drift. http://journals.plos.org/plosone/article?id=10.1371/journal.pone.0115203.

Greilhuber, J., Doležel, J., Lysák, M., Bennett, M.D., 2005. The origin, evolution and proposed stabilization of the terms 'genome size' and 'C-value' to describe nuclear DNA contents. Ann. Bot. 95 (1), 255–260.

Hao, M., Luo, J., Yang, M., Zhang, L., Yan, Z., 2011. Comparison of homoeologous chromosome pairing between hybrids of wheat genotypes Chinese Spring ph1b and Kaixian-luohanmai with rye. Genome 54 (12), 959–964.

Hardy, G.H., 1908. Mendelian proportions in a mixed population. Science 28 (706), 49–50.

He, Z., Pradhan, A.K., Harper, A.L., Bancroft, I., Parkin, I.A.P., Havlickova, L., Wang, L., 2017. Extensive homoeologous genome exchanges in allopolyploid crops revealed by mRNAseq-based visualization. Plant Biotechnol. J. 15 (5), 594–604.

Heslop-Harrison, J.S., 2000. Comparative genome organization in plants: from sequence and markers to chromatin and chromosomes. Plant Cell 12 (5), 617–635.

Hilu, K.W., 1993. Polyploidy and the evolution of domesticated plants. Am. J. Bot. 80 (12), 1494–1499.

Holland, J.B., Nyquist, W.E., Cervantes-Martinez, C.T., 2003. Estimating and interpreting heritability for plant breeding: an update. In: Plant Breeding Reviews. John Wiley & Sons, New York, NY, pp. 9–112.

Hoshmand, A.R., 1994. Experimental Research Design and Analysis: A Practical Approach for Agricultural and Natural Sciences. CRC Press, Boca Raton, FL (USA). 408 pp.

Huang, B.Q., Sheridan, W.F., 1996. Embryo sac development in the maize indeterminate gametophyte1 mutant: abnormal nuclear behavior and defective microtubule organization. Plant Cell 8 (8), 1391–1407.

Huskins, C.L., La-Cour, L., 1930. Chromosome numbers in *Capsicum*. Am. Nat. 64, 382–384.

Jassem, M., Sliwinska, E., Pilarczyk, W., 2000. Maternal inheritance of sugar concentration. J. Sugar Beet Res. 37 (2), 41–53.

Jiang, J., Gill, B.S., 1994. Nonisotopic in situ hybridization and plant genome mapping: the first 10 years. Genome 37 (5), 717–725.

Kanno, T., Habu, Y., 2011. siRNA-mediated chromatin maintenance and its function in Arabidopsis thaliana. Biochim. Biophys. Acta 1908 (8), 444–451.

Kawamura, K., Laurence, H.-S.O., Thouroude, T., Jeauffre, J., Foucher, F., 2015. Inheritance of garden rose architecture and its association with flowering behaviour. Tree Genet. Genomes 11 (2), 844.

Kubo, T., Kitazaki, K., Matsunaga, M., Kagami, H., Mikami, T., 2011. Male sterility-inducing mitochondrial genomes: how do they differ? Crit. Rev. Plant Sci. 30 (4), 378–400.

Ledbetter, C.A., Ramming, D.W., 1989. Seedlessness in grapes. Hort. Rev. 1989, 159–184.

Li, S.S., Khalid, N., Carlson, C., Zhao, L.P., 2003. Estimating haplotype frequencies and standard errors for multiple single nucleotide polymorphisms. Biostatistics 4 (4), 513–522.

Liu, L., Chu-Xin, C., Yang-Fan, Z., Xue, L., Qing-Wen, L., 2016. Maternal inheritance has impact on organic acid content in progeny of pear (*Pyrus* spp.) fruit. Euphytica 209 (2), 305–321.

Loomis, H., Weinberger, H., 1979. Inheritance studies of seedlessness in grapes. J. Am. Soc. Hort. Sci. 104, 181–184.

Maizonnier, D., 1984. Cytology. In: Sink, K.C. (Ed.), Monographs on Theoretical and Applied Genetics: Petunia. Springer-Verlag, Berlin, pp. 21–33.

Martis, M.M., Klemme, S., Banaei-Moghaddam, A.M., Blattner, F.R., Macas, J., 2012. Selfish supernumerary chromosome reveals its origin as a mosaic of host genome and organellar sequences. Proc. Natl. Acad. Sci. U. S. A. 109 (33), 13343–13346.

Matsuoka, Y., 2011. Evolution of polyploid *Triticum* wheats under cultivation: the role of domestication, natural hybridization and allopolyploid speciation in their diversification. Plant Cell Physiol. 52 (5), 750–764.

Maynard, D.N., Elmstrom, G.W., 1992. Triploid watermelon production practices and varieties. Acta Hortic. 318, 169–173.

Miller, J.S., Venable, D.L., 2000. Polyploidy and the evolution of gender dimorphism in plants. Science 289 (5488), 2335–2338.

Mogensen, H.L., 1996. The hows and whys of cytoplasmic inheritance in seed plants. Am. J. Bot. 83 (3), 383–404.

Morgan, T.H., Sturtevant, A.H., Muller, H.J., Bridges, G.B., 1915. The Mechanism of Mendelian Heredity. Henry Holt & Co., New York, NY. 372 pp.

Nagata, N., 2010. Mechanisms for independent cytoplasmic inheritance of mitochondria and plastids in angiosperms. J. Plant Res. 123 (2), 193–199.

Nicodemi, M., Pombo, A., 2014. Models of chromosome structure. Curr. Opin. Cell Biol. 28, 90–95.

Noher de Halac, I., Harte, C., 1985. Cell differentiation during megasporogenesis and megagametogenesis. Phytomorphology 35 (3/4), 189–200.

Oettler, G., 2005. The fortune of a botanical curiosity—triticale: past, present and future. J. Agric. Sci. 143, 329–346.

Orton, T.J., Steidl, R.P., 1980. Cytogenetic analysis of plants regenerated from colchicine-treated callus cultures of an interspecific *Hordeum* hybrid. Theor. Appl. Genet. 57, 89–95.

Petersen, R.G., 1994. Agricultural Field Experiments: Design and Analysis. Marcel Dekker, New York, NY (USA). 409 pp.

Pucha, H., Hohn, B., 2012. In planta somatic homologous recombination assay revisited: a successful and versatile, but delicate tool. The Plant Cell. https://doi.org/10.1105/tpc.112.101824.

Punnett, R.C., 1922. Mendelism. MacMillan Publishers, London (UK). 272 pp.

Quinn, G.P., Keough, M.J., 2002. Experimental Design and Data Analysis for Biologists. Cambridge University Press, Cambridge (UK). 537 pp.

Rajathy, T., Morrison, J.W., 1959. Cytogenetic studies in the genus *Hordeum* IV. Hybrids of *H. jubatum, H. brachyantherum, H. vulgare,* and a hexaploid *Hordeum* sp. Canadian J. Genet. Cytol. 1, 124–132.

Riley, R., Kimber, G., Chapman, V., 1961. Origin of genetic control of diploid-like behaviour of polyploid wheat. J. Hered. 52, 22–25.

Sattler, M.C., Carvalho, C.R., Clarindo, W.R., 2016. The polyploidy and its key role in plant breeding. Planta 243 (2), 281–296.

Sears, E.R., 1941. Chromosome pairing and fertility in hybrids and amphidiploids in the *Triticinae*. Missouri Agric. Exp. Station Res. Bull. 337, 1–20.

Sears, E.R., 1976. Genetic control of chromosome pairing in wheat. Ann. Rev. Genet. 10, 31–51.

Shaw, J., Lickey, E.B., Schilling, E.E., Small, R.L., 2007. Comparison of whole chloroplast genome sequences to choose noncoding regions for phylogenetic studies in angiosperms: the tortoise and the hare III. Am. J. Bot. 94 (3), 275–288.

Stebbins Jr., G.L., 1949. Types of polyploids: their classification and significance. Adv. Genet. 1, 403–409.

Talbert, L.E., Smith, L.Y., Blake, N.K., 1998. More than one origin of hexaploid wheat is indicated by sequence comparison of low-copy DNA. Genome 41 (3), 402–407.

Touzet, P., 2012. Mitochondrial genome evolution and gynodioecy. Adv. Bot. Res. 63, 71–98.

Visscher, P.M., 2008. Sizing up human height variation. Nat. Genet. 40, 489–490.

Weinberg, W., 1908. Über den nachweis der vererbung beim menschen. Jahreshefte des Vereins für vaterländische Naturkunde in Württemberg, 64, pp. 368–382.

Woodworth, C.M., Long, E.R., Jugenheimer, R.W., 1952. Fifty generations of selection for protein and oil in corn. Agron. J. 44, 60–65.

Wu F., Tanksley S.D., 2010. Chromosomal evolution in the plant family Solanaceae. BMC Genomics 11, Article number: 182.

Xie, S., Marasek-Ciolakowska, A., Ramanna, M.S., Arens, P., Visser, R.G.F., 2014. Characterization of B chromosomes in Lilium hybrids through GISH and FISH. Plant Syst. Evol. 300 (8), 1771–1777.

Xiong, Y.C., Li, F.M., Zhang, T., 2006. Performance of wheat crops with different chromosome ploidy: root-sourced signals, drought tolerance, and yield performance. Planta 224 (3), 710–718.

4

Engineered Population Structures

INTRODUCTION: THE NATURE OF POPULATIONS

The end products of plant breeding programs activities are *populations*. The biological definition of "population" is: (i) a group of organisms of one species that interbreed and live in the same place at the same time (e.g., deer *population*); (ii) a low-level taxonomic rank; (iii) a set of individuals, objects, or data from where a statistical sample can be drawn (Anon., 2016). The product of a plant breeding program is usually a set of seeds or asexual propagules that are used to produce a population that is a useful source of food, fiber, aesthetic enjoyment, environmental improvement, structural products, drugs, chemicals, or combinations of these elements for humans. The populations that are produced by plant breeders are not natural nor would they ever be expected to be found in nature. Populations developed by plant breeders are artificial, or *engineered*, for utility to humans. Engineered populations from plant breeding share many attributes with natural or feral populations and may be characterized using some of the same descriptors. Generally, engineered populations exhibit less genetic variability than natural populations, although many engineered populations are comprised of genetic variability that is not evident as phenotypic variability. Reduction in population phenotypic variability is accomplished by plant breeders through reduced V_G. This chapter will address the nature of engineered populations and how they are similar or different from natural populations. It will be assumed, unless stated otherwise, that all individuals in populations are diploid, the predominant genome configuration in angiosperms and gymnosperm sporophytes.

Plant populations in nature are a veritable snapshot in a long evolutionary movie. Plants look the way they do in a frame-by-frame (i.e., generation-by-generation) progression over time in this evolutionary movie analogy. The environment and competition for resources coupled with phenotypic (and underlying genotypic) variability create differences in fitness among individuals within populations. Those individuals that are adapted for the environment and ecosystem survive and reproduce while those that do not perish before they produce offspring. The environment might dictate changes in the fitness value of certain phenotypes, such as after a large meteorite impact, volcanic eruption, glaciation, massive flood, etc. Many species lack adequate genetic variability to cope with drastic environmental changes and may become extinct. If the populations of a species can adapt to incremental and gradual changes in the environment, they will change (or evolve) over time.

The adaptive consequences of self- and cross-pollination and asexual reproduction were discussed in Chapter 2. Generally, populations of cross-pollinating individuals, or outcrossers, have a genetic structure much like humans. Most individuals are similar, but also different in measurable ways. Likewise, humans are very similar, but nearly all (except identical twins raised together) can be distinguished from each other. Adult stature in humans ranges from about 1.25 to 2.50 m, and the distribution within genders resembles a normal curve. Interestingly, the mean of the curve has risen, although the gene pool has probably not changed nearly as dramatically (Dougherty, 2017). We attribute this to better diets and health regimen. In other words: *Phenotype = genotype + environment*. There are also rare alleles in the gene pool that condition statures outside of the normal range, such as achondroplastic dwarfism and pituitary gigantism.

Influenza A is a common human viral pathogen that is ubiquitous on earth, and always mutating to new virulent forms. Many human influenza A strains are, in fact, derived from mutations of those pathogenic on other animal species such as birds or other mammals (Anon., 2017). You have probably noticed that certain human strains seem to affect some people, but not others, including, perhaps, you. At least part of the difference in human susceptibility to flu virus has a genetic origin. If we were all identical, a human "monoculture," epidemics would be much more severe. This is but one of many examples of how intrinsic genetic variability within a population is beneficial.

New mutations that occur in outcrossing plant species are often not immediately manifested as a new and distinct phenotype. Most new mutant alleles are recessive to the predominant, or "wild-type" alleles, so are masked as heterozygotes after they first appear. It may be many generations and perhaps even hundreds of generations before a chance mating within the population involves two plants that carry the same rare, mutant allele. A mating of two heterozygous mutant carriers will result in 25% of progeny bearing the homozygous recessive mutant phenotype. If the mutant allele is highly deleterious, the haploid gametes that carry it may not survive long enough to undergo fertilization. If the mutant allele is marginally deleterious, or sub-lethal, gametes may fuse to form a zygote and an embryo may develop successfully from this zygote. Many sub-lethal or deleterious mutations result in death during early growth of the embryo. Mature plants that carry new recessive homozygous mutant genotypes may be affected positively, negatively or somewhere in between as compared to the wild-type genotype. A new flower color is seemingly inconsequential to fitness in plants, but it may be of critical importance to sexual propagation if corolla pigmentation is critical for the attraction of insect pollen vectors.

Individuals with new mutations that impart higher relative fitness than the wild-type phenotype will reproduce proportionately more, and, over time, the new mutant allele will come to predominate in the population. Mutations that improve the stress tolerance, disease resistance, or ability to repel herbivores may enhance fitness, but may also come at a cost if those factors (stress, pathogens, herbivores) cease to be present in the environment. If environmental or ecological stress factors persist, plants that possess the mutations for resistance or tolerance survive to reproduce, casting the new mutant allele forward to the next generation, while wild-type individuals with less fitness do not (Covert et al., 2013).

The mechanisms that create new alleles, gene combinations, and chromosomal structures appear to be more or less random, though evidence of patterns and "directed evolution" has been found (Stoltzfus and Yampolsky, 2009; Bloom and Arnold, 2009). Not all new genetic variation is beneficial, at least not in the immediate context in which it is first generated. In fact, most new mutant alleles are usually not desirable at all. The analogy "hitting a clock with a sledgehammer and expecting it to run better" comes to mind. The concept of lethality and sub lethality was discussed above. Most or even all individuals in a population of an outcrossing plant species carry several, perhaps many, potentially deleterious alleles that are usually masked in heterozygosity. The extent of masking of deleterious alleles by wild-type dominant alleles is thought to be responsible for the phenomenon of inbreeding depression (see Chapter 16; East, 1908).

Most new alleles are masked by dominant, partially dominant, or co-dominant counterparts that restore mostly normal function or impede the potential new advantage from being expressed (Ashri, 1989). Mates for reproduction are not chosen based on genomic DNA sequences but on overall phenotype. While most deleterious alleles in the genotype of one parent will be masked in progeny by others from the other parent, many deleterious alleles may be exposed according to probability and chance. This is a manifestation of genetic load, or the sum of all masked deleterious alleles that are carried in a population (Wallace, 1991; Barrett and Charlesworth, 1991; see Chapter 3). Without new alleles the population will eventually become extinct from the pressures wrought by an ever-changing environment. New mutations result, however, in desirable, undesirable and neutral alleles with regard to the environment and ecosystem. As the environment changes, undesirable alleles may be rendered neutral or even desirable (Allard, 1999).

Many plant species are genetically quite similar to humans with regard to sexually dimorphic mating system, such as hemp, holly, hops, juniper, spinach, pistachio, poplar, willow, ginkgo, and asparagus. In cases where both sexes occur on the same individual, such as with monoecy or hermaphrodity, other mechanisms have evolved to promote outcrossing (see Chapter 10). Humans are physiologically motivated by sex and sexual dimorphism, but outcrossing comes at a cost in plants. This cost is manifested by huge and showy flowers (to attract pollen vectors), physiological self-incompatibility systems, and copious pollen volumes. Outcrossing also comes at a tremendous long-term population cost due to genetic load. Inbreeding species, however, consume less energy for reproduction and experience lower accumulations of genetic load due to allele fixation. That outcrossing has persisted over time despite selective pressures for energy and genetic efficiencies attests to the long term importance of genetic variability (Wallace, 1991).

When the environment smiles on an individual so intensively that it is vastly more successful than its neighbors, the genes that it contains will be greatly increased in frequency in future populations. The mechanisms of mating and floral morphology are highly mutable and adaptable. Few natural species and populations are either 100% in-breeding or outcrossing, but usually somewhere in between, while favoring one end of the spectrum or the other (i.e., 90% outcrossing or inbreeding). The evolutionary tendency is for species to cast off the chains of outcrossing (e.g., the energy inefficiencies and the genetic load) in favor of genetically fixed individuals with high static relative fitness has been theorized especially in domesticated populations (Richards, 1986). Many examples of inbreeding crops domesticated from outcrossing derivatives in the wild have also been documented (Zizumbo-Villarreal et al., 2005; Beretta et al., 2015).

The alleles and gene combinations that condition high relative fitness are quickly fixed under inbreeding, and undesirable alleles are eliminated from the population (Ashri, 1989). Figs. 4.1 and 4.2 depict the drive to fixation under self-pollination for a single locus for eight generations. At a single locus, about 97% of individuals in the F_5 generation are homozygous, and approximately 68% are homozygous at ten loci. This phenomenon, the progression from heterozygosity to homozygosity during the course of recurrent self- or sib-pollination is referred to as *drive to fixation* under inbreeding (Allard, 1999).

Species that are propagated by sexual inbreeding are generally not robust over long periods of evolutionary time. They emerge from cross-pollinated progenitors through natural selection and successfully exploit a niche in the environmental and ecological continuum. As environmental and ecological fluxes occur, these inbreeding derivatives lack the genetic diversity necessary to cope with change and die out (Holsinger, 1992; Carr, 2013; Kariyat et al., 2013).

Though inbreeding species will ultimately die out in the course of evolutionary history, such species may persist for what is a long time in terms of human experience and perception, 100,000 to 10,000,000 years or more. Humans have selected and domesticated many of the most important crops under circumstances of inbreeding and genetic fixation. Examples include wheat, barley, oat, rice, soybean, tobacco, pea, green bean, lettuce, sunflower, and cotton. Others, such as tomato and peach, existed in the wild as predominantly outcrossers, and inbreeding derivatives were progressively selected during domestication (Taylor, 1986; Hegedus et al., 2006). Inbreeding is often associated with higher economic yield since the efficiency of pollination is higher than outcrossing and the associated energy costs are relatively lower (Takebayashi and Morrell, 2001).

Cultivars of crop species must be managed commercially as populations, at least at our current state of technical capabilities. Therefore, the plant breeder must always think in terms of the population as the end point, even though individuals may be clonally propagated for commerce. Before we dwell on biological and practical limitations, let the

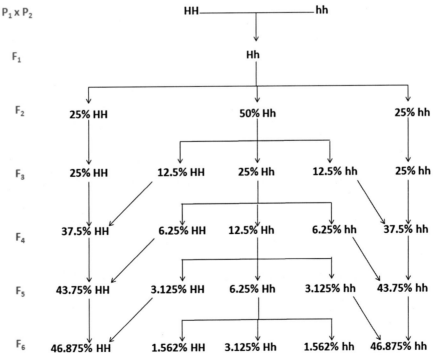

FIG. 4.1 Conceptual view of the drive to fixation following a monohybrid cross HH×hh and recurrent self-pollinations to produce successive filial generations.

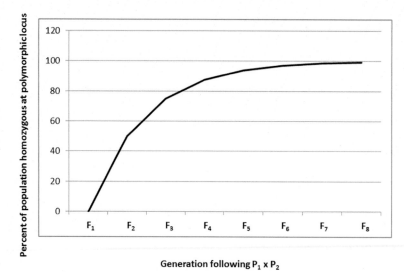

FIG. 4.2   The Y-axis denotes the % of individuals in the population (F_1, F_2, etc.) that are homozygous for two alleles at a single locus. The X-axis denotes the filial generations following a monohybrid cross (e.g., HH×hh).

imagination wander on the subject of possible genetic population structures. Most simplistically, a single genotype may be asexually replicated ad infinitum, also known as a clone. At the other extreme is random mating and rampant genetic variability, for example, the human population. While human populations exhibit more chaos than order, the mating is not totally random.

One can envisage a population wherein allele frequencies at infinite loci have been truncated from wild types, in a positive direction. All of the deleterious alleles (genetic load) have been selected out and, while some variation still exists, the range has been reduced such that individuals resemble each other much more than they would in the wild. An example of this scenario is the packet of wildflower seed purchased at the local garden center. The envelope description touts the manifestations of variability in a positive light: long flowering period, multiple flower colors, etc. Alternatively, each individual within the population could be identical, but heterozygous, not homozygous, at an infinite number of loci.

The possibility also exists for mixing genotypes in a prescribed manner, both in self and cross-pollinated species. The concepts of *multiline* (see below) and *composite* varieties will be addressed later in this text (Chapters 14 and 18), and these are both examples of engineered mixtures of prescribed genotypes (Helland and Holland, 2001). It is also possible to take the concept of genetic admixture even further, for example many different species in a crop ecosystem. While it would not be advisable to mix wheat and barley in the same field since most harvesting these days is done by machine, it might work in certain vegetables that are hand harvested and sorted in the field such as broccoli and cauliflower. "Multi-cropping" is a practice used extensively in agrarian societies where arable land is limiting. The espoused advantages include land- and resource-use efficiency, nutritional balance, and biodiversity to better protect the crop ecosystem against weed, disease, and arthropod pests.

PURE LINES AND MULTI-LINES

Johannsen (1903) first used the term "pure line" to explain his experimental results on recurrent self-pollination of garden bean (*Phaseolus vulgaris*). Following two generations of self-pollination from a single 800 mg "grandmother" bean, no differences among progeny classes from the parents were observed (Table 4.1). Selection pressure did not change the performance of the population. Johannsen termed such an unchanging population as a *pure line*. Examples of the many crop species that exist as populations of genetically identical individuals were listed above. Suffice it to say that the strategy of maintaining populations as pure lines has been extremely successful in these instances, especially since most of the agricultural consumption of humans depends on crops that still exist as pure line populations, for example small grains, soybean, cotton, and sunflower.

The pure line population is theoretically comprised of individuals all of the same panhomozygous genotype. Many so-called pure lines feature residual segregation at a finite number of polymorphic loci. Populations of inbreeding crop species that are not pure lines may be found in agricultural production in developing countries, but

TABLE 4.1 W.L. Johannsen Selected a Single Bean that Weighed 800 Mg then Selfed it for Two Generations

Weight class of mother bean (mg)	Offspring of mother bean		
	Average weight	#	SD
250–300	445.0	40	84.0
300–350	498.0	53	64.8
350–400	453.0	164	69.9
400–450	447.0	155	65.5
450–500	434.0	103	74.9
500–550	468.0	102	79.1
550–600	458.0	95	79.9
Average of all beans	454.4	712	74.0

The table excerpts some of the results he observed among progeny from this S_2 population.

these are increasingly rare, usually considered "landraces." Most modern cultivars of inbreeding crop species were developed using breeding methods that are predicated on the development of true pure lines (Chapters 13 and 14). Inbreeding and selection, the common features of all breeding methods for self-pollinating crop species, results in pure line populations even if practiced on genotype admixture landraces.

The primary advantages of the pure line population are phenotypic uniformity and fixed desirable genotypes. Population phenotypic variability is counterproductive to the goal of the grower to be as efficient as possible. If $V_G = 0$, then $V_P = V_E$. The objective of the grower is to minimize V_E and to maximize P (mean population phenotype). This will result in the greatest possible performance and uniformity. The reader has probably passed by countless fields of grain waving in the wind the seed heads all borne at the same height, and all of the same maturity, color, size, and shape. This enables the grower to adjust the harvesting equipment to a given set of specifications that will result in the highest possible yield.

Pure line populations are also extremely easy to maintain and to "increase" (make larger populations of) the seeds. One simply has to plant the seeds of the population and harvest the next generation, the genetic structure of of which will theoretically remain unchanged. Most predominantly self-pollinating crop species still exhibit outcrossing under certain circumstances, often promoted by extreme environmental conditions or developmental irregularities. Therefore, safeguards are still necessary to ensure the genetic, in addition to the biotic and abiotic, purity of pure line seed populations (see Chapter 11).

Not all plant species are readily adaptable to pure line population structures. Naturally cross-pollinated species will possess a genetic load, manifested as inbreeding depression as multiple heterozygous combinations are transformed to homozygosity (see Chapter 16). While it has been demonstrated in corn that recurrent selection will eventually purge genetic load from breeding populations, the procedure is tedious, expensive, and time consuming. Recurrent selection for phenotype and combining ability in corn began in the 1930s, and inbred performance has only reached commercially acceptable levels within the last 10 years. Of course the availability of powerful genomics tools and markers are accelerating this process tremendously (Gepts and Hancock, 2006).

The primary disadvantages of pure lines are genetic vulnerability and absence of heterosis, manifestations of the same forces that lead to the theoretical extinction of inbreeding species during the natural process of evolution. As the environment changes, and with no genetic variation within the population to adapt to the change, the species has no intrinsic means to achieve corresponding phenotypic change to cope with the new environment. The most striking examples of genetic vulnerability are disease epidemics or weed and insect pest imbalances. Not only are all of the plants within a monolithic population equally susceptible to any given pest or pathogen, they ultimately feed their demise by fostering blooms of populations of the pests that are eating or damaging them. Pure lines are particularly vulnerable to disease and pest epidemics in areas of the world where access to agrichemicals is lacking (Smithson and Lenne, 1996). Also, while genetic uniformity is consistent with the needs of the grower, it is not necessarily the best attribute for the consumer, particularly in developing countries where nutritional needs are met primarily or even exclusively by plants. Variation in food composition is essential to meet basic physiological and developmental requirements of humans under subsistence conditions (Lipton and Longhurst, 1985).

Heterosis (see Chapter 16) is defined as enhancement of individual and population performance based on polyheterozygosity. Pure lines are theoretically panhomozygous and lack genomic heterozygotes that impart heterosis. The phenomenon of heterosis has been demonstrated under limited circumstances in autogamous species such as wheat, cotton, and tomato, but the magnitude of the advantage of the hybrid over the inbred is usually relatively small. The mating system of autogamous species is notoriously difficult to subvert for mass production of F_1 hybrid progeny. Consequently, most autogamous species are bred as pure lines and the potential for greater crop performance from heterosis remains only speculative.

Multiline populations (see discussion of genotype admixtures above) represent a compromise between these opposing opportunities and problems posed by pure lines. In the most simplistic case, the multiline population is an admixture of pure lines that are identical to each other at most genetic loci but differ with regard to one or a few loci that control disease and pest vulnerability. The best example is vertical disease resistance (see Chapter 19). The multiline disease-resistant wheat population performs theoretically like a pure line but contains several resistance genes to different races of a prevalent pathogen (Wilson et al., 2001). Thus, if a certain race attacks the population only a proportion of the population will be affected, avoiding a total crop loss.

RANDOM MATING WITH TRUNCATED ALLELIC FREQUENCIES

A population with enhanced frequencies of desired alleles is the targeted endpoint of a plant breeding program, or more accurately the marketable intermediate, resulting from directional population selection in a cross-pollinated crop species (see Chapters 9 and 15). The mean of the progeny population moves in the direction of selection (see Chapters 3 and 9). The rate of desirable phenotype change over time depends on selection intensity, heritability, linkage, and mutation rate. The logical presumption by the novice is that increased selection pressure will lead to faster results. While this might be true in early generations, and with very large population sizes, this strategy alone can lead to an overall reduction of performance due to inbreeding of an outcrossing crop species, or the unintentional fixation of undesirable alleles due to linkage disequilibrium. Where mating is not controlled during the process, this is called *mass selection*. If mating is controlled it is called *recurrent selection*.

A balanced, sustained program of mass selection, or recurrent selection, featuring multiple environments and selection for multiple attributes, will culminate in populations that are improved as compared with the source populations or starting germplasm. Selected populations will exhibit reduced genetic variability, manifested as reduced phenotypic variability, but retain a degree of polymorphism for environmental buffering and to mask genetic load. Most cross-pollinated crop species consist of this type of population structure during early stages of domestication (referred to as *open-pollinated* or *"OP"* varieties), or when crop value is insufficient to drive higher-powered approaches, such as hybrids (see below). As deleterious alleles are removed by selection, the population is progressively more tolerant of inbreeding, and the longer-term endpoint is comparable to the pure line, even though the mating system may continue to promote outcrossing.

Humans survived for over 20,000 years using the directional phenotypic selection population enhancement strategy. New breeding methods developed over the past 150 years have mostly supplanted directional population selection and OP cultivars. Corn cultivars were open pollinated until the early 20th century when new hybrid strategies were employed to harness heterosis and minimize phenotypic variation due to V_G. Simultaneously, the population of humans on Earth has increased exponentially.

HYBRID AND SYNTHETIC POPULATIONS

Heterosis, or *hybrid vigor*, is the tendency for the progeny of parents to exceed the midparent (average of the two) or, in some exceptional cases, to exceed both parents (referred to as *heterobeltiosis*). While overdominance is well known for allelic interactions at certain loci, the phenomenon is quite remarkable when considering the cumulative effects of hundreds, or even thousands of heterozygous loci, on overall plant performance. Humans have known about heterosis for centuries, if not of the mechanisms that control it. In fact, we still do not understand completely how heterosis works. The debate between the hypotheses of collective dominance or overdominance has continued to the present day, and the endurance of the argument suggests that the truth lies somewhere in between (see Chapter 16). Heterosis is likely the result of many phenomena, including collective dominance, overdominance, and genetic buffering effects. After all, it is easy to imagine that possessing two different alleles might be preferable to having two copies of either one.

Early humans made crosses between plants from divergent populations and observed the immediate effects of heterosis, and certainly also noted that these effects on population performance diminished following subsequent cycles of uncontrolled mating. The theoretically pure hybrid population would consist of an infinite number of individuals, all genetically identical, and all heterozygous at all genomic loci. Thus, all individuals are simultaneously heterotic and identical, satisfying the most important criteria for the ideal crop population. In certain crop species, this possibility has come close to realty, for example, in F_1 hybrid corn varieties.

Experiments on the phenomena of hybrid vigor and inbreeding depression were conducted as early as the mid-19th century by Darwin and Beal (reviewed by Sprague, 1946). Groundbreaking experiments by East, Shull, and Jones in the early 20th century (Jones, 1917; East, 1936; Shull, 1948) clearly demonstrated the potential of hybrid over OP corn cultivars. Many independent breeding programs to develop hybrid corn and other cross-pollinated crop species were stimulated by these success stories. To achieve the goal of genetic uniformity and panheterozygosity, two individuals that are panhomozygous, or inbred, must be hybridized. The flip-side of heterosis, *inbreeding depression*, was also well documented before 1900 CE. Early entrepreneurs hoped that the value of the hybrid seed would compensate for the poor performance of the parental inbreds.

The entrepreneurs were wrong, at least initially, but the goal of developing hybrid varieties did not languish. The parents of the commercial hybrid populations were engineered to be slightly heterotic, thus overcoming the effects of inbreeding depression. This was accomplished by developing and combining several parental inbred populations. Two parents that were substantially similar, even closely related, could be hybridized to give rise to an intermediate population that was still mostly uniform genetically, but sufficiently heterozygous to give overall population vigor a boost. Then this population, or these populations, could be hybridized to deliver the commercial seed population, termed *3-way* and *double cross* hybrids (Fig. 4.3). The result was cultivars that featured a modicum of heterosis and were more uniform and higher yielding than were existing OP populations, but with a competitive seed cost. Eventually, breeders eliminated most of the deleterious alleles that caused inbreeding depression, thus making inbred parents sufficiently vigorous to produce cost-effective seed crops. The possibility of vigorous pure line inbreeds eliminated the need for three-way and double cross hybrids and they were ultimately replaced by single-cross or "F_1" hybrids starting in the 1950s. Later, male sterility was introduced to further reduce the costs associated with F_1 hybrid seed production (Duvick, 1984).

A single-cross or F_1 hybrid is theoretically a monolithic panheterozygote. This invokes the ecological concerns discussed above (pure lines) concerning genetic monocultures and vulnerability. The 1970 southern corn leaf blight (SCLB) epidemic is often cited as emblematic of such a problem created by a monoculture of hybrid plants bearing a single cytotype (Frankel and Bennett, 1970; Frankel and Hawkes, 1975), but there is little evidence that heterozygous individuals are as vulnerable as are their pure line, homozygous counterparts. While the SCLB epidemic did occur in hybrid corn varieties, it was a consequence of the monoculture of cytoplasmic, mitochondrial, genomes that caused the problem. Recall that the culprit was Texas cms, or "T" cytoplasm, used for efficient hybrid seed production. Seed companies immediately set about to produce the exact same hybrids in fertile cytoplasmic backgrounds using hand detasseling, and the SCLB problem was averted in ensuing years up to the present (Harlan, 1980).

The primary bottleneck for the economic viability of hybrid populations is the cost of development and seed production. Individual hybrid plants can be produced easily and inexpensively, but what about thousands or millions of them? That is the reason T-cms was used in the first place, to make large quantities of hybrid seed cost-efficiently.

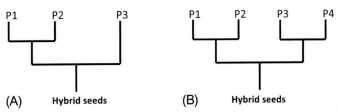

FIG. 4.3 Strategies to utilize a modicum of heterosis to overcome inbreeding depression in hybrid variety development programs: (A) three-way scheme and (B) double-cross scheme.

The use of male sterility and other mating control techniques to produce hybrid seed will be discussed in Chapters 11 and 16 since the necessary parental genotypes must be interwoven with the breeding methods. In some crop species where heterosis has been documented, no genetic strategies (e.g., male sterility or self-incompatibility) have been developed to facilitate hybrid seed production. This is particularly applicable in horticultural species where the value of the individual plant is high. The high value of individual plants has made it possible in some instances to produce large quantities of hybrid seeds by hand pollination.

Another strategy that has been espoused, although not as successfully as F_1 hybrids, is F_2 hybrid populations. To the marketer, this might sound like a viable alternative if the F_1 seed is too expensive, an analogous scenario to the logic for three-way and double-cross hybrids. Since heterosis is a quantitative concept related to the degree of heterozygosity, the F_1 will segregate 1:2:1 at each heterozygous locus, resulting in the release of phenotypic and genotypic variability in the F_2 hybrid population. In a crop cultivar, variability is usually not a desired attribute. In fact, this release of phenotypic variability is a major reason that seed companies prefer hybrid cultivars, the inability of the farmer to regenerate the cultivar by simply retaining seeds of the previous crop. In certain instances, however, differences in the performance of F_1 and F_2 populations have been found to be negligible (Bosland, 2005). If heterosis and V_A can be uncoupled to a sufficient degree, the F_2 hybrid strategy plausible.

ASEXUALLY PROPAGATED POPULATIONS (CLONES)

Asexual, or clonal, propagation is a popular alternative in both nature and plant breeding. Asexual propagules are usually multicellular organs (e.g., tubers, rhizomes, bulbils), but may also be apomictic seeds. The complete circumvention of sex has been accomplished successfully in many plant species. The outcome of clonal propagation is somewhat similar to self-pollination in that a genetically monolithic population is produced, but the underlying population genetics are very differentfor the two propagation strategies . With regard to asexual reproduction in natural habitats, individuals with the highest relative fitness values survive to contribute genotypes to the next generation. Since sex is absent, mitosis is the basis of reproduction. Genomic constitutions that are sexually dysfunctional, such as triploids and aneuploids, may be perfectly viable under recurrent asexual propagation cycles.

Like self-pollination, asexual propagation results in genetically uniform populations. Therefore, it is predicted that asexual propagation should be an evolutionary dead end, as is inbreeding, but this hypothesis has been found to be mostly incorrect. Many or most asexually-propagating species may also have a functional sexual system that generates adequate genetic variability to enhance the long-term fitness of the species. Moreover, mitotic cell divisions have been discovered to be imperfect and somatic recombinations and mutations are constantly creating new genetic variability, albeit to a lesser degree than meiosis (Friedt and Brune, 1987). Clonal tissues may often exist as a genetic mixture as opposed to a set of genetically identical cells. Over evolutionary time, therefore, the clone may generate and maintain sufficient genetic variability to adapt to changing environments with fluctuating genetic mixtures. Heritable information is also transmitted and expressed in progeny on small nuclear RNA molecules (snRNA) that do not obey Mendel's laws. As blasphemous as it might seem, perhaps Lamarck's view on acquired characteristics was at least partially correct in this respect (Suter and Martin, 2010).

Early plant breeders quickly discovered that populations regenerated from seeds often produced less desirable results than did populations regenerated from somatic tissues (Neiman et al., 2014). They knew, for example, that grapes could be propagated by excising scions (branches) and simply sticking them into moist soil, where they would take root and form new progeny plants. Not all plant species will root and regenerate whole functioning plants from cuttings or other tissue sources so readily, however (Hussey, 1984). The best plants could be selected and vegetatively propagated, and the ensuing generations mostly reflected the phenotypes of the original selections, unlike those propagated from seeds. We now know, of course, that the individuals must have been heterozygous following many cycles of cross-pollination and that seed-propagated populations would segregate at multiple loci, thus frustrating the early propagator, who had no knowledge of Mendelian inheritance.

The multiplication rate under asexual propagation is usually low, necessitating the use of a large number of source plants (Hussey, 1984). Therefore, early populations of domesticated crops propagated clonally were probably not genetically uniform. Technology gradually became more sophisticated, and populations of asexually propagated crop species almost always can be traced to a single individual that has been multiplied over many cycles and years. Since asexual population structure has few a priori genetic assumptions or requirements, the strategy for population increase tends to be popular among plant breeders. The fundamental limitation of clonal propagation is the ability to efficiently clone whole plants into large pangenetic populations. Therefore, this type of population structure is found where natural botanical systems are already in place (e.g., potato tubers, sugarcane stolons, garlic bulbils, and turfgrass apomictic seeds). If methods to clonally propagate a plant species are to be developed, the costs can only be justified in cases where single individuals have enormously high value (e.g., ornamental flowers).

In summary, domesticated crop species have been genetically honed into many different population structures. A number of factors come into play in the determination of the type of population structure that best suits each species: the natural mating system, the degree of heterosis and inbreeding depression, and the economic value. This is the first step taken by the plant breeder, unless circumstances have already made the decision.

A pictorial genetics contrast of the different population structures described in this chapter is provided below (Figs. 4.4–4.6):

Individual
1   AAbbccDDeeFFGGhhIIjjKKLLmmnnOOppQQRRssTTuuVV
2   AAbbccDDeeFFGGhhIIjjKKLLmmnnOOppQQRRssTTuuVV
3   AAbbccDDeeFFGGhhIIjjKKLLmmnnOOppQQRRssTTuuVV
4   AAbbccDDeeFFGGhhIIjjKKLLmmnnOOppQQRRssTTuuVV
5   AAbbccDDeeFFGGhhIIjjKKLLmmnnOOppQQRRssTTuuVV
6   AAbbccDDeeFFGGhhIIjjKKLLmmnnOOppQQRRssTTuuVV
7   AAbbccDDeeFFGGhhIIjjKKLLmmnnOOppQQRRssTTuuVV
8   AAbbccDDeeFFGGhhIIjjKKLLmmnnOOppQQRRssTTuuVV
9   AAbbccDDeeFFGGhhIIjjKKLLmmnnOOppQQRRssTTuuVV
10   AAbbccDDeeFFGGhhIIjjKKLLmmnnOOppQQRRssTTuuVV
(A)

Individual
1   AAbbccDDeeFfGGhhIIjjKKLLmmnnOOppQQRrssTTuuVV
2   AabbccDDeeFFGGhhIIjjKKLLmmnnOoppQQRRssTTuuVv
3   AABbccDDeeFFGGhhIIjjKkLLmmnnOOppQQRRssTTuuVV
4   AAbbccDdeeFFGGhhIIjjKKLLmmnnOOppQQRrssTTuuVV
5   AabbccDDeeFFGGhhIIjjKKLLmmnnOoppQQRRssTTuuVV
6   AAbbccDDeeFFGGhhIIjjKKLLmmNnOOppQQRRssTTuuVv
7   AAbbccDDeeFFGGhhIIjjKkLLmmnnOOppQQRRssTTuuVV
8   AabbccDdeeFFGGhhIIjjKKLLmmnnOOppQQRRssTTuuVV
9   AAbbccDDeeFFGGhhIIjjKKLLmmnnOoppQQRrssTTuuVV
10   AABbccDDeeFFGGhhIIjjKKLlmmnnOOppQQRRssTTuuVv
(B)

Individual
1   AabbccDDeeFfGGhhIIjjKkLLmmnnOOppQqRrssTTuuVv
2   AabbccDDeeFfGGhhIIjjKkLLmmnnOOppQqRrssTTuuVv
3   AabbccDDeeFfGGhhIIjjKkLLmmnnOOppQqRrssTTuuVv
4   AabbccDDeeFfGGhhIIjjKkLLmmnnOOppQqRrssTTuuVv
5   AabbccDDeeFfGGhhIIjjKkLLmmnnOOppQqRrssTTuuVv
6   AabbccDDeeFfGGhhIIjjKkLLmmnnOOppQqRrssTTuuVv
7   AabbccDDeeFfGGhhIIjjKkLLmmnnOOppQqRrssTTuuVv
8   AabbccDDeeFfGGhhIIjjKkLLmmnnOOppQqRrssTTuuVv
9   AabbccDDeeFfGGhhIIjjKkLLmmnnOOppQqRrssTTuuVv
10   AabbccDDeeFfGGhhIIjjKkLLmmnnOOppQqRrssTTuuVv
(C)

FIG. 4.4   Ten individuals from hypothetical pure line (A), OP (B), and F$_1$ hybrid or clonal (C) populations with regard to genotypes at 22 random loci. *Black*, homozygotes; *red*, heterozygotes.

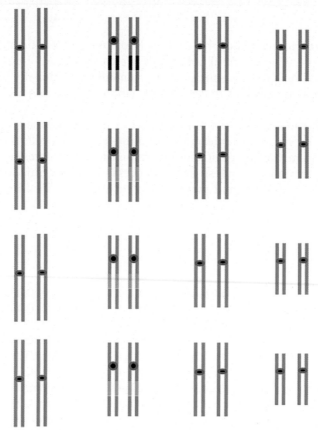

FIG. 4.5 Hypothetical karyotype of a species 2n = 2x = 8 representing a multiline population. The different colored chromosome segments represent different sources of DNA sequences that encode different phenotypes, for example resistance to different races of a pathogen. The rest of the genome is monolithic.

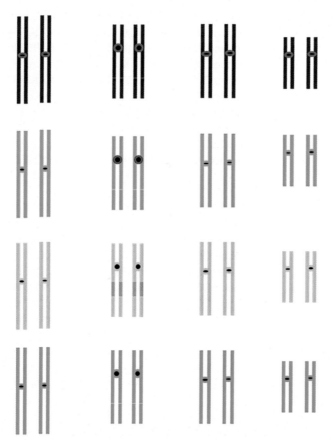

FIG. 4.6 Hypothetical karyotype of a species 2n = 2x = 8 representing a composite population. The different colored chromosomes represent different varieties that are mixed together to form the population.

References

Allard, R.W., 1999. Principles of Plant Breeding, second ed. Wiley & Sons, Inc., New York, NY. 254 pp.

Anon., 2016. Population—Definition. Biology On-Line Dictionary. Archived at: www.biology-online.org/dictionary/Population.

Anon., 2017. Transmission of Influenza Viruses From Animals to People. Archived at: https://www.cdc.gov/flu/about/viruses/transmission.htm.

Ashri, A., 1989. Major gene mutations and domestication of plants. In: Sijurbjörnsson, B. (Ed.), Plant Domestication by Induced Mutation. IAEA, Vienna, AUS, pp. 3–9.

Barrett, S.C.H., Charlesworth, D., 1991. Effects of a change in the level of inbreeding on the genetic load. Nature 352 (6335), 522–524.

Beretta, M., Donati, F., Ciriaci, T., Sabatini, E., 2015. GxE interaction in lettuce bolting phenomena. Acta Hortic. 1100, 141–144.

Bloom, J.D., Arnold, F.H., 2009. In the light of directed evolution: pathways of adaptive protein evolution. Proc. Natl. Acad. Sci. U. S. A. 106, 9995–10000.

Bosland, P.W., 2005. Second generation (F_2) hybrid cultivars for jalapeno production. HortScience 40 (6), 1679–1681.

Carr, D.E., 2013. A multidimensional approach to understanding floral function and form1. Am. J. Bot. 100 (6), 1102–1104.

Covert, A.W., Lenski, R.E., Wilke, C.O., Ofria, C., 2013. Experiments on the role of deleterious mutations as stepping stones in adaptive evolution. Proc. Nat. Acad. Sci. U. S. A. 110 (34), E3171–E3178.

Dougherty, M.J., 2017. Why are we getting taller as a species? Sci. Am. www.scientificamerican.com/article/why-are-we-getting-taller/.

Duvick, D.N., 1984. Genetic contributions to yield gains of U.S. hybrid maize, 1930 to 1980. In: Fehr, W.R. (Ed.), Proceedings of a Symposium Sponsored by Div. C-1, Crop Science Society of America, 2 December 1981, in Atlanta, Georgia. Crop Science Society of America and American Society of Agronomy, Atlanta, GA, pp. 15–47.

East, E.M., 1908. Inbreeding in corn. Rep. Conn. Agric. Exp. Sta. 1907, 419–429.

East, E.M., 1936. Heterosis. Genetics 21, 375–397.

Frankel, O.H., Bennett, E., 1970. Genetic Resources in Plants: Their Exploration and Conservation. Blackwell, Oxford, UK. 554 pp.

Frankel, O.H., Hawkes, J.G. (Eds.), 1975. Crop Genetic Resources for Today and Tomorrow. Cambridge Univ. Press, Cambridge, UK. 492 pp.

Friedt, W., Brune, U., 1987. Recombination: asexual recombination in higher plants. In: Progress in Botany. vol. 49. pp. 192–215.

Gepts, P., Hancock, J., 2006. The future of plant breeding. Crop Sci. 46 (4), 1630–1634.

Harlan, J.R., 1980. Crop monoculture and the future of American agriculture. In: Batie, S.S., Healy, R.G. (Eds.), Future of American Agriculture as a Strategic Resource. FAO/UN, Oslo, Norway, pp. 225–250.

Hegedus, A., Szabo, Z., Nyeki, J., Halasz, J., Pedryc, A., 2006. Molecular analysis of S-haplotypes in peach, a self-compatible *Prunus* species. J. Am. Soc. Hort. Sci. 131 (6), 738–743.

Helland, S.J., Holland, J.B., 2001. Blend response and stability and cultivar blending ability in oat. Crop Sci. 41 (6), 1689–1696.

Holsinger, K.E., 1992. Ecological models of plant mating systems and the evolutionary stability of mixed mating systems. In: Wyatt, R. (Ed.), Ecology and Evolution of Plant Reproduction. Chapman & Hall, New York, NY, pp. 169–191.

Hussey, G., 1984. Clonal propagation of plants from cells, tissues and meristems. In: Lea, P.J., Stewart, G.R. (Eds.), Genetic Manipulation of Plants and its Application to Agriculture. Clarendon Press, Oxford [Oxfordshire], UK, pp. 197–217.

Johannsen, W.L., 1903. Ueber Erblichkeit in Populationen Und in Reinen Linien. Gustav Fischer Verlag, Jena, Germany.

Jones, D.F., 1917. Dominance of linked factors as a means of accounting for heterosis. Proc. Natl. Acad. Sci. U. S. A. 3, 310–312.

Kariyat, R.R., Sinclair, J.P., Golenberg, E.M., 2013. Following Darwin's trail: interactions affecting the evolution of plant mating systems. Am. J. Bot. 100 (6), 999–1001.

Lipton, M., Longhurst, R., 1985. In: Leathers, H.D., Foster, P. (Eds.), The World Food Problem: Toward Ending Undernutrition in the Third World. Lynne Rienner Publishers, Boulder, CO. 433 pp.

Neiman, M., Sharbel, T.F., Schwander, T., 2014. Genetic causes of transitions from sexual reproduction to asexuality in plants and animals. J. Evol. Biol. 27 (7), 1346–1359.

Richards, A.J., 1986. Plant Breeding Systems. George Allen & Unwin, London, UK. 529 pp.

Shull, G.H., 1948. What is heterosis? Genetics 33, 439–446.

Smithson, J.B., Lenne, J.M., 1996. Varietal mixtures: a viable strategy for sustainable productivity in subsistence agriculture. Ann. Appl. Biol. 128 (1), 127–158.

Sprague, G.F., 1946. The experimental basis for hybrid maize. Biol. Rev. 21, 101–120.

Stoltzfus, A., Yampolsky, L.Y., 2009. Climbing mount probable: mutation as a cause of nonrandomness in evolution. J. Hered. 100 (5), 637–647.

Suter, C.M., Martin, D.I.K., 2010. Paramutation: the tip of an epigenetic iceberg. Trends Genet. 26 (1), 9–14.

Takebayashi, N., Morrell, P.L., 2001. Is self-fertilization an evolutionary dead end? Revisiting an old hypothesis with genetic theories and a macroevolutionary approach. Am. J. Bot. 88 (7), 1143–1150.

Taylor, I.B., 1986. Biosystematics of the tomato. In: Atherton, J.G., Rudich, J. (Eds.), The Tomato Crop. Chapman and Hall, New York, NY, pp. 1–34.

Wallace, B., 1991. Fifty Years of Genetic Load: An Odyssey. Cornell University Press, Ithaca, NY. 174 pp.

Wilson, J.P., Gates, R.N., Panwar, M.S., 2001. Dynamic multiline population approach to resistance gene management. Phytopathology 91 (3), 255–260.

Zizumbo-Villarreal, D., Colunga-GarciaMarin, P., Payro de la Cruz, E., Delgado-Valerio, P., Gepts, P., 2005. Population structure and evolutionary dynamics of wild-weedy-domesticated complexes of common bean in a Mesoamerican region. Crop Sci. 45 (3), 1073–1083.

References

5

Mass Selection and the Basic Plant Breeding Algorithm

INTRODUCTION

Plant breeding is defined as human actions that result in the permanent desirable genetic change of a population of plants. The journey by this book from theoretical and historical underpinnings through the actual processes now known as plant breeding thus begins. Mass selection is the simplest form of plant breeding. This primitive method was among the range of basic activities practiced in agriculture by early humans. Since the mid-19th century CE plant breeding had advanced and expanded to become more progressively more specialized and science-based. Following the consideration of the biological facets of plant breeding this textbook will present the various methods (Section 2; Chapters 13–19) according to the mating system and appropriate circumstances.

Mass selection is covered in Section 1 ("Elements and Underpinnings of Plant Breeding") as a prelude to the basic plant breeding algorithm and domestication. This will create a framework for subsequent, more complex strategies that will be covered in Section 2. The reader is referred to many other informative and enlightening textbooks to appreciate that the subject of plant breeding may be approached and treated in many different ways (Allard, 1999; Fehr, 1987; Poehlman and Sleper, 2013; Kuckuck et al., 1991; Borojevic, 1990; Chahal and Gosal, 2002; Acquaah, 2012; Fleury and Whitford, 2014; Singh, 2006; Brown et al., 2014).

MASS SELECTION AND PLANT DOMESTICATION

Most of our domesticated plant populations were passed down to us by the previous generations who carefully maintained seeds or asexual propagules from which the next year's bounties could be derived. A handful of crop species were developed relatively recently for example highbush blueberry and American cranberry (*Vaccinium corymbosum*, and *V. macrocarpon*) (Ehlenfeldt, 2009; Polashock and Vorsa, 2002). As was discussed earlier, this process of selecting individuals to contribute seeds to the next year's crop has only been documented in writing over the past

200 years. Prior to the 19th century, selected populations of seeds and agricultural know-how was passed along with other crucial life skills, as agrarian societies had little specialization. Everyone grew food crops and maintained their own seed stocks (Duvick, 1996).

Plant breeding is one of the oldest professions since it was practiced hand-in-hand with agriculture. Food acquisition or production has always been the most important of all human goals. It probably didn't take primitive humans very long to discover that seeds from the best plants tended to result in the best progeny plants. This primitive form of plant breeding they practiced in affecting desirable genetic changes is now called *mass selection*. This method consists of the best individual plants being selected and seeds from selected plants mixed together to comprise the population for the next production season. The rest of the seeds were eaten, used for medicinal purposes, made into rattles, or used some other way since little could be wasted under subsistence conditions. Mass selection is still the most widely used plant breeding method because it is easy to understand and communicate and makes no assumptions about reproductive behavior or the inheritance of the traits being addressed. The method is also very effective since it is predicated on the gradual increase of desirable alleles and corresponding decrease of undesirable ones (Duvick, 1996). A *panmictic* (randomly mating) population begets another panmictic population for outcrossing species while an admixture of pure lines begets another admixture of pure lines for self-pollinating species. In the latter example, very strong selection pressures can have the effect of eliminating most or all but one of the components of an admixture of genotypes resulting in a pure line (Ashri, 1989).

Uncontrolled mating, or mating without human intervention, is an important feature of mass selection. This point is mostly irrelevant for self-pollinated species since mating is, by definition, controlled. In Chapter 10, natural mating systems will be described further, including the notion that pure cross- and self-pollination are rarely realized. Most species exist in a continuum between pure cross- and self-pollination. Therefore, uncontrolled mating actually does affect the rate of progress in self-pollinated species as well as cross-pollinated species. By selecting the best individuals from which to harvest seed in outcrossing species, primitive humans were in fact only controlling half of the gametes, the female. The male gamete (pollen) could come from any individual in the population including those with poor performance.

Genetic progress under selection historically was (and still is) exceedingly slow, but over long periods of time, as long as 20,000 years, crafted populations that were markedly different from wild progenitors (Heiser, 1973). While this seems like an eternity with respect to the number of human generations, 20,000 years is extremely short within the context of the evolution of angiosperms (Wang et al., 1999). From the beginnings of agriculture (about 18,000 BCE) until about 1900 CE, populations progressed gradually from the wild to being wholly domesticated and completely dependent on humans for survival and reproductive capacity (Duvick, 1996). Generally, progress under mass selection is relatively slower for horticultural than agronomic crop species due to low heritability (h^2) (see Chapter 3) for many or most targeted traits. Horticultural crops are characterized by phenotypes that tend to have a relatively higher proportion of V_E and V_{GxE} than agronomic species.

Since mating is not controlled in mass selection, no assumption is made of the mating system employed by the targeted plant species. Mass selection was used successfully by early humans for both self- and cross-pollinating species. The rate of progress and intermediate- and long-term outcomes with respect to genetic population structures would vary with mating. Populations of self-pollinating species domesticated by mass selection would typically be comprised of a large number of geographically distinct sub-populations, each relatively fixed genetically.

Mass selection would likely have quickly depleted whatever genetic variability was available during the domestication process, particularly in self-pollinating species where both female and male parent were mostly controlled. Further progress would only have been possible from recombination of tightly linked alleles in repulsion, new mutations, and the influx of new genes from outside populations. There is evidence that mass selection and domestication have indirectly degraded certain phenotypes of agricultural value that were not directly selected during the domestication process. For example, a comparison of wild and domesticated cranberry (*V. macrocarpon*) showed that domesticates exhibited lower tolerance to gypsy moth herbivory and coincidentally expressed lower quantities of the natural defense hormone cis-jasmonic acid than wild ancestors (Rodriguez-Saona et al., 2011).

Carrot (*Daucus carota* L.) is an excellent example of a crop species that was domesticated over millennia via mass selection. Carrot is the most widely grown member of the Apiaceae (formerly Umbelliferae). This diverse and complex plant family includes several other vegetables, such as parsnip, fennel, celery and celeriac, parsley, arracacha, and many herbs and spices (Rubatzky et al., 1999). Like other plants of this family, carrot seeds are aromatic and consequently have long been used as a spice or herbal medicine. In fact, carrot seeds were found in early human habitation sites as long as 3000 to 5000 years ago in Switzerland and Germany (Laufer, 1919). These seeds are thought to be from wild carrot used for flavor or medicine (Simon et al., 2008).

The genus Daucus includes approximately 20 species scattered worldwide but the majority of species within the Apiaceae occur around the Mediterranean basin (Sáenz Laín, 1981). Domesticated carrot has been discovered to have originated in central Asia (Afghanistan), unlike most other species of *Daucus* and Apiaceae. The first evidence of carrot cultivation in central Asia dates to about 900 CE (Mackevic, 1929). Wild carrot (a.k.a. the common worldwide weed "Queen Anne's Lace") had spread beyond central Asia prior to domestication. The primitive Afghan carrot spread westward to Persia in the 900s CE, Middle East, and North Africa in the 1000s CE, Spain in the 1100s CE, and Northern Europe in the 1300s CE (Banga, 1957a, b). Turkey is regarded as a secondary center of diversity for carrot with evidence of diverse carrot landraces in the past and up to the present. The spread of cultivated carrot eastward is not as well documented as that to the west, but history records its use beginning in China in the 1300s and Japan in the 1700s CE (Simon et al., 2008).

The genetic improvement of carrot has been an ongoing effort throughout its cultivation and domestication. Before the 20th century CE, carrot production was small-scale in family or community gardens. A portion of the crop was likely protected in the field over winter with mulch or the best roots saved in cellars were replanted the subsequent spring to produce a seed crop. There is no written record of which traits were evaluated or any other detail of the selection process in this period, but all domesticated carrot cultivars differ from wild progenitors in forming larger, smoother storage roots, so it is clear that these traits also were improved through recurrent mass selection. Among the most important traits selected during domestication was "slow bolting", or the need for a longer period of cold vernalization before the plant is transformed from vegetative to reproductive growth (Simon et al., 2008).

Taproot color and flavor were among the most important early selection criteria since taproots of wild carrot are far too bitter to be consumed directly as food. Other evidence that color and flavor were targeted for selection is the existence of a broad spectrum of populations that exhibit a broad range of flavors and colors. Wild carrot roots are white or very pale yellow. Early plant breeders developed populations that featured purple and yellow tap roots, the only colors recorded historically until the 16th to 17th centuries CE. Thereafter, orange carrots were first described and soon came to be preferred in both the eastern and western production areas (Banga, 1957a, 1957b; Rubatzky et al., 1999; Simon, 2000). It is not known why 17th century CE carrot breeders shifted their preference to orange types, but this change in preference inadvertently provided a rich source of vitamin A to carrot consumers. Soon after orange carrots became popular, the first named carrot cultivars were described in terms of shape, size, color, and flavor, and the first commercially sold carrot seeds included reference to this growing list of distinguishing traits, all resulting from recurrent mass selection for these traits (Simon et al., 2008).

Human societies evolved in concert with the plants, animals, and microbes they domesticated. As populations of domesticates became more specialized and narrowly adapted, more human intervention was necessary to maintain and propagate them. Evidence has been advanced that the decline Amerindian populations in South America following European contact was at least in part due to the inability to maintain populations of domesticated plants and animals following disruption of societies by the Europeans (Clement, 1999).

As populations became more differentiated from those found in nature, they would assume intrinsic value to humans by virtue of utility. It is well known that seeds were disseminated as a consequence of trade among cultural centers (Heiser, 1973; Abbo et al., 2012). Thus, exploration for new populations that exhibited desired properties would have been highly successful, and likely took place in many locations on Earth during the transition from nomadic to sedentary human societies (Zohary and Hopf, 2000; Zeder, 2008). Subsequent progress under selection would have been limited following rapid initial gains, since no opportunities for recombination and segregation of extant variation exists within inbreeding species.

Populations of cross-pollinated species would, of course, contain a higher degree of genetic variability in the wild than self-pollinated species, and mass selection would have taken a longer time to hone the genetic structure into an entity that was both acceptably uniform and high relative performance. The introduction of new seeds from distant lands would have had a further long-term genetic benefit (with apologies to proponents of "native plants"): transfer (i.e., introgression) of new alleles into the population followed by recombination and segregation and resulting in new, resulting in potentially more desirable genotypes. By analyzing archeological and modern grape seed morphology, Terral et al. (2010) concluded that older clones in southern France originated from many diverse geographical regions. In another example, trade routes in Asia routinely featured rice from diverse cultures, mainly lowland (*japonica*) and upland (*indica*) types. Studies of genomic structure have shown that genes from the *japonica* and *indica* types were cross-introgressed (Gross and Zhao, 2014).

THE PLANT BREEDING ALGORITHM

All breeding programs have a theoretical beginning and ending, and progress through similar phases or stages. All of the different methods employed by plant breeders in the broad range of plant species and breeding objectives share the same fundamental steps:

BREEDING OBJECTIVES
(Chapter 6)
↓
GERMPLASM ACQUISITION AND EVALUATION
(Chapters 7, 8)
↓
RECURRENT CONTROLLED MATING AND SELECTION
AMONG LIMITED GENOTYPES FROM GERMPLASM
(Chapters 9-10, 13-19)
↓
NEW GENETICALLY DISTINCT POPULATIONS WITH
PROSPECTIVE IMPROVEMENTS
↓
TESTING AGAINST STANDARD POPULATIONS
(Chapter 11)
↓
PROPRIETARY PROTECTION (Chapter 12)
↓
INCREASE OF PROPAGULES/SEEDS (Chapter 11)
↓
RELEASE INTO COMMERCE
(Chapter 12)

Historically, the plant breeder was engaged in all of these phases. As human societies advanced, economies became more segmented, and occupations within economies grew progressively more specialized. This has made it possible to invent, manufacture, and mass-market very complex things at relatively low cost for example automobiles and computers. Likewise, the development of new plant genotypes has become progressively more segmented and specialized over time and with more technical sophistication.

The seed industry in the U.S., including the segment pertaining to horticultural crop species, is a good example of the segmentation and specialization wrought within the field of plant breeding. Each of the basic steps outlined above is now undertaken by separate teams of devoted specialists within the seed industry from germplasm and genetic refinement through seed production and sales and marketing. Each of the steps is further divided into more areas of specialization. Within the third step alone, the breeder may interact with cell and molecular biologists, plant pathologists, entomologists, and physiologists in the planning and execution of selection strategies.

In reality, most breeding programs have no defined beginning or end points. A newly posted plant breeder often inherits the germplasm, breeding lines, and client needs of a predecessor, and continues his/her work in a similar but altered direction. In this manner, plant breeding continues the tradition of continuity of germplasm over long periods of time. Some plant breeders do not readily yield to the next generation, however, and hoard the populations and methods they have crafted over their careers. This misguided sense of ownership can result in disruptions of continuity and loss of germplasm.

Plant breeding encompasses yet another facet that is not usually addressed in textbooks: scientific curiosity and discovery. Each new selection and mating, or testing of new germplasm, yields results never seen before. As a trained scientist, the plant breeder makes note of such new observations and alters his/her course to capture emerging opportunities for academic and commercial purposes. The breeder may set out on a certain course with clearly defined objectives and methods then progressively change course as scientific results dictate and in the direction of new alternatives or market fluctuations. Many horticultural crops, and particularly woody perennials, require extremely protracted time periods to run the gamut from germplasm to released variety; perhaps even decades. It is not reasonable to presume that science and market needs will remain static for that long. Breeding lines are subjected to shifting sets of selection criteria and mating schemes, and the path to the endpoint is not a straight one. Most breeders also maintain breeding lines for the sole purpose of gathering and recombining desirable alleles, also referred to as population improvement. This is analogous to an intermediate step between raw germplasm and populations that are proceeding toward imminent commercialization. Thus, most breeding programs are dynamic and deviate from the original plan, changing with new scientific evidence and market needs.

Each of these steps of the basic plant breeding algorithm will be addressed in detail in subsequent chapters. What follows below is a brief description of how they fit together in a continuum, and the general role played by the practicing plant breeder in each.

GERMPLASM

This subject will be covered in more detail in Chapters 7 and 8. Most plant breeders maintain a "working collection" of genetic variability, including various breeding lines, commercial varieties with different geographical and climatological ranges, and "wild" populations. Despite the general nature and geographic dispersion of germplasm, exploration and acquisition, organization, storage, and dissemination have become very specialized areas of study. An apt analogy for germplasm would be money and banking. Not so long ago, everyone kept most of their assets in a bank where a modest return would be provided to allow the bank to use the assets for higher-yielding loans. The bank was a physical entity with a safe and reserves of all sorts of currency denominations. If one needed hard currency or a loan they would physically visit the bank to file a request. The system for handling germplasm has historically been very similar to the monetary bank. In fact, locations, where germplasm is maintained, are often referred to as "banks".

The monetary banking industry has become much more complex. Most of the money a person makes never exists as physical currency at all but as vouchers that travel electronically from the employer's server to the employee then to vendors. Most consumers devise a complex monetary voucher routing system that fits their individual needs to pay for goods and services, maintain cash reserves, and contribute long-term interest-earning savings. The array of investment alternatives is truly bewildering, each coming with an abstraction of risk. The germplasm economy has likewise become very complicated, and global. But unlike currency, germplasm is a tangible thing.

Germplasm is a collection of genes from which individual genes or sets of genes may be withdrawn and used to further improve the phenotypes of populations of commercial varieties of plants. A gene, of course, is a basic unit of genetic function, the double-stranded DNA base sequence stretching from 5' to 3' ends, with all of the control, promoter, coding, and non-coding structural sequences included. The gene is a sequence of nucleotide bases that encodes a mRNA that, in turn, usually encodes a polypeptide. The corresponding polypeptide interacts with other polypeptides and macro-molecular units and the environment to affect an observable phenotype.

Is germplasm a soup containing a mixture of all of the genes, their allelic forms, etc.? Of course it is not. Germplasm is a general term that embodies all potential modes of gene reservoirs. In the overwhelming majority of instances, the basic unit of germplasm storage (or "currency") is in the form of seeds. Seeds are convenient biological packages that have evolved in angiosperms and gymnosperms to insulate new sporophytes against harsh environmental conditions until the next period of favorable environments arrives. Each seed represents a potentially unique combination of genes, most of little interest to the plant breeder, but a few are rough diamonds glistening in the piles of relatively worthless ore. It is the job of the breeder to find the gems and ignore the ore, and thereafter to be a genetic alchemist.

The vast majority of plant germplasm on planet Earth is in situ, or in natural living habitats. Over many millennia humans have prospected a small proportion of this in situ biodiversity and subjected it to primitive plant breeding (usually mass selection). This resulted in populations that are more useful to humans than the original populations in the wild. Only recently, within the last 200 years, have there been efforts to glean genes for potentially useful phenotypes from the wild into organized collections. Historically, most of these collections have been in the form of seeds collected in wild habitats, and maintained in repositories (or "banks") in the form of populations, or lots, of seeds (Fig. 5.1).

In many crop species, and particularly woody perennial horticultural plants, vegetative tissues or organs may be used to constitute germplasm. Asexual propagules are also utilized as germplasm and breeding intermediates in horticultural species that have evolved clonally, for example Liliaceae (e.g., lily, daffodil, iris, onion, garlic). Recent technological advances in plant cell and tissue culture have paved the way for units as small as individual cells to be stored indefinitely, then recultured and regenerated into whole plants on demand. For a multitude of reasons, the use of cultured cells and tissues to store and disseminate germplasm has not proven to be broadly practical. The technology is expensive and challenging to sustain due to the need for continuous energy and aseptic environments. It is even easier and cheaper to isolate and store individual genes in replicating bacterial plasmids than it is to develop and store cultured plant cells and tissues (Towill, 1989).

Since we have the capability to insert genes into plant genomes at will (Chapter 8), this mode of germplasm storage may well gain in importance in future years. Transgenic technology, and especially the use of *Agrobacterium tumefaciens* T-DNA, necessitates plant tissue culture to obtain transformed plants. Therefore, this strategy is wrought with technical hurdles and maintenance inputs, even more daunting than cell/tissue culture alone. Plant breeders prefer to maintain genes in the form of plants or units that will easily develop into plants unless a compelling reason

FIG. 5.1   Seed lots of germplasm accessions stored in a controlled environment room at the Uzbek Research Institute of Plant Industry (formerly the Vavilov Institute of Research).

to deal directly with individual genes at the DNA level arises. Maintaining genes as isolated DNA or RNA sequences is expensive and technically challenging. Genes mostly only have value in the context of expression within whole plants so it is convenient, efficient, cost-effective, and appropriate to maintain them within this context.

Most plant breeders typically maintain a compact and easily accessible germplasm collection that is pertinent to the goals and objectives of their programs for example adaptation to a specific geographical range or economic uses. If the constituency served by the breeder can be measured as such, there is no reason to maintain a large collection that includes populations that are adapted to other geographical regions or for other economic purposes, unless they carry other genes of interest to the targeted product concepts. Each population requires energy and space for storage and must be monitored constantly for viability. Depending on the length of time the seed or asexual propagule can remain viable the germplasm must be replenished periodically, a very time-consuming and potentially space-consuming endeavor. Since the plant breeder is in the business of creating new, exciting commercial populations, the maintenance of germplasm is usually viewed as a necessary chore, not the central focus of the program.

Seeds have been traded among cultures during the agricultural era since the advent of trade routes. Most significantly, the "Silk Road" from eastern Asia to Europe played an important role in the introduction of new crop species between Asia and Europe. Later, as cultures explored new continents nautically, the ship's crew would typically include a botanist charged with collecting and describing new plant forms (Perdue and Christenson, 1989).

The U.S. Department of Agriculture Office of Seed and Plant Introduction was established in 1898 and became the de facto coordinator of U.S. plant germplasm acquisition, storage, and dissemination (White et al., 1989; Stoner and Hummer, 2007). At that point in history, it was abundantly clear that few of the economic plant species upon which the young country was dependent upon for food were indigenous to North America, nearly all species having been brought here from Europe, Asia, and Africa during the 16th to 18th centuries CE.

The USDA/ARS/NPGS (Agricultural Research Service/National Plant Germplasm System) now houses and makes available seeds and asexual propagules to scientists worldwide, the current collection at about 460,000 accessions (Stoner and Hummer, 2007). Similar collections have been established around the world, most notably in the former Soviet Union, now St. Petersburg, Russia and Tashkent, Uzbekistan, where the famed botanist Nikolai I. Vavilov honed his theories on the Centers of Origin and Diversity of Species and Homologous Series of Variation (Vavilov, 1926; Harlan, 1992).

While germplasm was initially viewed as a curiosity and an underpinning to efforts in the budding field of plant breeding, the destruction of environments and habitats started to reach media headlines starting in the 1960s. The term "biodiversity" soon crept into our lexicon and has risen quickly to a level of broad awareness and socio-political concern. Simultaneously, nations began to regard plant and animal germplasm as a component of the portfolio of extant natural resources, similar to oil and strategic mineral deposits (Lawson, 2004; Chapter 2).

Consequently, plant germplasm has become a politically charged issue since most Centers of Origin and Diversity for economic plant species coincide with developing areas of the world. Plant breeding and the flight of native seeds from undeveloped or developing countries to North America, Europe, Japan, and others became emblematic of the pillage of these cultures by the west for capitalistic gains. Developing countries have banded together and formed an international agreement that ultimately provided for fair compensation to the sovereign nation of geographic origin (Barton and Siebeck, 1994). Consequently, the acquisition of seeds taken from the wild or of landrace populations is usually a quagmire, requiring mountains of paperwork to trace the origins of seeds and propagules that enable intellectual property (IP) professionals to assign ownership and impute royalties.

MATING AND SELECTION

A more thorough treatment of the topics of mating and selection in plant breeding will be presented in Chapters 9 and 10. A typical plant breeder devotes over 90% of their time, effort, and resources on the "recurrent controlled mating and selection" step (Step #3) of the algorithm. In many cases, however, such as economic plant species of relatively minor importance or where the "infrastructure" is lacking, the plant breeder must take on more of the load in other steps including germplasm acquisition and maintenance, breeding line evaluation, and seed production. This often extends to the promotion of the species for the intended economic use, a necessary prerequisite to driving financial support for breeding efforts.

Much of the challenge and allure of plant breeding is matching a species targeted for improvement with the best possible germplasm and selection and controlled mating scheme. While various general schemes will be described in this textbook, each species is truly unique, and adaptations of the established methods are always necessary. Each plant breeder does things a little differently, and for different reasons. Breeders of horticultural plant species tend to be even more varied than those of agronomic crops. There are often many different strategies that can ultimately be successful. There is no right or wrong answer, only some strategies that are demonstrably or arguably better than others.

Selection and controlled mating are the two major tools available to the plant breeder to affect genetic changes at the level of the individual and population. Selection enriches the pool of gametes for desired alleles (Fig. 5.2), and controlled mating affects how they are combined to generate populations where selection will be most fruitful (Fig. 5.3). The process of meiosis lies at the core of the plant breeder's toolbox. Once an enriched pool of gametes bearing desirable alleles is presented to the mating process, they are sliced and diced into a diverse array of recombinants that are sifted through by the plant breeder for the best phenotypes. The toolbox is currently being enhanced by new advancements in genomics, allowing plant breeders to begin to bypass the phenotype as the manifestation of genotype and select directly on DNA sequences (Kang et al., 2016).

Since selection and mating are both sensitive to population sizes, statistics and probability are employed to document factual differences among populations in each distinct breeding scheme. In practice, most plant breeders lack the infinite time, space, and money necessary to maximize the statistical probability of success, and must make strategic compromises at nearly every decision point. Much of the text describing individual methods will describe ways to enhance probabilities of success with the lowest concomitant expenditure of resources.

FIG. 5.2   Carrot breeder Dr. Philip Simon (left) of the USDA/ARS Madison, WI measuring foliage of carrot plants pursuant to selection of the most desirable individuals from the population.

I. ELEMENTS AND UNDERPINNINGS OF PLANT BREEDING

FIG. 5.3   Controlled mating by hand pollination of a cassava flower. *From https://www.flickr.com/photos/mreichwage/19325779232.*

NEW POPULATIONS WITH PROSPECTIVE IMPROVEMENTS

How does the plant breeder decide when a breakthrough has been accomplished; that a breeding objective has been reached; that a new population has what it takes to be elevated to the status of a commercial product? The plant breeder usually conducts performance trials of advanced selections to compare newly developed populations with examples of successful products that are already in commerce. The results of these trials prompt the plant breeder to take the next step to demonstrate the genetic veracity of these advanced selections. In the public sector, the transition from incipient commercial product to release isn't necessarily well defined or objective, depending on the size of the market and procedural demands. All too often, populations are released that lack merit, depressing the bottom line of commercial users. The process by which prospective products are advanced into the market in the private sector, and for major crop commodities, tends to be more defined and objective.

TESTING OF CANDIDATE POPULATIONS

Chapter 11 will present more details on post-breeding activities, including the testing of advanced breeding populations. It is quite common for the plant breeder to become excessively and sometimes inappropriately enamored of their breeding populations. They are often driven to craft seemingly endless improvements beyond what practical demand would dictate, and even to the point of perfection itself. Like one's child growing up to adulthood, the plant breeder often has difficulty learning how to let go; to say goodbye to the plant populations they have created. Eventually, however, the bills must be paid. Populations that are developed by breeding programs that exceed commercial cultivars with respect to overall performance are named new cultivars and offered into commerce. Perfection, after all, is a concept relative to time, space and opinion, and can never be truly attained.

The first step in the procession from breeding population to "finished cultivar" is the performance trial. This is a set of controlled experiments wherein the unknown populations are contrasted with known standards with regard to established performance criteria. These field experiments are best accomplished under conditions that most closely emulate those of actual commercial production. In most instances, effective methods to measure plant population performance under controlled laboratory or greenhouse conditions have not yet been developed. The performance trials program should be consistent with accepted statistical methods, allowing components of phenotypic variance (e.g., V_G, V_E, V_{GxE}) to be determined (Cooper and DeLacy, 1994).

Prior to cultivar release, the plant breeder has completed steps 1 (establishment of breeding objectives, 2 (germplasm acquisition) and 3 (recurrent mating and selection) of the overall program with a clear understanding of the state of the art and targeted market. The performance trials program is the acid test of how well he/she has done. For this reason, variety trials are often or even usually conducted by an objective third party. Each program is tailor-made to the crop species and geomarketing targets. Performance trialing is an excellent point to involve clients in "participatory plant breeding". This is a relatively new trend where prospective buyers of seeds of new varieties are invited to participate in the selection of new products (Chable et al., 2008).

A program of performance trials generally proceeds in stages. The entire program usually takes three years for temperate, annual crop species and longer for perennials or if the economic stakes are high. The performance trials step may require less time if generation time is short and multiple tests can be conducted within a year or if the economic stakes are low. There are usually several staggered performance trials in progress at any given time. Therefore, new cultivars may be advanced to commercial status continuously. The seed business is, of course, highly competitive and active development of new products is an important means by which to maintain remain viable.

Large seed companies generally employ a defined mechanism to move populations along in the pipeline. While it is important that populations be shepherded as rapidly as possible, there are limits to the rate of progress, and severe consequences for making decisions too hastily. It is also important that plant breeders be accorded contact with seed production specialists, the sales and marketing team, and any horticulturists or agronomists working on the company's behalf. These specialists participate in the development of the breeding goals and provide valuable insights into the practical advantages and disadvantages of each trait. In the most effective organizations, the specialists downstream of the breeding program become de facto members of the breeding team and contribute to programs in immeasurable ways.

Ideally, the end point is very few populations recommended for commercialization, the entries that have consistently outpaced the standards, or that bring something new that the farmer or another consumer will want. In some instances, a given breeding program does not produce any new commercial entities. The decision to commercialize is very costly and must be made with the confidence that performance attributes will make the new variety a hands-down winner.

LARGE-SCALE SEED PRODUCTION

Once the commitment has been made to commercialize a targeted breeding population, it passes into the realm of scale-up to provide propagules to the farmer or grower. As was discussed earlier, this function was until recently the providence of the farmer as well, who kept a portion of the harvest to plant for next year's crop. Production of seed or asexual propagules now primarily falls into the hands of specialists who are affiliated with seed companies or quasi-public seed certification organizations. Chapter 11 will go into more details about how propagules are produced, and the roles played by the plant breeder in these processes.

Producing seeds is analogous to making copies of an original document. Xerographic copy machines have been in the mainstream for over 50 years, so most everyone is familiar with what they can do. The quality of the copy is usually not the same (and not as desirable) as the original. The hue of black-gray-white may be different, and the subtle fine details are not as well defined. If one makes a copy of a copy, and so forth, the fidelity decreases progressively. With cheap copiers the fidelity is poor to begin with, and deteriorates rapidly with successive copying, while in more expensive models the fidelity, or fine resolution, is usually higher.

The seed production manager strives to accomplish a similar set of tasks as does the xerographic copier. He/she is presented with a population, or populations, from the plant breeder, and asked to make exact copies for large-scale production in agriculture. The two main parameters of interest are fidelity and cost: maximizing the former and minimizing the latter. In the simplest case, the "saved seed" model suffices, wherein the breeder's seed is used to plant a population from which a subsequent generation is harvested, and distributed to farmers for their operations. Situations may arise where it is not practical to proceed directly from breeder's (seeds produced by the plant breeder in connection with the breeding program) to commercial seed lots (e.g., low inventories, low seed viability, or high volume demand). Additional multiplication steps may be required under these circumstances.

The general operational model is one that corresponds to our analogy of the xerographic copier. A copy is made of the breeder's seed, usually by planting in a configuration that will permit matings to occur, either self- or cross-pollinations, and subsequent seeds to mature. For outcrossing species utilizing insect vectors, pollinating agents are often added to facilitate seed production. Population sizes must be adequate to recover identical allelic frequencies in subsequent generations, and the physical configuration of the plants used for seed production must also facilitate gene flow (e.g., square or ovoid vs. linear). For cross-pollinating species, such seed increases must be conducted with adequate "isolation distance", the minimum distance between two populations such that vectors will not significantly bridge pollen and egg.

If one generation of seed multiplication is not adequate to meet volume targets more generations may be employed. Genetic fidelity to the original breeder's seed lot is presumed to decrease with each generation of replication. The first generation of replication is referred to as "registered", the second as "certified" and third as "foundation". Foundation seed is generally used for the production of commercial seeds to farmers. The plant breeder responsible for the development of a given variety is usually involved with the seed production process, and especially with the monitoring of registered, certified, and foundation seed populations. He/she may even conduct additional selections within these populations to further ensure the best possible performance.

Private seed companies usually operate in-house seed production programs and maintain stocks of various populations of commercial populations. For large-volume U.S. crop species such as corn, soybean, wheat, cotton, and forage turfgrass, independent organizations have been established to oversee the seed certification process, usually on a state-by-state basis in cooperation with the land grant institution. These organizations are usually funded by farmers and often go beyond the assurance of genetic purity to the assurance of performance. Seed certification organizations uphold standards for varietal performance as a condition for approval to sell seeds to farmers.

POST-SEED MODIFICATIONS

The purveyors of propagules, and especially seeds for growers to use to establish crops, have devised many value-added services and enhancements. Seeds of many horticultural crops such as onion, basil, carrot, and many ornamental flowers are either very small or irregularly shaped. Small size and irregular shape makes it difficult for growers to use machines to separate and sow seeds. Inert coatings comprised of clay silt have been developed to transform the size and shape of seeds to permit the use of machinery. Fungicidal agents, Rhizobacteria, and mycorrhizae can sometimes be added to the colloidal mixture to promote beneficial microbial-plant interactions or reduce losses to pathogens that incite seedling diseases such as "damping off" (TeKrony, 2006).

Another popular seed modification is "osmotic priming". Many horticultural crop species exhibit seed germination times that are relatively slow and uneven for example onion, lettuce, and bell pepper. It is possible to improve the germination dynamics of seeds of some horticultural species by exposing them to a solution that balances seed water imbibition with limits on physiological and developmental activity. By using salts or other compounds such as polyethylene glycol, that impart osmotic pressure potential on the solution, seeds are hydrated to the point that the lag time to germination is alleviated, but seeds are prevented by the osmoticum from proceeding (Cantliffe, 2003). In effect they are "synchronized at the germination starting gate". After the osmotic agent is removed seeds proceed directly to germination. This treatment is irreversible and requires that the seeds be sown within a very short period of time, usually only a few days (Welbaum et al., 1998).

The horticultural industry makes extensive use of whole plants to establish commercial crops. While it is outside the scope of this textbook to address this vast area of technical material, there are implications to the realm of plant breeding. Nursery stock for woody perennials is often sold in the form of grafted units where the scion and rootstock were bred and commercialized independently. The scion and rootstock are usually genetically distinct and may even represent distantly related species that are not sexually compatible (Lewis and Alexander, 2008). Grafting technology has been applied to annual crops as well, such as vegetable species in Solanceae and Cucurbitaceae (King et al., 2010).

References

Abbo, S., Lev-Yadun, S., Gopher, A., 2012. Plant domestication and crop evolution in the near east: on events and processes. Crit. Rev. Plant Sci. 31 (3), 241–257.

Acquaah, G., 2012. Principles of Plant Genetics and Breeding, second ed. Wiley/Blackwell, New York, NY (USA). 756 pp.

Allard, R.W., 1999. Principles of Plant Breeding, second ed. Wiley & Sons, Inc., New York, NY (USA). 254 pp.

Ashri, A., 1989. Major gene mutations and domestication of plants. In: Sigurbjornsson, B. (Ed.), Plant Domestication by Induced Mutation. IAEA, Vienna, Australia, pp. 3–9.

Banga, O., 1957a. Origin of the European cultivated carrot. Euphytica 6, 54–63.

Banga, O., 1957b. The development of the original European carrot material. Euphytica 6, 64–76.

Barton, J.H., Siebeck, W.E., 1994. Material Transfer Agreements in Genetic Resources Exchange: The Case of the International Agricultural Research Centres. 61 International plant genetic resources institute, Rome (Italy).

Borojevic, S., 1990. Principles and Methods of Plant Breeding. Elsevier, Amsterdam, (NETH). 368 pp.

Brown, J., Caligari, P., Campos, H., 2014. Plant Breeding, second ed. Wiley-Blackwell, New York, NY (USA). 296 pp.

Cantliffe, D.J., 2003. Seed enhancements. Acta Hortic. (607), 53–59.

Chable, V., Conseil, M., Serpolay, E., Le Lagadec, F., 2008. Organic varieties for cauliflowers and cabbages in Brittany: from genetic resources to participatory plant breeding. Euphytica 164 (2), 521–529.

Chahal, G.S., Gosal, S.S., 2002. Principles and Procedures of Plant Breeding: Biotechnological and Conventional Approaches. Narosa Publ., New Delhi (India). 604 pp.

Clement, C.R., 1999. 1492 and the loss of Amazonian crop genetic resources. I. the relation between domestication and human population decline. Econ. Bot. 53 (2), 188–202.

Cooper, M., DeLacy, I.H., 1994. Relationships among analytical methods used to study genotypic variation and genotype-by-environment interaction in plant breeding multi-environment experiments. Theor. Appl. Genet. 88 (5), 561–572.

Duvick, D.N., 1996. Plant breeding, an evolutionary concept. Crop. Sci. 36 (3), 539–548.

Ehlenfeldt, M.K., 2009. Domestication of the highbush blueberry at Whitesbog, New Jersey, 1911-1916. Acta Hortic. (810)147–152.

Fehr, W.R., 1987. Principles of Cultivar Development Volume 1: Theory and Technique. MacMillan Publ., New York, NY. 536 pp.

Fleury, D., Whitford, R., 2014. Crop Breeding: Methods and Protocols. Humana Press, New York, NY (USA). 255 pp.

Gross, B.L., Zhao, Z., 2014. Archaeological and genetic insights into the origins of domesticated rice. Proc. Natl. Acad. Sci. U. S. A 111 (17), 6190–6197.

Harlan, J.R., 1992. Crops and Man, second ed. American Soc. Agronomy, Madison, WI (USA). 284 pp.

Heiser, C.B., 1973. Seed to Civilization: The Story of Man's Food. W. H. Freeman, San Francisco, CA (USA). 243 pp.

Kang, Y.J., Lee, T., Lee, J., Shim, S., Jeong, H., Satyawan, D., Kim, M.Y., Lee, S.H., 2016. Translational genomics for plant breeding with the genome sequence explosion. Plant Biotechnol. J. 14 (4), 1057–1069.

King, S.R., Davis, A.R., Zhang, X., Crosby, K., 2010. Genetics, breeding and selection of rootstocks for Solanaceae and Cucurbitaceae. Sci. Hortic. 127 (2), 106–111.

Kuckuck, J., Kobabe, G., Wenzel, B., 1991. Fundamentals of Plant Breeding. Springer-Verlag, Berlin (GER). 236 pp.

Laufer, B., 1919. Sino-Iranica. In: Chicago Field Museum of Natural Hist. Pub. 201. Anthropol. Ser., Vol. 15, pp. 451–454.

Lawson, C., 2004. Patents and the CGIAR system of international agricultural research Centres' germplasm collections under the international treaty on plant genetic resources for food and agriculture. Aust. J. Agric. Res. 55 (3), 307–313.

Lewis, W.J., Alexander, D.M., 2008. Grafting & Budding: A Practical Guide for Fruit and Nut Plants and Ornamentals. Landlinks, Collingwood, VIC (Australia). 102 pp.

Mackevic, V.I., 1929. The carrot of Afghanistan. Bull. Appl. Bot. Genet. Plant Breed. 20, 517–562.

Perdue, R.E., Christenson, G.M., 1989. Plant exploration. In: Janick, J. (Ed.), Plant Breeding Reviews. Vol. 7. Timber Press, Portland, OR (USA), pp. 67–94.

Poehlman, J.M., Sleper, D.A., 2013. Breeding Field Crops, third ed. Iowa State Univ. Press, Ames, IA (USA). 566 pp.

Polashock, J.J., Vorsa, N., 2002. Development of SCAR markers for DNA fingerprinting and germplasm analysis of American cranberry. J. Am. Soc. Hort. Sci. 127 (4), 677–684.

Rodriguez-Saona, C., Vorsa, N., Singh, A.P., Johnson-Cicalese, J., Szendrei, Z., Mescher, M.C., Frost, C.J., 2011. Tracing the history of plant traits under domestication in cranberries: potential consequences on anti-herbivore defences. J. Exp. Bot. 62 (8), 2633–2644.

Rubatzky, V.E., Quiros, C.F., Simon, P.W., 1999. Carrots and Related Vegetable Umbelliferae. CABI Publ., New York, NY (USA). 294 pp.

Sáenz Laín, C., 1981. Research on *Daucus* L. (Umbelliferae). An. Jard. Bot. Madr. 37, 481–533.

Simon, P.W., 2000. Domestication, historical development, and modern breeding of carrot. In: Janick, J. (Ed.), Plant Breeding Reviews. Vol. 18. Timber Press, Portland, OR, pp. 157–190.

Simon, P.W., Freeman, R.E., Viera, J.V., Boiteux, L.S., Briard, M., Nothnagel, T., Michalid, B., Kwon, Y.-S., 2008. Carrot. In: Prohens, J., Nuez, F. (Eds.), Vegetables II. Handbook of Plant Breeding. Vol. 2. Springer, New York, NY (USA), pp. 327–357.

Singh, B.D., 2006. Plant Breeding: Principles and Methods. Kalyani Publishers, New Delhi (INDIA). 1018 pp.

Stoner, A., Hummer, K., 2007. 19th and 20th century plant hunters. HortSci. 42 (2), 197–199.

TeKrony, D.M., 2006. Seeds: the delivery system for crop science. Crop. Sci. 46 (5), 2263–2269.

Terral, J.-F., Tabard, E., Bouby, L., Ivorra, S., Pastor, T., Figueiral, I., Picq, S., Chevance, J.-B., Jung, C., Fabre, L., Tardy, C., Compan, M., Bacilieri, R., Lacombe, T., This, P., 2010. Evolution and history of grapevine (*Vitis vinifera*) under domestication: new morphometric perspectives to understand seed domestication syndrome and reveal origins of ancient European cultivars. Ann. Bot. 105 (3), 443–455.

Towill, L.E., 1989. Biotechnology and germplasm preservation. In: Janick, J. (Ed.), Plant Breeding Reviews. Vol. 7. Timber Press, Portland, OR (USA), pp. 159–182.

Vavilov, N.I., 1926. Centers of origin of cultivated plants (transl.). In: Löve, D. (Ed.), Origin and Geography of Cultivated Plants. Cambridge University Press, Cambridge (UK).

Wang, R.L., Stec, A., Hey, J., Lukens, L., Doebley, J., 1999. The limits of selection during maize domestication. Nature 398 (6724), 236–239.

Welbaum, G.E., Shen, Z., Oluoch, M.O., Jett, L.W., 1998. The evolution and effects of priming vegetable seeds. Seed Technol. 20 (2), 209–235.

White, G.A., Shands, H.L., Lovell, G.R., 1989. History and operation of the National Plant Germplasm System. In: Janick, J. (Ed.), Plant Breeding Reviews. vol. 7. Timber Press, Portland, OR (USA), pp. 5–56.

Zeder, M.A., 2008. Domestication and early agriculture in the mediterranean basin: origins, diffusion, and impact. Proc. Natl. Acad. Sci. U. S. A. 105 (33), 11597–11604.

Zohary, D., Hopf, M., 2000. Domestication of Plants in the Old World: The Origin and Spread of Cultivated Plants in West Asia, Europe, and the Nile Valley. Oxford University Press, New York (USA). 316 pp.

CHAPTER

6

Breeding Objectives

INTRODUCTION

As with any acquired skill, plant breeding is an endeavor that consumes resources: Time, space, work, raw materials (water, nutrients), machines and devices. Although many would argue that the practice of plant breeding can be a lot of fun, few cases exist where programs are demonstrated to exist purely for the sake of pleasure derived. The basic plant breeding algorithm and a cursory introduction to the concept of germplasm were presented in Chapter 5. The core of the algorithm is cyclic, shuttling back and forth between selection and controlled mating. Successful plant breeding programs always have endpoints, indicating that the cyclic core of the algorithm culminates in a population that is deemed to be "finished." The finish line may be an abstract one, perhaps even unattainable, but it is always there, beckoning the plant breeder to keep striving forward. If the plant breeder is not concerned about the end product and associated expenditure of resources, it is likely that somebody else is in charge that does. This chapter will focus on the need for and issues surrounding the establishment of breeding objectives that will drive the choice of germplasm elements and the subsequent selection and mating scheme.

The plant breeder must establish clear breeding objectives to ensure that the program is designed and managed properly. The objectives paint a detailed picture of a theoretical population consisting of individuals that possess the desired combination of specific phenotypic attributes. An apt analogy would be the architectural plans, or "blueprints", for a building project. The blueprint defines the end product of the building project and specifies the finishes and materials, making it possible to estimate the total resource requirements. The breeding program is successful when a population matching or exceeding the goals set forth by the original breeding objective (or blueprint) is achieved.

Two fundamental issues must be addressed during the process of developing breeding objectives: technical feasibility and financial returns (Brown and Caligari, 2008). These authors listed the following considerations in developing breeding objectives:

- Future socio-political and economic factors that provide opportunities or threats to the status quo
- New opportunities for attaining or characterizing yield
- Opportunities for added value to the consumer or end user
- What diseases or pests are likely to be of greatest importance in the future?
- What agricultural system is the new variety developed for?
- Interactions of the above factors

Breeding objectives are comprised of the phenotypes that will be selected during the selection/controlled mating phase, how and when the selection will be applied, and most importantly what is the anticipated outcome at the culmination of the project. In this regard, it is crucial to define the population structure of the end product (i.e., pure line, hybrid, OP, clone, etc.; see Chapter 4) and how it will be intellectually protected (Brown and Caligari, 2008). The marker-assisted techniques described in Chapter 8 are, in some cases, lowering the barriers of technical and economic feasibility to the range of tolerance (Foolad and Panthee, 2012). One example is fruit size in sweet cherry (*Prunus avium* L.) a trait known to be quantitative with relatively low heritability. The discovery of effective QTLs has rendered it more efficient and cost-effective to mount a breeding program with the objective of larger fruit size in this species (Rosyara et al., 2013).

Similar findings were reported in common bean where the employment of molecular markers to introgress or pyramid multiple major genes and QTL for disease resistance led to the more rapid development of several cultivar releases than would have been possible using traditional breeding methods (Beaver and Osorno, 2009). The black raspberry industry in the United States has steadily declined since 1940 due to the lack of adapted and disease resistant cultivars. New technology tools will facilitate informed decisions regarding black raspberry germplasm value and usage, crossing, and selection through marker-assisted breeding, and will be useful for breeding programs across the United States (Bassil et al., 2014).

One example of a breeding objective is a population that is precisely identical to cultivar X but with resistance to disease pathogen P, to which cultivar X is now susceptible. It is easy to visualize how such a breeding objective will determine most of the elements of the program promulgated to achieve this goal. Cultivar X will be hybridized with individuals of a germplasm population found to be resistant to pathogen P. A controlled mating and selection algorithm will be devised to combine the phenotype of cultivar X with a gene or set of genes from the resistant parent that confers resistance to pathogen P.

The term "program" can be defined at least two different ways. In the narrow sense a whole program is a progression from raw germplasm to finished variety. In the broader sense "the program" never ends for the breeder. Each juncture during the execution of a defined project reveals new gene combinations and information that suggest new approaches for the existing project or new opportunities for the future. One result often, and usually, leads to the initiation of yet another pathway or feeds into one that is already in progress. Most plant breeders never retire, they only cease to continue. In this chapter the "narrow sense" definition of program will apply.

The formulation of breeding objectives might seem intuitive but is often the factor that is most responsible for a program succeeding or failing. Breeding objectives must tread a fine line between what is new and exciting and what is possible to achieve. The new cultivar must be demonstrably different from and better than anything that is already in the market. The target defined by breeding objectives is always in the mind's eye of the plant breeder; something that has never existed so must be imagined and aspired to.

Breeding objectives must be theoretically attainable and practically feasible. If the standard yield of a given fruit crop stands at 200 bushels per acre it is probably unrealistic and technically impossible to realize an objective of 400 bushels per acre within a reasonable period of time. If a tree species used for lumber takes 18 years to reach harvest maturity it may be possible to reduce generation time in small increments but it is probably not technically feasible for a reduction by major increments to 10 years or less. Remarkable technical achievements are evident, but advancements in plant breeding are generally incremental in terms of a 5–20% advance within the context of a single program. Generation time plays an important role since advancements are realized from one generation to the next. More incremental progress is usually attained during the productive lifetime of the breeder of annual as compared to woody perennial crop species.

It is often useful and effective to structure breeding objectives into subunits, milestones, or horizons. Since progress is usually measured in years, and sometimes decades, this affords the satisfaction of progress. It is also desirable where possible to distill complex phenotypes down to measurable components. Progress against a numerical target is easier to visualize than an abstract one. It also makes it easier to relate project status to stakeholders and to justify the continuation of a program in jeopardy.

Should the plant breeder surrender to compelling probabilities and shrink objectives accordingly? If the breeder of the above two hypothetical crop species sets, respectively, 225 bushels per acre or a 16 year generation time as the benchmarks, there is a high likelihood that success will eventually be achieved. Moreover, the breeder can bask in the glow of success, not in the dim light of failure to reach more ambitious goals. By setting 225 bushels or 16 years as the ultimate benchmarks for success, has the breeder set the bar too low? If the breeding objectives had been more ambitious, perhaps 235 bushels or 14 years respectively, would the program have been pursued with more energy and intensity and more progress realized?

How are breeding objectives prioritized and assembled into defined programs? Like any inventor, the breeder often has ideas that are based on serendipity, emerging scientific results, the novel combination of known factors, or just plain stupidity. The essence of the new product or breeding strategy idea is important; this drives the plant breeder to work relentlessly and without regard to time or physical/mental strain.

Breeding objectives must ultimately be grounded in some aspect of human utility since agriculture is applied biology in service of humans. Not surprisingly, the stakeholders, or clients, of the breeding program do not necessarily buy into the dreams of the visionary, and a compromise is necessary to sustain ongoing financial support. The plant breeder often conducts formal or informal referenda on industry needs the results of which are distilled into a weighted priority list. Armed with the list, the plant breeder knits clientele priorities together into an overall program that is practical, achievable in demonstrable increments and wins the support of stakeholders. If the clients want their plants to fly, the breeder must find a way to tell them it is impossible, yet also attempt to maintain their long-term support. Perhaps they will be pleased if the breeder can engineer the seeds to fly?

A theoretical example of such a "needs" list for hypothetical economic plant species "S", an ornamental cut flower, as extracted from growers, or primary producers: (i) resistance to stem canker, a fungal disease (all commercial varieties susceptible), (ii) concentrated flowering (currently five weeks, growers want 3.5 weeks), (iii) prolonged petal life (currently five days, growers want eight days), (iv) pure white petals (none currently exist), and (v) variegated red/pink flowers (only full red or pink currently exist). The first three imperatives are oriented towards cost-savings benefits to growers and distributors.

Of all the top five imperatives identified by growers, only priority #i (disease resistance) is a prospective candidate for a high likelihood of success via plant breeding. Based on hundreds of examples in a broad spectrum of higher plants, disease resistance is likely to be found within the gene pool accessible to the target species, and preponderance of reports suggests that the inheritance of resistance is likely to be simple (Ayliffe and Lagudah, 2004; Li et al., 2013; Zhang et al., 2013). All of the other prospective breeding targets have little or no technical feasibility due to either lack of genetic variability or low h^2. Since inheritance of disease resistance is probably simple, the corresponding breeding program targeting resistance to stem canker would be straightforward (e.g., backcross; see Chapter 18). It should be pointed out that breeding targets (iv) and (v) that involve corolla pigmentation could be addressed via non-traditional breeding strategies such as mutation breeding, transformation, or genome editing (see Chapter 8).

With regard to clientele priority #iii, prolonged petal life, perhaps information in the field of plant physiology can suggest agricultural management techniques to achieve the same outcome. Perhaps the topical or systemic application of plant growth regulators is found to promote prolonged petal life in the species? Is this approach cheaper, faster, or more effective than what is possible through a prolonged plant breeding strategy? Clientele priority #ii (concentration of flower set) may be a phenotype that is controlled by many interacting genes rendering a breeding approach more problematic than if genetic control is simple. This phenotype may also be amenable to modulation with exogenous plant growth regulators.

The consensus of growers is that the market would respond favorably to impart added value to cultivars that feature novel corolla pigments. Although added value to the consumer may translate into higher wholesale prices at the farm gate it is not a given that unit production costs will be comparable to those of existing cultivars. Further, retailers and distributors are usually the first layers of the product chain to reap benefits from breeding advancements aimed directly at consumers. It may take a long time, if ever, for the added value of the cultivar to reach the farm gate. Consequently, growers are usually more supportive of projects aimed at reducing unit production costs or abating risks.

It may be the case that the primary clientele entity is the retailer, distributor, or agent thereof. In that case, added consumer value such as new corolla pigmentation patterns may be accorded high priority. In considering whether a program whose breeding objective is a new corolla pigment the issue of technical feasibility is a crucial concern. Is there a genetic basis for white or variegated flower petals in this species? The envisioned combinations of corolla pigmentation (priorities #iv and #v) have been observed in other plant species, so are technically possible. The biochemical pathways for carotenoid and anthocyanin pigments have been elucidated, and many of the underlying genes have been identified and isolated (Gonzalez et al., 2008; Rodríguez-Villalón et al., 2009; Shumskaya and Wurtzel, 2013; Liu et al., 2014; Perrin et al., 2017).

This particular list pertains to one group of constituent clients: the growers. It is perhaps not surprising, therefore, that issues impinging on costs and risks associated with the growth phase are high on their list of imperatives. Resistance to a pathogen will obviate the need for prophylactic and/or curative pesticides, or foster higher overall quality and wholesale price. Concentrated flower set will reduce unit costs attributable to harvesting labor, and prolonged petal life may enable the grower to accumulate harvests and store under refrigeration. This capability will, in turn, contribute to reduced wholesale shipping costs.

The second group of clients is polled: retail marketers and sales outlets for cut flowers. Their list of priorities that need to be addressed is quite different from that of the growers: (i) pure white flowers, (ii) variegated red/pink flowers, (iii) prolonged petal life, (iv) lush foliage subtending blooms, and (v) larger bloom size.

The growers and retailers are usually at odds with each other with regard to business goals and models. While they ultimately depend on each other, relationships are often adversarial, particularly in the negotiation of fair wholesale prices at the farm gate. Industry sectors must ultimately form an alliance and compromise on research and development priorities for mutual benefit. Growers place highest priorities on characters that will reduce production costs and less emphasis on traits perceived to have value to end users, but also recognize that consumers of end products are the ultimate arbiters of value. Retailers seem not to care about the production problems faced by growers but gladly support any effort to reduce wholesale prices. In the narrow context, retailers prefer that research and development focus exclusively on consumer traits that will impart a competitive advantage directly to them. Complicating matters further, the total economic value of crop phenotype benefits aimed at growers is usually much less than those aimed at consumers. The conflict is more evident in certain crop species than in others. How will the breeder mitigate this conflict of priorities?

There are no absolute answers to this dilemma, except that everything should be done in balance, recognizing that both groups are essential for a vibrant industry and each has distinct and legitimate needs. If the growers provide funding to the plant breeder, they will have some influence in the choice of breeding objectives. If the breeder is working in academia, the choice of objectives will be made in favor of projects that garner the highest financial support, that have a high probability of practical success, that impart the greatest measure of public benefits, and will contribute to the knowledge base. If the breeder is working within the context of a private sector company, however, such esoteric byproducts are not usually rewarded, and may even be discouraged.

Ideally, the breeding objectives should combine as many distinct phenotypes as possible. In the case of a cut flower species independent breeding objectives may be combined, for example pure white large flowers, stem canker resistance, prolonged petal life, concentrated flower set, and lush subtending foliage. It may not be possible to combine all of these characters within an acceptable context of cost and time as defined by clients. In that case the program may be broken down into incremental segments, where traits are addressed sequentially. One example of this strategy is the development of a sequential recurrent selection strategy to develop new potato (*Solanum tuberosum*) varieties with disease resistance and improved tuber quality (Bradshaw et al., 2009).

Another example is the step-wise breeding of varieties with multiple disease resistance (Singh and Schwartz, 2010). This will take longer and require more patience and long-term financial support. Clients often prefer to mitigate costs and risks by pursuing a multi-faceted breeding program in segments and sequences. New information management software has been developed to assist breeders in sorting out the multitude of individual and interaction effects in a breeding program with multiple targets (Yan and Frégeau-Reid, 2008).

The order in which traits are addressed is subjective, but attention will inevitably be paid to those that are easier to manipulate than others, the easiest ones first, hardest ones last. Since plant breeding is a long-term project, it is necessary to demonstrate incremental progress. Therefore an annual cycle to evaluate the project followed by justification and reaffirmation of support is usually an effective compromise.

Plant breeding originated from the agricultural need to save seeds for the next year's crop. As more technologies were invented to enhance effectiveness and efficiency, farming skills such as plant breeding became more specialized and transitioned from farmers to technology centers such as the U.S. land-grant universities. Demand for seeds of improved cultivars increased to the point that plant breeding is promulgated progressively more by seed companies engaged in free enterprise.

U.S. seed companies alone attain annual sales of the magnitude of ~US$10 billion. Most of the value of these sales is based on proprietary varieties developed by plant breeders working within these seed companies. Plant breeding programs in the public sector have been eliminated or drastically reduced over the past 50 years while plant breeding in the private sector has expanded (Gepts and Hancock, 2006; Hancock and Stuber, 2008). Plant breeders in the public sector that remain are focused on crop species of marginal economic but significant non-economic importance. They are also engaged in basic research to elucidate biological mechanisms of phenogenesis or are developing new and improved methods of gene manipulation and selection. Many projects that originate in the public sector are usually assimilated by seed and agrichemical companies once a pathway to a business opportunity is clearly established.

Public plant breeding programs still exist in new crops or those of marginal or specialized value. Public breeding may continue to be focused on economically important crops for projects such as increasing the range of crop adaptation (Shinada et al., 2014). Another emerging class of objectives is to address vulnerability in economically important crops such as maize where the genetic base of commercial cultivars has narrowed over time (Carena, 2013). Plant breeders affiliated with educational or research institutes also tend to focus on training the next generation of plant breeders, germplasm collection, genetic studies of trait inheritance, and elucidation of new knowledge in the field of genetics (Duvick, 1996; Bliss, 2007; Stamp and Visser, 2012). If the public program is successful to the point that a significant market develops then a private enterprise will step in to assume the position. Not surprisingly, this transition is not always smooth.

Breeding objectives in the public sector tend to be relatively abstract and crude. The financial implications of project costs and the value of end products are often not fully understood or appreciated. If the goal is to develop germplasm that resembles commercial varieties but carries additional desirable genes, the need for rigorous varietal testing is greatly reduced. Breeding objectives in private seed companies, however, tend to be highly defined with respect to targeted phenotypes and markets. Breeding programs are also supported by a detailed business plan and financial analysis that clearly describes investments and returns to allow sound decisions to be made among competing opportunities (Fuglie and Walker, 2001).

Over the past 40 years, agricultural economists and theorists have conducted research into methods for the development of effective and efficient breeding objectives (Simmonds, 1979; Muller and Zeddies, 1988; Dhillon et al., 1993). Much of the theory underpinning breeding objectives was developed in long-lived woody perennial species where the financial consequences of ineffective and/or inefficient objectives are greatest. If poor planning leads to an ill-advised or defective breeding program in a pulp or lumber wood crop, the economic impacts can be absolutely devastating. Consequently, a tremendous body of knowledge on the decision-making process in formulating breeding objectives of tree species used to make pulp or wood has accumulated. For example, it was reported that among a broad range of parameters tested in radiata pine, a model that focused selection in a balance of 1.0 raw volume:1.5 wood density was optimal under forest conditions in Chile (Apiolaza and Garrick, 2001). Many similar examples have appeared in the scientific literature regarding studies in forest pulp and timber trees to correlate breeding objectives with the intended practical benefits (Borralho et al., 1993; Fries and Ericsson, 1998; Byram et al., 2005; Ivkovic et al., 2006).

In some cases, societal benefits will co-opt the necessity for detailed financial analysis to justify a breeding project in purely monetary terms. Projects of this ilk will likely be found in the public sector (Morris and Heisey, 2003). Competition among plant breeders in the public and private sectors in Switzerland is encouraged to broaden the national focus of agricultural research and development (Mann, 2013).

Different breeding objectives will prompt the development of different breeding strategies and diversity of technical approaches. For example, an array of strategies and approaches in eggplant (*Solanum melongena*) breeding was described by Hurtado et al. (2015). Grafting technology in *Citrus* is expanding to include the breeding of rootstock cultivars for resistance to soil-borne diseases (Grosser et al., 2015). Breeders are also aggressively targeting consumer-driven value-added traits such as seedlessness via triploidy/parthenocarpy (see Chapters 10 and 17). Similarly, rootstock breeding has become an area of breeding focus in peach (*Prunus persica*) with the goals of seed-propagation, disease resistance, and scion dwarfing (Reighard, 2002).

Yield and total biomass produced by annual legumes remain major objectives for breeders but other issues are becoming more important, such as environment-friendly, resource use efficiency including symbiotic performance, resilient production in the context of climate change, adaptation to new sustainable cropping systems, adaptation for a broader range of end uses, and new ecological services such as pollinator protection.

These trends translate into more complex and integrated objectives for breeders. A holistic approach to legume breeding is becoming more important for defining objectives with farmers, processors, and consumers. Consequently, cultivar structures are likely to be more multi-faceted and complex (Duc et al., 2015).

Environmental sustainability is an important area in which a new class of breeding objectives has emerged. One example of this trend is with invasive ornamentals such as purple loosestrife (*Lythrum salicaria*) and *Ruellia simplex*. Breeding programs have been established with the objective to develop and release sexually sterile varieties and populations that will interbreed with escapes to introduce cytogenetic sources of sterility into natural areas. This will, in turn, contribute to efforts to modulate the spread of invasive species (Freyre and Deng, 2013). In oat (*Avena sativa*) the concept of breeding for factors that impact the seed to consumer ("life cycle assessment") sustainability of agricultural practices has been proposed and preliminarily tested (McDevitt and Mila i Canals, 2011).

PLANT IDEOTYPES AND IDEOTYPE BREEDING

It is sometimes beneficial for the plant breeder to establish an abstract image of the ideal plant, or "holy grail" that is under pursuit. Such an image helps to keep the program on track when the program is plagued by a litany of diversions and unforeseen obstacles. An abstract image is also easy for everyone, including lay people and clients, to understand conceptually and to base notions for change.

The *plant ideotype* embodies such an abstract image. The entity, in essence, is a cartoon that is a composite of trait images. Fig. 6.1 is an example of non-specific general ideotypes for a dicot and a monocot crop species. These general ideotypes demonstrate how the overall plant phenotype is broken down into components such as tiller number, stature, axillary branching, internode length, and root system architecture for targeted selection. The ideotype is brandished into the mind's eye prior to the performance of selections, along with quantitative specifications for what is acceptable. This brandished image renders the decision-making process for large numbers of segregating progeny more tenable. The ideotype also serves as the basis for agreement and/or training if more than one person will be performing selections.

The notion of developing and using ideotypes as tools for plant breeding was first articulated by Donald (1968), but many breeders had undoubtedly hatched similar schemes prior to this publication. Later, Rasmusson (1984) described efforts to test the validity of the use of ideotypes as strict guides for defining breeding objectives and prescribing selection criteria. After ten years of research contrasting traditional and ideotype-driven breeding, he reported in that book chapter that there was no evidence to support the usefulness of ideotypes. Rasmussen did conclude that the ideotype breeding concept is important for the success of plant breeding because it forces the practitioner to think about components of cultivar performance and how they interact.

Fig. 6.2 depicts general ideotypes for pome (apple, pear) scion architecture. Each general ideotype has advantages and disadvantages pertaining to costs of management and gross yield (Laurens, 1998; Lauri and Costes, 2004). By defining the tree architecture phenotype precisely and as quantitatively as possible, the identification of correlated juvenile morphological markers has been enabled as a time-saving strategy (Bendokas et al., 2012). Both quantitative and qualitative characterization of the scion is essential for effective breeding of pome scions and rootstocks (Tworkoski and Fazio, 2016).

The ideotype may or may not be well grounded in reality or physiological, genetic, or developmental morphological reality. In the case of grass species, tiller number and stature may be manifestations of the same developmental process(es), so cannot be dealt with independently. It may not be possible to achieve low tiller number and short stature simultaneously, for example if low tiller number triggers longer stems and taller stature. The ideotype is, therefore, a dynamic concept that becomes more refined and grounded in reality with each iterative cycle.

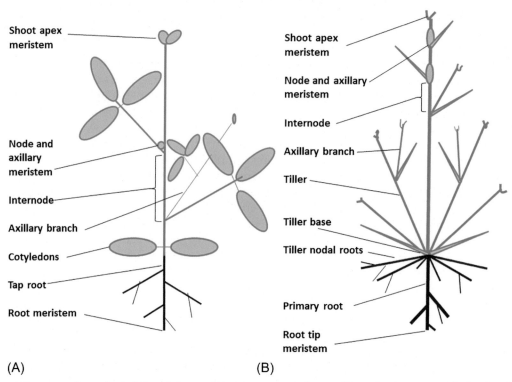

(A)　　　　　　　　　　　　　　　　　(B)

FIG. 6.1 General ideotypes of hypothetical dicotyledonous (A) and monocotyledonous (B) species that are specified with values for the plant breeder to use as a tool in aiding the selection process.

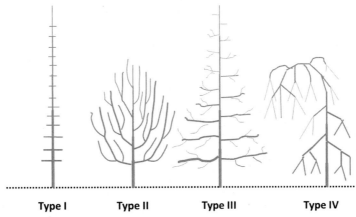

FIG. 6.2 General scion ideotypes for pome fruit architecture.

Ideotypes devised by the author in fresh market tomato breeding programs are presented in Figs. 6.3–6.5. In this case, ideotypes are often used to train technical and summer workers who may be assisting in conducting field, sorting/weighing, and lab evaluations. It is important for all members of the team to be on the same page with regard to the end targets. More detailed consumer-driven ideotypes for tomato were developed and tested by Lieshout (1993).

Research since the report of Rasmusson (1984) on the efficacy of ideotype breeding has led to results that confirm the limited utility of this strategy. Much of this research has invoked advance statistical modeling theory. For example, Picheny et al. (2017) showed that the multi-objective optimization formulation method successfully characterized key plant traits and identified a continuum of optimal solutions, ranging from the most feasible to the most efficient.

Ideotype breeding in common bean (*Phaseolus vulgaris*) has emphasized the selection of modified morphological traits that include upright, indeterminate growth habit. Progress in the medium-seeded pinto bean appeared to have been limited by negative linkages between small seed size and desired architectural traits. The ideal pinto bean ideotype differs from that of the related small-seeded navy bean ideotype by having fewer pods per plant and fewer

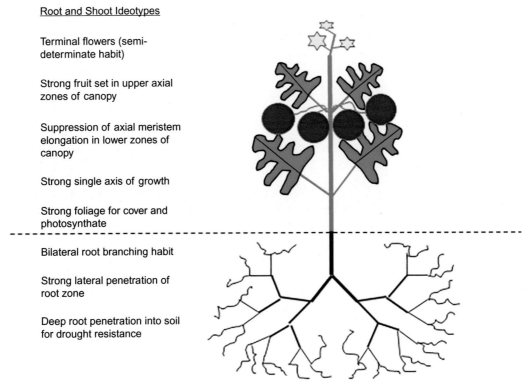

FIG. 6.3 Ideotype for general shoot and root architecture in a fresh market tomato breeding program.

<u>Fruit Ideotype</u>

Slightly flattened
globe profile

Consistent lateral
circumference
profile (not lobed)

Moderately firm

Deep red external
color

Consistent external
color

Smooth epidermis

Very small blossom
and stem scarring

FIG. 6.4   Fruit ideotype for a fresh market tomato breeding program.

<u>Internal Fruit
Ideotype</u>

Consistent deep red
color in pericarp,
placenta, and
locules

Absence of
inconsistencies in
tissue texture

Absence of voids
and hollowness

Intermediate
balance of pericarp,
placenta, and
locules

FIG. 6.5   Internal fruit ideotype for a fresh market tomato breeding program.

seeds per pod. Selection for medium-seeded, upright genotypes in other commercial classes would be greatly facilitated through the use of the pinto bean ideotype as a genetic bridge (Brothers and Kelly, 1993).

Several benefits could be derived for southern pine breeding programs by incorporating ideotypes, defined by the authors as "conceptual models which explicitly describe plant phenotypic characteristics that are hypothesized to produce greater yield" (Martin et al., 2001). This study concluded that the benefits of using ideotypes include improvement in trait heritabilities and genetic correlations, the higher genetic gain in diverse silvicultural environments, guidance for developing mating designs, and provision of a framework for synthesis of tree production physiology knowledge. Obstacles to the development of ideotypes for southern pines mostly related to the difficulty of linking traits and processes that operate at small spatial and temporal scales. Technologies that permit measures of component processes at relevant scales, the likely future importance of intensive clonal forestry and the movement towards large-scale genetic block plot experiments will obviate these obstacles in the future.

China's "super" rice breeding project produced many F_1 hybrid cultivars using a combination of the ideotype approach and intersubspecific heterosis (Peng et al., 2008). Ideotypes featured large panicle size, reduced tillering capacity, and improved lodging resistance emphasizing the top three leaves and panicle position within a canopy in order to meet the demand of heavy panicles for a large source supply. These authors concluded that the ideotype breeding objective strategy was effective for breaking the "yield ceiling" of irrigated rice cultivars.

Brunel-Muguet et al. (2011) proposed a technique of ideotype development that incorporated metric ranges of phenotypes extant in collections of germplasm for the targeted crop species. This group concluded that model-assisted analysis of the effects of genetic diversity demonstrated its usefulness in helping to identify the parameters which have most influence that could be improved by breeding programs.

Da Silva et al. (2014) studied the relationships of fruit tree architecture to the efficacy of an ideotype-driven breeding strategy. They concluded that the relationship of fruit tree architecture to performance is equivocal, especially with regard to the definition of varietal ideotypes and the selection of architectural traits in breeding programs. This study demonstrated that a light interception modeling approach could contribute to screening architectural traits and their relative impact on tree performance. This protocol modification opened up new perspectives for breeding and genetic selection to be assisted by varietal ideotype definition.

Van Oijen and Höglind (2016) presented a new procedure by which process-based models can help design breeding objective ideotypes, defined by them as virtual cultivars that optimally combine properties of existing cultivars. The procedure consisted of four steps: (i) Bayesian calibration of model parameters using data from cultivar trials, (ii) Estimating genetic variation for parameters from the combination of cultivar-specific calibrated parameter distributions, (iii) Identifying parameter combinations that meet breeding objectives, (iv) Translating model results to practice, i.e., interpreting parameters in terms of practical selection criteria.

Prospects for the success of an ideotype-driven breeding strategy can be enhanced when the morphological traits comprising the ideotype are buttressed by underlying QTL. For example, QTL analysis removes part of random errors of measured model input parameters, and that QTL information can successfully be coupled with crop models to replace measured parameters. Further, QTL-based modeling overcomes the limitations of designing ideotypes by using models that ignore the inheritance of model input traits (Yin et al., 2003). These authors proposed a strategy that integrates marker assisted selection (MAS; see Chapter 9) into model-based ideotype framework to support breeding for high crop yield. For this approach to be effective, there is a need to develop crop models that are capable of predicting yield differences among genotypes in a population under various environmental conditions.

In poplar, a segregating F_2 population derived from two highly divergent species, *Populus trichocarpa* and *Populus deltoides*, was used to evaluate the genetic basis of canopy structure and function in a clonally replicated plantation. QTL clustering was observed for morphological or developmental integration in poplar, i.e., traits with similar developmental origins are more strongly correlated with one another than traits with different developmental origins. There were important implications of these molecular genetic results for ideotype breeding of poplar (Wu, 1998).

Prospects for the success of an ideotype-driven breeding approach depend on the crop species, the range of germplasm, the breeding objectives, and the range of environments. For example, in plantain (*Musa* spp., AAB group) understanding the relationships between growth characteristics and yield potential is useful for developing an ideotype to guide breeding efforts. Few common pathways determining yield potential were found among the plantain landraces and a common ideotype for plantain breeding was not, therefore, possible. In this study, defined ideotypes differed for each landrace and according to the production system (Ortiz and Langie, 1997).

INTERGENIC INTERACTIONS AND PLEIOTROPY

Breeding objectives and representations thereof tend to conform to lists of attributes or composites of independent characters. Traits often interact in curious and unpredictable ways, and the corresponding breeding objectives should take this into account. Certain genes appear to exert effects on more than one phenotype, a phenomenon known as *pleiotropy*. Most examples of pleiotropy are for genes that have desirable effects on one phenotype but undesirable effects on others. If the objective is higher overall economic yield, another level of sub-objectives may be established that comprise components of yield, since they may be easier to identify and select than the composite trait (see Chapter 9). Components of yield, however, are often at odds with each other. Selection for higher seed number alone will inevitably result in lower average seed weight (Rasmusson, 1984). Simultaneous selection for higher grain seed number and high unit weight may result in a shift towards higher germ or bran ratio.

Moving one trait in a desirable direction over time by incremental selection may affect another trait in ways that are not so advantageous. Tomato fruit have been found to be an excellent source of many nutritious substances, including Vitamins A and C. Vitamin A, or retinoic acid, is derived from the compound β-carotene that contributes "yellow-orange" to the overall pigmentation of the fruits of tomatoes, and fruits and foliage of many other species as well. This pigment is involved in the light-harvesting complex, capturing energy within the incident wavelengths 375–525 nm. Lycopene is another carotenoid pigment commonly found in tomato fruits, imparting the red coloration to the epidermis and pericarp. This compound is a transitional product in the same biochemical synthesis pathway leading to β-carotene (Fig. 6.6; see also carotenoid biosynthetic pathway in Fig. 3.3). Recent human health studies have shown that lycopene is a powerful antioxidant that is effective in reducing the incidence of certain types of cancer, for example, early-stage prostate (Story et al., 2010).

Many mutations, such as Og^c (from the heirloom cultivar "Old Gold Crimson") have been discovered that alter the pigmentation of tomato fruit, including high β-carotene and high lycopene (Scott et al., 2013). Mutants that are high in the former tend to be low in the latter, and vice-versa. The mutant Og^c phenotype exhibits elevated lycopene and depressed β-carotene in tomato fruits. It seems that there are only so many isoprene skeletons available and

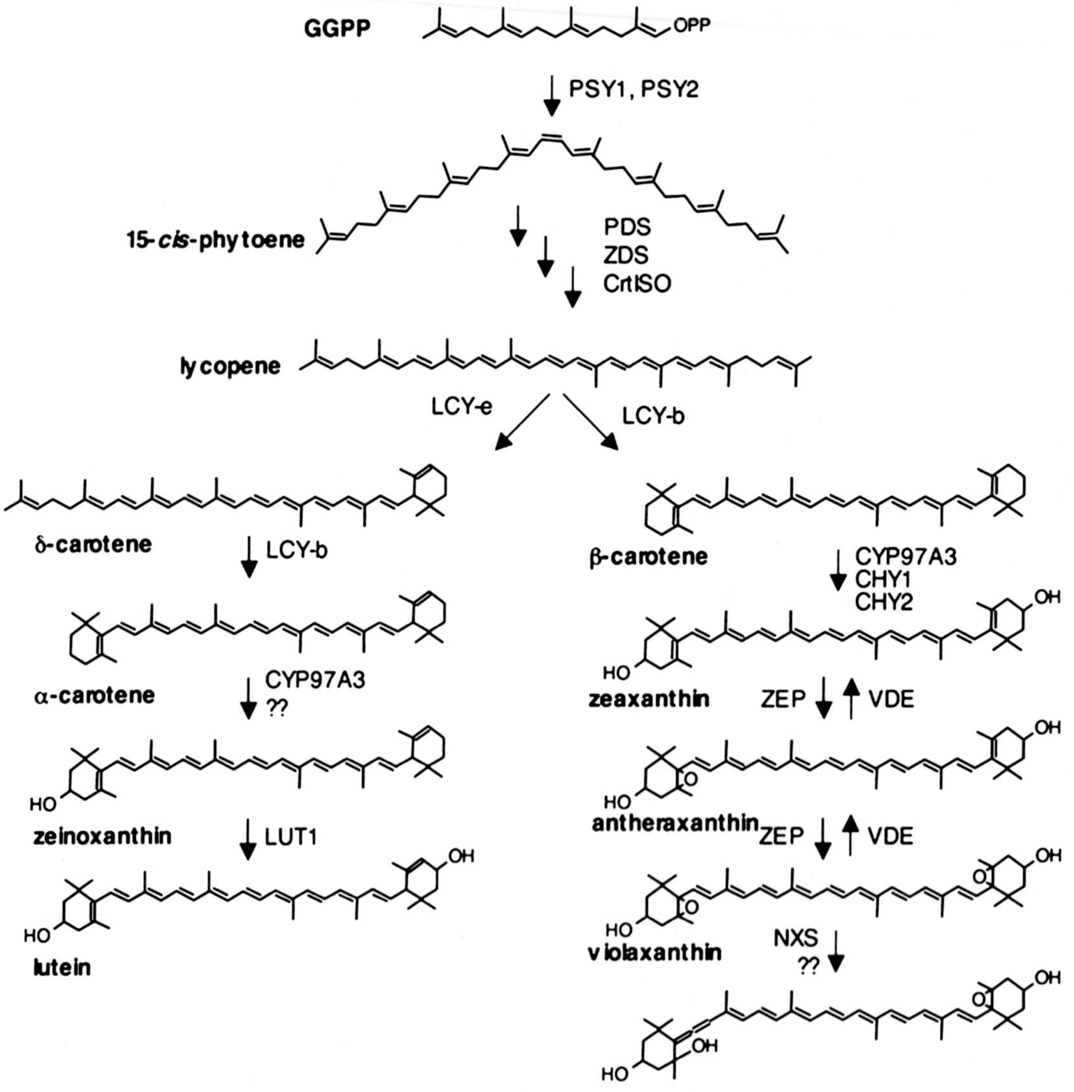

FIG. 6.6   Biosynthetic pathway of primary carotenoid pigments from GGPP. The wild-type pathway bifurcates at lycopene to either δ- and α-carotene or β-carotene. Selection for high β-carotene generally comes at the expense of lycopene precursors. Selection for high lycopene generally results in reduced lycopene cyclase (LCY) e and b activity, resulting in reduced β-carotene levels in fruits. *From https://bmcplantbiol.biomedcentral. com/articles/10.1186/1471-2229-6-13.*

that some precursors end up as β-carotene, some as lycopene while other precursors contribute to other carotenoid pigments or derivatives. All of the steps in the pathway are, of course, catalyzed by specific enzymes, the apparent control point for the mutations. By selecting for higher β-carotene, lower lycopene fruit levels are usually also inadvertently selected (Hirschberg et al., 1997). Another mutation in tomato, *Hp*, appears to increase the concentration of all carotenoid pigments, so operates at the level of isoprene skeleton pool size, not within the realm of the actual pathway (Yen et al., 1997). Both *Og^c* and *Hp* are associated with perceivable changes in foliar appearances and slight depression of overall fruit yield (Scott et al., 2013).

Genome engineering and gene editing are beginning to illuminate ways that metabolic pathways such as carotenoid biosynthesis may be altered without intractable pleiotropic gene interactions. For example, three transformants containing up-regulated *Lyc-b* construct show a significant increase in fruit beta-carotene content. The fruits from these plants displayed different epidermal color ranging from orange to orange-red depending on the lycopene/β-carotene ratio. Fruits from down-regulated *Lyc-b* transformants show up to 50% inhibition of Lyc-b expression, accompanied by a slight increase in lycopene content (Rosati et al., 2000). In another study with tomato, ripe fruits expressing a transgenic phytoene desaturase (*crtI*) gene showed significant increases in beta-carotene (threefold) but a reduction in the total carotenoid content (twofold) (Fraser et al., 2001). In potato, the first dedicated step in the beta-epsilon- branch of carotenoid biosynthesis, lycopene epsilon cyclase (*Lcy-e*), was silenced by introducing an antisense *Lcy-e* transgene under the control of the patatin promoter. qPCR measurements confirmed the tuber-specific silencing of *Lcy-e*. Antisense tubers showed significant increases in beta-beta-carotenoid levels, with beta-carotene showing the maximum increase (up to 14-fold) and total carotenoids increased up to 2.5-fold (Diretto et al., 2006).

Negative pleiotropy has also been observed in a cytoplasmic male sterility system of maize. Selection for "Texas" cytoplasmic male sterility (T-cms) for efficient production of F_1 hybrid corn seeds ultimately led to genetic vulnerability, mass susceptibility to *Cochliobolus heterostrophus*, the causal fungal pathogen of southern corn leaf blight (Levings and Siedow, 1992). In another example of negative pleiotropy, selection for the absence of trichomes (surface hairs) to eliminate worker irritation during hand harvesting of fruits and vegetables may inadvertently introduce susceptibility to herbivorous insects. Selection for the reduction or elimination of natural toxicants could also have the same result (Edwards and Singh, 2006). Alternatively, selection for insect tolerance alone may result in high plant toxicity (see Chapter 19).

INTERSPECIFIC INTERACTIONS

Resistance and tolerance to biotic factors that occur in the environment, e.g., pathogens, insect pests, animals, and weeds, is a major focus of plant breeding efforts (see Chapter 19). Of all the organisms that impinge on the health and productivity of cultivated plants, disease pathogens have received by far the most attention. Gene pools of nearly all domesticated plant species appear to include resistances to diseases and other pests, perhaps a vestige of the mechanisms that protected plants prior to domestication. Pathologists and entomologists are taught that the most effective parasites, pathogens and herbivores are the stealthy ones. To the herbivore or parasite the host is the goose that lays the golden eggs. If a pathogen, insect pest, mammalian herbivore, or weed is too aggressive the host will die along with the food source. Elements of ecosystems that are usually in exquisite balance in the wild often lack balance in monocultures.

By selecting for strong resistance to disease pathogens or insect pests, then growing resistant plants in monoculture, the stage is set for ecological imbalance. The pathogens and insects knock on the plant's doors, but can't get in. Bacteria, viruses, fungi, insects, and weeds are organisms that generally have extremely high reproductive rates. They also must be highly adaptable for survival. With such a strong selective pressure, it is not surprising that host genetic resistances to pathogens and parasites are usually fleeting. The plant breeder must, therefore, be cognizant of the effect that breeding objectives being pursued will have on other occupants of the agricultural ecosystem.

The breeding objective "disease resistance" will incorporate very different approaches depending on whether the aim is for long- or short-term solutions to the problem. "Vertical resistance" (also referred to as "race-specific resistance") is relatively easy to identify in germplasm and select for but has the ancillary effect of selecting for new pathogen races (Heath, 1996). The "horizontal resistance" (also referred to as "non-race-specific resistance") strategy (Chapter 19) was born out of the realization that reduced selection pressure on the biotic factor would lessen the probability that new and higher virulence would develop (Parlevliet and Zadoks, 1977; Young, 1996).

RETURN ON INVESTMENT

This chapter opened with an explanation of the need for breeding objectives to spell out the endpoint(s) clearly and to serve as the basis to estimate the cost to get to that endpoint in terms of time, labor, travel, space, equipment, etc. Since plant breeding is almost always a long-term, multi-year endeavor, it is instructive and beneficial to employ an adaptation of the cost-benefit analyses used in business to help to decide among competing project alternatives.

Among all preceding textbooks on the subject of plant breeding, the one that described this need most clearly and urgently was "Principles of Crop Improvement" (Simmonds, 1979). The reader is urged to consult with Section 3 of Chapter 10 of that textbook for a more thorough analysis of the costs and benefits of a plant breeding program. A cursory example is presented here to acquaint the reader with the process. Let us presume that the real cost of a prospective plant breeding project in present dollars will be $500,000. Too often in the public sector, the justification for proceeding is that the resulting cultivar "will garner increased demand for targeted clientele". Who are the customers, and how many actually will buy the new variety? How much will customers pay for the new cultivar? What proportion of existing or new customers will prefer the new cultivar over existing alternatives? How much does it cost to breed the new cultivar and to produce seeds or planting stock?

A better mechanism to decide whether a prospective project is economically feasible is clearly needed. One such mechanism is the *return on investment* (ROI) analysis used widely in many forms to make decisions among an array of research and development projects, including plant breeding (Simmonds, 1979). If hypothetical seed company X has $5 million per year to invest into R&D, they must make prudent decisions since breeding programs are not only long-term but are also risky, both technologically and strategically. One or more assumptions about the total net cost associated with the breeding strategy may be faulty. For such a long enterprise, it is, in fact, likely that changes in strategy (and associated costs) will occur during the course of the project. Moreover, market or competitive assumptions may shift or not be realized. For example, if company X invests into a breeding program for disease resistance, then the pathogen suddenly disappears, or a chemical company introduces an effective product to control it, the investment may be a total loss.

ROI for a plant breeding project is best estimated by an adaptation of a *discounted cash flow* (DCF) analysis. A discounted model should be used in deference to other methods because any given breeding program will take 5–10 years or more to reach fruition. As time progresses, the value of the initial investment increases, while inflation erodes net present value and the return on other indexed investments steadily builds. In other words, if $1 is invested into the breeding program and another $1 into a blue chip stock, the breeding investment will be worth less than the blue chip in year 1, 2, 3, etc. So why should the seed company continue to pour resources into the breeding program in lieu of the blue-chip stock? The answer, of course, is that the value of the breeding investment will exceed the blue chip in the longer term, perhaps 12, 15, or 20 years. In other words, the net present value (NPV) at an acceptable future date will exceed future value (FV) if the same investment was simply put into a bank account.

The DCF analysis permits one to calculate the net present value of the enterprise at any given point in time. This allows managers to compare actual financials with those that were projected at the outset of the project and thus keep the project on track, not spiraling out of fiscal control. For this purpose, there will inevitably be changes that occur in the DCF scenario as the project progresses from theoretical to actual. In the case of investment into a breeding program, the investor (Company X or a commodity group, for example) will be mainly interested in the "bottom line", how long until and under what sales assumptions the project reaches "break even" or net positive NPV at the end of the project.

DCF analyses always require assumptions, usually pertaining to the following:

- Up-front or non-recurring costs (e.g., equipment, land, supplies, labor recruitment)
- Ongoing or recurring costs (e.g., equipment depreciation, leases, supplies, labor, travel, energy, insurance, communication, recordkeeping, account management)
- Annual changes in recurring costs as the project progresses
- Length of time until R&D phase reaches fruition (e.g., the new variety is ready for sale)
- Total sales by year, ramping to market saturation, then stable? Declining? (new cultivars usually peak in market demand then decline)
- Costs of production and marketing (e.g., land, labor, equipment, transportation to make new, certified planting seed, trade, and consumer advertising)
- Investment (future value, or FV) and discount (present value, or PV) annualized rates

It is generally much easier to make valid assumptions about the costs and risks of the actual breeding program than to predict the returns from the sale of a product. Predicting the actual rates of market return and inflation are also tricky due to unforeseen events and trends. Table 6.1 provides a simplified illustration of a DCF analysis for a hypothetical breeding program that might be considered by Company X. The overall aim of the company is to maximize income for shareholders in the shortest possible time frame. Being in the seed business and specializing in proprietary plant varieties, the company is accustomed to and comfortable with projects that may take over ten years before a positive NPV is realized. In the case of the hypothetical project under consideration, break-even is not realized until after 13 years. The annualized simple rate of return (SRR) for this project is approximately 15%.

The company may choose to use an alternative parameter to evaluate the merits of the project, referred to as the internal rate of return (IRR). The *internal rate of return* (IRR) is defined as the discount rate that gives a *net present value* (NPV) of zero. This statistic is calculated using the same assumptions without the discount rate and a time duration in years from the inception of the project:

$$NPV = \left\{ a\left[1 - (1+i)^N \right] \right\} / i$$

where a = annual net cash flow, i = interest or discount rate, N = number of years.

Company X will most likely operate with some form of annual business cycle wherein new and existing projects are evaluated for approval or continuation. The plant breeder working for X will be teamed with financial specialists to investigate the parameters that govern each product. In large companies, there may be hundreds of projects proceeding simultaneously, each approved with a clear understanding of the expected financial parameters. Smaller concerns operate under similar constraints, but often take a softer, more intuitive approach to project selection. IRR, SRR, and break-even are used as a fundamental criterion to choose the top candidate projects from among many alternatives presented to the exercise. In some cases, the minimum thresholds are well established, and projects will not even receive any consideration unless financial statistics exceed them.

There are entire library sections devoted to business metrics, so this cursory treatment may seem out of place in a plant breeding textbook. The purpose for discussing the elementary nuts and bolts of ROI is that an understanding of the process imparts a greater appreciation for the impact of resource expenditures on the plant breeding process. By reducing up-front or early costs, or those incurred early in the program, the overall NPV and IRR of a given program are improved dramatically. The exercise also forces the plant breeder to break down breeding objectives into market assumptions. This is the driving force that drives the need for the program in the first place. Most potential breeding programs never make it past the drawing board after it is determined that R&D costs exceed anticipated

TABLE 6.1 DCF Analysis for a Hypothetical Plant Breeding Project with the Following Assumptions: $350 K Non-Recurring Costs; $400 K Annual Recurring R&D Costs for 7 years, Followed by $200 K, $150 K, and $100 K in Years 8, 9, and 10; Production Costs Ramping from $50 K in Year 9 to $500 K in Year 12; Sales Ramping from $500 K in Year 11 to $5000 K in Year 13; Annual Discount Rate = 0.96

Year	R&D costs	Production costs	Discounted Sum	Total sales	Discounted sales	Discounted net	Accumulative net (NPV)
1	(750)		(720)			(720)	(720)
2	(400)		(369)			(369)	(1087)
3	(400)		(354)			(354)	(1441)
4	(400)		(340)			(340)	(1781)
5	(400)		(326)			(326)	(2107)
6	(400)		(313)			(313)	(2420)
7	(400)		(301)			(301)	(2721)
8	(200)		(144)			(144)	(2865)
9	(150)	(50)	(139)			(139)	(3004)
10	(100)	(250)	(233)			(233)	(3237)
11		(400)	(256)	500	320	64	(3173)
12		(500)	(307)	2000	1228	921	(2252)
13		(500)	(295)	5000	2949	2654	402
14		(500)	(283)	5000	2831	2548	2950
15				5000	2718	2718	5668

sales. By taking the time to write down the assumptions that underpin a given breeding objective, a debate can ensue to strengthen the validity of the assumptions and to elicit a consensus among cohorts. This will make the going easier during the "dog days" of the middle years of the program when most of the investment has been made and several years still remain before sales can commence. Investors are usually excited at the outset of a project and enthusiasm wanes as the project goes more and more into the red. Documentation of the financial metrics can remind investors of the commitment they made and renew prospects for excitement about the culmination of the project.

The plant breeder working in the public domain also benefits by subjecting prospective breeding projects to such a financial analysis even when the financial context is not as well defined or urgent. While there may not be a cohort of business managers to convince of the veracity of a project before it is initiated, the public breeder still needs to have a credible story to tell. Why is this particular project being pursued? If it is not to capture a huge, untapped market, what other reasons may exist? Many breeding objectives are developed to deliver benefits that are of a more intangible nature, or to service a need that is difficult to put a price tag onto. Examples include human nutrition and general well-being or plant population attributes that will have a broad impact but no defined financial marketplace. Other public breeding programs exist to support the needs of agricultural segments that are not profitably addressed by the private sector. Small grains are one example. It is so easy to simply save seed from the previous harvest for next year's planting cycle that the cost of certified planting seed is barely above the futures price quoted by the Chicago Mercantile Exchange™. While it may be difficult to distill the impacts of new varieties down to mere numbers, it is instructive to employ some form of ROI analysis to help to decide among competing alternative projects.

References

Apiolaza, L.A., Garrick, D.J., 2001. Breeding objectives for three silvicultural regimes of radiata pine. Can. J. For. Res. 31 (4), 654–662.

Ayliffe, M.A., Lagudah, E.S., 2004. Molecular genetics of disease resistance in cereals. Ann. Bot. 94 (6), 765–773.

Bassil, N., Gilmore, B., Hummer, K., Weber, C., Dossett, M., Agunga, R., Rhodes, E., Mockler, T., Scheerens, J.C., Filichken, S., Lewers, K., Peterson, M., Finn, C.E., Graham, J., Lee, J., Fernandez-Fernandez, F., Fernandez, G., Yun, S.J., Perkins-Veazie, P., 2014. Genetic and developing genomic resources in black raspberry. Acta Hortic. 1048, 19–24.

Beaver, J.S., Osorno, J.M., 2009. Achievements and limitations of contemporary common bean breeding using conventional and molecular approaches. Euphytica 168 (2), 145–175.

Bendokas, V., Gelvonauskiene, D., Siksnianas, T., Staniene, G., Siksnianiene, J.B., Gelvonauskis, B., Stanys, V., 2012. Morphological traits of phytomers and shoots in the first year of growth as markers for predicting apple tree canopy architecture. Plant Breed. 131 (1), 180–185.

Bliss, F.A., 2007. Education and preparation of plant breeders for careers in global crop improvement. Crop Sci. 47 (3), S250–S261.

Borralho, N.M.G., Cotterill, P.P., Kanowski, P.J., 1993. Breeding objectives for pulp production of *Eucalyptus globulus* under different industrial cost structures. Can. J. For. Res. 23 (4), 648–656.

Bradshaw, J.E., Dale, M.F.B., Mackay, G.R., 2009. Improving the yield, processing quality and disease and pest resistance of potatoes by genotypic recurrent selection. Euphytica 170 (1–2), 215–227.

Brothers, M.E., Kelly, J.D., 1993. Interrelationship of plant architecture and yield components in the pinto bean ideotype. Crop Sci. 33 (6), 1234–1238.

Brown, J., Caligari, P.D.S., 2008. An Introduction to Plant Breeding. Blackwell Publishing, Ltd., Oxford (UK), pp. 18–33.

Brunel-Muguet, S., Aubertot, J.-N., Dürr, C., 2011. Simulating the impact of genetic diversity of *Medicago truncatula* on germination and emergence using a crop emergence model for ideotype breeding. Ann. Bot. 107 (8), 1367–1376.

Byram, T.D., Myszewski, J.H., Gwaze, D.P., Lowe, W.J., 2005. Improving wood quality in the western gulf forest tree improvement program: the problem of multiple breeding objectives. Tree Genet. Genomes 1 (3), 85–92.

Carena, M.J., 2013. Challenges and opportunities for developing maize cultivars in the public sector. Euphytica 191 (2), 165–171.

Da Silva, D., Han, L., Faivre, R., Costes, E., 2014. Influence of the variation of geometrical and topological traits on light interception efficiency of apple trees: sensitivity analysis and metamodelling for ideotype definition. Ann. Bot. 114 (4), 739–752.

Dhillon, S.S., Kumar, P.R., Gupta, N., 1993. Breeding objectives and methodologies. Monographs Theoret. Appl. Genet. 1993, 8–20.

Diretto, G., Tavazza, R., Welsch, R., Pizzichini, E., Mourgues, F., Papacchioli, V., Beyer, P., Giuliano, G., 2006. Metabolic engineering of potato tuber carotenoids through tuber-specific silencing of lycopene epsilon cyclase. BMC Plant Biol. 6 (1), 13.

Donald, C.M., 1968. The breeding of crop ideotypes. Euphytica 17, 385–403.

Duc, G., Agrama, H., Bao, S., Berger, J., Bourion, V., De Ron, A.M., Gowda, C.L.L., Mikic, A., Millot, D., Singh, K.B., Tullu, A., Vandenberg, A., Vaz Patto, M.C., Warkentin, T.D., Zong, X., 2015. Breeding annual grain legumes for sustainable agriculture: new methods to approach complex traits and target new cultivar ideotypes. Crit. Rev. Plant Sci. 34 (1–3), 381–411.

Duvick, D.N., 1996. Plant breeding, an evolutionary concept. Crop Sci. 36 (3), 539–548.

Edwards, O., Singh, K.B., 2006. Resistance to insect pests: what do legumes have to offer? Euphytica 147 (1–2), 273–285.

Foolad, M.R., Panthee, D.R., 2012. Marker-assisted selection in tomato breeding. Crit. Rev. Plant Sci. 31 (2), 93–123.

Fraser, P.D., Romer, S., Kiano, J.W., Shipton, C.A., Mills, P.B., Drake, R., Schuch, W., Bramley, P.M., 2001. Elevation of carotenoids in tomato by genetic manipulation. J. Sci. Food Agric. 81 (9), 822–827.

Freyre, R., Deng, Z., 2013. Breeding *Ruellia* and *Caladium* at the University of Florida. Acta Hortic. (1002), 223–229.

Fries, A., Ericsson, T., 1998. Genetic parameters in diallel-crossed scots pine favor heartwood formation breeding objectives. Can. J. For. Res. 28 (6), 937–941.

Fuglie, K.O., Walker, T.S., 2001. Economic incentives and resource allocation in U.S. public and private plant breeding. J. Agric. Appl. Econ. 33 (3), 459–473.

Gepts, P., Hancock, J., 2006. The future of plant breeding. Crop Sci. 46 (4), 1630–1634.

Gonzalez, A., Zhao, M., Leavitt, J.M., Lloyd, A.M., 2008. Regulation of the anthocyanin biosynthetic pathway by the TTG1/bHLH/Myb transcriptional complex in *Arabidopsis* seedlings. Plant J. 53 (5), 814–827.

Grosser, J.W., Gmitter, F.G., Dutt, M., Calovic, M., Ling, P., Castle, B., 2015. Highlights of the University of Florida Citrus Research and Education Center's comprehensive citrus breeding and genetics program. Acta Hortic. (1065), 405–413.

Hancock, J.F., Stuber, C., 2008. Sustaining public plant breeding to meet future national needs. HortScience 43 (2), 298–299.

Heath, M.C., 1996. Plant resistance to fungi. Can J. Plant Pathol. 18 (4), 469–475.

Hirschberg, J., Cohen, M., Harker, M., Lotan, T., Mann, V., Pecker, I., 1997. Molecular genetics of the carotenoid biosynthesis pathway in plants and algae. Pure Appl. Chem. 69, 2151–2158.

Hurtado, M., Vilanova, S., Plazas, M., Gramazio, P., Prohens, J., 2015. Development of breeding programmes in eggplant with different objectives and approaches: three examples of use of primary genepool diversity. Acta Hortic. (1099), 711–718.

Ivkovic, M., Wu, H.X., McRae, T.A., Powell, M.B., 2006. Developing breeding objectives for radiata pine structural wood production. I. Bioeconomic model and economic weights. Can. J. For. Res. 36 (11), 2920–2931.

Laurens, F., 1998. Review of the current apple breeding programmes in the world. Objectives for scion cultivar improvement. Acta Hortic. (484), 163–170.

Lauri, P.E., Costes, E., 2004. Progress in whole-tree architectural studies for apple cultivar characterization at INRA, France—contribution to the ideotype approach. Acta Hortic. (663), 357–362.

Levings, C.S.I.I.I., Siedow, J.N., 1992. Molecular basis of disease susceptibility in the Texas cytoplasm of maize. Plant Mol. Biol. 19 (1), 135–147.

Li, Y., Huang, F., Lu, Y., Shi, Y., Zhang, M., Fan, J., Wang, W., 2013. Mechanism of plant-microbe interaction and its utilization in disease-resistance breeding for modern agriculture. Physiol. Mol. Plant Pathol. 83, 51–58.

van Lieshout, O., 1993. Consumer-oriented quality improvement of tomatoes in Indonesia. How to construct an ideotype? How to assess quality problems? Euphytica 71 (3), 161–180.

Liu, Z., Ming-Zhu, S., De-Yu, X., 2014. Regulation of anthocyanin biosynthesis in *Arabidopsis thaliana* red pap1-D cells metabolically programmed by auxins. Planta 239 (4), 765–781.

Mann, S., 2013. Is "multifunctionality" a useful framework for plant breeding? A critical analysis of the institutional design in Switzerland. Agroecol. Sust. Food 37 (3), 363–378.

Martin, T.A., Johnsen, K.H., White, T.L., 2001. Ideotype development in southern pines: rationale and strategies for overcoming scale-related obstacles. For. Sci. 47 (1), 21–28.

McDevitt, J.E., Mila i Canals, L., 2011. Can life cycle assessment be used to evaluate plant breeding objectives to improve supply chain sustainability? A working example using porridge oats from the UK. Int. J. Agric. Sustain. 9 (4), 484–494.

Morris, M.L., Heisey, P.W., 2003. Estimating the benefits of plant breeding research: methodological issues and practical challenges. Agric. Econ. 29 (3), 241–252.

Muller, P., Zeddies, J., 1988. Economic valuation of varieties and breeding objectives. Z. Pflanzenzucht. 100 (1), 59–70.

Ortiz, R., Langie, H., 1997. Path analysis and ideotypes for plantain breeding. Agron. J. 89 (6), 988–994.

Parlevliet, J.E., Zadoks, J.C., 1977. The integrated concept of disease resistance: a new view including horizontal and vertical resistance in plants. Euphytica 26 (1), 5–21.

Peng, S., Khush, G.S., Virk, P., Tang, Q., Zou, Y., 2008. Progress in ideotype breeding to increase rice yield potential. Field Crops Res. 108 (1), 32–38.

Perrin, F., Hartmann, L., Dubois-Laurent, C., Welsch, R., Huet, S., Hamama, L., Briard, M., Peltier, D., Gagné, S., Geoffriau, E., 2017. Carotenoid gene expression explains the difference of carotenoid accumulation in carrot root tissues. Planta 245 (4), 737–747.

Picheny, V., Casadebaig, P., Trépos, R., Faivre, R., Da Silva, D., Vincourt, P., Costes, E., 2017. Using numerical plant models and phenotypic correlation space to design achievable ideotypes. Plant Cell Environ. 40 (9), 1926–1939.

Rasmusson, D.C., 1984. Ideotype research and plant breeding. In: Gustafson, J.P. (Ed.), Gene Manipulation in Plant Improvement, Proceedings of the 16th Stadler Genet. Symp. Plenum Press, New York, NY (USA), pp. 95–119.

Reighard, G.L., 2002. Current directions of peach rootstock programs worldwide. Acta Hortic. (592), 421–427.

Rodríguez-Villalón, A., Gas, E., Rodríguez-Concepción, M., 2009. Phytoene synthase activity controls the biosynthesis of carotenoids and the supply of their metabolic precursors in dark-grown *Arabidopsis* seedlings. Plant J. 60 (3), 424–435.

Rosati, C., Aquilani, R., Dharmapuri, S., Pallara, P., Marusic, C., Tavazza, R., Bouvier, F., Camara, B., Giuliano, G., 2000. Metabolic engineering of beta-carotene and lycopene content in tomato fruit. Plant J. 24 (3), 413–420.

Rosyara, U.R., Bink, M.C.A.M., van de Weg, E., Zhang, G., Wang, D., Sebolt, A., Dirlewanger, E., Quero-Garcia, A., Schuster, M., Iezzoni, A.F., 2013. Fruit size QTL identification and the prediction of parental QTL genotypes and breeding values in multiple pedigreed populations of sweet cherry. Mol. Breed. 32 (4), 875–887.

Scott, J.W., Myers, J.R., Boches, P.S., Nichols, C.G., Angel, F.F., 2013. Classical genetics and traditional breeding. In: Liedl, B.E., Labate, J.A., Stommel, J.R., Slade, A., Kole, C. (Eds.), Genetics, Genomics, and Breeding of Tomato. CRC Press, Boca Raton, FL (USA), pp. 37–73.

Shinada, H., Yamamoto, T., Yamamoto, E., Hori, K., Yonemaru, J., Matsuba, S., Fujino, K., 2014. Historical changes in population structure during rice breeding programs in the northern limits of rice cultivation. Theor. Appl. Genet. 127 (4), 995–1004.

Shumskaya, M., Wurtzel, E.T., 2013. The carotenoid biosynthetic pathway: thinking in all dimensions. Plant Sci. 208, 58–63.

Simmonds, N.W., 1979. Principles of Crop Improvement. Longman, London (UK). 408 pp.

Singh, S.P., Schwartz, H.F., 2010. Breeding common bean for resistance to diseases: a review. Crop Sci. 50 (6), 2199–2223.

Stamp, P., Visser, R., 2012. The twenty-first century, the century of plant breeding. Euphytica 186 (3), 585–591.

Story, E.N., Kopec, R.E., Schwartz, S.J., Harris, G.K., 2010. An update on the health effects of tomato lycopene. Annu. Rev. Food Sci. Technol. 1, 189–210.

Tworkoski, T., Fazio, G., 2016. Hormone and growth interactions of scions and size-controlling rootstocks of young apple trees. Plant Growth Regul. 78 (1), 105–119.

Van Oijen, M., Höglind, M., 2016. Toward a Bayesian procedure for using process-based models in plant breeding, with application to ideotype design. Euphytica 207 (3), 627–643.

I. ELEMENTS AND UNDERPINNINGS OF PLANT BREEDING

Wu, R.L., 1998. Genetic mapping of QTLs affecting tree growth and architecture in *Populus*: implication for ideotype breeding. Theor. Appl. Genet. 96 (3–4), 447–457.

Yan, W., Frégeau-Reid, J., 2008. Breeding line selection based on multiple traits. Crop Sci. 48 (2), 417–423.

Yen, H.C., Shelton, B.A., Howard, L.R., Lee, S., Vrebalov, J., Giovannoni, J.J., 1997. The tomato high-pigment (*hp*) locus maps to chromosome 2 and influences plastome copy number and fruit quality. Theor. Appl. Genet. 95 (7), 1069–1079.

Yin, X., Stam, P., Kropff, M.J., Schapendonk, A.H.C.M., 2003. Crop modeling, QTL mapping, and their complementary role in plant breeding. Agron. J. 95 (1), 90–98.

Young, N.D., 1996. QTL mapping and quantitative disease resistance in plants. Annu. Rev. Phytopathol. 34, 479–501.

Zhang, Y., Lubberstedt, T., Xu, M., 2013. The genetic and molecular basis of plant resistance to pathogens. J. Genet. Genomics 40, 23–35.

CHAPTER

7

Germplasm and Genetic Variability

OUTLINE

INTRODUCTION

In Chapter 5 *germplasm* was defined as a "collection of genes for use in the improvement of plants". These genes must be in a form that is accessible to the plant breeder, so are usually embedded into plant genomes and stored as seeds or other types of propagules. Alternatively, germplasm may take the form of cultured cells or tissues or may consist of a sequence of DNA in a transformation vector. Germplasm is regarded as "raw material" for breeding, a milieu from which desirable genes may be extracted and collected into the genomes of new commercial individuals and populations.

To be successful the plant breeder must ultimately target, acquire, and assemble the best possible combination of alleles for the targeted breeding objective. "Best" can, of course, have different meanings depending on the context. In a controlled experiment where values are placed on certain outcomes, "best" can be entirely objective. In other less defined situations, "best" is an arbitrary term, and subject to consensus. It can also be an abstract term, unattainable, but a concept to be strived for. No examples exist wherein the best combination of alleles was simply plucked from the wild and conveyed directly into a commercial context. "Best" is a concept in plant breeding where a plant performs in a manner consistent with highest expectations. Food and pharmaceuticals, fiber and structural materials, fuels, aesthetic beauty, environmental restoration are some examples of how plants have been made to serve humankind. Plants can and do provide food for humans in the wild, but our anthropomorphic view is that they would rather not, since they do put up a measure of resistance. Historically, most plant breeding efforts have been invested toward increased productivity or yield (Hoisington et al., 1999).

Plant genetic variability is a major component of Earth's total biodiversity. Without genetic variability plant breeding exists in a vacuum. Absence of genetic variability for plant breeding is analogous to house construction; possessing tools and a plan for a house, but having no building materials. In our building analogy needed raw materials are concrete, wooden lumber of many dimensions, metal fittings and fasteners, etc. No one raw material for building or gene for breeding can do the complete job. In house building the plan specifies the quality and quantity of raw materials, whereas in plant breeding, genetic variability is more cryptic and elusive. In essence, genetic variability is equivalent to proportion of polymorphic loci and numbers of alleles per locus (Frankel et al., 1999).

Horticultural Plant Breeding. https://doi.org/10.1016/B978-0-12-815396-3.00007-X

The genetic raw materials to be amassed must be relevant to breeding objectives (Chapter 6). What phenotypic attributes will be targeted (e.g., yield, quality, and disease resistance)? What will be the general genetic structure (e.g., OP vs. hybrid) of the finished commercial populations? Where will the finished commercial populations, or cultivars, be produced (e.g., upstate New York vs. South Florida or the Chilean coast during northern winters)? What production and harvesting methods will be employed (e.g., hand vs. machine harvested)? Answers to these questions provide the plant breeder with a starting point to build a working gene collection to fulfill the prescribed objectives (Plucknett et al., 1987).

The working germplasm collection of the plant breeder is arranged conceptually in concentric spheres, or "tiers" (Van Hintum, 1995; Swanson, 1998). The distinction between what is considered germplasm and breeding lines is blurred. Breeding lines are dynamic and unfinished, an intermediate between beginning and endpoints. The breeding populations most useful to plant breeders tend to be those that are closer to being finished. Genes may be extracted from more refined populations without excessive need for prolonged mating and selection to rid the pools of genes that condition unwanted phenotypes (Yonezawa et al., 1995). The further from commercial populations one goes toward the direction of raw germplasm, the more difficult it is to extract targeted genes using plant breeding methods. Examples of such populations include unadapted cultivars, very old populations adapted to different climates, wild, weedy relatives, etc.

The first, or most accessible, germplasm tier should always be the varieties against which the targeted commercial population will eventually compete (Fig. 7.1). It is of crucial importance that the plant breeder is well versed in the current state of the art and science and also aware of the performance of contemporary cultivars of the targeted crop species. The best way to accomplish this is to acquire top cultivars that are the top sellers or are otherwise at the forefront of the market. These cultivars should always be included in experimental plantings, making comparisons with as many other populations and in as many contexts as possible. With breeder's exemptions for using germplasm protected by patents, plant variety protection (PVP) certificates, and trade secrets (see Chapter 12), it is also common and acceptable to use commercial cultivars as a source of desirable genes in the breeding program. The plant breeder must affect heritable changes to this germplasm that are deemed to be sufficient that the new cultivar stands apart phenotypically. Care must be taken, however, to abstain from the use of germplasm, cultivars, genomic constructs, genes, and genetic applications that are bound by broad and specific patent protection (Ghijsen, 2009).

The second germplasm layer, or tier, consists of fundamentally strong populations upon which the plant breeder wishes to build (Fig. 7.1). These are usually older and proven cultivars that still perform relatively well but have fallen out of favor with the market. The first and second layers may overlap, but there are pitfalls to the simple enhancement or combination of popular cultivars developed by others. The laws that protect against genetic piracy are

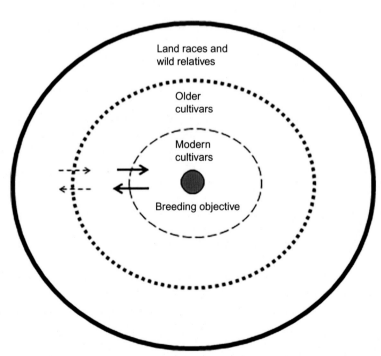

FIG. 7.1 The concept of concentric tiers, or layers, of population sources of useful genes for crop species improvement.

becoming stronger and more favorable to the inventors, including the addition of simply-inherited traits by back-crossing (Ghijsen, 2009; Chapter 18).

The third tier of working germplasm is comprised of older historic cultivars, perhaps 30–100 years since their respective period of commercial popularity (Fig. 7.1). Most of these cultivars were based on even more ancestral varieties that may not be available in extant germplasm collections. If such historic populations can be acquired, they may be valuable sources of genes and gene combinations that impart stress tolerance and disease resistance (Chapter 19).

The "outer" layers of accessible working germplasm include landrace, or primitive, cultivars that retain much of the character of wild populations but are demonstrably more useful than wild ancesters. Also included in the "outer layers" are feral (domesticates that became re-established in the wild) or wild progenitor populations if they still exist as such (the forces of evolution may have rendered them reproductively isolated from cultivated forms). The concepts articulated by Nikolai Vavilov apply here and will be described and discussed below. For many crop species such as corn the chain of progenitor populations from the wild to contemporary cultivars does not exist. Until recently, the origin of modern corn was the subject of a heated intellectual debate promulgated by theorists Paul Mangelsdorf and George Beadle during the mid-20th century (described in Smith, 2004; see for more recent findings on the evolution of maize in Eubanks, 2001; Matsuoka et al., 2002; Jaenicke-Despres et al., 2003; Beissinger et al., 2016; see also Chapter 15, Brief history of corn breeding).

The plant breeder may also wish to include closely related species in the working germplasm collection. The concept of species is defined mostly by reproductive isolation but the boundary that defines the limits of gene flow is a fuzzy one. In some families, disparate plant groups, and even some taxa with different chromosome numbers may be hybridized. Sexual fertility in these amphiploids may be restored with a chromosome doubling agent such as colchicine. It is also true that plant taxa that are nearly indistinguishable may not hybridize. The inclusion of related species in the germplasm collection is pointless unless some method exists by which flow between gene pools of interest can be developed. Breeders of long-lived woody perennial crop species generally narrow germplasm collections to include only "adapted" breeding lines within the targeted species. An elapsed time equivalent to many human lifetimes are necessary to complete a traditional breeding program in woody perennial species that includes "unadapted" populations.

The plant species targeted by the plant breeder may already be highly studied and have a substantial body of genetic knowledge already in place. With the advent of accessible molecular and genomic research methods the list of plant species about which the genome has been studied is growing rapidly. For species that are well-studied model systems in biology, such as tomato, genetic stock populations that are useful for plant breeders will likely also be available in the public domain. At least some of these genetic stocks may contain genes of direct commercial interest, such as disease resistance, novel pigmentation, desirable growth habit or stature, enhanced fruit quality, improved seed endosperm/embryo composition, or better floral attributes. If no such information or genetic stocks exist for the targeted species, an excellent opportunity is presented to the plant breeder and geneticist to initiate a program on basic inheritance. Any new information on the genetics of an unknown plant species is always welcome in the scientific community. This is especially germane for new plant breeders in the public sector, such as university faculty who need to publish original research results to fulfill the academic expectations of their appointments.

As cellular and molecular technologies have expanded in plant biology and more species' genomes have been sequenced, an age is approaching when genes may be simply "pulled off the shelf" and plugged into chosen plants. Most traits in horticultural crop species tend to be multigenic or quantitative, and affected by subtle interactions between genes, environment, and genetic x environment interactions. Therefore, genomic technology will not completely supplant what has been termed "classical plant breeding", or selection and controlled sexual mating of whole plants in populations, for a very long time.

EXTANT GENE POOLS

The definition of *gene pool* is the total range of germplasm available for genetic improvement of a given species by sexual hybridization (Simpson and Sedjo, 1998). The overwhelming majority of genetic variability is locked up in living plants, seeds, and asexual propagules, or *extant gene pools* (Frankel and Bennett, 1970). Other sources of genetic variability, such as DNA in fossilized samples (e.g. "Jurassic Park") and genes accessible via new technologies, are available to a very limited extent. It is clear that applications of genomic technology in plants will accelerate in the future. It is conceivable that all natural and synthetic genes will someday be considered to be in one unified gene pool (Briggs and Walters, 1997).

Primitive human cultures were aware that certain types of plants came from certain geographical areas, or climatic zones (Flodin, 1999). As humans explored the planet during the 16th and later centuries, it likely became apparent that different geographical areas were characterized by different flora and fauna. Charles Darwin first articulated the theory in 1859 that species arose by geographic isolation of population subsets, using Galapagos Island finches as one of many examples (Darwin, 1859; Grant, 1991). Later, Nikolai Vavilov of Russia and the former Union of Soviet Socialist Republics (USSR) traveled the world to collect seeds and observations led to his advancement of the Theory of Centers of Origin and Variation (1926; Vavilov, 1951). Based on his experiences, Vavilov proposed seven major Centers (Fig. 7.2), based on the high density of landraces and wild progenitors of domesticated crop species. These Centers of Origin were scattered among the continents, but a preponderance of species was concluded to have originated in the "old world", including Europe, Africa, Asia Minor, Central Asia, and Australo-Asia (Oldfield, 1989).

The primary economic species that originated in the "new world" were of the families Solanaceae (potato, pepper, tomato), Fabaceae (Lima bean, runner bean), Cucurbitaceae (winter squashes and gourds), and Asteraceae (sunflower, Jerusalem artichoke). Of these, only the sunflower (*Helianthus annuus*) and pepper (*Capsicum annuum*) are native to North America (Harlan, 1951; Basu and De, 2003). While the ornamental value of native plants from the western hemisphere is on the rise, spurred by concerns over adverse effects on non-native species, most of the inhabitants of our gardens and landscapes are still from Europe and Asia. Thus, the overwhelming proportion of the foods and aesthetic fulfillment of North Americans originated from elsewhere on the planet Earth.

The concept of Centers of Origin is of crucial importance to plant breeders because it provides the best estimation of where untapped genetic variability may be discovered (Harlan, 1976). If no desirable genes are found among existing germplasm sources, the search proceeds in a stepwise fashion and ultimately to additional collections in the species' Center of Origin, conspecific populations, and sexually compatible wild species. In many instances, the Center of Origin for an economic plant species has been unavailable for collecting due to adverse geopolitical factors. Central Asia encompasses the region circumscribed by the Tien Shan, Himalayan, and Caucasus Mountain Ranges. Much of this area fell within the boundaries of the former Soviet Union until the individual republics became independent in 1990. Subsequently, world collections have become more accessible to western scientists, but recent geopolitical developments may reverse this trend. While collections of economic plant species were made by Soviet scientists including Vavilov, who later established a second branch of the St. Petersburg (then Leningrad) based Lenin All-Union Academy of Agricultural Sciences in Tashkent, Uzbekistan, it is still unclear exactly how comprehensive these were (Loskutov, 1999). During World War II Leningrad was besieged and dedicated All-Union Academy scientists famously died of starvation rather than to consume the seeds they were tasked with protecting. The St.

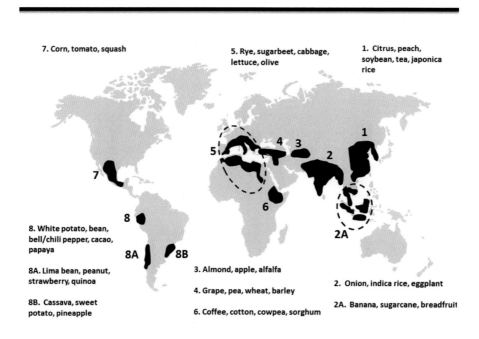

FIG. 7.2   Centers of origin and diversity for the major crop species as identified by Nikolai I. Vavilov and modified by Harlan (1951).

Petersburg facility is still active and now known as the All-Russian Research Institute of Plant Industry whereas the Tashkent facility operates independently as the Uzbek Research Institute of Plant Industry (Mavlyanova et al., 2005).

In most cases, the Center of Origin for a crop species coincides with the geographical locus of the first domestication, known as the Center of Diversity. In some instances, however, the Centers of Origin and Diversity do not occur together. Such is the situation in *Brassica* species where certain species experienced the bulk of domestication geographically distant from the Center of Origin (the Mediterranean basin; Guo et al., 2014).

BIODIVERSITY: GENETIC VARIABILITY IN NATURAL ECOSYSTEMS

The majority of all genetic variability is currently and historically stored in natural ecosystems. This germplasm in wild animal, plant, and microbial populations, it is at peril for becoming lost (Dasmann, 1991). Each wild population likely embodies unique alleles and allele/gene combinations that arose and were selected for over immense periods of evolutionary time. Humans currently lack the means to recover or reconstruct germplasm lost due to habitat destruction. Earth is changing all the time both climatically and geologically, and old species become extinct as new species are emerging. Mass extinctions have been documented by the fossil evidence at several points in earth's history. Such catastrophic biological calamities are thought to have occurred in response to large asteroid impacts or massive volcanic eruptions. Habitats on earth are in a constant state of flux, and over 99% of all species that have ever existed are now extinct (Frankel et al., 1999). Therefore, extinction is a natural process, unless it is wrought by the inadvertent consequences of human activities.

Humans are also culpable for both irreversible losses of species and potentially valuable genes through CO_2-induced climate change (Van den Berg and Feinstein, 2009) and physical destruction of fragile wild habitats (Lin and Liu, 2006; Riordan et al., 2015; Hooftman et al., 2016). Most plant species diversity exists within the tropics, the band that encompasses earth between the north and south latitudes of the Tropics of Cancer and Capricorn. The tropical belt is considered mostly primitive with regard to human socio-political development. Human cultures in the tropics have historically and persistently remained agrarian and subsistence, and the worst impact on total biodiversity has been a "slash and burn" strategy toward land use (Brown et al., 2007). Economic pressures wrought by globalization are hastening the destruction of natural tropical ecosystems and habitats by pandering to short-lived demand for commodities and natural resources. These pressures are not necessarily limited to the tropics, of course, and biodiversity in general is threatened (Cox and Wood, 1999).

GENETIC VARIABILITY MAINTAINED IN SITU

The next level of genetic conservation lies at within the realm of biological preserves such as parks and botanical gardens. These are also referred to as in situ *collections*. Many of the benefits of retaining natural habitats are realized with in situ collections, but this comes at a substantial cost. National, state, and local parks require public resources for maintenance (Wilkes, 1991). Botanical gardens are often quasi-public organizations or private foundations that derive support from a combination of fees and donations. The altruistic among us would encourage the transfer of proportionately more natural habitats into public trusts, but even the most fervent would acknowledge that limits exist. As specific tracts are scrutinized for use as in situ reserves, it is crucial that ecologists and population geneticists be routinely consulted to ensure the maximum preservation of biodiversity per available resources (Gomez et al., 2005; Schlottfeldt et al., 2015). In situ collections are important sources of germplasm for certain types of domesticates such as ornamental and medicinal plant species, but not as important thus far in most horticultural food crop species (McFerson, 1998).

GERMPLASM REPOSITORIES

Our populations of economically important cultivated species were passed down from ancestors, usually in the form of seeds collected from harvests and retained for future crop establishment. Seeds are living entities that require specific environmental conditions to promote long-term viability. Seeds do not live forever. Since seeds give rise to plants that beget more seeds that are mostly faithful genetic copies, loss of seeds to mortality implies a loss of genetic variability. Many publications have appeared that address the issue of preservation of plant genetic resources in the form of seeds (e.g., Ford-Lloyd and Jackson, 1986; Frankel et al., 1999; Razdan and Cocking, 1997).

It is unclear at what point in time and space a collection of seeds became the first collective germplasm repository, or bank. European monarchies sanctioned maritime expeditions during the period 1500–1700 aimed at assessing economic opportunities in distant continents, and explorers are known to have brought back seeds and other plant propagules to Europe. Little or no documentation exists, however, on how these propagules were managed and maintained after they reached Europe (Thompson and Harris, 2010).

The U.S. Department of Agriculture Office of Seed and Plant Introduction was established in 1898 and was for a period the de facto coordinator of U.S. plant germplasm acquisition, storage, and dissemination. Subsequently, this office became a part of the Agricultural Research Service and was named the National Plant Germplasm System (NPGS), headquartered at the Beltsville Agricultural Research Center (BARC), Beltsville, MD (USA) where it resides to the present (White et al., 1989; Stoner and Hummer, 2007). Under the auspices of USDA the collective functions of germplasm acquisition, organization, storage, and dissemination have expanded substantially, now encompassing a large number of facilities, and professional staff that oversees the various activities associated with plant germplasm. These facilities and staff are scattered throughout North America, the Caribbean basin, and the Pacific Islands.

USDA/ARS/NPGS is charged with the following:

- Germplasm Acquisition
- Germplasm Storage
- Germplasm Information Management
- Germplasm Dissemination
- Research on germplasm storage technologies

The NPGS organizational structure consists of a main coordinating laboratory and office staff in Beltsville, MD (Beltsville Agricultural Research Center, BARC), a National Center for Genetic Resources Preservation in Fort Collins, CO, four regional Plant Introduction Stations (Geneva, NY; Tifton, GA; Ames, IA, Pullman, WA), and eight clonal repositories that specialize in the storage of vegetative plant material of woody perennials (Fig. 7.3). NPGS is modeled after the land-grant system wherein USDA/ARS partners with individual state Agricultural Experiment Stations (AESs) and Extension Services and with county governments. NPGS includes primarily state AESs, who collaborate primarily in the establishment of policies and germplasm evaluation and utilization (White et al., 1989).

Each major crop species in the U.S. has an associated Crop Advisory Committee that consists of scientists from many research and teaching institutions, primarily land grant universities (White et al., 1989). The working group is charged with the development of descriptors pertaining to the economic performance of the targeted crop species, and the identification of germplasm gaps. NPGS uses the descriptors to fashion the "Passport" database for each targeted species (see below). With knowledge of all perceived gaps in the germplasm collection, NPGS formulates priorities that are applied against subsequent requests for exploration or acquisition.

With regard to acquisition of new germplasm either uncollected or collected but not in the NPGS system two programs are advanced: germplasm exchange with international peer institutions and geo-global germplasm exploration. The importance of exploration is waning as germplasm has progressively become regarded more as a natural resource than as a global resource. Funding or sanctioning expeditions to foreign lands with the expressed

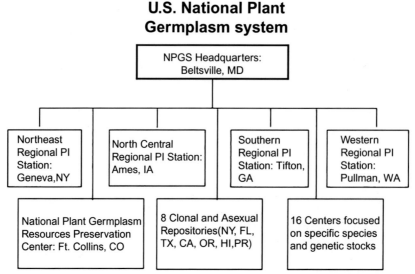

FIG. 7.3 Basic organizational structure of USDA/ARS/NPGS.

purpose of finding germplasm in the wild or amongst primitive societies have become increasingly problematic. Reciprocal agreements mandating a paper trail to trace gene ownership, geographic origins, destinations, and commercial applications are typical. The Consultative Group for International Agricultural Research (CGIAR) has led the way in this quest to repatriate genetic resources to the geopolitical source. CGIAR is the umbrella non-profit international organization that assumed control in 1971 of the International Agriculture Research Centers that were established after World War II by the Rockefeller Foundation. The Center where Dr. Norman Borlaug conducted his groundbreaking work on short/stiff-stemmed grains was CYMMYT in Mexico, one of the original Rockefeller International Crop Centers (see below).

CGIAR has formulated strong policies on germplasm exploration and exchange that have been co-opted by governments. While it seems fair that the country of origin that may include the Center of Origin and/or diversity for a given crop species is compensated for the removal of germplasm, the process has had the effect of crimping the free flow of germplasm across political borders. The CGIAR-FAO convention is that intellectual property protection is not permitted on any germplasm exchanged or derived entities under a sanctioned agreement. Therefore, for-profit concerns such as biotechnology and seed companies tend to avoid this cache of genetic variability. As was stated earlier, most Centers of Origin coincide with developing countries, or governments that are not necessarily on good political terms with nations from the "west". Exploration within developed countries, or those on very friendly terms with the west, is usually not associated with such difficulty (Fowler et al., 1998).

Germplasm exchange has, therefore, grown to be more important than in situ exploration. Institutions within countries that encompass the geographic area endemic to the germplasm of interest may be contacted and an agreement established that defines what each party expects to gain from the germplasm exchange (or quid pro quo. Any germplasm that is acquired in this manner may be excluded from finished cultivars targeted for patent or PVP protection (Chapter 12). This requirement will, of course, force private industry to confine their efforts to germplasm that is already available, or can be acquired exclusively of this requirement.

Each new population, seed lot, or set of clones that is acquired by NPGS enters into a highly defined procedure by which the germplasm entity is identified, described, and relegated for storage and dissemination. Individual scientists may obtain genetic entities independent of NPGS, but cooperation is encouraged, especially for populations that appear to have broad significance. The actual materials are first inspected by the USDA Animal and Plant Health Inspection Service (APHIS) to ensure freedom from infestation by unwanted pests prior to being transferred to NPGS at BARC. In some instances cryptic pests such as viruses and prions are a concern and further quarantine is warranted. Such materials are usually subjected to a "grow-out" and scrutiny under confinement before release into the general NPGS system. New molecular technologies to identify and quantify pests are being adopted that render tedious, inaccurate, and time-consuming "grow-outs" unnecessary.

Each population is assigned a *Plant Introduction* (PI) number that will accompany it everywhere within NPGS and this number is usually retained by recipients of germplasm to aid in the value of the NPGS database. The PI population and its descendants are thereafter referred to as an *accession*. PI numbers are assigned sequentially without regard to species, and had reached 655,520 as of 2008, the last published NPGS inventory. Many accessions are discarded or abandoned over the years as they are found to be duplicates of other populations or are otherwise not needed. NPGS currently oversees over 460,000 accessions and germplasm facilities for maintenance, introduction, and study in all regions of the U.S. (Fig. 7.4).

At the time of accession establishment a *passport* form is completed, summarizing key attributes that will be entered into the NPGS database known as GRIN (Germplasm Resources Information Network). Passport categories are based on descriptors developed by Crop Advisory Committees that exist to support germplasm issues pertaining to specific crop species with representatives from ARS, state AESs, and industry. The passport data record includes the global positioning system (GPS) coordinates where the sample originated or was collected, if available (Stoner and Hummer, 2007).

Each new accession is subsequently transferred to the National Laboratory for Genetic Resources Preservation (NLGRP) in Fort Collins, CO (USA) and one of the four PI stations or many clonal repositories where most of the daily operations of germplasm storage, maintenance, and dissemination is conducted. Each of the four PI stations and repositories operates a controlled-environment enclosure wherein seeds or asexual plant propagules are stored. The environmental conditions are maintained at levels that have been shown to prolong viability and preserve genetic purity. For seeds, viability is adversely affected by high temperatures and humidity (Krishnan et al., 2004). Therefore, the environment of the seed storage vault is usually controlled at $\leq 10\,°C$ and $\leq 30\%$ RH. The storage enclosure of the clonal repository is, conversely, operated at low temperature and high relative humidity since live plant tissues are not well preserved under desiccation (Roos, 1989).

Seeds are the fundamental storage unit for most economic crop species. Under the prescribed storage conditions, viability can be maintained for up to many years, perhaps over 10–15. Certain species produce seeds that

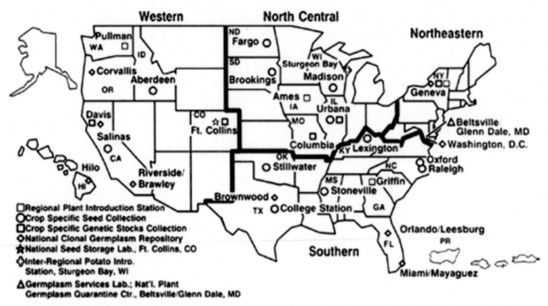

FIG. 7.4   The U.S. NPGS Regions and locations of affiliated repositories for seeds and asexual propagules. *From Shands, H.L. and G.A. White.*
1990. New crops in the U.S. national plant germplasm system. p. 70-75. In: J. Janick and J.E. Simon (eds.), Advances in new crops. Timber Press, Portland, OR,
Retrieved from https://hort.purdue.edu/newcrop/proceedings1990/V1-070.html.

are notoriously short-lived. Even under the best possible storage conditions, viability may only be maintained for 1–3 years. Population replenishment, therefore, is an important function of the PI station or repository. Technical staff must frequently assess the viability of propagules, especially those known to be short-lived. Viability assessment of seeds is usually accomplished by germination tests that measure both rate and vigor. When the proportion of germinating vs. non-germinating seeds drops below a designated threshold level, for example 70–80%, the decision to replenish the stock is triggered (Desai, 2004).

The PI stations usually maintain storage and working stocks of each accession, both of which need to be replenished periodically due to loss of vigor. Frequent replenishment of working stocks is necessary for accessions that attract high demand for samples among practitioners. In cases of high priority or "emergency", the PI station may dispense seed from the storage stock to a practitioner, but if the working stock is depleted, clientele usually must wait until the stocks are replenished, sometimes taking two years or more White et al., 1989).

Each species has a specific procedure for replenishment, depending on mating type, the genetic structure of individuals and populations, the generation time, growing methods, etc. The procedure will be different for self- vs. cross-pollinating species. The latter must either be accomplished within enclosures, usually screen- or mesh-covered "cages" or through adequate geographical distance from other sexually-compatible flowering populations to prevent vectors from introducing contaminating pollen. This is not as necessary, of course, for obligate inbreeding species that do not usually disperse pollen (Ward et al., 2008).

Each accession also has a theoretical minimum population size for replenishment of the accession. Cross-pollinated populations are generally associated with polymorphic alleles and Hardy-Weinberg equilibria and larger population sizes are needed to assure that allelic frequencies are accurately conveyed to the new derivative population. Inadequate population sizes will lead to changes in allelic frequencies due to drift and founder's effects. Self-pollinating species are usually characterized by allelic fixation and population size is not as critical except where the original population consists of a mixture of many fixed genotypes such as a composite or multiline (Frankel et al., 1999).

The NPGS PI stations and repositories entertain requests from practitioners for samples of accessions maintained by the system. The Germplasm Resources Information Network (GRIN) is an interactive online database that permits users to search NPGS collection and request seeds or asexual propagules of specific accessions. NPGS clients are expected to limit accession requests only to those that are needed for established breeding programs. NPGS clients consist primarily of plant breeders in the U.S., but also include scientists in other fields, scientists in other countries, educators, government policy-makers, and regular citizens. Considerable discretion and latitude are granted to the PI stations and clonal repositories in responding to requests for germplasm. While no monetary charges are levied against requests, each one necessitates a significant cost in terms of compliance and the eventual need for replenishment that is supported by tax revenues. Each request for germplasm is, therefore, weighed against potential benefits. Requests by domestic scientists are usually granted unless they are not adequately specific or the need is not well-documented.

While no money changes hands in these transactions tacit agreement exists wherein the scientist will provide NPGS with any information gained pertaining to the performance of accessions. The U.S. Land Grant University System is a partner with the USDA/ARS/NPGS in the evaluation of germplasm (White et al., 1989) especially since the NPGS is not charged with the use of germplasm to produce cultivars. In cases of exceptional need NPGS may either conduct evaluations internally or contract with outside agents to evaluate a limited set of accessions. Any information pertaining to accession performance is entered into GRIN for use by subsequent clients seeking germplasm for breeding programs.

The NLGRP in cooperation with Colorado State University provides a program comprehensive redundancy of all U.S. germplasm accessions (Table 7.1). NLGRP does not entertain requests from germplasm users. These requests are coordinated with BARC staff and the PI stations and clonal repositories. Seed and clonal germplasm storage facility failures at PI stations and clonal repositories occur rarely and NLGRP is used as a backup, providing a new source of germinal populations for the re-establishment of inventories and working stocks. NLGRP also conducts research on emerging technologies for germplasm storage such as cryopreservation of plant organs, tissues, and cells and the isolation and storage of DNA and RNA sequences. Research is also promulgated on new techniques for propagule viability monitoring, reducing costs, and enhancing population management effectiveness.

A separate government unit has been established within the USDA with charged with keeping unwanted germplasm out of the United States. The Federal Insecticide, Fungicide, and Rodenticide Act (FIFRA) provides funding for establishing and operating the APHIS under the auspices of USDA and in close cooperation with Departments of Agriculture and Environmental Protection affiliated with individual states. They are charged with preventing plants, animals, and microbes from entering the U.S. that could adversely affect agriculture and human health. Most everyone has heard of Dutch elm disease, chestnut blight, plum pox virus, kudzu, mile-a-minute weed, Mediterranean fruit fly, purple loosestrife, hemlock wooly adelgid, water hyacinth, Asian longhorn beetle, Japanese knotweed, Asian tiger mosquito, Emerald ash borer, brown marmorated stink bug and, more recently, red-spotted lantern fly. With the rapid expansion of global trade, incidences of unintentional introduction of invasive, non-native species of weeds, insects, and pathogens that threaten agriculture have increased tremendously. For example, the Asian longhorn beetle and emerald ash borer were thought to have arrived in North America in the wooden packing materials used to protect large consumer goods during shipment from Asia (Cappaert et al., 2005; Haack et al., 2009).

Plant breeders must interact with APHIS when bringing germplasm into the U.S., especially for seeds and clones from geographical locations that are infested with diseases and insects not already in the U.S. APHIS sometimes establishes strict moratoria on the movement of plant materials across the border. One example is of plum pox virus on *Prunus* in southeastern Pennsylvania and elsewhere in North America. APHIS responded by prohibiting the importation of virtually all *Prunus* clones to protect the industry within the continental U.S. Unfortunately, this regulation also stifled U.S. *Prunus* breeders.

The germplasm collections in St. Petersburg, Russia and Tashkent, Uzbekistan, initiated in the 1920s by Nikolai I. Vavilov, were mentioned above. The Ministry of Agriculture and/or Institute of Agricultural Research of virtually every country on Earth sponsors a unit that oversees plant germplasm, the size and range of activities of which vary tremendously (Plucknett et al., 1987). Most significant, however, are the germplasm collections and activities housed within the International Centers of the CGIAR, a global non-profit consortium aimed at alleviating global hunger and environmental destruction (Fuccillo et al., 1997). These centers were established as a philanthropic attempt to

TABLE 7.1   NLGRP Inventory Held in the Base Collection as of February 16, 2017

Number of NPGS safety backup samples	445,722
Number of NPGS unique accessions	566,345
Number of Non-NPGS (black box) accessions	324,507
Number of genera	1893
Number of species	9279
Percent NPGS seed accessions backed-up	85%
Percent NPGS vegetatively-propagated accessions backed-up	14%

apply advances in agriculture initially to the alleviation of world hunger, culminating in the "Green Revolution" of the 1960s. The focus of international centers has been broadened to include all forms of "sustainable" agriculture. The centers were established in host countries wherein problems of human starvation were acute and also where major food crops were produced. Currently, CGIAR operates 15 centers, the following 12 of which deal directly or indirectly with plant genetic resources:

- International Rice Research Institute (**IRRI**), Los Banos, Philippines
- Center for International Forestry Research (**CIFOR**), Bogor, Indonesia
- Centro Internacional de Mejoramiento de Maize y Trigo (**CIMMYT**), El Batan, Texcoco, Mexico
- International Crops Research Institute for the Semi-Arid Tropics (**ICRISAT**), Hyderabad, India
- International Center for Agricultural Research in the Dry Areas (**ICARDA**), Beirut, Lebanon (recently moved from Aleppo, Syria)
- World Agroforestry Centre (**ICRAF**), Nairobi, Kenya
- International Institute of Tropical Agriculture (**IITA**), Ibadan, Nigeria
- West Africa Rice Development Association (**AfricaRice**), Bouake, Ivory Coast
- Centro Internacional de Agricultura Tropical (**CIAT**), Cali, Columbia
- Center Internacional de la Papa (**CIP**), Lima, Peru
- International Food Policy Research Institute (**IFPRI**), Washington DC, USA
- International Plant Genetic Resources Institute (**IPGRI**), Rome, Italy

The World Vegetable Center (WorldVeg; previously known as the Asian Vegetable Research and Development Center [AVRDC]) in Shanhua, Taiwan was founded in a similar fashion as the CGIAR centers in 1971. WorldVeg is a member of the Association of International Research and Development Centers for Agriculture, AIRCA, very similar in charter and activities to CGIAR. AIRCA includes many other research organizations that sponsor genebanks, including the Tropical Agricultural Research and Higher Education Center (CATIE; Cartago, Costa Rica), Crops For the Future (CFF; Kuala Lumpur, Malaysia), the International Center for Biosaline Agriculture (ICBA; Dubai, United Arab Emirates), and the International Network for Bamboo and Rattan (INBAR; Beijing, China).

Through most of the 20th century worldwide germplasm exploration was accomplished without hindrance. During the reconstruction period following WWII countries progressively fell into economic tiers: "developed" (U.S., Canada, Europe, Japan, Australia), "developing" (India, China, USSR/Russia, Brazil, Argentina), and "underdeveloped" (most other countries in South America, Africa, Asia). Developed countries came to be known as resource users, taking from the developing and underdeveloped countries, using them as a source of cheap labor for consumer goods. The oil-producing countries banded together to form a cartel that would control the return to countries for that natural resource. Germplasm ultimately came to be considered as a natural resource in the same vein as fossil fuels and minerals but has not historically been treated as a commodity. It is difficult to assign a monetary value to germplasm since the utility lies in subtle changes in the underlying genomic DNA sequences. The germplasm holder tends to overvalue while those seeking it tend to discount germplasm value since most of the actual value of germplasm lies in successful applications and development. For example, natural resources and commodities such as metals, plastics, and rubber have much less value than the automobile that is manufactured from them.

Developing and underdeveloped countries came together in 1971 with financial support by the World Bank to form the CGIAR. The tenets of germplasm valuation and rules for international engagement with CGIAR and other international agricultural center consortia were subsequently established an International Treaty on Plant Genetic Resources for Food and Agriculture sanctioned by the Food and Agriculture Organization (FAO) (Guarino et al., 1995). The treaty essentially banned the practice of collecting germplasm without a material transfer agreement mandating that no intellectual property could result. While this treaty was expedient on a world political level, the flow of germplasm from Centers of Origin and Diversity into breeding programs and basic research has been shuttered.

Any estimate of the total germplasm holdings in the many worldwide facilities is difficult to document and is also constantly changing. A very rough estimate of the number of holdings is summarized in Table 7.2. The actual number of holdings is much higher since Table 7.2 only addresses major organizations. Every plant breeder maintains substantial and important individual germplasm collections that are mostly inestimable. It should be emphasized that many of these accessions are duplicated in one or more other facilities. For example, all or nearly all of the collection maintained at the Global Seed Vault on Svalbard Island, Norway consists of redundancies from other facilities. Technology for the precise characterization of genetic identity (see below) will be essential to render these collections as representative and comprehensive as possible.

Preservation of biodiversity has become an issue of broad social concern, especially with the widely accepted tenet that human consumption of fossil fuels is driving atmospheric changes, including elevated CO_2 levels, culminating

TABLE 7.2 A Crude Estimate of the Number of Accessions Held in Major Germplasm Collections Around the World

Country	Center or facility	Species focus	Estimated total # accessions
Norway	GSB/Svalbard	All economic	1,000,000
USA	NPGS/USDA	All economic	460,000
China	Ministry of Ag	All economic	450,000
Russia	Ministry of Ag	All economic	335,000
India	Ministry of Ag	All economic	80,000
Phillipines	IRRI/CGIAR	Rice	86,000
India	ICRISAT/CGIAR	Chickpea, millet sorghum, peanut	86,000
Lebanon[a]	ICARDIA/CGIAR	Cereal grains, forages, legumes	77,000
Mexico	CYMMYT/CGIAR	Wheat, maize	75,000
Columbia	CIAT/CGIAR	Cassava, legumes, forages	66,000
Cameroon	IITA/CGIAR	Cowpea, rice, root crops	40,000
Peru	CIP/CGIAR	Potato, sweet potato	12,000
Total			2,767,000

[a] *ICARDIA temporarily relocated from Aleppo, Syria to Beirut, Lebanon.*

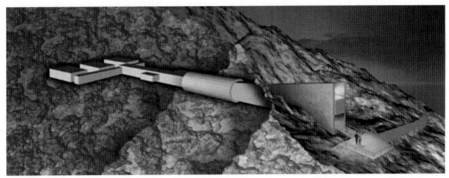

FIG. 7.5 Geographic location and conceptual design of the Global Seed Storage Vault on Svalbard Island, Norway. *Source: Global Crop Diversity Trust.*

in rising global temperature and sea level (Clark et al., 2016). A change in planetary climate caused by the rapid consumption of hydrocarbon-based energy sources has spawned many initiatives about how to alter human activities and cope with prospective irreversible environmental changes. One idea that has already been realized is the Global Seed Vault on Svalbard Island, Norway (Charles, 2006). This facility is designed to utilize the naturally cool and dry environmental conditions that exist deep within mountain terrain in the arctic zone to store germplasm (Fig. 7.5). As of 2017, the Svalbard facility inventory consisted of nearly 1.0 million germplasm accessions, most of which are duplicates of collections held elsewhere on Earth.

BIOTECHNOLOGY TO FOSTER AND CHARACTERIZE PHENOTYPIC DIVERSITY AND GENETIC VARIABILITY

The use of technology to regenerate lost genetic variability (Brown et al., 1997) and the growth in size and species diversity of germplasm collections held around the world has led to logistical challenges. The concept of "core collection" offers one strategy to deal with the ever-increasing theoretical size of germplasm collections (Chavarriaga-Aguirre et al., 1999; Odong et al., 2013). However, the problems of managing increasing species diversity are only just emerging, particularly in the germplasm collections of wild species. Wild collections encompass forest trees, forage plants, medicinal and industrial plants, the wild relatives of cultivated plants, and plant species designated as endangered. Regenerating such diverse collections presents a myriad of problems. Accessions of wild species are generally more costly to regenerate than those of cultivated species. This is due to more complex life histories, multi-faceted breeding systems, ecological interactions, and impeded accessibility. It is worth asking: "why bother to regenerate wild accessions at all?", and instead rely on resampling conserved sources in situ, frozen DNA libraries, or both. Modern empirical studies of the population genetics of wild populations focus on host-pathogen co-evolution, spatial subpopulation structure and dynamics, restricted sampling strategies, breeding system variation, and colonizing history and molecular divergence and phylogeny. Each of these topics has important lessons for the optimum prioritizing of accessions, or the best methods of their regeneration. Optimum regeneration of wild species requires a clear definition of objectives and priorities among accessions, monitoring of mating system and genetic structure, maintenance of accession purity and associated passport data, and biologically realistic and flexible guidelines for sample size.

Technology will continue to impact the nature of germplasm and its utilization by humans. New findings in cell biology and cryobiology will enable longer storage of smaller quantities of materials. Advances in molecular biology, particularly DNA sequencing and quantification of sequence homology, have made it possible to measure genetic variability accurately at the level of informational code. Ultimately, tools will continue to be developed that will make the practice of germplasm acquisition and information management more efficient and effective. For example, existing collections are replete with redundancies that DNA sequence comparisons will readily identify and weed out.

Lewontin and Hartl (1991) were among the first to propose the use of DNA sequence information to measure the degree of genetic variability in the spectrum of biodiversity. Evolution has proceeded through stepwise changes in genomic DNA sequences; hence DNA sequencing has been adopted as a powerful tool to discern evolutionary relationships among species and the underlying mechanisms of evolution (Kubis et al., 1998; Keller and Yi, 2014; Biscotti et al., 2015; Charlesworth and Charlesworth, 2016). Frankham et al. (2010) have presented an excellent primer on the use of molecular markers to measure genetic variability at the level of the individual and population.

This notion was advanced further (Sherwin et al., 2006; Bretting and Wildrlechner, 2010). For example, He et al. (2003) used several measures of allelic polymorphism to characterize *Solanum lycopersicum* (tomato) germplasm, including alleles per locus and average polymorphism information content (PIC; Fig. 7.6). A subsequent study extended this analysis to wild species of *Solanum* (*S. pennellii*, *S. pimpinellifolium*, *S. habrochaites*) related to cultivated tomato (i.e., at least partially cross-compatible), and up to 13 alleles per locus were identified (Frary et al., 2005). Similar studies have been conducted on many other horticultural plant species such as grape *Vitis* spp. (Salmaso et al., 2004), crucifer vegetables *Brassica* sp. (Tonguc and Griffiths, 2004), coffee *Coffea* spp. (Poncet et al., 2004), cassava *Manihot esculenta* (Xia et al., 2005), snap bean *Phaseolus vulgaris* (Blair et al., 2006), kiwifruit *Actinidia* spp. (Fraser et al., 2007), almond *Prunus dulcis* (Sorkheh et al., 2007), carrot *Daucus carota* (Just et al., 2007), cucumber *Cucumis sativus* (Hu et al., 2010), *Rubus* brambles (Castillo et al., 2010), croton *Codiaeum variegatum* (Deng et al., 2010), hops *Humulus lupulus* (Howard et al., 2011), *Citrus* spp. (Amar et al., 2011), rubber *Hevea brasiliensis* (Li et al., 2012), cowpea *Vigna unguiculata* (Gupta et al., 2012), and cypress *Cupressus* spp. (Yang et al., 2016).

Table 7.3 summarizes findings on DNA sequence polymorphism among a diverse group of horticultural crop species. These studies employed a broad range of different marker strategies and targeted different regions of the genome but the results are still remarkably similar. Since genomic DNA sequence variability is easy to find and measure and has obvious utility in quantifying and qualifying germplasm and biodiversity, these studies are having a significant impact on the field and discipline of plant breeding.

Concepts of the determination of phenotype are rapidly changing as well. The traditional paradigm was gene→mRNA transcript→polypeptide→→→phenotype. Inquiry on the origins of phenotypes has spawned the notions of not only genome, but also transcriptome, proteome, metabolome, phenome, and other orders of macrointeractions (Pagel and Pomiankowski, 2008). The phenotype is the key factor that drives evolutionary fitness and economic value. It is certain that the old paradigm that dictates a focus only on genomic DNA sequence is antiquated.

$$PIC = 1 - \sum_{i=1}^{l} p_i^2 - 2\sum_{i=1}^{l} \sum_{j=i+1}^{l} p_i^2 p_j^2$$

FIG. 7.6 The definition of polymorphism information content (PIC); where *l* is the total number of fragments (bands) for an SSR and p_i and p_j are the frequencies of the ith and jth fragment in the populations investigated. A PIC value of 1 indicates that the marker can differentiate each line, and 0 indicates a monomorphic marker. *From Gaitán-Solís, E., Choi, I.-Y., Quigley, C., Cregan, P., Tohme, J., 2008. Single nucleotide polymorphisms in common bean: their discovery and genotyping using a multiplex detection system. Plant Genome 1 (2), 125–134.*

TABLE 7.3 Genomic DNA Sequence Polymorphism Among a Diverse Sample of Horticultural Crop Species

Species	Common name	% polymorphic	Avg PIC	Range PIC	Reference
Asimina triloba	Pawpaw	100.0	0.69	0.29–0.92	Lu et al. (2011)
Brassica oleracea	Cole vegetables	36.1	0.56	0.25–0.86	Tonguc and Griffiths (2004)
Capsicum spp.	Pepper	51.9	0.76	0.31–0.91	Lee et al. (2004)
Carica papaya	Papaya	73.0	0.47	0.08–0.81	de Oliveira et al. (2010)
Chrysanthemum morifolium	Chrysanthemum	90.0	0.99	0.95–0.99	Feng et al. (2016)
Citrullus lanatus	Watermelon	78.3	0.55	0.04–0.91	Liu et al. (2016)
Citrus spp.	Citrus	99.0	0.95	0.89–0.98	Amar et al. (2011)
Codiaeum variegatum	Croton	81.0	0.22	0.15–0.29	Deng et al. (2010)
Corylus avellana	Hazelnut	88.2	0.73	0.29–0.90	Gürcan et al. (2010)
Cucumis sativus	Cucumber	92.9	0.39	0.09–0.75	Hu et al. (2010)
Manihot esculenta	Cassava	73.8	0.55	0.19–0.75	Raji et al. (2009)
Momordica charantia	Bitter gourd	77.8	0.23	0.20–0.25	Gaikwad et al. (2008)
Phaseolus vulgaris	Snap bean	53.0	0.59		Blair et al. (2006)
Prunus dulcis	Almond	96.2	0.71	0.56–0.86	Sorkheh et al. (2007)
Punica granatum	Pomegranate	54.1	0.40		Moslemi et al. (2010)
Rubus spp.	Brambles	88.5	0.55	0.29–0.82	Castillo et al. (2010)
Solanum lycopersicum	Tomato	41.1	0.38	0.09–0.67	He et al. (2003)
Solanum spp.	Wild tomato	92.0[a]	0.64	0.00–0.91	Frary et al. (2005)
Vicia sativa	Vetch	82.1	0.63	0.12–0.96	Cil and Tiryaki (2016)
Vigna unguiculata	Cowpea	36.0	0.34	0.18–0.64	Gupta et al. (2012)

[a] *Wild Solanum spp. as compared to S. lycopersicum.*

The phenotype is determined by a complex interaction of transcribed and untranscribed genomic DNA, RNA transcripts, polypeptides, and metabolites, the latter classes of which may also be trans-generational (i.e., have a heritable component). These new vistas into how phenotypes originate will drive new strategies in germplasm acquisition, evaluation, and utilization by plant breeders (Tanksley and McCouch, 1997).

Can we afford to sacrifice genetic variability that is continually lost to human activity and "natural" causes? Intuitively, it appears that low-stringency systems for the storage of materials that contain information that is both accessible and genetically significant will have the greatest impact on the challenge of germplasm preservation and maintenance. Currently, there is an emphasis on readily accessible forms of genetic variability, such as seeds and sexual propagules. Storage requirements and rapid deterioration make this mode of germplasm handling excessively expensive. Dried tissue samples can be stored for relatively long periods of time without deterioration, as "ancient DNA" has demonstrated. The genetic information contained therein, however, may be adulterated. The additional steps of re-naturation or gene isolation, followed by introduction into the context of a functioning organism are needed. These steps are currently also expensive, but technological advances have made substantial inroads into cost reduction and the lowering of technical barriers.

Will these emerging technologies render our established systems of germplasm, storage of seeds and vegetative organs, perpetual need for revitalization, obsolete? Will the NPGS and World Center systems as they are currently structured go the way of the slide rule and horse and carriage? Yes, but not in the "foreseeable" future. Despite the incredibly rapid advance of the cutting edge of science, the demand for "classical" plant breeding methodologies is likely to persist for decades to come. Large numbers of genes and their primary, tertiary, quaternary, etc. interactions that condition whole-plant phenotypes of economic interest are not yet amenable to elucidation by molecular biology. When we can feed the DNA sequence of a whole genome into a computer that will spit out a functioning hologram of the resulting organism, we will be getting close.

References

Amar, M.H., Biswas, M.K., Zhang, Z., Guo, W., 2011. Exploitation of SSR, SRAP and CAPS-SNP markers for genetic diversity of *Citrus* germplasm collection. Sci. Hortic. 128 (3), 220–227.

Basu, S.K., De, A.K., 2003. *Capsicum*: historical and botanical perspectives. In: De, A.K. (Ed.), The Genus *Capsicum*. Taylor and Francis, London and New York (UK and USA), pp. 1–15.

Beissinger, T.M., Wang, L., Crosby, K., Durvasula, A., Hufford, M.B., Ross-Ibarra, J., 2016. Recent demography drives changes in linked selection across the maize genome. Nat. Plants 2 (16084), 1–7.

Biscotti, M.A., Olmo, E., Heslop-Harrison, J.S., 2015. Repetitive DNA in eukaryotic genomes. Chromosome Res. 23 (3), 415–420.

Blair, M.W., Giraldo, M.C., Buendía, H.F., Tovar, E., Duque, M.C., Beebe, S.E., 2006. Microsatellite marker diversity in common bean (*Phaseolus vulgaris* L.). Theor. Appl. Genet. 113 (1), 100–109.

Bretting, P., Wildrlechner, M., 2010. Genetic markers and plant genetic resource management. Plant Breed. Rev. 13, 11–86.

Briggs, D., Walters, S.M., 1997. Plant Variation and Evolution. Cambridge Univ. Press, Cambridge, UK. 512 pp.

Brown, A.H.D., Brubaker, C.L., Grace, J.P., 1997. Regeneration of germplasm samples: wild versus cultivated plant species. Crop. Sci. 37 (1), 7–13.

Brown, S., Hall, M., Andrasko, K., Ruiz, F., Marzoli, W., Guerrero, G., Masura, O., Dushku, A., DeJong, B., Cornell, J., 2007. Baselines for land-use change in the tropics: application to avoided deforestation projects. Mitig. Adapt. Strat. Glob. Chang. 12 (6), 1001–1026.

Cappaert, D., McCullough, D.G., Poland, T.M., Siegert, N.W., 2005. Emerald ash borer in North America: a research and regulatory challenge. Am. Entomol. 51 (3), 152–165.

Castillo, N.R.F., Reed, B.M., Graham, J., Fernández-Fernández, F., Bassil, N.V., 2010. Microsatellite markers for raspberry and blackberry. J. Am. Soc. Hort. Sci. 135 (3), 271–278.

Charles, D., 2006. A 'forever' seed bank takes root in the arctic. Science 312 (5781), 1730–1731.

Charlesworth, B., Charlesworth, D., 2016. Elements of Evolutionary Genetics. MacMillan Learning. ISBN-10: 0-9815194-2-3. 768 pp.

Chavarriaga-Aguirre, P., Maya, M.M., Tohme, J., Duque, M.C., Iglesias, C., Bonierbale, M.W., Kresovich, S., Kochert, G., 1999. Using microsatellites, isozymes and AFLPs to evaluate genetic diversity and redundancy in the cassava core collection and to assess the usefulness of DNA-based markers to maintain germplasm collections. Mol. Breed. 5 (3), 263–273.

Cil, A., Tiryaki, I., 2016. Sequence-related amplified polymorphism and inter-simple sequence repeat marker-based genetic diversity and nuclear DNA content variation in common vetch (*Vicia sativa* L.). Plant Genet. Resour. 14 (3), 183–191.

Clark, P.U., Shakun, J.D., Marcott, S.A., Mix, A.C., Eby, M., Kulp, S., Levermann, A., Milne, G.A., Pfister, P.L., Santer, B.D., Schrag, D.P., Solomon, S., Stocker, T.F., Strauss, B.H., Weaver, A.J., Winkelmann, R., Archer, D., Bard, E., Goldner, A., Lambeck, K., Pierrehumbert, R.T., Plattner, G., 2016. Consequences of twenty-first-century policy for multi-millennial climate and sea-level change. Nat. Clim. Chang. 6, 360–369.

Cox, T.S., Wood, D., 1999. The nature and role of crop biodiversity. In: Wood, D., Lenné, J.M. (Eds.), Agrobiodiversity: Characterization, Utilization, and Management. CABI Publishing, Cambridge, MA (USA), pp. 35–57.

Darwin, C.R., 1859. On the Origin of Species by Means of Natural Selection, or the Preservation of Favoured Races in the Struggle for Life. John Murray, London, UK. 502 pp.

Dasmann, R.F., 1991. The importance of cultural and biological diversity. In: Oldfield, M.L., Alcorn, J.B. (Eds.), Biodiversity: Culture, Conservation, and Eco-Development. Westview Press, Boulder, CO (USA), pp. 7–15.

Deng, M., Chen, J., Henny, R.J., Li, Q., 2010. Genetic relationships of *Codiaeum variegatum* cultivars analyzed by amplified fragment length polymorphism markers. HortScience 45 (6), 868–874.

Desai, B., 2004. Seeds Handbook: Biology, Production, Processing, and Storage, second ed. Marcel Dekker, Inc., New York, NY, pp. 549–568.

Eubanks, M.W., 2001. The mysterious origin of maize. Econ. Bot. 55 (4), 492–514.

Feng, S., Ren-Feng, H., Meng-Ying, J., Jiang-Jie, L., Xiao-Xia, S., Jun-Jun, L., Zhi-An, W., Hui-Zhong, W., 2016. Genetic diversity and relationships of medicinal *Chrysanthemum morifolium* revealed by start codon targeted (SCoT) markers. Sci. Hortic. 201, 118–123.

Flodin, N.W., 1999. Nutritional influences in the geographic dispersal of Pleistocene man. Ecol. Food Nutr. 38 (1), 71–99.

Ford-Lloyd, B., Jackson, M., 1986. Plant Genetic Resources: An Introduction to Their Conservation and Use. Edward Arnold, Northampton, UK. 162 pp.

Fowler, C., Smale, M., Gaiji, S., 1998. Germplasm Flows Between Developing Countries and the CGIAR: An Initial Assessment. FAO Publication, Vienna, AUS. 25 pp. http://www.fao.org/docs/eims/upload/206939/gfar0065.PDF.

Frankel, O.H., Bennett, E., 1970. Genetic Resources in Plants: Their Exploration and Conservation. Blackwell, Oxford UK. 554 pp.

Frankel, O.H., Brown, A.H.D., Burdon, J.J., 1999. The Conservation of Plant Biodiversity. Cambridge Univ. Press, Cambridge, UK. 299 pp.

Frankham, R., Ballou, J.D., Briscoe, D.A., 2010. Introduction to Conservation Genetics. vol. 2 Cambridge Univ. Press, Cambridge (UK), pp. 41–114.

Frary, A., Xu, Y., Liu, J., Mitchell, S., Tedeschi, E., Tanksley, S., 2005. Development of a set of PCR-based anchor markers encompassing the tomato genome and evaluation of their usefulness for genetics and breeding experiments. Theor. Appl. Genet. 111 (2), 291–312.

Fraser, L.G., McNeilage, M.A., Tsang, G.K., De Silva, H.N., MacRae, E.A., 2007. The use of EST-derived microsatellites as markers in the development of a genetic map in kiwifruit. Acta Hortic. 753, 169–176.

Fuccillo, D., Sears, L., Stapleton, P. (Eds.), 1997. Biodiversity in Trust. Cambridge Univ. Press, Cambridge, UK. 371 pp.

Gaikwad, A.B., Behera, T.K., Singh, A.K., Chandel, D., Karihaloo, J.L., Staub, J.E., 2008. Amplified fragment length polymorphism analysis provides strategies for improvement of bitter gourd (*Momordica charantia* L.). HortScience 43 (1), 127–133.

Ghijsen, H., 2009. Intellectual property rights and access rules for germplasm: benefit or straitjacket? Euphytica 170 (1–2), 229–234.

Gomez, O.J., Blair, M.W., Frankow-Lindberg, B.E., Gullberg, U., 2005. Comparative study of common bean (*Phaseolus vulgaris* L.) landraces conserved *ex situ* in genebanks and *in situ* by farmers. Genet. Resour. Crop. Evol. 52 (4), 371–380.

Grant, P.R., 1991. Natural selection and Darwin's finches. Sci. Am. 265, 82–87 (October 1999).

Guarino, L., Ramantha Rao, V., Reid, R., 1995. Collecting Plant Genetic Diversity: Technical Guidelines and Legal Issues in Plant Germplasm Collecting. CAB International, Rome, Italy, pp. 13–30.

Guo, Y., Chen, S., Li, Z., Cowling, W.A., 2014. Center of origin and centers of diversity in an ancient crop, *Brassica rapa* (turnip rape). J. Hered. 105 (4), 555–565.

Gupta, S.K., Bansal, R., Gopalakrishna, T., 2012. Development of intron length polymorphism markers in cowpea [*Vigna unguiculata* (L.) Walp.] and their transferability to other *Vigna* species. Mol. Breed. 30 (3), 1363–1370.

Gürcan, K., Mehlenbacher, S.A., Botta, R., Boccacci, P., 2010. Development, characterization, segregation, and mapping of microsatellite markers for European hazelnut (*Corylus avellana* L.) from enriched genomic libraries and usefulness in genetic diversity studies. Tree Genet. Genomes 6 (4), 513–531.

Haack, R.A., Hérard, F., Sun, J., Turgeon, J.J., 2009. Managing invasive populations of Asian long-horned beetle and citrus long-horned beetle: a worldwide perspective. Annu. Rev. Entomol. 55, 521–546.

Harlan, J.R., 1951. The anatomy of gene centers. Am. Nat. 85, 97–103.

Harlan, J.R., 1976. The plants and animals that nourish man. Sci. Am. 235, 88–97.

He, C., Poysa, V., Yu, K., 2003. Development and characterization of simple sequence repeat (SSR) markers and their use in determining relationships among *Lycopersicon esculentum* cultivars. Theor. Appl. Genet. 106 (2), 363–373.

Hoisington, D., Khairallah, M., Reeves, T., Ribaut, J.-M., Skovmand, B., Taba, S., Warburton, M., 1999. Plant genetic resources: what can they contribute toward increased crop productivity? Proc. Natl. Acad. Sci. U. S. A. 96 (11), 5937–5943.

Hooftman, D.A.P., Edwards, B., Bullock, J.M., 2016. Reductions in connectivity and habitat quality drive local extinctions in a plant diversity hotspot. Ecography 39 (6), 583–592.

Howard, E.L., Whittock, S.P., Jakše, J., Carling, J., Matthews, P.D., Probasco, G., Henning, J.A., Darby, P., Cerenak, A., Javornik, B., Kilian, A., Koutoulis, A., 2011. High-throughput genotyping of hop (*Humulus lupulus* L.) utilising diversity arrays technology (DArT). Theor. Appl. Genet. 122 (7), 1265–1280.

Hu, J., Zhou, X., Li, J., 2010. Development of novel EST-SSR markers for cucumber (*Cucumis sativus*) and their transferability to related species. Sci. Hortic. 125 (3), 534–538.

Jaenicke-Despres, V., Buckler, E.S., Smith, B.D., Gilbert, M.T.P., Cooper, A., Doebly, J., Paabo, S., 2003. Early allelic selection in maize as revealed by ancient DNA. Science 302 (5648), 1206–1208.

Just, B.J., Santos, C.A.F., Fonseca, M.E.N., Boiteux, L.S., Oloizia, B.B., Simon, P.W., 2007. Carotenoid biosynthesis structural genes in carrot (*Daucus carota*): isolation, sequence-characterization, single nucleotide polymorphism (SNP) markers and genome mapping. Theor. Appl. Genet. 114 (4), 693–704.

Keller, T.E., Yi, S.V., 2014. DNA methylation and evolution of duplicate genes. Proc. Natl. Acad. Sci. U. S. A. 111 (16), 5932–5937.

Krishnan, P., Nagarajan, S., Moharir, A.V., 2004. Thermodynamic characterisation of seed deterioration during storage under accelerated ageing conditions. Biosyst. Eng. 89 (4), 425–433.

Kubis, S., Schmidt, T., Heslop-Harrison, J.S., 1998. Repetitive DNA elements as a major component of plant genomes. Ann. Bot. 82, 45–55.

Lee, J.M., Nahm, S.H., Kim, Y.M., Kim, B.D., 2004. Characterization and molecular genetic mapping of microsatellite loci in pepper. Theor. Appl. Genet. 108 (4), 619–627.

Lewontin, R.C., Hartl, D.L., 1991. Population genetics in forensic DNA typing. Science 254 (5039), 1745.

Li, D., Xia, Z., Deng, Z., Liu, X., Dong, J., Feng, F., 2012. Development and characterization of intron-flanking EST-PCR markers in rubber tree (*Hevea brasiliensis* Muell. Arg.). Mol. Biotechnol. 51 (2), 148–159.

Lin, Z., Liu, H., 2006. How species diversity responds to different kinds of human-caused habitat destruction. Ecol. Res. 21 (1), 100–106.

Liu, G., Xu, J.H., Zhang, M., Li, P.F., Yao, X.F., Hou, Q., Zhu, L.L., Ren, R.S., Yang, X.P., 2016. Exploiting Illumina sequencing for the development of InDel markers in watermelon (*Citrullus lanatus*). J. Hortic. Sci. Biotechnol. 91 (3), 220–226.

Loskutov, I.G., 1999. Vavilov and His Institute: A History of the World Collection of Plant Genetic Resources in Russia. International Plant Genetic Resources Institute, Rome, Italy. 188 pp.

Lu, L., Pomper, K.W., Lowe, J.D., Crabtree, S.B., 2011. Genetic variation in pawpaw cultivars using microsatellite analysis. J. Am. Soc. Hort. Sci. 136 (6), 415–421.

Matsuoka, Y., Vigouroux, Y., Goodman, M.M., 2002. A single domestication for maize shown by multilocus microsatellite genotyping. Proc. Natl. Acad. Sci. U. S. A. 99 (9), 6080–6084.

Mavlyanova, R.F., Abdullaev, F.K., Khodjiev, P., Zaurov, D.E., Molnar, T.J., Goffreda, J.C., Orton, T.J., Funk, C.R., 2005. Plant genetic resources and scientific activities of the Uzbek Research Institute of Plant Industry. HortScience 40 (1), 10–14.

McFerson, J.R., 1998. From in situ to ex situ and back: the importance of characterizing germplasm collections. HortScience 33 (7), 1134–1135.

Moslemi, M., Zahravi, M., Khaniki, G.B., 2010. Genetic diversity and population genetic structure of pomegranate (*Punica granatum* L.) in Iran using AFLP markers. Sci. Hortic. 126 (4), 441–447.

Odong, T.L., Jansen, J., van Eeuwijk, F.A., van Hintum, T.J.L., 2013. Quality of core collections for effective utilisation of genetic resources review, discussion and interpretation. Theor. Appl. Genet. 126 (2), 289–305.

Oldfield, M.L., 1989. The Value of Conserving Genetic Resources. Sinauer Assoc, Inc., Sunderland, MA (USA), pp. 12–52.

de Oliveira, E.J., Amorim, V.B.O., Matos, E.L.S., Costa, J.L., da Silva Castellen, M., Padua, J.G., Dantas, J.L.L., 2010. Polymorphism of microsatellite markers in papaya (*Carica papaya* L.). Plant Mol. Biol. Report 28 (3), 519–530.

Pagel, M., Pomiankowski, A., 2008. The organismal prospect. In: Pagel, M., Pmiankowski, A. (Eds.), Evolutionary Genomics and Proteomics. Sinauer Assoc. Inc. Publ., Sunderland, MA (USA), pp. 1–10.

Plucknett, D.L., Smith, N.J.H., Williams, J.T., Anishetty, N.M., 1987. Gene Banks and the World's Food. Princeton Univ. Press, Princeton, NJ (USA). 247 pp.

Poncet, V., Hamon, P., Minier, J., Carasco, C., Hamon, S., Noirot, M., 2004. SSR cross-amplification and variation within coffee trees (*Coffea* spp.). Genome 47 (6), 1071–1081.

Raji, A.A.J., Anderson, J.V., Kolade, O.A., Ugwu, C.D., Dixon, A.G.O., Ingelbrecht, I.L., 2009. Gene-based microsatellites for cassava (*Manihot esculenta* Crantz): prevalence, polymorphisms, and cross-taxa utility. BMC Plant Biol. 9 (118), 1–11.

Razdan, M.K., Cocking, E.C., 1997. Biotechnology in conservation of genetic resources. In: Razdan, M.K., Cocking, E.C. (Eds.), Conservation of Plant Genetic Resources In Vitro. General Aspects, vol. 1. Science Publ., Inc., Enfield, NH (USA), pp. 3–25.

Riordan, E.C., Gillespie, T.W., Pitcher, L., Pincetl, S.S., Jenerette, G.D., Pataki, D.E., 2015. Threats of future climate change and land use to vulnerable tree species native to Southern California. Environ. Conserv. 42 (2), 127–138.

Roos, E.E., 1989. Long-term seed storage. In: Janick, J. (Ed.), Plant Breeding Reviews. vol. 7. Timber Press, Portland, OR (USA), pp. 129–158.

I. ELEMENTS AND UNDERPINNINGS OF PLANT BREEDING

Salmaso, M., Faes, G., Segala, C., Stefanini, M., Salakhutdinov, I., Zypirian, E., Toepfer, R., Grando, M.S., Velasco, R., 2004. Genome diversity and gene haplotypes in the grapevine (*Vitis vinifera* L.), as revealed by single nucleotide polymorphisms. Mol. Breed. 14 (4), 385–395.

Schlottfeldt, S., Walter, M., Emília, M.T., de Carvalho, A.C.P.L.F., Soares, T.N., Telles, M.P.C., Loyola, R.D., Diniz-Filho, J.A.F., 2015. Multi-objective optimization for plant germplasm collection conservation of genetic resources based on molecular variability. Tree Genet. Genomes 11 (2), 836.

Sherwin, W.B., Jabot, F., Rush, R., Rossetto, M., 2006. Measurement of biological information with applications from genes to landscapes. Mol. Ecol. 15 (10), 2857–2869.

Simpson, R.D., Sedjo, R.A., 1998. The value of genetic resources for use in agricultural improvement. In: Evenson, R.E., Gollin, D., Santaniello, V. (Eds.), Agricultural Values of Plant Genetic Resources. CABI Publ., New York, NY (USA), pp. 55–66.

Smith, C.W., 2004. Corn: Origin, History, Technology, and Production. John Wiley & Sons, New York. 949 pp.

Sorkheh, K., Shiran, B., Gradziel, T.M., Epperson, B.K., Martínez-Gómez, P., Asadi, E., 2007. Amplified fragment length polymorphism as a tool for molecular characterization of almond germplasm: genetic diversity among cultivated genotypes and related wild species of almond, and its relationships with agronomic traits. Euphytica 156 (3), 327–344.

Stoner, A., Hummer, K., 2007. 19th and 20th century plant hunters. HortScience 42 (2), 197–199.

Swanson, T., 1998. The source of genetic resource values and the reasons for their management. In: Evenson, R.E., Gollin, D., Santaniello, V. (Eds.), Agricultural Values of Plant Genetic Resources. CABI Publ., New York, NY (USA), pp. 67–81.

Tanksley, S.D., McCouch, S.R., 1997. Seed banks and molecular maps: unlocking genetic potential from the wild. Science 277 (5329), 1063–1066.

Thompson, P., Harris, S., 2010. Seeds, Sex and Civilization: How the Hidden Life of Plants has Shaped Our World. Thames & Hudson, New York, NY (USA). 272 pp.

Tonguc, M., Griffiths, P.D., 2004. Genetic relationships of *Brassica* vegetables determined using database derived simple sequence repeats. Euphytica 137 (2), 193–201.

Van den Berg, R.D., Feinstein, O.N., 2009. Evaluating Climate Change and Development. Transaction Publishers, New Brunswick, NJ (USA). 438 pp.

Van Hintum, T.J.L., 1995. Hierarchical approaches to the analysis of genetic diversity in crop plants. In: Hodgkin, T., Brown, A.H.D., Van Hintum, T.J.L., Morales, E.A.V. (Eds.), Core Collections of Plant Genetic Resources. John Wiley & Sons, New York, NY (USA), pp. 23–34.

Vavilov, N.I., 1951. The Origin, Variation, Immunity, and Breeding of Cultivated Plants. Transl. by K. Starr Chester, Selected Writings, Ronald Press, New York, NY (USA). 364 pp.

Ward, K., Gisler, M., Fiegener, R., Young, A., 2008. The Willamette Valley seed increase program developing genetically diverse germplasm using an ecoregion approach. Nativ. Plants J. 9 (3), 334–350.

White, G.A., Shands, H.L., Lovell, G.R., 1989. History and operation of the National Plant Germplasm System. In: Janick, J. (Ed.), Plant Breeding Reviews. vol. 7. Timber Press, Portland, OR (USA), pp. 5–56.

Wilkes, G., 1991. In situ conservation of agricultural systems. In: Oldfield, M.L., Alcorn, J.B. (Eds.), Biodiversity: Culture, Conservation, and Eco-Development. Westview Press, Boulder, CO (USA), pp. 86–101.

Xia, L., Peng, K., Yang, S., Wenzl, P., de Vicente, M.C., Fregene, M., Kilian, A., 2005. DArT for high-throughput genotyping of cassava (*Manihot esculenta*) and its wild relatives. Theor. Appl. Genet. 110 (6), 1092–1098.

Yang, H., Zhang, R., Jin, G., Feng, Z., Zhou, Z., 2016. Assessing the genetic diversity and genealogical reconstruction of cypress (*Cupressus funebris* Endl.) breeding parents using SSR markers. Forests 7 (160), 1–14.

Yonezawa, K., Nomura, T., Morishima, H., 1995. Sampling strategies for use in stratified germplasm collections. In: Hodgkin, T., Brown, A.H.D., Van Hintum, T.J.L., Morales, E.A.V. (Eds.), Core Collections of Plant Genetic Resources. John Wiley & Sons, New York, NY (USA), pp. 35–53.

Further Reading

Anon., 1972. Genetic Vulnerability of Major Crops. National Academy of Sciences, USA, Washington, DC (USA). 307 pp.

8

Enhancement of Germplasm

INTRODUCTION

If nature does not provide enough of what humans need it is human nature to develop ways to find or manufacture more. Demand spawns efforts to create supply. Such is the case with plant germplasm. Most of biota on planet Earth has been scrutinized intensively by humans over the past 20,000+ years for species that may be domesticated and used as sources of food, drugs, fibers, fuel, and aesthetic pleasure. The earliest plant breeders, the farmers, eventually exhausted available genetic variability within the populations they cultivated and selected. Considering all of the potential germplasm available on Earth to present-day plant breeders, and over 3 billion years of evolutionary history that created and selected for trillions of mutations, it is hard to believe that the range of extant variability currently on the planet would not include genes of interest for all potential breeding objectives. The cumulative forces of recombination, segregation, and selection have pushed the gene pools of economic plant populations to the limits of performance, especially in the major grain crops. Species are constantly adapting to an ever-changing environment, and new strains of biotic enemies constantly appear.

After all accessible germplasm has been thoroughly tested for the presence of genes that might enhance plant performance, but none are found, what if anything can be done? Naturally-occurring genetic variability appears at a relatively slow rate relative to the needs of plant breeders. One approach is to accelerate the rate of appearance of genetic variability. Technological advances have also made it possible to explore entirely novel ways to expand gene pools. Molecular biologists have discovered phenomena wherein events trigger the induction of new genetic or phenotypic variability or, partly due to increased rates of transposon activity, but the underlying mechanisms are ill-understood (Pace and Feschotte, 2007). We are only beginning to understand how newly appearing functional genes have evolved under natural conditions (Fan et al., 2008). Methods to enhance genetic variability at present are relatively crude and limited in scope. New discoveries in plant molecular biology are, however, accelerating the pace of improved tools and methods to change the expression and function of genes.

Ideally, the plant breeder can expand the pool of genes accessible for plant improvement to include all species, not limited only to the range of reproductive compatibility. All eukaryotes have a common evolutionary ancestor and a consequence of this process is that genes with similar genomic DNA sequences tend to exhibit analogous functions across a broad spectrum of diverse species (Koonin, 2005). If a species targeted for improvement lacks an attribute that another species possesses it is logical, therefore, to presume that the genes responsible for the phenotype could be moved from one species to the other and still give rise to the same desirable phenotype.

PLOIDY CHANGES AND CHROMOSOME ENGINEERING

Natural processes that mitigate the structure and function of higher plant nuclear genomes were reviewed in Chapter 3. The structure and number of chromosomes are in a constant state of flux. The presumed "mistakes" that lead to partial genomic duplications, deletions, translocations and inversions, and the heritable chromosome structural changes they contribute to, may be manifestations of the natural evolutionary phenomena that generate new variability. This may also be the case for chromosome *endo*-reduplication and altered spindle orientations that lead to polyploidization, and interspecific hybridizations that lead to multiple spindles and subsequent aneuploidy, amphiploidy, and haploidy (Darlington and Thomas, 1937; Elliott, 1958).

Chromosome structural change is a major factor involved with the speciation process (Lagercrantz, 1998; Schubert, 2007; Schubert and Lysak, 2011). Speciation can be viewed as a way to generate and perpetuate overall genetic variability. For example, humans have become habituated to an omnivorous diet. Today, we consume wheat, eggs, tuna fish, tomatoes, lettuce, canola oil; tomorrow we consume corn, beans, chicken, peppers, avocado, and onions. These are all different species that were domesticated to fulfill different human food needs. Plant breeding is currently practiced almost entirely within and among existing species. Could the processes that drive speciation be accelerated to the point that new species is a viable strategy for attaining better economic plant populations? If that is too ambitious, perhaps intermediate approaches such as polyploids, aneuploids, and amphiploids could contribute to the fulfillment of breeding objectives?

Polyploids appear frequently in the plant kingdom but do not persist in evolutionary time, and are perhaps intermediate stages in the evolutionary process (Luo et al., 2009; Lysak, 2014). Eukaryotes aspire to diploid genomic behavior and polyploids usually become "diploidized" over evolutionary time (Mandáková et al., 2010). For example, chromosome instability and rearrangements were observed in an artificial allotetraploid followed by progressively more stability and fewer rearrangements in subsequent generations (Xiong et al., 2011).

The question "if two gene copies enhances a targeted phenotype, why wouldn't three, four, or more (copies) be as good or even better?" was posed in Chapter 3. One answer to this question is provided by the accumulated observations of the comparative performance of diploids and corresponding autopolyploids induced artificially by the microtubulin-binding alkaloid colchicine. The performance of induced polyploids is usually less than the corresponding ancestral diploids, but there are many exceptions to this generalization (Sybenga, 1992).

The preponderance of evidence supports the conclusion that, despite extra copies of expressed and regulatory genes, autotetraploids are not necessarily larger or stronger than their diploid counterparts (Renny-Byfield and Wendel, 2014). Are there any other characteristics imparted by polyploidy that lead us to conclude that the condition is comparatively superior to diploidy? Are polyploids, for example, more tolerant to pests, diseases, or environmental stresses? Polyploids have, in fact, been found to be generally more tolerant of biotic and abiotic insults than are corresponding diploids (Sattler et al., 2016). While most individual systems operate as would be predicted from an understanding of the diploid, self-incompatibility (see earlier discussion) is a possible exception. Sporophytic self-incompatibility may operate differently in diploids and polyploids even though the underlying alleles are still present (Mable et al., 2004). Self-incompatibility apparently depends not only on SI genotype but also gene dosage and modifiers (Chen, 2007). With regard to economic crops that depend on reproduction for their value, such as the major grain and fruit crops, polyploids are generally undesirable because normal sexual functions leading to seed yield is impaired (see Chapter 3).

Are there any possible circumstances at all wherein the plant breeder would want to utilize polyploidy? One example would be the exceptional cases where polyploids perform at a higher level than the corresponding diploids. Plant organs in polyploids are often larger than those of diploids, and this would be desirable in cases where the organ has economic value, for example, fruit (Zeldin and McCown, 2002) and root (Hahn et al., 1994). Polyploids offer the possibility of engineering new genomic configurations with multiple alleles at specific loci, culminating in "fixed heterosis" (Ortiz, 1997). Another beneficial use of polyploids is to provide a "bridge" for the transition of desirable genes from sexually-compatible alloploid wild species to a cultivated diploid (Pertuze et al., 2003; Sattler et al., 2016). He et al. (2017) reported that genome structural rearrangement occurs frequently in polyploid crop species, and the transfer of genome sequences between ancestral parents is part of this process.

The most prominent example of the practical use of autopolyploidy in horticultural crop species is seedless fruit. From a human culinary standpoint, seeds often get in the way of a pleasurable eating experience. For example, the fruit of diploid banana is full of hard seeds as compared with the contemporary seedless varieties that most of the world is familiar with. Seedless bananas are sexually-sterile triploids. Since three genomes will not segregate evenly during meiosis the developing gametophytes do not receive full haploid genomes leading to seed abortion and absence of seeds in banana fruits (Heslop-Harrison, 2011).

The fruit organ was not meant to be devoid of seeds since it evolved mainly to attract seed dispersal vectors and to protect seeds from environmental fluxes. Development of fruit and seed is strongly coupled to the extent that fruits do not usually complete development unless sexual reproduction has been consummated resulting in the next sporophyte generation, that is encapsulated in seeds as dormant embryos. In certain species or mutants, however, fruit and seed development are uncoupled, a condition known as *parthenocarpy*.

Triploid plants can usually grow and develop normally but cannot generate functional gametes. If a triploid plant is also parthenocarpic, it is possible that seedless fruits will develop and ripen. As a further complication, developing seeds often secrete phytohormones that stimulate fruit growth. Thus, larger numbers of seeds will drive the development of larger fruits and the complete lack of seeds will lead to correspondingly smaller fruits (Bolmgren and Eriksson, 2010). It is possible to counteract small fruit size in seedless cultivars by supplying the phytohormone exogenously, for example, cytokinins and gibberellic acid (Hayata et al., 1995; Chao et al., 2011).

How can one obtain a triploid? Answer - by simple mathematics: 2 + 1; or a 2× gamete fused with a 1× gamete. Theoretically, an autotetraploid will give rise to a 2× gamete that, when fertilized with a 1× = 1n gamete from a diploid, will produce a triploid. The plant breeder uses colchicine to obtain a 4× plant from a 2× then crosses the 4× with a 2× to get a 3× that is sterile and does not produce seeds, but will in some cases produce parthenocarpic fruit. This has been accomplished commercially in banana (Heslop-Harrison, 2011), navel orange (Davies, 1986), and watermelon (Mohr, 1986). Seedlessness has been reported in many other horticultural crop species (*Cucumis melo*), but this is usually due to a mutation affecting seed development following pollination (Hayata et al., 2001). Seedless grapes exemplify an alternative genetic strategy: mutations that result in seed abortion (e.g., stenospermocarpic) prior to the deposition of the seed coat (Karaagac et al., 2012). Recent research results have implicated the MADS-box *AGL11* gene in seed development, and speculated that manipulating the expression of orthologs of this gene could lead to seedless fruits in a broad spectrum of horticultural species (Ocarez and Mejía, 2016).

Other uses for induced polyploidy exist within the realm of plant breeding methodology, for example, the combination and bridging of distinct gene pools (Tamayo-Ordóñez et al., 2016). Classical cytogenetic experiments were conducted wherein wheat was hybridized with different diploid *Triticum* species (Sears, 1969). Hybridization between species, also referred to as *interspecific hybridization* or *wide crosses*, has been demonstrated to be more easily accomplished in the family Gramineae than in most other plant families. Sometimes it is necessary, however, to "rescue" the embryo from a developing seed that is aborting due to endosperm incompatibility. Embryo rescue is accomplished by culturing the aborting embryo on nutrient medium allowing growth and development to continue (Mii, 2012).

When one entire homologous pair in the economic parent is supplanted by the homologous pair of homeologues from the wild parent, the entity is called a *substitution line* (Fig. 8.1). When pairs of chromosomes are found along with a whole genomic set from the economic parent, the entity is called an *addition line* (Ladizinsky, 1992; Dhaliwal, 1992). Both addition and substitution lines are examples of aneuploidy. While these genetic entities are interesting and can yield information on gene mapping and function, they rarely contribute anything of practical significance (Ji and Chetelat, 2003; Schauer et al., 2008). Adding or substituting chromosomes is sort of like switching or adding body parts between a human and an ape; the nerves and endocrine functions would not necessarily connect the same way. Substitution lines likely have duplicated or missing genetic information, while addition lines will have extra copies of some genes, but not most others (Hermsen, 1992; Kuckuck et al., 1991).

The biological processes of reproductive isolation and speciation were addressed earlier in this textbook (Chapter 2). These processes present intractable barriers to the unrestricted transfer of genes among plants, animals, and microbes. Undaunted, many plant breeders continue to try to make wider and wider crosses and sometimes succeed in spite of the textbooks (Khush and Brar, 1992). If sexual hybridization is unsuccessful, there are several alternative technical approaches to achieving gene flow. One example is the *somatic hybrid*, where the rigid polysaccharide walls are enzymatically removed from cells of different plant species then fused by chemical or electrical forces (Fig. 8.2). This approach has not led to any commercial successes to date, for example, the ill-fated pomato (potato + tomato; Melchers et al., 1978) or eggato (eggplant + tomato; Samoylov et al., 1996).

If an *amphiploid* (syn *hemizygous; Fig. 8.2*) hybrid can be successfully synthesized and grown to sexual maturity, it will most likely be sterile, unable to produce functioning gametes. The reason for infertility is likely chromosome pairing abnormalities during meiosis. It may be possible to restore fertility by simply doubling the chromosome number of the interspecific hybrid (i.e., an *amphidiploid*; see first three steps of Fig. 8.1). Thus, each distinct chromosome will now have a homologue with which to pair, and meiosis I can, theoretically, proceed to the formation of viable gametophytes.

A good example is the case of Triticale (x*Triticosecale Wittmack*), a new allopolyploid (2n = 6× = 42; AABBRR; see Fig. 8.3) crop species that was developed from the interspecific hybrid of 4× durum wheat (*T. turgidum* 2n = 28);

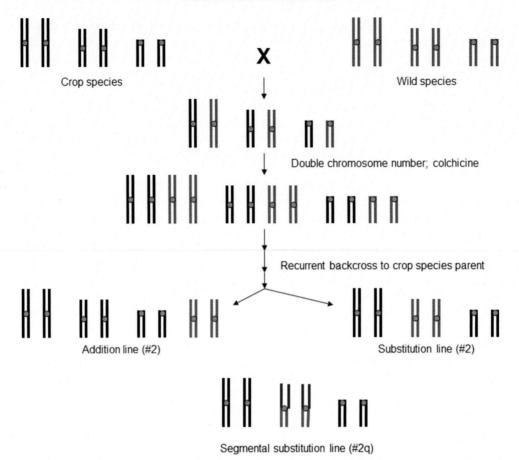

FIG. 8.1   Use of interspecific hybridization to create chromosome addition and substitution lines.

Plant Protoplast Production and Fusion

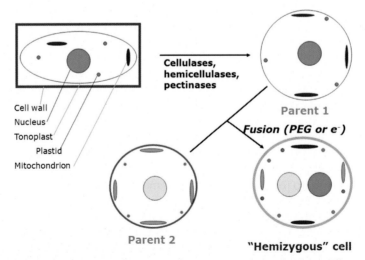

FIG. 8.2   The process of somatic hybridization: Protoplasts are produced from somatic cells of two different plant species then induced to fuse by disrupting cell membranes chemically (polyethylene glycol or PEG) or with an electric field (e⁻).

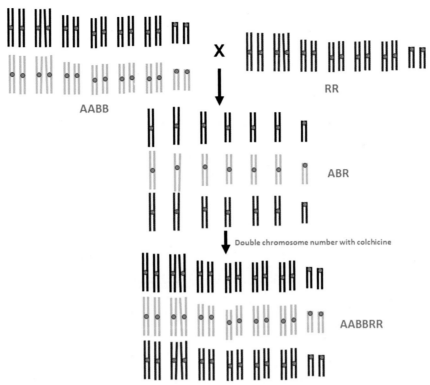

FIG. 8.3 Synthesis of x*Triticosecale* by crossing durum wheat (AABB) with rye (RR) then doubling the chromosome number of the amphiploid hybrid (ABR) to achieve a sexually fertile genotype (AABBRR).

AABB) and rye (*Secale cereale* 2n = 2x = 14; RR) (Tsen, 1974). The original goal in the development of triticale was to combine the endosperm characteristics of durum wheat with the cold-hardiness of rye, such that glutinous flour could be produced in subarctic latitudes (Rajathy, 1977). The resulting entity has met with limited success following its introduction in the 1950s amid sensational fanfare, and the crop is still grown in northern Canada. A small sustained market has also emerged for triticale flour that has certain unique flavor, texture, and baking attributes.

Most plant breeders are understandably daunted by the prospects and challenges that attend the development of an entirely new crop species, such as triticale. Interspecific hybridization, however, has also proven to be an effective intermediate for the transfer of genes between distinct gene pools. Such a program generally begins with a cross between an economic species and a wild relative that possesses the desired attribute not present in the domesticated gene pool. The resulting amphidiploid may be crossed recurrently to the economic plant species and the resulting population regains acceptable sexual fertility (*backcross*; the subject of Chapter 18). Cytogenetic observations of metaphase I in the amphidiploid usually show that bivalent pairings are present and that viable gametes are produced, but these compound genomes are often unstable and individual chromosomes are sometimes lost (Liu and Li, 2007; Tu et al., 2009). With each successive backcross to the economic parent, more fertility and characteristics of the economic parent are usually recovered.

If the recurrent backcross program is successful the economic parent is reconstituted with the simple addition of the desired attribute from the wild relative (see Fig. 8.1). It is not surprising that simply inherited traits tend to be the most readily adaptable to this approach, since the meiotic transfer of small homeologous chromosome segments from the wild species to chromosomes of the economic species must occur (segmental substitution line; Fig. 8.1).

The use of colchicine to increase ploidy levels in eukaryotes is very common. Ploidy reduction, however, is not so easily accomplished. Two primary methods have been developed, both of which incorporate a subversion of natural processes. The first strategy is to trick gametophytes into behaving like sporophytes and the second is to synthesize a genetically-unstable interspecific hybrid (Chen and Hayes, 1992). Unstable interspecific hybrids sometimes culminate in haploid plants of one of the parents, not usually either or both. Biological barriers and technical challenges have limited the availability of haploid plants to relatively few economic plant species: wheat, rice, canola, cabbage, barley, tobacco, peppers (Germanà, 2011). Plant breeders use haploidy and dihaploidy mainly as tools to facilitate plant breeding, and not to directly enhance gene pools. Methods to produce haploids and uses of haploids in plant breeding will be covered in more detail in a later chapter (Chapter 14).

INDUCED MUTATIONS

A *mutation* is a naturally-occurring or induced change to an organism that is inherited by its progeny. A mutation may result in a corresponding measurable phenotypic change or may be *silent* (i.e., no observable phenotype). Mutations represent new sources of genetic variability and may contribute to the enhancement of germplasm. The notion has been advanced, therefore, that mutations may be utilized as germplasm in plant breeding programs. Several excellent monographs on the subject of mutations and ways to produce them appeared during the 1950s and 1960s, but they are wrought with details on technical aspects of mutagenic agents and exposure. Much of the research on mutagenesis during the early-mid 20th century was focused on ionizing radiation, whereas more recent efforts have preferentially involved chemical mutagens. A brief and cogent review of mutagenesis and mutation breeding is provided in the textbook by Mayo (1987).

Naturally-occurring mutations were first addressed in Chapter 3 within the context of population genetics. One important class is referred to as "point mutations" including genomic DNA nucleotide base changes, duplications, and deletions, but also multi-base changes that alter gene function, but not gross chromosome structure and number (Fig. 8.4). About 99% of all point mutations result in recessive alleles so are not directly observed in the diploid organisms in which they occur (Gottschalk and Wolff, 1983). They are only observed in homozygous recessive individuals that occur by inbreeding or chance in subsequent generations.

"Natural" mutations occur spontaneously, induced by environmental factors that we generally consider to be "normal" (e.g., subatomic particles and incident electromagnetic radiation; naturally-occurring and synthetic chemicals and substances). Moreover, natural, or spontaneous, mutations occur at a mostly consistent rate per locus and cell division cycle: 10^{-6} to 10^{-9} (Borojević, 1990). Different phenotypic traits are sometimes characterized by different mutation rates. For example in corn, the spontaneous *mutation rate* for colorless aleurone is $\sim 5 \times 10^{-4}$, whereas shrunken endosperm is $\sim 1 \times 10^{-6}$ (Stadler, 1942).

A given individual of a multicellular eukaryotic species, therefore, inevitably contains many mutations and also many cells that contain new mutations. Animal immune systems eliminate those that are identified as foreign due to new epitopes, and most mutant cells are not present in the germline, or somatic cell line that leads to gametes, so are not passed along to progeny (Fig. 8.5). It is incorrect to consider naturally-occurring mutations as "mistakes", since mitosis and meiosis would have much higher genetic fidelity if they were so deleterious. Rather, it is more appropriate to regard mutations as a natural mechanism by which new genetic variability is generated, and that may contribute positively to the gene pools of organisms both under cultivation and in the wild.

Somatic mutations are sometimes visible as "sectors" in a "normal" background of cells, the combination of which is called a *chimera* (Fig. 8.6). Depending on where and when the mutation occurs, distinct types of chimeras may be distinguished: periclinal, mericlinal, sectoral (Fig. 8.7). This is easily seen for pigment genes in plants with showy flowers, where sectoral chimeras often have commercial value, with intricate displays of alternating colors (Konzak, 1984). In species where techniques of asexual propagation are well developed, it may be possible to excise and propagate the mutant sector from the chimera and obtain uniform populations from a mutant plant, called a *sport*.

Certain of these floral sectors occur at a much higher frequency than most mutations and have been demonstrated to be a consequence of mobile DNA, or transposable, elements, that move about the genome when stimulated by

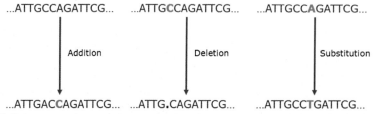

Types of Mutations

- **Point:** Nucleotide base additions, deletions, substitutions
- **Generally recessive** (loss of function): B → b allelic form of gene

...ATTGCCAGATTCG...   ...ATTGCCAGATTCG...   ...ATTGCCAGATTCG...

Addition   Deletion   Substitution

...ATTGACCAGATTCG...   ...ATTG.CAGATTCG...   ...ATTGCCTGATTCG...

FIG. 8.4   Types of point mutations: nucleotide base addition, deletion and substitution.

Developmental dynamics of mutations

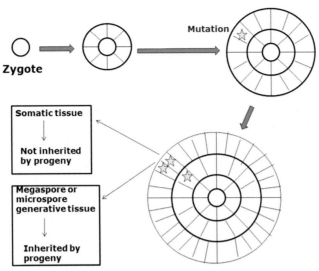

FIG. 8.5 The concept of "germline" (the somatic cell lineage leading to gametes) and "somatic line" (somatic cell lineages not leading to gametes) and the consequences to the inheritance of new mutations by progeny.

FIG. 8.6 An example of a sectoral chimera for flower color in *Azalea*.

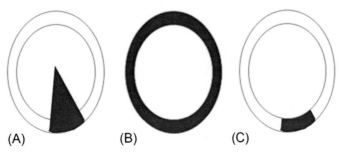

FIG. 8.7 (A) Sectoral chimera (trans L1 L2 L3 layers); (B) Periclinal chimera (L1); (C) Mericlinal chimera (L1).

stress or developmental stage, and consequently alter gene function (Bennetzen, 2000). Early plant breeders propagated and perpetuated these sports as whole plants and populations. It is probable that somatic mutations played an important role in the domestication of woody perennial plants such as tree and vineyard fruits, where the sport may be selected, excised, and propagated directly.

Mutations lead directly to the expansion of gene pools for the species in which they occur. Most naturally-occurring point mutations (Fig. 8.4) have been shown to convert functional into non-functioning genes (Muller, 1927; Stadler, 1954). Consequently, most new mutations also behave recessively in genetic studies. Mutation processes, driven by the environmental insult to DNA and/or mistakes in the mechanisms of DNA replication, are considered to be mostly random (Gustafsson, 1947). Base duplications or deletions that do not alter the sequence by a multiple of 3 nucleotide base pairs are frame-shift mutations that usually truncate gene activity altogether due to shortened or nonsense transcripts (Gustafsson and Tedin, 1954). Base changes may or may not change the amino acid at the corresponding position in the polypeptide, but most changes to amino acids having different biochemical properties would likely not enhance gene function in the short term. Rarely, a point mutation results in demonstrably enhanced performance; the amino acid change makes the protein function better, or in a different way that benefits the plant (or human perception of plant performance). Alternatively, the point mutation may occur in the associated control sequences of a gene, and the resulting change in gene expression may alter phenotype in the desired fashion.

In between gross chromosome structural changes and point mutations lies a realm of heritable fluxes that defies classification: changes in the number of tandem repeat sequences in non-transcribed regions, simple sequence repeats (SSRs), and a host of sequence alterations involving more than one nucleotide base in the DNA and too small to see microscopically. The direct connection between such mutations and phenotypic change is not well established with these examples, such as with polyploids.

Can the spontaneous mutation rate be altered? The answer to that question came after the experimental, wartime, and peaceful uses of radioisotopes. As it turns out, there are no reliable or consistent ways that the natural or spontaneous mutation rate may be reduced, but there are many agents that can effectively increase it. Following the discovery of unstable radioisotopes and certain light waves in the non-visible spectrum experiments were conducted with laboratory organisms that demonstrated the powerful mutation-inducing properties of ionizing radiation (e.g., $\lambda = 0.03$ to $3.00\,nm$, or X-rays) and ultraviolet (UV; $\lambda = 10$ to $300\,nm$) within the electromagnetic spectrum visible to humans (Fig. 8.8). The warnings about human exposure (up to 10^5 REM) were realized following the thermonuclear bomb detonations at Hiroshima and Nagasaki, Japan in 1945 when tens of thousands perished from acute radiation exposure (Listwa, 2012). Three generations later, the rate at which new mutations is appearing is greatly elevated over the background (i.e., mutation rate of unexposed cohorts), and several hundred deaths have been directly attributed to DNA damage from ionizing radiation in 1945 (Heidenreich et al., 2007). Most of these new mutations were deleterious, killing the exposed or appearing in later generations, the main effect being elevated leukemia and urothelial carcinoma rates (Listwa, 2012). A recurrence of these tragedies was observed downwind of the failed nuclear power plant at Chernobyl, Ukraine in 1986.

Subsequent scientific studies demonstrated that energy in these wavelengths resonates with DNA to cause breakage and other biochemical alterations, most of which are corrected by existing DNA polymerase repair mechanisms.

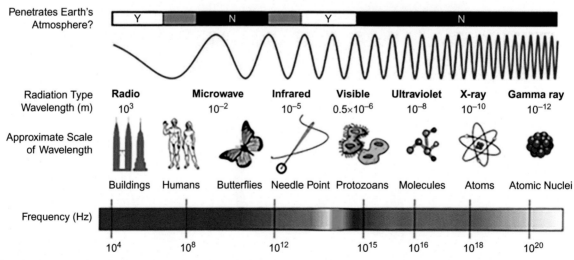

FIG. 8.8   The electromagnetic frequency spectrum; the mutagenic range is $<10^{-7}$ m. *From https://en.wikipedia.org/wiki/Electromagnetic_spectrum#/media/File:EM_Spectrum_Properties_edit.svg.*

The double-stranded structure of DNA provides natural buffering against mutagenic agents in the environment by these repair mechanisms (Kimura et al., 2004). Further exploratory experiments showed that certain chemicals in solution were also able to interact with DNA to incite mutations. Examples include nucleotide base analogues and alkylating agents such as ethyl methanesulfonate (EMS; Fehr, 1987). Both types of chemical mutagens essentially "trick" the DNA replicating machinery by methylating nucleotide bases, resulting in permanent changes, mainly base transitions (e.g., $A \rightarrow T$) and transversions (e.g., $A \rightarrow G$ or C).

Thus, plant breeders can expand the gene pool by accelerating the appearance of new mutations with mutagenic chemicals or ionizing electromagnetic radiation. Since mutations appear randomly, and most are deleterious, the gene pool is not necessarily expanded in a manner useful or desirable for breeding. The targeted gene may be only one in 10^5 in the genome, mandating prohibitively large populations to ensure a reasonable probability of recovering the targeted mutation (Fehr, 1987).

Since DNA and RNA are the universal genetic code molecules on Earth, and physiological and biochemical mechanisms of life are similar or orthologous among all organisms, mutagens that work on plants also work on humans and other organisms. Many mutagenic agents are also carcinogenic in humans and other mammals (Ames, 1986). Therefore, stringent safety precautions for shipping, handling, storage, and disposal of chemical mutagens are absolutely essential. Similarly, radioisotopes used to generate ionizing electromagnetic radiation must be addressed with care. Exposure to radiation by humans must be minimized and monitored. Finally, many mutagens are expensive or require expensive machines to generate them. The plant breeder is generally encouraged or persuaded, therefore, to explore and exhaust other sources of desirable genetic variability before resorting to a mutation breeding strategy (see below).

The expansion of the germplasm pool via induced mutagenesis can, however, be an effective strategy in asexually-propagated horticultural crop species, and especially ornamentals where important phenotypes (flower or foliar color, color patterns, shape, etc.) are often simply-inherited (Schum, 2003; Kleynhans, 2011). A population with otherwise desirable performance attributes is mutagenized then screened directly for desirable phenotypic variability. If such phenotypic variants are found and are stable over time and following asexual propagation, they may be amplified into large populations and sold as a finished product (Schum, 2003).

A generalized procedure is usually employed to expand crop species' gene pools via induced mutations, known as *mutation breeding* (Fig. 8.9). A population is assembled to which the mutagen will be administered, referred to as the M_0 generation. This population is usually the commercial variety or inbred parent of a commercial hybrid variety. The population must be synchronized to effectively receive the mutagen pulse, since they work best on actively growing and dividing tissues. Seeds are a convenient choice because they are naturally synchronized, small, actively

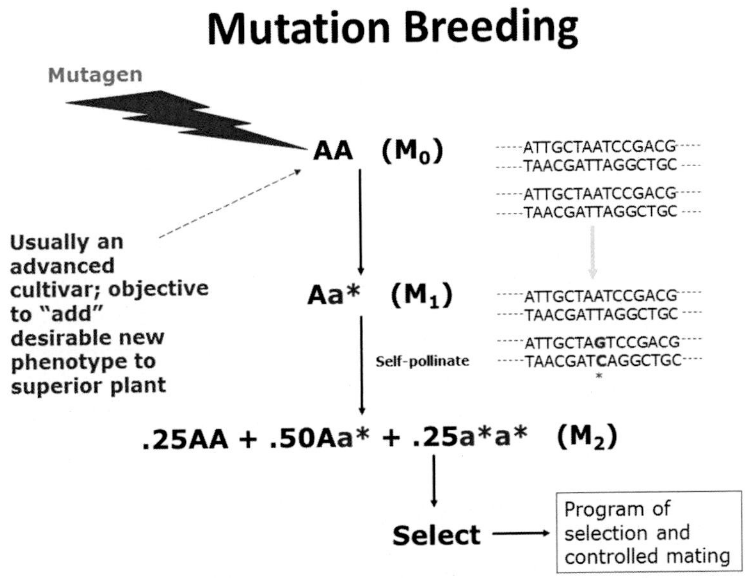

Mutation Breeding

FIG. 8.9 The process of gene pool expansion by induced mutations, or *mutation breeding*. M_0 is the starting population, usually imbibed seeds or young shoots. The M_1 is the mutagenized population that is self-pollinated to generate the M_2 population. The M_2 will segregate for new mutations that were induced in the M_0 and present in a heterozygous configuration in the M_1. The new desirable phenotype (a*a*) is selected within the M_2 population then used as a starting point in a more comprehensive plant breeding program aimed at combining the new mutation with a background of other desirable genotypes. The nature of the mutation depicted ($AT \rightarrow GC$) is only one of many different possibilities (base substitutions, additions, deletions, insertions, translocations, inversions, etc.).

growing (following hydration), and a cell lineage within the apical meristem will give rise to the reproductive organs.

The mutagen is administered in a pulse, the concentration and duration of which are highly controlled. Each mutagen has a characteristic lethality curve, usually quantified by LD_{50} or the dose at which half the population is killed. The LD_{50} is a function of both the concentration and intensity of and duration of the mutagen pulse. A confusing lexicon has been developed to describe the absorbance of ionizing radiation by humans and the resulting mutagenic effects, and these terms are sometimes also used in connection with plant mutagenesis (Yonezawa and Yamagata, 1977; Schum, 2003).

The mutagenized M_0 population is referred to as the M_1. If the mutagen pulse was too strong or lengthy, near-term viability is adversely affected due to adverse physiological effects, known analogously as radiation sickness in humans. If the mutagen pulse was too weak, no new mutations would be observed either in the M_1 or subsequent generations. Most new induced mutations will behave recessively, so will be masked by the dominant (functional) allele not be observed in the M_1. The M_1 must be self-pollinated to produce the M_2 generation, wherein recessive mutations will be first observed among the 25% homozygous recessive segregants.

Because multicellular sporophyte organisms are almost always used for mutation breeding, and each constituent cell reacts independently to the mutagen, it is most likely that the M_1 generation will consist of chimeric plants (Fig. 8.5) comprised of many different mutant genotypes. Physiological and/or indirect effects of mutagen exposure are not well understood in plants, but mutagenized plants (M_1) often exhibit transient phenotypes. For these reasons, the M_1 is not generally used for anything other than the production of the M_2 (Yonezawa and Yamagata, 1977).

The M_2 generation is the first opportunity to select for the appearance of new, desirable mutations. If the original mutagen pulse was completely effective, mutant phenotypes should appear frequently among M_2 individuals, including phenotypes of interest to the plant breeder. Ideally, M_2 plants with desirable mutations do not also have additional mutations that are undesirable, but this possibility is difficult to preclude (Yonezawa and Yamagata, 1977). The selected M_2 plants must usually be submitted into additional breeding programs, the likes of which will be described later in this text (Chapters 13–19). Undesirable mutations will be eliminated during the course of controlled mating and selection within the context of these programs.

It has been estimated that over 500 released varieties owe their enhanced performance or novelty to induced mutations (Gottschalk and Wolff, 1983). Despite this impressive number mutation breeding is not particularly popular. Mutagenic agents embody highly invasive threats to human health. Consequently, access, use, storage, and disposal of mutagens are difficult, expensive, dangerous, and highly regulated. In an increasingly risk-averse world, it is unlikely that mutation breeding as it was originally envisioned in the mid-20th century will become a more attractive option in the future. Collective understanding of the forces that induce plant retrotransposons to become active is rapidly expanding (Paszkowski, 2015). Perhaps retrotransposons will be employed as targeted mutagens to expand germplasm pools in the future?

CELL CULTURE STRATEGIES

Working in the 1950s, Folke Skoog and other plant biologists developed the capabilities to culture plant organs, tissues, and cells independently of whole plants, or in vitro (Latin for "in glass"; Armstrong, 2002; Dhaliwal, 2002; Sharma, 2015). As compared to bacteria and fungi, cultured plant tissues and cells require growth media and environments that are comparatively much more complex and demanding. Approximately 20 major and minor salt compounds, a few organic compounds, a carbon/energy source (usually sucrose) and phytohormones are mixed in a typical plant tissue/cell growth medium. Habituated plant tissues and cells can live and prosper for years in culture, and some may be manipulated to exhibit desirable attributes such as overproduction of valuable phytochemical substances. To maintain vigor, cultured plant tissues or cells must be fed continuously and waste products must be removed. This is usually accomplished by frequent serial transfers into fresh medium.

Certain cultured plant organs can readily be regenerated into whole plants, such as meristems, flowers, leaves (Read and Preece, 2009). After exhaustive empirical studies, the conditions were developed by which whole plants could be regenerated first from tissues and later individual cells (Sugiyama, 2015). This is referred to as *totipotency* or the retention of information and know-how by the single cell to make a whole plant with phenotypic attributes of the somatic-celled ancestor (Chawla, 2002). Until relatively recently plants were presumed to possess this possibility while mammals and other Animalia did not. With the success of recent cloning experiments on domestic mammals, it is apparent that animals are totipotent as well, although requiring much more stringent conditions to proceed from single somatic cells to whole organisms (e.g., "Dolly" the sheep; Brem and Kuhholzer, 2002).

Many potential uses have been projected for plant tissue and cell culture technologies. Among them is genetic improvement. Theorists and futurists in the 1970s foresaw plant breeding taking place in a test tube or Petri dish instead of an open field or greenhouse. We only needed to figure out how the phenotype could be reduced to a simple chemical challenge, akin to popular approaches in bacterial genetics. New hybrids could be synthesized by fusing protoplasts, cells with the walls stripped off. The selected cells or somatic hybrids could then be regenerated into whole plants again that would go directly into agricultural commerce (Kuckuck et al., 1991). Some even envisioned "artificial seeds", packaged somatic embryos regenerated from a chemically-selected cell culture.

This vision did not unfold in the anticipated way. Most whole-plant phenotypes could not readily be adapted to allow for selection in the environment of the cell and culture medium. Somatic hybrids did not ultimately lead to results that could not have been obtained using sexual crosses. It is not surprising that the "wide crosses" that could not be synthesized sexually would lead to dubious outcomes. When the "pomato" (potato + tomato) somatic hybrid was regenerated into a whole plant, the root system resembled the tomato and foliage resembled the potato, the opposite of what was originally targeted. Since pomato plants were completely sterile it was problematic to devise the next steps beyond the hybrid. As unlikely as the pomato was, even less probable hybrid combinations were attempted, for example angiosperm plant (albino) + prokaryotic blue-green algae.

Another unexpected observation of plant cell and tissue culture during the mid-20th century was de novo genetic variability. While cultured plant organs tended to result in genetically clonal entities, cultured somatic tissues and cells that were undifferentiated were often genetically unstable. This instability was manifested by cytogenetic changes from wild type with regard to chromosome number and structure (Orton, 1984b) and molecular changes in polypeptides (Lassner and Orton, 1983). Phenotypic variants that differed from the original source plant were often, but not always, observed among populations of plants regenerated from cell and tissue cultures. Among plants regenerated from plant cell and tissue cultures albinism, altered leaf shape and size, altered floral structure and function, and pigmentation were reported. Most of these phenotypic variants were later shown to be heritable (Ahloowalia, 1986). Comparisons of cell and tissue cultures with plants regenerated from them suggested that the range of genetic variability had been much reduced during the regeneration process (Orton, 1980). Analyses of the types and frequencies of new mutants demonstrated that the underlying processes responsible for the new mutants were not random. Specifically, high frequencies of specific types of mutants are usually observed and few or none of most others (Orton, 1984a).

Larkin and Scowcroft (1981) first coined the term *somaclonal variation* (Fig. 8.10) to describe the phenomenon of phenotypic and genotypic variability from cell and tissue cultures. Evans et al. (1984) later extended the concept and coined the term *gametoclonal variation* to describe variation from cultured microspores and megaspores. It was speculated that many of the new mutant alleles derived from cultures would be useful for plant improvement, an extension of the use of sectoral chimeras (sports) in woody perennial breeding (Kuckuck et al., 1991). Many examples of somaclonal mutants that are useful for plant breeding have been developed, such as male sterility for hybrid seed production, new pigments, elevated vitamin/cofactor levels, and increased fruit soluble solids (Bajaj, 1990).

Is somaclonal variation an effective method for the expansion of the gene pool? Conventional wisdom would accord this tool the same level of significance as mutation breeding. If the term "cell/tissue culture" is substituted for "mutagen", the procedure for utilizing somaclonal variability is essentially the same as mutation breeding. In effect, cell culture is used as a mutagenic agent, although it has not been proven that the process is inherently mutagenic. Researchers developed new nomenclatural systems to describe somaclonal variants and their descendants, but it was inconsistent with the "filial" (F_1, F_2, etc.) and "mutated" (M_0, M_1, etc.) nomenclature already established in literature (Chawla, 2002). The process of using somaclonal variation as a tool to enhance plant germplasm entails the introduction of an explant from a targeted individual/population into cell/tissue culture followed by the regeneration of whole plants. The population of regenerated plants is then screened for variation of interest and also self-pollinated to unmask any new recessive alleles.

An alternative explanation of the basis of somaclonal variation is that mitosis is inherently mistake-prone and whole-plant selective forces continually weed out variants in a manner similar to the animal immune system (Orton, 1984a; Bairu et al., 2011; Wang and Wang, 2012). Cells in culture lack these whole-plant selective forces provided by a developmental context allowing variants to survive and, sometimes, prosper. Observations that more disorganized and less associated cells and tissues tend to exhibit more genetic variability are consistent with this hypothesis (Orton, 1983; Orton, 1984a, b; Wang and Wang, 2012).

The following differences are applicable in the comparison of somaclonal variation vs. mutation breeding with respect to the enhancement of pools of genetic variability (Henry, 1998):

- Higher frequencies of certain mutants, lower of others
- Low human risk of exposure to cell/tissue culture process

FIG. 8.10   The origin of somaclonal variation during the course of culturing plant tissues and cells.

- High cost and technical prowess for cell/tissue culture
- Some plant species are easier to manipulate as cultured cells and tissues than are others

Advances in genomics and nucleic acid sequencing technologies will provide better phenotypes to study the sources of de novo genetic variability in plant tissue and cell cultures. For example, Campbell et al. (2011) devised a retrotransposon-based marker system to study somaclonal variation in cell cultures of barley (*Hordeum vulgare*). This strategy identified 29 polymorphisms of which 12 were novel non-parental bands among regenerated plants.

GENETIC TRANSFORMATION

Most consumers have been exposed to the concepts of biotechnology and GMO foods and drugs, especially the fact that they present potential or perceived risks to humans and other mammals (Devos et al., 2014). Biotechnology is a relatively new term that describes any technology based on the utility of living organisms or components thereof, including plant breeding. The term has, however, been narrowly applied to include only the molecular and cellular manipulation of organisms. The ownership and manipulation of life are politically-charged issues that date back to the landmark Diamond v Chakrabarty U.S. Supreme Court decision in 1980 (Kevles, 1994).

New technologies, "discovery", and product development are expensive and time consuming, but can yield large financial rewards when they lead to valuable products, processes, and intellectual properties. The Diamond v Chakrabarty decision and its derivatives have led to a liberal extension of U.S. patent and copyright laws to life forms and their derivatives (e.g., drugs and other bioactive compounds). Many small biotechnology companies were formed in the 1980s to capitalize on biological intellectual property opportunities, but most of these have disappeared or been acquired by larger multinational corporations. Patented life forms and the tools to manipulate them genetically have come to be owned, consequently, by a relatively few global organizations. Opponents of such liberal extensions of capitalism have targeted biotechnology as an example of what can go wrong (Bernauer, 2003).

The ability to synthesize any gene or DNA sequence or to isolate it, alter it, add desirable expression controls, and re-insert the altered sequence into the genome of a plant is one of the most powerful methods of germplasm enhancement. Unlike the techniques described previously in this chapter transformation has had, and will continue to have, meaningful and significant impacts on plant breeding. Therefore, the technology will be described and discussed in greater detail.

Genetic transformation was first demonstrated in prokaryotes. If a bacterial strain that lacks a specific gene is cultured in the presence of exogenous DNA including the gene, it may be taken up and become part of the bacterial genome by the natural process of transformation (Avery et al., 1944; Hershey and Chase, 1952). The term "transformation" is also used in plants but the process probably never occurs in nature in the manner that is seen in prokaryotes. The tools to cut and splice DNA were also discovered in bacteria. In the late 1960s bacterial *restriction endonucleases* (REs), a rudimentary form of immunity to pathogenic viruses (phage), were discovered. REs recognize short oligonucleotide sequences in invading viral particles that are not present in the host bacterial genomes. REs then bind to and cut viral DNA (or RNA) into non-infective fragments. Different REs were isolated from different species of bacteria, each one specific to different characteristic nucleotide sequences and number of nucleotides. The number of nucleotides in the sequence generally ranges from four to seven. The "four-cutters" usually produce many more fragments than "seven-cutters" for any given genome. Over 60 distinct REs are currently available from commercial sources.

Many of the different REs made staggered cuts, leaving "sticky" single-stranded ends on the resulting DNA fragments. Enterprising scientists immediately envisioned the possibility of recombinant DNA by annealing two sources of DNA both cut by the same RE. In the early 1970s Stanley Cohen of Stanford University and Herbert Boyer of the University of California San Francisco (see Chapter 1 for short biographies) developed the concept of the recombinant plasmid. They obtained a series of U.S. patents (US4363877 B1, US4237224 A, US4468464 A, and US4740470 A) that described the process of using REs to produce chimeric (recombinant DNA) plasmids. These patents were licensed to E. I. Lilly and Genentech, Inc., and led to the first biopharmaceutical products based on recombinant DNA (Buchholz and Collins, 2010).

Plasmids are found in many bacteria, existing as small independently replicating entities separate from the main genome. They are often associated with transient functions, such as drug resistance. By splicing foreign DNA into a bacterial plasmid, and re-introducing it into bacteria, the resulting recombinant plasmid would replicate to produce millions of exact copies of the spliced DNA sequence (Thackray, 1998).

By the late 1970s, it was routine to isolate plant genes and to splice them into bacterial plasmids. The resulting copies could be studied intensively, later including the actual DNA sequence thereof. They could also be translated in vitro, and the corresponding polypeptide isolated and studied. Methods were developed by which isolated genes could be modified, nucleotide bases inserted, deleted, and substituted. Finally, techniques were forged by which DNA sequences could be synthesized, initially only short segments, but improvements have made longer sequences attainable (Buchholz and Collins, 2010).

The isolated DNA was of only esoteric interest unless it could be re-inserted into the replicating genome of the whole plant. The crown gall disease of woody perennial plant species has proven to be an excellent platform for the genetic transformation of angiosperms. The pathogenic crown gall bacterium, *Agrobacterium tumefaciens* (and other close relatives in the genus *Agrobacterium)* was found to incite tumors consisting of rapidly dividing and expanding plant cells formed at a wound site. The tumor continued to grow even after the bacteria were removed with antibiotics. Tumor tissues exhibited altered metabolism characterized by a peculiar group of amino acids not normally seen in plants. *A. tumefaciens* was found to express specific enzymes for the catalysis of these amino acids and also for the use of resulting breakdown products for energy and carbon skeletons for metabolism, growth, and cell division. Scientists were surprised to discover that host crown gall tumor cells contained a piece of bacterial DNA in their nuclear genomes that was responsible for all of these physiological and biochemical changes (Murphy, 2007; Vasil, 2008).

Subsequent experiments showed that *A. tumefaciens* contained a plasmid that encoded genes for the synthesis of the peculiar amino acids. The plasmid also contained genes that encoded enzymes that mediated steps of the DNA transfer phenomenon: cell recognition and attachment, plasmid injection, DNA integration. The actual DNA sequences that ended up in the chromosomes of the tumor were referred to as "transferred" or T-DNA. The nucleotide sequence of the borders of this region was critical to the integration process. It appeared that it didn't much matter what DNA sequences were in between these borders, as long as it met certain length requirements (Murphy, 2007).

With the successful development of directed angiosperm plant transformation by Rob Horsch and Rob Fraley of Monsanto Corp. in 1984, the concept of gene pool was radically changed (Horsch et al., 1984). Thereafter, it was possible to insert any gene into a plant genome regardless of phylogenetic (or synthetic) origin. The fundamental

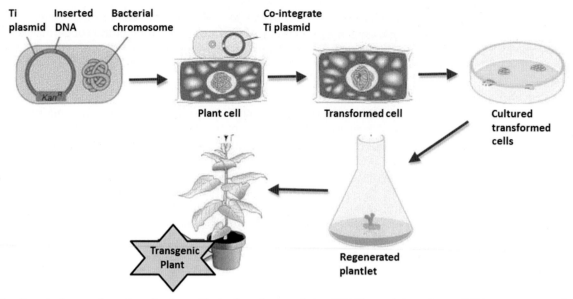

FIG. 8.11 Steps in the transformation of a plant with *Agrobacterium tumefaciens* T-DNA containing a foreign DNA sequence. *From http://2010. igem.org/Team:Nevada/Agrobacterium_Transformations.*

steps involved with *A. tumefaciens*-mediated plant transformation are depicted in Fig. 8.11. Crop species like tomato could be transformed with genes from corn, papaya, monkeys, lobsters, bacteria, or artificially synthesized DNA sequences. Advances in forensic sciences have made it possible to isolate genes from our evolutionary past, frozen into fossils for millions of years (Helgason et al., 2007; Ottoni et al., 2009). DNA sequence synthesis technologies have enabled the design and construction of artificial genes in test tubes. Derivatives from fossil and artificial genes may be transformed into genomes of plants. The possibilities are endless and speculations of potential uses for transformation are as boundless as the imagination. Frightening and unlikely scenarios have been speculated that have misled the public into a blanket distrust of biotechnology (Harlander, 2002).

Thousands of angiosperm plant transformations have been successfully performed since 1984 and incremental improvements have been made in all steps relevant to the process (Galun and Breiman, 1997). Most notable among these was the expansion of the range of plant species amenable to the use of *A. tumefaciens* for the introduction of foreign DNA. The host range of the pathogen was limited to a small range of woody perennial plant species. Therefore, it was thought that applications of *A. tumefaciens* for transformation would be highly species-specific. By tinkering with the T-DNA borders of the TI plasmid, however, strains of *A. tumefaciens* have been developed that are effective on monocots as well, including the major food grain crops. A parallel technology for foreign gene introduction was developed to circumvent species applications: the process of *biolistics*, physically introducing DNA on a projectile shot from a *gene gun*. Biolistics is plagued by low numbers of transformants but the technique is necessary if *A. tumefaciens* cannot be made to work (Murphy, 2007). Biolistics is also used extensively in conjunction with genome editing systems like CRISPR-cas9 (see below).

Different modes of altered gene function have been achieved with transformation (Chawla, 2002). Existing genes may be silenced, accentuated, or altered to produce entirely new phenotypes. Further, it is possible to add new genes that impart new functions not previously observed in a given species. Gene expression may be modulated by number of copies or altered control sequences (e.g., promoters). Silencing of undesirable genes was first thought to be a consequence of antisense (sequence complementary to mRNA) or sense (sequence homologous to mRNA) inactivation of mRNA. Gene silencing appears, rather, to be a consequence of the RNAi pathway in the nucleus of eukaryotes. The modulation of gene expression is an important new capability that is difficult if not impossible in traditional plant breeding (Ditt et al., 2001).

Scientists proceeded blindly with transformation technology in the 1990s, assuming that if the technology worked it would be adopted and would revolutionize the fields of plant and animal breeding. In the race to bring biotechnology products to market, Calgene, a small California biotechnology start-up company, produced the Flavr Savr® tomato, wherein the synthesis of the ripening hormone gas ethylene was inhibited by a gene introduced by *A. tumefacien*- mediated transformation. The tomato variety behind Flavr Savr® was submitted to the FDA for GRAS approval and a debate on the risks posed by genetically modified organisms (GMOs) used for food and drugs ensued. The FDA concluded in 1994 that risk was measured by the interaction of introduced DNA sequence and host organism, not the technology used to accomplish the entity. Thus, the regulation of transgenic/GMO foods and drugs was applied on a case-by-case and not a wholesale basis (Kramer and Redenbaugh, 1994).

Assuming that other countries would follow suit and thinking that the GMO food/drug issue had been resolved, corporate efforts were expanded for the development and introduction of GMO plant cultivars. Almost immediately, European consumers voiced opposition to GMO foods and drugs and political opposition surged when it was discovered that American companies had already co-mingled GMO and non-GMO food ingredients in food products exported from the U.S. to Europe. The backlash contributed to a major public relations debacle, the aftermath of which severely hampered the continued expansion of the technology (Montpetit et al., 2007).

The first wave of GMO cultivars was aimed at capturing added value at the farm gate, not from value to end users. "Roundup-Ready"® and "Liberty-Link"® soybean, cotton, and corn contained bacterial genes that conferred resistance respectively to the herbicides glyphosate and glufosinate, thus rendering those broad-spectrum and environmentally friendly (relative to most other herbicides) products to be selective. Both compounds bind and inactivate enzymes in the biosynthetic pathways of essential amino acids not produced by animals. The bacterial genes provide an active enzyme replacement to transformants, so they are resistant to the herbicide (Funke et al., 2006).

These new cultivars also contributed to sales of "Roundup"® and "Liberty"®. Antagonists protested that the varieties would lead to increased dependence on the herbicides, purportedly inconsistent with high standards of environmental stewardship. Both herbicides exhibit low toxicity to animals, however, and are purportedly degraded rapidly in soil as compared to more environmentally persistent selective herbicides used in non-GMO crops. Subsequently, mutant weeds have been discovered following the widespread cultivation of Roundup-Ready® and Liberty-Link® cultivars that are resistant to the broad-spectrum herbicide, adding fuel to the argument against the new GMO varieties. Despite the perceived negatives, GMO herbicide-resistant soybeans occupy most of the cultivated areas of the U.S., South America, and China at the time of the publication of this book (Shaner, 2014).

The *cry* genes of the soil-borne bacterium *Bacillus thuringiensis* (Bt) encode a toxin effective against specific insect pests but that is harmless to most other organisms. These genes have been successfully isolated and transformed into corn, potatoes, and cotton and conferring cultivars with excellent resistance to corn rootworm and earworm, Colorado potato beetle, and cotton pink bollworm (Gassmann et al., 2009). The environmental impact realized by the corresponding reduction in chemical pesticide use was touted by the agricultural biotechnology industry, but a report appeared on the collateral toxic effects of Bt on the monarch butterfly (Sears et al., 2001). Bt resistance also appeared in certain pest populations (Tabashnik et al., 2009). The EPA mandated that any crop plant expressing a gene with pesticidal properties must be labeled as a pesticide, antithetical to the perception of Bt cultivars as safe and wholesome foods. These factors were enough to turn the tide against Bt corn and potato, although Bt cotton still enjoys a substantial market (Murphy, 2007). Apparently, it is worse to ingest foreign genes than to wear them.

Transformation methods are not yet robust or comprehensive enough to accommodate traits with quantitative or complex inheritance. The upper limit would appear to be fixed at characters determined by a maximum of 2–3 distinct genes. If the major genes responsible for the expression of a complex trait are known there is no reason that transformation techniques could not be applied in a stepwise fashion to "pyramid" them in a single plant. Another concern is how to coordinate the expression of newly introduced genes with the rest of the genome.

Ownership of intellectual property presents a further complication that will be addressed in greater detail in Chapter 12. While hybrid corn varieties have an inherent protection against plant-back or outright piracy of varieties, pure line OP varieties of soybean and cotton do not. The Plant Variety Protection Act that governs intellectual property in sexually-propagated plant species exempts farmers from saving seed to plant future crops. Since the cost of the new technology is high and recaptured through sales of seed, companies employed a new approach using contractual law. Each container of seed of GMO cultivars bears a written contract that, in effect, constitutes an agreement not to propagate the product for additional seed for any purpose. The enforcement of these contracts has surely presented a huge challenge to the biotechnology industry.

At present, there are few barriers to the production of GMO crops in the U.S., but major hurdles still exist in Europe, Canada, Australia, and Japan (Montpetit et al., 2007; Teng, 2008). Other countries seem mostly indifferent to the environmental threats posed by GMO crops or the theological and socio-economic arguments against biotechnology. "Golden" rice that is GMO and high in β-carotene has replaced conventional varieties in India, Malaysia, Indonesia, and African and Southeast Asian nations (Dawe and Unnevehr, 2007). This provides a valuable source of vitamin A for populations that lack other dietary sources of this nutrient. China has embraced all forms of biotechnology, seeing it as one of many tools to help become self-reliant in food production.

Most scientists, including plant breeders, agree that plant transformation and attendant technologies are inherently safe and will ultimately be embraced by society and become fully integrated into the crop improvement "tool box". Most new ideas faced such opposition during the course of history and are now fully integrated into most societies, for example telecommunications, automobiles, and personal computers.

GENOME EDITING

Genome editing in organisms via site-directed mutagenesis is an alteration of a naturally occurring process (Songstad et al., 2017). This process involves designed *sequence-specific nucleases* to incite targeted genomic DNA nucleotide sequence changes. Depending on the locus and type of nucleotide change, the phenotypic effects range from *knock-out* or *knock-in* of coding gene functions or *oligonucleotide directed mutagenesis* where specific custom *single nucleotide polymorphisms* (SNPs) may be targeted.

The **C**lustered **R**egularly **I**nterspaced **S**hort **P**alindromic **R**epeats (CRISPR)/Cas9 system is a simple and efficient tool for genome editing (Fig. 8.12; Mei et al., 2016). CRISPR/Cas is a nucleotide sequence-driven nuclease system that evolved in bacteria as a defense mechanism against viral attack, in a similar manner as restriction endonucleases (Barrangou et al., 2007; Gasiunas and Siksnys, 2013). By delivering the Cas9 nuclease complexed with a synthetic guide RNA (sgRNA) into a cell, the cell's genome can be cut at the desired location, allowing existing genes to be removed and/or new ones added (Ledford, 2015). Techniques described earlier (transformation, above) are used to incorporate CRISPR-Cas9 into plant cells including *A. tumefaciens* T-DNA plasmids and biolistics (Fauser et al., 2014).

The Cas9 endonuclease is a four-component system that includes two small RNA molecules named CRISPR RNA (crRNA) and trans-activating CRISPR RNA (tracrRNA). The native Cas9 endonuclease was engineered into a more manageable two-component system by fusing the two RNA molecules into a sgRNA that, when combined with Cas9, could find and cut the DNA target specified by the sgRNA (Jinek et al., 2012). The simplicity of the type II CRISPR nuclease, with only three required components (Cas9 along with the crRNA and trRNA) makes this system amenable to adaptation for genome editing. Based on the type II CRISPR system described previously, this group developed a simplified two-component system by combining trRNA and crRNA into a single synthetic single guide RNA (sgRNA). sgRNA-programmed Cas9 was shown to be as effective as Cas9 programmed with separate trRNA and crRNA in guiding targeted gene alterations (Jinek et al., 2012).

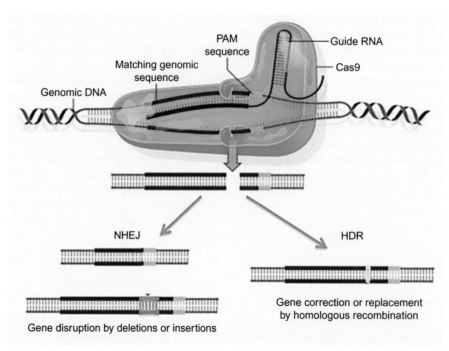

FIG. 8.12   The CRISPR-Cas9 complex depicting the repair of an induced sequence-specific break in genomic DNA to insert targeted donor DNA. PAM = protospacer-associated motif; HDR = homology-directed repair; NHEJ = nonhomologous end joining. *From Charpentier, E., Doudna, J.A., 2013. Biotechnology: rewriting a genome. Nature. 495 (7439), 50–51.*

Higher organisms, including plants, continuously encounter DNA breaks caused by external sources such as sunlight as well as internal processes such as those that release free radical molecules. To endure, these organisms have developed efficient mechanisms for repairing the multitude of DNA breaks that occur in each cell every day. DNA repairs can be generally classified in two ways: (1) non-homologous end joining and (2) homology-directed repair. Repair without a template occurs through the nonhomologous end joining (NHEJ) pathway and can be used to disrupt the function of a gene, effectively deleting it. Repair using a template through the homology-directed repair (HDR) pathway can enable precise alterations and insertions from the template DNA sequence into the genome (Sander and Joung, 2014).

Non-homologous end-joining (NHEJ) is the dominant DNA repair pathway in plants. It does not use a DNA template for the repair process and instead functions by simply identifying two broken ends of DNA and bonding them back together. This DNA repair process can often result in the insertion or deletion of random DNA sequences at the repair site. If the broken DNA sequence represented a plant's gene, the function of this gene becomes disrupted and it is "knocked out".

Homology-directed repair, often referred to as HDR, requires a second unbroken strand of DNA that harbors sequence that is identical to that flanking the broken DNA as well as the desired change (edit, specific and targeted alteration of the sequence, or additional genetic material to insert). It uses that unbroken DNA strand as a template to repair the DNA. Genes can be deleted by targeting Cas9 to cut the desired gene. Repair by the NHEJ pathway can be used either to disrupt the DNA sequence that codes for genes or, in the case of two cuts flanking the gene, the entire gene can be removed. An example of such an application is the next generation of high extractable starch and waxy corn hybrids by Bayer/DuPont/Pioneer.

Plant breeders and geneticists are just beginning to apply this gene editing technology for crop trait improvement as this textbook is being published. CRISPR/Cas9 has already been demonstrated to be equally effective for all topographical regions of the eukaryotic genome (euchromatin, heterochromatin, repeated regions, coding genes, introns/exons, cis/trans-expression-control sequences; Feng et al., 2016). Before rational strategies can be designed to implement this technology broadly, however, there is a need to expand the availability of crop-specific vectors, genome resources, and transformation protocols. These challenges will certainly be overcome along with the continued evolution of the CRISPR/Cas9 system particularly in the areas of manipulation of large genomic regions, transgene-free genetic modification, development of breeding resources, the discovery of gene function, and improvements upon CRISPR/Cas9 components. The CRISPR/Cas9 editing system appears poised to transform crop trait improvement well into the future (Schaeffer and Nakata, 2015).

Despite the use of *Agrobacterium tumefaciens* and biolistics to introduce CRISPR-Cas9 vectors, regulatory agencies in the U.S. have concluded that entities arising from this technology will not be regulated as transgenic derivatives because the DNA introduced into plant cells is present only transiently (Butler and Douches, 2016). For example, the U.S. Department of Agriculture (USDA) viewed Cibus' SU Canola™, the first commercial product arising from plant genome editing and had its test launch in 2014, as non-genetically modified (non-GM) (Songstad et al., 2017). This will render the use of CRISPR-Cas9 and other homology-based nuclease systems as more acceptable than direct transformation with foreign genes resulting in GMO crop plants.

The potential to beneficially expand gene pools with CRISPR-cas9 and other genome editing tools appears to be boundless. Traditional plant breeding is constrained by the limits of sexual reproduction and bridging germplasm through wholesale hybridization. Transformation and genome editing transcend these barriers and enable the plant breeder to target specific loci within the genome for change. Since many of the resource requirements of traditional breeding programs are mandated by the presence of the overwhelming majority of non-targeted genes, new molecular methods will revolutionize methodology in addition to providing better raw materials for cultivar development.

References

Ahloowalia, B.S., 1986. Limitations to the use of somaclonal variation in crop improvement. In: Semal, J. (Ed.), Somaclonal Variation and Crop Improvement. Martinus Nijhoff Publisher, Hingham, MA (USA), pp. 14–27.

Ames, B.N., 1986. Food constituents as a source of mutagens, carcinogens, and anticarcinogens. In: Knudsen, I.B., Ames, B.N. (Eds.), Genetic Toxicology of the Diet: Proceedings of a Satellite Symposium of the Fourth International Conference on Environmental Mutagens. A.R. Liss, New York, NY (USA), pp. 3–32.

Armstrong, D.J., 2002. Folke K. Skoog: in memory and tribute. J. Plant Growth Regul. 21 (1), 3–16.

Avery, O.T., MacLeod, C.M., McCarty, M., 1944. Studies on the chemical nature of the substance inducing transformation of pneumococcal types: induction of transformation by a deoxyribonucleic acid fraction isolated from *Pneumococcus* type III. J. Exp. Med. 79 (2), 137–158.

Bairu, M.W., Aremu, A.O., Van Staden, J., 2011. Somaclonal variation in plants: causes and detection methods. Plant Growth Regul. 63 (2), 147–173.

Bajaj, Y.P.S., 1990. Somaclonal variation—origin, induction, cryopreservation, and implications in plant breeding. In: Bajaj, Y.P.S. (Ed.), Biotechnology in Agriculture and Forestry. vol. 11. Springer-Verlag, Berlin (GER), pp. 3–48.

Barrangou, R., Fremaux, C., Deveau, H., Richards, M., Boyaval, P., Moineau, S., Romero, D.A., Horvath, P., 2007. CRISPR provides acquired resistance against viruses in prokaryotes. Science 315, 1709–1712.

Bennetzen, J.L., 2000. Transposable element contributions to plant gene and genome evolution. Plant Mol. Biol. 42 (1), 251–269.

Bernauer, T., 2003. Genes, Trade, and Regulation: The Seeds of Conflict in Food Biotechnology. Princeton University Press, Princeton, NJ (USA). 229 pp.

Bolmgren, K., Eriksson, O., 2010. Seed mass and the evolution of fleshy fruits in angiosperms. Oikos 119 (4), 707–718.

Borojević, S., 1990. Principles and Methods of Plant Breeding. Elsevier, Amsterdam (NE). 236 pp.

Brem, G., Kuhholzer, B., 2002. The recent history of somatic cloning in mammals. Cloning Stem Cells 4 (1), 57–63.

Buchholz, K., Collins, J., 2010. Concepts in Biotechnology: History, Science and Business. John Wiley, Chichester (UK). 471 pp.

Butler, N.M., Douches, D.S., 2016. Sequence-specific nucleases for genetic improvement of potato. Am. J. Potato Res. 93 (4), 303–320.

Campbell, B.C., LeMare, S., Piperidis, G., Godwin, I.D., 2011. IRAP, a retrotransposon-based marker system for the detection of somaclonal variation in barley. Mol. Breed. 27 (2), 193–206.

Chao, C.T., Khuong, T., Zheng, Y., Lovatt, C.J., 2011. Response of evergreen perennial tree crops to gibberellic acid is crop load-dependent. I: GA$_3$ increases the yield of commercially valuable 'Nules' Clementine Mandarin fruit only in the off-crop year of an alternate bearing orchard. Sci. Hortic. 130 (4), 743–752.

Chawla, H.S., 2002. Introduction to Plant Biotechnology, second ed. Science Publishers, Enfield, NH (USA). 538 pp.

Chen, Z.J., 2007. Genetic and epigenetic mechanisms for gene expression and phenotypic variation in plant polyploids. Annu. Rev. Plant Biol. 2007, 377–406.

Chen, F.Q., Hayes, P.M., 1992. The genetic basis of seed set in barley genotypes varying in compatibility with *Hordeum bulbosum*. Genome 35 (5), 799–805.

Darlington, C.D., Thomas, P.T., 1937. The breakdown of cell division in a *Festuca-Lolium* derivative. Ann. Bot. 1, 747–761.

Davies, F.S., 1986. The navel orange. Hortic. Rev. 1986, 129–180.

Dawe, D., Unnevehr, L., 2007. Crop case study: GMO golden rice in Asia with enhanced vitamin A benefits for consumers. AgBioforum 10 (3). www.agbioforum.org/v10n3/v10n3a04-unnevehr.html.

Devos, Y., Aguilera, J., Diveki, Z., Gomes, A., Liu, Y., Paoletti, C., du Jardin, P., Herman, L., Perry, J.N., Waigmann, E., 2014. EFSA's scientific activities and achievements on the risk assessment of genetically modified organisms (GMOs) during its first decade of existence: looking back and ahead. Transgenic Res. 23 (1), 1–25.

Dhaliwal, H.S., 1992. Unilateral incompatibility. In: Kalloo, G., Chowdury, J.B. (Eds.), Distant Hybridization in Crop Plants. Springer-Verlag, Berlin (GER), pp. 32–40.

Dhaliwal, H.S., 2002. Progress and prospects of plant tissue culture and genetic engineering for crop improvement. In: Arora, J.K., Marwaha, S.S., Grover, R. (Eds.), Biotechnology in Agriculture and Environment. Asiatech Publisher, Inc., New Delhi (India), pp. 11–28.

Ditt, R.F., Nester, E.W., Comai, L., 2001. Plant gene expression response to *Agrobacterium tumefaciens*. Proc. Natl. Acad. Sci. U. S. A. 98 (19), 10954–10959.

Elliott, F.C., 1958. Plant Breeding and Cytogenetics. McGraw-Hill Book Co., New York, NY (USA). 395 pp.

Evans, D.A., Sharp, W.R., Medina-Filho, H.P., 1984. Somaclonal and gametoclonal variation. Am. J. Bot. 71, 759–774.

Fan, C., Emerson, J.J., Long, M., 2008. The origin of new genes. In: Pagel, M., Pomiankowski, A. (Eds.), Evolutionary Genomics and Proteomics. Sinauer Association Inc. Publisher, Sunderland, MA (USA), pp. 27–44.

Fauser, F., Schiml, S., Puchta, H., 2014. Both CRISPR/Cas-based nucleases and nickases can be used efficiently for genome engineering in *Arabidopsis thaliana*. Plant J. 79 (2), 348–359.

Fehr, W.R., 1987. Mutation breeding. In: Principles of Cultivar Development Volume 1: Theory and Technique. MacMillan Publ. Co., New York, NY (USA), pp. 287–303.

Feng, C., Yuan, J., Wang, R., Liu, Y., Birchler, J.A., Han, F., 2016. Efficient targeted genome modification in maize using CRISPR/Cas9 system. J. Genet. Genomics 43 (1), 37–43.

Funke, T., Han, H., Healy-Fried, M.L., Fischer, M., Schönbrunn, E., 2006. Molecular basis of the herbicide resistance of Roundup-Ready crops. Proc. Natl. Acad. Sci. U. S. A. 103 (35), 13010–13015.

Galun, E., Breiman, A., 1997. Transgenic Plants. Imperial College Press, London (UK). 376 pp.

Gasiunas, G., Siksnys, V., 2013. RNA-dependent DNA endonuclease Cas9 of the CRISPR system: Holy Grail of genome editing? Trends Microbiol. 21 (11), 562–567.

Gassmann, A.J., Carrière, Y., Tabashnik, B.E., 2009. Fitness costs of insect resistance to *Bacillus thuringiensis*. Annu. Rev. Entomol. 2009, 147–163.

Germanà, M.A., 2011. Anther culture for haploid and doubled haploid production. Plant Cell Tiss. Org. Cult. 104 (3), 283–300.

Gottschalk, W., Wolff, G., 1983. Induced Mutations in Plant Breeding. Springer-Verlag, Berlin (GER), pp. 1–9.

Gustafsson, A., 1947. Mutations in agricultural plants. Hereditas 33, 1–100.

Gustafsson, A., Tedin, O., 1954. Plant breeding and mutations. Acta Agric. Scand. 4, 633–639.

Hahn, S.K., Bai, K.V., Chukwuma, Asiedu, R., Dixion, A., Ng, S.Y., 1994. Polyploid breeding of cassava. Acta Hortic. 380, 102–109.

Harlander, S.K., 2002. The evolution of modern agriculture and its future with biotechnology. J. Am. Coll. Nutr. 21 (3S), 161S–165S.

Hayata, Y., Li, X.X., Osajima, Y., 2001. Sucrose accumulation and related metabolizing enzyme activities in seeded and induced parthenocarpic muskmelons. J. Am. Soc. Hortic. Sci. 126 (6), 676–680.

Hayata, Y., Niimi, Y., Iwasaki, N., 1995. Synthetic cytokinin--1-(2-chloro-4-pyridyl)-3-phenylurea (CPPU)--promotes fruit set and induces parthenocarpy in watermelon. J. Am. Soc. Hortic. Sci. 120 (6), 997–1000.

He, Z., Pradhan, A.K., Harper, A.L., Bancroft, I., Parkin, I.A.P., Havlickova, L., Wang, L., 2017. Extensive homoeologous genome exchanges in allopolyploid crops revealed by mRNAseq-based visualization. Plant Biotechnol. J. 15 (5), 594–604.

Heidenreich, W.F., Cullings, H.M., Funamoto, S., Paretzke, H.G., 2007. Promoting action of radiation in the atomic bomb survivor carcinogenesis data? Radiat. Res. 168 (6), 750–756.

Helgason, A., Pálsson, S., Lalueza-Fox, C., Ghosh, S., Sigurðardóttir, S., Baker, A., Hrafnkellson, B., Arnodattir, L., Porsteinsdottir, U., Stefansson, K., 2007. A statistical approach to identify ancient template DNA. J. Mol. Evol. 65 (1), 92–102.

Henry, R.J., 1998. Molecular and biochemical characterization of somaclonal variation. In: Henry, R.J. (Ed.), Somaclonal Variation and Induced Mutations in Crop Improvement. Kluwer Academic Publishers, Dordrecht (Germany), pp. 485–499.

Hermsen, J.G.T., 1992. Introductory considerations on distant hybridization. In: Kalloo, G., Chowdury, J.B. (Eds.), Distant Hybridization in Crop Plants. Springer-Verlag, Berlin (Germany), pp. 1–14.

Hershey, A., Chase, M., 1952. Independent functions of viral protein and nucleic acid in growth of bacteriophage. J. Gen. Physiol. 36 (1), 39–56.

Heslop-Harrison, J.S., 2011. Genomics, banana breeding and superdomestication. Acta Hortic. 897, 55–62.

Horsch, R.B., Fraley, R.T., Rogers, S.G., Sanders, P.R., 1984. Inheritance of functional foreign genes in plants. Science 223, 496–499.

Ji, Y., Chetelat, R.T., 2003. Homoeologous pairing and recombination in *Solanum lycopersicoides* monosomic addition and substitution lines of tomato. Theor. Appl. Genet. 106 (6), 979–989.

Jinek, M., Chylinski, K., Fonfara, I., Hauer, M., Doudna, J.A., Charpentier, E., 2012. A programmable dual-RNA-guided DNA endonuclease in adaptive bacterial immunity. Science 337 (6096), 816–821.

Karaagac, E., Vargas, A.M., de Andrés, M.T., Carreño, I., Ibáñez, J., Carreno, J., Martinez-Zapater, J.M., Cabezas, J.A., 2012. Marker assisted selection for seedlessness in table grape breeding. Tree Genet. Genomes 8 (5), 1003–1015.

Kevles, D.J., 1994. Ananda Chakrabarty wins a patent: biotechnology, law, and society. Hist. Stud. Phys. Biol. Sci. 25, 111–135.

Khush, G.S., Brar, D.S., 1992. Overcoming barriers in hybridization. In: Kalloo, G., Chowdury, J.B. (Eds.), Distant Hybridization in Crop Plants. Springer-Verlag, Berlin (Germany), pp. 47–61.

Kimura, S., Tahira, Y., Ishibashi, T., Mori, Y., Mori, T., Hashimoto, J., Sakaguchi, K., 2004. DNA repair in higher plants; photoreactivation is the major DNA repair pathway in non-proliferating cells while excision repair (nucleotide excision repair and base excision repair) is active in proliferating cells. Nucleic Acids Res. 32 (9), 2760–2767.

Kleynhans, R., 2011. Potential new lines in the Hyacinthaceae. Acta Hortic. 886, 139–145.

Konzak, C.F., 1984. Role of induced mutations. In: Vose, P.B., Blixt, S.G. (Eds.), Crop Breeding: A Contemporary Basis. Pergamon Press, New York, NY (USA), pp. 216–292.

Koonin, E.V., 2005. Orthologs, paralogs, and evolutionary genomics. Annu. Rev. Genet. 39, 309–338.

Kramer, M.G., Redenbaugh, K., 1994. Commercialization of a tomato with an antisense polygalacturonase gene: the FLAVR SAVR tomato story. Euphytica 79 (3), 293–297.

Kuckuck, H., Kobabe, G., Wenzel, G., 1991. Fundamentals of Plant Breeding. Springer-Verlag, Berlin (GER). 236 pp.

Ladizinsky, G., 1992. Crossability relations. In: Kalloo, G., Chowdury, J.B. (Eds.), Distant Hybridization in Crop Plants. Springer-Verlag, Berlin (GER), pp. 15–31.

Lagercrantz, U., 1998. Comparative mapping between *Arabidopsis thaliana* and *Brassica nigra* indicates that *Brassica* genomes have evolved through extensive genome replication accompanied by chromosome fusions and frequent rearrangements. Genetics 150 (3), 1217–1228.

Larkin, P.J., Scowcroft, W.R., 1981. Somaclonal variation, a novel source of variability from cell cultures for plant improvement. Theor. Appl. Genet. 60, 197–214.

Lassner, M.W., Orton, T.J., 1983. Detection of somatic variation. In: Tanksley, S.D., Orton, T.J. (Eds.), Isozymes in Plant Genetics and Breeding, Part A. Elsevier, Amsterdam, pp. 209–218.

Ledford, H., 2015. CRISPR, the disruptor. Nature 522 (7554), 20–24.

Listwa, D., 2012. Hiroshima and Nagasaki: The Long Term Health Effects. K-1 Project Report. Center for Nuclear Studies, Columbia University. Archived at https://k1project.columbia.edu/news/hiroshima-and-nagasaki.

Liu, M., Li, Z.Y., 2007. Genome doubling and chromosome elimination with fragment recombination leading to the formation of *Brassica rapa*-type plants with genomic alterations in crosses with *Orychophragmus violaceus*. Genome 50 (11), 985–993.

Luo, M.C., Deal, K.R., Akhunov, E.D., Akhunova, A.R., Anderson, O.D., Anderson, J.A., Blake, N., Clegg, M.T., Coleman-Derr, D., Conley, E.J., Crossman, C.C., Dubkovsky, J., Gill, B.S., Gu, Y.Q., Hadam, J., Heo, H.Y., Huo, N., Lazo, G., Ma, Y., Matthews, D.E., McGuire, P.E., Morel, P.L., Qualset, C.O., Renfro, J., Tabanao, D., Talbert, L.E., Tian, C., Toleno, D.N., Warburton, M.L., You, F.M., Zhang, W., Dvorak, J., 2009. Genome comparisons reveal a dominant mechanism of chromosome number reduction in grasses and accelerated genome evolution in Triticeae. Proc. Natl. Acad. Sci. U. S. A. 106 (37), 15780–15785.

Lysak, M.A., 2014. Live and let die: centromere loss during evolution of plant chromosomes. New Phytol. 203 (4), 1082–1089.

Mable, B.K., Beland, J., Di Berardo, C., 2004. Inheritance and dominance of self-incompatibility alleles in polyploid *Arabidopsis lyrata*. Heredity 93 (5), 476–486.

Mandáková, T., Joly, S., Krzywinski, M., Mummenhoff, K., Lysak, M.A., 2010. Fast diploidization in close mesopolyploid relatives of *Arabidopsis*. Plant Cell 22 (7), 2277–2290.

Mayo, O., 1987. The Theory of Plant Breeding. Clarendon Press, Oxford, pp. 195–202.

Mei, Y., Wang, Y., Chen, H., Zhong, S.S., Xing-Da, J., 2016. Recent progress in CRISPR/Cas9 technology. J. Genet. Genomics 43 (2), 63–75.

Melchers, G., Sacristan, M.D., Holder, A.A., 1978. Somatic hybrid plants of potato and tomato regenerated from fused protoplasts. Carlsberg Res. Commun. 43, 201–218.

Mii, M., 2012. Ornamental plant breeding through interspecific hybridization, somatic hybridization and genetic transformation. Acta Hortic. 953, 43–54.

Mohr, H.C., 1986. Watermelon breeding. In: Bassett, M.J. (Ed.), Breeding Vegetable Crops. AVI Publisher & Co., Inc., Westport, CT (USA), pp. 37–66.

Montpetit, E., Rothmayr, C., Varone, F., 2007. The Politics of Biotechnology in North America and Europe: Policy Networks, Institutions, and Internationalization. Lexington Books, Lanham, MD (USA). 295 pp.

Muller, H.J., 1927. Artificial transmutation of the gene. Science 66, 84–87.

Murphy, D.J., 2007. Plant Breeding and Biotechnology: Societal Context and the Future of Agriculture. Cambridge University Press, Cambridge (UK). 423 pp.

Ocarez, N., Mejía, N., 2016. Suppression of the D-class MADS-box *AGL11* gene triggers seedlessness in fleshy fruits. Plant Cell Rep. 35 (1), 239–254.

Ortiz, R., 1997. Secondary polyploids, heterosis, and evolutionary crop breeding for further improvement of the plantain and banana (*Musa* spp. L) genome. Theor. Appl. Genet. 94 (8), 1113–1120.

Orton, T.J., 1980. Chromosomal variability in tissue cultures and regenerated plants of *Hordeum*. Theor. Appl. Genet. 56, 101–112.

Orton, T.J., 1983. Experimental approaches to the study of somaclonal variation. Plant Mol. Biol. Report. 1, 67–76.

Orton, T.J., 1984a. Genetic variation in somatic tissues: method or madness? In: Ingram, D.H.S. (Ed.), Advances in Plant Pathology. vol. 2. Academic Press, London, pp. 153–189.

Orton, T.J., 1984b. Somaclonal variation: theoretical and practical considerations. In: Gustafson, J.P. (Ed.), Gene Manipulation in Plant Improvement. Plenum, New York, pp. 427–468.

Ottoni, C., Koon, H.E.C., Collins, M.J., Penkman, K.E.H., Rickards, O., Craig, O.E., 2009. Preservation of ancient DNA in thermally damaged archaeological bone. Naturwissenschaften 96 (2), 267–278.

Pace II, J.K., Feschotte, C., 2007. The evolutionary history of human DNA transposons: evidence for intense activity in the primate lineage. Genet. Res. 17, 422–432.

Paszkowski, J., 2015. Controlled activation of retrotransposition for plant breeding. Curr. Opin. Biotechnol. 32, 200–206.

Pertuze, R.A., Ji, Y., Chetelat, R.T., 2003. Transmission and recombination of homeologous *Solanum sitiens* chromosomes in tomato. Theor. Appl. Genet. 107 (8), 1391–1401.

Rajathy, T., 1977. Plant breeding evolving. Can. J. Genet. Cytol. 19, 595–602.

Read, P.E., Preece, J.E., 2009. Micropropagation of ornamentals: the wave of the future? Acta Hortic. 812, 51–61.

Renny-Byfield, S., Wendel, J.F., 2014. Doubling down on genomes: polyploidy and crop plants. Am. J. Bot. 101 (10), 1711–1725.

Samoylov, V.M., Izhar, S., Sink, K.C., 1996. Donor chromosome elimination and organelle composition of asymmetric somatic hybrid plants between an interspecific tomato hybrid and eggplant. Theor. Appl. Genet. 93 (1–2), 268–274.

Sander, J., Joung, J.K., 2014. CRISPR-Cas systems for editing, regulating, and targeting genomes. Nat. Biotechnol. 32, 347–355.

Sattler, M.C., Carvalho, C.R., Clarindo, W.R., 2016. The polyploidy and its key role in plant breeding. Planta 243 (2), 281–296.

Schaeffer, S.M., Nakata, P.A., 2015. CRISPR/Cas9-mediated genome editing and gene replacement in plants: transitioning from lab to field. Plant Sci. 240, 130–142.

Schauer, N., Semel, Y., Balbo, I., Steinfath, M., Repsilber, D., Selbig, J., Pleban, T., Zamir, D., Fernie, A.R., 2008. Mode of inheritance of primary metabolic traits in tomato. Plant Cell 20 (3), 509–523.

Schubert, I., 2007. Chromosome evolution. Curr. Opin. Plant Biol. 10 (2), 109–115.

Schubert, I., Lysak, M.A., 2011. Interpretation of karyotype evolution should consider chromosome structural constraints. Trends Genet. 27 (6), 207–216.

Schum, A., 2003. Mutation breeding in ornamentals: an efficient breeding method? Acta Hortic. 612, 47–60.

Sears, E.R., 1969. Wheat cytogenetics. Annu. Rev. Genet. 3, 451–468.

Sears, M.K., Hellmich, R.L., Stanley-Horn, D.E., Oberhauser, K.S., Pleasants, J.M., Mattila, H.R., Siegfried, B.D., Dively, G.P., 2001. Impact of Bt corn pollen on monarch butterfly populations: a risk assessment. Proc. Natl. Acad. Sci. U. S. A. 98 (21), 11937–11942.

Shaner, D.L., 2014. Lessons learned from the history of herbicide resistance. Weed Sci. 62 (2), 427–431.

Sharma, D.P., 2015. Plant tissue culture in crop improvement. Acta Hortic. 1085, 273–284.

Songstad, D.D., Petolino, J.F., Voytas, D.F., Reichert, N.A., 2017. Genome editing of plants. Crit. Rev. Plant Sci. https://doi.org/10.1080/07352689.2017.1281663:1-23.

Stadler, L.J., 1942. Some observations on gene variability and spontaneous mutation. In: The Spragg Memorial Lectures on Plant Breeding, third ed. Michigan State College Press, East Lansing, MI (USA).

Stadler, L.J., 1954. The gene. Science 120, 811–819.

Sugiyama, M., 2015. Historical review of research on plant cell dedifferentiation. J. Plant Res. 128 (3), 349–359.

Sybenga, J., 1992. Manipulation of genome composition. B. Gene dose: duplication, polyploidy and gametic chromosome number. In: Sybenga, J. (Ed.), Cytogenetics in Plant Breeding. Springer-Verlag, Berlin-Heidelberg, pp. 327–371.

Tabashnik, B.E., Van Rensburg, J.B.J., Carriere, Y., 2009. Field-evolved insect resistance to Bt crops: definition, theory, and data. J. Econ. Entomol. 102 (6), 2011–2025.

Tamayo-Ordóñez, M.C., Espinosa-Barrera, L.A., Tamayo-Ordóñez, Y.J., Ayil-Gutiérrez, B., Sánchez-Teyer, L.F., 2016. Advances and perspectives in the generation of polyploid plant species. Euphytica 209 (1), 1–22.

Teng, P.S., 2008. Bioscience Entrepreneurship in Asia: Creating Value With Biology. World Scientific, Hackensack, NJ (USA). 335 pp.

Thackray, A., 1998. Private Science: Biotechnology and the Rise of the Molecular Sciences. University of Pennsylvania Press, Philadelphia, PA (USA). 268 pp.

Tsen, C.C., 1974. Triticale: First Man-Made Cereal. American Association of Cereal Chemists, St. Paul, MN (USA). 291 pp.

Tu, Y., Sun, J., Ge, X., Li, Z., 2009. Chromosome elimination, addition and introgression in intertribal partial hybrids between *Brassica rapa* and *Isatis indigotica*. Ann. Bot. 103 (7), 1039–1048.

Vasil, I.K., 2008. A history of plant biotechnology: from the cell theory of Schleiden and Schwann to biotech crops. Plant Cell Rep. 27 (9), 1423–1440.

Wang, Q.M., Wang, L., 2012. An evolutionary view of plant tissue culture: somaclonal variation and selection. Plant Cell Rep. 31 (9), 1535–1547.

Xiong, Z., Gaeta, R.T., Pires, J.C., 2011. Homoeologous shuffling and chromosome compensation maintain genome balance in resynthesized allopolyploid *Brassica napus*. Proc. Natl. Acad. Sci. U. S. A. 108 (19), 7908–7913.

Yonezawa, K., Yamagata, H., 1977. On the optimum mutation rate and optimum dose for practical mutation breeding. Euphytica 131 (2), 413–426.

Zeldin, E.L., McCown, B.H., 2002. Towards the development of a highly fertile polyploid cranberry. Acta Hortic. 574, 175–180.

Further Reading

Maluszynski, M., van Zanten, L., Ashri, A., Brunner, H., Ahloowalia, B., Zapata, F.J., Weck, E., 1995. Mutation techniques in plant breeding. In: Sigurbjörnsson, B., Machi, S., Dargie, J.D., Gustafson, J.P. (Eds.), Induced Mutations and Molecular Techniques for Crop Improvement. IAEA, Vienna (AUS), pp. 489–504.

Improvement of Selection Effectiveness

INTRODUCTION

Selection is the most powerful tool available to the plant breeder to affect changes in allelic frequencies within populations. Since the worlds of agriculture and consumers of agricultural products deal only with actual phenotypes and not genomic DNA sequences from which phenotypes are partly derived, most selection is accomplished by a decision either to include or not include an individual in the next mating cycle based solely on phenotype. The genes of the selected individual will be included in the new gene pool, and those of excluded individuals will not. The new population will have different allelic frequencies at loci that govern the economic phenotype: Desirable alleles will be enriched and undesirable alleles will be depressed in relative frequency. The classical view of the process and outcomes of selection is depicted in Fig. 9.1.

Let us return to the fundamental principle that **phenotype = genotype + environment** (Chapter 3). From this fundamental, the following was extrapolated to account for sources of phenotypic variability within populations (Chapter 3):

$$V_P = V_G + V_E, \text{expanded to } V_P = V_A + V_D + V_I + V_E + V_{GxE}$$

Further, let us consider the effectiveness of selection under the following scenarios of relative variances (Wricke and Weber, 1986; Fig. 9.2):

1. $V_P = 0$: All individuals in a given population are phenotypically identical; selection, therefore, is impossible.
2. $V_P = V_E$ (i.e. $V_G = 0$): All phenotypic variability in a given population is determined by effects of the environment. Selection will not alter allelic frequencies in succeeding generations.
3. V_P/V_A = narrow sense $h^2 \approx 0.1$: Response to selection should be "slow and steady"
4. Narrow sense $h^2 \approx 0.9$: Response to selection should be rapid.

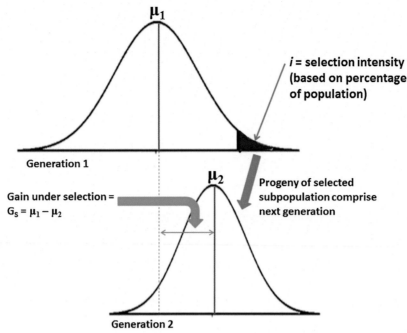

FIG. 9.1   The classical model of selection on V_P practiced on a hypothetical population where $V_G > 0$.

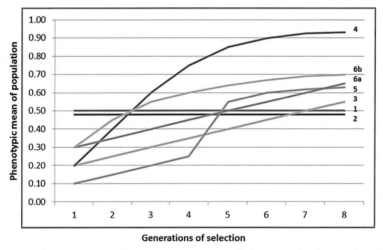

FIG. 9.2   Mean of targeted phenotype in populations of successive generations of selection under the scenarios described in the text (see above).

5. V_P/V_G = broad sense $h^2 \approx 0.9$; $H^2_{NS} \approx 0.1$: Most genetic variability for a given trait is explained by non-additive effects; therefore, response to selection will be unpredictable, possibly including periods of stagnation followed by rapid changes.
6. V_{GxE} is large relative to V_A, V_D, V_I: Response to selection depends on environment.; may be rapid in some environments (e.g. 6a), slower in others (e.g. 6b).

   The ability of the plant breeder to observe the phenotype of an individual and to "see its genes" is the essence of the craft. This skill often becomes so obtuse that the breeder is at a loss to explain the reasons behind his/her conclusions, and often chalks it up to intuition. Many years and locations and individuals selected and tested are necessary to achieve this level of intuition. The human eye (and sometimes nose, hand, palette, and even ear) and brain are the most sophisticated scanning and integrating devices for complex morphological and sensory patterns, at least at this point in time. As this book is published, research into the development of robotics to measure and evaluate complex morphological phenotypes is proceeding. When conducting phenotypic selection on a population of 10,000 or more individuals, however, no machine is yet available to match or replace the human senses and brain to determine what is desirable and what is not. This "breeding intuition" is often what sets apart a good from an exceptional plant breeder. Extraordinary intuition is also an important factor that made "Bob" such an effective plant breeder (Chapter 3).

Recall that the equation $V_P = V_G + V_E$ and derivatives is always subject to a set of assumptions pertaining to specific populations and environments. Each combination of genotypes and environments will lead to a unique mathematical result. Some populations are already largely devoid of genetic variability, so V_P will be determined mostly by V_E. Other populations may be replete with polymorphisms and the corresponding V_G is relatively higher. When a plant breeder states that "heritability of X trait is Y", "Y" is probably a composite or average of many observations. Alternatively, "Y" may be the most optimistic estimate given ideal parameters of population and environment. The savvy response would be to inquire about how this particular heritability estimate was determined, i.e. "… under what set of genotypes and under which environments was "Y" estimated?"

HERITABILITY AND RESPONSE TO SELECTION

The theoretical relationships of parents and offspring with regard to genotype and phenotype were first described by Wright (1921). He elucidated the relationships between the crucial selection criteria (phenotypic variance, selection intensity, heritability) and expected gain under selection, assuming additive gene action and random assortment of alleles:

$$Gs = (i)\left(\sqrt{Vp}\right)\left(h^2\right)$$

- Gs = gain from selection
- i = constant based on selection intensity in standard deviation units
- Vp = total phenotypic variation
- h^2 = narrow sense heritability

In the following example (Fig. 9.3), selection is taking place for shorter internode length in a hypothetical crop species. It is assumed that $i = 2.063$, or 5% of the population with the shortest internodes. V_P will change as selection is practiced recurrently due to progressive depletion of V_G. In this example, V_P is 3.75 in the starting population then decreases incrementally (3.50, 3.25, 3.00) in subsequent selected populations. h^2 is varied in the example to show the effect if the other parameters are constant. This is, obviously, a highly contrived example only for the sake of illustrating the effect of h^2.

Conversely, by knowing the gain under selection and other parameters, it is possible to impute narrow sense h^2 by applying the same equation. It is important to note that the assumptions implicit in the validity of these theoretical relationships are rarely satisfied. This set of logical relationships is, at best, a crude estimate of reality.

The progression of gain under selection is predictable in this example, but reality is rarely this simple or clear. V_E changes constantly due to fluctuating climate, light intensity, ecological interactions, etc. Changes in V_E affect V_P and,

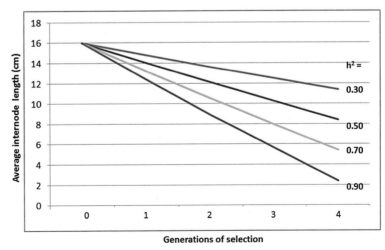

FIG. 9.3 Illustration of the effect of heritability (h^2) when starting from the same population; 4 cycles of selection for short internode length.

therefore, h^2. V_P is a neutral factor in this relationship. The plant breeder must screen available germplasm to identify the starting material that is the most beneficial to the planned breeding program, but otherwise has little control over V_P. The starting population will also define the phenotype that will be altered by changes in allelic frequencies, driven by selection.

The plant breeder may increase or decrease the value of i but this factor is directly linked to the number of individuals in the selected sample. Increasing i will have the effect of increasing GS as long as the basic assumptions hold. As i increases, the population size of the sample decreases to the point that allelic frequencies may be affected by genetic drift (Falconer and MacKay, 1996).

It is very clear that the fundamental factor that the plant breeder may vary in the quest to change phenotypes in a desirable way is h^2. As was stated earlier, h^2 is a conditional value, determined by the set of genotypes present in the population and the environments under which they are grown and evaluated. Therefore, the plant breeder may alter h^2 by altering population genetic structure and/or environment(s). We shall see later in the chapter that the plant breeder may also alter h^2 by selecting for surrogate phenotypes that are genetically tied to the targeted phenotype of interest. The most effective surrogate phenotype is comprised of the DNA sequences that are associated with desirable alleles, or *quantitative trait loci* (QTL; Paterson et al., 1988; Bernardo, 1998; Da et al., 2000).

ENHANCED HERITABILITY: OPEN FIELD PRODUCTION

Humans are experts at changing and controlling environments. The weather we are confronted with changes constantly and often does not suit us. To survive, we may either relocate to a place where the weather is better or we may build dwellings that have walls and roofs to keep us dry, protected from "elements", and at the desired air temperature. We invent devices to safely illuminate, heat, and cool these enclosed dwellings. It was logical, therefore, to apply principles of controlled environments to agriculture to attain better performance of domesticated animals and plants.

The V_E term of phenotypic variability is rooted in the degree of heterogeneity and sequences of environmental parameters experienced by a population during a period of time when the phenotype is being determined. If identical twins are separated and reared in disparate environments, they may end up exhibiting an array of different phenotypes even though they are genetically identical. Contrasting identical twins separated at birth is a very popular way to gauge the relative effects of genes and environment on different aspects of human morphology and behavior (e.g. Stunkard et al., 1990).

It is impossible to control environments completely, and phenotypic variability is always present even among sets of genetically identical individuals. If environmental parameters are controlled more progressively, V_E is usually reduced accordingly. If V_E is reduced while V_G is not changed, V_P should decrease and h^2 should increase with the increased proportion determined genetically as opposed to environmentally. Since h^2 is elevated, response to selection should also increase.

How can environmental parameters be controlled in the open plots of land in which breeding populations of plants are grown? The plant breeder must be vigilant and meticulous in choosing where plants are to be produced. The field should be characterized by environmental parameters at or near the mid-point of the range wherein the targeted cultivar will be grown. This will help ensure that results obtained will be relevant to program clients, since V_{GxE} terms increase in relative magnitude in marginal environments (Fig. 9.1). It also helps to demonstrate to clients and sponsors that plants appear similar to what they would be under the management programs of targeted users, adding credibility to the results and outcomes.

The open field chosen for breeding populations should be flat (unless grades are the norm as is the case with some orchard crops) and have a uniform soil type. The degree of compaction and drainage should be consistent and appropriate to accepted management practices. Levels of nutrients should also be evenly distributed throughout the plot. Geographical areas characterized by deep alluvial soils with high organic matter, and especially table land at the bottom of undulating valleys, tend to be the most uniform.

Most management practices include soil tillage and some type of grading, such as raised beds, and these should be conducted in an consistent fashion. Intrusions by structures and wooded areas that introduce differences in shading and wind patterns are discouraged. Water and supplemental nutrients should be administered to all plots and plants in an equitable manner. All other management practices, such as weed, insect and animal pest, disease abatement should correspond to accepted commercial practices and be uniformly applied. Juxtaposition with adjacent plots that might be the source of phytoactive chemical drifts in air or water should be avoided. The author recalls one instance where a poinsettia performance trial was located adjacent to an asphalt parking lot that was resurfaced during the experiment. The border of the trial exhibited peculiar growth patterns that mimicked phytohormone effects, later determined to be volatile substances emanating from the pavement resurfacing agent.

Population establishment should also conform to accepted standards and densities should be uniform within the open field plot. If plots are established from seeds, they should have been stored and handled according to a defined and accepted procedure. Spacing and depth of sown seeds should be equivalent. Transplants, if used, can be a source of undesirable environmental variation. They should be produced in a manner that ensures a uniform population of plants in the subsequent field, even when genetic differences are present, for the observation of genetic differences is a good indication that everything is being done correctly. Following stand establishment, all management practices pertaining to individual plants must be administered equitably, for example shearing, pruning, mowing, thatching, topping, staking, trellising, or grafting.

Each management practice comes with a cost. Efforts to minimize environmental variation within a field plot are also associated with incremental costs. Most plant breeders endeavor to work with the largest possible populations to maximize the probability of observing rare desirable genetic events. The combined cost of the plot and necessary inputs ultimately imposes practical limits. Armed with knowledge pertaining to the crop species, populations, and traits of interest, the breeder can make educated compromises that maximize the probabilities of success and impact considering these limitations.

ENHANCED HERITABILITY: ENCLOSURES

The next increment of environmental control is *greenhouse production*, often referred to as *controlled environment agriculture*. Growing plants in greenhouses is associated with higher unit expenses than open fields due to the costs of the structures, transparent material maintenance, and increased labor and energy inputs. Notwithstanding economic challenges, many high-value perishable plant products are routinely produced in greenhouses throughout the world including indoor foliage plants, cut flowers, and high-value food crops such as herbs and certain fruits and vegetables. Transplants of vegetables and ornamental bedding annuals and perennials are also routinely produced in greenhouses. Such structures conceptually provide a barrier against biotic factors such as weeds, pests, and diseases. With the advent of open-roof designs to achieve more energy-efficient air temperature management, unwanted organisms can more easily gain entry into the structure. The artificial greenhouse environment can also favor certain introduced arthropod pests and disease pathogens. More powerful pest management strategies than those developed for open field culture are often necessary.

Most plant breeders located in temperate climates use greenhouses extensively in their respective programs. The overwhelming majority of greenhouse uses are associated with protection of plant populations during prolonged periods of adverse weather, primarily winter in extreme northern and southern latitudes. Thus, the plant breeder can make progress during winter months without the need for expensive travel to geographical areas that offer favorable climates. Many plant breeders of high-value crop species in the northern hemisphere spend proportionately larger amounts of time during the winter in the tropics or southern hemisphere.

The greenhouse, if properly designed and equipped, can provide narrower environmental parameters than the best managed open field (Fig. 9.4). Stringent sanitation measures can exclude unwanted and irrelevant pests and diseases. The temperature can be tightly controlled with forced air conventional heating and cooling, floor heating or evaporative cooling, and varied according to diurnal cycles or plant developmental stage. Lighting can also be supplemented or administered selectively as is appropriate for special needs such as flower forcing. With the advent of light-emitting diodes (LEDs), artificial lighting is poised to offer more desirable spectral quality at cheaper prices.

The soil medium is usually constituted from defined, controlled components and dispensed into uniform containers or beds. Alternatively, plants may be produced hydroponically without soil medium, though this may not be consistent with the product development needs of the plant breeder targeting a crop that is produced under conditions of open-field agriculture. Water may also be purified and disinfested to established standards.

All of these factors may combine to reduce V_E to an even lower increment than can be achieved in the open field. While reduced V_E will portend higher h^2 and response to selection, cost is an important factor in determining whether a greenhouse environment may be used in a breeding program. Even more important, however, is the relevance of results to the production context at which genetic improvements are aimed. If V_{GxE} is relatively high for the traits under selection, breeding populations that emerge from cycles of selection in greenhouses may not perform predictably in open fields. As environments become more marginal or exclusive, V_{GxE} tends to increase proportionately (Fig. 9.2).

FIG. 9.4 Producing crop populations under controlled environmental conditions where the surface is comprised of high-transmission polymers, the interior and all surfaces are sanitized, water is purified and purged of pathogens, temperature and humidity are controlled, and day length is controlled with artificial lighting of high intensity and spectral quality.

ENHANCED HERITABILITY: ENVIRONMENTAL CHAMBERS

The *growth chamber* is the next increment of environmental control, a small box in which all parameters are managed to a much greater extent than even the greenhouse. Such devices have not proven to be of much use to plant breeders except for certain phenotypes that require highly defined conditions, such as certain diseases or performance parameters. The costs per individual plant maintained in controlled environment chambers are simply too high to accommodate meaningful population sizes and relevance to agriculture is too low to ensure that composite plant selections will culminate in useful entities.

Controlled environment chambers have high applicability in breeding programs involving the development of genetic resistance or tolerance to disease pathogens or insect pests. Many disease phenotypes are highly sensitive to environmental ranges, and controlled environment chambers are essential to discern genetically resistant from susceptible individuals. This subject will be addressed in further detail in Chapter 19.

SELECTION BASED ON PROGENY TESTS

Direct selection on individual phenotypes within populations is conceptually straightforward. Recall that one method of estimating h^2 is to regress progeny performance onto that of the parents. A related concept is the selection of individuals based on the performance of their respective progeny. This strategy is particularly effective in cases of low to intermediate heritability, amenable to species that are outcrossing, and breeding for hybrid cultivars. Progeny tests are used in cross-pollinated plant species to estimate combining ability (Chapter 14), a useful parameter for the selection of inbred parents in a hybrid breeding program (Han et al., 2006; Luan et al., 2010).

Consider a population that is segregating for a complex trait under selection by the plant breeder. Each individual exhibits a phenotype that may or may not be an accurate reflection of its underlying genotype. If h^2 is relatively high, the association between phenotype and genotype will be stronger than if h^2 is lower. Each individual considered for selection is used to generate derivative populations of progeny. Progeny populations are then measured for the targeted phenotype and contrasted. Based on the mean and variance of progeny performance, the original individual is selected for continuation in the mainline breeding program (Gallais, 1991).

The use of progeny tests to select individuals is an extension of the notion that offspring inherit their genes from parents. It is often presumed that children will end up much as their parents did. Is individual X a smart guy? He must be, since all of his children are brilliant; and they got it from somewhere (probably from mom!)." The acorn doesn't drop far from the tree." Perhaps we should choose our leaders based on how their children have fared? Or would they be too old?

SELECTION ON TRAIT COMPONENTS

The notion that most measurable or observable characteristics of whole organisms have been found to be controlled by many independent and/or interacting genes was advanced in Chapter 3. Inheritance patterns that are more continuous than discrete, appearing to defy characterization by Mendel's Laws, were defined as complex, multigenic, or quantitative. Not surprisingly, most traits of economic significance fall into this category: for example, yield, stature, growth habit, fruit or flower quality, stress tolerance, maturity date, and a score of other attributes. Such traits are often characterized by low h^2 (Mayo, 1987). Complex traits can frequently be partitioned into constituent component "subtraits" that contribute to the whole phenotype, for example yield components (number of units, size of individual units, etc.). By directing selection on one or more trait components that have higher heritability, it has been demonstrated that faster progress in the composite phenotype can sometimes be realized (Sane and Amla, 1990).

Fruit yield is a good example of such a complex quantitatively inherited trait. Component subtraits include number of fruit clusters, numbers of fruits set per cluster, average fruit weight, packout (percent marketable), premature fruit drop, shelf life, and many others (Iezzoni and Mulinix, 1992; Cramer and Wehner, 2000; Stephens et al., 2012). Cultivar A may be low yielding, but have a large number of fruit clusters; perhaps yield is sacrificed to small fruit size or low fruit per cluster ratio? In this case, selection would be appropriately directed at larger fruit mass or number of fruit per cluster instead of total harvest yield. In cultivar B, the factor most limiting to yield may be low number of fruit clusters; cultivar C premature fruit drop, and so forth.

One potential problem in selecting on partitioned trait components is the impacts on V_P due to interactions of phenotypic components are ignored. It is obvious that V_P will include many significant interactions such as number of clusters and fruit size. By selecting for larger fruit mass, the number of fruit clusters may drop proportionately and fruit drop may become a problem. By selecting for one component of yield partitioning, other components are often negatively affected. It is likely that intense selection for increased fruit size alone will also lead to reduced numbers of clusters or fruit per cluster (or both). Therefore, any selection on components must be conducted with simultaneous counterbalancing selection on other components (Zalapa et al., 2006).

The aim of plant physiologists is to understand how the internal machinery of plants works. Results often emerge from research in plant physiology that may be directly applicable in plant breeding. Using the fruit yield example cited above, plant physiological results have illuminated all of the key rate-limiting steps in the fixation of carbon and final transformation and transportation into fruit and seed storage reserves. By selecting for higher enzymatic rates at these metabolic choke points, either higher unit activity or larger enzyme quantities, it may be possible to positively influence the flow of fixed carbon into fruit yield (Trentacoste et al., 2012; Shekar et al., 2013; Rogers et al., 2014; Munawar et al., 2015).

SELECTION BASED ON COMPOSITE PHENOTYPIC SCORE

All successful cultivars are comprised of a balance of many desirable traits. Agronomic crop species tend to be relatively simple, net yield and harvest quality being the primary factors defining success. The successful horticultural crop species cultivar, in contrast, is driven by a larger number of performance factors, and quality of the finished product is much more critical than it is for agronomic crop species. Several multiple trait selection schemes have been developed, including independent culling levels, tandem selection, and composite index selection. These schemes can result in improvement even for traits with unfavorable associations (Henning and Teuber, 1996; Yan and Frégeau-Reid, 2008; Luby and Shaw, 2009).

Yield in horticultural crops is not solely tonnage of biomass produced in the field. Rather, it is the proportion of the crop that can be harvested and brought to market in a condition and at a price acceptable to the consumer. Quality in a horticultural food crop species may include flavor, color, shape, size, degree of damage, nutrient levels, and traits that permit greater perceived food safety or environmental sustainability. Quality in a horticultural ornamental or nursery crop species is equally multi-factorial. Traits with unfavorable associations will be of concern to the breeder if the cause is unfavorably correlated genetic effects, especially those resulting from pleiotropy (Luby and Shaw, 2009). As QTL are discovered and verified that may provide better methods to isolate the heritable components of economically-valuable phenotypes, they may be substituted in the resulting multi-factorial model (Slater et al., 2016; Shimomura et al., 2017)

Composite index selection is based on a composite score that is an imputed numerical value that is based on the sum of weighted criteria evaluation subscores. Some of these subscores may be genuine measurements (such as weight, height, volume, or fruit pulp solids or pH) and others may be subjective values based on assessment relative to a standard. A scale is often devised to emulate the degree to which the trait being evaluated attains an ideal maximum, such as 1–5 or 1–10. In a 1–5 subjective scale, a "1" would be attributed to a breeding population that exhibits the lowest or least desirable end of the range and a "5" would represent the highest or most desirable end of the phenotypic continuum.

For example, in a processing tomato (*Solanum lycopersicum*) breeding program, the following traits are to be assigned subscores:

1. Maturity date (1 = earliest; 5 = latest)
2. Plant growth habit (1 = compact determinate; 5 = open indeterminate)
3. Plant vigor (1 = least vigorous; 5 = most vigorous)
4. Foliar disease damage (1 = least damage; 5 = most damage)
5. Presence of jointlessness (1 = most jointed; 5 = least jointed)
6. Concentration of fruit set (1 = least concentrated; 5 = most concentrated)
7. Gross fruit yield (in avg. kg per 5M plot)
8. Packout (raw proportion of marketable fruits)
9. Fruit shape (1 = least desirable; 5 = most desirable)
10. Fruit firmness (1 = least firm; 5 = most firm)
11. Average fruit size (g)
12. External fruit color (1 = least desirable; 5 = most desirable)
13. External fruit color consistency (1 = least consistent; 5 = most consistent)
14. Internal fruit color (1 = least desirable; 5 = most desirable)
15. Internal fruit color consistency (1 = least consistent; 5 = most consistent)
16. Fruit stem scar (1 = smallest; 5 = largest)
17. Fruit blossom scar (1 = smallest; 5 = largest)
18. Fruit pulp titratable acidity (in ml NaOH equivalent)
19. Fruit pulp soluble solids (in °brix)
20. Fruit pulp furaneol (ng/g pulp)
21. Fruit pulp geranial (ng/g pulp)
22. Fruit pulp decadianol (ng/g pulp)

In this example, 20 tomato breeding lines under consideration for advancement to the next cycle are evaluated independently for all of these phenotypes, and the results are depicted in Table 9.1. Each trait is first considered independently with regard to its contribution to the overall value of the breeding line. In the case of processing tomato, mid-season maturity is desired, so raw values for maturity are transformed into a truncating scale (1 = 1; 2 = 2; 3 = 3; 4 = 2; 5 = 1)(see column 1a). The rest of the subscore values are concluded to be directly proportional, one end of the objective scale more desirable than the other.

The model used in this hypothetical example is CS = *1a* + *2* + *3* + (*6*×*4*) + *5* + *6* + ((*7*/10)-3) + ((*8*×10)-5) + *9* + *10* + ((*11*/100)x2) + *12* + *13* + *14* + *15* + *16* + *17* + (*18*×2) + (*19*×2) + (*20*×10) + (*21*×2) + (*22*×3). If we apply a selection pressure of 25% to this set of candidate breeding lines based on overall composite performance scores, ABC009, ABC014, ABC015, ABC016, and ABC019 are advanced to the next generation, and the unselected entries are discarded or recycled in the breeding program.

Models for the imputation of composite scores are devised intuitively from breeding objectives (Chapter 6) and experience with the traits being evaluated with respect to breeding objectives. The model should evolve over time as imputed scores are regressed onto the performance scores of progeny. Weighting and skewing of raw data to more effectively develop new populations that approximate breeding objectives should be re-examined at every cycle. Breaking down breeding objectives into a composite score model for selection, driven by economic returns into discrete components of performance that may be treated independently, is an "art" that takes time and experience for the plant breeder to master.

SELECTION OF LINKED MOLECULAR MARKERS: MARKER-ASSISTED SELECTION (MAS)

What are the factors that are responsible for low h^2? Heritability is affect by ascending orders of interactions between gene promoters and transcription factors, gene products (transcription, transcript processing, translation,

TABLE 9.1 Imputation of Composite Breeding Value Scores Based on 22 Constituent Trait Subscores

Breeding line	Subscore value																							CS[a]	Select
	1	1a	2	3	4	5	6	7	8	9	10	11	12	13	14	15	16	17	18	19	20	21	22		
ABC001	3	3	3	2	2	1	4	68.2	0.74	4	4	253.1	3	3	4	4	3	3	2.65	1.94	0.41	2.44	0.93	77.232	
ABC002	4	2	3	3	3	4	3	78.4	0.78	3	4	226.8	2	2	4	4	2	3	2.41	2.23	0.29	2.12	1.12	73.956	
ABC003	3	3	2	3	3	1	4	73.5	0.79	3	2	289.7	1	2	3	2	3	3	2.54	2.57	0.25	3.09	0.92	71.704	
ABC004	4	2	3	3	3	1	4	81.8	0.68	4	2	265.5	3	4	4	3	4	3	2.49	2.40	0.22	2.54	0.85	74.900	
ABC005	4	2	4	4	4	2	3	55.0	0.55	5	2	238.4	4	3	3	4	4	4	2.77	2.39	0.30	3.23	0.95	74.398	
ABC006	2	2	3	3	4	1	5	53.6	0.57	3	4	190.6	5	4	5	5	3	2	3.26	2.95	0.38	3.11	1.15	79.762	
ABC007	2	2	2	3	3	5	5	67.1	0.80	4	3	336.3	3	3	4	4	3	3	3.17	2.60	0.23	2.29	0.82	83.316	
ABC008	1	1	2	3	3	5	5	64.3	0.82	3	4	122.5	4	4	5	5	1	1	2.62	2.51	0.27	1.87	1.04	76.900	
ABC009	3	3	2	3	2	5	4	75.5	0.85	4	5	301.8	3	3	4	1	3	4	3.23	2.79	0.22	2.10	0.80	84.926	*
ABC010	3	3	3	3	2	4	4	80.7	0.81	2	3	321.0	3	2	2	2	4	4	1.96	2.41	0.25	2.45	0.91	76.460	
ABC011	4	2	5	4	4	1	2	68.2	0.71	2	3	228.7	4	3	3	4	2	2	2.95	3.24	0.53	3.18	1.75	74.784	
ABC012	5	1	5	3	4	2	1	38.1	0.51	3	3	249.0	3	4	4	5	3	2	2.72	2.80	0.37	3.46	1.62	69.410	
ABC013	4	2	5	5	5	1	3	77.6	0.66	2	2	318.6	4	3	3	3	3	2	2.38	2.75	0.21	3.14	1.88	72.012	
ABC014	3	3	3	4	4	4	3	88.9	0.82	4	4	366.9	4	3	2	3	5	3	2.29	2.37	0.18	2.96	1.35	84.518	*
ABC015	3	3	4	4	2	5	4	84.3	0.84	5	5	358.2	4	4	3	3	4	4	3.02	2.99	0.55	3.33	1.71	99.304	*
ABC016	4	2	4	3	2	5	4	71.8	0.85	5	5	321.4	3	5	3	4	5	3	2.63	2.43	0.29	2.30	0.99	87.698	*
ABC017	3	3	3	2	3	1	4	29.6	0.50	3	4	160.1	5	4	5	5	2	2	2.95	3.07	0.33	2.48	0.82	71.922	
ABC018	2	2	2	3	1	1	5	66.0	0.74	3	2	264.9	2	1	3	3	2	2	2.54	2.84	0.38	2.62	0.86	71.678	
ABC019	4	2	3	4	3	4	4	82.1	0.79	3	2	333.7	3	2	3	2	3	5	3.11	3.16	0.78	3.12	1.56	89.044	*
ABC020	5	1	5	2	2	5	2	48.7	0.63	1	1	166.1	5	5	4	5	2	2	4.12	5.12	0.75	3.90	0.47	81.682	

a CS = Composite Breeding Line Score.

post-translational processing), and the environment. Simply-inherited traits are characterized by little or no interaction of gene expression or translation with the environment. For example, the gene that results in black seed coat is likely transcribed according to a distinct developmental or environmental signal and the transcript is translated into a gene product that acts on metabolism to condition the synthesis and deposition of black pigment in the seed coat. The white gene follows a similar pathway leading to white seed coat. If the seed coat color phenotype is manifested in a thousand shades of gray, however, and many genes are involved, each interacting with each other and intra- and extracellular environmental cues, the connection between phenotype and corresponding genotype is more equivocal and occluded than for simply inherited phenotypes. It becomes more difficult to predict phenotype based on genotype alone and h^2, therefore, is reduced.

The skill of the plant breeder is to examine an individual and assess the phenotype and, indirectly, surmise the underlying genotype. The opportunity to test breeding intuition against a hard standard, such as DNA sequence of the coding frames and corresponding control domains of all the responsible genes is one welcomed by all plant breeders. By selecting directly on genomic DNA sequences responsible for the phenotype, the interactions of gene products and environments that obscure the relationship between genotype and phenotype are bypassed, and h^2 is 1.00 (Khanna, 1990). The widespread ability of plant breeders to exercise selection directly on genomic DNA sequence in lieu of whole-plant phenotype is fast approaching. It is likely that complex phenotypes that were historically characterized as quantitatively inherited will be understood at the DNA sequence level within the foreseeable future (see Functional genomics section below).

While molecular biologists unravel the remaining secrets of the eukaryotic genome, an effective selection strategy combining phenotype and DNA selection has been developed, known as *marker assisted selection* (MAS). Marker loci have been used since the early 20th century in the study of the structure of genomes (Lewontin, 1974; Langridge and Chalmers, 2005). Examination of the cosegregation ratios of pairs of genes controlling simply inherited phenotypes revealed instances where parental combinations clearly outnumbered recombinants. Ultimately, this led to the discovery of linkage, then linkage groups, and followed by the realization that linkage groups and chromosomes were different manifestations of the same genome organizational feature (Morgan et al., 1925; Sinnott and Dunn, 1939).

In the 1950s, a new adaptation of liquid chromatography was applied to polypeptides wherein an electric field was applied to a mixture in a salt buffer solution. Polypeptides tend to be positively charged at neutral (biological) pH ranges, so migrate toward the cathode pole in this electric field. When a porous matrix, or gel, was overlaid on the electric field, the proteins could be physically fixed in place long enough to visualize differences in the rate of migration. Large, neutrally charged polypeptides moved more slowly in the charged matrix, while small, positively charged species moved more quickly. The technique was modified to the extent that the migration of nearly identical proteins differing in a single amino acid subunit could be captured, appearing as bands emanating from the original point source. Various types of stains and dyes were developed to visualize the polypeptides. Certain of these were specific to all polypeptides, while others were linked to enzymatic activity. In the latter, the appearance of dye in a region of gel matrix was evidence that a protein with specific enzymatic activity was localized there. The technique of electrophoretic separation in a gel matrix followed by non-specific or activity staining came to be known as gel electrophoresis (McMillan, 1984).

Isozymes, or *isoenzymes*, are proteins that have the same catalytic function but that differ in amino acid sequence. These proteins are usually related evolutionarily, so differ in relatively few amino acid units. Mutations in the coding sequence for such enzymes often results in isozymes within the same individual that have different electrophoretic mobility. Thus, isozymes viewed by gel electrophoresis may be subjected to classical genetic studies, and have been shown to obey Mendel's laws (Weeden, 1984).

Isozymes constitute excellent marker genes since they are easily detected and absolutely diagnostic (Stuber, 1990). Plant geneticists seized on the new treasure trove of markers in the 1960s and 1970s, adding isozyme markers to already existing linkage maps, or using isozymes as a starting point (McMillan, 1984). The notion of using molecular markers to find constituent genes underlying quantitative traits and using markers to select for desirable alleles (later alleles at quantitative trait loci) was first advanced by Tanksley et al. (1982).

For example, the aps-1[1] allele of tomato (*Solanum lycospersicum*) was shown to be either very tightly linked to the locus for resistance to root knot nematode resistance (*Mi*) or synonymous with that locus (Medina-Filho and Stevens, 1980; Rick, 1984; Stevens and Rick, 1986). Commercial varieties grown in the western U.S. during the mid-20th century were plagued by root knot nematode (*Meloidogyne incognita*), a microscopic soil-borne roundworm that is parasitic on many plant species. Tests of available germplasm within the confines of the species *S. lycospersicum* (syn *Lycopersicon esculentum*) revealed no sources of resistance/tolerance but certain populations of a related wild species *S. peruvianum*, however, appeared to express resistance (Medina-Filho and Stevens, 1980). The interspecific *S. lycopersicum x S. peruvianum* hybrids were successfully produced, and a recurrent backcross program was promulgated to transfer the prospective resistance gene into cultivated tomato. Subsequent linkage studies determined that the resistance gene *Mi* was closely linked to a molecular marker *Aps-1* that encoded an isozyme of the enzyme acid phosphatase. These two loci were found to be so tightly linked that no recombinant gametes were ever observed among ~50,000 segregating progeny. Some scientists even theorized that resistance was conditioned by the *S. peruvianum* APS enzyme but the linkage hypothesis was ultimately proven to be correct (Messeguer et al., 1991). Nematode resistance is extremely difficult and expensive to screen for, and the possibility that APS could be used as a surrogate marker was proposed (Medina-Filho and Stevens, 1980). Nearly 40 years later, this technique is still routinely used to select for root knot nematode resistance in tomato due to accuracy, low cost, and simplicity, forging the path for MAS to become a powerful mainstream breeding tool.

Surrogate marker systems based on nucleotide base polymorphisms in genomic DNA have been developed since 1990 and now predominate over isozymes due to the unlimited number of polymorphisms possible, relatively high efficiency, and low (and still decreasing) costs. DNA markers may be broadly classified according to the method by which they are detected: (i) hybridization-based; (ii) polymerase chain reaction (PCR)-based and (iii) DNA sequence-based (Gupta et al., 1999; Jones et al., 1997; Joshi et al., 1999; Winter and Kahl, 1995). DNA fragments can be separated electrophoretically in a gel matrix and visualized by staining with ethidium bromide or a non-toxic alternative, such as GelRed®, or hybridizing with tagged (colorimetric or ^{32}P) DNA probes. DNA markers are particularly useful if they reveal polymorphisms or if sequence differences exist at the same locus. Markers that do not discriminate between genotypes are called *monomorphic*. Polymorphic markers may also be described as codominant or dominant depending on whether they can discriminate between homozygotes and heterozygotes. In general, dominant markers are either present or absent whereas codominant polymorphic markers exhibit differences in fragment size. The different forms of a DNA marker (e.g. different sized bands on gels) are marker "alleles" in the same manner that whole-plant phenotypic genes may have different alleles.

Following the discovery of restriction enzymes (REs) in the 1960s (Meselson and Yuan, 1968), research ultimately led to a new class of molecular markers known as *restriction fragment length polymorphisms* (RFLPs). These markers are based on the DNA sequence specificity of individual REs, usually 4–6 nucleotide bases in length, a few longer or shorter. The 4-cutters (REs that recognize a sequence of four specific nucleotide bases then cuts and repairs the sugar-PO_4 backbone) tend to make larger numbers of cuts in the genome than do the 6-cutters, due to simple probability. If a given restriction is used to digest the entire genomes of an individual plant, thousands of fragments are released into the mixture, corresponding to the genomic DNA fragments between the sites recognized by the RE used.

DNA fragments, like polypeptides, have a net electrical charge at slightly acidic buffer pH, and will, therefore, migrate in an electrical field. Since there are only four different nucleotide bases and the bases are interspersed more or less randomly in the genome, DNA fragments tend to exhibit different rates of migration based mainly on length, or number of nucleotide bases in the fragment.

If a mixture of DNA fragments following digesting with a restriction enzyme is subjected to gel electrophoresis fragments will migrate according to length, shorter fragments migrating further from the origin than longer fragments. If the gel containing the migrating DNA fragments is stained with a DNA-specific stain, the individual bands may be discerned but it is impossible to discern fragment identity since there is likely to be thousands of them. What is needed is a way to identify specific fragments based on a meaningful criterion such as nucleotide base sequence.

During the time period that restriction enzymes were under intense study experimental work was also proceeding on a parallel course, *affinity association kinetics*. The DNA code itself is based on chemical affinity, of adenine for thymine, and cytosine for guanine, thus driving the fidelity of strand replication. Scientists discovered that DNA strands could be physically separated and would re-anneal in solution, based on collective base affinities (Buchholz and Collins, 2010). A strand 5′-AATTGGCC-3′ has tremendous affinity for 3′-TTAACCGG-5′, but much less so if the base sequences are not 100% complementary.

Affinity of DNA fragments based on nucleotide base sequence, or binding, was developed as a way to identify specific fragments in background resulting from digestion of a genome with a restriction enzyme. If the electrophoresed gel was blotted onto a matrix, initially nitrocellulose paper, DNA fragments would tend to be absorbed and bound to the matrix. Next, the DNA "probe" was prepared, a specific sequence fragment containing ^{32}P in lieu of the predominant stable phosphorus isotope. The probe and blotted filter were then immersed in a buffer solution that encouraged DNA re-annealing, or *hybridization*. After the solution was removed, a piece of X-ray photographic film was overlaid over the hybridized filter. Emulsion precipitates appeared in the same physical vicinity as the ^{32}P. This procedure was named *Southern blotting* after the original developer (Southern, 1975).

If the probe hybridized to fragments from two individuals that had migrated to different positions, and were therefore of different lengths, the corresponding DNA sequence relative to the site recognized by the restriction enzyme must have been different, or polymorphic (Fig. 8.4). Therefore, REs were useful for the discovery of DNA sequence polymorphisms that exist in genomic DNA. RE polymorphisms do not necessarily reside in regions of the genome that are transcribed into gene products. The phenotypic mutants that were used by early geneticists in fruit flies, maize, and ascomycetes were often lethal or sub-lethal, complicating the experiments due to skewing of observed segregation ratios. Since RFLPs do not usually affect fitness of individuals or have a negligible effect, there is no limit to the number that can be characterized in any given individual. There is a limit, however, on the number of REs that have been made available to scientists and plant breeders as tools to visualize markers, currently about 600.

RFLPs are tedious to visualize due to the large number of rigorous steps involved, and the need for a probe such as ^{32}P or monoclonal antibodies to visualize the DNA fragments. Advances in the understanding of the mechanisms of DNA synthesis by many researchers ultimately led to a further development in 1988 known as *polymerase chain reaction* (Saiki et al., 1988). If one had a purified DNA oligomer, a fragment of defined sequence and length, it was possible to hybridize it with genomic DNA. Components of the DNA synthesis and repair apparatus could be combined to incite the synthesis of a genomic fragment that bridged two sites that were homologous to the oligomer. By altering, or cycling, the temperature of the mixture, such that DNA would thermally anneal, then denature, then re-anneal, large numbers of fragments could be obtained, enough to visualize on a gel matrix with DNA-specific dyes. Kary Mullis, the corresponding author on the seminal paper and inventor of the PCR process, was awarded the Nobel Prize in Medicine in 1993 in honor of this monumental discovery and invention.

If an individual plant harbors a mutation in the sequence homologous to the oligomer, it will no longer hybridize, thus altering the length of the corresponding amplified PCR fragment. Such polymorphisms came to be known as *random amplified polymorphic DNA* (RAPD). Empirical evidence determined that oligomers in the range of 8–12 nucleotide bases in length work best, yielding discernible polymorphisms in reliably detectable numbers. The beauty of the technique lies in the oligomers that can easily be synthesized, then replicated and maintained in bacterial plasmids or other suitable vectors. The cost and prevalence of molecular polymorphisms have both been improved incrementally, and RAPDs were added to nascent linkage maps that already contained morphological, isozyme, and RFLP markers. In some cases, RAPD markers soon came to predominate in research aimed at defining genome organization and function (Kalinski, 1993).

RAPDs do have drawbacks as a marker system of choice. The inheritance of RAPD markers is usually totally dominant, although codominance is sometimes observed. This obscures the analysis of results somewhat since the heterozygote cannot be readily distinguished from one of the parental homozygotes. RAPD results are sometimes difficult to repeat from lab to lab due to differences in protocols. Further, RAPD results are sometimes difficult to interpret due to minor mismatches between the primer and the template (Kalinski, 1993).

Methods have been developed that enable the plant breeder or researcher to observe PCR phenotypic results in a very short period of time, known as *real-time PCR* (rtPCR) or *quantitative PCR* (qPCR). Sensitive detectors have been developed that measure primer amplification during PCR thermal cycles at very small titers, allowing results and conclusions to be drawn in very short periods of time compared to conventional PCR that is taken to an end point then the amplification products are separated electrophoretically in an agarose gel matrix. Either non-specific probes (fluorescent dyes) or specific probes (oligonucleotides) can be used for detection of amplified PCR primer products during qPCR. Sophisticated equipment that allows reaction mixtures [DNA + primer(s)] with combines the thermo-cylcing, amplification product detection, and computer-driven analyses are available to plant breeders, greatly accelerating the time required to obtain targeted results. Decreasing costs and faster availability of custom-synthesized oligonucleotide primers has also made this technology very attractive for plant breeders using MAS (Logan et al., 2009).

With time and additional efforts, new marker systems were added to the plant breeding toolbox, for example, amplified fragment length polymorphisms (AFLPs), simple sequence repeats (SSRs), and single nucleotide polymorphisms (SNPs). All of these molecular marker systems are based fundamentally on detection of DNA sequence homology, or lack thereof, the comparison leading to demonstration of polymorphism. The effectiveness of different marker systems for detecting targeted polymorphisms varies, and the plant breeder is challenged to identify the marker system that will be best for a specific application. For example, patterns of polymorphism revealed by the two marker systems were found by Sood et al. (2016) to vary in *Avena*. The level of polymorphism was higher for SSR (100%) than RAPD (85.82%). Differences in marker effectiveness were also reported in *Oryza* by Choi et al. (2016).

The costs of molecular markers have dropped dramatically, reproducibility has improved markedly, and overall accessibility of marker technology has increased monumentally since the early 1980s. Most practicing plant breeders currently have at least some access to the technology and many possess the know-how and equipment in their own programs (Logemann and Schell, 1993; Lörz and Wenzel, 2005; Malik et al., 2009; Henry, 2013; Prentis et al., 2013). As new marker systems are developed, the plant breeder is challenged to adopt the most cost-effective platform for the application at hand. For example, kompetitive allele-specific PCR (KASP®), a proprietary technology of LGC Ltd. that can distinguish alleles at variant loci, has been espoused as a cost-effective single-step genotyping technology that is cheaper for GS than SSRs and more flexible than genotyping by sequencing (GBS) or array-based genotyping (Steele et al., 2018).

Kordrostami and Rahimi (2015) presented a table that compared the characteristics, advantages, and disadvantages of the predominant marker systems used by plant breeders. This table is reproduced below with minor modifications (Table 9.2). Each situation is unique for species, phenotypes studied, and institutional limitations, and different markers will be more applicable in some instances over others. In general, all marker systems may be used for marker assisted breeding (see below), but some may be better than others depending on the situation.

Many other marker systems have been developed that are useful under specific circumstances or for targeted applications. Multilocus markers such as sequence-related amplified polymorphism (SRAP), inter-simple sequence repeat (ISSR), selectively amplified microsatellite polymorphic loci (SAMPL), AFLP and RAPD are suitable for estimating diversity and relationships among a given set of individuals, but for marker-assisted selection, locus specific markers such as sequence characterized amplified region (SCAR) and cleaved amplified polymorphic sequences (CAPS) markers are more appropriate (Paran and Michelmore, 1993; Biswas et al., 2011; Gulsen, 2016).

Schulman (2007) described a marker system based on genomic retrotransposons, sequence-specific amplification polymorphism (SSAP). Retrotransposons are ubiquitous, active, and abundant in plant genomes, and this marker system was developed based on the insertional polymorphism created upon element replication. The long terminal repeat (LTR) retrotransposons are well suited as molecular markers. As dispersed and ubiquitous transposable elements, their "copy and paste" life cycle of replicative transposition leads to new genome insertions without excision of the original element (Kalendar et al., 2010). Doubled haploid populations through gametophytic embryogenesis (see Chapter 14) are valuable for the efficient use of these markers. SSAP showed about four- to nine-fold more

TABLE 9.2 Comparison of the Characteristics, Advantages, and Disadvantages of Popular Dna Sequence-Based Marker Systems. RFLP: Beckmann and Soller (1983); Tanksley et al. (1989). RAPD: Welsh and McClelland (1990); Williams et al. (1990). SSR (also known as microsatellites): Powell et al. (1996); Taramino and Tingey (1996). AFLP: Vos et al. (1995). SNP: Collard et al. (2005)

Factor	RFLP	RAPD	AFLP	SSR	SNP
Genomic coverage	Low copy coding region	Whole genome	Whole genome	Whole genome	Whole genome
Relative quantity of DNA required	10–50 μg	1–100 ng	1–100 ng	50–120 ng	≥50 ng
Relative quality of DNA required	High	Low	High	Intermediate to high	High
Types of polymorphism identified	Single base changes, indels	Single base changes, indels	Single base changes, indels	Changes in repeat length	Single base changes, indels
Polymorphism sensitivity[a]	Intermediate	High	High	High	High
Effective multiplex ratio	Low	Intermediate	High	High	Intermediate to high
Inheritance	Co-dominant	Dominant	Dominant or codominant	Codominant	Codominant
Type of probes/primers	Low copy DNA or cDNA	10 bp random oligonucleotides	Specific sequence	Specific sequence	Allele-specific PCR primers
Demands of technical rigor	High	Low	Intermediate	Low	High
Health risks (e.g. radioactivity, toxic dyes)	Yes and no	No	Yes and no	No	No
Reproducibility	High	Low to intermediate	High	High	High
Time demands	High	Low	Intermediate	Low	Low
Amenability to automation	Low	Intermediate	High	High	High
Hardware and supply costs per output	High	Low	Intermediate	High	High
IP licensing requirement?	No	Yes	Yes	Yes	Yes
Range of application	Mendelian	Polymorphism	Mendelian and polymorphism	All purposes	All purposes

a Relative ability to distinguish alleles or DNA nucleotide polymorphisms.

diversity than AFLP and had the highest number of polymorphic bands per assay ratio and the highest marker index (Tam et al., 2005). SSAP markers were highly efficient in detecting genetic similarity in *Citrus*, while SSR markers may be more useful for segregation studies and genome mapping (Biswas et al., 2011).

Marker systems have also been developed for the study and manipulation of epigenetic changes that affect phenotypes. For example, methylation sensitive amplification polymorphism (MSAP) is a commonly used method for assessing DNA methylation changes in plants. This method involves gel-based visualization of PCR fragments from selectively amplified DNA that are cleaved using methylation-sensitive restriction enzymes (Chwialkowska et al., 2017).

Other novel marker systems, including single feature polymorphisms, diversity array technology and restriction site-associated DNA markers, have also been developed, where array-based assays have been utilized to provide for the desired ultra-high throughput and low cost. Microarray-based markers are the markers of choice for the future and are already being used for construction of high-density maps, quantitative trait loci (QTL) mapping (including expression QTLs) and genetic diversity analysis with a limited expense in terms of time and money (Gupta et al., 2008).

Plant breeders have used this form of analysis for contrasting germplasm, particularly where the goal is to breed hybrid cultivars. If the ultimate objective is to maximize heterosis (see Chapter 16), one approach is to choose source germplasm that differs with regard to as many alleles as possible. Many studies have appeared that correlate degree of molecular polymorphism with heterosis (horticultural species examples: Riaz et al., 2001; Sreekala and Raghava, 2003; Geleta et al., 2004; Steinfath et al., 2010; Jagosz, 2011).

Molecular markers are also powerful tools to enhance selection efficiency. This is by virtue of both absolute heritability ($h^2 = 1.00$) and "tight" linkage. If a molecular locus is linked to another locus that conditions a known phenotype A, it may be possible to select for A by using the molecular locus, even though function is probably unrelated. This strategy became known as MAS (Paterson et al., 1991; Ribaut and Hoisington, 1998; Xie and Xu, 1998) and is substantially similar for all molecular marker systems and simply-inherited traits. While studying joint inheritance patterns, linkages inevitably are discovered. To be useful as surrogate selection markers, the linkage must be on the order of ≤5 cM, or even ≤1 cM. At 5 cM, 10% of the gametes from double heterozygotes will be recombinant gametes, and will yield an undesirable result (selection for a susceptible, against a resistant, type). The problem of undesirable recombinants can mostly be eliminated if two surrogate molecular marker loci are used, one flanking each side of the gene of interest. If a gene is flanked by two molecular markers, each 10 cM distant, and selection is based on the presence of both, the probability of undesirable recombinants is reduced from 20% to 1% (Tanksley et al., 1981; Orton, 1983).

Most traits of practice economic significance are not simply inherited. Despite the challenges posed by systems involving many interacting genes and environments, plant breeders have forged ahead to develop molecular tools to improve selection efficiency and effectiveness. The fields of genetics and breeding have made tremendous strides by basing all observations on the unit gene. Thus, it was reasoned that progress on traits exhibiting complex patterns of inheritance should also be broken down to individual QTL. Molecular markers have been inextricably involved with these experimental studies, and ultimately in selection systems that have emerged from the work (Tanksley, 1993; Weeden et al., 1994; Collard et al., 2005).

QTL have been very useful as surrogates for selecting genes that underlie quantitatively inherited phenotypes. If one gene conditions one discrete phenotype, the relationship between the two is easily discerned. In fact, the gene is usually named after the phenotype with which it is associated: the *T* gene for tallness, the *R* gene for resistance, the *Ms* gene for male sterility, etc. The same sort of relationship may be extended to cases where a phenotype is conditioned by 2 or 3 genes, but usually breaks down if $n \geq 4$. The identity of individual genes is progressively more difficult to follow as more are involved in the phenotype, and interact with each other, and also with environmental factors (Hospital, 2003). QTL allow the plant breeder to select directly on genomic regions that contain desirable alleles at many loci that determine quantitatively-inherited phenotypes, bypassing the obfuscation of the phenotype by epistasis and the environment.

QTL are easy to understand if all intra- and intergenic interactions are nil (i.e., $V_G = V_A$) or if individual genes have incrementally similar effects, as with the example cited in Chapter 3. In this case, individuals with the extremes of the phenotype may be hybridized to give an intermediate F_1 population that, when selfed, yields a normal F_2 curve relative to this phenotype. The inheritance patterns of most traits, unfortunately, do not conform to such a picture (Knapp, 1994). MAS is most useful, in fact, when the phenotype of interest has low h^2.

A MAS breeding program begins with the identification of parental genotypes that represent the extremes of the quantitatively inherited trait being targeted; for example tall vs. short individual plant stature (or internode length). The next step is to identify and verify genomic DNA polymorphisms that distinguish the tall from the short parents. It is important to fully *saturate* the genome with markers such that all linkage groups are covered and that no gaps of more than ~10 cM are evident. The third step is to conduct a joint inheritance experiment, usually involving the generation of a segregating population such as F_2 or backcross (F_1 x tall parent or F_1 x short parent). Individuals of the segregating population are evaluated for marker and stature phenotypes. Customized statistical software and powerful computers are used to impute linkage relationships from the resulting data sets. Linkages are detected as they were defined in Chapter 3; a departure from independent assortment of alleles. The higher the percentage of parental associations, the more tightly linked the markers and whole-plant phenotypes are. If sufficiently large populations are employed, and h^2 is not hopelessly minute, the results will yield an understanding of the number of genes involved and their relative effects on the whole-plant phenotype. Xu (2002) provides an excellent summary of the mapping procedure and statistical methods employed QTL mapping and analyses of relative effects.

The example of breeding for stature in a hypothetical plant species will be developed further. Let us assume that this hypothetical species has a genome that consists of 5 linkage groups (i.e. 2n = 2x = 10). These linkage groups are named by Roman numerals I, II, III, IV, and V. A population of tall individuals and another of short individuals are assembled; the plant breeder will select these populations carefully and with regard to many other attributes than stature, but we will only focus on stature for now.

Next, the plant breeder employs one or more marker systems to illuminate polymorphisms that distinguish the tall and the short population (or individual). This may have already been done; comprehensive maps of marker loci exist for dozens of crop species, and the plant breeder should start with what is already known and available. Many crop species are associated with a group of scientists who have established a consensus marker map, and keep it updated, for example Solanaceae (e.g. tomato, potato, eggplant, tobacco, petunia, pepper; see "SolCap web site:

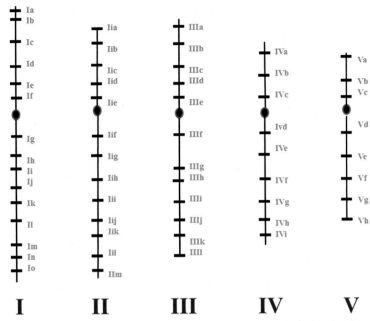

FIG. 9.5   Linkage map of the hypothetical crop species described in the text with 57 hypothetical molecular markers (Ia, Ib, …., Vh) mapped to the five linkage groups.

http://solcap.msu.edu/index.shtml). SolCap makes updated maps and databases with marker polymorphisms across a broad spectrum of germplasm available to scientists as a public resource (Scott, 2010).

Using available resources and by conducting mapping studies of molecular markers, we find 57 hypothetical markers that are dispersed broadly across the five linkage groups of our hypothetical species (Fig. 9.5). It is advisable to locate and employ more markers than this to achieve genome saturation, since it is a challenge to fully represent all regions of the genome. Frequently, some regions end up with either poor or no representation by marker loci.

Let us assume that it is possible that all 57 markers are polymorphic between the tall and short populations. Next, a segregating population for the polymorphic markers and stature phenotype is generated, in this case F_2 (Fig. 9.6). We will assume that stature in our hypothetical species behaves in an additive fashion and that all constituent genes exert more-or-less equal effects. This will result in a normal curve in the F_2 for stature. The 57 qualitative marker loci will segregate 1:2:1 for parental and heterozygous configurations and cosegregate according to the established map (Fig. 9.5). The genes underlying stature also segregate and these phenotypes will be analyzed in the next step.

Cosegregation ratios of stature and marker alleles will be distorted toward overrepresentation of parental associations where there are linkages. The degree of the distortion will be highest where linkages are most tight (or closest). We are observing the stature phenotype, however, that is determined by many constituent genes and also has an

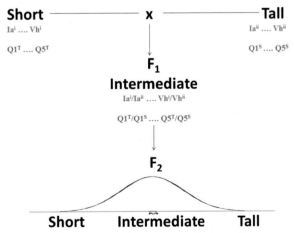

FIG. 9.6   Generation of a F_2 population segregating jointly for marker polymorphisms and stature.

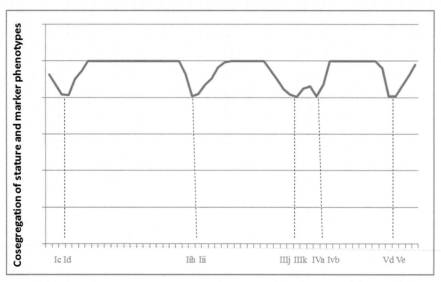

FIG. 9.7 Graph depicting cosegregation of stature and marker phenotypes along the axis of the genome, or five linkage groups of the hypothetical species. Five peaks of coseqregation distortion are detected in this example, indicating five QTL of approximately equal effects on stature.

environmental component (V_E). Therefore, the mapping software uses an interval mapping algorithm that employs maximum likelihood theory and generates parameters *logarithmic of odds* (LOD) and *likelihood ratio statistic* (LRS) scores. These scores are used to impute the most likely location of the underlying QTL with respect to marker loci on the genomic map (Fig. 9.7). Before permutation tests were widely accepted as an appropriate method to determine significance thresholds, a LOD score of between 2.0 and 3.0 (most commonly 3.0) was usually chosen as the significance threshold (Collard et al., 2005).

Thus, the linkage map with the QTL for stature in this hypothetical species located at the positions indicated in the cosegregation distortion analysis is as shown in Fig. 9.8. In this instance, there are no markers that appear to cosegregate absolutely with stature. Ideally, there will be one or more marker loci that cosegregate with the QTL phenotype, providing a starting point for studies to isolate the actual genes that underlie the phenotype.

It is now possible to use the linked molecular markers as surrogates in a breeding program for stature in this hypothetical species. If shorter stature plants are targeted, the plant breeder selects for the marker alleles linked with QTL for the short parent (Ic^i.Id^i; IIh^i.IIi^i; IIIj^i.IIIk^i; IVa^i.IVb^i; and Vd^i.Ve^i). By selecting for tightly linked flanking markers, assurance of including the QTL for short stature is virtually absolute. Fig. 9.9 illustrates the use of MAS in this example to select for stature (internode length) as compared to selection under a range of h^2 depicted in Fig. 9.3.

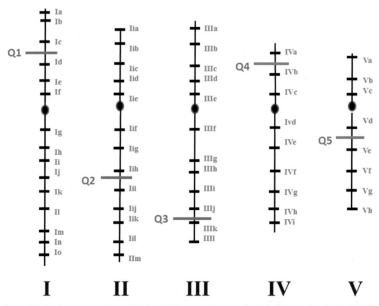

FIG. 9.8 Linkage map of the hypothetical crop species with the QTL superimposed with the marker loci; 5 QTL of approximately equal effects.

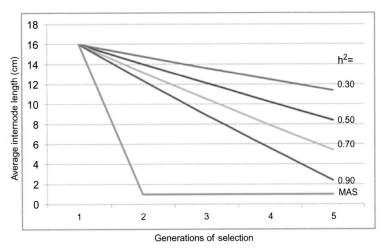

FIG. 9.9 Illustration of the effect of heritability (h^2) vs. MAS using the text example when starting from the same population; 4 cycles of selection for short internode length.

It is important to note that these markers may not be universal as surrogates for stature in this hypothetical (or any) species. If other tall/short populations are employed, they may be characterized by a different QTL set that determines stature. Collective results of many studies on quantitative traits, however, suggest that analogous sets of loci operate across a broad spectrum of germplasm (Pereira and Lee, 1995; Chaim et al., 2001; Frary et al., 2003).

Every phenotype, species, and environmental context is different. Most quantitatively inherited traits are not as straightforward as is the example outlined herein. Complications include larger number of QTL, unequal effects, dominance and epistatic intra- and intergenic effects, developmentally regulated gene expression, and GxE interactions. All of these factors greatly complicate the QTL detection and mapping program, and may diminish the veracity of linkage conclusions. If linkage relationships are equivocal, the power of constituent markers to select for QTL is greatly hampered.

QTL mapping and MAS are rooted deeply in theoretical statistical science. Mathematical underpinnings of these tools are beyond the scope of this textbook, but practitioners who are using these techniques extensively should acquire a working knowledge of the mathematical tenets on which they are built. A short list of references on the theoretical statistical underpinnings of QTL and MAS is as follows: (Lande and Thompson, 1990; Zhang and Smith, 1993; Gimelfarb and Lande, 1995; Hospital et al., 1997; van Berloo and Stam, 1998; Ribaut and Hoisington, 1998; Xie and Xu, 1998; Ribaut and Betran, 1999; Xu, 2002; Hospital, 2003; Collard et al., 2005; Goddard, 2009).

ALTERNATIVE METHODS FOR THE IMPUTATION OF MARKER BREEDING VALUES

Specialized plant genetic stocks, such as bi-parental and multi-parent mapping populations, mutant populations, and immortalized collections of recombinant lines have been generated to facilitate mapping and gene function analysis via association studies and QTL mapping. Alternatively, genotypic and phenotypic datasets on training populations can be used to develop models to predict the breeding value of lines. Varshney et al. (2014) have provided a useful description of the different types of populations and approaches that have been developed and are being used to find and map QTL and other useful DNA sequences for enhancing the effectiveness and efficiency of MAS:

Bulked segregant analysis (BSA): This approach identifies molecular markers associated with a trait of interest by genotyping DNA extracted from bulked samples of individuals at the trait's phenotypic extremes (Michelmore et al., 1991; Chen et al., 2011).

Double digest restriction-site associated DNA sequencing technology (ddRAD-seq): This is a reduced representation sequencing technology that samples genome-wide enzyme loci by next-generation sequencing. The ddRAD strategy is economical, time-saving, and requires little technical expertise or investment in laboratory equipment (Yang et al., 2017).

Genome-wide association studies (GWAS): These studies utilize collections of diverse, unrelated lines that are genotyped and phenotyped for traits of interest, and statistical associations are established between DNA polymorphisms and trait variation to identify genomic regions where genes governing traits of interest are located (George and Cavanagh, 2015; Ogura and Busch, 2015; Pascual et al., 2016; He et al., 2017; Viana et al., 2017).

Genotyping-by-sequencing (GBS): A highly multiplexed genotyping system involving DNA digestion with different enzymes and the construction of a reduced representation library, which is sequenced using a next-generation sequencing platform. It enables the detection of thousands of SNPs in large populations or collections of lines that can be used for mapping, genetic diversity analysis, characterizing polymorphism and orthologs in polyploids, and evolutionary studies (Limborg et al., 2016; Chung et al., 2017; Lee et al., 2017).

Marker-assisted back-crossing (MABC): In this form of marker-assisted selection, a genomic locus (gene or QTL) associated with a desired trait is introduced into the genetic background of an elite breeding line through several generations of backcrossing (Varshney and Dubey, 2009).

Multi-parent advanced generation inter-cross (MAGIC): A type of multi-parent population developed from four to eight diverse founder lines, generated to increase the precision and resolution of QTL mapping because of the larger number of alleles and recombination events compared to bi-parental mapping populations (Cavanagh et al., 2008; Verbyla et al., 2014; Pascual et al., 2016).

Nested association mapping (NAM): NAM combines advantages of linkage and association mapping and eliminates disadvantages of both; it takes into consideration recent and historical recombination events, facilitating high resolution mapping (Yu et al., 2008; Guo and Beavis, 2011; Rincent et al., 2017).

Quantitative trait locus (QTL): A genomic region encompassing one or more genes that accounts for a portion of the variation of a complex quantitative trait, identified by phenotyping and genotyping a segregating population followed by statistical analysis (Darvasi and Soller, 1992; Collard et al., 2005; Cooper et al., 2009).

Recombinant inbred line (RIL): An immortal mapping population consisting of fixed (inbred) lines in which recombination events between chromosomes inherited from two inbred strains are preserved. RILs are generated by crossing two divergent parents followed by several generations of inbreeding to achieve homozygosity (van Berloo and Stam, 1998; Calvo-Polanco et al., 2016).

Sequence-based mapping (SbM): An approach requiring deep sequencing (5× to 8× genome coverage) of two DNA pools derived from individuals from the phenotypic extremes of a segregating population, to identify candidate genes associated with a phenotype of interest (Paux et al., 2012).

Target Enrichment Sequencing (TES): Target enrichment sequencing (TES) is a powerful method to enrich genomic regions of interest and to identify sequence variations. TES is useful for SNP identification in non-model species where a genome reference not available (Peng et al., 2017).

Training population (TP): A genotyped and phenotyped reference breeding population used to develop a model to predict genomic-estimated estimate breeding values for genome selection (Zhao et al., 2012; Riedelsheimer and Melchinger, 2013; Guo et al., 2014; Isidro et al., 2015).

Whole genome re-sequencing (WGRS): This is a strategy to sequence an individual genome where short sequence reads generated by NGS are aligned to a reference genome for the species, providing information on variants, mutations, structural variations, copy number variation, and rearrangements between and among individuals, based on comparison to the reference genome (Hazzouri et al., 2015; Wu et al., 2016).

Targeting-induced local lesions in genomes (TILLING): A reverse genetics approach for the rapid discovery and mapping of induced causal mutation responsible for traits of interest (Comai and Henikoff, 2006; Raghavan et al., 2007; Tsai et al., 2011; Kumar et al., 2017).

GENOME SELECTION

Understanding the complex relationship between genotypic and phenotypic variation lies at the heart of genetics and is also absolutely crucial to applications in plant breeding. In traditional plant breeding, selection is practiced on the phenotype with resulting changes in population allelic frequencies in proportion to the relationship of genotype to phenotype. *Genomic selection* (GS), the ability to select for complex, quantitative traits based on marker data alone, has been developed by many researchers worldwide since about 2010 from the intersection of next-generation sequencing (NGS), new high-throughput marker technologies, and new statistical methods needed to analyze the data. GS works by estimating the effects of many loci spread across the genome. For typical crops, the requirements range from at least 200 to at most 10,000 markers and observations. GS can greatly accelerate the breeding cycle while also using marker information to maintain genetic diversity and potentially prolong gain beyond what is possible with phenotypic selection (Lorenz et al., 2011; Nakaya and Isobe, 2012; Varshney et al., 2017).

Until about 2010, marker data were expensive and laborious to generate, and MAS strategies were limited by the number of markers that could be assayed cost effectively (Rajsic et al., 2016). Only markers mapping to targeted genomic regions were utilized, consequently, to predict the presence or absence of agriculturally valuable traits. The availability of advanced DNA sequencing technologies (see below) provided genome-wide marker coverage at a drastically reduced cost per data point. This led to the ability to estimate overall breeding value of the entire genome with single nucleotide base-level precision. Advanced sequencing technologies have great promise for the identification of molecular markers that are powerful tools in plant breeding (Yang et al., 2015).

In addition to classical sequencing methodology (Sanger et al., 1977), a range of new, more powerful sequencing technologies have become available in recent years. These technologies are currently being used to sequence the genomes of a number of crops. New sequencing strategies are increasing sequencing rates, system throughput, and read lengths resulting in decreasing sequencing costs and reduced complexity, rigor, and cost of sample preparation (Pennisi, 2010; Schneider and Dekker, 2012; Chin et al., 2013). Consequently, geneticists and breeders are utilizing more sophisticated tools for sequencing-based mapping and genome-wide selection for the development of new breeding lines (Varshney et al., 2016). Advanced DNA sequencing methods are currently being applied to gene discovery in diverse species and populations, and as a foundation for large-scale modeling in both basic plant genetics and applied plant breeding. This technology is providing progressively more insights into the relationships of genotype and QTL to phenotype and how this information can be used to accelerate plant breeding (Garrido-Cardenas et al., 2017).

Advances in DNA sequencing provide tools for efficient large-scale discovery of markers for use in plants. Henry et al. (2012) published an excellent review of the plant breeding implications of NGS, 3rd-generation, and advanced-sequencing technologies. New options available to the plant breeder include large-scale amplicon sequencing, transcriptome sequencing, gene-enriched genome sequencing and whole genome sequencing. Sequencing the whole genome of parents identifies all the polymorphisms available for analysis in their progeny. Sequencing PCR amplicons of sets of candidate genes from DNA bulks can be used to define the available variation in these genes that might be exploited in a population or germplasm collection. Sequencing of the transcriptomes of genotypes that co-vary with the trait of interest may identify genes with patterns of expression that could explain the phenotypic variation. Sequencing genomic DNA enriched for genes by hybridization with probes for all or some of the known genes simplifies sequencing and analysis of differences in gene sequences between large numbers of genotypes and genes especially when working with complex genomes. The main challenges facing plant breeders is how to choose an appropriate genotyping method and how to integrate genotyping data sets obtained from various sources into the overall breeding strategy.

The broad objective of GS is to evaluate the effects of directional selection based on estimated genomic breeding values (GEBVs) for quantitative traits. In comparison to MAS, GS uses all available marker data for a population as predictors of breeding value, not just a targeted trait. GS integrates marker data from a training population with phenotypic and, when available, pedigree data collected on the same population to generate a statistical prediction model that imputes GEBVs for all genotyped individuals within a breeding population. GEBVs serves as a predictor of how well a plant will perform as a parent for crossing and generation advance in a breeding pipeline, based on the similarity of its genomic profile to other plants in the training population. Before the prediction model can be applied to a breeding population, the accuracy of the model is generally tested using cross-validation on subsets of the training population. The validated model can be applied to a breeding population where GEBVs are calculated for all lines for which genotypic information is available, and their phenotypic performance is predicted solely on the basis of that genotypic information (Barabaschi et al., 2016; Varshney et al., 2016).

Numerous studies have explored how selfing can be deployed to maximal benefit in the context of traditional plant breeding programs. McClosky et al. (2013) examined how selfing impacts the two key aspects of genomic selection GEBV prediction (training) and selection response in biparental populations. Selfing increases genomic selection (GS) gains by >70%. Gains in genomic selection response attributable to selfing hold over a wide range population sizes (100–500), h2 (0.2–0.8), and selection intensities (0.01–0.1) if the number of QTL is >20. The major cause of the improved response to genomic selection with selfing is through an increase in the occurrence of superior genotypes and not through improved GEBV predictions.

It is important to employ the most applicable and valid statistical and analytical methods for effective GS and genomics-assisted breeding (GAB). Advances in DNA sequencing technologies have prompted geneticists and breeders to utilize more sophisticated models for sequencing-based mapping and genome-wide selection for the development of new breeding lines (Montesinos-López et al., 2017). The availability of open-source and vertically integrated platforms will facilitate the modernization of crop breeding programs. Although phenotyping remains expensive and time consuming, prediction of allelic effects on phenotypes opens new doors to enhance genetic gain across crop cycles, building on reliable phenotyping approaches and good crop information systems (Varshney et al., 2014, Crossa et al., 2017).

The estimation of marker-target gene associations in MAS ignores genes with small effects that trigger underpinning quantitative traits. GS includes these minor genes by estimating marker effects across the whole genome on the target population based on a prediction model developed in the training population. Whole-genome prediction models estimate all marker effects in all loci and capture small QTL (Desta and Ortiz, 2014). Selection affects GEBV prediction accuracy as well as genetic architecture via changes in allelic frequencies and linkage disequilibria (LD), and the resulting changes are different from those in the absence of selection. Selection response tends to reach a plateau, but at higher marker density both the magnitude and duration of the response increase. Selection changes QTL allele frequencies and generates new but unfavorable LD for prediction (Long et al., 2011).

When combined with precise phenotyping methods, GS provides a powerful and rapid tool for identifying the genetic basis of agriculturally important traits and for predicting the breeding value of individuals in a plant breeding population. Phenotyping is a major operational bottleneck that limits the power and resolution of many kinds of genetic analysis. Scientists are developing new and more efficient breeding strategies that integrate genomic technologies and high throughput phenotyping to better utilize natural and induced genetic variation. Rapid developments in DNA sequencing technologies over the last decade have opened up many new opportunities to explore the relationship between genotype and phenotype with greater resolution than ever before.

As the cost of sequencing has decreased, breeders have begun to sequence large populations of plants, increasing the resolution of gene and QTL discovery and providing the basis for modeling complex genotype-phenotype relationships at the whole-genome level. Specialized plant genetic stocks, such as bi-parental and multi-parent mapping populations, mutant populations, and immortalized collections of recombinant lines have been generated to facilitate mapping and gene function analysis via association studies and QTL mapping in several crop species. Knowledge about the identity and map location of agriculturally important genes and QTL provides the basis for parental selection and MAS in plant breeding.

The trend for sequence-based genotyping to replace the use of fixed marker arrays seems realistic, particularly as the cost of sequencing continues to fall, and is already occurring for diploid crops with relatively small genome sizes (≤1 GB). For polyploids and crops with larger genomes, fixed SNP arrays will continue to be useful, particularly where they assay gene-specific or genome-specific markers that facilitate accurate mapping (Varshney et al., 2014).

GS and genome-assisted strategies will have an enormous impact on advances and breeding protocols in horticultural crops with long generation times such as biennials and woody perennials (Gemenet and Khan, 2017). The selection component of plant breeding will, progressively, shift from phenotype to genotype. The degree to which emphasis is shifted and the specific elements of phenotype vs. genomic sequences will be different in each distinct breeding program. The plant breeder will conduct two parallel efforts, one that is analogous to the current "traditional" breeding program (described in Chapters 13–18) and another that evaluates the degree of correlation ("training") of genomic DNA sequences with phenotypes (Fig. 9.10). Over time, the models that relate genotype to phenotype converge on absolute predictability until, theoretically, phenotype is irrelevant. Complexities of intra- and intergenic and GxE interactions that modulate phenotype will present a challenge for the strategy of pure GS.

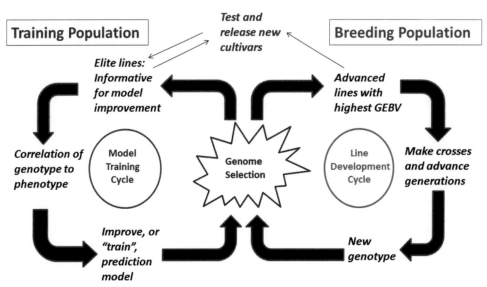

FIG. 9.10 Genomic selection training and breeding populations and their interactions in a plant breeding program. *Adapted from Sorrells, M.E., 2015. Genomic selection in plants: Empirical results and implications for wheat breeding. In: Ogihara, Y., Takumi, S., Handa, H. (Eds.) Advances in Wheat Genetics: From Genome to Field. Springer, Tokyo (Japan), pp. 401–409.*

FUNCTIONAL GENOMICS

The study of the organization and workings of the genetic apparatus of organisms is called *genomics*. This field has emerged from the explosion of molecular biology from the 1980s to the present starting with recombinant DNA and proceeding to large-scale rapid and efficient DNA, RNA, and polypeptide sequencing and biosynthesis protocols. Genomics is focused on how genetic information is organized, how it changes over time with evolutionary forces, how and when it is transcribed or otherwise expressed, and the relationship between the genome and higher order phenotypes. *Functional genomics* is the study of how genomes are structured and expressed to facilitate known biological processes in biochemistry and biophysics, physiology, reproduction, development, growth, energy transduction, transpiration, etc. (Leister, 2005). The overlap of functional genomics with plant breeding is substantial and new discoveries have already suggested new selection techniques for economically important phenotypes that either enhance heritability or reduce resource inputs (e.g., time or dollars).

Advanced DNA sequencing technologies that enable plant breeders to discern the organization of entire nuclear and organelle genomes and *transcriptomes* (the sequence, organization, and function of RNA transcripts) are now relatively inexpensive and accessible. Rapid polynuceotide sequencing technologies can be applied broadly among plant species for the development of genomic resources such as gene sequence and function databases (Bartoszewski and Malepszy, 2012). If a researcher discovers and sequences an open reading frame (ORF; transcribed gene of unknown context or function), the sequence can readily be used to find analogues in other species in which the function of the gene may be known (Kang et al., 2016).

While mapping QTL and employing markers as surrogate selection criteria (addressed above) are aimed at increasing h², functional genomics goes further to dissect the phenotype into molecular components (Delseny et al., 2010). Understanding the relationship between genomic DNA blueprints and the phenotypes they encode will shed enormous light on how to further enhance targeted phenotypes in a desirable fashion (Kang et al., 2016). Functional genomics is also starting to shed more light on adaptive advantages and disadvantages of polyploidy during evolution and within plant breeding programs (Yang et al., 2011).

QTL mapping is often a starting point for functional genomics. Close linkage associations will ultimately lead to the genes that actually encode the targeted phenotype. As with the *Mi—Aps-1* linkage in tomato cited above, the resulting linkage cassettes may also prove to be useful in the isolation of interesting and potentially valuable gene sequences. The molecular marker system used will generally predicate the possession of a DNA fragment that could be used to find and isolate the corresponding genomic fragment via affinity capture.

Recent developments suggest that rapid DNA sequencing technology will supplant currently popular DNA marker systems. GBS is now being used to mine polymorphisms, open reading frames (functional genes), control sequences, and transposons (Elshire et al., 2011; Davey et al., 2011). TILLING (see above) has also been viewed as an effective strategy to discern functional genomics systems because (i) it can be applied to any species, regardless of its genome size and ploidy level, and (ii) it provides a high frequency of point mutations distributed randomly. The mutagenic potential of chemical agents to generate a high rate of nucleotide substitutions has been proven by the high density of mutations reported for TILLING populations in various plant species. High-throughput TILLING permits the rapid and low-cost discovery of new alleles that are induced in plants. Recent trends in TILLING procedures rely on the diversification of bioinformatic tools, new methods of mutation detection, including mismatch-specific and sensitive endonucleases, but also various alternatives for SNP discovery using advanced sequencing technologies (Kurowska et al., 2011).

The nature of molecular technologies applied to plant breeding is changing far too quickly to make it useful to cover the cutting edge in great detail or to speculate about what is coming next. Suffice it to say that the successful plant breeder of the future must be intimate with not only the phenotype manifested in whole plants and populations, but also with the genome(s) (and other –omes) from which the phenotype is derived.

References

Barabaschi, D., Tondelli, A., Desiderio, F., Volante, A., Vaccino, P., Giampiero, V., Cattivelli, L., 2016. Next generation breeding. Plant Sci. 242, 3–13.

Bartoszewski, G., Malepszy, S., 2012. Plant genomics—its emerging role in crop improvement. Acta Hortic. (953), 23–29.

Beckmann, J., Soller, M., 1983. Restriction fragment length polymorphisms in plant genetic improvement: methodologies, mapping and costs. Theor. Appl. Genet. 67 (1), 35–43.

van Berloo, R., Stam, P., 1998. Marker-assisted selection in autogamous RIL populations: a simulation study. Theor. Appl. Genet. 96 (1), 147–154.

Bernardo, R., 1998. A model for marker-assisted selection among single crosses with multiple genetic markers. Theor. Appl. Genet. 97 (3), 473–478.

Biswas, M.K., Chai, L., Amar, M.H., Zhang, X., Deng, X.-X., 2011. Comparative analysis of genetic diversity in citrus germplasm collection using AFLP, SSAP, SAMPL and SSR markers. Sci. Hortic. 129 (4), 798–803.

Buchholz, K., Collins, J., 2010. Concepts in Biotechnology: History, Science and Business. Wiley-VCH; John Wiley, Distributor, Weinheim (Ger); Chichester (UK), p. 471.

Calvo-Polanco, M., Sánchez-Romera, B., Aroca, R., Asins, M.J., Declerck, S., Dodd, I.C., Martínez-Andújar, C., Albacete, A., Ruiz-Lozano, J.M., 2016. Exploring the use of recombinant inbred lines in combination with beneficial microbial inoculants (AM fungus and PGPR) to improve drought stress tolerance in tomato. Environ. Exp. Bot. 131, 47–57.

Cavanagh, C., Morell, M., Mackay, I., Powell, W., 2008. From mutations to MAGIC: resources for gene discovery, validation and delivery in crop plants. Curr. Opin. Plant Biol. 11 (2), 215–221.

Chaim, A.B., Paran, I., Grube, R.C., Jahn, M., van Wijk, R., Peleman, J., 2001. QTL mapping of fruit-related traits in pepper (*Capsicum annuum*). Theor. Appl. Genet. 102 (6–7), 1016–1028.

Chen, X., Hedley, P.E., Morris, J., Liu, H., Niks, R.E., Waugh, R., 2011. Combining genetical genomics and bulked segregant analysis-based differential expression: an approach to gene localization. Theor. Appl. Genet. 122 (7), 1375–1383.

Chin, C.-S., Alexander, D.H., Marks, P., Klammer, A.A., Drake, J., Heiner, C., Clum, A., Copeland, A., Huddleston, J., Eichler, E.E., Turner, S.W., Korlach, J., 2013. Non-hybrid, finished microbial genome assemblies from long-read SMRT sequencing data. Nat. Methods 10, 563–569.

Choi, J.-Y., Roy, N.S., Park, K.-C., Nam-Soo, K., 2016. Comparison of molecular genetic utilities of TD, AFLP, and MSAP among the accessions of japonica, indica, and tongil of *Oryza sativa* L. Genes Genomics 38 (9), 819–830.

Chung, Y.S., Kim, C., Choi, S.C., Tae-Hwan, J., 2017. Genotyping-by-sequencing: a promising tool for plant genetics research and breeding. Hortic. Environ. Biotechnol. 58 (5), 425–431.

Chwialkowska, K., Korotko, U., Kosinska, J., Szarejko, I., Kwasniewski, M., 2017. Methylation sensitive amplification polymorphism sequencing (MSAP-Seq)—a method for high-throughput analysis of differentially methylated CCGG sites in plants with large genomes. Front. Plant Sci. 8, 2056.

Collard, B.C.Y., Jahufer, M.Z.Z., Brouwer, J.B., Pang, E.C.K., 2005. An introduction to markers, quantitative trait loci (QTL) mapping and marker-assisted selection for crop improvement: the basic concepts. Euphytica 142 (1–2), 169–196.

Comai, L., Henikoff, S., 2006. TILLING: practical single-nucleotide mutation discovery. Plant J. 45 (4), 684–694.

Cooper, M., van Eeuwijk, F.A., Hammer, G.L., Podlich, D.W., Messina, C., 2009. Modeling QTL for complex traits: detection and context for plant breeding. Curr. Opin. Plant Biol. 12 (2), 231–240.

Cramer, C.S., Wehner, T.C., 2000. Path analysis of the correlation between fruit number and plant traits of cucumber populations. HortSci. 35 (4), 708–711.

Crossa, J., Pérez-Rodríguez, P., Cuevas, J., Montesinos-López, O., Jarquín, D., los Campos, G., Burgueño, J., González-Camacho, J.M., Pérez-Elizalde, S., Beyene, Y., Dreisigacker, S., Singh, R., Zhang, X., Gowda, M., Roorkiwal, M., Rutkoski, J., Varshney, R.K., 2017. Genomic selection in plant breeding: methods, models, and perspectives. Trends Plant Sci. 22 (11), 961–975.

Da, Y., VanRaden, P.M., Schook, L.B., 2000. Detection and parameter estimation for quantitative trait loci using regression models and multiple markers. Genet. Sel. Evol. 32 (4), 357–381.

Darvasi, A., Soller, M., 1992. Selective genotyping for determination of linkage between a marker locus and a quantitative trait locus. Theor. Appl. Genet. 85 (2/3), 353–359.

Davey, J.W., Hohenlohe, P.A., Etter, P.D., Boone, J.Q., Catchen, J.M., Blaxter, M.L., 2011. Genome-wide genetic marker discovery and genotyping using next-generation sequencing. Nat. Rev. Genet. 12 (7), 499–510.

Delseny, M., Han, B., Hsing, Y.I., 2010. High throughput DNA sequencing: the new sequencing revolution. Plant Sci. 179 (5), 407–422.

Desta, Z.A., Ortiz, R., 2014. Genomic selection: genome-wide prediction in plant improvement. Trends Plant Sci. 19, 592–601.

Elshire, R.J., Glaubitz, J.C., Sun, Q., Poland, J.A., Kawamoto, K., Buckler, E.S., Mitchell, S.E., 2011. A robust, simple genotyping-by-sequencing (GBS) approach for high diversity species. PLoS One 6 (5), e19379. https://doi.org/10.1371/journal.pone.0019379.

Falconer, D.S., MacKay, T.F.C., 1996. Introduction to Quantitative Genetics. vol. 4 Longman Green, Harlow, Essex, p. 480.

Frary, A., Doganlar, S., Daunay, M.C., Tanksley, S.D., 2003. QTL analysis of morphological traits in eggplant and implications for conservation of gene function during evolution of solanaceous species. Theor. Appl. Genet. 107 (2), 359–370.

Gallais, A., 1991. A general approach for the study of a population of testcross progenies and consequences for recurrent selection. Theor. Appl. Genet. 81 (4), 493–503.

Garrido-Cardenas, J.A., Mesa-Valle, C., Manzano-Agugliaro, F., 2017. Trends in plant research using molecular markers. Planta 247 (3), 543–557.

Geleta, L.F., Labuschagne, M.T., Viljoen, C.D., 2004. Relationship between heterosis and genetic distance based on morphological traits and AFLP markers in pepper. Z. Pflanzenzuchtung 123 (5), 467–473.

Gemenet, D.C., Khan, A., 2017. Opportunities and challenges to implementing genomic selection in clonally propagated crops. In: Varshney, R.K., Roorkiwal, M., Sorrells, M.E. (Eds.), Genomic Selection for Crop Improvement. Springer International Publishing, New York, NY, pp. 185–198.

George, A.W., Cavanagh, C., 2015. Genome-wide association mapping in plants. Theor. Appl. Genet. 128 (6), 1163–1174.

Gimelfarb, A., Lande, R., 1995. Marker-assisted selection and marker-QTL associations in hybrid populations. Theor. Appl. Genet. 91 (3), 522–528.

Goddard, M., 2009. Genomic selection: prediction of accuracy and maximisation of long term response. Genetica 136 (2), 245–257.

Gulsen, O., 2016. Applications of molecular markers in vegetable seed industry. Acta Hortic. (1142), 201–208.

Guo, B., Beavis, W.D., 2011. In silico genotyping of the maize nested association mapping population. Mol. Breed. 27 (1), 107–113.

Guo, Z., Tucker, D.M., Basten, C.J., Gandhi, H., Ersoz, E., Guo, B., Xu, Z., Wang, D., Gay, G., 2014. The impact of population structure on genomic prediction in stratified populations. Theor. Appl. Genet. 127 (3), 749–762.

Gupta, P., Varshney, R., Sharma, P., Ramesh, B., 1999. Molecular markers and their applications in wheat breeding. Plant Breed. 118, 369–390.

Gupta, P.K., Rustgi, S., Mir, R.R., 2008. Array-based high-throughput DNA markers for crop improvement. Heredity 101 (1), 5–18.

Han, Y., Bonos, S.A., Clarke, B.B., Meyer, W.A., 2006. Inheritance of resistance to gray leaf spot disease in perennial ryegrass. Crop. Sci. 46 (3), 1143–1148.

Hazzouri, K.M., Flowers, J.M., Visser, H.J., Hussam, S.M.K., Rosas, U., Pham, G.M., Meyer, R.S., Johansen, C.K., Zoã, A.A.F., Masmoudi, K., Haider, N., El Kadri, N., Youssef Idaghdour, M., Joel, A.A., Thirkhill, D., Markhand, G.S., Krueger, R.R., Zaid, A., Purugganan, M.D., 2015. Whole genome re-sequencing of date palms yields insights into diversification of a fruit tree crop. Nat. Commun. 6, 8824.

He, J., Yang, B., Xing, G., Lu, J., Gai, J., Xia, Q., Guan, R., Meng, S., Yang, S., Zhao, T., Li, Y., Wang, Y., 2017. An innovative procedure of genome-wide association analysis fits studies on germplasm population and plant breeding. Theor. Appl. Genet. 130 (11), 2327–2343.

Henning, J.A., Teuber, L.R., 1996. Modified convergent improvement: a breeding method for multiple trait selection. Crop. Sci. 36 (1), 1–8.

Henry, R.J., 2013. Evolution of DNA marker technology in plants. In: Henry, R.J. (Ed.), Molecular Markers in Plants. Wiley On-line Library. https://doi.org/10.1002/9781118473023.ch1.

Henry, R.J., Edwards, M., Waters, D.L.E., Gopala, K.S., Bundock, P., Sexton, T.R., Masouleh, A.K., Nock, C.J., Pattemore, J., 2012. Application of large-scale sequencing to marker discovery in plants. J. Biosci. 37 (5), 829–841.

Hospital, F., 2003. Marker-assisted breeding. In: Newbury, H.J. (Ed.), Plant Molecular Breeding. CRC Press, Boca Raton, FL, pp. 30–59.

Hospital, F., Moreau, L., Lacoudre, F., Charcosset, A., Gallais, A., 1997. More on the efficiency of marker-assisted selection. Theor. Appl. Genet. 95 (8), 1181–1189.

Iezzoni, A.F., Mulinix, C.A., 1992. Yield components among sour cherry seedlings. J. Am. Soc. Hortic. Sci. 117 (3), 380–383.

Isidro, J., Jannink, J.-L., Akdemir, D., Poland, J., Heslot, N., Sorrells, M.E., 2015. Training set optimization under population structure in genomic selection. Theor. Appl. Genet. 128 (1), 145–158.

Jagosz, B., 2011. The relationship between heterosis and genetic distances based on RAPD and AFLP markers in carrot. Plant Breed. 130 (5), 574–579.

Jones, N., Ougham, H., Thomas, H., 1997. Markers and mapping: we are all geneticists now. New Phytol. 137, 165–177.

Joshi, S.P., Renjekar, P.K., Gupta, V.S., 1999. Molecular markers in plant genome analysis. Curr. Sci. 77 (2), 230–240.

Kalendar, R., Antonius, K., Smýkal, P., Schulman, A.H., 2010. iPBS: a universal method for DNA fingerprinting and retrotransposon isolation. Theor. Appl. Genet. 121 (8), 1419–1430.

Kalinski, A., 1993. Application of RFLP and RAPD technologies to plant breeding. In: Plant Genome Data and Information Center. National Agricultural Library, Plant Genome Data and Information Center, Beltsville, MD.

Kang, Y.J., Lee, T., Lee, J., Shim, S., Jeong, H., Satyawan, D., Kim, M.W., Lee, S.H., 2016. Translational genomics for plant breeding with the genome sequence explosion. Plant Biotechnol. J. 14 (4), 1057–1069.

Khanna, K.R., 1990. Crop improvement in the perspective of agricultural advancement and the necessity for investigations at the molecular level. In: Khanna, K.R. (Ed.), Biochemical Aspects of Crop Improvement. CRC Press, Boca Raton, FL, pp. 3–33.

Kordrostami, M., Rahimi, M., 2015. Molecular Markers in Plants: Concepts and Applications. Research Gate. https://www.researchgate.net/publication/282954774_Molecular_markers_in_plants_Concepts_and_applications.

Kumar, A.P.K., McKeown, P.C., Boualem, A., Ryder, P., Brychkova, G., Bendahmane, A., Sarkar, A., Chatterjee, M., Spillane, C., 2017. TILLING by sequencing (TbyS) for targeted genome mutagenesis in crops. Mol. Breed. 37 (2), 14.

Kurowska, M., Daszkowska-Golec, A., Gruszka, D., Marzec, M., Szurman, M., Szarejko, I., Maluszynski, M., 2011. TILLING—a shortcut in functional genomics. J. Appl. Genet. 52 (4), 371–390.

Lande, R., Thompson, R., 1990. Efficiency of marker-assisted selection in the improvement of quantitative traits. Genetics 124 (3), 743–756.

Langridge, P., Chalmers, K., 2005. The principle: identification and application of molecular markers. In: Lörz, H., Widholm, J.M. (Eds.), Biotechnology in Agriculture and Forestry. vol. 55. Springer-Verlag, Berlin (Germany), pp. 3–22.

Lee, S.J., Choi, C., Kim, G.H., Kim, J.H., Ban, S.H., Kwon, S.I., 2017. Identification of potential gene-associated major traits using GBS-GWAS for Korean apple germplasm collections. Plant Breed. 136 (6), 977–986.

Leister, D. (Ed.), 2005. Plant Functional Genomics. Food Products Press, New York, NY, p. 677.

Lewontin, R.C., 1974. The Genetic Basis of Evolutionary Change. Columbia University Press, New York, NY, p. 352.

Limborg, M.T., Seeb, L.W., Seeb, J.E., 2016. Sorting duplicated loci disentangles complexities of polyploid genomes masked by genotyping by sequencing. Mol. Ecol. 25 (10), 2117–2129.

Logan, J., Edwards, K., Saunders, N. (Eds.), 2009. Real-Time PCR: Current Technology and Applications. Caister Academic Press, London, p. 284.

Logemann, J., Schell, J., 1993. The impact of biotechnology on plant breeding, or how to combine increases in agricultural productivity with an improved protection of the environment. In: Chet, I., Logemann, J., Schell, J. (Eds.), Biotechnology in Plant Disease Control. Wiley-Liss, New York, NY, pp. 1–14.

Long, N., Gianola, D., Guilherme, R.J.M., Weigel, K.A., 2011. Long-term impacts of genome-enabled selection. J. Appl. Genet. 52 (4), 467–480.

Lorenz, A.J., Chao, S., Asoro, F.G., Heffner, E.L., Hayashi, T., Iwata, H., Smith, K.P., Sorrells, M.E., Jannink, J.-L., 2011. Genomic selection in plant breeding: knowledge and prospects. Adv. Agron. 110, 77–123.

Lörz, H., Wenzel, G., 2005. Molecular Marker Systems in Plant Breeding and Crop Improvement. Springer, Berlin (Germany), p. 476.

Luan, F., Sheng, Y., Wang, Y., Staub, J.E., 2010. Performance of melon hybrids derived from parents of diverse geographic origins. Euphytica 173 (1), 1–16.

Luby, J.J., Shaw, D.V., 2009. Plant breeders' perspectives on improving yield and quality traits in horticultural food crops. HortSci. 44 (1), 20–22.

Malik, C.P., Wadhwani, C., Kaur, B., 2009. Crop Breeding and Biotechnology. Pointer Publishers: Distributed by Aavishkar Publishers, Distributors, Jaipur, Rajastan (India), p. 278.

Mayo, O., 1987. The Theory of Plant Breeding. Clarendon Press, Oxford. 334 pp.

McClosky, B., LaCombe, J., Tanksley, S.D., 2013. Selfing for the design of genomic selection experiments in biparental plant populations. Theor. Appl. Genet. 126 (11), 2907–2920.

McMillan, D.E., 1984. Plant isozymes: a historical perspective. In: Tanksley, S.D., Orton, T.J. (Eds.), Isozymes in Plant Breeding and Genetics, Part A. Elsevier Science Publisher, Amsterdam (Netherlands), pp. 3–14.

Medina-Filho, H.F., Stevens, M.A., 1980. Tomato breeding for nematode resistance: survey of resistant varieties for horticultural characteristics and genotype of acid phosphatase. Acta Hortic. 100, 383–393.

Meselson, M., Yuan, R., 1968. DNA restriction enzyme from *E. coli*. Nature 217 (5134), 1110–1114.

Messeguer, R., Ganal, M., de Vicente, M.C., Young, N.D., Bolkan, H., Tanksley, S.D., 1991. High resolution RFLP map around the root knot nematode resistance gene (*Mi*) in tomato. Theor. Appl. Genet. 82 (5), 529–536.

Michelmore, R.W., Paran, I., Kesseli, R.V., 1991. Identification of markers linked to disease-resistance genes by bulked segregant analysis: a rapid method to detect markers in specific genomic regions by using segregating populations. Proc. Natl. Acad. Sci. U. S. A. 88 (21), 9828–9832.

Montesinos-López, O.A., Montesinos-López, A., Crossa, J., 2017. Bayesian genomic-enabled prediction models for ordinal and count data. In: Varshney, R.K., Roorkiwal, M., Sorrells, M.E. (Eds.), Genomic Selection for Crop Improvement. Springer International Publishing, New York, NY, pp. 55–98.

Morgan, T.H., Bridges, C.B., Sturtevant, A.H., 1925. The genetics of *Drosophila*; bibliography. Genetica 2, 1–262.

Munawar, M., Hammad, G., Nadeem, K., Raza, M.M., Saleem, M., 2015. Assessment of genetic diversity in tinda gourd through multivariate analysis. Int. J. Veg. Sci. 21 (2), 157–166.

Nakaya, A., Isobe, S.N., 2012. Will genomic selection be a practical method for plant breeding? Ann. Bot. 110 (6), 1303–1316.

Ogura, T., Busch, W., 2015. From phenotypes to causal sequences: using genome wide association studies to dissect the sequence basis for variation of plant development. Curr. Opin. Plant Biol. 23, 98–108.

I. ELEMENTS AND UNDERPINNINGS OF PLANT BREEDING

Orton, T.J., 1983. Applications of isozyme technology in breeding cross-pollinated crops. In: Tanksley, S.D., Orton, T.J. (Eds.), Isozymes in Plant Breeding and Genetics, Part A. Elsevier Science Publisher, Amsterdam (Netherlands), pp. 363–376.

Paran, I., Michelmore, R.W., 1993. Development of reliable PCR-based markers linked to downy mildew resistance genes in lettuce. Theor. Appl. Genet. 85, 985–993.

Pascual, L., Albert, E., Sauvage, C., Duangjit, J., Bouchet, J.-P., Bitton, F., Desplat, N., Brunel, D., Le Paslier, M.-C., Ranc, N., Bruguier, L., Chauchard, B., Verschave, P., Causse, M., 2016. Dissecting quantitative trait variation in the resequencing era: complementarity of bi-parental, multi-parental and association panels. Plant Sci. 242, 120–130.

Paterson, A.H., Lander, E.S., Hewitt, J.D., Peterson, S., Lincoln, S.E., Tanksley, S.D., 1988. Resolution of quantitative traits into Mendelian factors by using a complete RFLP linkage map. Nature 335, 721–726.

Paterson, A., Tanksley, S.D., Sorrels, M.E., 1991. DNA markers in plant improvement. Adv. Agron. 44, 39–90.

Paux, E., Sourdille, P., Mackay, I., Feuillet, C., 2012. Sequence-based marker development in wheat: advances and applications to breeding. Biotechnol. Adv. 30 (5), 1071–1088.

Peng, Z., Fan, W., Wang, L., Paudel, D., Leventini, D., Tillman, B.L., Wang, J., 2017. Target enrichment sequencing in cultivated peanut (Arachis hypogaea L.) using probes designed from transcript sequences. Mol. Genet. Genomics 292 (5), 955–965.

Pennisi, E., 2010. Semiconductors inspire new sequencing technologies. Science 327 (5970), 1190.

Pereira, M.,.G., Lee, M., 1995. Identification of genomic regions affecting plant height in sorghum and maize. Theor. Appl. Genet. 90 (3–4), 380–388.

Knapp, S.J., 1994. Mapping quantitative trait loci. In: Phillips, R.L., Vasil, I.K. (Eds.), DNA-Based Markers in Plants. Kluwer Academic Publishers, Dordrect (Netherlands), pp. 58–96.

Powell, W.;.G., Machray, G., Provan, J., 1996. Polymorphism revealed by simple sequence repeats. Trends Plant Sci. 1, 215–222.

Prentis, P.J., Gilding, E.K., Pavasovic, A., Frere, C.H., Godwin, I.D., 2013. Molecular markers in plant improvement. In: Henry, R.J. (Ed.), Molecular Markers in Plants. Wiley On-line Library. https://doi.org/10.1002/9781118473023.ch5.

Raghavan, C., Naredo, M.E.B., Wang, H., Atienza, G., Liu, B., Qiu, F., McNally, K.L., Leung, H., 2007. Rapid method for detecting SNPs on agarose gels and its application in candidate gene mapping. Mol. Breed. 19 (2), 87–101.

Rajsic, P., Weersink, A., Navabi, A., Pauls, K.P., 2016. Economics of genomic selection: the role of prediction accuracy and relative genotyping costs. Euphytica 210 (2), 259–276.

Riaz, A., Li, G., Quresh, Z., Swati, M.S., Quiros, C.F., 2001. Genetic diversity of oilseed Brassica napus inbred lines based on sequence-related amplified polymorphism and its relation to hybrid performance. Z. Pflanzenzüchtung 120 (5), 411–415.

Ribaut, J.M., Betran, J., 1999. Single large-scale marker-assisted selection (SLS-MAS). Mol. Breed. 5 (6), 531–541.

Ribaut, J.M., Hoisington, D., 1998. Marker-assisted selection: new tools and strategies. Trends Plant Sci. 3, 236–239.

Rick, C.M., 1984. Tomato. In: Tanksley, S.D., Orton, T.J. (Eds.), Isozymes in Plant Breeding and Genetics, Part B. Elsevier Publisher, Amsterdam (Netherlands), pp. 147–166.

Riedelsheimer, C., Melchinger, A.E., 2013. Optimizing the allocation of resources for genomic selection in one breeding cycle. Theor. Appl. Genet. 126 (11), 2835–2848.

Rincent, R., Charcosset, A., Moreau, L., 2017. Predicting genomic selection efficiency to optimize calibration set and to assess prediction accuracy in highly structured populations. Theor. Appl. Genet. 130 (11), 2231–2247.

Rogers, G.S., Jobling, J.J., Weerakkody, P., 2014. Fruit growth and bioactive development in pomegranate fruit. Acta Hortic. 1040 (2014), 269–276.

Saiki, R., Gelfand, D., Stoffel, S., Scharf, S., Higuchi, R., Horn, G., Mullis, K., Erlich, H., 1988. Primer-directed enzymatic amplification of DNA with a thermostable DNA polymerase. Science 239 (4839), 487–491.

Sane, P.V., Amla, D.V., 1990. Genetics of photosynthetic components in relation to productivity. In: Khanna, K.R. (Ed.), Biochemical Aspects of Crop Improvement. CRC Press, Boca Raton, FL, pp. 109–151.

Sanger, F., Nicklen, S., Coulson, A.R., 1977. DNA sequencing with chain-terminating inhibitors. Proc. Natl. Acad. Sci. U. S. A. 74 (12), 5463–5467.

Schneider, G.F., Dekker, C., 2012. DNA sequencing with nanopores. Nat. Biotechnol. 30 (4), 326–328.

Schulman, A.H., 2007. Molecular markers to assess genetic diversity. Euphytica 158 (3), 313–321.

Scott, J.W., 2010. Phenotyping of tomato for SolCAP and onward into the void. HortScience 45 (9), 1314–1316.

Shekar, K.C., Ashok, P., Sasikala, K., 2013. Characterization, character association, and path coefficient analyses in eggplant. Int. J. Veg. Sci. 19 (1), 45–57.

Shimomura, K., Fukino, N., Sugiyama, M., Kawazu, Y., Sakata, Y., Yoshioka, Y., 2017. Quantitative trait locus analysis of cucumber fruit morphological traits based on image analysis. Euphytica 213 (7), 138.

Sinnott, E.W., Dunn, L.C., 1939. Principles of Genetics. McGraw-Hill, Inc., New York, NY, p. 408.

Slater, A.T., Cogan, N.O.I., Rodoni, B.C., Hayes, B.J., Forster, J.W., 2016. Improving the selection efficiency in potato breeding. Acta Hortic. (1127), 237–242.

Sood, V.K., Rana, I., Hussain, W., Chaudhary, H.K., 2016. Genetic diversity of genus Avena from north western-Himalayas using molecular markers. Proc. Nat. Acad. Sci. India, Sect. B: Biol. Sci. 86 (1), 151–158.

Southern, E.M., 1975. Detection of specific sequences among DNA fragments separated by gel electrophoresis. J. Mol. Biol. 98 (3), 503–517.

Sreekala, C., Raghava, S.P.S., 2003. Exploitation of heterosis for carotenoid content in African marigold (Tagetes erecta L.) and its correlation with esterase polymorphism. Theoret. Appl. Genet. 106 (4), 771–776.

Steele, K.A., Vyas, D., Witcombe, J.R., Quinton-Tulloch, M.J., Heine, M., Dhakal, R., Amgai, R.B., Khatiwada, S.P., 2018. Accelerating public sector rice breeding with high-density KASP markers derived from whole genome sequencing of indica rice. Mol. Breed. 38 (4), 38.

Steinfath, M., Gärtner, T., Lisec, J., Meyer, R.C., Altmann, T., Willmitzer, L., Selbig, J., 2010. Prediction of hybrid biomass in Arabidopsis thaliana by selected parental SNP and metabolic markers. Theor. Appl. Genet. 120 (2), 239–247.

Stephens, M.J., Alspach, P.A., Winefield, C., 2012. Genetic parameters associated with yield and yield components in red raspberry. Acta Hortic. (946), 37–42.

Stevens, M.A., Rick, C.M., 1986. Genetics and breeding. In: Atherton, J.G., Rudich, J. (Eds.), The Tomato Crop; A Scientific Basis for Improvement. Chapman and Hall, Ltd., London, pp. 35–109.

Stuber, C.W., 1990. Isozyme markers and their significance in crop improvement. In: Khanna, K.R. (Ed.), Biochemical Aspects of Crop Improvement. CRC Press, Boca Raton, FL, pp. 59–78.

Stunkard, A.J., Harris, J.R., Pedersen, N.L., McClearn, G.E., 1990. The body-mass index of twins who have been reared apart. N. Engl. J. Med. 322 (21), 1483–1493.

I. ELEMENTS AND UNDERPINNINGS OF PLANT BREEDING

Tam, S.M., Mhiri, C., Vogelaar, A., Kerkveld, M., Pearce, S.R., Grandbastien, M.-A., 2005. Comparative analyses of genetic diversities within tomato and pepper collections detected by retrotransposon-based SSAP, AFLP and SSR. Theor. Appl. Genet. 110 (5), 819–831.

Tanksley, S.D., 1993. Mapping polygenes. Annu. Rev. Genet. 27, 205–233.

Tanksley, S.D., Medina-Filho, H., Rick, C.M., 1981. The effect of isozyme selection on metric characters in an interspecific backcross of tomato - basis of an early screening procedure. Theor. Appl. Genet. 60, 291–296.

Tanksley, S.D., Medina-Filho, H., Rick, C.M., 1982. Use of naturally-occurring enzyme variation to detect and map genes controlling quantitative traits in an interspecific backcross of tomato. Heredity 49, 11–25.

Tanksley, S.D., Young, N.D., Paterson, A.H., Bonierbale, M., 1989. RFLP mapping in plant breeding: new tools for an old science. Biotechnology 7, 257–264.

Taramino, G., Tingey, S., 1996. Simple sequence repeats for germplasm analysis and mapping in maize. Genome 39, 277–287.

Torkamaneh, D., Boyle, B., François, B., 2018. Efficient genome-wide genotyping strategies and data integration in crop plants. Theor. Appl. Genet. 131 (3), 499–511.

Trentacoste, E.R., Puertas, C.M., Sadras, V.O., 2012. Modeling the intraspecific variation in the dynamics of fruit growth, oil and water concentration in olive (Olea europaea L.). Eur. J. Agron. 2012, 83–93.

Tsai, H., Howell, T., Nitcher, R., Missirian, V., Watson, B., Ngo, K.J., Lieberman, M., Fass, J., Uauy, C., Tran, R.K., Khan, A.A., Filkov, V., Tai, T.H., Dubcovsky, J., Comai, L., 2011. Discovery of rare mutations in populations: TILLING by sequencing. Plant Physiol. 156 (3), 1257–1268.

Varshney, R.K., Dubey, A., 2009. Novel genomic tools and modern genetic and breeding approaches for crop improvement. J. Plant Biochem. Biotechnol. 18 (2), 127–138.

Varshney, R.K., Terauchi, R., McCouch, S.R., 2014. Harvesting the promising fruits of genomics: applying genome sequencing technologies to crop breeding. PLoS Biol. 12 (6), e1001883. https://doi.org/10.1371/journal.pbio.1001883.

Varshney, R.K., Singh, V.K., Hickey, J.M., Xun, X., Marshall, D.F., Wang, J., Edwards, D., Ribaut, J.-M., 2016. Analytical and decision support tools for genomics-assisted breeding. Trends Plant Sci. 21 (4), 354–363.

Varshney, R.K., Roorkiwal, M., Sorrells, M.E., 2017. Genomic selection for crop improvement: an introduction. In: Varshney, R.K., Roorkiwal, M., Sorrells, M.E. (Eds.), Genomic Selection for Crop Improvement. Springer International Publishing, New York, NY, pp. 1–6.

Verbyla, A.P., George, A.W., Cavanagh, C.R., Verbyla, K.L., 2014. Whole-genome QTL analysis for MAGIC. Theor. Appl. Genet. 127 (8), 1753–1770.

Viana, J.M.S., Mundim, G.B., Pereira, H.D., Bastos Andrade, A.C., Fonseca e Silva, F., 2017. Efficiency of genome-wide association studies in random cross populations. Mol. Breed. 37 (8), 102.

Vos, P., Hogers, R., Bleeker, M., Reijans, M., van de Lee, T., Hoernes, M., Frijters, A., Pot, J., Peleman, J., Kuiper, M., Zabeau, M., 1995. AFLP: a new technique for DNA fingerprinting. Nucleic Acids Res. 23, 4407–4414.

Weeden, N.F., 1984. Evolution of plant isozymes. In: Tanksley, S.D., Orton, T.J. (Eds.), Isozymes in Plant Breeding and Genetics, Part A. Elsevier Science Publisher, Amsterdam (Netherland), pp. 177–208.

Weeden, N., Timmerman, G., Lu, J., 1994. Identifying and mapping genes of economic significance. Euphytica 73, 191–198.

Welsh, J., McClelland, M., 1990. Fingerprinting genomes using PCR with arbitrary primers. Nucleic Acids Res. 18, 7213–7218.

Williams, J., Kubelik, A., Livak, K., Rafalski, J., Tingey, S., 1990. DNA polymorphisms amplified by arbitrary primers are useful as genetic markers. Nucleic Acids Res. 18, 6531–6535.

Winter, P., Kahl, G., 1995. Molecular marker technologies for plant improvement. World J. Microbiol. Biotechnol. 11, 438–448.

Wricke, G., Weber, W.E., 1986. Quantitative Genetics and Selection in Plant Breeding. Waltude Gruyter, Berlin (Germany), p. 406.

Wright, S., 1921. Systems of mating 1. The biometric relations between parent and offspring. II. The effects of inbreeding on the genetic composition of a population. III. Assortative mating based on somatic resemblance. IV. The effects of selection. Genetics 6, 111–178.

Wu, Z., Zhang, T., Li, L., Xu, J., Qin, X., Zhang, T., Cui, L., Qunfeng, L., Li, J., Chen, J., 2016. Identification of a stable major-effect QTL (Parth 2.1) controlling parthenocarpy in cucumber and associated candidate gene analysis via whole genome re-sequencing. BMC Plant Biol. 16 (1), 182.

Xie, C., Xu, S., 1998. Efficiency of multistage marker-assisted selection in the improvement of multiple quantitative traits. Heredity 80, 489–498.

Xu, S., 2002. QTL analysis in plants. In: Camp, N.J., Cox, A. (Eds.), Quantitative Trait Loci. Humana Press, Totowa, NJ, pp. 283–310.

Yan, W., Frégeau-Reid, J., 2008. Breeding line selection based on multiple traits. Crop. Sci. 48 (2), 417–423.

Yang, X., Ye, C.-Y., Cheng, Z.-M., Tschaplinski, T.J., Wullschleger, S.D., Yin, W., Xia, X., Tuskan, G.A., 2011. Genomic aspects of research involving polyploid plants. Plant Cell Tissue Organ Cult. 104 (3), 387–397.

Yang, H., Li, C., Lam, H.-M., Clements, J., Yan, G., Zhao, S., 2015. Sequencing consolidates molecular markers with plant breeding practice. Theor. Appl. Genet. 128 (5), 779–795.

Yang, G.-Q., Chen, Y.-M., Wang, J.-P., Guo, C., Zhao, L., Wang, X.-Y., Ying, G., Li, L., Li, D.-Z., Guo, Z.-H., 2017. Development of a universal and simplified ddRAD library preparation approach for SNP discovery and genotyping in angiosperm plants. Plant Methods 12 (1), 39.

Yu, J., Holland, J.B., McMullen, M.D., Buckler, E.S., 2008. Genetic design and statistical power of nested association mapping in maize. Genetics 178 (1), 539–551.

Zalapa, J.E., Staub, J.E., McCreight, J.D., 2006. Generation means analysis of plant architectural traits and fruit yield in melon. Plant Breed. 125 (5), 482–487.

Zhang, W., Smith, C., 1993. Simulation of marker-assisted selection utilizing linkage disequilibrium: the effects of several additional factors. Theor. Appl. Genet. 86 (4), 492–496.

Zhao, Y., Gowda, M., Longin, F.H., Würschum, T., Ranc, N., Reif, J.C., 2012. Impact of selective genotyping in the training population on accuracy and bias of genomic selection. Theor. Appl. Genet. 125 (4), 707–713.

Further Reading

Sorrells, M.E., 2015. Genomic selection in plants: Empirical results and implications for wheat breeding. In: Ogihara, Y., Takumi, S., Handa, H. (Eds.), Advances in Wheat Genetics: From Genome to Field. Springer, Tokyo (Japan), pp. 401–409.

10

Natural Mating Systems and Controlled Mating

INTRODUCTION

Another important tool available to the plant breeder for affecting desirable changes in population allelic frequencies in addition to selection is controlled mating. Naturally occurring mating systems were presented in Chapters 1–4, including basic biology and genetic consequences of different mating systems. Natural mating systems are indelibly intertwined with the genetic structure of the starting population, with the breeding strategy, and final outcome of the program. Nearly all plant breeding programs incorporate some form of mating that is altered from the natural state to achieve breeding goals, usually in combination with selection. How and why do plant breeders change and subvert mating systems to develop genetically improved plant populations? This chapter will describe the basic features of natural mating systems then explain how plant breeders either use these features or alter them to control mating in breeding programs.

NATURAL PLANT MATING SYSTEMS

Professor A. J. Richards starts his fascinating textbook on plant mating systems with the statement: "Most animals, and all vertebrates, behave. They are able to choose their mates, and they unconsciously influence the genetic makeup and evolutionary patterns of their populations through this behavior." Richards (1986) continued: "Plants do not behave. Unlike most animals, higher plants are essentially stationary within a generation, although some plants have considerable powers of vegetative dispersal and pollen and seed dispersal between generations can also be sizable. They are not conscious, and so they cannot choose a mate."

It is a defining feature of plants that they are mostly stationary, incapable of motor functions including locomotion. Plants do not "behave" in the sense of having definable animal-like behavior. Plants do, however, distinguish their mates within habitats that are replete with mating alternatives. At the level of species or closely-related species, plants can discern which classes of gametes, among the mixture of thousands of gametes (pollen grains) will be selected as suitable mates.

While animals employ complicated behavioral patterns to choose mates and the gametes they bear, plants seem to be more passive. The drive to produce offspring for the next generation is equally strong, however, among plants and animals. The diversity of ways that plants produce offspring is truly astonishing. Flowering behavior in plants is among the most mutable and plastic of all traits, allowing species to quickly adapt to changes in environment, habitat, and ecosystem (Devaux et al., 2014).

The genetic outcomes of mating are generalized into classes such as autogamy (gametes from the same individual), heterogamy (gametes from different individuals), and anisogamy (no gametes at all; the embryo formed from maternal somatic cells). There are mitotic and meiotic "mistakes" that may result in a change of ploidy (haploidy, polyploidy, and aneuploidy; Chapter 3) and interspecific hybrids (alloploids) that also occur sporadically and can have a major impact on the course of evolution (Richards, 1986).

Theoretically, asexuality and autogamy are intrinsically more advantageous to selected individuals and populations than heterogamy over short periods of evolutionary history. Heterogamy is prevalent among a broad spectrum of plant species in nature and, therefore, other evolutionary forces must be present that prevent all species from becoming purely self-perpetuating. Charlesworth and Charlesworth (2012) proposed that inbreeding depression and heterosis (covered in Chapter 16) are the evolutionary forces responsible for the observed balance of autogamy and heterogamy in nature. To achieve autogamy or heterogamy, plants deploy an amazingly diverse array of floral modifications that promote self-pollination or attract animal (usually insect) vectors to move pollen from one plant to another (Kariyat et al., 2013). Alternatively, plants produce copious volumes of pollen that is carried to prospective female parents by wind currents. The flowers of these wind-pollinated females are adapted to effectively filter pollen out of the air much like a coral polyp filters microorganisms from seawater for food.

The process of evolution requires the constant generation of genetic variability that leads to the phenotypic variability upon which natural selection acts. Heterogamy is analogous to the mating system of most animal species on Earth. Plant species that are characterized by a mating system that includes heterogamy tend to generate nascent genetic variability that is hidden by dominant allelic interactions, and released slowly to the forces of natural selection as homozygotes. In contrast, recurrent cycles of autogamy drive all recessive alleles to fixation quickly, where natural selection may favor or extinguish them. Almost universally, plant species that feature mainly heterogamy are more genetically variable than those that feature mainly autogamy (Richards, 1986; Honnay and Jacquemyn, 2008).

It is widely accepted that the first terrestrial vascular plants were outcrossing hermaphrodites (Richards, 1986). It is certain that self-pollination was a mating system strategy that arose independently in a large proportion of plant taxa during the course of evolution. Fossil evidence indicates that the evolution of self- from cross-pollinated evolutionary intermediates is irreversible, but experimental results demonstrated examples of cross-pollinated individuals derived from self-pollinating species (Richards, 1986; Takebayashi and Morrell, 2001). Fossil evidence and advances in evolutionary theory prompted genetic theorist G. Ledyard Stebbins to speculate that self-pollination is not a sustainable strategy; rather it is an evolutionary "dead end" (Stebbins, 1950). Results of more recent and powerful genomics experiments have, to date, supported Stebbins' theory (Takebayashi and Morrell, 2001).

Recent molecular evidence has redefined plant phylogenetics and systematics (Fig. 10.1). Molecular biology is also having a tremendous impact on the field of plant reproductive biology (Milligan, 1996; Alström-Rapaport, 1997; Rostoks et al., 2006; Karron et al., 2012; Kim et al., 2016). Fig. 10.1 shows that the mating system is not a progressive evolutionary characteristic that is a distinguishing feature of taxa. Rather, plants appear to adapt reproductive systems to fit the needs of the ecological niche at a time in evolutionary history. Fig. 10.1 also reflects the impacts that humans have had on plants through artificial selection. For example, most wild *Solanum* spp. relatives of tomato (*S. lycopersicum*) are cross-pollinated. Strong recurrent selection for high fruit set and yield has probably resulted in tomato cultivars that are self-pollinated though the flowers retain many features typical of cross-pollination. There are many other examples of this phenomenon (e.g., *Prunus*).

Molecular evidence supports the conclusion that Magnoliaceae is the most primitive family of terrestrial angiosperm plants on Earth (Richards, 1986). The flowers of this family are perfect (hermaphroditic). It is presumed that other angiosperm families evolved from Magnoliaceae so it is not surprising that the overwhelming majority of angiosperm flowers are hermaphroditic (Table 10.1). While modifications in the mating system ranging to dioecy are not as common as hermaphrodity, floral development and function tend to be variable and androecy and gynoecy are common in some predominantly hermaphroditic species. Mutations or natural genes associated with male sterility and gynoecy are particularly common.

Reproductive biology and floral structure and function are, however, mostly consistent within plant taxa. Floral structure has, historically, been the defining attribute in plant systematics dating to the original works of Linnaeus (Richards, 1986). With a few exceptions, the taxonomic trees that were devised by contrasting floral structures agree with phylogenetic trees based on molecular polymorphisms (Scutt et al., 2006). Evidence points to the ability of plants to alter the

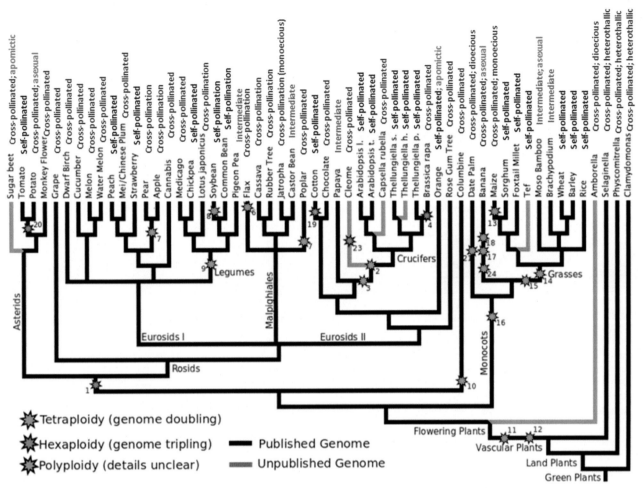

FIG. 10.1 Consensus phylogeny of plants ranging from unicellular algae (*Chlamydomonas*) to angiosperms based on imputed evolutionary relationships from genomic polymorphic molecular marker data. The "star" notations indicate where polyploidization events have occurred during species divergence. The script colors indicate basic mating system employed by the species or group of species: purple = cross-pollination; red = self-pollination; blue = intermediate (both self- and cross-pollination); green = mode of asexual reproduction. *Based on the cladogram of sequenced plant genomes (up to April 2013) generated by James Schnable at CoGe (http://genomevolution.org) and used with permission.*

TABLE 10.1 Prevalence of Floral and Sexual Configurations Among Terrestrial Angiosperm Species

Floral system	Sexual constitution	Sex of individual	% of angiosperm species
Hermaphrodity	Hermaphrodite	♀ and ♂	72
Monoecy	Female and Male flowers	♀ and ♂	5
Gynomonoecy	Hermaphrodite and female flowers	♀ and ♂	3
Dioecy	Female and male flowers	♀ or ♂	4
Gynodioecy	Hermaphrodite and female or male flowers	♀ and ♂	7
Other (asexual)			9

After Richards, A.J., 1986. Plant Breeding Systems. George Allen & Unwin, London (UK), 529 pp.

course of reproductive biology in response to the environment (Fig. 10.1). In general, if environmental parameters exceed the range for a typically cross-pollinated species, flowers change to become more receptive to self-pollination. Examples include temperature (Hiratsuka et al., 1989; Holsinger, 1996; Westwood et al., 1997), drought (Jorgensen and Arathi, 2013; Devaux et al., 2014; Waser and Price, 2016), and herbivory (Ivey and Carr, 2005; Carr, 2013; Campbell, 2014).

Floral development and structure are fundamental determinants of mating system that is driven by floral modifications that favor either self- or cross-pollination. Gymnosperms are predominantly monoecious, while primitive

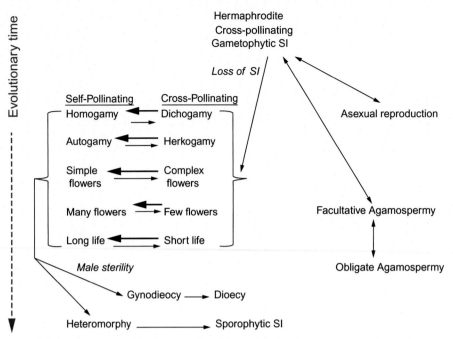

FIG. 10.2   Evolutionary trends in mating systems, floral structural, and floral functional modifications. SI = self-incompatibility; agamospermy = formation of seeds asexually; dichogamy = separation of anther dehiscence and stigma receptivity of a flower in time; homogamy = simultaneous anther dehiscence and stigma receptivity of a flower; herkogamy = separation of dehiscing anther and receptive stigma in space; autogamy = self-pollination within a flower; gynodioecy = hermaphrodite and pistillate flowers on an individual; dioecy = separation of staminate and pistillate flowers on different individuals; heteromorphy = the coexistence of two or more hermaphrodite floral classes among a group of individuals (e.g., pin and thrum; cob and papillate; tristyly). *Adapted from Richards, A.J., 1986. Plant Breeding Systems. George Allen & Unwin, London, UK, 529 pp.*

angiosperms were mostly hermaphrodites. A broad array of floral modifications is evident in hermaphrodites that drive cross- versus self-pollination (Fig. 10.2). Experimental and empirical evidence suggests that the pathways from dichogamous hermaphrodites led to homogamous hermaphrodites followed evolutionarily by the spatial separation of sexes on distinct floral structures (staminate and pistillate flowers).

It is apparent that the forces of natural selection operate differently on the mating systems of short-lived annual as compared to long-lived perennial plant species. Specifically, self-pollination is much more prevalent in annuals than in perennials (Fig. 10.3; Barrett et al., 1996).

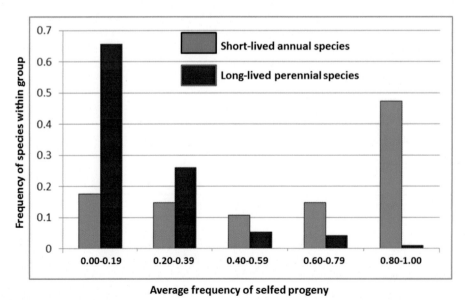

FIG. 10.3   Comparison of 74 short-lived annual vs. 96 long-lived perennial plant species with regard to the relative frequency of self-pollination (Barrett et al., 1996).

GENERALIZED FLOWER STRUCTURE AND FUNCTION

The flower is the structure in angiosperms that houses the plant's sexual organs, and it develops and is manifested in similar ways across the entire plant kingdom. Floral structure is a fundamental descriptor used to classify plant taxa, numbers of sepals, petals, stamens, carpels, etc. In the idealized flower depicted in Fig. 10.4, multiple purposes are fulfilled. The petals are showy and fragrant to attract insect pollinators. They sometimes even mimic the insect vector, inviting it to come down from the sky for sex. Frequently, special organs called nectaries produce and display sweet or musky nectar, a further inducement to hungry, sex-starved insects to visit the flower and introduce or carry away pollen.

The idealized flower is perfect, hermaphroditic, containing both functional male (androecium) and female (gynoecium) units. The androecium is comprised of the whorl of stamens, each a filament and anther. The filament is of variable length, and serves to position the anther precisely within the flower to encourage or discourage pollen dispersal within or outside the flower, and to make pollen available to prospective vectors. The anther houses the mature male gametophyte, the pollen grains, each usually binucleate or trinucleate. One of the pollen generative nuclei will fuse with the egg to form the zygote, another with the polar nuclei to form the endosperm (Brewbaker, 1967). The anther remains intact until a critical moment when it bursts or splits, termed dehiscence, releasing the payload of pollen grains to the vector or directly to the stigma.

The gynoecium (Fig. 10.5) has undergone a parallel developmental transformation to the androecium, and consists at maturity of the integument, the micropyle (opening for pollen tubes), the nucellus, and the female gametophyte (egg nucleus, polar nuclei, antipodals). The ovary may be singular, or may be composed of multiple units, ovules or carpels. It is easier for some to visualize the gynoecium by tracing the development of familiar fruits in reverse, for example apples or tomatoes. Each seed arose from an individual female gametophyte. This reality strikes at the heart of the efficiency of the breeding program, where a large proportion of the required labor is invested into making crosses. Singular ovules translate into one seed per pollination, while multicarpellate fruits produce multiple seeds per pollination.

Receptive sexual organs and gametes exist for fleeting moments and in tiny spaces, so the breeder must know where they are and when they are viable or receptive. Many plant species exhibit diurnal behavior, responding to cues such as dawn or dusk, dew, and temperature fluxes to "synchronize" their respective biological clocks (Nozue and Maloof, 2006). Thus, the plant breeder must respond to the same cues to be successful in controlled mating. Surgery is often practiced on flowers, anther-ectomies or castration to prevent self-pollination from taking place when crosses are desired. Devices have been devised to either use or discourage pollen vectors, also to control crossing. Examples include "bagging" flowers to keep away insects, and planting flowering plants in "cages" with insect pollinators thrown in (see Chapter 9). The plant breeder must be aware of how long pollen can be maintained in a viable state and the optimal conditions necessary for long-term storage. In many cases, pollen can be frozen for prolonged periods, even many years, and used later, much like bovine semen.

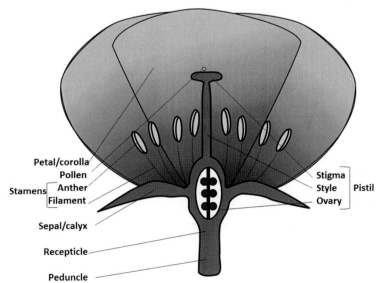

FIG. 10.4 Anatomy of the generalized angiosperm dicotyledonous hermaphrodite flower. *Source: https://pixabay.com/vectors/flower-dissection-dissected-31439/.*

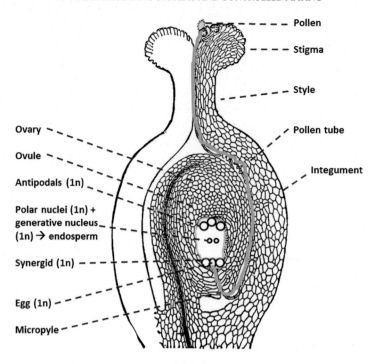

FIG. 10.5   Anatomy of the angiosperm pistil, or gynoecium. *From Goodale, G.L., 1885. Physiological Botany; I. Outlines of the Histology of Phænogamous Plants. II. Vegetable Physiology. American Book Co.*

The development of the zygote into the embryo and mature plant is accomplished by a precise interplay of many genes, the expression of which is controlled both temporally and spatially (Quatrano, 1986; Takayama and Isogai, 2005). The symphony of expressed genes into a functioning organism is crafted over long periods of evolutionary history. A union of pollen and egg from different species only rarely leads to subsequent signs of embryo development. To dispense with the developmental incompatibilities associated with unions of unrelated gametes, a system of recognition has arisen in plants. Pollen grains do not germinate and pollen tubes do not grow unless the appropriate chain of chemical events takes place, much like the "lock and key" analogy of enzymatic reactions or antigen-antibody recognition. Thus, with thousands of different plant species contributing to air- and insect-borne pollen, plants on which they land possess molecular and physiological systems that select the correct pollen.

FLORAL DEVELOPMENT AND TRANSCRIPTION FACTORS (MADS-BOX)

Molecular studies on gene expression and post-transcriptional activities in *Saccharomyces cerevisiae, Arabidopsis thaliana, Antirrhinum majus* (snapdragon), and mammals including humans led to the discovery of a generalized gene expression modulation system that is of broad significance in developmental pathways of all multicellular eukaryotes (Dubois et al., 1987; Meyerowitz et al., 1989; Schwarz-Sommer et al., 1990). This genomic domain binds to DNA sequences of high similarity to the motif $CC[A/T]_6GG$ termed the "CArG-box". It was concluded that the gene products encoded by this genomic region were DNA binding proteins that modulated transcription, or "transcription factors". Due to the obvious sequence homology of this region across these species and humans, the term "MADS-box" was applied: **M**CM1 from *S. cerevisiae*; **A**GAMOUS from *A. thaliana*; **D**EFICIENS from *A. majus*; **S**RF from *H. sapiens*. All multicellular eukaryotes have many MADS-box regions embedded in their genomes, each controlling different developmental and environmentally-modulated physiological processes and phenotypes (Theissen et al., 1996; Saedler et al., 2001; Hu and Liu, 2012).

Historically, floral development in eukaryotic plants was the first general characteristic to be intensively studied with regard to the structure, function, and evolution of MADS-box genes. It turns out that MADS-box function also controls fruit and seed development, encompassing an enormous portion of all topics in the field of plant breeding, including mating system and phenotypes of high economic value (Rounsley et al., 1995). MADS-box sequences are generally only 165–180 nucleotide base pairs in length translating to peptides 55–60 amino acids in length, but their importance overshadows the diminutive length (de Bodt et al., 2003).

Results of elegant experiments with flowering behavior in *A. thaliana* led to the development of the "ABC" model of floral development (Bowman et al., 1991). This simple model provides a conceptual framework for explaining how the individual and combined activities of the *ABC* genes produce the four organ types of the typical eudicot flower (Pelaz et al., 2000). Different MADS-box peptides form homo- and heterodimers that collectively bind with genomic DNA to turn on the genes necessary for the development of specific organs such as sepals, petals, stamens, and pistils, the four "whorls" of modified leaves we know as flowers (Coen and Meyerowitz, 1991; Sieburth et al., 1995; Rounsley et al., 1995; Riechmann et al., 1996; Immink et al., 2002; Busi et al., 2003). For example, 505 binding sites for the *A. thaliana* MADS-box gene product FLC were identified, mostly located in the promoter regions of genes and containing at least one CArG box. This is the motif known to be associated with MADS-box proteins such as FLC (Deng et al., 2011).

Given the requirement of *E*-class genes for floral organ specification, the ABC model is now often referred to as the ABCE model (Colombo et al., 1995; Pelaz et al., 2000; Honma and Goto, 2001; Ma, 2005). In *Arabidopsis*, the A-function genes are APETALA1 (API) and APETALA2 (AP2), B function is provided by APETALA3 (AP3) and PISTILLATA (PI), C function by AGAMOUS (AG), and E function by multiple SEPALLATA genes (SEP1–4). All but one (AP2) of the ABCE genes are members of the MADS-box gene family (Jofuku et al., 1994; Ma and de Pamphilis, 2000; Becker and Theissen, 2003).

MADS-box genes regulate the early step of specifying floral meristem identity as well as the later step of determining the fate of floral organ primordia (Mandel and Yanofsky, 1998). Fig. 10.6 illustrates how MADS-box gene products interact to coordinate this developmental process. Modifications of MADS-box genes due to forces of natural selection over evolutionary time have led to all of the modifications in mating systems depicted in Fig. 10.2.

Plant MADS-box proteins contain a DNA-binding (M), an intervening (I), a Keratin-like (K) and a C-terminal C-domain (Saedler et al., 2001). Evolutionary forces have powered a divergence of MADS-box genomic sequences to take on new roles in development and environmental responses. Many different versions of MADS-box organization have been elucidated (de Bodt et al., 2003). Plant MADS-domain sequences diverged much faster than those of animals. Gene duplication and sequence diversification were extensively involved in the creation of new genes during plant evolution and correlated with the evolution of increasingly complex body plans (Theissen et al., 1996; Theissen et al., 2000; Vandenbussche et al., 2004). In *A. thaliana*, approximately 100 MADS-box orthologs or derivatives have been identified to date (Airoldi and Davies, 2012). Expression patterns of floral MADS-box genes in primitive (or basal) angiosperms are more diverse than those of their counterparts in more recent eudicots and monocots (Kim et al., 2005).

Activities of MADS-box genes underlie all of the floral modifications discussed below that have an enormous impact on plant breeding strategies and the phenotypes under selection. Efforts to further understand how MADS-box genes and other transcription factors control developmental processes will elucidate opportunities to manipulate these genes in a desirable way. One example is the selective activation of male and female floral functions at will to facilitate targeted individual and population mating dynamics. Findings have also recently appeared showing that

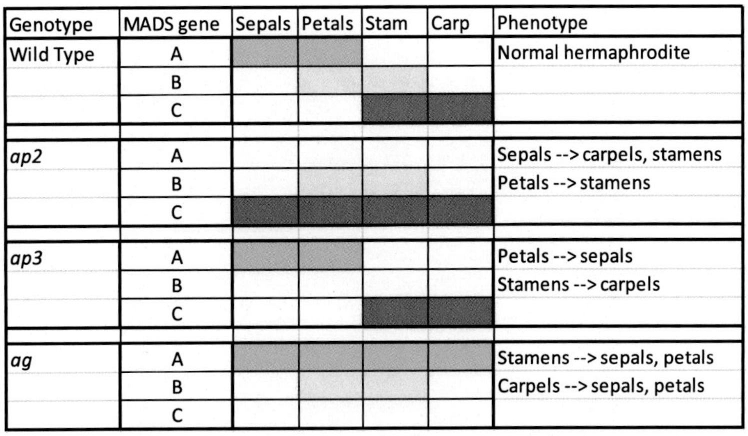

Genotype	MADS gene	Sepals	Petals	Stam	Carp	Phenotype
Wild Type	A					Normal hermaphrodite
	B					
	C					
ap2	A					Sepals --> carpels, stamens
	B					Petals --> stamens
	C					
ap3	A					Petals --> sepals
	B					Stamens --> carpels
	C					
ag	A					Stamens --> sepals, petals
	B					Carpels --> sepals, petals
	C					

FIG. 10.6 A model of the interaction of MADS-box genes in the determination of eukaryotic floral organs. According to this model, the identity of the different floral organs (sepals, petals, stamens and carpels) is determined by four combinations of floral homeotic proteins known as MADS-box proteins. Protein quartets, which are transcription factors, operate by binding to the promoter regions of target genes, that are activated or repressed for the development of the different floral organs. The indicated mutants in *Arabidopsis thaliana* (Genotype column) knock out one of the MADS-box genes, resulting in a corresponding change in floral phenotype that is explained by the ABC Model. Stam = stamens; Carp = carpels. Sources: Goto and Meyerowitz (1994); Jack et al. (1994); Liu and Meyerowitz (1995).

small ncRNAs such as microRNAs (miRNAs) and long ncRNAs (lncRNAs) have important roles in a wide range of biological processes such as the regulation of reproduction and sex determination (Li et al., 2015).

SELF-INCOMPATIBILITY

The separation of male and female gametes in time and space are macro-biological strategies utilized by plants to encourage cross-pollination and discourage self-pollination events under natural conditions. Physiological systems have also evolved in hermaphrodite flowers to promote cross-pollination, known as *self-incompatibility* (SI; de Nettancourt, 1972). SI is the ability of stigmatic or style cells of the pistil of a hermaphrodite flower to distinguish between pollen and pollen tubes from self or non-self. If the pollen source is determined by the detection mechanism to be "self", a chemical response is deposited that prevents or slows the germination/growth of the pollen tube while pollen tubes from non-self pollen grains continue to grow and accomplish gamete (and polar nuclei with tube nucleus) fusion.

SI has been observed and studied since the beginning of the field of genetics; first described in 1763 (Kolreuter, 1763). Attempts were made in the early 20th to explain segregation patterns in self-incompatible species soon after the rediscovery of Mendel's seminal experiments (De Vries, 1907; Correns, 1913). SI is relatively simple to identify. The species that bears SI will have the following characteristics: flowers are hermaphroditic and devoid of agamospermy, dichogamy or herkogamy; the anthers produce copious pollen; in the absence of pollen vectors (e.g., arthropods or wind), individual plants fail to self-pollinate; crosses to other individuals may or may not result in progeny; certain individuals may undergo apparently successful self-pollination (Richards, 1986).

Researchers conducted classical Mendelian genetic studies to unravel the inheritance of the self-infertile phenotype starting in the early 20th century (East, 1919a, b, c). The first in-depth description of SI was for the ancestral single-locus gametophytic form (Fig. 10.2; East and Mangelsdorf, 1925). Different forms of SI have been discovered based on the number of alleles at the SI locus and correspondence of the phenotype of the gamete to genotype or that of the parental sporophyte (Nasrallah et al., 1969; Richards, 1986). Among many alleles at the SI locus are often those associated with self-compatibility, or the ability to self-pollinate. Fig. 10.7 contrasts a *Brassica rapa* individual with

FIG. 10.7   Racemes of two distinct *Brassica rapa* individuals grown in the absence of pollen vectors: A. self-compatible (abundant seed set); and B. self-incompatible (no seed set).

strong sporophytic SI (on the right) as compared with one that bears a genotype with SC alleles (on the left; setting seeds in silique fruits despite the absence of pollen vectors).

Most SI systems are either gametophytic or sporophytic with multiple alleles at the S locus, but di-allelic forms of SI have been reported. The latter is usually associated with heteromorphy, and not commonly encountered in species of economic importance (Richards, 1986). Therefore, this textbook will focus on the more common forms of multi-allelic SI. The simplest case is the single locus gametophytic form where self-sterile and self-fertile interactions of pollen and pistil are governed by gametophyte genotype at the S locus (Table 10.2). The resulting inheritance patterns of self-sterility or -fertility for the same crosses are given in Table 10.3 and the genotypes of progeny from the parents in Tables 10.2 and 10.3 are presented in Table 10.4.

Pictorially, Fig. 10.8 exemplifies the various reactions of pollen and stigma with regard to S-locus (SI) genotype. S-alleles exist in certain species that impart either undetectable or weak SI. Let us presume that SF is an allele that is associated with lack of SI, and $S_FS_F \times S_FS_F$ is completely fertile. The S_XS_F genotype would likely result in a self-sterile phenotype since S_X is usually dominant over S_F.

The following method is used to estimate the number of S-locus alleles in a species with gametophytic self-incompatibility (Lewis, 1955):

1. Grow a large population of individuals of the species;
2. Perform a full diallel of pairwise crosses of these individuals;
3. Examine pollen germination, tube growth, or seed set on the female parents of these crosses to determine which crosses are compatible (fertile) or incompatible (sterile);
4. Determine the number of different cross-incompatible groups that are determined by the resulting data (this is presumed to be equal to the number of S-genotypes);
5. Apply the following formula:

$$N(S-genotypes) = n(n-1)/2.$$

where n = number of S-alleles.

In certain species, a large number of S-alleles have been found, for example *Oenothera organensis* where 45 alleles were reported (Emerson, 1939). An extrapolation of estimation methods led Lewis (1949) to conclude that *Trifolium*

TABLE 10.2 Interactions of pollen and pistil in a single S locus gametophytic SI system. Geno = S locus genotype of sporophyte that bears the male gametes; Pheno = SI phenotype of male gametes (pollen) from the corresponding genotype

Sporophyte Genotype	Male gametophyte genotypes											
	Geno		Geno			Geno			Geno			
	S_XS_X*	Pheno	S_XS_Y	Pheno		S_XS_Z	Pheno		S_WS_Z	Pheno		
S_XS_X*		S_X		S_X	S_Y		S_X	S_Z		S_W	S_Z	
S_XS_Y		S_X		S_X	S_Y		S_X	S_Z		S_W	S_Z	
S_XS_Z		S_X		S_X	S_Y		S_X	S_Z		S_W	S_Z	
S_WS_Z		S_X		S_X	S_Y		S_X	S_Z		S_W	S_Z	

No pollen germination or tube growth.

TABLE 10.3 Appearance of self-sterility (= S; no seeds in the absence of vectors) vs. self-fertility (= F; seeds in the absence of vectors) among crosses presented in Table 10.2

♀ Genotype	♂ Genotype			
	S_XS_X[a]	S_XS_Y	S_XS_Z	S_WS_Z
S_XS_X[a]	S	F[b]	F[b]	F
S_XS_Y	S	S	F[b]	F
S_XS_Z	S	F[b]	S	F[b]
S_WS_Z	F	F	F[b]	S

[a]*Homozygous genotypes at the S locus not common in nature due to SI and cross-pollination.*
[b]*Phenotype is "fertile" but not all pollen will produce progeny.*

TABLE 10.4   Expected genotypes of progeny from crosses of the SI genotypes presented in Tables 10.2 and 10.3

♀ Genotype	♂ Genotype			
	S_XS_X[a]	S_XS_Y	S_XS_Z	S_WS_Z
S_XS_X[a]	None	S_XS_Y	S_XS_Z	S_XS_W
				S_XS_Z
S_XS_Y	None	None	S_XS_Y	S_XS_W
			S_YS_Z	S_XS_Z
				S_YS_W
				S_YS_Z
S_XS_Z	None	S_XS_Y	None	S_XS_W
		S_ZS_Y		S_ZS_W
S_WS_Z	S_WS_X	S_WS_X	S_WS_X	None
	S_ZS_X	S_WS_Y	S_ZS_X	
		S_ZS_X		
		S_ZS_Y		

[a]*Homozygous genotypes at the S locus not common in nature due to SI and cross-pollination.*

Pollen produced by S_1S_2 sporophyte

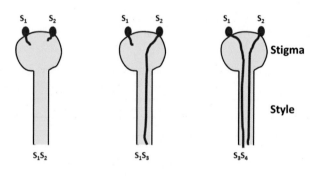

Genotype of pistil

FIG. 10.8   Pictogram of pollen (male gametophyte) interacting with the pistil (female sporophyte) in the case of gametophytic SI and multiple alleles at the S-locus.

pretense and *T. repens* have as many as 400 different S-alleles. Emerson theorized that a larger number of S-alleles endows the population with higher evolutionary fitness than if the population had a lower number of S-alleles since a higher proportion of random matings will produce offspring.

Sporophytic SI is much less common than gametophytic SI, and appeared later during the evolution of angiosperms (Richards, 1986; Fig. 10.2). Multi-allelic sporophytic SI is limited to two main families: Asteraceae and Brassicaceae (Richards, 1986). From the standpoint of impact on plant breeding, sporophytic SI is mainly of importance in the breeding of hybrid varieties of crucifer vegetables (cabbage/kale, turnip) and canola. In sporophytic SI, the SI phenotype of the pollen and pollen tube are determined by the sporophyte bearing the male gametes (as contrasted with gametophytic SI where the SI phenotype of the pollen and pollen tube are determined by the gametophyte). Therefore, an individual S_XS_Y will give rise to S_X and S_Y pollen, but they will both exhibit the same SI phenotype (S_XS_Y) despite having different haplotypes.

Tables 10.5–10.7 use the same hypothetical crosses as Tables 10.2–10.4 to provide a comparison of gametophytic vs. sporophytic SI:

TABLE 10.5 Interactions of pollen and pistil in a single S locus sporophytic SI system. Geno = S locus genotype of sporophyte that bears the male gametes; Pheno = SI phenotype of male gametes (pollen) from the corresponding genotype

Sporophyte Genotype	Male gametophyte genotypes							
	Geno		Geno		Geno		Geno	
	$S_xS_x^*$	Pheno	S_xS_Y	Pheno	S_xS_Z	Pheno	S_WS_Z	Pheno
$S_xS_x^*$		S_xS_x		S_xS_Y		S_xS_Z		S_WS_Z
S_xS_Y		S_xS_x		S_xS_Y		S_xS_Z		S_WS_Z
S_xS_Z		S_xS_x		S_xS_Y		S_xS_Z		S_WS_Z
S_WS_Z		S_xS_x		S_xS_Y		S_xS_Z		S_WS_Z

*No pollen germination or tube growth.

TABLE 10.6 Appearance of self-sterility (= S; no seeds in the absence of vectors) vs. self-fertility (= F; seeds in the absence of vectors) among crosses presented in Table 10.5

♀ Genotype	♂ Genotype			
	$S_xS_x^a$	S_xS_Y	S_xS_Z	S_WS_Z
$S_xS_x^a$	S	S	S	F
S_xS_Y	S	S	S	F
S_xS_Z	S	S	S	S
S_WS_Z	F	F	S	S

[a]Homozygous genotypes at the S locus not common in nature due to SI and cross-pollination.

TABLE 10.7 Expected genotypes of progeny from crosses of the SI genotypes presented in Tables 10.5 and 10.6

♀ Genotype	♂ Genotype			
	$S_xS_x^a$	S_xS_Y	S_xS_Z	S_WS_Z
$S_xS_x^a$	None	None	None	S_xS_W
				S_xS_Z
S_xS_Y	None	None	None	S_xS_W
				S_xS_Z
				S_YS_W
				S_YS_Z
S_xS_Z	None	None	None	None
S_WS_Z	S_WS_x	S_WS_x	None	None
	S_ZS_x	S_WS_Y		
		S_ZS_x		
		S_ZS_Y		

[a]Homozygous genotypes at the S locus not common in nature due to SI and cross-pollination.

Sporophytic SI provides a further opportunity (or complication) of interallelic interactions in determining pollen phenotypes. The most common interallelic interaction observed is dominance. If the genotype of the sporophyte bearing the male gamete is S_xS_Y and S_x is dominant over S_Y, the phenotype of the pollen will be S_x. Let us suppose that S_x is dominant over S_Y and that S_Z is dominant over S_x, while S_W and S_Z interact codominantly. Table 10.5 will be changed as follows (Table 10.8):

As was illustrated for gametophytic SI, Fig. 10.9 shows the various reactions of pollen and stigma with regard to S-locus (SI) genotype for sporophytic SI where all alleles interact codominantly.

TABLE 10.8 Table 10.5 reconfigured to illustrate the effect of dominance in sporophytic SI (see text)

Sporophyte Genotype	Male gametophyte genotypes							
	Geno $S_xS_x^*$	Pheno	Geno S_xS_y	Pheno	Geno S_xS_z	Pheno	Geno S_wS_z	Pheno
$S_xS_x^*$		S_x		S_x		S_z		S_wS_z
S_xS_y		S_x		S_x		S_z		S_wS_z
S_xS_z		S_x		S_x		S_z		S_wS_z
S_wS_z		S_x		S_x		S_z		S_wS_z

No pollen germination or tube growth.

Pollen produced by S_1S_2 sporophyte

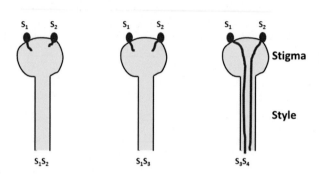

Genotype of pistil

FIG. 10.9 Pictogram of pollen (male gametophyte) interacting with the pistil (female sporophyte) in the case of sporophytic SI and multiple alleles at the S-locus.

Increasingly powerful molecular tools and knowledge of genomics have led to a better understanding of the intracellular and cellular mechanisms underlying SI (Stone and Goring, 2001). The coding genes responsible for the cellular SI signal response and pathogen recognition in hypersensitive disease resistance have been found to share sequence homology and likely evolved from a common ancestor (Matton et al., 1994; Sanabria et al., 2008). Interspecific pollen incompatibility in *Brassica* sp. is not the same as the sporophytic S-allele self-incompatibility (Udagawa et al., 2010). In many SI plant species the S-RNase gene encodes the pistil determinant, and the previously unidentified S-gene encodes the pollen determinant. S-RNases interact with pollen S-allele products to inhibit the growth of self-pollen tubes in the style. Pollen-expressed "F-box" genes showing allelic sequence polymorphism have recently been identified in close proximity to the S-RNase gene in members of the Rosaceae and Scrophulariaceae (Sijacic et al., 2004). This same group showed that transformation of S_1S_1, S_1S_2 and S_2S_3 *Petunia inflata* plants with the S_2-allele of F-box gene *PiSLF* causes breakdown of their pollen function in self-incompatibility, probably due to sense-driven transcript inactivation. They concluded that *PiSLF* encodes the pollen self-incompatibility determinant in *P. inflata*.

Recent discoveries have shed more light on the biology of self- and non-self recognition. The self-recognition system, as expressed in Brassicaceae and Papaveraceae, depends on a specific interaction between male and female S-determinants derived from the same S-haplotype. In contrast, the non-self-recognition system found in Solanaceae depends on non-self (different S-haplotype)-specific interaction between male and female S-determinant, and the male S-determinant genes are duplicated to recognize diverse non-self female S-determinants (Iwano and Takayama, 2012).

AGAMOSPERMY AND APOMIXIS

Natural selection has obviously worked in peculiar ways throughout evolutionary history as is evident from the range of phenotypes that have existed in specific times and places. It is clear that the existence of binary genomes and mating systems that feature combinations of disparate individuals and segregation of polymorphic alleles in successive generations are prevalent among eukaryotes, especially in animals where sexual dimorphism and heterogamy predominate.

In plants and certain animals, however, evolutionary forces have culminated in mating systems that circumvent sex entirely. Manifestations of sexual systems such as functional gonad or floral androecium or gynoecium are often present in individuals or populations that exhibit a functional "end around" the sexual process resulting in clonal, or asexual, progeny. The production of diploid asexual progeny is known as *apomixis*, while the more restrictive term *agamospermy* refers to apomixis manifested as sporophytic embryos within seeds. Species that exhibit apomixis nearly always involve asexual progeny of female, or maternal, origin (Richards, 1986).

The notion of asexuality evolving from sexuality has been considered a unidirectional process since the generation of new genetic variability during clonal reproduction is severely crimped as compared to sexual reproduction (Richards, 1986), but several phenomena are inconsistent with this conclusion. Apomixis occurs sporadically in different taxa and is observed in both early and late branching lineages, with several evolutionary reversals from apomixis to obligate sex. Adventitious embryony was the most frequent form of apomixis (148 genera) followed by apospory (110) and diplospory (68) (Hojsgaard et al., 2014). Across apomictic-containing orders and families, numbers of apomict-containing genera were positively correlated with total numbers of genera. In general, apomict-containing orders, families, and subfamilies of Asteraceae, Poaceae, and Orchidaceae were larger, i.e., they possessed more families or genera, than non-apomict-containing orders, families or subfamilies. Apomixis may facilitate diversification of polyploid complexes and evolution in angiosperms (Hörandl and Hojsgaard, 2012; Hojsgaard et al., 2014). Research results have shown that facultative apomixis may not have the predicted depressing effect on the continuous generation of new genetic variability. For example, Adolfsson and Bengtsson (2007) observed no measurable loss of genetic variability over evolutionary time in species that exhibit both apomixis and sexual function as compared to purely sexual populations.

Apomixis is a bane to the plant breeder seeking to achieve the recombination of desirable alleles from two or more individuals. The essential first step in this process is the development of hybrids via dichogamy. Species that feature apomixis may interfere with the production of sexual hybrids. Plant species that are apomictic may be classified as *obligate* or *facultative*. In obligate apomicts, all progenies are asexual. Facultative apomicts, however, bear asexual and under other circumstances bear sexual progeny (Richards, 1986).

Facultative apomixis that is under genetic or strict environmental control is a phenomenon of intense interest to plant breeders, particularly agamospermy. It is desirable during the breeding program to have access to sexual mechanisms that combine and sort massive numbers of polymorphic genes. Once the targeted genotype is achieved, invoking agamospermy as a tool for efficient cloning is a potentially powerful strategy to amplify the desirable genotype as a finished product for the farmer (Garcia et al., 1999; Pupilli and Barcaccia, 2012).

A consensus picture of the genetic control of the developmental processes that lead to sexual vs. apomictic seeds is coming into focus. The understanding of this developmental process will ultimately allow plant breeders to switch individual plants from sexual to apomictic and back at will. In angiosperms, two fundamental pathways of reproduction through seed exist: sexual or amphimictic (largely exploited by seed companies for breeding new varieties), and asexual or apomictic (for clonal seed production and thus enable efficient and consistent yields of high-quality seeds) (Barcaccia and Albertini, 2013). Initial experimental results demonstrated that sexual and apomictic embryos exhibited distinct patterns of genes and gene expression (Vielle-Calzada et al., 1996; Carman, 1997; Koltunow and Grossniklaus, 2003).

Apomixis has been observed frequently in phylogenetic branches that feature transient polyploidization. It is not clear yet what the relationship is between the transition from sexual to asexual reproduction and polyploidy, but the involvement of deactivated duplicated pseudogenes has been speculated. In support of this hypothesis, Martínez et al. (2007) concluded that apomixis was only present in polyploid plants of the genus *Paspalum* because of a pleiotropic lethal effect associated with monoploid gametes. In at least one case hybridization, and not polyploidy, was found to be the fundamental genomic feature associated with apomixis (Lovell et al., 2013).

Apospory, agamospermy, and apomixis are evolutionarily associated with genomic regions exhibiting similar phenomena as has been shown for dimorphic sex chromosomes of animals and some plants (see Heterogamy and dioecy below). Parallels exist between the dimorphic Y-chromosome and apomixis-bearing chromosomes, such as repression of recombination events, accumulation of transposable elements, and degeneration of genes (Pupilli and Barcaccia, 2012). Partial sequence analysis of the apomixis locus in *Paspalum* spp. revealed structural features of heterochromatin, namely the presence of repetitive elements, gene degeneration, and deregulation (Podio et al., 2014). The apospory-specific genomic region (ASGR) of *Pennisetum squamulatum* is sufficient for the inheritance of apomixis in this species. The ASGR is physically large (>50 Mb), highly heterochromatic, hemizygous, and recombinationally suppressed, similar to mammalian Y chromosomes (Huo et al., 2009). These properties have made it difficult to generate pools of segregants via sexual recombination to molecularly dissect these orthologous regions.

Arabidopsis thaliana is normally a dichogamous hermaphrodite, but there have been many mutants of floral function that have been helpful in illuminating the genetic basis of apomixis, especially when comparing genes and gene expression between *A. thaliana* and typically apomictic species. The *Arabidopsis f644* mutation allows for replication of the central cell and subsequent endosperm development without fertilization. When mutant *f644* egg and central cells are fertilized by wild-type pollen generative nuclei, embryo development is inhibited, and endosperm is overproduced. It was found that the *F644* gene encodes a "polycomb group" (PcG) protein identical to MEDEA (*MEA*), a gene whose maternally-derived allele is required for embryogenesis (Kiyosue et al., 1999). Similar findings pertaining to the appearance of apomixis in a broad array of plant taxa suggest that the polycomb model is a common theme. For example, the FIS-class gene *MhFIE* from apple encodes a predicted protein highly similar to polycomb group (PcG) protein fertilization-independent endosperm (FIE). Results suggested that *MhFIE* has evolved into the regulation of flower development and apomixis in *Malus* sp. (Liu et al., 2012).

Certain endosperm genes are expressed differentially depending on their parental origin, and this genomic imbalance is required for proper seed formation. This asymmetric parental effect on gene expression, or *genomic imprinting*, is controlled epigenetically through genomic histone modifications and DNA methylation. The fertilization-independent PcG genes control genomic imprinting by specifically silencing maternal or paternal target alleles through histone modifications. The function of some PcG components is required for viable seed formation in seeds formed via sexual and asexual processes (apomixis) in *Hieracium*, suggesting a conservation of the seed viability function in some eudicots (Rodrigues et al., 2010). In this species, apomixis occurs when sporophytic cells termed *aposporous initial* (AI) enlarge near sexual cells undergoing meiosis. AI cells in *Hieracium* subgenus *Pilosella* displace the gametogenic cells and divide by mitosis to form unreduced embryo sac(s) without meiosis (apomeiosis) that initiate fertilization-independent embryo and endosperm development. These events are controlled by the dominant *Loss of apomeiosis* (*LOA*) and *Loss of parthenogenesis* (*LOP*) loci (Tucker et al., 2012; Ogawa et al., 2013). A DNA methylation pathway active during reproduction is also essential for gametophyte development in the monocot *Zea mays* and likely plays a critical role in the differentiation between apomictic and sexual reproduction (Garcia-Aguilar et al., 2010).

A large body of experimental results has accumulated regarding the genetic and developmental mechanisms of facultative apomixis in *Poa pratensis* (Kentucky bluegrass). Five major genes modulate the switch from sexual to asexual embryo formation in this highly polyploid species, and combinations of these genes result in different manifestations of apomixis. Phenotypic fluxes best explained by variable expressivity and incomplete penetrance were documented (Matzk et al., 2005). The expression of two of these genes *PpSERK* and *APOSTART* were found to be different in apomictic and sexual individuals. It was shown that these genes are involved in the signal transduction of sexual vs. apomictic development of *P. pratensis* nucellus cells (Albertini et al., 2005). Obligate apomicts of *P. pratensis* do not bear seeds with developed endosperms whereas seeds of facultative apomicts may or may not have endosperm. Synthesis and excretion of auxins were shown to be the cause of observed differences (Niemann et al., 2012).

The current model of facultative apomixis in *P. pratensis* is instructive to plant breeders since the mechanism of the developmental pathway switch appears to be extremely similar across a broad phylogenetic range of angiosperm plant species. The five genes involved in the switch from a sexual to an asexual function that have been characterized in *P. pratensis* include *Apv*, *Ait*, *Ppv*, *Pit*, and *Mdv*. Dominant and recessive alleles of these genes interact to affect the developmental fate of cells originating in the female ancestral nucellus. *Apv- ait/ait* promotes the transition of archegonial cells toward meiosis, whereas *apv/apv Ait-* promotes differentiation into an aposporous initial cell that does not undergo meiosis (Pupilli and Barcaccia, 2012; Fig. 10.10).

Subsequently, the *Ppv- pit/pit* genotype promotes the development of the normal sexual haploid egg cell within the mature female gametophyte (see *** in Fig. 10.11). Further growth is arrested in this case until fertilization by a haploid sperm nucleus occurs. Conversely, the *ppv/ppv Pit-* genotype promotes the continued growth and development of the maternally-derived diploid egg cell into the sporophyte embryo. Generative nuclei of the male gametophyte fuse with the two central cells in the female gametophyte to form the endosperm, 3× in the sexual pathway and 5× in the apomictic pathway.

By using genome editing tools (e.g., CRISPR-cas9; see Chapter 8), it is theoretically possible to manipulate individuals to be either sexual or apomictic by turning on and off these five underlying genes. Thus, the plant breeder of the future will be able to engineer sexual breeding populations and induce apomixis for clonal propagation after the desired genomic construct is achieved, an extremely powerful and exciting possibility.

I. ELEMENTS AND UNDERPINNINGS OF PLANT BREEDING

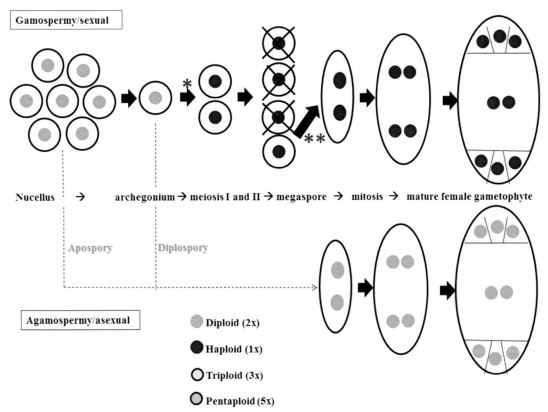

FIG. 10.10 The process of female gametophyte development in *Poa pratensis* starting with nucellus tissue. *The developmental point at which the *Apv- ait/ait* genotype leads to normal or the *apv/apv Ait-* genotype leads to asexual gametophyte development. **The developmental point at which the *Mdv-* genotype leads to further normal sexual development and mdv/mdv leads to the deterioration of any leaked sexual derivative cells.

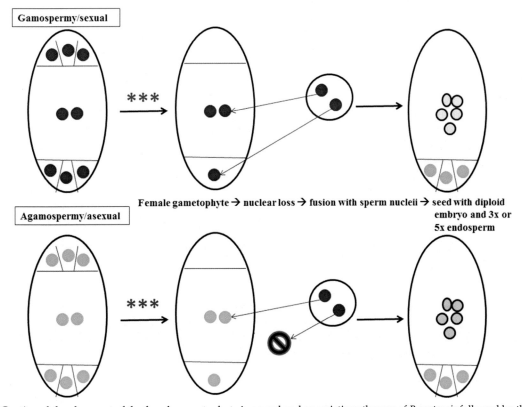

FIG. 10.11 Continued development of the female gametophyte in sexual and apomictic pathways of *P. pratensis* followed by the introduction of sperm and generative nuclei to form the new sporophyte and endosperm of the incipient seed. Antipodal and synergid nuclei are degraded in both cases. ***The *Ppv- pit/pit* genotype promotes the presence of a functional haploid egg cell within the female gametophyte that does not develop further until it fuses with a sperm nucleus from a male gametophyte. In the apomictic *ppv/ppv Pit-* genotype, the asexual 2n diplosporous egg cell continues to grow and develop in the absence of fusion with a sperm nucleus. Fusion of the male generative nucleus with the two central nuclei of the female gametophyte is similar in the two genotypes, resulting in a 3× endosperm in the sexual seed and a 5× endosperm in the apomictic seed.

HETEROGAMY AND DIOECY

It is widely accepted among botanists that terrestrial angiosperms evolved from a primitive ancestor that was a dichogamous hermaphrodite (Fig. 10.2; Richards, 1986). From this common ancestor, an amazing array of mating systems and morphological and physiological phenotypes, such as dioecy and dicliny, arose to promote a mechanism of gene flow that was driven by fluctuating forces of natural selection. This section will focus on dioecy while a later section will address other variations of dicliny.

Heterogamy implies that cross-pollination will predominate among individuals within the population. Dicliny involves the evolution of structural and genetic mechanisms that tend to enforce heterogamy. While monoecy and dioecy are common among gymnosperms, physical separation of sexes is less common in angiosperms than in gymnosperms. Only 4–6% of angiosperm species are dioecious (Alström-Rapaport, 1997; Guttman and Charlesworth, 1998; Renner, 2014; Vyskot and Hobza, 2015). Among these, about 31.6% of the dioecious species are wind-pollinated, compared with 5.5–6.4% of nondioecious angiosperms (Renner, 2014). Sex-determining loci and sex-linked regions evolved independently in many plant lineages, sometimes in closely related dioecious species, and often within the past few million years according to DNA sequencing evidence (Charlesworth, 2015). Sex specificity has evolved in 75% of plant families by male sterile or female sterile mutations but suppression of recombination at the sex determination locus and its neighboring regions leading to functionally and physically dimorphic sex chromosomes is lacking in most dioecious species (Ming et al., 2007). Some disagreement exists on the role of monoecy in the evolution of dioecy. Golenberg and West (2013) argued that most dioecious species evolved through monoecious intermediates.

Among these species of dioecious plants, similar genetic mechanisms have evolved regarding individual sex expression that mirrors the phenomenon of sexual dimorphism in animals. Starting from MADS-box genes controlling the differentiation of whorls of hermaphroditic flowers, those that control the development of stamens and pistils become dominant in the species such that one or more dominant loci condition male and female individuals. In almost all cases the heterozygote Mm is male and the homozygote mm is female. Over evolutionary time only Mm♂ x mm♀ matings persisted, leading to a perpetual regeneration of the sexes in equal proportions. The male is the heterozygote in the overwhelming majority of dioecious angiosperm plant species, but on occasion the female evolves as the heterozygote, for example, *Fragaria elator*, *Potentilla fruticosa*, and *Cotula* spp. (Richards, 1986). In stinging nettle (*Urtica dioica*), the sex determination locus has at least four alleles, some of which are dominant and others are additive (Glawe and de Jong, 2009).

Charlesworth and Charlesworth (2012) described a multi-step model of evolution from hermaphrodity to irreversible dioecy. The first step is the appearance of an M→m male sterile allele at a MADS-box gene. In this model, male sterile segregants are gynoecious, with no functional male gametes. Next, a second mutation f→F occurs in another MADS-box gene that promotes maleness in hermaphrodite flowers. If these two loci are linked, as is common with MADS-box genes, linkage cassettes evolve that prevent the simultaneous catastrophic occurrence of male and female sterility (Ming et al., 2011). These male sterile/female fertile + male fertile/female sterile linkage cassettes have higher theoretical fitness than the linkages in repulsion and come to predominate, therefore, over evolutionary time. Secondary mechanisms ultimately appear that reinforce the lack of recombination between M and F. A Y-chromosome evolves over evolutionary time from the original autosome/incipient X-chromosome (Fig. 10.12). The incipient Y-linked genes homologous to functional X-linked genes may be rendered degenerate during this process (Guttman and Charlesworth, 1998). Once recombination is suppressed around the sex determination region, an incipient Y chromosome starts to differentiate by accumulating deleterious mutations, transposable element insertions, chromosomal rearrangements, and selection for male-specific alleles (Charlesworth, 2002; Ming et al., 2007). While the general process of evolution of dimorphic sex chromosomes is likely similar to what occurred independently in animals, comparison of genomic sequence evidence demonstrates clear differences (Vyskot and Hobza, 2015).

Thus far, distinct dimorphic sex chromosomes that are very common in animals have only been observed in 40 plant species (Renner, 2014; Heikrujam et al., 2015). White campion (*Silene latifolia* = *Melandrium album*) appears to be at an intermediate evolutionary state of this process. The white campion Y-chromosome is rich in repetitive DNA, similar to mammalian Y-chromosomes. Unlike heterochromatic mammalian Y-chromosomes, however, that of white campion is mainly euchromatic (Grant et al., 1994). Papaya (*Carica papaya*) also appears to be at an intermediate point in the evolution of dimorphic sex chromosomes. The papaya male and hermaphrodite phenotypes are controlled by two different types of sex chromosomes: Y and Yh (Aryal and Ming, 2014).

Harkess et al. (2015) identified 570 differentially expressed genes among female (*mm*), male (*Mm*), and supermale (*MM*) asparagus (*Asparagus officianalis*) individuals and showed that significantly more genes exhibited male-biased than female-biased expression. Reduced recombination frequency was discovered within the sex-determining M-region of asparagus. Molecular cytogenetic and sequence analysis of bacterial artificial chromosomes (BACs) flanking the M-locus indicated that the BACs contain highly repetitive sequences localized to centromeric and

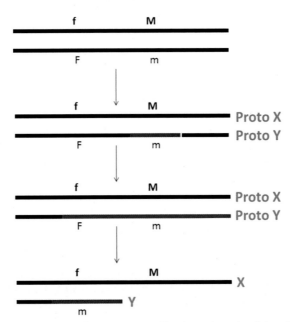

FIG. 10.12 Model for the evolution of dimorphic sex chromosomes in dioecious plants involving the appearance and fixation of MADS box mutations for gynoecious (mm) and androecious (F-) flowers from hermaphrodity. Black regions signify euchromatic, transcribed, non-repetitive sequences devoid of retrotransposons; red regions signify untranscribed heterochromatic, repetitive sequences rife with retrotransposons and absent of homology-based recombination (Charlesworth and Charlesworth, 2012).

pericentromeric locations on all asparagus chromosomes, obscuring mapping of the M-locus to the single pair of sex chromosomes (Telgmann-Rauber et al., 2007).

GYNOECY AND MALE STERILITY

Phenotypic plasticity with regard to floral development and function is extremely common in angiosperms (Richards, 1986). A range of mating systems is evident in plants including gynomonoecy, andromonoecy, monoecy, gynodioecy, and androdioecy. It is surmised that the evolutionary progression was from hermaphrodity to systems that feature physical separation of sexes (dicliny) in response to selective forces favoring cross-pollination. Mating system fluxes are modulated by both genetic and environmental factors. In most cases, plant species that exhibit dicliny are not monolithic with regard to mating system.

Plant breeders must be aware of both the genetic and environmental factors that underlie sex expression. Sex determination in the androdioecious species *Datisca glomerata* (Datiscaceae), for example, is controlled by at least two loci. Males are homozygous recessive at both loci and hermaphrodites have at least one dominant allele at each locus (Wolf et al., 1997).

Domesticated species of the family Cucurbitaceae have been extensively studied to characterize the factors responsible for the observed range of floral types and mating systems. Sex determination in cucumber (*Cucumis sativus* L.) plants is genetically controlled by the *F*, *M*, and *A* loci (Trebitsh et al., 1997). These loci interact to produce three different sex-expression phenotypes: gynoecious (*M-F-*), monoecious (*M-ff*), and andromonoecious (*mmff*). Gynoecious cucumber plants produce more ethylene than do monoecious plants (Yamasaki et al., 2001). The *M/m* gene in the dominant condition suppresses stamina development and thus leads to female flowers. The *F/f* gene in the dominant condition shifts the monoecious sex pattern downwards and promotes femaleness by causing a higher level of ethylene in the plant (Mibus and Tatlioglu, 2004). It has also been shown that MADS-box class C homeotic function is required for the position-dependent arrest of reproductive organs that is observed in cucumber (Kater et al., 2001).

Gene ontology studies revealed that the differentially expressed sex expression genes were derived from genes involved in the biogenesis, transport, and organization of cellular component, macromolecular and cellular biosynthesis, and the establishment of localization, translation, and other processes. For example, a cDNA fragment that encodes a putative GTP binding tyrosine phosphorylated protein A (*CsTypA1*) is developmentally regulated between monoecious and gynoecious genotypes (Barak and Trebitsh, 2007). Differential expression of genes involved in plant hormone signaling pathways, such as *ACS*, *Asr1*, *CsIAA2*, *CS-AUX1*, and *TLP*, indicates that phytohormones play a critical role in the sex determination (Wu et al., 2010). The ethylene biosynthetic gene *Cs-ACS1* (ACC synthase) is present in a single copy in monoecious (*ffMM*) plants whereas gynoecious plants (*FFMM*) contain an additional copy *Cs-ACS1G* that was mapped to the *F* locus (Trebitsh et al., 1997; Knopf and Trebitsh, 2006). Ethylene inhibited stamen development in gynoecious cucumbers but not in andromonoecious individuals. Ethylene responses in andromonoecious cucumber plants are generally reduced from those in maniacs and gynoecious plants (Yamasaki et al., 2001). Auxins and gibberellins have also been shown to influence sex expression in cucurbits (Papadopoulou and Grumet, 2005). Gibberellic acid was associated with male sex expression whereas cytokinins found to be associated with female sex expression. The cytokinin 6-benzyl adenine was shown to induce hermaphrodite flowers (Adhikari et al., 2012). Plant breeders have used these findings to devise ways to control sex expression and to efficiently produce large volumes of F_1 hybrid seeds (Kumar and Wehner, 2012).

Mutations from hermaphrodity to gynoecy, or the loss of male function and structure resulting in gynoecy or gynodioecy, are extremely common in angiosperms (Richards, 1986). Such mutations are referred to collectively by plant breeders as *male sterility*. They fall into one of three general categories: genic/nuclear (gms), cytoplasmic (cms), and genic/nuclear-cytoplasmic (also referred to as cms; see below). Numerous schemes for mechanisms controlled by nuclear genes (gms) have been devised to identify the timing and location of the inception of sterility. These schemes are divided into structural (gross organ changes) and functional. The latter may be divided into abnormal changes occurring in the male cells or surrounding anther tissues during microsporogenesis (Horner and Palmer, 1995).

Inheritance patterns of these distinct forms of male sterility are illustrated in Figs. 10.13–10.15. Evolutionarily, it has been hypothesized that cms arose independently of gms, then naturally-occurring mechanisms of the interaction of cytoplasmic and nuclear genomes resulted in cytoplasmic-genic male sterility (cgms; usually also referred to as cytoplasmic-nuclear male sterility and denoted cnms or cms in scientific literature) (Mulyantoro et al., 2009).

Cytoplasmic-nuclear male sterility (hereinafter cms) has proven to be the most important system for plant breeding applications due to the ratio of male sterile (female) plants among progeny of crosses with male fertile (male) plants. Only 50% of S x F progeny are sterile in gms systems, while 100% of S x F progeny are sterile in cms. Therefore,

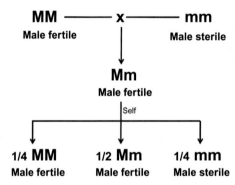

FIG. 10.13   Typical inheritance patterns for genic or nuclear male sterility (gms).

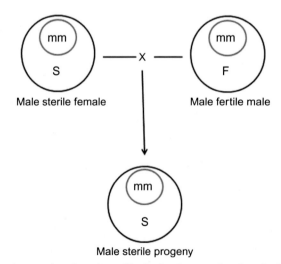

FIG. 10.14   Typical inheritance patterns for cytoplasmic male sterility in the absence of nuclear fertility restoration.

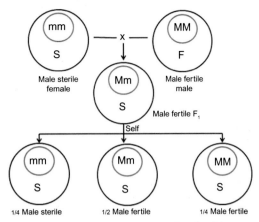

FIG. 10.15 Typical inheritance patterns for cytoplasmic male sterility with nuclear fertility restoration.

FIG. 10.16 (A) Normal hermaphrodite carrot umbel; (B) Petalloid cms carrot umbel.

cms is highly preferred as a tool for F_1 hybrid seed production and other population management applications (Chase, 2006; Engelke et al., 2010). Cms has been reported in over 150 distinct plant species (Bentolila et al., 2002). While cms systems within and among species are similar with regard to function and mode of inheritance, phenotypic manifestations are diverse. In carrot (*Daucus carota*), two main cms systems are described, *petaloid* (Fig. 10.16; stamens converted to a second whorl of petals) and *brown anther* (stamens discolored and devoid of pollen). As more studies have been published in a broader array of systems among diverse plant species, it is clear that there are some common themes and also an impressive array of different pathways to the cms phenotype (Chaumont et al., 1995; Landgren et al., 1996; Bentolila et al., 2002; Sandhu et al., 2007; Mulyantoro et al., 2009; Nizampatnam et al., 2009; Deng et al., 2016).

A useful form of cms was first reported in onion (*Allium cepa*; Jones and Emsweller, 1934). Henry A. Jones (Chapter 1) and coworkers were able to transfer cms to desirable genetic backgrounds via backcrossing (Chapter 18). Most of the basic strategies and techniques to develop A/S, B/M, and C/R lines for the breeding and production of F_1 hybrid cultivars were developed by Jones during the 1930s. Since then, the scientific and practical significance of male sterility has expanded tremendously and research into the causes and uses of male sterility has intensified.

Plant cms systems fall into two broad categories: developmental, or homeotic, and physiological. Homeotic mutants such as those described above (MADS-box) culminate in the conversion of the stamen whorl to a different whorl type in the hermaphrodite (Horner and Palmer, 1995; Leino et al., 2003; Linke et al., 2003). Male germ line cells are developmentally diverted to somatic growth and no gametes (pollen) are produced. Regarding the physiological category of cms interaction of nuclear and mitochondrial genes in energy transduction pathways is essential for the development of viable microspores. A disruption in this interaction can result in inviable microspores and pollen, manifested as cms. Strategies have been developed to mitigate male sterility in both homeotic and physiological cms systems by providing missing substrates or by altering the expression of underlying genes (Leino et al., 2003; Chase, 2006; Li et al., 2007; Ribarits et al., 2007; Sandhu et al., 2007; Konagaya et al., 2008; Engelke et al., 2010; Singh et al., 2010; Wang et al., 2012; Millwood et al., 2016).

Observations of cms in *Capsicum* sp. have demonstrated that flower development and cms phenotypes have a high GxE component (Shifriss and Guri, 1979). Cms in *C. annuum* is restored by one major dominant nuclear gene, restorer-of-fertility (*Rf*), together with some modifier genes and is also affected by temperature. Male fertility is manifested in a range of phenotypes including partial restoration of fertility and plants that simultaneously produce normal and aborted and non-dispersed pollen grains, resulting in low seed set per fruit (Min et al., 2009). Recent experimental results have revealed a multigenic nuclear restoration system in *Capsicum*, rendering the classical cms model depicted in Fig. 10.13 to be overly simplistic (Min et al., 2008). Most angiosperms for which extensive genomic sequence data exist contain multiple *Rf*-related genes that show a number of characteristic features including chromosomal clustering and unique patterns of evolution (Fujii et al., 2011; Barchenger et al., 2018).

Since the male sterile phenotype in cms systems results from an interaction of genes located in the nucleus and mitochondria, experimental studies have delved into mechanisms of mitochondrial inheritance and the transfer of DNA between the nucleus and cytoplasmic organelles. Molecular evidence has shown that the classical characterization of organelle genomes as static and nuclear genomes as dynamic by virtue of meiosis is naïve. Chloroplast genomes have been discovered to be relatively static while mitochondrial genomes are much more are much more mutable than previously thought. It is clear, for example, that there is a DNA base sequence homology-based recombination system at work in mitochondria of angiosperms (Conklin and Hanson, 1993). High levels of mtDNA genomic variability were observed among plants regenerated from tissue cultures of *Coffea arabica* with little or no nuclear or cpDNA variability (Rani et al., 2000). Cybrids of *Nicotiana tabacum* and *Hyoscyamus niger* resulted in a remarkable array of new interspecies recombinants (or genome chimeras) in mtDNA that were driven by DNA homology-driven recombination (Sanchez-Puerta et al., 2015). Sequence similarity with a retrotransposon element suggests a possibility that a transposon-like event transferred a nuclear sequence into the plastid genome of carrot (*Daucus carota*; Iorizzo et al., 2012). The transfer of genes between cytoplasmic organelles over evolutionary time has also been described (Straub et al., 2013).

In many cases, DNA sequence polymorphisms and changes in mtDNA sequence organization have been shown to be associated with cms. Evidence of transfer of DNA sequences between the nuclear and organelle genomes is widespread. The transformation of genes for desirable phenotypes into organelle instead of nuclear genomes is a viable and even preferable option in certain instances (Chase, 2006).

On the basis of the sexual phenotype of wild carrot (*D. carota* L.) plants and their progeny, two of 25 different mitotypes from geographically diverse populations were found to be highly associated with cms (Ronfort et al., 1995). Stability of the mitochondrial genome in species of Solanaceae is controlled by nuclear genes that suppress mitochondrial DNA rearrangements during development (Sandhu et al., 2007). A broad group of proteins called PPR (pentatricopeptide repeat) proteins has also shown to hold great promise for engineering male sterility in crop plants as most of the restorers belong to this category (Sofi et al., 2007). Using in silico techniques, 552 PPR domains were identified throughout the pepper (*Capsicum annuum*) genome (Barchenger et al., 2018). PPR proteins are involved in RNA processing, and mapping the selection data to a predicted consensus structure of an array of PPR motifs suggests that these residues are likely to form base-specific contacts to the RNA ligand (Fujii et al., 2011).

In radish (*Raphanus sativus* L.), for example, plants with the Ogura cms system nuclear genes encoding pentatricopeptide repeat (PPR) proteins are responsible for restoring fertility (Wang et al., 2013). In another study on this same species and cms system, a preliminary microarray experiment revealed that several nuclear genes concerned with flavonoid biosynthesis were inhibited in cms plants (Yang et al., 2008). Following interspecific somatic hybridization, restriction endonuclease profiles of mitochondrial DNA and molecular hybridization with specific genes of the mitochondrial genome used as probes indicated that mitochondrial DNA rearrangement had occurred between sunflower and chicory. The intensity of the rearrangements correlated with the degree of sterility of the different plants (Rambaud et al., 1993).

CONTROLLED MATING IN PLANT BREEDING PROGRAMS

Mating system is among the most important features of the crop species under development by a plant breeder. The mating system determines, in large part, the degree of genetic variability and how it is partitioned among populations, thus influencing germplasm acquisition and utilization strategies. Further, the natural mating system is the most useful and accessible tool for channeling desirable genetic information into individuals and populations. In many or most instances, the mating system is directly responsible for the economic value of the crop, for example, immature flowers (broccoli, cauliflower), flowers (nasturtium), flower parts (spices), fruits, nuts, and seeds. From the standpoint of plant breeding, the natural mating system is of fundamental importance for determining the population structure of the finished product. The targeted population structure will also be most consistent with a defined range of breeding protocols. Historically and practically, these are organized according to self- and cross-pollinated crop species (Chapters 13–17).

The choice of parents in a breeding program is the first step. The plant breeder has already concluded that the parents, as good as they may be, can be forged into something better as a consequence of a prescribed program of germplasm enhancement, selective mating, and selection. Chapters 13–18 will deal with the choice of parents from the range of available germplasm in more detail. In general, the parents should possess alleles that impart the range of phenotypes targeted in the finished product, or reasonably expected following the program of germplasm enhancement, controlled mating, and selection.

The "why" of mating comes down to the development of genotypes that deliver the desired performance at the level of both the individual and corresponding collective population. There are four general classes of theoretical mating classes based on the genotype or phenotype of prospective parents (Allard, 1999), although in practice situations conform to more of a continuum:

- Genetic assortative
- Genetic disassortative
- Phenotypic assortative
- Phenotypic disassortative

The reader may already have supposed that genetic and phenotypic criteria are not necessarily mutually exclusive though it is feasible to mate individuals that are phenotypically assortative but genetically disassortative. It is useful to simplify Allard's classification from four to two broad classes: assortative and disassortative. During the planning and execution of programs, plant breeders most often employ assortative matings with minor disassortative components. In other words, the parents of a mating are selected with a specific result in mind; a targeted amalgam of phenotypes.

Secondly, parents of matings are chosen in an effort to recombine desirable attributes. For example, if the plant breeder wishes to develop a new strawberry cultivar that has better fruit color, indeterminate flowering habit, and minimal stolon growth, one seminal parent will be chosen that contributes genes for generally horticultural plant type and economic yield (usually a prominent commercial cultivar) the other parent that contributes genes for enhanced fruit color, flowering habit, and stolon growth. The second ("donor") parent will also be chosen to possess phenotypes that are as consistent with the final outcome as possible.

A system of standardized mating designs has been devised to allow plant breeders to communicate across programs and species and to underpin quantitative genetic experiments with common elements. A mating design is a procedure by which progenies are generated during the course of plant breeding programs and in experimental studies. Breeders and geneticists use different mating designs for targeted purposes. The choice of a mating design for estimating genetic variances should be dictated by the objectives of the study, genetic parameters of the species, and practical limitations of time, space, cost, and labor. A composite of several independent studies described and contrasted different mating designs: bi-parental progenies (Type I; BIP), poly cross, top cross, North Carolina (Types I, II, III), Diallels (Types I, II, III, IV) and Line X tester design (Griffing, 1956; Kearsey and Pooni, 2004; Hallauer et al., 2010; Acquaah, 2012). In all mating designs, the individuals are selected randomly and crossed to produce progenies which are related to each other as either half-sibs or full-sibs. Multivariate analyses or analyses of variance (ANOVA) can be adapted to each mating design to estimate the components of variances (Nduwumuremyi et al., 2013).

MATERNAL INHERITANCE

Does it matter which individual is chosen as the female or male parent in a bimodal cross? In most cases, it does not. Correns and Baur first reported on the non-Mendelian inheritance of certain phenotypes in the early 1909 (described in Hagemann, 2010). Since then, it is increasingly clear that the process of gamete fusion or fertilization usually includes so little cytoplasm from the male gametophyte that in the overwhelming majority of cases the zygotic organelles were derived from the cytoplasm of the egg, or female parent (Lersten, 1980). How frequent is the paternal inheritance of organelle-encoded traits? Approximately one-third of angiosperm plant genera investigated display biparental cytoplasmic organelle inheritance to some degree (Mogensen, 1996). Paternal plastids were transmitted to the seed progeny in *Arabidopsis* at a low (3.9×10^{-5}) frequency (Azhagiri and Maliga, 2007). Similar findings have been reported in *Antirrhinum majus, Epilobium hirsutum, Nicotiana tabacum, Petunia hybrida*, and the grain crop species *Setaria italic*. It appears from studies of the fertilization process that the replication or digestion of organellar DNA in young generative cells just after pollen mitosis I is a critical point determining the degree of paternal inheritance (Nagata, 2010).

Traits that are encoded by genes residing in cytoplasmic organelles are broadly presumed to be maternally inherited. Genetic factors in cytoplasmic organelles were described in Chapter 3, generally either of plastid or mitochondrial origin. If the phenotypes of interest are in any way affected by genes that reside in cytoplasmic organelles, therefore, it may matter whether a plant is chosen to be the female or male parent in a cross. If one is unsure whether the trait(s) of interest have a cytoplasmic or maternal component to their inheritance, the cross will be conducted in both pairwise configurations: A♀ x B♂ and B♀ x A♂, also known as *reciprocals*. The convention is that the female parent of a given cross is identified first, followed by the male/pollen donor. In other words, in the cross A x B, A is presumed to be the female parent.

MATING OF INDIVIDUALS

Controlled mating is the hybridization or cross of two designated individuals. In the case of self-pollination, both parents are the same individual. All eukaryotes on Earth have a binary genomic structure with regard to sexual function. They possess two gametic sets of genes (haplotypes), bear haploid gametes, and produce progeny derived from the production of a new diploid (binary) individual from the fusion of haploid gametes. This axiom holds for polyploids since normal sexual function is based on diploid, and not polyploid, behavior.

One of two broad mating strategies is usually prevalent within any given species: cross- or self-pollination. The genetic outcomes of these diametric options are quite different, but some species may include subspecies or populations that range from cross- to self-pollination and everything in between (Table 10.9). Determination of natural mating system in many species appears to be intransient, but in other species, there may be a very strong environmental influence on the mating system (Richards, 1986). The proportion of self- and cross-matings for a specified individual or population may be highly variable according to genetic, developmental, and environmental factors.

The cross is a controlled sexual union of two chosen individuals (different or the same). One individual is selected to donate the egg nucleus (female parent), the other the sperm nucleus (male parent). Each species chosen for genetic improvement has a particular set of innate characteristics and challenges that impact on one's ability to perform

TABLE 10.9 Reported ranges of outcrossing from a sample of diverse plant species to exemplify the continuum from autogamy to heterogamy

Species	Reported range of outcrossing
Festuca microstachys	0.00–0.01
Spergula arvensis	0.00–3.00
Galeopsis tetrahit	0.00–16.00
Solanum pimpinellifolium	0.00–84.00
Hordeum vulgare	1.00–2.00
Avena barbata	1.00–8.00
Trifolium hirtum	1.00–10.00
Clarkia temborlensis	8.00–83.00
Lupinus bicolor	13.00–50.00
Limnanthes alba	43.00–97.00
Piectritus congesta	48.00–80.00
Helianthus annuus	60.00–91.00
Eucalyptus obliqua	64.00–84.00
Clarkia unguiculata	96.00

Excerpted from Richards, A.J., 1986. Plant Breeding Systems. George Allen & Unwin, London (UK), 529 pp. (P. 343).

crosses. Issues that impact the ability to affect controlled matings include the ability to promote flowering, the availability of compatible females and males, technical ability to compel the mating to occur, access to environmental ranges (thermal, moisture, light, etc.) to permit gametes and zygotes to perform as needed, and many other unique factors applicable to species and situations. Plant species typically undergo mating as a part of the reproduction process in nature, but it is sometimes difficult to obtain hybrids under artificial conditions. The successful plant breeder must become very skillful with regard to the steps necessary to successfully produce hybrids including the growing of plants that are predisposed to be parents in artificial mating, identifying and preparing flowers that can function as females and males, overcoming problems due to flower morphological factors or small size, and excelling at the efficiency, speed and accuracy of the individual steps in the process.

While corn (*Zea mays*) is not considered to be a horticultural crop species (even sweet corn is generally considered to be an agronomic crop), it does provide an excellent example of controlled mating in a monoecious plant species. The primary vector for corn pollen is wind. Once the stigmas ("silks") begin to emerge, all airborne *Z. mays* pollen that lands on them will fertilize one of the egg cells borne on the spike, or "cob". If a breeder desires to cross corn plant A with plant B, and A is the female, an impervious barrier must be interceded between potentially contaminating pollen and the stigma. The most useful and inexpensive device is the paper bag, usually about the size that held your school lunch (Fig. 10.17). Bags fulfill the most important criteria for the plant breeder: they are cheap and easy to find. The paper bag also possesses desirable properties that it filters pollen, "breathes" to allow for dissipation of moisture, and reflects light and heat. When the time comes to perform the pollination, the bag is simply loosened and removed briefly to introduce the desired pollen onto the stigmas then replaced and secured to prevent any future pollen contamination. The bag may be removed approximately five days or more after the corn cross is made, since the stigmas are no longer receptive (Russell and Hallauer, 1980). Adaptations of this strategy are applied across a broad spectrum of cross-pollinated species, altering the size and shape of the paper enclosure to fit the scale and position of the female.

Making crosses with hermaphroditic flowers is usually more challenging than either dioecious or monoecious flowers. The male and female parts are often closely situated, and in autogamous, self-pollinating species, the union of gametes may be affected well before the flower appears mature or is even open (i.e., cleistogamy). Generally, the plant breeder visits both the male and female parents in advance of a given cross to collect or prepare the bearers of the gametes. For the male, pollen is collected, often in whole flowers or anthers. For example in asparagus, buoyant pollen is aspirated manually from small male flowers into microfuge tubes that may be maintained as a source of male gametes directly or frozen for later use (Fig. 10.18).

The length of time that pollen remains viable varies among species (Stanley and Linskens, 1974; Stone et al., 1995). Pollen of some species may be stored for prolonged periods under cold, dry conditions. Pollen viability has been maintained for months or even years when stored at freezing temperatures (usually <10 °C) under desiccation (Hanna and Woeill, 1995; Towill, 2004). The ability to store pollen for prolonged periods is an asset to the plant breeder when the time of stigma receptivity of the chosen female is outside of the normal range of pollen viability. The author was involved with one such example, wherein two species of heather (*Erica* spp.) differed in flowering time by about six months. Pollen was collected from one species and stored in ampules with desiccant in a freezer. Later, the pollen was thawed and applied to stigmas of the other species, resulting in successful crosses. As it turns

FIG. 10.17 Female flowers ("silk") of maize (*Zea mays*) covered with paper bags following controlled crosses (introduction of pollen). *From Farmsaat.de https://www.farmsaat.de/en/breeding/.*

FIG. 10.18 A vacuum is applied to the end of a small tube to aspirate pollen out of male asparagus flowers; the pollen drops into a microfuge tube where large volumes may be collected for making crosses or frozen indefinitely for later use.

Emasculation of a Hermaphrodite Tomato Flower

Gently excise calyx, corolla, and anthers with jeweler's forceps

FIG. 10.19   Emasculation of a hermaphrodite tomato flower to produce a female for pollination in a controlled cross.

out, the two *Erica* populations were probably of the same species, but following geographical isolation in South Africa, different reproductive behaviors involving floral induction were manifested. The resulting hybrid population consisted of individuals that flowered at the times of both parents, and also at times in between.

In a species with hermaphrodite flowers, the female parent in the cross is prepared by emasculation or castration (Fig. 10.19) to ensure that any self-pollen within the flower will be eliminated. In many cases of heterogamy, such as SI, self-pollination is completely obviated and emasculation may be unnecessary. Extreme care must be taken during emasculation not to harm the gynoecium. Any wound site may result in the production of ethylene gas that will contribute to the abortion the developing embryo. The anthers also must be removed before they dehisce and pollen is shed, but there are also limitations as to minimum developmental stage. Care must also be taken during the performance of the cross to exclude any foreign pollen. Following emasculation, an impervious barrier is usually inserted between the stigma and any potentially desirable source of foreign pollen, such as the bag with corn. For physically small flowers, ingenious contraptions and processes are often devised by the plant breeder for this purpose.

Naturally-occurring dichogamy (protandry or protogyny) is frequently used to enhance the efficiency and effectiveness of making crosses. Protandry is a common floral feature in the family Apiaceae where consistent patterns of pollen viability and stigmatic receptivity are evident within flowers, umbellets, and umbels. The outer whorls of florets within umbels mature earlier than do the inner whorls and the time gap is predictable if environmental parameters (temperature; light) are consistent. By physically removing the inner whorls of flowers from developing umbellets before pollen is shed, controlled crosses may be affected by pollinating and covering the remaining outer-whorl flowers. In this manner, large numbers of crosses can be made with relatively little effort or physical damage due to contact with fragile flowers. Skilled plant breeders may consistently achieve a high proportion of crosses to self-pollinations (90–99%) with this strategy. Since flower size in most species of the family Apiaceae is relatively small and the number of seeds per pollination low (1–2), this strategy is necessary to obtain adequate numbers of hybrids in a meaningful breeding program.

Mechanisms that enforce outcrossing in hermaphrodites such as SI may interfere with the ability to perform a given cross. SI may also be subverted by the plant breeder to obtain large populations of pure hybrid individuals. In situations where the selected male and female parents are separate individuals but have incompatible genotypes at SI loci (see above), empirical research has demonstrated that certain "tricks" can be employed to overcome SI mechanisms. For example in *Brassica* species, sporophytic SI glycoproteins are not present and active on stigmatic surfaces until a point in flower development shortly before the corolla opens. Emasculation of immature flowers followed by pollination may result in successful self- or incompatible matings; known as *bud pollination* (Shivanna et al., 1978). It may also be possible to physically remove the stigma where SI recognition glycoproteins are deposited to allow germinating pollen tubes to grow directly into and through the style to the ovary. Certain chemical compounds such as CO_2 have also been shown to depress the overall effectiveness of the SI system (Lao et al., 2014).

The distinction between species defines the limits of individual gene pools. Crosses are considered to be compatible or plausible within species and often among species within certain genera (see broader discussion in Chapter 2). In certain instances, crosses within species may be subject to barriers that are usually post-zygotic in nature. Certain cultivated species of Orchidaceae are characterized by the failure of seeds to develop following the fusion of sperm and egg nuclei. This phenomenon is caused by an incompatibility between the embryo and endosperm (Arditti, 1984). Successful orchid breeders are adept at rescuing embryos from dying seeds and culturing them on artificial media under sterile conditions. These expensive efforts are abetted by the fact that individual hybrid plants may be worth hundreds or even thousands of dollars.

A "self" or self-pollination is a form of a cross that is associated with certain conveniences and potential pitfalls. Species that are naturally self-pollinating and exhibit physical or physiological mechanism that enforces self-pollination are generally allowed to undergo this process without any outside help. Mechanical devices are sometimes employed to enhance the frequencies of successful self-mating. For example in tomato (*S. lycopersicum*) a flower shaker or vibrator is often used to agitate pollen within flowers on greenhouse-grown plants, thus maximizing fertilization (Martin-Closas et al., 2007). This is presumed to emulate natural vibrations under outdoor field conditions from wind, insects, and birds that promote pollination. Insects such as bumblebees may be used to enhance self-pollination in genetically monolithic populations of tomato, but may also culminate in unwanted cross-pollinations in mixed plant populations.

The breeder of an autogamous species such as tomato, pea, and peach often presumes that no outcrossing ever occurs. Within the gene pool of autogamous species may reside alleles that promote cross-pollination. One example is style length in tomato, where certain uncommon alleles result in exposed stigmas that are accessible to visiting insects (Tikoo and Anand, 1980; Orton et al., 2016). If the outcrossing rate is significant, on the order of 2–5% or more, measures may be necessary to ensure that self-pollination is affected. If the species is adapted to insect pollen vectors, a simple bag or cage may be sufficient to exclude unwanted pollen. If wind or water vectors are involved, more rigorous or aggressive steps are warranted to exclude unwanted pollen from female parents.

For cross-pollinating species in which physiological or physical mechanisms operate to enforce outcrossing, it may be necessary to overcome them using similar or identical approaches to those described earlier to accomplish self-pollination. In outcrossing species that exhibit a high degree of inbreeding depression, it may be advantageous to pursue an alternative strategy for achieving homozygosity, such as *sib pollination* or *half-sib pollination*, crosses between two individuals that share two (full-sib) or one (half-sib) parents (see Chapter 16).

Where a number of individuals or populations are involved in a given breeding program, the plant breeder may wish to perform a defined set or pattern of crosses between them. The most common example is *diallel* wherein all pairwise crosses of individuals or populations are performed. Thus, for n individuals, there are n(n-1) possible crosses since the self-matings are excluded. If the phenotype of interest in the program has no cytoplasmic inheritance component, then both reciprocal hybrids are equivalent (e.g., AxB=BxA) and the number of combinations reduces by half n(n-1)/2 possible crosses. Diallels and half diallels are often performed in hybrid breeding programs for the estimation of combining ability of prospective inbred parents (see Chapter 15).

MATINGS AMONG AND WITHIN POPULATIONS

The plant breeder often bulks individuals into populations and manages genes at the level of populations to achieve efficiencies and also to foster natural processes leading to gene recombination and segregation. Self-pollinating plants may be handled as populations because genes are already locked into lineages defined by F_2 individuals (Chapter 13), such as with the "bulk population" breeding method (Chapter 14).

A form of multiple pairwise crosses within a population known as the *intercross* is similar to the diallel except that the outcome is not as strictly confined. The intercross is usually employed, rather, to impose a step that is analogous to random mating. In the intercross all individuals in the population are hybridized, and the progeny are bulked and carried to the next generation.

Two fundamental types of controlled mating that pertain strictly to outcrossing species are *open pollination* and *hybrid*. Open pollination is another emulation of random mating. Although the term is sometimes used in connection with self-pollinating species, it is a misnomer since the pollinations are usually "self" and not "open". "Open pollination" is generally used to describe situations where reproduction in a population of plants is unmanaged or where mating occurs without human intervention. The only controls exercised by the breeder in situations involving open pollination are the choice of individuals constituting the population and planting configuration. With regard to the latter factor, the type of mating that takes place may be affected by planting configuration. If 100 plants are situated in a row along a narrow canyon, the number of different matings will be lower than if the physical planting configuration is more two-dimensional. Genotypes should be randomly dispersed throughout the planting site used to accomplish population mating by open pollination. The overseer may augment naturally occurring vectors, such as beehives for insect-pollination or wind-generating machines. These will foster greater dispersion of pollen among the population resulting in progeny that will be more closely representative of true random mating, also known as *panmixis*.

F_1 hybrid cultivars are based on controlled hybridizations of two inbreds, or genetically fixed, parental breeding lines (Chapter 16). Specifically, the cross of these two inbred lines must be repeated many times to produce the volume of seeds to meet farmer demand. The genetic structure of the F_1 hybrid cultivar is a multiply heterozygous genetically monolithic population. Thus, any attempt by the farmer to sexually propagate the cultivar will culminate in mass segregations and corresponding phenotypic disarray within populations of progeny. The farmer must return to the producers of seeds or vegetative stocks at the beginning of each cycle to acquire more F_1 hybrid seeds, paying whatever prices the market will bear.

The seeds of a F_1 hybrid cultivar are derived, ideally, from the fewest number of crosses as possible since the costs associated with producing hybrids are high. Most crop species do not exhibit ranges of fecundity that approach this possibility. Some species produce only a single seed per individual cross, some as high as 100,000 (Copeland and McDonald, 2012). The seed company prefers that the ratio of seeds per individual cross is as high as possible. If the fecundity of the crop species is very low, investment of resources and time into the increase of available seeds of the inbred parent breeding lines may be necessary. Seed production always implies the involvement of sexual reproduction, and the possibility that undesirable genetic variability will be generated from uncontrolled environmental factors such as vector-mediated outcrossing.

Maize has been the archetype for the success of hybrid crop cultivars. Since *Z. mays* is monoecious, seed companies have historically used human labor (usually summer high school students), with or without machinery, to emasculate large populations of female plants rapidly. No machine or high school student has been yet invented that can effectively and efficiently separate the gynoecium and androecium from a hermaphroditic flower as is possible with corn. One approach that is used to produce large volumes of hybrid seeds of species with hermaphrodite flowers, however, is hand-pollination. Hand pollination is feasible only where and when the value of the hybrid seed is extremely high. Hand-pollination is currently performed for F_1 hybrid seed production of certain high-value vegetable crops such as tomato and bell pepper. The same technique is used as was described earlier in this chapter for making individual crosses, but adapted to a much larger scale. Human labor is, in this instance, a fundamental determinant of the overall cost of F_1 hybrid seed production. Large-scale F_1 hybrid seed production by hand-pollinations was initially conducted in the U.S. but operations were quickly moved to other countries with lower labor costs such as Mexico, Chile, Japan, and Taiwan. As the cost of labor increased in these countries, seed companies relocated hand-pollinated seed production to Eastern Asian countries such as Korea, The Phillipines, Malaysia, Thailand, India, and China. Seed companies continue to search for a source of human labor the purest hybrid seed at the lowest possible cost. As the standard of living and wages increase in Eastern Asia, where will the seed companies go next? Hand-pollination is only economically and technically feasible for certain high-value crops. This hybrid seed production strategy is also only justified where the performance of hybrid varieties clearly and consistently exceeds that of open-pollinated populations. Profit margins for hybrids are characteristically higher than OPs, and hybrids offer inherent protection against genetic piracy.

Self-incompatibility was discussed above as a possible tool to eliminate the need for physical emasculation of hermaphroditic flowers in producing hybrid individuals. The same strategy may be extended to entire populations. If two self-incompatible, but cross-compatible, populations are planted in the same location and allowed to intermate, the only seeds that should theoretically result will be F_1 hybrids. This strategy has been successfully employed for the large-scale production of F_1 hybrid seeds of *Brassica oleracea, B. rapa,* and *B. napus,* including cabbage, broccoli, cauliflower, kale, kohlrabi, Brussels sprouts, canola, and rutabaga (Ruffio-Chable et al., 1999; Ripley and Beversdorf, 2003). In practice, self-incompatibility tends to break down under stress and as the plant ages, so some level of self-contaminants is usually inevitable and is widely tolerated. Since flying insect vectors are involved in the transfer of *Brassica* pollen, strict isolation procedures must be observed.

Floral structure and function are highly mutable in plants. Plants behave differently according to environmental cues and floral morphology can be highly variable within a species, or even within populations of individuals. Since evolutionary fitness is measured by forces that perpetuate the species, the plasticity of reproductive systems is understandable. Plants must be able to reproduce even if their survival is threatened. For example, a typically outcrossing species may suddenly revert to autogamy during periods of stress, presumably a mechanism for survival (Glémin et al., 2008; Horisaki and Niikura, 2008).

The interaction of cytoplasmic and nuclear genes to condition a change from hermaphrodity and gynoecy (male sterility) has been used extensively to accomplish large-scale population crosses, most commonly to produce hybrid seeds (Havey, 2004). Male sterility systems are currently widely used in the production of hybrid onion, carrot, sunflower, and sorghum. A male sterile plant can be considered to be genetically castrated. When planted en masse with a sexually fertile male population, all seeds harvested from the male sterile (female) counterpart will be F_1 hybrid. The corresponding breeding program is oriented such that male sterile female and male fertile male inbred parental populations are the endpoint. Plant breeders use cms with nuclear restorer genes to facility F_1 hybrid breeding programs in ways that will be described in Chapter 16.

Male sterility played a prominent role in one of the most dramatic events in the annals of modern plant breeding, the southern corn leaf blight (SCLB) epidemic of 1970. Cytoplasmic male sterility mutants have been known in *Z. mays* for decades, and three distinctly different sources have been identified: "S", "C", and "T" (Havey, 2004). Each is distinguished by both morphology and the nuclear genes that restore them to sexual fertility. Seed companies decided in the 1950s and 1960s to pursue male sterility as a way to reduce the cost of hybrid corn seed, and programs were established based on both S and T cytoplasm. As it happened in T cytoplasm mitochondrial DNA, the apparent locus of susceptibility to the pathogen *Bipolaris* (*Helminthosporium*) *maydis* was apparently the same gene that controlled T-cms. In 1970, most of the hybrid corn in the main production areas of the U.S. corn belt had been produced using T cytoplasm, and was therefore susceptible to the SCLB pathogen. The corn harvest was devasted that year, and the "rest of the story" is now etched in history (Bruns, 2017). Fortunately for corn seed companies, male sterility is not strictly required for large-scale hybrid seed production, and programs were rapidly retooled to eliminate T cytoplasm and cms from stocks of female inbred parent populations. The production of hybrid corn seed returned to the use of human labor to detassel the female plants. In the process, a crucial lesson was learned (if not already learned from similar historical instances of genetic vulnerability, such as the Irish potato famine) about the potential perils of genetic uniformity.

Possible uses of gynoecious mutations in Cucurbitaceae (e.g., *Cucumis sativus*) for hybrid seed production are similar to male sterility in other species, but these mutations are notoriously difficult to manipulate in breeding programs and have not, therefore, been widely used. It has been discovered, alternatively, that certain plant growth regulators (e.g., gibberellic acid or GA) and certain inorganic salts can be applied as exogenous solutions to floral primordial that will be converted from staminate to pistillate flowers, or vice versa. Chemical modulation of sex expression in Cucurbitaceae has proven to be a highly useful method where a prospective female inbred is developed then sprayed with GA to convert it to gynoecy and interplanted with the male parent for pollination (Zhang et al., 1994).

Hybrid cultivars in certain species require a source of pollen to produce fruits or seeds. For example, in self-incompatible species where reproduction is required for fruit set, such as with cherries, crop production must be conducted in the vicinity of a source of self-compatible pollen. In another example, seedless triploid watermelons require a source of viable pollen from fertile diploid plants to obtain fruit set, even though the seeds within the triploids quickly abort. Seeds of triploid watermelon varieties are generally marketed in tandom with a diploid "pollenizer" line that are interplanted (Dittmar et al., 2010). If the grower or customer prefers an old-fashioned seeded watermelon, fruits from the pollenizer can fill this demand.

References

Acquaah, G., 2012. Principles of Plant Genetics and Breeding, second ed. Wiley/Blackwell, New York, NY (USA). 756 pp.

Adhikari, S., Bandyopadhyay, T.K., Ghosh, P., 2012. Hormonal control of sex expression of cucumber (*Cucumis sativus* L.) with the identification of sex linked molecular marker. Nucleus 55 (2), 115–122.

Adolfsson, S., Bengtsson, B.O., 2007. The spread of apomixis and its effect on resident genetic variation. J. Evol. Biol. 20 (5), 1933–1940.

Airoldi, C.A., Davies, B., 2012. Gene duplication and the evolution of plant MADS-box transcription factors. J. Genet. Genomics 39 (4), 157–165.

Albertini, E., Marconi, G., Reale, L., Barcaccia, G., Porceddu, A., Ferranti, F., Falcinelli, M., 2005. *SERK* and *APOSTART*: candidate genes for apomixis in *Poa pratensis*. Plant Physiol. 138 (4), 2185–2199.

Allard, R.W., 1999. Principles of Plant Breeding, second ed. John Wiley & Sons, New York, NY (USA). 264 pp.

Alström-Rapaport, C., 1997. On the Sex Determination and Evolution of Mating Systems in Plants. Uppsala Univ. Press, Uppsala (Sweden).

Arditti, J., 1984. An history of orchid hybridization, seed germination and tissue culture. Bot. J. Linn. Soc. 89 (4), 359–381.

Aryal, R., Ming, R., 2014. Sex determination in flowering plants: papaya as a model system. Plant Sci. 217, 56–62.

Azhagiri, A.K., Maliga, P., 2007. Exceptional paternal inheritance of plastids in *Arabidopsis* suggests that low-frequency leakage of plastids via pollen may be universal in plants. Plant J. 52 (5), 817–823.

Barak, M., Trebitsh, T., 2007. A developmentally regulated GTP binding tyrosine phosphorylated protein A-like cDNA in cucumber (*Cucumis sativus* L.). Plant Mol. Biol. 65 (6), 829–837.

Barcaccia, G., Albertini, E., 2013. Apomixis in plant reproduction: a novel perspective on an old dilemma. Plant Reprod. 26 (3), 159–179.

Barchenger, D.W., Said, J.I., Yang Zhang, Y., Song, M., Ortega, F.A., Ha, Y., Kang, B.-C., Bosland, P.W., 2018. Genome-wide identification of chile pepper pentatricopeptide repeat domains provides insight into fertility restoration. J. Am. Soc. Hort. Sci. 143 (6), 418–429.

Barrett, S.C.H., Harder, L.D., Worley, A.C., 1996. Comparative biology of plant reproductive traits. Phil. Trans. Royal Lond. B. 351, 1272–1280.

Becker, A., Theissen, G., 2003. The major clades of MADS-box genes and their role in the development and evolution of flowering plants. Mol. Phylogenet. Evol. 29, 464–489.

Bentolila, S., Alfonso, A.A., Hanson, M.R., 2002. A pentatricopeptide repeat-containing gene restores fertility to cytoplasmic male-sterile plants. Proc. Natl. Acad. Sci. U. S. A. 99 (16), 10887–10892.

de Bodt, S., Raes, J., Florquin, K., Rombauts, S., Rouze, P., Thiessen, G., de Peer, Y.V., 2003. Genomewide structural annotation and evolutionary analysis of the type I MADS-box genes in plants. J. Mol. Evol. 56 (5), 573–586.

Bowman, J.L., Smyth, D.R., Meyerowitz, E.M., 1991. Genetic interactions among floral homeotic genes of *Arabidopsis*. Development 112, 1–20.

Brewbaker, J.L., 1967. The distribution and phylogenetic significance of binucleate and trinucleate pollen grains in the angiosperms. Am. J. Bot. 54 (9), 1069–1083.

Bruns, H.A., 2017. Southern corn leaf blight: a story worth retelling. Agron. J. 109 (4), 1–7.

Busi, M.V., Bustamante, C., D'Angelo, C., Hidalgo-Cuevas, M., Boggio, S.B., Valle, E.M., Zabaleta, E., 2003. MADS-box genes expressed during tomato seed and fruit development. Plant Mol. Biol. 52 (4), 801–815.

Campbell, S.A., 2014. Ecological mechanisms for the coevolution of mating systems and defence. New Phytol. 205 (3), 1047–1053.

Carman, J.G., 1997. Asynchronous expression of duplicate genes in angiosperms may cause apomixis, bispory, tetraspory, and polyembryony. Biol. J. Linn. Soc. 61 (1), 51–94.

Carr, D.E., 2013. A multidimensional approach to understanding floral function and form. Am. J. Bot. 100 (6), 1102–1104.

Charlesworth, D., 2002. Plant sex determination and sex chromosomes. Heredity 88 (2), 94–101.

Charlesworth, D., 2015. Plant contributions to our understanding of sex chromosome evolution. New Phytol. 208 (1), 52–65.

Charlesworth, B., Charlesworth, D., 2012. Elements of Evolutionary Genetics. Roberts & Co. Publ, Greenwood Village, CO (USA). 734 pp.

Chase, C.D., 2006. Genetically engineered cytoplasmic male sterility. Trends Plant Sci. 11 (1), 7–9.

Chaumont, F., Bernier, B., Buxant, R., Williams, M.E., Levings, C.S.I.I.I., Boutry, M., 1995. Targeting the maize T-urf13 product into tobacco mitochondria confers methomyl sensitivity to mitochondrial respiration. Proc. Natl. Acad. Sci. U. S. A. 92 (4), 1167–1171.

Coen, E.S., Meyerowitz, E.M., 1991. The war of the whorls: genetic interactions controlling flower development. Nature 353, 31–37.

Colombo, L., Franken, J., Koetje, E., van Went, J., Dons, H.J., Angenent, G.C., van Tunen, A.J., 1995. The petunia MADS box gene FBP11 determines ovule identity. Plant Cell 7, 1859–1868.

Conklin, P.L., Hanson, M.R., 1993. A truncated recombination repeat in the mitochondrial genome of a *Petunia* CMS line. Curr. Genet. 23 (5–6), 477–482.

Copeland, L.O., McDonald, M.B., 2012. Seed Science and Technology, fourth ed. Springer Science, New York, NY (USA). 466 pp.

Correns, C., 1913. Selbststerilität und individualstoffe. Biol. Centr. 33, 389–423.

De Vries, H., 1907. Plant Breeding. Open Court Publishing Co., Chicago, IL (USA).

Deng, Z., Li, X., Wang, Z., Jiang, Y., Wan, L., Faming, D., Fengxiang, C., Dengfeng, H., Guangsheng, Y., 2016. Map-based cloning reveals the complex organization of the *BnRf* locus and leads to the identification of *BnRf*b, a male sterility gene, in *Brassica napus*. Theor. Appl. Genet. 129 (1), 53–64.

Deng, W., Ying, H., Helliwell, C.A., Taylor, J.M., Peacock, W.J., Dennis, E.S., 2011. FLOWERING LOCUS C (FLC) regulates development pathways throughout the life cycle of *Arabidopsis*. Proc. Natl. Acad. Sci. U. S. A. 108 (16), 6680–6685.

Devaux, C., Lepers, C., Porcher, E., 2014. Constraints imposed by pollinator behaviour on the ecology and evolution of plant mating systems. J. Evol. Biol. 27 (7), 1413–1430.

Dittmar, P.J., Monks, D.W., Schultheis, J.R., 2010. Use of commercially available pollenizers for optimizing triploid watermelon production. HortScience 45 (4), 541–545.

Dubois, E., Bercy, J., Descamps, F., Messenguy, F., 1987. Characterization of two new genes essential for vegetative growth in *Saccharomyces cerevisiae*: nucleotide sequence determination and chromosome mapping. Gene 55, 265–275.

East, E.M., 1919a. Studies on self-sterility III. The relation between self-fertile and self-sterile plants. Genetics 4, 341–345.

East, E.M., 1919b. Studies on self-sterility IV. Selective fertilization. Genetics 4, 346–355.

East, E.M., 1919c. Studies on self-sterility V. A family of self-sterile plants, wholly cross-sterile *inter se*. Genetics 4, 356–363.

East, E.M., Mangelsdorf, A.J., 1925. A new interpretation of the hereditary behavior of self-sterile plants. Proc. Natl. Acad. Sci. U. S. A. 11, 166–171.

Emerson, S., 1939. A preliminary survey of the *Oenothera organensis* population. Genetics 24, 524–537.

Engelke, T., Hirsche, J., Roitsch, T., 2010. Anther-specific carbohydrate supply and restoration of metabolically engineered male sterility. J. Exp. Bot. 61 (10), 2693–2706.

Fujii, S., Bond, C.S., Small, I.D., 2011. Selection patterns on restorer-like genes reveal a conflict between nuclear and mitochondrial genomes throughout angiosperm evolution. Proc. Natl. Acad. Sci. U. S. A. 108, 1723–1728.

Garcia, R., Asins, M.J., Forner, J., Carbonell, E.A., 1999. Genetic analysis of apomixis in *Citrus* and *Poncitrus* by molecular markers. Theor. Appl. Genet. 99 (3–4), 511–518.

Garcia-Aguilar, M., Michaud, C., Leblanc, O., Grimanelli, D., 2010. Inactivation of a DNA methylation pathway in maize reproductive organs results in apomixis-like phenotypes. Plant Cell 22 (10), 3249–3267.

Glawe, G.A., de Jong, T.J., 2009. Complex sex determination in the stinging nettle *Urtica dioica*. Evol. Ecol. 23 (4), 635–649.

Glémin, S., Petit, C., Maurice, S., Mignot, A., 2008. Consequences of low mate availability in the rare self-incompatible species *Brassica insularis*. Conserv. Biol. 22 (1), 216–221.

Golenberg, E.M., West, N.W., 2013. Hormonal interactions and gene regulation can link monoecy and environmental plasticity to the evolution of dioecy in plants. Am. J. Bot. 100 (6), 1022–1037.

Goto, K., Meyerowitz, E.M., 1994. Function and regulation of the *Arabidopsis* floral homeotic gene PISTILLATA. Genes Dev. 8, 1548–1560.

Grant, S., Houben, A., Vyskot, B., Siroky, J., Pan, W.H., Macas, J., Saedler, H., 1994. Genetics of sex determination in flowering plants. Dev. Genet. 15 (3), 214–230.

Griffing, B., 1956. Concept of general and specific combining ability in relation to diallel crossing systems. Aust. J. Biol. Sci. 9, 463–493.

Guttman, D.S., Charlesworth, D., 1998. An X-linked gene with a degenerate Y-linked homologue in a dioecious plant. Nature 393 (6682), 263–266.

Hagemann, R., 2010. The foundation of extranuclear inheritance: plastid and mitochondrial genetics. Mol. Gen. Genomics 283 (3), 199–209.

Hallauer, A.R., Carena, M.J., Filho, J.B.M., 2010. Quantitative Genetics in Maize Breeding, third ed. Springer, Berlin (GER). 664 pp.

Hanna, W.W., Woeill, L.W., 1995. Long-term pollen storage. In: Janick, J. (Ed.), Plant Breeding Reviews. John Wiley & Sons, Chichester, MA (USA), pp. 179–207.

Harkess, A., Mercati, F., Shan, H., Sunseri, F., Falavigna, A., Leebens-Mack, J., 2015. Sex-biased gene expression in dioecious garden asparagus (*Asparagus officinalis*). New Phytol. 207 (3), 883–892.

Havey, M.J., 2004. The use of cytoplasmic male sterility for hybrid seed production. In: Daniell, H., Chase, C. (Eds.), Molecular Biology and Biotechnology of Plant Organelles. Springer, Amsterdam (NETH), pp. 617–628.

Heikrujam, M., Sharma, K., Prasad, M., Agrawal, V., 2015. Review on different mechanisms of sex determination and sex-linked molecular markers in dioecious crops: a current update. Euphytica 201 (2), 161–194.

Hiratsuka, S., Tezuka, T., Yamamoto, Y., 1989. Analysis of self-incompatibility reaction in Easter lily by using heat treatments. J. Amer. Soc. Hort. Sci. 114 (3), 505–508.

Hojsgaard, D., Klatt, S., Baier, R., Carman, J.G., Hörandl, E., 2014. Taxonomy and biogeography of apomixis in angiosperms and associated biodiversity characteristics. Crit. Rev. Plant Sci. 33 (5), 414–427.

Holsinger, K.E., 1996. Pollination biology and the evolution of mating systems in flowering plants. Evol. Biol. 29, 107–149.

Honma, T., Goto, K., 2001. Complexes of MADS-box proteins are sufficient to convert leaves into floral organs. Nature 409, 525–529.

Honnay, O., Jacquemyn, H., 2008. A meta-analysis of the relation between mating system, growth form and genotypic diversity in clonal plant species. Evol. Ecol. 22 (3), 299–312.

Hörandl, E., Hojsgaard, D., 2012. The evolution of apomixis in angiosperms: a reappraisal. Plant Biosys. 146 (3), 681–693.

Horisaki, A., Niikura, S., 2008. Developmental and environmental factors affecting level of self-incompatibility response in *Brassica rapa* L. Sex. Plant Reprod. 21 (2), 123–132.

Horner, H.T., Palmer, R.G., 1995. Mechanisms of genic male sterility. Crop Sci. 35 (6), 1527–1535.

Hu, L., Liu, S., 2012. Genome-wide analysis of the MADS-box gene family in cucumber. Genome 55 (3), 245–256.

Huo, H., Conner, J.A., Ozias-Akins, P., 2009. Genetic mapping of the apospory-specific genomic region in *Pennisetum squamulatum* using retrotransposon-based molecular markers. Theor. Appl. Genet. 119 (2), 199–212.

Immink, R.G.H., Gadella Jr., T.W.J., Ferrario, S., Busscher, M., Angenent, G.C., 2002. Analysis of MADS box protein-protein interactions in living plant cells. Proc. Natl. Acad. Sci. U. S. A. 99 (4), 2416–2421.

Iorizzo, M., Senalik, D., Szklarczyk, M., Grzeblus, D., Spooner, D., Simon, P.W., 2012. De novo assembly of the carrot mitochondrial genome using next generation sequencing of whole genomic DNA provides first evidence of DNA transfer into an angiosperm plastid genome. Biomed Central Plant Biol. 2012, 1–17.

Ivey, C.T., Carr, D.E., 2005. Effects of herbivory and inbreeding on the pollinators and mating system of *Mimulus guttatus* (Phrymaceae). Amer. J. Bot. 92 (10), 1641–1649.

Iwano, M., Takayama, S., 2012. Self/non-self discrimination in angiosperm self-incompatibility. Curr. Opin. Plant Biol. 15 (1), 78–83.

Jack, T., Fox, G.L., Meyerowitz, E.M., 1994. *Arabidopsis* homeotic gene APETALA3 ectopic expression. Transcriptional and post-transcriptional regulation determines floral organ identity. Cell 76 (4), 703–716.

Jofuku, K.D., den Boer, B.G., Van Montagu, M., Okamuro, J.K., 1994. Control of Arabidopsis flower and seed development by the homeotic gene APETALA2. Plant Cell 6, 1211–1225.

Jones, H.A., Emsweller, S.I., 1934. A male-sterile onion. Proc. Am. Soc. Hort. Sci. 34, 582–585.

Jorgensen, R., Arathi, H.S., 2013. Floral longevity and autonomous selfing are altered by pollination and water availability in *Collinsia heterophylla*. Ann. Bot. 112 (5), 821–828.

Kariyat, R.R., Sinclair, J.P., Golenberg, E.M., 2013. Following Darwin's trail: interactions affecting the evolution of plant mating systems. Amer. J. Bot. 100 (6), 999–1001.

Karron, J.D., Ivey, C.T., Mitchell, R.J., Whitehead, M.R., Peakall, R., Case, A.L., 2012. New perspectives on the evolution of plant mating systems. Ann. Bot. 109 (3), 493–503.

Kater, M.M., Franken, J., Carney, K.J., Colombo, L., Angenent, G.C., 2001. Sex determination in the monoecious species cucumber is confined to specific floral whorls. Plant Cell 13 (3), 481–493.

Kearsey, M.J., Pooni, H.S., 2004. The Genetical Analysis of Quantitative Traits, first ed. Garland Science, London (UK). 396 pp.

I. ELEMENTS AND UNDERPINNINGS OF PLANT BREEDING

Kim, C., Guo, H., Kong, W., Chandnani, R., Lan-Shuan, S., Paterson, A.H., 2016. Application of genotyping by sequencing technology to a variety of crop breeding programs. Plant Sci. 242, 14–22.

Kim, S., Koh, J., Yoo, M.J., Kong, H., Ma, H., Soltis, P.S., Soltis, D.E., 2005. Expression of floral MADS-box genes in basal angiosperms: implications for the evolution of floral regulators. Plant J. 43 (5), 724–744.

Kiyosue, T., Ohad, N., Yadegari, R., Hannon, M., Dinneny, J., Wells, D., Katz, A., Margossian, L., Harada, J.J., Goldberg, R.B., 1999. Control of fertilization-independent endosperm development by the MEDEA polycomb gene in *Arabidopsis*. Proc. Natl. Acad. Sci. U. S. A. 96 (7), 4186–4191.

Knopf, R.R., Trebitsh, T., 2006. The female-specific *Cs-ACS1G* gene of cucumber. A case of gene duplication and recombination between the non-sex-specific 1-aminocyclopropane-1-carboxylate synthase gene and a branched-chain amino acid transaminase gene. Plant Cell Physiol. 47 (9), 1217–1228.

Kolreuter, J.G., 1763. Vorläufige nachricht von einigen das gaschlecht der pflanzen betreffenden versuchen und beobachtungen, nebst Fortsetzungen 1, 2v. 3:266 (Ostwald's Kassiker 41 Leipzig, GER).

Koltunow, A.M., Grossniklaus, U., 2003. Apomixis: a developmental perspective. Annu. Rev. Plant Biol. 2003, 547–574.

Konagaya, K., Ando, S., Kamachi, S., Tsuda, M., Tabei, Y., 2008. Efficient production of genetically engineered, male-sterile *Arabidopsis thaliana* using anther-specific promoters and genes derived from *Brassica oleracea* and *B. rapa*. Plant Cell Rep. 27 (11), 1741–1754.

Kumar, R., Wehner, T.C., 2012. Growth regulators improve the intercrossing rate of cucumber families for recurrent selection. Crop Sci. 52 (5), 2115–2120.

Landgren, M., Zetterstrand, M., Sundberg, E., Glimelius, K., 1996. Alloplasmic male-sterile *Brassica* lines containing *B. tournefortii* mitochondria express an ORF 3′ of the *atp6* gene and a 32 kDa protein. Plant Mol. Biol. 32 (5), 879–890.

Lao, X., Suwabe, K., Niikura, S., Kakita, M., Iwano, M., Takayama, S., 2014. Physiological and genetic analysis of CO_2-induced breakdown of self-incompatibility in *Brassica rapa*. J. Exp. Bot. 65 (4), 939–951.

Leino, M., Teixeira, R., Landgren, M., Glimelius, K., 2003. *Brassica napus* lines with rearranged *Arabidopsis* mitochondria display CMS and a range of developmental aberrations. Theor. Appl. Genet. 106 (7), 1156–1163.

Lersten, N.R., 1980. Reproduction and seed development. In: Fehr, W.R., Hadley, H.H. (Eds.), Hybridization of Crop Plants. Amer. Soc. Agron/ Crop Sci. Soc. Amer. Publishers, Madison, WI (USA), pp. 17–43.

Lewis, D., 1949. Incompatibility in flowering plants. Biol. Rev. 24, 427–469.

Lewis, D., 1955. Sexual incompatibility. Sci. Prog. 172, 593–605.

Li, S.F., Iacuone, S., Parish, R.W., 2007. Suppression and restoration of male fertility using a transcription factor. Plant Biotechnol. J. 5 (2), 297–312.

Li, Z., Yu-Chan, Z., Yue-Qin, C., 2015. miRNAs and lncRNAs in reproductive development. Plant Sci. 238, 46–52.

Linke, B., Nothnagel, T., Borner, T., 2003. Flower development in carrot CMS plants: mitochondria affect the expression of MADS box genes homologous to GLOBOSA and DEFICIENS. Plant J. 34 (1), 27–37.

Liu, D., Dong, Q., Sun, C., Wang, Q., You, C., Yao, Y., Hao, Y., 2012. Functional characterization of an apple apomixis-related MhFIE gene in reproduction development. Plant Sci. 185–186, 105–111.

Liu, Z., Meyerowitz, E.M., 1995. LEUNIG regulates AGAMOUS expression in *Arabidopsis* flowers. Development 121, 975–991.

Lovell, J.T., Aliyu, O.M., Mau, M., Schranz, M.E., Koch, M., Kiefer, C., Song, B., Mitchell-Olds, T., Sharbel, T.F., 2013. On the origin and evolution of apomixis in *Boechera*. Plant Reprod. 26 (4), 309–315.

Ma, H., 2005. Molecular genetic analyses of microsporogenesis and microgametogenesis in flowering plants. Annu. Rev. Plant Biol. 56, 393–434.

Ma, H., de Pamphilis, C., 2000. The ABCs of floral evolution. Cell 101, 5–8.

Mandel, M.A., Yanofsky, M.F., 1998. The *Arabidopsis* AGL9 MADS box gene is expressed in young flower primordia. Sex. Plant Reprod. 11 (1), 22–28.

Martin-Closas, L., Puigdomenech, P., Pelacho, A.M., 2007. Pollination techniques for the improvement of greenhouse tomato production in two crop cycles. Acta Hortic. 761, 327–332.

Martínez, E.J., Acuña, C.A., Hojsgaard, D.H., Tcach, M.A., Quarin, C.L., 2007. Segregation for sexual seed production in *Paspalum* as directed by male gametes of apomictic triploid plants. Ann. Bot. 100 (6), 1239–1247.

Matton, D.P., Nass, N., Clarke, A.E., Newbigin, E., 1994. Self-incompatibility: how plants avoid illegitimate offspring. Proc. Natl. Acad. Sci. U. S. A. 91 (6), 1992–1997.

Matzk, F., Prodanovic, S., Bäumlein, H., Schubert, I., 2005. The inheritance of apomixis in *Poa pratensis* confirms a five locus model with differences in gene expressivity and penetrance. Plant Cell 17 (1), 13–24.

Meyerowitz, E.M., Smyth, D.R., Bowman, J.L., 1989. Abnormal flowers and pattern formation in floral development. Development 106, 209–217.

Mibus, H., Tatlioglu, T., 2004. Molecular characterization and isolation of the F/f gene for femaleness in cucumber (*Cucumis sativus* L.). Theor. Appl. Genet. 109 (8), 1669–1676.

Milligan, B.G., 1996. Estimating long-term mating systems using DNA sequences. Genetics 142 (2), 619–627.

Millwood, R.J., Moon, H.S., Poovaiah, C.R., Muthukumar, B., Rice, J.H., Abercrombie, J.M., Abercrombie, L.L., Green, W.D., Stewart, C.N., 2016. Engineered selective plant male sterility through pollen-specific expression of the *EcoRI* restriction endonuclease. Plant Biotechnol. J. 14 (5), 1281–1290.

Min, W.K., Kim, S., Sung, S.L., Kim, B.D., Lee, S., 2009. Allelic discrimination of the restorer-of-fertility gene and its inheritance in peppers (*Capsicum annuum* L.). Theor. Appl. Genet. 119, 1289–1299.

Min, W.K., Lim, H., Lee, Y.P., Sung, S.K., Kim, B.D., Kim, S., 2008. Identification of a third haplotype of the sequence linked to the restorer-of-fertility (*Rf*) gene and its implications for male sterility phenotypes in peppers (*Capsicum annuum* L.). Mol. Cell 25, 20–29.

Ming, R., Bendahmane, A., Renner, S.S., 2011. Sex chromosomes in land plants. Annu. Rev. Plant Biol. 2011, 485–514.

Ming, R., Wang, J., Moore, P.H., Paterson, A.H., 2007. Sex chromosomes in flowering plants. Am. J. Bot. 94 (2), 141–150.

Mogensen, H.L., 1996. The hows and whys of cytoplasmic inheritance in seed plants. Amer. J. Bot. 83 (3), 383–404.

Mulyantoro, Chen, S., Wahyono, A., Ku, H., 2009. Modified complementation test of male sterility mutants in pepper (*Capsicum annuum* L.): preliminary study to convert male sterility system from GMS to CMS. Euphytica 169 (3), 353–361.

Nagata, N., 2010. Mechanisms for independent cytoplasmic inheritance of mitochondria and plastids in angiosperms. J. Plant Res. 123 (2), 193–199.

Nasrallah, M.E., Burber, J.T., Wallace, D.H., 1969. Self-incompatibility proteins in plants: detection, genetics, and possible mode of action. Heredity 24, 23–27.

Nduwumuremyi, A., Tongoona, P., Habimana, S., 2013. Mating designs: helpful tool for quantitative plant breeding analysis. J. Plant Breeding Genet. 1 (3), 117–129.

Nettancourt, D.D., 1972. Self-incompatibility in basic and applied research with higher plants. Genet. Agraria 26, 163–216.

Niemann, J., Wojciechowski, A., Janowicz, J., 2012. Identification of apomixis in the Kentucky bluegrass (*Poa pratensis* L.) using auxin test. Acta Soc. Bot. Pol. 81 (3), 217–221.

Nizampatnam, N.R., Doodhi, H., Kalinati Narasimhan, Y., Mulpuri, S., Viswanathaswamy, D.K., 2009. Expression of sunflower cytoplasmic male sterility-associated open reading frame, orfH522 induces male sterility in transgenic tobacco plants. Planta 229 (4), 987–1001.

Nozue, K., Maloof, J.N., 2006. Diurnal regulation of plant growth. Plant Cell Environ. 29 (3), 396–408.

Ogawa, D., Johnson, S.D., Henderson, S.T., Koltunow, A.M.G., 2013. Genetic separation of autonomous endosperm formation (*AutE*) from the two other components of apomixis in *Hieracium*. Plant Reprod. 26 (2), 113–123.

Orton, T., Nitzsche, P., Honig, J., Donderalp, V., 2016. Genomics to detect and measure departures from autogamy in domesticated tomato. Adv. Crop Sci. Technol. 4 (3), 74.

Papadopoulou, E., Grumet, R., 2005. Brassinosteriod-induced femaleness in cucumber and relationship to ethylene production. HortScience 40 (6), 1763–1767.

Pelaz, S., Ditta, G.S., Baumann, E., Wisman, E., Yanofsky, M.F., 2000. B and C floral organ identity functions require *SEPALLATA* MADS-box genes. Nature 405, 200–203.

Podio, M., Cáceres, M.E., Samoluk, S.S., Seijo, J.G., Pessino, S.C., Ortiz, J.P.A., Pupilli, F., 2014. A methylation status analysis of the apomixis-specific region in *Paspalum* spp. suggests an epigenetic control of parthenogenesis. J. Exp. Bot. 65 (22), 6411–6424.

Pupilli, F., Barcaccia, G., 2012. Cloning plants by seeds: inheritance models and candidate genes to increase fundamental knowledge for engineering apomixis in sexual crops. J. Biotechnol. 159 (4), 291–311.

Quatrano, R.S., 1986. Regulation of gene expression by abscisic acid during angiosperm embryo development. Oxford Surv. Plant Cell Mol. Biol. 2, 467–477.

Rambaud, C., Dubois, J., Vasseur, J., 1993. Male-sterile chicory cybrids obtained by intergeneric protoplast fusion. Theor. Appl. Genet. 87 (3), 347–352.

Rani, V., Singh, K.P., Shiran, B., Nandy, S., Goel, S., Devarumath, R.M., Sreenath, H.L., Raina, S.N., 2000. Evidence for new nuclear and mitochondrial organizations among high-frequency somatic embryogenesis-derived plants of allotetraploid *Coffea arabica* L. (Rubiaceae). Plant Cell Rep. 19 (10), 1013–1020.

Renner, S.S., 2014. The relative and absolute frequencies of angiosperm sexual systems: dioecy, monoecy, gynodioecy, and an updated online database. Am. J. Bot. 101 (10), 1588–1596.

Ribarits, A., Mamun, A.N.K., Li, S., Resch, T., Fiers, M., Herberle-Bors, E., Liu, C., Touraev, A., 2007. Combination of reversible male sterility and doubled haploid production by targeted inactivation of cytoplasmic glutamine synthetase in developing anthers and pollen. Plant Biotechnol. J. 5 (4), 483–494.

Richards, A.J., 1986. Plant Breeding Systems. George Allen & Unwin, London (UK). 529 pp.

Riechmann, J.L., Krizek, B.A., Meyerowitz, E.M., 1996. Dimerization specificity of *Arabidopsis* MADS domain homeotic proteins APETALA1, APETALA3, PISTILLATA, and AGAMOUS. Proc. Natl. Acad. Sci. U. S. A. 93 (10), 4793–4798.

Ripley, V.L., Beversdorf, W.D., 2003. Development of self-incompatible *Brassica napus*. II. *Brassica oleracea* S-allele expression in *B. napus*. Z. Pflanzenzüchtung 122 (1), 6–11.

Rodrigues, J.C.M., Luo, M., Berger, F., Koltunow, A.M.G., 2010. Polycomb group gene function in sexual and asexual seed development in angiosperms. Sex. Plant Reprod. 23 (2), 123–133.

Ronfort, J., Saumitou-Laprade, P., Cuguen, J., 1995. Mitochondrial DNA diversity and male sterility in natural populations of *Daucus carota* ssp *carota*. Theor. Appl. Genet. 91 (1), 150–159.

Rostoks, N., Ramsay, L., MacKenzie, K., Cardle, L., Bhat, P.R., Roose, M.L., Svensson, J.T., Stein, M., Varshney, R.K., Marshall, D.F., 2006. Recent history of artificial outcrossing facilitates whole-genome association mapping in elite inbred crop varieties. Proc. Natl. Acad. Sci. U. S. A. 103 (49), 18656–18661.

Rounsley, S.D., Ditta, G.S., Yanofsky, M.F., 1995. Diverse roles for MADS box genes in *Arabidopsis* development. Plant Cell 7 (8), 1259–1269.

Ruffio-Chable, V., Le Saint, J.P., Gaude, T., 1999. Distribution of S-haplotypes and relationship with self-incompatibility in *Brassica oleracea*. 2. In varieties of broccoli and romanesco. Theor. Appl. Genet. 98 (3–4), 541–550.

Russell, W.A., Hallauer, A.R., 1980. Corn. In: Fehr, W.R., Hadley, H.H. (Eds.), Hybridization of Crop Plants. Amer. Soc. Agron/Crop Sci. Soc. Amer. Publishers, Madison, WI (USA), pp. 299–312.

Saedler, H., Becker, A., Winter, K.U., Kirchner, C., Theissen, G., 2001. MADS-box genes are involved in floral development and evolution. Acta Biochim. Pol. 48 (2), 351–358.

Sanabria, N., Goring, D., Nürnberger, T., Dubery, I., 2008. Self/nonself perception and recognition mechanisms in plants: a comparison of self-incompatibility and innate immunity. New Phytol. 178 (3), 503–514.

Sanchez-Puerta, M.V., Zubko, M.K., Palmer, J.D., 2015. Homologous recombination and retention of a single form of most genes shape the highly chimeric mitochondrial genome of a cybrid plant. New Phytol. 206 (1), 381–396.

Sandhu, A.P.S., Abdelnoor, R.V., Mackenzie, S.A., 2007. Transgenic induction of mitochondrial rearrangements for cytoplasmic male sterility in crop plants. Proc. Natl. Acad. Sci. U. S. A. 104 (6), 1766–1770.

Schwarz-Sommer, Z., Huijser, P., Nacken, W., Saedler, H., Sommer, H., 1990. Genetic control of flower development: homeotic genes in *Antirrhinum majus*. Science 250, 931–936.

Scutt, C.P., Vinauger-Douard, M., Fourquin, C., Finet, C., Dumas, C., 2006. An evolutionary perspective on the regulation of carpel development. J. Exp. Bot. 57 (10), 2143–2152.

Shifriss, C., Guri, A., 1979. Variation in stability of cytoplasmic male sterility in *Capsicum annuum*. J. Am. Soc. Hort. Sci. 104, 94–96.

Shivanna, K.R., Heslop-Harrison, Y., Heslop-Harrison, J., 1978. The pollen-stigma interaction: bud pollination in the Cruciferae. Acta Bot. Neerl. 27 (2), 107–119.

Sieburth, L.E., Running, M.P., Meyerowitz, E.M., 1995. Genetic separation of third and fourth whorl functions of AGAMOUS. Plant Cell 7 (8), 1249–1258.

I. ELEMENTS AND UNDERPINNINGS OF PLANT BREEDING

Sijacic, P., Wang, X., Skirpan, A.L., Wang, Y., Dowd, P.E., McCubbin, A.G., Huang, S., Kao, T.H., 2004. Identification of the pollen determinant of S-RNase-mediated self-incompatibility. Nature 429 (6989), 302–305.

Singh, S.P., Pandey, T., Srivastava, R., Verma, P.C., Singh, P.K., Tuli, R., Sawant, S.V., 2010. BECLIN1 from *Arabidopsis thaliana* under the generic control of regulated expression systems, a strategy for developing male sterile plants. Plant Biotechnol. J. 8 (9), 1005–1022.

Sofi, P.A., Rather, A.G., Wani, S.A., 2007. Genetic and molecular basis of cytoplasmic male sterility in maize. Commun. Biometry Crop Sci. 2 (1), 49–60.

Stanley, R.G., Linskens, A.R., 1974. Pollen Biology, Biochemistry, and Management. Springer-Verlag, Berlin (GER). 258 pp.

Stebbins, G.L., 1950. Variation and Evolution in Plants. Columbia Univ, Press, New York, NY (USA). 643 pp.

Stone, S.L., Goring, D.R., 2001. The molecular biology of self-incompatibility systems in flowering plants. Plant Cell Tissue Organ Cult. 67 (2), 93–114.

Stone, J.L., Thomson, J.D., Dent-Acosta, S.J., 1995. Assessment of pollen viability in hand pollination experiments: a review. Am. J. Bot. 82, 1186–1197.

Straub, S.C.K., Cronn, R.C., Edwards, C., Fishbein, M., Liston, A., 2013. Horizontal transfer of DNA from the mitochondrial to the pastid genome and its subsequent evolution in milkweeds (Apocynaceae). Genome Biol. Evol. 5 (10), 1872–1885.

Takayama, S., Isogai, A., 2005. Self-incompatibility in plants. Ann. Rev. Plant Biol. 56, 467–489.

Takebayashi, N., Morrell, P.L., 2001. Is self-fertilization an evolutionary dead end? Revisiting an old hypothesis with genetic theories and a macroevolutionary approach. Am. J. Bot. 889 (7), 1143–1150.

Telgmann-Rauber, A., Jamsari, A., Kinney, M.S., Pires, J.C., Jung, C., 2007. Genetic and physical maps around the sex-determining M-locus of the dioecious plant asparagus. Mol. Gen. Genomics 278 (3), 221–234.

Theissen, G., Becker, A., Di Rosa, A., Kanno, A., Kim, J.T., Munster, T., Winter, K.U., Saedler, H., 2000. A short history of MADS-box genes in plants. Plant Mol. Biol. 42 (1), 115–149.

Theissen, G., Kim, J.T., Saedler, H., 1996. Classification and phylogeny of the MADS-box multigene family suggest defined roles of MADS-box gene subfamilies in the morphological evolution of eukaryotes. J. Mol. Evol. 43 (5), 484–516.

Tikoo, S.K., Anand, N., 1980. Development of tomato genotypes with exserted stigma and a seedling marker for use as female parents to exploit heterosis. Curr. Sci. 49 (8), 326–327.

Towill, L., 2004. Pollen storage as a conservation tool. In: Guerrent, E., Havens, K., Maunder, M. (Eds.), Ex Situ Plant Conservation: Supporting Species in the Wild. Island Press, Covela, CA (USA), pp. 180–188.

Trebitsh, T., Staub, J.E., O'Neill, S.D., 1997. Identification of a 1-aminocyclopropane-1-carboxylic acid synthase gene linked to the female (*F*) locus that enhances female sex expression in cucumber. Plant Physiol. 113 (3), 987–995.

Tucker, M.R., Okada, T., Johnson, S.D., Takaiwa, F., Koltunow, A.M.G., 2012. Sporophytic ovule tissues modulate the initiation and progression of apomixis in *Hieracium*. J. Exp. Bot. 63 (8), 3229–3241.

Udagawa, H., Ishimaru, Y., Li, F., Sato, Y., Kitashiba, H., Nishio, T., 2010. Genetic analysis of interspecific incompatibility in *Brassica rapa*. Theor. Appl. Genet. 121 (4), 689–696.

Vandenbussche, M., Zethof, J., Royaert, S., Weterings, K., Gerats, T., 2004. The duplicated B-class heterodimer model: whorl-specific effects and complex genetic interactions in *Petunia hybrida* flower development. Plant Cell 16 (3), 741–754.

Vielle-Calzada, J.P., Nuccio, M.L., Budiman, M.A., Thomas, T.L., Burson, B.L., Hussey, M.A., Wing, R.A., 1996. Comparative gene expression in sexual and apomictic ovaries of *Pennisetum ciliare* (L.) link. Plant Mol. Biol. 32 (6), 1088–1092.

Vyskot, B., Hobza, R., 2015. The genomics of plant sex chromosomes. Plant Sci. 236, 126–135.

Wang, X., Singer, S.D., Liu, Z., 2012. Silencing of meiosis-critical genes for engineering male sterility in plants. Plant Cell Rep. 31 (4), 747–756.

Wang, Z.W., Wang, C., Gao, L., Mei, S.Y., Zhou, Y., Xiang, C.P., Wang, T., 2013. Heterozygous alleles restore male fertility to cytoplasmic male-sterile radish (*Raphanus sativus* L.): a case of overdominance. J. Exp. Bot. 64 (7), 2041–2048.

Waser, N.M., Price, M.V., 2016. Drought, pollen and nectar availability, and pollination success. Ecology 97 (6), 1400–1409.

Westwood, J.H., Tominaga, T., Weller, S.C., 1997. Characterization and breakdown of self-incompatibility in field bindweed (*Convolvulus arvensis* L.). J. Hered. 88 (6), 459–465.

Wolf, D.E., Rieseberg, L.H., Spencer, S.C., 1997. The genetic mechanism of sex determination in the androdioecious flowering plant, *Datisca glomerata* (Datiscaceae). Heredity 78, 190–204.

Wu, T., Qin, Z., Zhou, X., Feng, Z., Du, Y., 2010. Transcriptome profile analysis of floral sex determination in cucumber. J. Plant Physiol. 167 (11), 905–913.

Yamasaki, S., Fujii, N., Matsuura, S., Mizusawa, H., Takahashi, H., 2001. The *M* locus and ethylene-controlled sex determination in andromonoecious cucumber plants. Plant Cell Physiol. 42 (6), 608–619.

Yang, S., Terachi, T., Yamagishi, H., 2008. Inhibition of chalcone synthase expression in anthers of *Raphanus sativus* with Ogura male sterile cytoplasm. Ann. Bot. 102 (4), 483–489.

Zhang, Q., Gabert, A.C., Baggett, J.R., 1994. Characterizing a cucumber pollen sterile mutant: inheritance, allelism, and response to chemical and environmental factors. J. Am. Soc. Hort. Sci. 119 (4), 804–807.

Further Reading

Benz, B.F., 2005. Archeological evidence of teosinte domestication from Guilá Naquitz. Oaxaca. Proc. Nat. Acad. Sci. U. S. A. 98 (4), 2104–2106.

C H A P T E R

11

Cultivar Testing and Seed Production

INTRODUCTION

Both the testing of cultivars and production of propagules for dissemination to producers following the end of the plant breeding phase are generally beyond the purview of the plant breeder. These subjects will be accorded cursory coverage in this chapter because plant breeders are usually involved in the underlying processes. Coverage will be superficial since cultivar performance trials and large-scale seed production are peripheral to plant breeding, and abundant reference materials are available to scientists and practitioners that focus specifically on field trial design and analysis and the applied science of population replication.

The plant breeding program is technically complete after a set of candidate cultivar populations are deemed to be ready for final testing and production of seeds or asexual propagules. In the private sector, the organizational chart for the visualization, research and development, verification, and production of new plant cultivars is well defined and articulated. The plant breeder is a crucial member of the team that establishes the vision and takes it from idea and germplasm to finished product status. In a vertically integrated seed company, the "performance trials department" is charged with documenting the performance of the new cultivar as compared to designated established product standards. The "seed stock department" inherits the seminal finished populations from the plant breeder and applies principles of population biology and other sciences to produce large volumes of physically and genetically pure seeds or other propagules to be sold to farmers.

The plant breeder is not a disengaged bystander in these downstream activities. Extensive interaction and iteration with other departments are crucially important to successfully test, increase, and market the new product. The plant breeder plays an essential role in the development of new product concepts and assessing technical and business feasibility. With regard to seed stock increases, the plant breeder plays the crucial role of verifying the genetic integrity in newly produced propagule lots as compared to standards of phenotype metrics and qualities, now including the extensive use of molecular markers to assess genetic purity and quality attributes.

The relationship of the plant breeder with the leader of the performance trials department, or "trials coordinator," is often more challenging. Inevitably, the plant breeder develops hard-held beliefs about the performance of breeding lines as compared to commercial standard cultivars during the population refinement phase. This bias renders the plant breeder hopelessly ineffective as the person that verifies the comparative performance of the prospective new cultivar. The trials coordinator must be inherently unbiased to be effective. It is not unusual for the plant breeder and trials coordinator to be at odds on conclusions about candidate cultivar performance. Constructive debate is an important feature of any multi-faceted business group. It is incumbent on plant breeding organizations, therefore, to foster affirmative interfaces between the plant breeder and other members of the product development team.

The author once hired a professional who specialized in the design and management of a performance trials testing program. She was initially assigned the responsibility of rigorously testing incipient cultivars that were emerging from a linked biotechnology-plant breeding effort. The breeding program lacked resources, however, and the new trials coordinator was ultimately asked to expand the range of responsibilities to include the development of inbreds for a hybrid breeding program. Her training and experiences in a large, multinational seed company had taught her that this was not a good idea. She warned her supervisors about the dangers of assigning breeding and testing responsibilities to the same person. Why was she so concerned? From the standpoint of efficiency and familiarity with the intricacies of the cultivars under development, it would seem to make sense to combine these functions.

The reason for her concern was rooted in the allure of the plant breeding process. As the plant breeder weaves the fibers of raw germplasm into a finished tapestry it becomes increasing difficult to extricate him/herself from the incremental and obscure value of each fiber. Any selected entity, individual or population, was chosen for a purpose, and the compelling reasons for the decision are not usually apparent to the outsider. The plant breeder often develops an anthropomorphic attitude, a kind of kinship, about the fledgling populations, especially with regard to comparisons with competing materials. Objectivity may be blurred when making comparisons of populations from the program vs. those developed by others. Since overall performance is a composite phenotype governed by dozens of subjective and objective measurements, it is relatively easy to tip the scales in favor of the home team.

The farmer does not necessarily share the plant breeder's enthusiasm for his/her wares or appreciate all the underlying efforts and insights. For the producer, the function of the cultivar is to fulfill an expectation for a finished product with a minimum of risk. There will be plenty of other risks to contend with: weather, labor, diseases and pests, fickle markets, and many other challenges. Little concern exists that the finished cultivar was developed using this special source of germplasm or that unique and stupendous method for selection or mating. The expected result is what is expected, over and over. If the cultivar does not perform as expected, the farmer will quickly cease planting it, and "word will get out". Worse, if the cultivar exhibits an attribute that is antagonistic to the economic outcome, the plant breeder may be branded with a reputation that will dampen future efforts.

As the responsibility for the development of new cultivars of economic plants was passed on from farmers to seed companies, expectations have been substantially elevated. Seed companies and certifying organizations are charged with demonstrating that the new cultivar is significantly better than peers and that the seeds or planting stocks are viable, genetically pure, free of pathogens, and devoid of other organic and inorganic contaminants. The farmer is already faced with a bewildering multitude of risk factors, and the infusion of any more by virtue of unsubstantiated cultivar performance is certainly unwelcome.

As seed companies were formed and became more capable and sophisticated, it was inevitable that seed production would become a highly technical and specialized process. Seed production was historically the providence of the farmer who simply retained a portion of the previous season's harvest for next year's plantings. While saving seeds, farmers practiced mass selection, the algorithm that ultimately gave humanity the legacy of gene pools enhanced for uses as food and medicine, fiber and structure, aesthetics, etc. In certain instances, farmers still engage in the practice of saving seed, especially in self pollinated pure-line grain species such as wheat, rice, soybeans, and sunflowers. Where the varieties under cultivation are of commercial origin, and especially if they are protected under intellectual property statutes (Chapter 12), this possibility is rapidly disappearing.

Cultivar testing and seed production are not functionally or theoretically linked activities but have been grouped as the next steps following the conclusion of the plant breeding program. Plant breeders inevitably engage in a modicum of both candidate cultivar testing and seed production. Engagement with these processes also gives the plant breeder insights into the status of the market and cultivar attributes that might render the breeding process or seed production to be more efficient or precise. Since plant breeders store and deploy germplasm in the form of seeds and other types of propagules, they are continually producing more of them or replenishing populations that are experiencing loss of viability. Successful populations that are submitted for seed production must exhibit certain fundamental characteristics, and care must be exercised during the breeding process that future seed production needs are attended to.

CULTIVAR TESTING

A credible cultivar testing program is a prerequisite for all breeding programs, especially within commercial seed companies. If the cultivar testing program is doing its job properly, the feelings of the plant breeder about the relative performance of candidate cultivars do not matter. The reputation of the organization in assuring that finished cultivars have been thoroughly and adequately tested prior to commercial release is paramount. Every effort will be exerted to ensure that no cultivars are released that may fail and undermine this reputation. The memories of farmers are much more iron-clad than are those of elephants. For example, a prominent seed company sold a specific eggplant variety to eastern U.S. growers that were later proven to be mislabeled. While eggplants were indeed harvested by farmers, who bought this mislabeled seed, the fruits were sufficiently different that conventional wholesale markets did not accept them. Lawsuits and regulatory actions ensued. The farmers still talk about it decades later and still refuse to buy seeds from this company.

Public domain plant breeding programs also undergo a process of performance trialing prior to cultivar release, but the organization of operations is usually entirely different than the private sector process. For obscure species of small economic value such as minor vegetables and herbs, plant breeding is usually a one-stop-shop. The plant breeder is expected to do it all: Germplasm maintenance and evaluation, breeding, pathology and entomology, variety testing, and seed production. He/she is often also expected to teach, develop extension programs, and conduct basic research. Such a person is well acquainted with the changing of hats and the ability to simultaneously breed candidate cultivars then assess their relative level of commercial value may come naturally.

Most small grain and soybean breeding in the U.S. is conducted in the public domain, primarily at land-grant universities and USDA/ARS, although this situation is changing with the advent of GMO and other value-added product strategies. The predominance of public-sector breeding of these species is prevalent because the commercial potential of self pollinated grain crop seeds is limited. Commercial populations of these species are pure lines (Chapter 13) propagated by self-pollination. Consequently, it is relatively easy for farmers to save seed from the grain harvests if the cost of buying seed becomes inordinately more expensive than the labor for them to do so. Individual states have responded to the need for supplies of pure, vigorous seed of varieties that meet performance specifications by forming seed certification agencies. These are cooperatives comprised of government and farmers. The seed production functions of such organizations will be addressed below.

Seed Certification agencies are entrusted, among other things, with the testing of candidate cultivars against current industry standards. In this role, they operate independently of the breeding programs and as an unbiased service to agriculture in general. Most of the state governments have organized seed quality assurance standards that are enforced by agencies or surrogates, and most states also play host of trade organizations that promulgate planting seed quality and standards of product performance (Poehlman and Sleper, 2013).

New candidate populations are constantly being entered into the testing process and, at the other end of the process, chosen cultivars are submitted for seed production while unchosen ones are discarded or kicked back for further breeding inputs. The duration of time from the point of candidate submission to final decision varies according to crop species and circumstances, but a "rule of thumb" for annual crop species is three years. A negative decision about any given candidate cultivar, however, may be made in as little as one to two years. In rare instances, the entire program may be completed in two years or less, or extend for more than four years. For woody perennial species that feature long generation times and gaps until performance can be measured, the candidate cultivar testing process must be tailored substantially to provide results in the shortest possible time frame while remaining relevant to the realities of commercial agriculture. Time frames are expanded beyond three to four years for woody perennial species with longer generation times.

Several other conventions also apply to candidate cultivar testing programs. Any given cultivar testing program has a maximum capacity that is limited by available resources (attributable to labor, space, production costs, evaluation services, etc.). The cultivar testing program will usually be divided into phases or steps (1–3). Step 1, 2, and 3 trials are conducted every year. The step 2 trials for the current year consist of candidates that were advanced from the step 1 trials from the previous cycle, then step 3 from step 2.

Let us assume that 30 candidate populations are submitted for rigorous testing in the performance trials program. The plant breeder struggles incessantly to keep this number within a reasonable range since there is never a shortage of populations worthy of further evaluation. Adequate quantities of seed or other propagules are usually provided by the plant breeder to the performance trials coordinator when the evaluation of a candidate cultivar begins. The trials coordinator withdraws seeds, therefore, from the same source population, and observed differences may be concluded to be a consequence only of the environment. If different seed lots are used it is possible that genetic differences in the populations will confound results.

By year/step 2, the number of candidates may be reduced from about 30 to 5–10. The other 20–25 candidates are discontinued due to relatively low or marginal performance. After three years (step 3), the target for the number of populations selected for commercialization may be on the order of zero to two, and rarely as many as three, that will be recommended for release and submission into large-scale seed production. Seed production costs and physical capacity possessed by the organization present severe limitations to the number of populations that may achieve cultivar status during any given cycle.

The trials coordinator must be an accomplished and respected agronomist or horticulturist. Credibility is crucial since the often competing views of the R&D and marketing entities are subjugated to the results of the trialing program. The business plan for introducing and marketing the product depend largely on the results of the performance trialing program.

The objective is to gather sufficient information about the performance of candidate populations such that confident decisions can be rendered about their relative value as a product or source of germplasm. The company, or seed certification agency, does not want to recommend commercialization unless the facts truly support the expectation of ongoing and consistently superior performance of the candidate population. It is expensive to commercialize a cultivar due to seed production and marketing/sales costs. The drag on short- and long-term (due to loss of grower confidence) revenues as a consequence of "fielding a loser" is also substantial.

Trial locations and production methods should replicate or emulate those of targeted customers as much as possible. Extreme care must be taken to select, prepare, and manage experimental field sites within defined ranges that conform to the targeted area of introduction. For example, the soils should be uniform and representative of the targeted eco-geographical range of the new cultivars. Cultural practices should be administered uniformly and also closely mirror those that are employed in a production setting. Finally, the measurement of performance must accurately reflect economic benefits imparted by the new product to the grower or end user. A common practice is to enter into an agreement with a high-quality commercial grower that exemplifies the target customer for the execution of these trials. This is strategy is economical, ensures that appropriate methods are being employed, and may also help to get the word out to potential customers.

The trials coordinator (or trials program) must also be experienced and skilled at statistics, experimental design, and data interpretation. Large planting seed organizations usually employ expert statisticians to ensure the appropriateness and soundness of experimental designs and analytic procedures. Statistical science provides a clear framework for the quantification of the probability of specific decisions, and specifically that "population A is better than population B." If the trials are well planned and executed and an appropriate statistical design is employed, credible conclusions about the relative performance of all trial entries are possible. If requisite experimental distribution assumptions are adequately satisfied, the probability that phenotypic differences are due to chance alone can be set at any prescribed level, usually <5%. A probability >5% is deemed by scientists and for business applications to be too risky, and <1% portends the possibility that excessive risk avoidance will come with a sacrifice of legitimate opportunities.

A robust design always embraces replication: the plot size (number of individuals), the number of plots per experiment (usually 3–4), and the number of experiments (locations, years). Plot size is an important consideration. The larger the plot, the more that results will tend to represent large-scale performance in real production populations. Costs multiply quickly as plot size is expanded, and a compromise is always necessary. Each crop species tends to be characterized by a convention in the trade for minimum trial plot size. The experimental design also mandates randomization to minimize effects on performance attributable to physical location. Popular experimental designs in agricultural research are "completely randomized", "randomized complete block", and "Latin squares." The reader is referred to one of many excellent references on biological and agricultural statistics for background information on experimental design (Bender et al., 1982; Haaland, 1989; Zolman, 1993; Hoshmand, 1994; Petersen, 1994; Neter et al., 1996; Clewer and Scarisbrick, 2001; Quinn and Keough, 2002; Glass, 2007).

It is crucial to test candidate cultivars over diverse locations and multiple years to ascertain the degree of GxE interactions with regard to targeted phenotypes that are characteristic of the new population. The locations should circumvent the geographical range to which the new cultivar is targeted. Ideally, the range will be as broad as possible to avoid a proliferation of cultivars with specific adaptation. Low relative GxE proportion of V_P is typical for cultivars with broad adaptation, while high relative GxE proportion of V_P is more typical for specific adaptation. Locations (soil, microclimates, water quality, etc.) can be controlled, while environmental and GxE effects due to unique climatic patterns attributable to growing seasons cannot, but the effects of both locations and years must be addressed in the testing program. The reader is referred to one or more of the following excellent references to assess GxE interaction terms in horticultural crop species (Moreno-Gonzalez and Crossa, 1998; Manrique and Hermann, 2002; Nichols et al., 2002; Lacaze and Roumet, 2004).

An example of one entire 3-year candidate cultivar testing cycle is as follows (for a typical horticultural crop species such as an annual vegetable or small fruit):

- Step 1: 30 candidate populations, 10 checks (commercial standards).
    Trials observational (not replicated)
    3 locations, 20 plants per plot

Performance evaluated and ranked
Top ≤12 candidates that are ≥checks advanced to Step 2

- Step 2: ≤12 candidate populations, 10 checks (same as step 1)
  Trials replicated 3 times
  6 locations, 20 plants per plot
  Performance evaluated, means calculated and statistically contrasted
  Top ≤6 candidates that are >checks advanced to Step 3

- Step 3: ≤6 candidate populations, 5 checks (top commercial standards)
  Trials replicated 3 times
  3 locations, 10,000 plants per plot
  Performance evaluated, means calculated and statistically contrasted
  Top 0–2 candidates advanced to cultivar status and large-scale seed production.

The plant breeder is forewarned that only the best candidates should be submitted into the rigorous performance testing program. Many seed companies devise organizational decision matrices that incorporate input from not only the breeder(s), but also the trials coordinator, seed production staff, and sales and marketing. Such a decision matrix is applied at each step through large-scale seed production and sales. In this manner, the likelihood that expensive mistakes occur is greatly reduced.

Organizations that promulgate plant breeding efforts in the private sector, for example seed companies, frequently collaborate with researchers in the public domain for purposes of variety testing (Cantliffe, 2002). A fundamental goal of the U.S. land grant university system is to identify new cultivars that will enhance the profitability of growers in specific geographic regions or hardiness zones. The public performance trial program may be initiated by either party. Private companies are sometimes solicited for their new products and requested to submit seeds of prospective entries for such trials. Alternatively, one or more companies approach one or more public land grant institutions to establish this service. The seed company usually bears the cost of performance trials that incorporate their entries, usually on a per-entry basis. The prospective entries for cooperative land grant university trials are usually at the commercial of "advanced experimental" (usually step 2 or 3; see above) populations that are poised for commercialization.

Cooperative performance trials are both economical and enlightening for the private companies, public universities, and local growers that are given access to the trials. Unfortunately, most public U.S. land-grant universities no longer accord a high priority to such trials under the guise of scholarship or extension recognized for rewarding faculty and staff performance (Cantliffe, 2002). This shift in priorities has created a major void since public land-grant universities are needed as a source of unbiased information for growers and other agricultural producers (Williams and Roberts, 2002).

The performance trials program is a source of valuable information to the plant breeder. Unbiased trials provide the breeder with the measure of relative performance that is essential to keep programs relevant and competitive. The process can, however, foment considerable anxiety as well since clear, quantitative results are generated on the relative performance of the populations with which the breeder has labored for years. Even if entries do not measure up, it is possible to employ the results for choosing germplasm with which to mount new efforts and improve future results. The fundamental goal of the cultivar performance trials program is to clearly and accurately document the performance of a prospective commercial crop population with a minimum of risk imparted to the clientele that will engage in significant risk in adopting the new cultivar.

CULTIVAR RELEASE

Following the verification of superior performance in the rigorous unbiased performance testing program, the population that was molded by the breeding program is now ready to go out into the real world. The launching of the new cultivar is generally referred to as a *release*. "Release" is an accurate descriptor for what actually happens since the new cultivar was not previously available, kept close to the vest by the breeder and seed company during the development and testing processes. Following rigorous testing and statistical contrasting, the new population clears the final hurdle and has earned the chance to be offered to prospective customers.

The release is a proclamation comprised of an announcement of the new cultivar, a summary of key attributes including how it is better than what is already available, and where and how seeds or planting stock will be available for distribution. Releases from public programs are usually announced in scientific journals such as *HortScience*, while those from private programs are more likely to appear in promotional venues such as seed catalogs, internet web pages, and social media.

The interplay of plant breeding programs in the public and private sectors has often sparked debate over the appropriate roles that should be played by each (Morris et al., 2006). In general, public breeding programs serve the needs of a defined clientele that is not otherwise being met by the private sector. If public breeding programs are successful entire industries may arise from the commercial opportunities made possible by the advent of cultivars that occupy a new agricultural or economic niche. Inevitably, private companies may be established to engage in business in connection with the new opportunity, including incremental plant breeding (Innes, 1984).

Where both public and private breeding programs involving a specific crop species coexist, the private companies may view the development of finished cultivars by the public entity as an infringement by a subsidized competitor on the marketplace (Bliss, 2006). Conversely, public entities may regard the private companies with some suspicion, believing that they may simply take their cultivar releases and re-release them under a different name, leaving the public breeder without any avenue to demonstrate industry impact. Formal mechanisms to address and dispatch these incidences of overlap and redundancy are lacking. As a general rule, it is advisable that the private sector is allowed to address market opportunities for new horticultural cultivars. If the needs of the industry are not satisfied by the private sector or for opportunities that are deemed to be too small or narrow, the public university will fill these gaps as needed and feasible. The public sector also has providence over longer-term projects that are not adequately appealing to the private sector, and the development of germplasm that exhibits desirable socio-political but not necessarily economic impacts (Fuglie and Walker, 2001; Delmer, 2005).

The compromise that is usually forged between the public and private sectors with regard to the development of new varieties/cultivars amounts to a division of responsibilities. The public program conducts basic research on the inheritance of economically important traits and administers breeding to shepherd genes into general phenotypic classes, then the private programs take over and complete the breeding process to the finished varieties. The "unfinished" populations that are made available to the private companies for their subsequent breeding are called *germplasm releases*.

TRANSIENT AND DURABLE POPULATION NAMES

Plant breeders and, ultimately, marketing departments give populations of plant names to distinguish them and to create excitement and demand among prospective customers. Throughout the breeding program intermediary populations are usually given a numerical nameplate derived from the record-keeping system developed or adopted by the breeder. An example would be XY-146, population #146 of perhaps a thousand or more that was selected or mated during the calendar year 20XY. Most plant breeders maintain a *pedigree database* that documents all pertinent information about breeding populations including genetic interrelationships. The history of population XY-146 should be explained in detail in the database along with the identity of ancestors and specific selection and mating schemes and the known phenotypes or other genetic/physiological attributes it currently possesses. Breeding populations are ephemeral and dynamic, always in a state of flux. As the population is altered over time and with actions that alter genetic makeup, new names and database entries are forged. Populations may be retained for long periods of time as genetic "backups" or to use as germplasm in future breeding efforts.

The rules and conventions for botanical plant nomenclature were described in Chapter 2. The "variety" category was used historically as a botanical taxon, and still appears in older literature, now replaced largely by "subspecies". To avoid confusion, the term "cultivar" (cultivated variety) was coined and was used more or less synonymously with the current definition of "variety": a population that exhibits heritable attributes of commercial value. As such, it has little botanical value, and is used almost entirely as a part of the product marketing strategy.

While Latin species epithets are often chosen to describe the appearance or growth habit, cultivar names are not intended to be unambiguously correct or unbiased. Instead, they are chosen to reflect the best attributes of the population or to convey a positive outlook. In this manner, they are more similar to consumer product names like "Corvette"® or "Pop-Tart"®. In many instances, the population is one of a series of populations from the same program or with the same general attributes, and a common nameplate is minted to draw the attention of customers to that fact.

The names of the populations that are being worked on change from year to year. These utilitarian names do not necessarily appeal to the targeted customer of the seed of the cultivar. The finished cultivar is a biological entity, a population that embodies a unique and valuable collection of genes, so why not give it a catchy name? Who gets to name cultivars? It depends entirely on the context. Rules and conventions vary with the private vs. public sector and economic size of the market. Within the private sector, the breeder and management in the "one-stop-shop" or sole proprietorship usually have the absolute power to make this decision. If the breeder is employed by a larger organization that includes sophisticated sales and marketing the process of naming of new plant cultivars is incrementally more complicated. At the very least a committee that includes representation from sales and marketing is charged with the assignment approval of product names. In larger seed companies, sophisticated marketing studies are done to contrast the perceptions from sampled customers, resulting in catchy names like "Dominator," "Cash Register," or "KaChink." The voice of the plant breeder in this discussion is often drowned out.

SEED PRODUCTION

The plant breeding and performance testing processes conclude with the identification of one or more populations that are targeted for mass distribution to farmers and other agricultural producers. The context of this distribution may be in the public domain, through a land grant institution or USDA/ARS or a seed certification agency, or more likely from private enterprise. The beauty of genetics is that it not only provides for the generation of new variability for potential capture but that natural mechanisms of reproduction also provide for methods by which desired genotypes may be preserved and amplified. Ultimately, the ability to consistently replicate populations traces back to the fidelity of DNA replication and cell division.

If a plant breeder possesses 1000 seeds that represent a commercially promising population of a given crop species, it is possible, and even relatively straightforward, to convert this population of 1000 seeds into billions of seeds that possess equivalent genetic properties. The increased seeds will give rise to populations that exhibit identical phenotypic properties as the ancestral seeds. The seeds from this amplification process may then be distributed to producers, who insert them into their production system, and achieve similar results across the set of environmental conditions prescribed by the breeder, trials coordinator, and Stock Seed Manager or Coordinator.

Seed production is the art and science of using natural plant reproductive systems to amplify genotypes and conserve population genetic structures accurately while maximizing parameters of seed quality. This is accomplished most directly by planting seeds of the original population and harvesting self- and/or cross-pollinated progeny. If random mating is a fundamental property of the crop species, proper management of pollen vectors must be applied. Large-scale seed production of F_1 hybrid cultivars involves many successful and accurate female x male matings (Clayton et al., 2009). The development of large numbers of propagules of a clonally propagated species or cultivar carries an entirely different set of requirements (McDonald, 1995).

A pure line cultivar, bred to be fixed at all genomic loci, may be increased by planting seeds provided by the breeder and harvesting S_1, S_2, S_3, etc. progeny. Theoretically, the S_1, S_2, S_3, etc. populations should be genetically identical. As was intimated earlier, farmers still engage in the practice of "saving seed", akin to performing seed production on their own behalf. Pure line varieties, such as most small grains, soybeans, and certain vegetables (e.g., tomatoes) are readily adapted to such a scheme, although seed extraction and purification carry some additional requirements above and beyond the mere harvesting for agricultural products (Clayton et al., 2009). Moreover, mating systems are only rarely absolute. A small frequency of outcrossing is typically observed in self-pollinating species and, likewise, a small frequency of self-pollination is observed in outcrossing species. Care must often be taken to prevent the flow of unwanted pollen into a self-pollinating population, especially if the product specifications are very precise. For example, if pollen from a GMO population invades a non-GMO population, a miniscule frequency of hybrids in the progeny may still taint the cultivar.

Seed certification as a process and science-based farmer-university-government cooperative came into existence in the U.S. during the mid-20th century (Frolik and Lewis, 1944; Cooke, 2002). In this process, the Stock Seed Manager receives populations of *foundation* or *breeder* seeds from the breeder, and is expected to convert these small populations into the final product, large populations of genetically equivalent seeds. Breeder's seed represents the most accurate representation of the genetic basis of the corresponding cultivar. The individual and population are in the absolute correct genomic state, analogous to the "master" recording of music or photography. As copies are made of the master, followed by copies made of the copies, fidelity of reproduction of the original product is sacrificed due to resolution limitations in equipment or recording media. In seed production, the skill of the stock seed manager modulates the loss of fidelity. The higher the skill of the manager, the better will be the fidelity of genotype and population genetic reproduction.

Biological reproductive mechanisms operate with much, much greater fidelity than those used in sound or photographic reproduction. Nonetheless, the quest for perfection is of a similar order. Any populations that are derived from breeders' seed by any form of reproduction, either sexual or asexual, are assigned secondary status. A convention exists that pertains to the progression from primary → secondary → tertiary → quarternary, etc.:

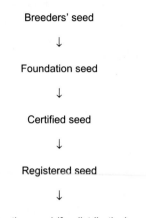

Breeders' seed

↓

Foundation seed

↓

Certified seed

↓

Registered seed

↓

Planting seed (for distribution)

In practice, all of these steps may not be necessary to achieve adequate seed amounts for sale or distribution. It may be possible in rare instances to proceed directly from breeders to planting seed, for this is theoretically the most preferable. Immense fecundity would be required, on the order of >1.0 million seeds per plant. Usually, it takes several steps to reach populations of that magnitude. Where possible, the intervening steps are simply skipped (e.g., breeders' seed → planting seed, foundation seed → planting seed). With each reproductive cycle removed from breeder's seed, the genetic value is presumed to be progressively reduced with each cycle of increase even though the populations may be virtually identical.

Population structures other than pure line require more strict procedures for amplification, according to the genetic consequences that accompany inadequacies. For OP populations, where foundation seed will be produced from breeders seed by cross-pollination or a mixture of cross- and self-pollination, and certified seed will be produced from foundation seed by intermating, etc., the main factors at play are population size and pollen flow. The fidelity of genetic reproduction of the population is maximized as $n \to \infty$, so all populations for the next seed production step should be as large as can be sustained. As $n \to 0$ allelic frequencies are not reproduced accurately and parental population sizes of <100 individuals should be avoided.

The seed stock manager must be intimately attuned to the mechanisms that affect pollen flow in the targeted plant species. Most OP populations will be subject to natural forces and mechanisms that promote outcrossing, so attention must be directed at protocols and inputs that will affect them. The most important among these is isolation from potential sources of contaminating pollen. Genetic and botanical research studies have led to the development of accepted standards of minimum isolation distance for most important crop species (Navazio, 2013). Such distances may vary from several hundred meters to several kilometers. Isolation distances are standards that are subject to adjustment to address prevailing environmental conditions. Reproductively compatible wild species and cultivated escape plants that occur in the hedgerows and roadsides are always a concern. They may contribute pollen that fertilizes females in the seed production field, manifested later by the appearance of "off types" in the farmer's field (Ellstrand, 1992).

Starting in the 1990s, the introduction of genetically modified (GMO) populations into open environments prompted more research on the dynamics of temporal and spacial pollen flow (Morris et al., 1994; Lavigne et al., 2008). The introgression of genetically modified genomes from GMO cultivars into conventional populations presents a challenge for labeling and characterizing plants. Some interests are worried about GMO genomes ending up in endemic and weed populations where they may upset natural ecological systems (Lavigne et al., 2008).

Where the employment of large distances between populations is impossible or impractical, other strategies may be employed to prevent unwanted pollen flows. Prominent among these is the use of covered frames or "cages" (Bosland, 1993; Navazio, 2013; Fig. 11.1). The impervious covering must be complete since arthropod vectors are attracted by floral volatiles and reflected light wavelengths. All openings should be secured with laps or zippers, and the bottom edge covered with soil heaped over the perimeter. The covering material must be inspected for tears and holes that might allow contaminating vectors to enter the enclosure. The composition of the covering material

FIG. 11.1 Typical cage structure for enclosure of flowering plants to exclude contaminating pollen from targeted matings. Cages are built inexpensively from rigid wood, plastic, or metal materials and a covering matrix that allows for light penetration and gas exchange, but excludes vectored pollen.

is specialized to be effective for this application. The covering material must transmit adequate light and gases to foster plant growth, while excluding the pollen vector. This is especially challenging for wind-pollinated species. Cages vary in size from <10 to over 100 m in length and height must be great enough to accommodate both plant and human stature. The framing and covering materials must be strong enough to withstand periodic strong wind forces and intense precipitation events.

Inputs may also be used to maximize the flow of pollen within the population used for seed increase. If the species is wind-pollinated, external fan devices or helicopters may be helpful to achieve random mating. For insect-pollinated species, vectors are often introduced, such as beehives of European honeybees. Beekeepers and seed companies frequently intertwine their operations for mutual benefit.

The seed stock manager is charged with the generation of new populations that are characterized by several standards of purity. Genetic purity is only one category, the absence of individuals considered to be off-types, usually resulting from unwanted pollen or self-pollinations during the course of hybrid seed production. Additional standards of biotic and abiotic purity are applied for adulteration of the lot with weed seeds, insects, pathogens, plant debris, and soil particles. These latter standards have been established industry-wide by the various national seed trade associations, and also by the International Seed Trade Association.

With increasing specialization of stock seed population structure comes corresponding complications and details for the seed stock manager. For example, if the cultivar population structure is a synthetic population, the stock seed manager is provided with a set of inbreds for intercrossing. The plant breeder must specify whether the end product will be a first, second, third, etc. generation synthetic. If the commercial product must be first generation synthetic (consisting of the first intercross of inbreds), the seed stock manager must determine the appropriate parameters to achieve targeted results including maximization of intercrossing and minimization of self-matings. In some instances it may be necessary to increase seed stock populations of the inbreds before the actual commercial seed increase may proceed.

The seed stock manager is challenged most by the production of seed of hybrid cultivars. The breeder submits small seed packets of two or more inbreds and the seed stock manager is expected to convert these into massive populations of pure hybrid seed usually accomplished in phases (Fig. 11.2). The inbred lines must be amplified to numbers deemed adequate for the production of a projected quantity of hybrid seed. For autogamous species, this may be relatively straightforward by simply conducting an OP seed increase. Since the lines are already considered to be inbred, it matters little whether the within-population matings are self- or cross-pollinations. If spurious or intentional changes to mating exist that prevent the inbred from perpetuating itself by simple open pollination, such as male sterility or self-incompatibility, additional measures must be taken to ensure success.

If the targeted crop species is dioecious, the male and female parents of the hybrid will, quite literally, be male and female. It will, therefore, be impossible to increase inbred populations by self- or open-pollination. It may be possible

Hybrid Seed Production

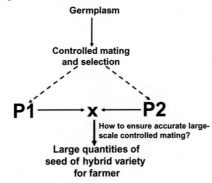

FIG. 11.2 Conceptual process of production of large quantities of pure F₁ hybrid seeds.

to induce the formation of male flowers in a genetically female population or vice-versa with growth regulators and then accomplish self-pollination as in a hermaphrodite (Khryanin, 1987). Sex conversion is often not an efficient or absolute process, or it is too expensive to justify in a business model. Alternatively, it may be feasible to increase populations of the males and females asexually using tissue culture/micropropagation techniques. Cloning inbred parental lines is currently used in certain horticultural crops for seed production in addition to the production of plants for direct sale to consumers, for example, asparagus (Javouhey, 1990).

In situations where male sterility is used for hybrid seed production the male will retain the ability to be perpetuated by self- and open- pollination, but the female will not. The reader is referred to Chapter 16 for a description of the breeding plans aimed at the incorporation of cytoplasmic male sterility and nuclear fertility restoration genes into parental inbreds for hybrid seed production. After the breeding phase is completed, the seed stock manager will obtain three populations from the breeder: the male (often referred to as the" R" or "C" line), the female (often termed the "S" or "A" line), and the maintainer (called the "M" or "B" line). The R/C population is propagated or increased by self/open pollination since it is a functional hermaphrodite. The S/A line is male sterile, so must be propagated via matings with the M/B line that is, ideally, isogenic for all genes other than those that control male sterility/fertility. For hybrid seed production, the S/A line is mated with the R/C line (Fig. 11.3).

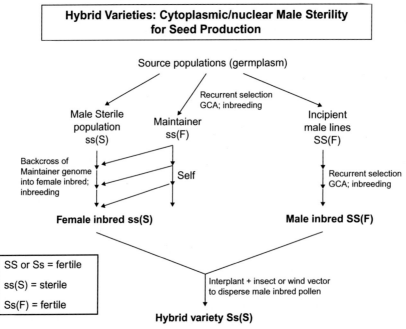

FIG. 11.3 The utilization of cytoplasmic/nuclear male sterility to produce large quantities of genetically pure F₁ hybrid seeds. The male sterile ss(S) population corresponds to the S/A line, the maintainer ss(F) to the M/B line, and the incipient male SS(F) lines to the F/C line described in the text.

Certain hybrid population structures a synthesized using three or more inbred lines. Examples include 3-way, double cross, six-way, and synthetic varieties. The same general strategies are used as those described above for F_1 hybrid varieties, but the complications presented by an increase in the number of inbreds are immense. More cages, more isolation plots, and more recordkeeping ultimately consume more resources, making these population structures progressively more expensive and time-consuming to produce.

Each crop plant species will present the seed stock manager with a unique set of challenges to the quest to produce large populations consisting of individuals of the proper genotype. Continuing with special cases in hybrid seed production, heterogamy presents a barrier to self and open pollination of an inbred line, for example, self-incompatibility. In *Brassica* crop species, this is overcome by employing "bud pollination". The self-incompatibility mechanisms in this family, especially the substances that govern recognition and inhibition, are not fully expressed until flowers reach maturation (Takasaki et al., 2000). Thus, it is possible, although resource consumptive, to affect self-pollination by introducing pollen to immature stigmata (Sun, 1938).

For autogamous species, the amplification of inbred lines presents much less challenge than for heterogamous species. The performance of mass cross-pollinations necessary in the production of hybrid seed, however, is an enormous challenge. Difficulty in forcing autogamous species to cross-pollinate is, in fact, the fundamental barrier to the breeding of hybrid varieties of certain of the world's most important crops, including the small grains, soybeans, and cotton. If the market is adequately robust to support the additional cost of hand pollination, hybrid varieties of autogamous crop species may be commercially feasible. For example, tomato male sterile mutants and transformants have been extensively described in tomato (Sandhu et al., 2007), but seed of F_1 hybrid cultivars is still produced by hand pollination due to the high value of the seeds (Watson, 2008). The efficiency and effectiveness of pollen transfer from fertile to sterile tomato populations are poor and male sterile systems have not yet proven to be useful for F_1 hybrid seed production.

The commercial success of F_1 hybrid tomato cultivars is a tribute to stock seed production managers in the private sector. The cost of labor must be sufficiently low accuracy of the hand pollinations sufficiently high that only certain labor pools are appropriate for the job. The geographical site of hybrid tomato seed production has shifted over the decades from the U.S. to Mexico and South America, to Japan, then Taiwan, Korea, Thailand and, most recently, mainland China and India. The seed stock manager must develop a skilled, dedicated, efficient, and effective workforce and oversee a process by which stock seeds are transported to the hybrid seed production site. Pollinations and seed extractions from mature fruits are then accomplished, followed by packaging and transport back to repackaging and distribution facilities. The import-export of biological materials mandates that phytosanitation statutes and commerce certificates regulations are adhered to. As the standard and cost of living in China and India will most certainly rise in the coming years, hybrid tomato seed will either rise accordingly or production will again be relocated where costs are low but the quality is still achievable. At some point, logic dictates an end to the wanderlust of this enterprise.

TESTING THE GENETIC PURITY OF SEED AND CLONAL POPULATIONS

Many types of impurities can adulterate seed lots. The Association of Official Seed Analysts (AOSA) and the International Seed Trade Organization (ISTA) mostly agree on the classes of seed purity tests, and what ranges of impurities are acceptable in commerce, but the methods to calculate numbers vary. The most general test is the Total Purity Test or the percentage by weight of a given lot that is attributable to pure seeds. Impurities include biological (genetic impurities, weed seeds, plant debris, and other biological material) and mineral (rocks, soil, paper products). Depending on the needs of commercial vendors, many other specific tests have been developed that focus on other sources of seed contamination.

The process of increasing the size of populations by sexual propagation is not analogous to photocopying documents. The biological "pixel" in the cloning process is the nucleotide base in DNA and RNA that has been found to be highly accurate; much more accurate than a photocopying machine. Notwithstanding the technical challenges described earlier in the chapter, many factors can contribute to an unacceptable result. Examples include inadequate isolation distances to other conspecific seed increases, sexually compatible weeds or flowering crops, and mixups of seed or clonal product lots before and after the actual increase step.

Before the discovery of highly heritable and easily visualized molecular markers, a standard test for genetic purity was the *grow-out*. This crude test consisted of the scrutiny of the gross phenotype of the population in question, often replicated over location and time. The result was presented as a proportion of "off-types" that were presumed to represent genetic contaminants. The grow-out is inefficient, inaccurate, and expensive, but many seed companies still perform these tests as a final quality assurance step in the product release procedure. Methods employed in conventional seed industry grow-out tests vary with the species, genetic structure of the commercial population and other circumstances (e.g., size of plants). Approximately 50–100 plants space-planted (1.5–3× standard spacing) plants side-by-side with a known control.

Molecular markers have been employed since the 1980s as a more accurate, efficient, and inexpensive alternative to the grow-out (Arús et al., 1982; Arús, 1984). The most effective and simplistic scenario involves populations with $V_G = 0$, for example, clones, pure lines, and F_1 hybrids. In cases where $V_G > 0$, such as open-pollinated populations, top cross hybrids, and synthetics, the use of molecular markers to assess the genetic purity of the population or divergence from genetic expectations, is more challenging.

The simplest case is the F_1 hybrid population where the two parents are inbreds. Only one definitive dimorphic marker is needed for which the parents and hybrid can be distinguished (Fig. 11.4; Arús et al., 1985). Finding effective dimorphic markers is a relatively straightforward process using qPCR (Ballester and Vicente, 1998). The two parents are challenged with a panel of candidate markers to find dimorphism than the use of the marker to distinguish the two homozygotes and the heterozygote is confirmed before the execution of the purity test (Crockett et al., 2000).

Testing inbred pure lines (or inbred parents of hybrid cultivars) for genetic purity is slightly more subjective. The targeted inbred population must be genotyped with respect to an exhaustive panel of markers known to be polymorphic among the domesticate subset of varieties. If the population in question is distinguished by rare (<0.10) alleles, fewer markers are needed for confirmation of purity. Conversely, if the population genotype consists of prevalent (>0.50) alleles, more markers are usually needed. Essentially, the decision regarding "pure" vs. "impure" individuals is buttressed by a low probability of a recurring genotype. For example, if the genotype of a population is $A_1A_1B_4B_4C_3C_3D_2D_2E_7E_7$, and the frequencies of A_1, B_4, C_3, D_2, and E_7 are all >0.90, the joint probability of observing this genotype among all possible genotypes is $0.90^5 = 0.59$; whereas if the frequencies are <0.10, the joint probability is $0.10^5 < 0.00001$.

Molecular Markers for Hybrid Seed Purity Testing

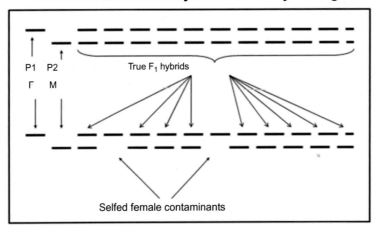

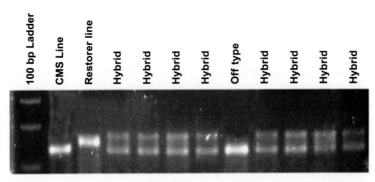

FIG. 11.4   A simple molecular test for genetic purity of F_1 hybrid seed progeny is presented where the parents are inbred and polymorphic for alleles of a marker gene that can be detected using an analytical technique such as electrophoresis. The top panel is a graphic representation of the bands that would be seen on a gel matrix. The bottom panel is a photo of an actual gel that illustrates banding patterns associated with inbred parents, the F_1 hybrid, and on "off-type" (selfed female) contaminant (Tamilkumar et al., 2009).

In cases where $V_G = 0$, or where this is the valid presumption, the agent or client must first fix the targeted level of resolution for detection of genetically impure individuals. For most horticultural applications, a resolution of 0.01 (1 in 100) is acceptable. When considering agronomic crops such as corn or sunflowers, more accuracy (0.001) may be required. Ideally, three or four independent samples of 100 seeds/individuals from the population in question are adequate, the level of genetic purity expressed as the average percent of the replications.

If $V_G > 0$ many assumptions pertaining to randomness and the nature of the frequency distribution must be considered when imputing statistical inferences. The archetype population is fully characterized with regard to allelic frequencies at 10 or more polymorphic marker loci. In cases where allelic frequencies are relatively low (<0.10), it is necessary to employ large sample populations (up to 1000 individuals) to achieve adequate accurately. Statistical inference is invoked to determine whether allelic frequencies in test populations are the same or different than the archetype (Remund et al., 2001).

References

Arús, P., 1984. Genetic purity of commercial seed lots. In: Tanksley, S.D., Orton, T.J. (Eds.), Isozymes in Plant Genetics and Breeding Part a. Elsevier Press, Amsterdam, NE, pp. 415–423.

Arús, P., Tanksley, S.D., Orton, T.J., Jones, R.A., 1982. Electrophoretic variation as a tool for determining seed purity and for breeding hybrid varieties of Brassica oleracea. Euphytica 31, 417–428.

Arús, P., Shields, C.R., Orton, T.J., 1985. Application of isozyme electrophoresis for purity testing and cultivar identification of F_1 hybrids of Brassica oleracea. Euphytica 34, 651–657.

Ballester, J., Vicente, M.C.d., 1998. Determination of F_1 hybrid seed purity in pepper using PCR-based markers. Euphytica 103 (2), 223–226.

Bender, F.E., Douglass, L.W., Kramer, A., 1982. Statistical Methods for Food and Agriculture. AVI Pub. Co, Westport, CT (USA). 345 pp.

Bliss, F., 2006. Plant breeding in the U.S. private sector. HortScience 41 (1), 45–47.

Bosland, P.W., 1993. An effective plant field cage to increase the production of genetically pure chili (Capsicum spp.) seed. HortScience 28 (10), 1053.

Cantliffe, D.J., 2002. Vegetable variety trials: an administrative point of view. HortTechnology 12 (4), 584–586.

Clayton, G.W., Brandt, S., Johnson, E.N., O'Donovan, J.T., Harker, K.N., Blackshaw, R.E., Smith, E.G., Kutcher, H.R., Vera, C., Hartman, M., 2009. Comparison of certified and farm-saved seed on yield and quality characteristics of canola. Agron. J. 101 (6), 1581–1588.

Clewer, A.G., Scarisbrick, D.H., 2001. Practical Statistics and Experimental Design for Plant and Crop Science. J. Wiley & Sons, Chichester NY (USA). 332 pp.

Cooke, K.J., 2002. Expertise, book farming, and government agriculture: the origins of agricultural seed certification in the United States. Agric. Hist. 76 (3), 524–545.

Crockett, P.A., Bhalla, P.L., Lee, C.K., Singh, M.B., 2000. RAPD analysis of seed purity in a commercial hybrid cabbage (Brassica oleracea var. capitata) cultivar. Genome 43, 317–321.

Delmer, D.P., 2005. Agriculture in the developing world: connecting innovations in plant research to downstream applications. Proc. Natl. Acad. Sci. U. S. A. 102 (44), 15739–15746.

Ellstrand, N.C., 1992. Gene flow among seed plant populations. New For. 6 (1/4), 241–256.

Frolik, E.F., Lewis, R.D., 1944. Seed certification in the United States and Canada. J. Am. Soc. Agron. 36 (3), 183–193.

Fuglie, K.O., Walker, T.S., 2001. Economic incentives and resource allocation in U.S. public and private plant breeding. J. Agric. Appl. Econ. 33 (3), 459–473.

Glass, D.J., 2007. Experimental Design for Biologists. Cold Spring Harbor Laboratory Press, New York, NY (USA). 206 pp.

Haaland, P.D., 1989. Experimental Design in Biotechnology. Marcel Dekker, New York, NY (USA). 259 pp.

Hoshmand, A.R., 1994. Experimental Research Design and Analysis: A Practical Approach for Agricultural and Natural Sciences. CRC Press, Boca Raton, FL (USA). 408 pp.

Innes, N.L., 1984. Public and private breeding of horticultural food crops in Western Europe. HortScience 19 (6), 803–808.

Javouhey, M., 1990. Fifty years of asparagus breeding valorized through twelve years of vitroculture. II. Future prospects for micropropagation of the best selected clones. Acta Hortic. 271, 129–133.

Khryanin, V.N., 1987. Hormonal regulation of sex expression in plants. Adv. Agric. Biotechnol. 21, 115–150.

Lacaze, X., Roumet, P., 2004. Environment characterisation for the interpretation of environmental effect and genotype x environment interaction. Theor. Appl. Genet. 109 (8), 1632–1640.

Lavigne, C., Klein, E.K., Mari, J.-F., Ber, F.L., Adamczyk, K., Monod, H., Angevin, H., 2008. How do genetically modified (GM) crops contribute to background levels of GM pollen in an agricultural landscape. J. Appl. Ecol. 45 (4), 1104–1113.

Manrique, K., Hermann, M., 2002. Comparative study to determine stable performance in sweetpotato (Ipomoea batatas [L.] lam.) regional trials. Acta Hortic. 583, 87–94.

McDonald, J.G., 1995. Disease control through crop certification: herbaceous crops. Can. J. Plant Pathol. 17 (3), 267–273.

Moreno-Gonzalez, J., Crossa, J., 1998. Combining genotype, environment and attribute variables in regression models for predicting the cell-means of multi-environment cultivar trials. Theor. Appl. Genet. 96 (6/7), 803–811.

Morris, W.F., Kareiva, P.M., Raymer, P.L., 1994. Do barren zones and pollen traps reduce gene escape from transgenic crops? Ecol. Appl. 4 (1), 157–165.

Morris, M., Edmeades, G., Pehu, E., 2006. The global need for plant breeding capacity: what roles for the public and private sectors? HortScience 41 (1), 30–39.

Navazio, J., 2013. The Organic Seed Grower: A Farmer's Guide to Vegetable Seed Production. Chelsea Green Publ., White River Junction, VT (USA). 388 pp.

Neter, J., Wasserman, W., Kutner, M.H., Nachtsheim, C., 1996. Applied Linear Statistical Models. Vol. 4. WCB/McGraw-Hill, Boston, MA (USA). 1408 pp.

Nichols, M.A., Godfrey, A.J.R., Wood, G.R., Qiao, C.G., Ganesalingam, S., 2002. An improved imputation method for incomplete GxE trial data for asparagus. Acta Hortic. (589),111–116.

Petersen, R.G., 1994. Agricultural Field Experiments: Design and Analysis. Marcel Dekker, New York, NY (USA). 409 pp.

Poehlman, J.M., Sleper, D.A., 2013. Breeding Field Crops. Blackwell Publishing Professional, Ames, IA (USA). 538 pp.

Quinn, G.P., Keough, M.J., 2002. Experimental Design and Data Analysis for Biologists. Cambridge University Press, Cambridge (UK). 537 pp.

Remund, K.M., Dixon, D.A., Wright, D.L., Holden, L.R., 2001. Statistical considerations in seed purity testing for transgenic traits. Seed Sci. Res. 11, 101–120.

Sandhu, A.P.S., Abdelnoor, R.V., Mackenzie, S.A., 2007. Transgenic induction of mitochondrial rearrangements for cytoplasmic male sterility in crop plants. Proc. Natl. Acad. Sci. U. S. A. 104 (6), 1766–1770.

Sun, V.G., 1938. Self-pollination in rape. J. Am. Soc. Agron. 30 (9), 760–762.

Takasaki, T., Hatakeyama, K., Suzuki, G., Watanabe, M., Isogal, A., Hinata, K., 2000. The S receptor kinase determines self-incompatibility in *Brassica* stigma. Nature 403 (6772), 913–916.

Tamilkumar, P., Jerlin, R., Senthil, N., Ganesan, K.N., Jeevan, R.J., Raveendran, M., 2009. Fingerprinting of rice hybrids and their parental lines using microsatellite markers and their utilization in genetic purity assessment of hybrid rice. Res. J. Seed Sci. 2 (3), 40–47.

Watson, B., 2008. Hybrid or open-pollinated? Nat. Gardening Assoc. 032017, 1–6.

Williams, T.V., Roberts, W., 2002. Is vegetable variety evaluation and reporting becoming a lost art? An industry perspective. HortTechnology 12 (4), 553–559.

Zolman, J.F., 1993. Biostatistics: Experimental Design and Statistical Inference. Oxford University Press, New York, NY (USA). 343 pp.

CHAPTER

12

Protection of Proprietary Plant Germplasm

INTRODUCTION

The biological properties of cell division allow us to perpetuate cultivars that exhibit a predictable set of performance characteristics. Sexual and asexual propagation has been used for both saving seed for the subsequent year's production and for genetic population improvement. The conservative nature of mitosis and assortative properties of meiosis, however, also present a clear pathway to the potential theft of unique plant genotypes without the permission of the breeders and organizations that own them. This chapter will present the concept of ownership of a plant genotype or set of plant genotypes and describe how to obtain legal title to this relatively new form of property. We must first deal with the concept of germplasm ownership. When did plants cease to be a shared natural resource and become personal or corporate property?

The notion that sexual seeds or asexual propagules and the hereditary material within them are subject to human ownership is an ancient one. The advent of agriculture quickly led to stores of selected seeds, bulbs, and cuttings for the following season. Each human tribes trove of germplasm was unique and embodied features that addressed specific needs for food, fiber, fragrances, medicines, and nurturing of domesticated animals. As tribes began to trade, seeds of unique selections were one of the most valuable barters. No phenomenon exemplifies the perceived value of plant germplasm better than the so-called "tulipomania" in Holland during the 1630s. Demand for bulbs of selections with unique and rare floral characters soared to absurd proportions until rampant speculation splintered the economy, resulting in a spectacular market crash by 1640 (Pollan, 2002).

Given that plant genotypes may be "owned", instances of piracy are surprisingly common. The imperfect association between genotype and phenotype and the ease of committing the larceny contribute to the relative ease of proprietary germplasm theft. All new plant breeders are quickly acquainted by sages with accounts of stolen germplasm, whether factual, mythical or somewhere in between. Such accounts often are laced with aspects of mystery, intrigue, and misdirected genius. Typical examples of impropriety (true or fanciful) include posing as a trusted colleague, dumpster diving, conspiring with the janitor who sweeps the floor of the research facility, and walking through a field trial with glue-ladened gloves to snag seeds or special shoes with soles that pinch seeds from the ground. The author has heard accounts of all of these strategies to steal germplasm. Prior to recent advances in biochemistry and molecular biology, genotype identity was based mostly on morphological phenotypes that were plagued by low heritability, rendering the identification of allegedly stolen property to be tenuous. The ability to police germplasm precisely with DNA sequencing has greatly improved the precision of genotype identification and, consequently, strengthened the ability to police intellectual property and proprietary germplasm (see "Enforcement of PVP and Patents" below).

Patents are legal contracts that are sanctioned, supported, and policed by governments or groups of governments. They are issued to the inventors of objects or processes to give them a head start in the marketplace for a limited time period, thus promoting innovation and economic opportunities. The original period of protection was 17 years, but the U.S. Patent Office has extended the term to 20 years. Another form of intellectual property protection that is used extensively by plant breeders is Plant Variety Protection (PVP), similar but with distinct differences (see below).

Copyrights and trademarks are intended to apply, respectively, to written or recorded compositions and names of products or processes. The period of protection extends for the entire life of the author or composer (and heirs) or trademark holder. Copyrights and trademarks are used only infrequently by plant breeders, so most of this chapter will focus on plant patents and PVP. Recent rulings and legal opinions have rendered copyrights to be more attractive to plant breeders as a form of proprietary protection of germplasm, but applications are only beginning (Burk, 2017). The laws and conventions of the U.S. will presented as an example, but those of most other countries or consortia are similar, and are/were often drafted to emulate U.S. codes.

Plant breeders are, as a group, suspicious of competitors and tend to possess an inflated sense of the value of their cache of germplasm. From that predisposition, most plant breeders have welcomed interventions by governments to provide a means to protect biological inventions. Most plant breeders are also notoriously averse to paperwork and regulatory processes and especially to the intrusion into their affairs by government bureaucrats. As plant genotypes have been invented that exhibit higher relative value, new offices of "patenting and technology transfer" within public and private sector organizations have appeared to protect biological intellectual. These offices are staffed by lawyers trained in contract law. Very few have more than a cursory knowledge of biology. The individuals charged with the protection of intellectual property often evolve from passive service providers to playing an active role in the planning of research activities. If an end product cannot be protected, many private sector organizations will not support the enabling research and development (Zullow and Karmas, 2008).

PLANT PATENTS

The U.S. Constitution spells out the right of inventors to secure legal property protection for creative efforts leading to new objects and processes to make them (Ihnen and Jondle, 1989). Before 1930, however, the applications of patents, trademarks, and copyrights did not extend to living organisms. While plant breeding had been practiced along with agriculture for millenia, it wasn't until the late 19th century that the commercial potential of genetically improved crop plant populations was demonstrated. As the industrial revolution portended larger and more vertically-integrated farms, private companies began to spring up in the 19th century that focused on the sale of seeds of unique populations of horticultural and agronomic crops to farmers (Butler and Marion, 1985).

The Plant Patent Act (PPA) of 1930 provided this new industry with the protection it needed to be confident in the large investments of labor and time associated with plant breeding. The commercial successes of improved plant cultivars achieved by Luther Burbank by the early 20th century were evident, and were cited as evidence that plant breeding should be accorded the same protection as other forms of invention. The Act led to Section 161 of Title 35, with oversight and enforcement by the U.S. Department of Commerce Office of Patents and Trade Names (Butler and Marion, 1985).

The PPA of 1930 provided, and continues to provide, for the following:

> Whoever invents or discovers and asexually reproduces any distinct and new variety of plant, including sports, mutants, hybrids and newly found seedlings, other than tuber propagated plant or plant found in an uncultivated state, may obtain a patent therefor, subject to the conditions and requirements of this title.

In addition to "tuber propagated plants", bacteria were also excluded from the act. In the late 1920s, when the wording of the legislation was under development, the understanding of bacteria was very limited, and it was even debated whether these organisms could be defined as "living" in the same sense as plants and animals. The exclusion of tuberous plants was politically motivated by potato producing states. The act was also intended only for asexually propagated plants since sexual reproduction was presumed to affect plant population phenotype uniformity adversely. Virus contamination that affected plant performance was common in potatoes and other asexually-propagated crops during the early 20th century, and virus-free planting stocks were virtually nonexistent in 1930.

The provisions that had previously applied to utility patents were extended to plants in the PPA of 1930 (Anon., 2017):

- Novelty
- Utility
- Non-obvious to those skilled in the art

Novelty may be demonstrated by any single unique phenotypic attribute or combinations of attributes, including but not limited to:

- Growth habit
- Resistance to disease or soil
- Flower color
- Leaf color
- Fruit color
- Flavor
- Productivity
- Perfume or fragrance
- Ease of sexual reproduction

In addition, certain requirements or conditions must be satisfied. One important area pertains to disclosures of the object or process prior to the filing of the patent application. For the purposes of our discussion, if the plant for which patent protection is being sought had been distributed previously to others freely and without restriction, then the application is voided. Secondly, the concept of enablement must be satisfactorily demonstrated. Enablement is access to the necessary materials and instructions such that someone skilled in the art can re-create the invention. The PPA of 1930 contains an exemption from enablement compliance provided that a description of the variety is as complete as is reasonably possible, but this was later modified (Butler and Marion, 1985).

Each plant patent pertains specifically to a single variety, or cultivar. Groups of cultivars that share a common feature are not patentable as such. The PPA of 1930 gave the holder of a plant patent the right to exclude others from asexually propagating the plant or selling or even using the plant so propagated. The act also required patent holders to document and demonstrate any acceptable phenotypic variability, or the range of phenotypes being claimed, for use in future patent infringement legal actions. Interestingly, the Act also requires the patentee to assign a name to the cultivar. No plants collected directly from wild habitats may be awarded patent protection (Butler and Marion, 1985).

PLANT VARIETY PROTECTION

Plant breeders employed by seed companies are involved primarily with sexually-propagated plant species that are not included in the PPA of 1930. Specialists in patent law were adamant that only asexual propagation could be made to fit the original intent of precise replication of the invention. Since sexual reproduction usually introduced genetic variability and corresponding phenotypic variability, it was concluded that populations derived from seeds did not constitute an entity that could be effectively protected from piracy (Staub et al., 1996).

The International Union for Protection of New Varieties of Plants (UPOV) is an organization headquartered in Geneva, Switzerland whose constituency is primarily European, but has enjoyed progressively more participation from countries from North and South America and Asia. UPOV was first organized in 1961 and charged with the development of an effective system for the protection of sexually propagated plant varieties. A code for plant variety protection (PVP) quickly came into existence (Anon., 2015). The U.S. Congress passed the Plant Variety Protection Act (PVPA) in 1970 based on most of the elements of the UPOV PVP system (Evans and Taylor, 1989).

The PVPA of 1970 was substantially similar to the PPA of 1930, but with certain notable exceptions. Rather than establishing a precise standard for varietal phenotypes, PVPA provided for the application of scientifically-established statistical tests to determine whether two population entities were the same or different. The phenotypes that could be used as bases for establishing novelty and utility were broad, including morphological, physiological, or biochemical attributes. Most sexually-propagated crop species were included, with some notable (and politically-motivated) exceptions (Evans and Taylor, 1989).

The procedure for obtaining a PVP certificate for a plant variety incorporates most of the same steps as those in Patents and Trademarks, and the PPA of 1930 (Blakeney, 2012). While patents are the purview of the U.S. Department of Commerce, an overview of PVP is also assigned to the U.S. Department of Agriculture. The PVP application includes an analysis of "prior art", or a summation of the history of public domain information and previous inventions pertinent to the current invention. The PVP application also includes a list of specific claims for which the inventor is seeking protection. The PVP examiner, who is trained in a specific area or discipline, is assigned and charged with the verification of assumptions and claims. The process is designed to be simple enough that the inventor may complete the application without legal assistance, but in practice specialized IP attorneys now play a prominent role in the application process. The entire process usually takes 2–3 years and can cost $50,000 or more per application (Smith, 2008).

After a PVP Certificate is issued, the holder is entitled to protection from the following forms of infringement (Ihnen and Jondle, 1989):

- Unauthorized sale or other distribution of the variety, or cultivar
- Unauthorized importation or exportation
- Unauthorized sexual propagation of the variety, or cultivar
- Using the variety, or cultivar in the direct production of another variety or cultivar (e.g., as a parent in a hybrid combination)
- Using derivatives for the asexual propagation of the variety, or cultivar

The PVPA of 1970 does carry certain exemptions. The original legislation excluded certain "soup vegetables", granted following political maneuvering by a large consumer products company, but this was deemed to be purely arbitrary and was rescinded in 1980 (Zullow and Karmas, 2008). Protection is also excluded from the following:

- Fungi, bacteria, and first generation (e.g., F_1) hybrids
- Seed used for research purposes
- Seed saved by farmers for the establishment of next year's crop

With regard to the exemption of F_1 hybrids, it is still possible to obtain PVP certificates for the inbred parent populations. Hybrid varieties already enjoy intrinsic protection from piracy by virtue of the population structure. The only way to faithfully propagate a multiply heterozygous genotype is by asexual propagation, a strategy that is technically or economically impractical for most crop species.

Fungi and bacteria that had been excluded by the PPA of 1930 were covered by a later extension of patent law (see below). The latter two exemptions have, in retrospect, been the source of considerable consternation and debate. The definition of "research purposes" is vague, but assumed to be the incorporation of the protected variety into a breeding program without recourse or compensation to the inventor. How much "plant breeding" is necessary and sufficient to qualify for this exemption? This question has been argued extensively in the courts, the rule of thumb being more than two distinct populations and greater than or equal to two generations of mating/selection (Staub et al., 1996).

The farmers' exemption was included to preserve a very long-standing tradition, but has proven to be a major impediment to long-term investments into plant breeding programs. This exemption applies almost entirely to self-pollinated agronomic crop species such as small grains, soybeans, and cotton. It is presently very uncommon for farmers of horticultural crop species to save seeds unless there is no alternative to obtain targeted genotypes. Seed companies have circumvented this exemption by imposing sales contracts that prohibit buyers of seed from engaging in seed saving practices (Chen, 2005).

The PVPA of 1970 further presents the prospective certificate holder with certain requirements. The examiner may request original data and other additional information to substantiate the factual basis for the claims. Before the certificate is issued, a seed sample must be deposited with an approved third party for purposes of enablement. The act carries a mandatory licensing clause that encourages the beneficial use of the invention and also discourages the filing of applications for "defensive" purposes. For example, Company A might devise a PVPA application with the sole purpose to disrupt the business operations of competing Company B, but the mandatory licensing clause negates the application of the legislation for this purpose (Butler and Marion, 1985).

The U.S. PVPA was amended in 1991 to incorporate provisions developed by UPOV, rendering the U.S. PVP codes synonymous with those of UPOV (Smith, 2008). These newly adopted provisions took effect in 1995. They provided for:

- Increase of the duration of protection from 17 to 20 years (25 years for trees and vines)
- Extended protection to include harvested plant organs (e.g., cut flowers)
- Included the explicit prohibition of "over-the-fence" transactions
- Protection extends to apomictic seeds (e.g., Kentucky bluegrass)
- Insertion of DNA sequences (e.g., transformation) does not override varietal infringement
- *Most importantly*: provided for protection from "essentially derived varieties", or sale of seed of a variety, or cultivar, expressing the essential characteristics of a protected variety, or cultivar

Additional "fine tuning" to the PVPA of 1970 was also applied as follows:

- Seeds provided to outside parties for testing after the filing of an application are protected
- The inventor may begin to sell seed before the application for PVP protection is filed, but no more than one year

After the 20 year period of protection expires, the certificate is no longer in force, and all parties are free to produce and sell the cultivar. For OP populations and pure lines, one would simply propagate the cultivar as it was originally sold. For hybrid cultivars where the inbreds are held by the original holder of PVP certificates, the inbreds revert to yet another form of proprietary protection, trade secrets (see below).

TRADE SECRETS

The tenet that you cannot steal what you have no knowledge of or cannot gain access to is the basis of proprietary protection by means of a trade secret. Theft of a trade secret may be prosecuted, but only for the physical object of the larceny, not for the underlying intellectual value such as genotype. Many individuals or organizations resort to trade secrets as the default mode of protection because the resources to pursue more formal means are lacking or if the proprietary entity is worth less than the costs associated with pursuing formal proprietary protection. These scenarios are common in cases of cultivars of specialty horticultural crop species where market size may be quite small.

In many cases, trade secrets are even preferable to patents and PVP certificates, since competitors are not provided with any information whatsoever about the program; objectives, germplasm, breeding methods, marketing strategies, etc. It may even be possible and preferable to maintain the existence of a program as secret so as not to alert potential competitors of a business opportunity.

Unlike patents or PVP, trade secrets are protected without legal registration or procedural formalities and can be protected for an unlimited time period. For these reasons, the protection of trade secrets may appear to be particularly attractive due to the absence of expenses and lack of disclosure. There are some conditions for the information to be regarded as a trade secret by legal authorities that vary from country to country:

- The information must be secret.
- It must have commercial value because it is a secret.
- It must have been subject to reasonable steps by the rightful holder of the information to keep it secret.

Hybrid varieties are mostly protected by trade secrets since little incremental competitive advantage is gained by obtaining PVP protection. Bags of seed of hybrid varieties often contain a small proportion of self-pollinated contaminants, the products of self-pollination of the female parent, one of the inbred populations that are proprietary to the company selling the hybrid seed. In rarer instances, the male parent may also be included as a contaminant in the hybrid seed lot. In a landmark 1993–1994 case, the Pioneer Hi-Bred Company brought successful litigation against Holden Foundation Seeds for infringement of a trade secret in such a manner (Anon., 1994). Holden was ruled to be an "unlawful appropriator" of the inbreds maintained by Pioneer as trade secrets, and a substantial financial award ($46,703,230.00) was levied against Holden. This decision conveyed an important message to the industry about the potential penalties associated with misappropriated genotypes.

COPYRIGHTS, TRADEMARKS, AND SERVICE MARKS

A *copyright* is a form of property ownership that protects original works of authorship including literary, dramatic, musical, and artistic works, such as poetry, novels, movies, songs, computer software, and architecture. The duration of copyright protection depends on several factors. For works created by an individual, protection lasts for up to 70 years beyond the life of the author or owner. For works created anonymously, pseudonymously, and for hire, protection lasts 95 years from the date of publication or 120 years from the date of creation, whichever is shorter (Anon., 2019).

A *trademark* is defined as a symbol, word, or words legally registered or established by use as representing a company or product that is a form of proprietary protection indicating source or origin (Moore, 1993). A *service mark* is a word, phrase, symbol, and/or design that identifies and distinguishes the source of a service rather than goods. Some examples include: brand names, slogans, and logos. The term "trademark" is often used in a general sense to refer to both trademarks and service marks. Unlike patents and copyrights, trademarks (and service marks) do not expire after a set term of years. Trademark rights come from actual "use" and can last indefinitely while the owner continues to use the mark in commerce to indicate the source of goods and services within a business context (Anon., 2019). Since the 1980s, we are seeing progressively more trademarks have been used in the marketing of many horticultural products such as nursery and landscape plant products, perennial fruit cultivars, and certain specialty vegetables and herbs (Elliott, 1991; Hutton, 1991; Moore, 1993; Bester, 2013).

In recent years, trademarks have been used in association with cultivars as a strategy to protect intellectual property rights (Hutton, 1991). In cases where a targeted population is protected by both a plant patent and a copyright, the US Patent and Copyright Office requests that the copyright be included in the patent application. After a plant patent expires, the inventor loses control over who may grow, use, or sell the patented cultivars. The inventor may, however, continue to control the use of the trademarks that are firmly entrenched in association with the cultivar. Thus, the combined application of both a patent and trademark to a new plant cultivar can often strengthen the breadth and duration of protection without substantial additional costs.

UTILITY PATENTS

U.S. patent statutes did not apply for inventions involving living organisms until 1980, except as provided for by the PPA of 1930 (e.g., asexually propagated plants; see above). In 1980, the U.S. Supreme Court agreed to hear a case that argued in favor of the extension of patent protection to include all living organisms. That case, Diamond v. Chakrabarty (1980), ultimately provided the cornerstone upon which the entire biotechnology industry was built (Kevles, 1994; Kieff and Olin, 2003). The researcher, Ananda Chakrabarty, had developed a new strain of bacteria that was capable of breaking down long-chain hydrocarbons for the clean-up of oil spills in water environments. The U.S. Supreme Court ruled that the new bacterial strain could be patented since it would not have existed without a substantial intercession of human intervention. The U.S. Patent Office was compelled by the Court to expand the organizational purview to include living organisms and portions derived therefrom (e.g., cells, organs, tissues, cells, organelles, plasmids, and DNA sequences, molecules, etc.).

The extension of utility patents to plant variety protection came largely as a consequence of a challenge brought to the Patent Office Board of Appeals in 1985 on corn varieties that were high in the amino acid tryptophan: Ex Parte Hibberd (227 U.S.P.Q.443) (Hodgins, 1989). As a result of this decision, the U.S. Patent Office began to accept applications for plant material including sexually reproduced plant cultivars provided that the criteria for utility patents have been fulfilled. Utility patents are now granted broadly for biological products and processes, including plant materials and uses. A fundamental feature of utility patents is that they may be issued to cover many genotypes that are subject to a similar set of novel uses. To qualify for such extended protection, utility patent applications carry additional requirements to prove novelty/distinction and non-obviousness (Plant et al., 1982):

- The scope and content of the prior art
- Ascertainment of the differences between prior art and the included claims
- Assessment of the level of skill necessary for the independent development of the invention

Additionally, if a patent is sought for a specific functioning gene, some degree of novelty regarding the altering of naturally-occurring alleles is essential. This additional requirement resulted from a 2013 decision of the U.S. Supreme Court regarding patents issued on the naturally-occurring human alleles BRCA1 and BRCA2 (Marshall, 2013). If John or Joanne Q. Public could easily have devised the invention, then the requirement for non-obviousness is not satisfied. This standard is, therefore, very subjective.

Until about 2000 CE, the proprietary genotype of an individual or population was defined entirely by the phenotype it gave rise to. Because phenotype = genotype + environment, the genotype was not defined, therefore, very precisely unless the trait of interest was highly heritable. It was relatively easy to misappropriate and perpetuate somebody else's patented genotypes and elude detection because a measure of doubt existed about the precise constitution of the genotype. With the advent of efficient and effective DNA sequencing methods, the ordering of nucleotide bases is used more and more to define genotypes in biological inventions (Barton, 1997). Molecular biology is increasingly providing accurate insights into the genetic and physiological mechanisms of phenogenesis, such as gene suppression (Chi-Ham et al., 2010). As of this writing, however, composite plant phenotypes are still used prominently to satisfy statutory requirements of uniqueness and utility. It must be proven that the actual gene responsible for the phenotype has been isolated and sequenced to be included in the patent script. The sequence of flanking genomic segments and QTL may be used to help to detect and prosecute genotype thefts.

MATERIAL TRANSFER AGREEMENTS

In instances where trade secrets are not sufficient to achieve adequate protection, but PVP or patent laws are either inappropriate, too unwieldy, or specifically exclude the form of protection being sought, material transfer agreements (MTAs) are often used in plant breeding to preserve property ownership rights. These are written documents that spell out the terms to which signed parties agree to in sharing proprietary information or materials. The enforcement of MTAs is by torte or contract law. The advantages are that they are relatively fast and inexpensive to develop and institute. While government entities recognize MTAs in civil courts, they are written and executed by the appropriate private parties without government involvement. The MTA may be as simple as the implied contract printed on the bag of seed that denies the right to propagate to the buyer. The main disadvantage lies in the enforcement that is uneven, unpredictable, and may take an inordinate time periods, up to several years (Rodriguez, 2005).

A new term has come into existence as the myriad of intellectual property (IP) protection laws have evolved and pervaded the industries in which biological materials and processes are invented: *Freedom to operate* (Kimpel, 1999). If a new product or service is developed that uses any patented materials or processes for which necessary arrangements have been forged with patent-holders, the product or service may not necessarily be free to sell without a license to enable the use of those materials or processes. In the private sector, FTO is a formalized process that generates a formatted report that articulates the ownership and licensing status of critical technologies for a targeted product or service. Specialized companies offer services to conduct a thorough assessment of the IP landscape to (1) identify all third party IP rights; (2) assess the levels of the risks posed; (3) consider how the risks can be managed; and (4) communicate conclusions to the business. Extensive negotiations are often necessary to ensure that a targeted project enjoys the freedom to operate before further resources are invested (Le Buanec, 2005).

INTELLECTUAL PROPERTY RIGHTS IN THE PUBLIC SECTOR

The treatment of strategies to protect intellectual property rights (IPR) for proprietary plant genotypes presented above pertains primarily to research and development in the private sector. While plant breeders the private and public sectors share the same goal of developing improved cultivars, there are important differences that must be considered when developing appropriate IPR for cultivars developed in the public sector. Plant breeding at public universities and foundations often focuses on marginal, specialty, or longer-term crop species with high social returns on investment but with less emphasis on monetary gains. Following product development and proof of concept by public sector organizations, commercialization is accomplished by the private sector with little or no return on investment to the public entity. The public breeding entity often collaborates with the private counterpart to commercialize public cultivars, and considerations must be made to facilitate this technology transfer (Tracy et al., 2016).

The costs of operating an effective plant breeding program have escalated as new, more powerful molecular tools have been developed. Concomitantly, funding of public organizations such as land grant universities and government-sponsored research stations has dropped while tax-supported grant programs have become more multi-dimensional and ambitious in scope. The commercial value of improved plant germplasm is, increasingly, regarded as a prospective strategy to garner funds to support research activities in public institutions.

The basic strategy for public sector research organizations is to obtain patent or PVP protection for improved germplasm and to develop licensing relationships with outside organizations based on the anticipated commercial value of the genotype. As an incentive to public breeders and other researchers, a portion of derived royalties is returned to support programs. Depending on the context in which agreements are struck, the terms can become very equivocal and complicated.

The IPR and licensing strategies of two large public organizations involved in plant breeding are contrasted to illustrate some of these issues and complications: The University of Florida (UF)/Institute of Food and Agricultural Sciences (IFAS) and the University of Wisconsin (UW). At UF, new cultivars are most commonly protected by either PVP or Plant Patent and released directly to a separate entity, the Florida Foundation Seed Producers (FFSP). FFSP then applies for intellectual property protection, develops licenses and distributes royalties. In contrast to the UF Office of Technology Licensing (OTL), royalty distribution through UF FFSP is weighted toward the inventor's program when total royalty amounts are lower and divides them more equitably across units and the Florida Agricultural Experiment Station when royalties increase. In the FFSP system, 70% of the royalties will return to the inventor's program. The vast majority of UF-IFAS cultivars earn less than $50,000 in annual royalties. Despite the trend that plant breeding activities are migrating from the public to the private sectors, the number of germplasm and cultivar releases by IFAS has grown from under 15 prior to 1930 to nearly 260 between 2000 and 2010 (Fig. 12.1). Since 2000, these modest royalties have allowed a broad spectrum of UF plant breeding programs to grow and thrive (Tracy et al., 2016).

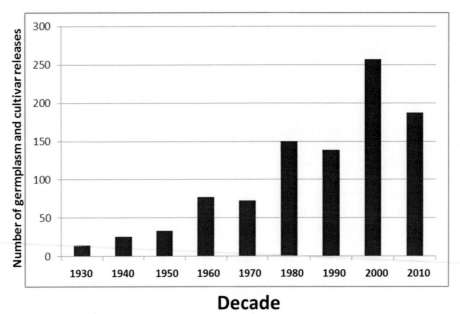

FIG. 12.1   The numbers of plant population releases (germplasm and cultivars) by the Florida Institute of Food and Agricultural Sciences from before 1930 to 2016, depicted by decade. *Adapted from Tracy, W.F., Dawson, J.C., Moore, V.M., Fisch, J., 2016. Intellectual Property Rights and Public Plant Breeding. College of Agricultural and Life Sciences, University of Wisconsin, Madison, WI, 70 pp., archived at: http://sbc.ucdavis.edu/files/271437.pdf.*

At UW cultivars have historically been released through the Wisconsin Crop Improvement Association (WCIA) that maintains seed inspection and quality programs. All new cultivars created by UW faculty are handled by the Wisconsin Alumni Research Foundation (WARF) with a standard distribution system for royalties to WARF and UW units (administration and academic units). Because this standard distribution system returns no revenue to the program that created the invention, plant breeders developed an alternative with WARF and WCIA. When WARF helps faculty members start small companies to commercialize a new cultivar or plant genotype concept, the main asset of that start-up is the intellectual property developed by the faculty member. WARF allows these businesses to keep some of the revenue and distributes the remainder as royalties under their standard distribution system (Tracy et al., 2016).

The halcyon times during the early- and mid-20th century, when public land grant universities and tax-funded institutions developed and dispersed genetically-improved plant germplasm to the private sector without any attribution or financial quid pro quo, are over. It is expected that efforts to develop, capture ownership of, and capitalize on plant-based inventions by public organizations will further intensify as technologies advance and public resources for research and education are static or declining.

SUMMARY

The plant breeder has two general strategies by which formal protection may be obtained for novel germplasm or cultivars: PVP Certificate or Patent. The patent category may be divided further into plant patents (PP) and utility patents (UP). Trade secrets, trademarks, and copyrights are also an option. What factors must be considered in deciding which mode of protection will be most suited to any given situation?

In PVP and PP, the varietal population genotype and imparted phenotypic attributes are protected, whereas in UP the uses of the variety or varieties are accorded protection. Most experts conclude that UP is effective, therefore, for protecting both the genotype and its uses. PVP, PP, and UP provide for 20 years of protection, but PP extends the duration to 25 years for long-lived perennials (Table 12.1).

ENFORCEMENT OF PVP AND PATENTS

When a plant breeder moves from one organization to another, the whereabouts of proprietary germplasm is often a concern. A new cultivar appears shortly after the plant breeder leaves Company X to join up with Company Y, and

TABLE 12.1 Comparison of the Alternatives with Respect to the Key Criteria that Distinguish them

Criterion	PVP	PP	UP
Cost	~$4500	$5000–8000	$10,000–30,000
Asexual reproduction?	Yes	Yes	Yes
Sexual reproduction?	Yes	No	Yes
Research/breeding exemption?	No (since 1998)	No	No
Training of assigned examiner	Plant scientist	Lawyer	Lawyer
Number of varieties/decision	1	1	As many as possible

Company X is suspicious that proprietary germplasm has been pirated. Alternatively, a mysterious prowler is seen running from a breeding nursery in the dead of night and a new cultivar suddenly appears that bears an uncanny resemblance to one that was being tested in the nursery. The reader may suspect indulgence in feigned drama, but might be surprised to learn that such scenarios have occurred.

Molecular markers and DNA sequences have proven to be the best and most unequivocal tools to protect and enforce intellectual property rights for biological inventions. Early uses of protein and DNA markers were in genetic identity cases (see discussion of Pioneer v. Holden above). DNA markers are currently ubiquitous in human parental identity litigations and exploration of genealogical (e.g., paternity) origins.

This application relies heavily on adaptations of population genetic principles (Kjeldgaard and Marsh, 1994). An individual DNA polymorphism is manifested by two or more alleles, the frequencies p, q, etc. of which can be measured. By making specific assumptions regarding the population from which two individuals were extracted, the degree of genetic identity may be estimated based on the incidences of like and unlike alleles. Thus, the likelihood that two individuals are related is greater if they share uncommon alleles. No assumptions have generally been invoked for linkage or linkage disequilibrium, each polymorphism treated independently, but this is changing with the availability of powerful new DNA sequencing methods and mapping software (Fister et al., 2017). For independently segregating markers, joint probability estimates increase concomitantly with the number of polymorphisms used in the comparison. DNA marker technology has been used to deduce germplasm piracy in the world of commercial plant breeding (e.g., Pioneer v Holden; Blakeney, 2009).

It is much more difficult to prove statistically that two individuals are identical than it is to prove they are different. The author has served in litigations in examples of both scenarios. In the first case, germplasm theft was suspected. A system of molecular markers was developed and applied against a panel of populations, unknown to the researchers who were charged with measuring the degree of genetic similarities/differences. The results were so compelling that the accused party quickly settled out of court. The fundamental parameter is estimated probability, in this case of being identical. For plants and animals, the courts seem to be convinced if the order is of the magnitude of 10^6. For humans, where the stakes are understandably higher, the acceptable order is more like 10^7 or even 10^8 (Blakeney, 2009).

In the second case, tests were conducted to determine whether two populations were different. While molecular marker analyses showed the populations to be substantially similar, they differed with regard to several alleles, and the probability of being identical was well below 10^6, so the jury quickly decided that they were indeed different. The costs and efforts associated with the first case (demonstrating sameness) were at least 10-fold greater than in the second (demonstrating differences).

To avoid the complications and expense of using the naturally-occurring genome to prove or disprove the genetic identity of an organism or population, the strategy of implanting unique DNA oligonucleotide sequences via *Agrobacterium tumefaciens* or biolistics mediated transformation has been proposed and demonstrated (Palazzoli et al., 2010; Fister et al., 2017). Genome editing technology can potentially also serve this purpose. While the presence of unique DNA in the protected entity by transformation makes it easier to protect, the footprint also mandates a GMO label in many jurisdictions.

Acknowledgment

The contents of this chapter were derived substantially from information provided by Professor William A. Meyer, Ph.D., Department of Plant Biology, Rutgers University.

References

Anon., 1994. United States Court of Appeals, Eighth Circuit. Docket Numbers 92-3292, 92-3556, archived at https://openjurist.org/35/f3d/1226/pioneer-hi-bred-international-v-holden-foundation-seeds-inc-pioneer-hi-bred-international.

Anon., 2015. The UPOV Convention, Farmers' Rights and Human Rights, German Federal Ministry for Economic Cooperation and Development. Deutsche Gesellschaft für Internationale Zusammenarbeit, Bonn (Germany), p. 104.

Anon., 2017. Intellectual Property and Genomics. National Institute of Health, National Human Genome Research Institute. archived at https://www.genome.gov/19016590/intellectual-property/.

Anon., 2019. Basic Facts: Trademarks, Patents, and Copyrights. US Patent and Trademark Office. archived at https://www.uspto.gov/trademarks-getting-started/trademark-basics/trademark-patent-or-copyright.

Barton, J., 1997. Intellectual property and regulatory requirements affecting the commercialization of transgenic plants. Appendix, In: Galun, E., Breiman, A. (Eds.), Transgenic Plants. Imperial College Press, London, UK, pp. 254–280.

Bester, C.A., 2013. Model for commercialisation of honeybush tea, an indigenous crop. Acta Hortic. 1007, 889–894.

Blakeney, M., 2009. Intellectual Property Rights and Food Security. CABI Publishing, Wallingford, OX,p.266. see p. 53.

Blakeney, M., 2012. Patenting of plant varieties and plant breeding methods. J. Exp. Bot. 63 (3), 1069–1074.

Burk, D.L., 2017. DNA copyright in the administrative state. Univ. Calif. Law Rev. 51, 1297–1350.

Butler, L.J., Marion, B.W., 1985. The Impacts of Patent Protection on the U.S. Seed Industry and Public Plant Breeding. University of Wisconsin-Madison, College of Agricultural and Life Sciences, Madison, WI. 128 pp.

Chen, J., 2005. The parable of the seeds: interpreting the plant variety protection act in furtherance of innovation policy. Notre Dame Law Rev. 81, 105–166.

Chi-Ham, C.L., Clark, K.L., Bennett, A.B., 2010. The intellectual property landscape for gene suppression technologies in plants. Nature Biotechnol. 28 (1), 32–36.

Elliott Jr., W.H., 1991. Property rights and plant germplasm. HortScience 26, 364–365.

Evans, K.H., Taylor, E.E., 1989. Issues and challenges in the administration of the plant variety protection act. Am. Soc. Agron. 52, 157–159. ASA Special Publ.

Fister, K., Fister Jr., I., Murovec, J., Borut, B., 2017. DNA labeling of varieties covered by patent protection: a new solution for managing intellectual property rights in the seed industry. Transgenic Res. 26 (1), 87–95.

Hodgins, D.S., 1989. Life forms protectable as subjects of U.S. patents—microbes to animals (perhaps). Biomater. Artif. Cells Artif. Organs 17 (2), 205–224.

Hutton, R.J., 1991. New funds for plant breeding. HortScience 26 (4), 361–362.

Ihnen, J.L., Jondle, R.J., 1989. Protecting plant germplasm: alternatives to patent and plant variety protection. Am. Soc. Agron. 52, 123–143. ASA Special Publ.

Kevles, D.J., 1994. Ananda Chakrabarty wins a patent: biotechnology, law, and society. Hist. Stud. Phys. Biol. Sci. 25, 111–135.

Kieff, F.S., Olin, J.M., 2003. Perusing property rights in DNA. Adv. Genet. 50, 125–151.

Kimpel, J.A., 1999. Freedom to operate: intellectual property protection in plant biology and its implications for the conduct of research. Annu. Rev. Phytopathol. 37, 29–51.

Kjeldgaard, R.H., Marsh, D.R., 1994. Intellectual property rights for plants. Plant Cell 6 (11), 1524–1528.

Le Buanec, B., 2005. Plant genetic resources and freedom to operate. Euphytica 146 (1–2), 1–8.

Marshall, E., 2013. Supreme court rules out patents on 'natural' genes. Science 340 (6139), 1387–1388.

Moore, J.N., 1993. Plant patenting: a public fruit breeder's assessment. HortTechnology 3 (3), 262–266.

Palazzoli, F., Testu, F., Merly, F., Bigot, Y., 2010. Transposon tools: worldwide landscape of intellectual property and technological developments. Genetica 138 (3), 285–299.

Plant, D.W., Reimers, N.J., Zinder, N.D., 1982. Patenting of Life Forms. Cold Spring Harbor Laboratory Publication, Cold Spring Harbor, NY. 337 pp.

Pollan, M., 2002. Desire: beauty/plant: the tulip. In: The Botany of Desire. Random House, New York, NY, pp. 59–110.

Rodriguez, V., 2005. Material transfer agreements: open science vs. proprietary claims. Nat. Biotechnol. 23 (4), 489–491.

Smith, S., 2008. Intellectual property protection for plant varieties in the 21st century. Crop Sci. 48 (4), 1277–1290.

Staub, J.E., Gabert, A., Wehner, T.C., 1996. Plant variety of protection: a consideration of genetic relationships. HortScience 31 (7), 1086–1091.

Tracy, W.F., Dawson, J.C., Moore, V.M., Fisch, J., 2016. Intellectual Property Rights and Public Plant Breeding. College of Agricultural and Life Sciences, University of Wisconsin, Madison, WI. 70 pp. archived at http://sbc.ucdavis.edu/files/271437.pdf.

Zullow, K.A., Karmas, R.A., 2008. Protecting intellectual property in plants and seeds. Cereal Foods World 53 (6), 319–321.

BREEDING METHODS

Introduction to Section II

The fundamental plant breeding algorithm was first introduced in Chapter 5. The figure used to illustrate the algorithm will be repeated to remind the reader of the basic elements of the plant breeding program (Fig. I2.1). All of these elements were introduced and described in great detail in Section 1. Breeding objectives and the processes leading to their formulation were presented in Chapter 6. The acquisition, maintenance, and enhancement of germplasm were covered in Chapters 7 and 8. Selection and how to improve efficiency and effectiveness of selection and controlled mating were presented in Chapters 9 and 10. A cursory treatment pertinent to plant breeding of methods for cultivar testing, seed production, and release was provided in Chapter 11. Finally, protecting intellectual property emanating from the project, including new discoveries in germplasm enhancement or selection effectiveness, or the resulting unique cultivar, were presented in Chapter 12. The progression of Section 1 may have seemed to be disjointed, but was, in reality, carefully crafted.

Section 2 will delve into specific examples of breeding strategies that have been developed and promulgated to breed new cultivars of crop plants. The method used to develop a new cultivar is developed or adapted is based on the biology of the crop species and properties of the targeted end product. Fundamentally, the structure of the engineered population (Chapter 4) will be based on the most efficient and precise use of biological processes to construct or perpetuate it. The natural mating system of the featured crop species is inevitably involved in this process. If a crop species is autogamous, the population structure of the finished cultivar will generally be a pure line, or poly-homozygote. Self-pollination is used extensively in the development and propagation of the cultivar because to do otherwise would introduce costs. Chapters 13 and 14 will cover breeding methods used predominantly to develop pure line populations in self-pollinated crop species.

In the case of heterogamous crop species, the use of natural mating systems is more problematic. Cross-pollination has the potential to disperse collections of desirable alleles that have been concentrated by controlled mating and

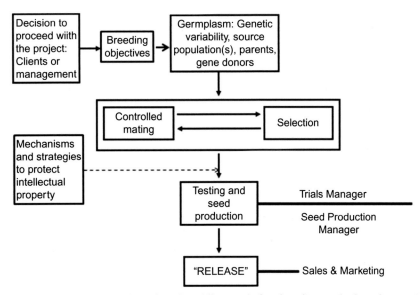

The Fundamental Plant Breaeding Algorithm

FIG. I2.1   Elements of the fundamental plant breeding algorithm. All general plant breeding methods and strategies are organized according to this scheme.

selection. Primitive breeding methods used for domestication (e.g., mass selection) of cross-pollinated species were stymied by uncontrolled mating and the constant segregation of selected genotypes. Consequently, inbreeding is used extensively in the breeding of cross-pollinated crop species to control mating and to constrain the process of concentration of desired alleles. Controlled hybridizations are subsequently employed to combine the resulting inbred lines that contain the selected desirable alleles.

Many crop species are clonally propagated. Clonal propagation may be applied for both self- and cross-pollinated crop species, but is more important for the latter because it provides the power to select and fix specific desirable genotypes from a broad spectrum of unstable genotypes. Since clonal propagation is usually more expensive (per plant/harvested unit) than seed-propagation, clonally propagation is more common as a breeding strategy in horticultural rather than agronomic crop species. Autogamous crop species are easily propagated by seed production from genetically fixed foundation population. Seed-propagation of finished cultivars of heterogamous species, as shall be presented in Chapters 15–17, is more challenging than for autogamous crop species.

The genetic differences between self- and cross-pollinated crop species are much more pervasive than simply the direct consequences of the mating system. Over time the forces of allele unmasking in homozygotes that are prevalent in self-pollinated crop species have tended to eliminate deleterious and sublethal alleles, also known as *genetic load* (Muller, 1950; Anderson et al., 1992; Crow, 1997). Conversely, genetic load tends to be much higher in cross-pollinated crop species (Barrett and Charlesworth, 1991; Couvet and Ronfort, 1996; Marriage and Orive, 2012; Böhm et al., 2017). Ultimately, genetic load is manifested in inbreeding depression and heterosis (Chapters 3 and 4) that is not usually associated with self-pollinated crop species. Breeding methods have been crafted to capitalize on these fundamental and innate genetic forces. Methods applied to self-pollinated crop species are predicated on the synthesis of heterozygotes by mating followed by the systematic winnowing of desirable from undesirable genotypes by recurrent self-pollination and selection, culminating in a pure line. In cross-pollinated crop species, methods are aimed at the simultaneous enrichment of desirable alleles, homogenization of extraneous alleles to achieve uniformity, and selection of alleles that enhance performance as polyheterozygotes. The most advanced archetype of population structure in cross-pollinated crop species, the F_1 hybrid cultivar, is ostensibly heterozygous at a multitude of loci (polyheterozygous), thus maximizing heterosis. In contrast, the open-pollinated population is a compromise of heterosis and uniformity achieved by panmixis instead of controlled hybridization (da Silva Dias, 2010).

Since the end product of the algorithm for a self-pollinated crop species is a fixed (panhomozygous) genotype, the V_P term is devoid of V_D; hence $V_P = V_A + V_I + V_E + V_{GxE}$. Therefore, any selection for V_D during the breeding process is inconsequential and potentially spurious. Methods are crafted to minimize the possibility that V_D will be inadvertently selected. In cross-pollinated crop species, V_D is potentially a significant source of desirable (or undesirable) phenotypic variability and must be strongly considered in any effective breeding strategy. Fig. I2.2 illustrates the distinctions between self- and cross-pollinated crop species in genetic terms.

Self Pollinated (=inbreeding) vs.Cross Pollinated (=outcrossing) Species

Self Pollinated	Cross Pollinated
Genetic structure of finished individual/population:	Genetic structure of finished individual/population:
AAbbccDDEEffGGHHiiJJkk etc.	*Open Pollinated:*
	AabbCcDDEeffGgHHiiJjkk
	AAbbCcDdEEffGgHhiiJjKk
	AabbCcDDEEFfGGHHiiJjkk etc.
= Pure line	*Hybrid:*
$V_T = V_A + V_D + V_I + V_E + V_{GXE}$	AaBbCCDdEeFfGghhIiJjKk etc.
No heterozygotes in pure line	$V_T = V_A + V_D + V_I + V_E + V_{GXE}$
Therefore:	
Select for $V_A + V_I + V_{GXE}$	Select for $V_A + V_D + V_I + V_{GXE}$

FIG. I2.2 The contrast between population genetic structures and sources of phenotypic variability in self- vs. cross-pollinated crop species.

The existence of a continuum of outcrossing rates in natural plant populations/species was presented in Chapter 10 (Clegg, 1980). How can we discern the best general strategy (i.e., fixation vs. heterozygosity) for the species that fall somewhere in the middle of the range and not at the extremes (i.e., absolute self- and cross-pollination)? Mathematical models have been invoked to study gene flow and genetic load, but the implications of these models have not been extended into the realm of practical plant breeding (Shaw et al., 1981; Cheliak et al., 1983; Kelly, 2005; Bradshaw, 2017). In general, the arbitrary dividing line is in the range of 20% outcrossing (=80% self-pollination).

PRELIMINARY STEPS

Prior to embarking on a breeding program certain milestones should already have been accomplished. Most of these benchmarks also apply to all other breeding methods as well. A thorough understanding of the botany of the species of interest, such as mating system and developmental and adaptive plasticity, is essential. The principles of the pedigree method are adaptable to self-pollinating species, where pure line breeding theories were first hypothesized and scientifically verified. The breeding objectives (see Chapter 6) will be established, based on a survey of clientele needs and the attributes amenable to breeding ranked according to a composite score of economic importance and technical feasibility.

An exhaustive search of the literature is important to take advantage of any information that might be used to render the program more effective such as inheritance patterns and available germplasm. If the goal of the breeding program is to combine two traits that are linked in repulsion, the program will be designed and executed differently than if the alleles are unlinked or linked in coupling. The literature may also impart knowledge of the inheritance of the characters that bear on population sizes and mating and selection strategies.

Breeding objectives should define the population structure of the finished product and combination of targeted attributes it will possess. Standards for selection, such as for size, color, shape, etc. are often helpful so that data are comparable from year to year, plot to plot, etc. Finally, a clear understanding of the resource needs of the project should be developed including the time, space, labor, and equipment and supplies needed. In the private sector, this information is used to compute the return on investment, and the project may or may not go forward based on the outcome of this analysis (Chapter 6; Simmonds, 1979).

Technology applications, such as computer hardware and software and molecular tools, tend to proceed very quickly, but plant breeding programs are limited by generation times and the seasonality of the geographical areas in which the crop will be produced. Impatience is a common source of concern or even anxiety during the course of the breeding program. Therefore, special attention should be paid to time management issues such as off-season advancement of generations where two or more generations may be obtained in one year. Such nurseries may require artificial environments such as greenhouses or open fields in far-away places. Arrangements for such needs may take time.

Most importantly, the germplasm that will be used for the breeding program must be carefully and thoroughly considered. The success of the program depends tremendously on the choice of parents used in the initial crosses. Ideally, one parent will be a generally accepted standard variety that lacks a key attribute. The other parent will usually possess the gene for this attribute and is not strikingly deficient in other aspects of economic performance. The program will proceed most expeditiously if the second parent is as similar phenotypically to the first as possible. The best possible plan and execution cannot compensate for the poor choice of initial germplasm.

Among the first steps in devising an effective breeding strategy for a targeted species is consideration of the mating system. For self-pollinated species in which finished cultivars will be pure lines, one of several alternative general methods are used, often in combination: Pedigree, bulk population, single seed descent, or doubled haploid. Some breeders have developed F_1 hybrid varieties of self-pollinated crop species, but the performance of these often does not surpass the pure line enough to justify the extra work and expense. For cross-pollinated species in which finished cultivars will be enriched for both desirable alleles and also heterozygotes, recurrent selection (an embellishment of mass selection; see Chapters 15 and 16) has proven to be most effective (Moll et al., 1978; Gallais, 1991). Enriched populations may be released directly in the form of open-pollinated populations or clones, or inbred to parental lines that are selectively combined to generate hybrids.

The backcross method (covered in Chapter 18), predicated on the transfer of one or a few genes from one population to another, is equally effective for both self- and cross-pollinated species. Backcrossing is also used extensively to introgress genes among phylogenetically-related crop and wild species, a process sometimes referred to as "pre-breeding" to distinguish it from actual cultivar development. In this book, introgression is covered primarily

Natural Mating System and Breeding Methods

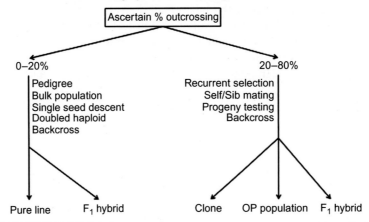

FIG. I2.3   General method to distinguish self- and cross-pollinated crop species, and a summary of broadly-based strategies that are effective for each.

in Chapter 8, and the backcross method is addressed in Chapter 18 primarily as a method to develop finished cultivars. The reader is apprised that the method is the same whether introgressing genes among populations or species, except that sexual dysfunction may disrupt gene flow among species.

Fig. I2.3 illustrates a hypothetical decision process to select the best breeding method for specific mating circumstances. Pure lines and OP populations tend to be more common for low-value crops, and clones and F_1 hybrids are more prevalent for higher-value crops.

References

Anderson, N.O., Ascher, P.D., Widmer, R.E., 1992. Lethal equivalents and genetic load. Plant Breed. Rev. 10, 93–127.

Barrett, S.C.H., Charlesworth, D., 1991. Effects of a change in the level of inbreeding on the genetic load. Nature 352 (6335), 522–524.

Böhm, J., Schipprack, W., Friedrich, H.U., Melchinger, A.E., 2017. Tapping the genetic diversity of landraces in allogamous crops with doubled haploid lines: a case study from European flint maize. Theor. Appl. Genet. 130 (5), 861–873.

Bradshaw, J.E., 2017. Plant breeding: past, present and future. Euphytica 213 (3), 60.

Cheliak, W.M., Morgan, K., Strobeck, C., Yeh, F.C.H., Dancik, B.P., 1983. Estimation of mating system parameters in plant populations using the EM algorithm. Theor. Appl. Genet. 65 (2), 157–161.

Clegg, M.T., 1980. Measuring plant mating systems. BioScience 30 (12), 814–818.

Couvet, D., Ronfort, J., 1996. Relationship between inbreeding depression and selfing: the case of intrafamily selection. Heredity 76, 561–568. Pt 6.

Crow, J.F., 1997. The high spontaneous mutation rate: is it a health risk? Proc. Natl. Acad. Sci. U. S. A. 94 (16), 8380–8386.

da Silva Dias, J.C., 2010. Impact of improved vegetable cultivars in overcoming food insecurity. Euphytica 176 (1), 125–136.

Gallais, A., 1991. Three-way and four-way recurrent selection in plant breeding. Plant Breed. 107 (4), 265–274.

Kelly, J.K., 2005. Family level inbreeding depression and the evolution of plant mating systems. New Phytol. 165 (1), 55–62.

Marriage, T.N., Orive, M.E., 2012. Mutation-selection balance and mixed mating with asexual reproduction. J. Theor. Biol. 308, 25–35.

Moll, R.H., Cockerham, C.C., Stuber, C.W., Williams, W.P., 1978. Selection responses, genetic environmental interactions, and heterosis with recurrent selection for yield in maize. Crop Sci. 18 (4), 641–645.

Muller, H.J., 1950. Our load of mutations. Am. J. Human Genet. 2 (2), 111–176.

Shaw, D.V., Kahler, A.L., Allard, R.W., 1981. A multilocus estimator of mating system parameters in plant populations. Proc. Natl. Acad. Sci. U. S. A. 78 (2), 1298–1302.

Simmonds, N.W., 1979. Principles of Crop Improvement. Longman Group Ltd., New York, NY. 408 pp.

The Pedigree Method

REVIEW OF THE GENETIC IMPLICATIONS OF SELF-POLLINATION OR OTHER ASSORTATIVE MATING SCHEMES

The fundamental biology of autogamy was covered in Chapters 4 and 10. The distinctions in genetic outcomes between self- and cross-pollination were covered in the Section 2 Introduction. In summary, recurrent self-pollination of a heterzygote results in a "drive to fixation" with regard to all genomic loci that are heterozygous in the F_1. The rate of drive to fixation among individuals within filial populations per self-pollinated generation is 50% (or 25% for each parental homozygote), a geometric progression. Fig. 13.1 illustrates the progression in terms of % homozygosity at a single locus (left side) and the change of a hypothetical set of loci that are heterozygous in the F_1 (right side).

The sources of phenotypic variation within filial generation populations change dramatically during the process of inbreeding and allele fixation. In the F_1, where all polymorphic loci are heterozygous, the proportion of V_P that is attributable to V_D is at its highest point. With each succeeding generation, the proportion of V_P attributable to V_D decreases geometrically as the loci that exert genetic effects on V_P are progressively driven from heterozygosity to homozygosity. At the endpoint of the fixation process, the population is theoretically comprised of panhomozygous individuals, and $V_D = 0$ (Fig. 13.2). Therefore, all the V_D that is present in the early generations of the fixation process (F_2, F_3, F_4, F_5) is a distraction to the plant breeder since V_D will not contribute to P in the cultivar. Consequently, all breeding methods for self-pollinated crop species minimize phenotypic selection during early filial generations where higher V_D is expected to be present. Alternatively, selection may be practiced for the presence of desirable alleles based on linked QTL (see "Use of QTL and MAS to enhance pedigree method effectiveness" below).

The essence of the early steps of breeding protocols for self-pollinated crop species is illustrated in Fig. 13.3. In this example, *Capsicum annuum* parent 1 that bears small elongated red fruits is hybridized with parent 2 that bears large blocky yellow fruits. The corresponding F_1 bears fruits that are intermediate in size, elongated, and red; indicating that fruit size is inherited additively, elongated is dominant over blocky, and red is dominant over yellow. The F_2 individuals segregate for fruit size, shape, and pigmentation. The breeder may, therefore, retrieve novel recombinants of all these characteristics.

Breeding Methods for Self Pollinated Species: Drive to Allele Fixation

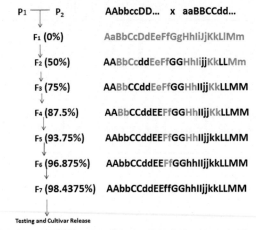

P₁ ⊤ P₂   AAbbccDD... x aaBBCCdd...

F_1 (0%)   AaBbCcDdEeFfGgHhIiJjKkLlMm

F_2 (50%)   AABbCcddEeFfGGHhIijjKkLLMm

F_3 (75%)   AABbCCddEeFfGGHhIIjjKkLLMM

F_4 (87.5%)   AABbCCddEEFfGGHhIIjjKkLLMM

F_5 (93.75%)   AAbbCCddEEFfGGHhIIjjkkLLMM

F_6 (96.875%)   AAbbCCddEEFfGGhhIIjjkkLLMM

F_7 (98.4375%)   AAbbCCddEEffGGhhIIjjkkLLMM

Testing and Cultivar Release

FIG. 13.1   Left side: Drive to allele fixation under recurrent self-pollination at a locus that is polymorphic for two alleles at a single locus among original parents: Filial generation with the % of individuals in the population homozygous at a single locus. Right side: Hypothetical allele configuration of many unlinked polymorphic loci or genomic sequences during the inbreeding process (F_1 to F_7).

Self Pollinated Crops: Selection Strategy

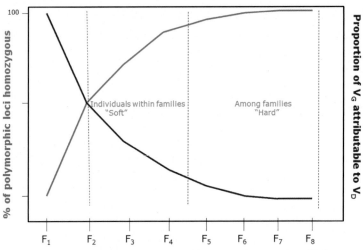

FIG. 13.2   The theoretical decrease of V_D concomitant with the increase of fixation with succeeding filial generations from a polyheterozygote F_1.

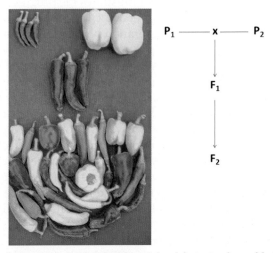

FIG. 13.3   A small, elongated, red fruited inbred is hybridized with an inbred featuring large, blocky, yellow fruits. The F_1 has intermediate sized, elongated, and red fruits. The F_2 segregates for fruit size, shape, and color, including new phenotypes (e.g., tangerine) not present in the parents or F_1.

All the breeding methods germane to self-pollinated crop species described in Chapters 13 and 14 start and end the same way. The starting point is a bilateral mating of parents that embody polymorphisms at targeted loci expected to affect P in a desirable way. The mating would be described as assortative with some disassortative properties (see Chapter 10). Ideally, the breeding methods for self-pollinated crop species allow the breeder to combine desirable alleles from both parents into a F_1 then select for new genomic combinations of alleles in a stepwise manner while fixation proceeds a generation at a time (Fig. 13.4).

The alternative breeding strategies presented in Chapters 13 and 14 are variations on the same theme. The main distinctions between them pertain to the size and maintenance of $F_1 \rightarrow F_8$ populations, selection pressures exerted during this progression, number of generations per unit time, and the size of the database that is generated during the program. Different breeding methods were devised to contend with a broader array of practical limitations and special parameters faced by plant breeders. The main factors that favor one strategy over another are budget, available time, other resources (e.g., space, staff, often exclusive of budget), and the accessibility of haploid plant technology (Fig. 13.5).

Fundamental Genetic Approach for Breeding Self Pollinated Plant Species*

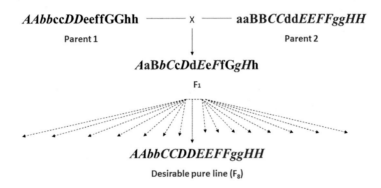

AAbbccDDeeffGGhh —— X —— *aaBBCCddEEFFggHH*

Parent 1   Parent 2

AaB*b*CcDd*Ee*Ff*GgHh*

F₁

AAbbCCDDEEFFggHH

Desirable pure line (F₈)

*Desirable genotypes in italics

FIG. 13.4   The general and fundamental process for recombining and selecting new desirable fixed genotypes from less desirable parental populations in breeding strategies for self-pollinated crop species.

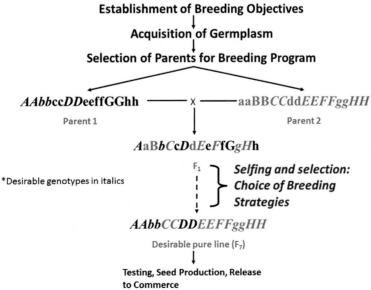

Establishment of Breeding Objectives

Acquisition of Germplasm

Selection of Parents for Breeding Program

AAbbccDDeeffGGhh —— X —— *aaBBCCddEEFFggHH*

Parent 1   Parent 2

*Desirable genotypes in italics

A*aBbCcDdEe*Ff*GgHh*

F₁   *Selfing and selection:*
          Choice of Breeding
          Strategies

AAbbCCDDEEFFggHH

Desirable pure line (F₇)

Testing, Seed Production, Release to Commerce

FIG. 13.5   Similarities and differences between breeding strategies for self-pollinated crop species. The different strategies are applied during the $F_1 \rightarrow F_8$ fixation progression (in red).

Selection and Family Relations Under Self Pollination

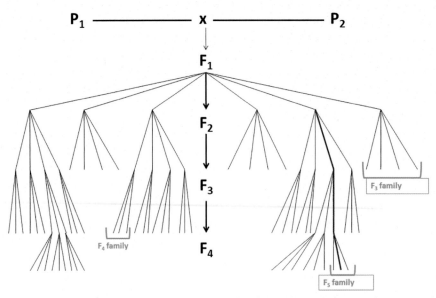

FIG. 13.6   The "family" concept that is helpful to plant breeders in comprehending and communicating the genetic relationships of individuals in the $F_1 \rightarrow F_8$.

Fig. 13.6 illustrates the "family" concept to comprehend the relationships of individuals as the fixation progression proceeds under recurrent self-pollination. From one F_1 polyheterozygous genotype, the theoretical number of distinct polyhomozygous individuals increases exponentially with successive filial generations. As the number of polymorphic loci in the F_1 increases, the theoretical number of unique genotypes in the F_2 also increases exponentially (Table 13.1). Hence, it is easy to visualize why the breeding program can quickly escalate in resource and time requirements, necessitating the need for alternative strategies to economize on these resources without undue sacrifice of breeding effectiveness.

Another important consideration in breeding methods for self-pollinated crop species is linkage of loci that impact the targeted phenotype, or phenotypes. The numbers presented in Table 13.1 presume that loci are unlinked. These numbers can change dramatically in cases of linkage, particularly for close linkages and alleles that are locked in undesirable configurations in the original parents. Desirable haplotypes are referred to as *coupling* and undesirable haplotypes as *repulsion* (Fig. 13.7). Using natural mating systems and in cases of tight linkages (e.g., < 5 cM), very large populations are often necessary to recover rare recombination events that convert repulsion to coupling haplotype configurations (Table 13.2).

All breeding strategies that have been developed for self-pollinated crop species are based on the assumption of 100% autogamy. As was described in Chapter 10, it is not typical for plant mating systems to be exclusively self- or cross-pollinating. Plants have devised natural physiological and anatomical mechanisms to modulate mating

TABLE 13.1   The Effects of Number of Polymorphic Loci between P_1 and P_2 and the Theoretical Numbers of Distinct F_1 Gametes, F_2 Genotypes, and Theoretical "Perfect" (Not Considering Random Effects) F_2 Population Size

Number of allelic pairs	Different F_1 gametes	Different F_2 genotypes	Smallest "perfect" F_2 population size
1	2	3	4
2	4	9	16
3	8	27	64
4	16	81	256
10	1024	59,049	10,84,576
n	2^n	3^n	4^n

Adapted from Allard, R.W., 1960. Principles of Plant Breeding. Wiley & Sons, New York, 485 pp.

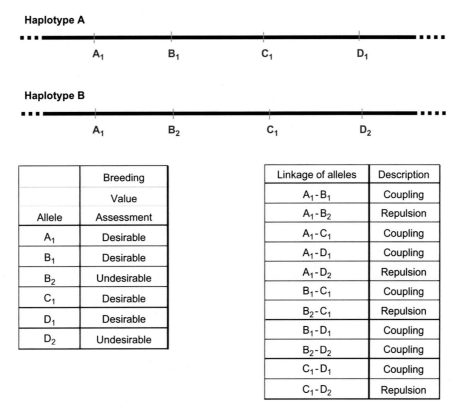

FIG. 13.7 Definitions and examples of coupling and repulsion linkage associations.

TABLE 13.2 The effect of linkage on the proportion of parental and recombinant F_2 individuals

Distance between A and B loci (cM)	% of F_2 AB/AB if F_1 is	
	AB/ab	Ab/aB
50	6.25	6.25
25	14.06	1.56
10	20.25	0.25
2	24.01	0.01
1	24.50	0.0025
p	$\frac{1}{4}(1-p)^2$	$\frac{1}{4}p^2$

As the proportion of desired associations decreases, the corresponding size of populations needed to identify recombinants increases accordingly. 50 cM is synonymous with independent assortment, or unlinked.

Adapted from Allard, R.W., 1960. Principles of Plant Breeding. Wiley & Sons, New York, 485 pp.

system based on both genetic and environmental factors to ensure the survival and fitness of the next generation. Heterogamy and autogamy are mating strategies that both have potential advantages and disadvantages, so it is not surprising that plant species exhibit inherent plasticities. While many species are predominantly autogamous or heterogamous, most are somewhere in between. Even the cereal grain crop species on which most of the most of the self-pollinated breeding theories are based are not exclusively autogamous. In general, cross-pollination will erode the intensity of the drive to fixation by self-pollination, as exemplified for the $F_3 \rightarrow F_5$ interval in Table 13.3. Therefore, progressively higher levels of cross-pollination degrade the theoretical assumptions upon which self-pollinating crop species breeding strategies are based.

Lastly, the drive to fixation may be modulated by altering the mating scheme. All the examples cited above have presumed self-pollination, the default mating system since it is the natural condition and requires no additional labor

TABLE 13.3   The effect of outcrossing on the drive to fixation in predominantly self-pollinated crop species

Percent outcrossing	Frequency in F$_3$			Frequency in F$_4$			Frequency in F$_5$		
	AA	Aa	aa	AA	Aa	aa	AA	Aa	aa
0	0.375	0.250	0.375	0.438	0.125	0.438	0.469	0.063	0.469
10	0.368	0.275	0.368	0.419	0.163	0.419	0.447	0.106	0.447
30	0.338	0.324	0.338	0.381	0.238	0.381	0.403	0.193	0.403
50	0.313	0.374	0.313	0.344	0.313	0.344	0.360	0.281	0.360
70	0.288	0.424	0.288	0.306	0.388	0.306	0.316	0.369	0.316
90	0.263	0.474	0.263	0.269	0.463	0.269	0.272	0.456	0.272
100	0.250	0.500	0.250	0.250	0.500	0.250	0.250	0.500	0.250

TABLE 13.4   The rates of disappearance of heterozygotes at a single polymorphic locus under different mating systems

Generation[a]	% of population heterozygous			
	DH	Self	Full sib	Half sib
1	100.00	100.00	100.00	100.00
2	0.00	50.00	50.00	50.00
3	0.00	25.00	37.50	43.75
4	0.00	12.50	28.13	38.28
5	0.00	6.25	21.09	33.50
6	0.00	3.13	15.82	29.31
7	0.00	1.56	11.87	25.65

[a] *Assuming P$_1$ (AA) × P$_2$ (aa); for DH, generation 1 is F$_1$ and generation 2 is DH$_1$; for self pollination generation 1 is F$_1$, generation 2 is F$_2$, etc.; for full sib and half sib generations are consecutive under these mating schemes.*
Fixation is negatively correlated with the proportion of heterozygotes.

inputs. The rate of drive to fixation can be accelerated by employing a doubled haploid step in the F$_1$, or decelerated by applying relaxed genetic assortative mating strategies such as sib- or half-sib (Table 13.4). Depending on circumstances, the breeder may wish to speed up the breeding process (to address a time-sensitive objective) or slow down the inbreeding process (to allow more opportunities for repulsion linkages to be converted to coupling).

THE PEDIGREE METHOD INTRODUCTION

The *pedigree* breeding method is considered the archetype for self-pollinated plant species (Allard, 1960; Poehlman and Sleper, 2013). This method is widely practiced by plant breeders in one form or another, and provides ample opportunity for elements of traditional and advanced sciences with intuition/art to contribute to project success. The pedigree method, or derivatives of it, is also widely used in cross-pollinated species for inbred parent development (Acquaah, 2012). Since the method is a cornerstone for the breeding of self-pollinated crop species, it will be covered here and cited in the later chapters on breeding methods for cross-pollinated crop species. This strategy is a good place for a student to begin since the strengths and weaknesses of alternative approaches may be fully appreciated with full knowledge of how an entire pedigree program works, and what information and materials it delivers (Allard, 1960; Briggs and Knowles, 1967).

A pedigree is a system of familial relationships, or genealogical record. A dog or horse is said to have a "good pedigree" if all ancestors adhered to accepted standards of the represented breed. The study of human pedigrees historically led to a greater understanding of inherited characters and disorders such as blood antigen groups, hemophilia, sickle cell anemia, type 1 diabetes, amyotrophic lateral sclerosis ("Lou Gehrig's Disease"), dwarfism, albinism, and certain cancers and mental dysfunctions (e.g. Alzheimer's Disease). This information regarding the presence and absence in ancestors of alleles that control key human phenotypes is allows prospective parents to consider the risks of transmitting undesirable alleles to their children (Thompson, 1985).

The pedigree breeding was first developed in Sweden during the early 20th century (Newman, 1912 cited in Fehr, 1987). The method takes its name from the fact that records are usually kept for individuals of all matings, selections, and performance evaluations (Fehr, 1987). Typically, the plant breeder selects two parents from which he/she wishes to extract and combine desirable attributes, or alleles. For self-pollinating plant species, the parents are presumed to be genetically fixed, or pure lines. The parents are crossed to generate a polyheterozygous F_1 hybrid, the derivatives of which are repetitively self-pollinated until total homozygosity, or an acceptable degree of consistent phenotypic uniformity (fixation), is reached. For practical purposes, fixation is generally considered to be adequate for commercialization by the F_7 or F_8.

We learned in Chapter 4 that if parents X and Y differ with regard to alleles at a single locus (A and a), then nearly 94% of F_5 individuals will be homozygous at that locus, 47% homozygous for each allele (AA and aa; Fig. 4.1; Allard, 1960). If ten genetic loci are polymorphic between the two parents, approximately 49% of corresponding F_5 individuals will be completely homozygous at all ten loci. The parents chosen for any given breeding project, including those involving the pedigree method, would typically be polymorphic with respect to many more than ten loci. The number of polymorphisms would more likely be in the range of hundreds or more.

The pedigree method, like other methods developed for inbreeding species, is aimed at the extraction of desirable alleles from both parents and their combination into a single genotype. Likewise, the method is predicated on the removal of the undesirable alleles of both parents, replacement with the more desirable counterpart from the other parent. The analogy of the poker game, invoked to exemplify the concept of heritability in Chapter 3, is illustrative of this concept. In a hypothetical variation on the poker theme, two players are dealt five cards each and instructed to work together as a team to fashion the best possible five-card hand. Player 1 holds 10H, 3C, AH, 8S, 6D and player 2 holds JH, QH, KH, 2D, 9C. By taking cards from each hand, both of which are nearly worthless, we can achieve a royal flush. In this analogy, the plant breeder seeks to take one hand, very good but lacking one to two key cards, and make it better by extracting the one to two needed cards from another hand, the result being better than either of the two hands alone. While the poker analogy may be instructive, rarely would this enormous degree of genetic complementation and epistasis be encountered in the biological world.

As the name "pedigree" suggests, records are kept of genetic lineages including matings and selections at all points in the breeding program (Fehr, 1987; Poehlman and Sleper, 2013). The parents are chosen following careful consideration of performance and relatedness from known germplasm records and, more recently, genomic sequence information. Ideally, and theoretically, each individual in the network of descendants from a hybrid combination comprises a point in a potentially massive database. As the program proceeds and descendants become more and more inbred, genetic differences between individuals within filial families diminish to zero. Individuals are then bulked into pure-breeding populations. The advent of computers, software, robotic phenotyping, and customized data input/output hardware devices has made pedigree breeding programs much easier to analyze and promulgate (Brown et al., 2014; Fleury and Whitford, 2014).

Data may be entered directly at the experimental plot site onto portable recorders, or downloaded from computerized combines that automatically encode entries from GPS satellite signals. Specialized agricultural database computer software (e.g., Agrobase®, Doriane®, Phenome®, Peditree®, E-Brida®, Plabsoft®, NOAH®, GGT 2.0®, and others) is widely used that prints bar-coded field stakes. A hand-held scanner rapidly and easily records plot entries along with relevant data. The database software also automatically performs all relevant statistical analyses. For the reader who has never known life without computers, it is difficult to fathom the monumental volume of plant breeding work that has been supplanted by advances in information technology (Bliss, 2006). Unfortunately, the resulting time savings have often not benefited the plant breeder directly but instead reduced resource requirements and the project workforce.

Has information technology (IT) made it easier to evaluate plant and population performance? For measurable physical traits such as yield, size, color, and even flavor and aroma, machines now exist that can do the job faster than humans. Molecular markers such as QTL are even more amenable to analyses by computers and software (Collard et al., 2005). In most cases, electronic sensors interfaced with computer hardware and software can evaluate phenotypes more accurately than humans, but many physical traits still need the human eye and brain to impute a value. Agronomic grain crops breeders often have access to sophisticated plot combines that clean and weigh samples, automatically recording the data according to GPS coordinates. Horticultural plant breeders usually do not have access to the full range of tools that are available to agronomic crop species breeders since horticultural phenotypes tend to be more specialized and qualitative. No IT machine can yet match or exceed the speed and accuracy of the human senses and brain for evaluation of subtle, qualitative attributes.

HISTORY OF THE PEDIGREE METHOD

The pedigree method is derived from two basic concepts: hybridization and pure-line selection. Plant breeders have long used hybridization for the improvement of self-pollinated grain crops such as wheat. William Farrer of Australia developed many important wheat varieties during the late 19th century using a precursor of the pedigree method that featured hybridization to combine desirable characters of two cultivars (Hayes et al., 1955). Simultaneously, several plant breeders in Europe reported similar results circa 1890 (Newman, 1912 cited in Fehr, 1987; Jensen, 1988). The term "pedigree" is one alternative strategy that incorporates the segregation and independent evaluation of a set of descendants following hybridization (the other being the bulking of populations; Jensen, 1988).

The genetic basis of pure lines was first articulated by Johannsen (1903, 1909; see Chapter 1) who attributed true-breeding to homozygosity. The veracity of the Pure Line Theory was most convincingly demonstrated by East (1935a, b) in his experiments on corolla tube length in the self-pollinating species *Nicotiana rustica*. The pedigree method was highly praised in the seminal plant breeding textbook first published by Allard (1960). Therein, the method was claimed to "allow the greatest exercise of skill of any method used for self-fertilizing crops". This endorsement and many reports of successful applications undoubtedly influenced many subsequent plant breeders to accord high credibility to the pedigree method.

CHOICE OF PARENTS

Most breeding methods designed for self-pollinated crop species, including the pedigree breeding program, begin with a cross of two parents. The parents are chosen based on breeding objectives, knowledge base, and available germplasm (Chapter 7; Allard, 1960; Simmonds, 1979). The pedigree program is most appropriate in cases where the genetic determinants for two or more characteristics from separate germplasm sources are combined into a single genome and adequate resources are available to support associated time, space, and labor requirements. In many or most cases, the targeted traits are quantitatively inherited. Additionally, the traits may have low to moderate heritability (e.g., 0.3–0.7). If the aim is to transfer 1–3 simply-inherited traits from one population to another, the backcross method (Chapter 18) is more appropriate than the pedigree method.

The parents are chosen not only based on the targeted traits, but particularly on the "background" of other characteristics that render them of commercial value (Jensen, 1988). The range of acceptable phenotypes is extremely narrow in the realm of contemporary crop cultivars, especially for highly evolved crops such as cotton, wheat, corn, and soybeans. If the end product of a breeding program does not possess the fundamental attributes that are acceptable to established markets, it will not matter whether success in combining targeted new traits of interest is achieved.

Ideally, the genetic "background" of both parents is close to acceptable commercial type. The background is the general phenotype exclusive of the traits targeted by the breeding program, although such traits (or associated QTL) are selected for and against during the execution of the program. In practice, one parent often possesses the best available combination of background attributes while the other parent is not quite of commercial quality but contributes genes for a characteristic that has been identified as critically important in the needs assessment to establish the demand and business viability of the project (Brown et al., 2014).

The following two examples will be used to illustrate the features of the Pedigree Method and how it contrasts with other strategies (see Chapter 14). *Example 1*: The best peach cultivar being grown commercially is subject to excessive fruit drop under cool, moist conditions. A breeding line has been identified that is mostly acceptable, and also has significantly reduced cool/moist fruit drop. *Example 2*: The best tomato cultivar is high yielding and great tasting, but too soft to be shipped for long distances. Another tomato cultivar or breeding population is close to commercial type in terms of plant habit and fruit size and shape, but not as high yielding or of as high quality. The second population, however, exhibits firm fruit that can withstand the rigors of long-distance shipping. In these hypothetical examples, both fruit drop in peach and fruit firmness in tomato are quantitatively inherited. The pedigree program strives to combine the best peach genetic background with low fruit drop, or the best tomato genetic background with firm fruit.

THE METHOD

The starting point of a pedigree breeding program is the hybridization of two genetically complementing parents followed by generations of recurrent self-pollination of individuals until genetic fixation is achieved (Fig. 13.8). Desirable alleles contributed by each parent are selected in segregating individuals and families along the way.

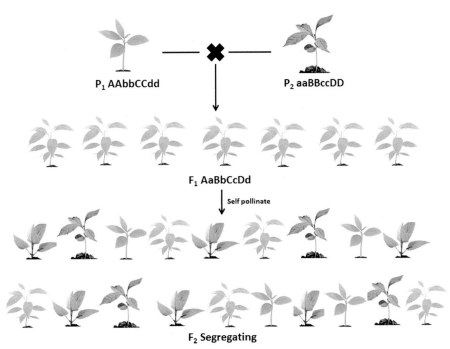

FIG. 13.8 The first steps of the pedigree breeding program consist of the hybridization of complementing polyhomozygous parents to generate a polyheterozygous F_1. A small population of F_1 plants is grown principally to observe the interaction of alleles (dominance effects) and produce adequate quantities of F_2 seeds. A relatively large F_2 population is then grown to observe the initial segregation of targeted traits and to generate seeds of F_3 families.

Fixation is presumed to have been reached by the F_7–F_9 depending on the crop species and circumstance. The duration of each selfing cycle depends on the mating system of the targeted species and the measures taken to economize on elapsed time within and between generations. Seed and plant dormancy often stymies efforts by the plant breeder to squeeze in as many generations over time as possible.

A breeding cycle or generation may take as long as a year for an annual crop species and up to 10 years for a woody perennial crop species. Thus, the entire program may take as long as 10–40 years or more to the point of producing a candidate new cultivar, followed by additional time for rigorous testing. An enterprising plant breeder may be able to shorten the generation time, but this usually requires resource-draining inputs, such as controlled environments (e.g., greenhouses) or nurseries in distant lands that feature suitable climates and growing factors (soil, available labor, etc.). Care must be exercised in exerting selection on populations growing under conditions not representative of the area of targeted production. Selection under these circumstances should be limited to traits of high heritability (including QTL for complex quantitatively inherited traits).

The two parents are crossed to produce a F_1 generation (Fig. 13.8). The original cross usually is accomplished many times, since a relatively large F_2 population is often essential for success. In cases of low pollination efficiency, as low as one seed per individual pollination, the cross may need to be repeated hundreds of times to produce adequate quantities of seeds to meet the needs of the project. An example of such as case is peach, where one pollination results in only one F_1 progeny. A breeding program may call for a population of 100–200 hybrid individuals, necessitating the same number of crosses. For tomato, where a single pollination yields up to several hundred seeds, the cross may only need to be performed once.

If both parents are genetically fixed, all F_1 plants from within and among crosses should be identical. While it is important to note the phenotype of the F_1 as compared to the parents to discern intra- and inter-genic interactions and to understand modes of inheritance of the traits under pursuit and/or degree of heterosis, the primary function of the F_1 generation is to combine all of the genes of interest into one diploid genome. Selection is never applied to F_1 populations unless the original parents were not fixed resulting in the appearance of undesirable segregants.

The F_1 is selfed to give rise to an F_2 population. Both peach and tomato are primarily autogamous, so the plant breeder only needs to harvest seeds from F_1 plants to obtain the F_2. If the original parents were genetically fixed, and all F_1 individuals from different crosses are identical, it does not matter whether some of the matings result from outcrossing. If the original parents were not genetically fixed, however, F_1 individuals will be different from one another, and the outcome of self-pollinations and outcrosses from those F_1 individuals will also be different.

The F_2 generation is a major point of genetic divergence, or segregation (Simmonds, 1979). Generally, a relatively large population of F_2 plants must be observed to identify and select those with the highest potential to deliver the desired improved population. For a single polymorphic locus A, the parents are AA and aa, the F_1 Aa, and the F_2 segregates 1AA:2Aa:1aa. For two loci, there are nine genotypic classes, for three there are 27, etc. (=3^n; n = number of polymorphic loci; Table 13.1). If ten or more polymorphisms are involved, the number of F_2 genotypic classes is greater than 60,000. If a population of 100 F_2 individuals is scrutinized, therefore, joint segregation of only a maximum of three to four polymorphic genes may be fully recovered. Most parents in a pedigree breeding program will be polymorphic at a much larger number of loci, so 100 F_2 individuals is a woefully inadequate sample of the potential range of genetic recombinants. Hence, the need for large populations is obvious. How large? The population should be as large as possible, up to the theoretical maximum (plus the statistical interval of certainty; Allard, 1960; MacKey, 1963; Kang and Namkoong, 1988; Kuckuck et al., 1991; Singh, 2006). What separates successful from unsuccessful plant breeders is the ability to forge effective compromises on resource needs such as population size, space, and labor. Will it be better to scrutinize 100 F_2 individuals from each of 50 different crosses, 1000 each from 5 crosses, or 5000 from one? The answer to this question is: "It depends".

Depending on the complexity of inheritance patterns, the selection pressures on F_2 populations should be moderate, weak, or absent. Pedigree breeding theory generally specifies little or no selection on F_2 populations to allow for the drive to fixation to diminish V_D and linkage disequilibria (Reed, 2009). While these types of intra- and intergenic interactions may ultimately be used beneficially in commercial populations, they tend to occlude the ability to surmise genotype based on phenotype (i.e., erode narrow sense heritability; Simmonds, 1979). In our examples, the inexperienced breeder may be tempted to select F_2 plants that exhibit desirable combinations of attributes but is disappointed when the results of these selections do not deliver desirable individuals in later generations. The experienced plant breeder, however, is always driven to seek efficiencies, and will inevitably practice selection where and to what extent it may enrich breeding populations for allelic combinations of highest potential.

Each F_2 plant is a gateway to a line of genetically unique descendants (Figs. 13.6 and 13.8). If any potentially unique genotype in the F_2 is not included or is eliminated by spurious selection, an entire set of potentially valuable derivative genotypes is also eliminated. Therefore, each F_2 individual is treated as a source of potentially valuable pure lines, even if it is phenotypically undesirable. All F_2 descendants (e.g., F_3, F_4, etc.) will thereafter be treated not only as individuals, but also as families. Each F_3 individual is related to other F_3 individuals by virtue of F_2 parent, the F_4 to the F_3 parent, etc. (Fig. 13.6). As the drive to fixation progresses, emphasis gradually shifts from individuals to families, since the siblings that comprise them will be more and more similar, until the F_{7-9} when individuals within families tend to become phenotypically indistinguishable.

Each selected F_2 plant is selfed to produce a F_3 family (Fig. 13.9). The pedigree method, along with others that pertain primarily to self-pollinated crops, was first developed mainly by breeders of small grains such as wheat, barley, rice, and oats. The floral structure in these economic species is botanically a spike consisting of individual flowers borne on a central rachis, but is often referred to as a "head." When a F_2 plant, or any small grain plant, is allowed to self pollinate, the seeds on the spike constitute the F_3. The entire F_3 family is often subsequently planted in a common row, historically called a "head row." This strategy also adapts well for self-pollinated horticultural crop species. By planting families of individuals together, it is easier to spot genetic tendencies. If the heritability of a trait under selection is low, it may be possible to discern the presence of underlying genes by examining progeny families. Recall that one method of estimating heritability is the regression of progeny and parent performance (Chapter 3).

The F_3 generation is the next step towards the recurrent recombination and segregation of genes in the F_2 parent that will ultimately become fixed. For each polymorphic locus in the original parents, the F_2 segregates 1:1 homozygotes:heterozygotes. All of the selected F_2 individuals that are already homozygous at a given locus will not change from the F_2 to the F_3. If the selected F_2 individual is heterozygous, then the F_3 will segregate 1:2:1, or 25% P_1P_1, 50% P_1P_2, and 25% P_2P_2. On average, for each of all loci that are heterozygous in the F_1, an F_3 individual will be 75% homozygous.

Most experienced plant breeders resist the inclination to apply strong selective pressures in the F_3. Since 25% of each polymorphic locus is still heterozygous, a considerable degree of recombination and segregation is yet to come in subsequent generations, and V_D is still a considerably large proportion of V_P. Linkages in repulsion need more time and opportunities (i.e., larger populations) to be converted to linkages in coupling. Ideally, no selection at all would be practiced, but in reality, some "intermediate" selection (<50%) is inevitably necessary to reduce the size of the overall population being carried (Fig. 13.6). With experience, the plant breeder instinctively recognizes individuals and families that have little or no potential to deliver the targeted phenotype or ideotype. Selection is practiced both among and within F_3 families since individuals are still segregating within families.

Selected F_3 individuals are then selfed to give the F_4 (Fig. 13.10). The precise steps that are used in this regard vary between programs. The total number of potential F_4 families based on the original F_2 population increases

F₃ Families : "Soft" selection within families

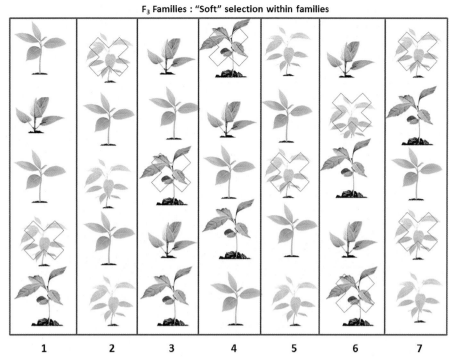

1   2   3   4   5   6   7

FIG. 13.9   The F₃ is generated from self-matings of F₂ individuals. Each F₂ individual gives rise to an F₃ "family." "Soft" selection (<20%) on targeted breeding objective traits may be practiced, but selection should be kept to a minimum to avoid the elimination of potentially valuable individuals due to V_D or repulsion linkages.

F₄ families: Slightly harder selection within families

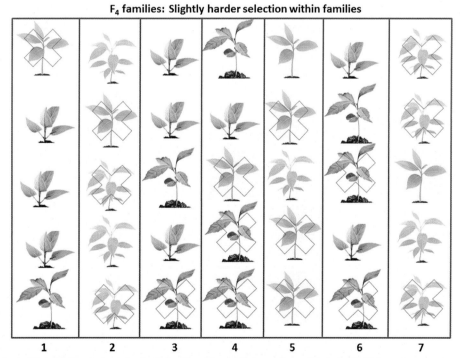

1   2   3   4   5   6   7

FIG. 13.10   Each selected F₃ individual gives rise to F₄ families. V_D and linkage disequilibria in the F₄ have decreased as compared to the F₃ so that selection pressures may increase accordingly, both within and, to a lesser extent, among F₄ families.

exponentially. Each of a theoretical F₂ population of 100 could give rise to 100 F₃ individuals, or 10,000 total. If 100 F₄ individuals are to be observed from each F₃ individual, the total number climbs to 1,000,000. From there, we proceed to the F₅, F₆, F₇ and beyond in a similar progression (Figs. 13.11–13.13). Progressively stronger selection pressures may be applied, and more emphasis shifts from within to among families as individuals approach fixation and

F₅ families: Slightly stronger selection within and among families

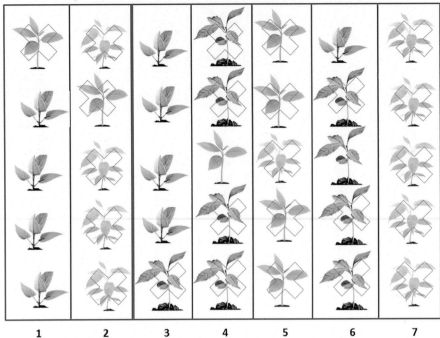

FIG. 13.11   Each selected F₄ individual gives rise to F₅ families. V_D and linkage disequilibria in the F₅ have decreased as compared to the F₄ so that selection pressures may increase and be shifted from "within" to "among" F₅ families.

F₆ Families: Strong selection within and among families

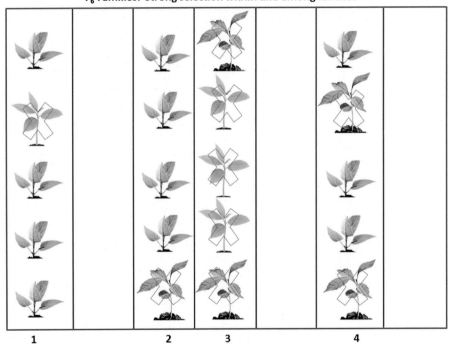

FIG. 13.12   Each selected F₅ individual gives rise to F₆ families. V_D and linkage disequilibria in the F₆ have decreased further as compared to the F₅ so that selection pressures may increase and be shifted progressively more from "within" to "among" F₆ families.

individuals within families are more and more similar. It is evident that limited population sizes and increasingly stronger performance selection are necessary to achieve affordable and feasible program parameters.

To reiterate, the pedigree method is associated with comprehensive and consistent records of the relationships of individuals and populations and their performance. The amount of work necessitated by the pedigree strategy varies by program and according to the individual breeder's style but is almost always more information-intensive than

F$_7$ Families: Strong selection among familes

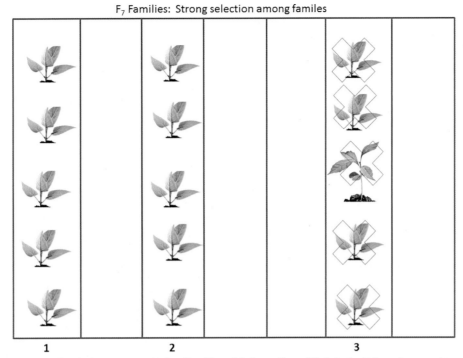

1   2   3

FIG. 13.13 Each selected F$_6$ individual gives rise to F$_7$ families. V$_D$ and linkage disequilibria in the F$_7$ have decreased to nearly 0, so all selection pressures may increase.

other strategies for self-pollinated crop species. What are the benefits of this supplemental information? These data provide the breeder with elevated confidence in applying hard selection pressures. The difficulty with any breeding program for inbreeding species is to dispatch unworthy genotypes quickly and to establish genetic fixation with the best genotypes as quickly as possible. The pedigree method makes use of a historical biological paradigm: the performance of progeny is the best measure of the potential of the corresponding parent. If most progeny from a certain lineage perform poorly, it is highly likely that the ancestors did not contribute worthy genes. Therefore, most or all of the populations or individuals in that lineage should be eliminated from further consideration.

What is the relative importance of parental selection, the number of breeding populations, and the size of each population in a pedigree breeding program? Also, what are the optimum combinations between the number and size of breeding populations? It was demonstrated that selection response was largest when the maximum number of breeding populations was used (Kang and Namkoong, 1988). The effect of the number of breeding populations was minor, however, when selection was practiced among the parents or when heritability was less than 1.0. Large selection responses could be obtained with a wide range of combinations between number and size of breeding populations (Bernardo, 2003).

PEDIGREE BREEDING EXAMPLES

Let us return to the examples described earlier pertaining to (i) fruit drop resistant peach; and (ii) increased tomato fruit firmness. In example i, time frames and population sizes are significantly skewed within the pedigree breeding program to provide for practical realities of plant size and generation time. Resource limitations dictate that population sizes must be modulated since commercial densities for fruit trees may be up to 1000 times less than that for annual crops. Each generation may take up to 5 or more years to complete, resulting in relaxed standards for phenotypic uniformity; or genetic fixation. The F$_4$ or F$_5$ may be judged to exhibit acceptable levels of phenotypic uniformity where the F$_7$ or F$_8$ is a requirement for most annual crop species. Most importantly, the pedigree strategy may be short-circuited if a superior phenotype is selected and propagated asexually (see Chapter 17).

In our example, the standard peach cultivar, susceptible to fruit drop, was crossed with an otherwise mediocre parent that exhibited fruit drop resistance. Among 50–100 F$_2$ individuals from this cross, V$_P$ and phenotypic values would generally vary tremendously, ranging from the phenotype of one parent to the other (and sometimes exceeding this range). The 25–50 F$_2$ individuals that embody the best combination of overall performance and fruit

TABLE 13.5 Comparison of population sizes and selection pressures for hypothetical pedigree breeding programs of peach (example 1; absence of cold-induced fruit drop) and tomato (example 2; enhanced fruit firmness)

	Peach					Tomato				
		Families and indiv.		Selection pressure			Families and indiv.		Selection pressure	
	Population[a]	No. families	Indiv. per family	Within families	Among families	Population[a]	No. families	Indiv. per family	Within families	Among families
P_X	5	NA	1–5	NA	NA	10	NA	10	NA	NA
F_1	1–5	1	1–5	1.00	NA	10	1	10	1.00	NA
F_2	50–100	1	50–100	0.50	NA	500–1000	1	500–1000	0.50	NA
F_3	125–250	25–50	5	0.20	0.25	2500–5000	250–500	10	0.15	0.15
F_4	100–200	20–40	5	0.20	0.20	3750–7500	375–750	10	0.10	0.10
F_5	50–100	4–8	5	0.20	0.20	380–750	38–75	10	0.10	0.05
F_6	5–15	1–2	5	0.00	0.50	20–40	2–4	10	0.10	0.50
F_7	5	1	5			10	1–2	10	0.00	0.5

[a] No. families × no. plants per family.

drop resistance are selected (Table 13.5). From each of the 25–50 selected F_2 individuals, five F_3 individuals are harvested. Each of the 25–50 families of five F_3 individuals (125–250 trees total), related to each other by a F_2 parent, is grown in the same physical vicinity, usually in the same orchard. It is crucial to grow breeding populations and perform selections in the area of adaptation with regard to climate, soil type, endemic pests, etc. to ensure that the end result will be adapted to the targeted geographical region.

The plant breeder scrutinizes both F_3 families and individuals for the best combination of general horticultural attributes and fruit drop resistance. Since V_D and linkage disequilibria are still relatively high in the F_3, a "soft" selection comprised of the top 20% of individuals is applied from within F_3 families. Hypothetically, this equates to approximately 20–40 individuals advanced to the F_4. If five F_4 individuals are obtained from each selected F_3 parent, the total number of prospective trees in the program is reduced to 100–200 (Table 13.5).

In example ii (enhanced tomato fruit firmness), project parameters pertaining to population sizes and generation times are more comparable to the small grain examples presented in many previously published plant breeding textbooks. Commercial population densities for market tomato are in the range of 5000–10,000 per acre, an order of magnitude less than the planting densities for small grains. Similarly, as for the peach fruit drop example, the standard tomato variety with soft fruits is hybridized with a parent that produces firmer fruits. The firmness donor parent will, ideally, be as close to the general background, fruit type, maturity, etc. as the standard tomato variety.

A F_2 population size in the range of 500–1000 is feasible and practical for a pedigree breeding program involving large-fruited, or market tomato. Assuming a selection intensity of 50%, 250–500 F_2 individuals are selected to contribute selfed progeny to the F_3. Since the F_3 is still segregating from the F_2 (50% heterozygous per locus per F_3 individual), larger populations of F_3 families are desirable, perhaps 20–50 plants per family. For reasons of cost feasibility, market tomato breeding population, or plot sizes are usually maintained at ten (Table 13.5) giving a total number of 2500–5000 F_3 individuals. Modulation of population sizes to reduce resource needs in tomato breeding programs is, however, very common.

The market tomato breeder scrutinizes both the F_3 families and individuals within F_3 families for the best combination of general horticultural attributes and fruit firmness. Analogous to the peach example, V_D and linkage disequilibria are still relatively prominent in the F_3, the top 25% of individuals are selected from within the top 15% of F_3 families. Hypothetically, this is approximately 375–750 individuals advanced to the F_4. If ten individuals per F_4 family are derived from each selected F_3 parent, the total number of prospective F_4 plants will be 3750–7500.

Returning to example i, 20–40 selected F_4 families of five individual trees each are planted together in a breeding orchard and scrutinized for performance. The F_4 generation is on the cusp of homozygosity (87.5% per polymorphic locus). A modicum of heterozygosity persists in the F_4 and selection should be applied both among individuals within families and among families. The F_3 families may also be traced back to F_2 ancestors. If the selections performed on the F_2 and F_3 generations result in subpar phenotypes, the collective performance of all F_4 individuals that trace back to a F_2 ancestor may reveal the source of undesirable genes, and permit the breeder to eliminate the entire lineage. This is an example of useful collateral information that may be obtained during the execution of the pedigree breeding method.

If 20% of peach trees are selected from 20 to 40 F_4 families, the total number of F_4 individuals selected to comprise the F_5 is 20–40 (Table 13.5; see Fig. 13.11 for an illustration of the transition from the F_4 to the F_5). The F_5 isapproaching fixation and selection will shift progressively from "within families" to "among families". Empirically, five F_5 individuals are obtained from each of the four to eight selected F_4 plants, or 20–40 total. The total number of individuals in the program decreases while the proportion of desirable genotypes increases.

In example ii, the 375–750 selected market tomato F_4 families of ten individual plants apiece are planted in a field nursery and evaluated for plant and fruit characteristics. F_4 plants are then selected from within and among families at an intensity of approximately 10% (i.e., 10% of families; 10% of individuals within families), the total number of F_4 individuals selected to continue to the F_5 is about four to eight (Table 13.5). The F_5 is closer to fixation than the F_4, so selection will shift progressively from "within families" to "among families." The 38–75 F_5 market tomato families are, once again, evaluated in a breeding nursery located in the targeted area of cultivar adaptation. The selection pressure intensifies in the F_5 to 5% among families. Selection will result in two to four families, one plant per family, advanced to the F_6 (Table 13.5).

Returning to example i, the four to eight F_5 peach families are planted in a breeding orchard and scrutinized for performance. Were the F_4, F_3, and F_2 selections that led to these selected lineages good choices? Did the ancestral selections lead to descendants that exhibit consistently excellent performance? If not, the entire F_2 lineage may be discarded. Selection in the F_5 is a continuation of the trend from previous generations: more severe among families and less severe within families. 20%, or one to two, of F_5 families are selected in our hypothetical example, and 20% of individuals within these selected families (i.e., one per family) are also selected to serve as parents for the F_6, giving rise to five to ten F_6 individuals (Table 13.5).

The two to four selected market tomato (example ii) F_6 families of ten individual plants apiece are, once again, planted together in a breeding nursery and evaluated. Plants are then selected from within and among F_6 families at 50% intensity among families and 10% within families. The total number of F_6 individuals selected to continue to the F_7 is one to two (Fig. 13.13). The F_7 is, under most circumstances, considered to be fixed, and is advanced directly into cultivar trial testing and seed production.

The F_7 is, on average, 98.4375% homozygous per polymorphic locus per individual. Most pedigree breeding programs would be considered finished at this juncture, ending with the identification of candidates for rigorous varietal testing (see Chapter 11). In example i (peach fruit drop resistance) let us surmise that five F_7 progenies are obtained from each selected F_6 tree. The five to ten F_7 individuals (five each of one to two selected F_6 parents) constitute a mother block for the generation of clonal populations from cuttings. Populations of clonally-propagated derivatives from F_7 individuals are evaluated in orchards replicated over time and space.

In example ii, seeds from self-pollinations are harvested from selected firm-fruited tomato F_7 individuals selected for advancement into the replicated testing phase.

USE OF QTL AND MAS TO ENHANCE PEDIGREE METHOD EFFECTIVENESS

The development and employment of molecular markers as surrogates for linked or pleiotropic genes associated with desirable whole plant phenotypes was discussed in detail in Chapter 8. Marker-assisted selection (MAS) is of particular significance as an adjuvant to "conventional" pedigree breeding when the markers (QTL) are co-dominantly inherited. For tightly linked co-dominant markers that have been comprehensively and accurately mapped to the point of genome saturation (i.e., QTL account for virtually 100% of V_P), MAS has a profound impact on the design and execution of a pedigree breeding program.

Co-dominant QTL may be used to select not only for the linked genomic determinants of targeted traits, but also for genomic configuration attributes such as fixation (Breseghello and Coelho, 2013; Rosyara et al., 2013). In other words, both controlled mating and selection may contribute to the drive to fixation and selection may be used concomitantly for both phenotype and genotype fixation. Integration of QTL may begin as early as the F_2 where rare, desirable genotypes may be identified while avoiding the effects of V_D and linkage disequilibrium. Theoretically, it is possible to obtain desirable homozygous candidates in the F_3 through the intensive and effective integration of MAS and multiple linked QTL.

The availability of a broad range of segregating populations of defined familial relationships, such as is generated by a pedigree breeding program, can also be of immense value in the identification and mapping of QTL (Jannink et al., 2001; Bink et al., 2002; Wu et al., 2003; Crepieux et al., 2004; Arbelbide et al., 2006). The innate design of the pedigree strategy, including the detailed records of genetic relationships of individuals and phenotypic breeding values, has been very useful for this purpose. Database catalogs of QTL and trait prediction coefficients will lead to the widespread use of QTL for the rapid and efficient breeding of all major crop species.

PEDIGREE METHOD: OTHER CONSIDERATIONS

Accurate records of the performance of parents and F_1 and all selected descendants from each individual cross will not only provide crucial information for devising aggressive and effective selection strategies but will likely reveal data that pertain to the genetic basis of the traits under selection. For public domain plant breeders, where publication of results is a requirement, collected data and any resulting inferences are important for expanding the base of knowledge in addition to career advancement. A better understanding of the heritable basis of horticultural traits and relationships of whole-plant phenotypes to genomic DNA sequences imparts a solid foundation for breeding programs. Often, the knowledge and breeding lines developed in one breeding program are used as the basis for the initiation of subsequent projects.

While one breeding program is in progress there are likely many other projects that are proceeding in parallel. Programs at the F1, F2, F3, etc. generations are in progress at any given point in time. Many different crosses are often accomplished leading to independent pedigree programs aimed at the same general result (e.g., improved peach tree architecture, firmer shipping tomatoes). The relationships of the parents may also be taken into consideration, and intermediate progeny compared between competing programs. It would not be unusual for an entire program to be abandoned if a competing, parallel program were generating more promising results.

It is also likely that other projects are scattered in time, especially considering the long-term nature of the work and the need for a constant flow of results and improved varieties. In any given year, 50–5000 new crosses may be done, 50–5000 F_1 populations increased to F_2 status, 50–5000 F_2 arrays evaluated, and various F_3, F_4, etc. generations evaluated and selected. If all of the programs are strictly pedigree programs, the plant breeder must surely be truly blessed with a multitude of resources: space, labor, equipment, and, of course, money. It is more likely that a few problems are being addressed that are most amenable to the pedigree strategy and other, less intense strategies (see Chapter 14), are utilized for other objectives.

Where and when is the pedigree method most appropriate and effective? Alternative strategies for self-pollinated crop species will be described and analyzed in Chapter 14 along with the relative strengths and weaknesses of each as compared with the pedigree method. The pedigree method is generally preferable to alternative methods because it is the most grounded in science and also generates more information as a byproduct of the program. The data serve as the informational bedrock on which to base rational decisions.

Selection of desirable genotypes begins at the F_2 and is promulgated during each generation until fixation is reached, and selection is most effective if applied under the same conditions under which the crop will be grown commercially. If the traits involved are simply inherited or have high heritability under diverse environments, other selection methods such as controlled environments or distant winter nurseries may offer better efficiencies. Sufficient resources must be available to accommodate targeted population sizes and growing costs, and information management requirements, including software licenses.

In reality, each pedigree program is adapted to fit the particular and unique needs and challenges of crop species and breeding objectives. Further adjustments are inevitable to address unanticipated issues that are encountered along the way. Decisions must be made on the fly, and the plant breeder never quite knows how and when a given breeding program will be concluded. Resources counted on when a program is begun may dry up midstream, and the plant breeder must quickly change directions if a finished product is to come of all the hard work invested. Elements of several different strategies are often combined into a single program. Alternatively, a program may begin as a pedigree program and is then adapted to a different approach in later stages, or may start with a less resource-intense strategy and switched to a pedigree program later on. The "proof is in the pudding" as the saying goes. If the pudding does not turn out well, it is often advisable to have the recipe written down to show to detractors, and to use in forging future improvements in the method.

References

Acquaah, G., 2012. Principles of Plant Genetics and Breeding, second ed. Wiley/Blackwell, New York, NY. 756 pp.

Allard, R.W., 1960. Principles of Plant Breeding. Wiley & Sons, New York. 485 pp.

Arbelbide, M., Yu, J., Bernardo, R., 2006. Power of mixed-model QTL mapping from phenotypic, pedigree and marker data in self-pollinated crops. Theor. Appl. Genet. 112 (5), 876–884.

Bernardo, R., 2003. Parental selection, number of breeding populations, and size of each population in inbred development. Theor. Appl. Genet. 107 (7), 1252–1256.

Bink, M.C.A.M., Uimari, P., Sillanpaa, M.J., Janss, L.L.G., Jansen, R.C., 2002. Multiple QTL mapping in related plant populations via a pedigree-analysis approach. Theor. Appl. Genet. 104 (5), 751–762.

Bliss, F., 2006. Plant breeding in the U.S. private sector. HortScience 41 (1), 45–47.

Breseghello, F., Coelho, A.S.G., 2013. Traditional and modern plant breeding methods with examples in rice (*Oryza sativa* L.). J. Agric. Food Chem. 61 (35), 8277–8286.

Briggs, F.N., Knowles, P.F., 1967. Introduction to Plant Breeding. Reinhold Publ. Co., New York, NY. 426 pp.

Brown, J., Caligari, P., Campos, H., 2014. Plant Breeding, second ed. Wiley-Blackwell, New York, NY. 296 pp.

Collard, B.C.Y., Jahufer, M.Z.Z., Brouwer, J.B., Pang, E.C.K., 2005. An introduction to markers, quantitative trait loci (QTL) mapping and marker-assisted selection for crop improvement: the basic concepts. Euphytica 142 (1–2), 169–196.

Crepieux, S., Lebreton, C., Servin, B., Charmet, G., 2004. IBD-based QTL detection in inbred pedigrees: a case study of cereal breeding programs. Euphytica 137 (1), 101–109.

East, E.M., 1935a. Genetic reactions in *Nicotiana*. I. Compatibility. Genetics 20, 403–413.

East, E.M., 1935b. Genetic reactions in *Nicotiana*. II. Phenotypic reaction patterns. Genetics 20, 414–442.

Fehr, W.R., 1987. Principles of Cultivar Development Volume 1: Theory and Technique. MacMillan Publ. Co., New York, NY. 536 pp.

Fleury, D., Whitford, R., 2014. Crop Breeding: Methods and Protocols. Humana Press, New York, NY. 255 pp.

Hayes, H.K., Immer, F.R., Smith, D.C., 1955. Methods of Plant Breeding. McGraw-Hill, New York. 551 pp.

Jannink, J.L., Bink, M.C.A.M., Jansen, R.C., 2001. Using complex plant pedigrees to map valuable genes. Trends Plant Sci. 6 (8), 337–342.

Jensen, N.F., 1988. Plant Breeding Methodology. John Wiley & Sons, New York, NY. 676 pp.

Johannsen, W., 1903. Ueber Erblichkeit in Populationen und in reinen Linien. G. Fischer Verlag, Jena.

Johannsen, W., 1909. Elemente der exakten erblichkeitslehre. G. Fischer Verlag, Jena.

Kang, H., Namkoong, G., 1988. Inbreeding effective population size under some artificial selection schemes. 1. Linear distribution of breeding values. Theor. Appl. Genet. 75 (2), 333–339.

Kuckuck, J., Kobabe, G., Wenzel, B., 1991. Fundamentals of Plant Breeding. Springer-Verlag, Berlin GER. 236 pp.

MacKey, J., 1963. Autogamous plant breeding based on already highbred material. In: Akerberg, E., Hagberg, A. (Eds.), Recent Plant Breeding Research, Svalof, 1946–1961. John Wiley & Son, New York, NY.

Poehlman, J.M., Sleper, D.A., 2013. Breeding Field Crops, third ed. Iowa State Univ. Press, Ames, IA. 566 pp.

Reed, D.H., 2009. When it comes to inbreeding: slower is better. Mol. Ecol. 18 (22), 4521–4522.

Rosyara, U.R., Marco, C.A., Bink, M., van de Weg, E., Zhang, G., Wang, D., Sebolt, A., Dirlewanger, E., Quéro-Garcia, J., Schuster, M., Iezonni, A.F., 2013. Fruit size QTL identification and the prediction of parental QTL genotypes and breeding values in multiple pedigreed populations of sweet cherry. Mol. Breed. 32 (4), 875–887.

Simmonds, N.W., 1979. Principles of Crop Improvement. Longman Group Ltd., New York, NY. 408 pp.

Singh, B.D., 2006. Plant Breeding: Principles and Methods. Kalyani Publishers, New Delhi (INDIA). 1018 pp.

Thompson, E.A., 1985. Pedigree Analysis in Human Genetics. Johns Hopkins Univ. Press, Baltimore, MD. 240 pp.

Wu, J., Jenkins, J.N., Zhu, J., McCarty Jr., J.C., Watson, C.E., 2003. Comparisons of quantitative trait locus mapping properties between two methods of recombinant inbred line development. Euphytica 132 (2), 159–166.

Further Reading

Chahal, G.S., Gosal, S.S., 2002. Principles and Procedures of Plant Breeding: Biotechnological and Conventional Approaches. Narosa Publishing, New Delhi. 604 pp.

14

Other Breeding Methods for Self Pollinated Plant Species

INTRODUCTION

The pedigree method is considered the archetype for breeding self-pollinated crop species. Other strategies that offer potential advantages for certain situations will be presented in this chapter. Every perceived advantage can also be considered a disadvantage depending on the context of the application. It is judicious to consider a prospective breeding program for a self-pollinated crop species first with the pedigree strategy in mind and subsequently make adjustments or employ alternative strategies as the situation dictates. The three alternative strategies that will be addressed in this chapter are the *bulk population method*, the *single seed descent method*, and the *doubled haploid method*. The first two methods were developed in the early 20th century CE for agronomic species and have historically been taught and practiced for over 50 years (Allard, 1960), but the doubled haploid method is more recent; developed and practiced since the mid-1970s (Borojevic, 1990; Pickering and Devaux, 1992; Baenziger, 1996). These strategies all share one important attribute: they are all less resource consumptive than is the pedigree method in one way or another. The reduction of resource inputs, of course, always portends a corresponding sacrifice, the consequences of which will be estimated in terms of precision, risks, opportunities for integration of technologies such as MAS, the genetic potential of the end product, and collateral information capture.

THE BULK POPULATION METHOD

The *bulk population* strategy is similar to the pedigree method in structure and time requirements but without detailed records or familial performance data. The method begins and ends the same ways as for the pedigree method. Two parents, chosen for the potential for complementation of genotypes, are hybridized, and the resulting F_1 hybrids are grown to produce the F_2 population. At this point in the program, the bulk population and pedigree methods diverge. A similar number of F_2 individuals may be produced for the bulk population as in the pedigree method.

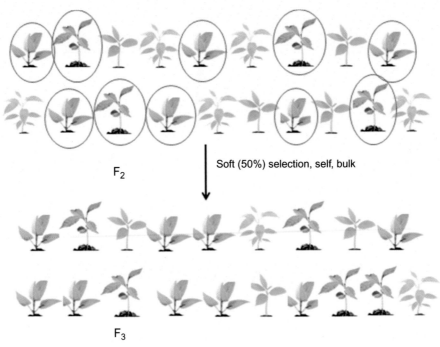

FIG. 14.1　Advancement of F_2 selections to the F_3 under the bulk population method.

The F_2 population for both the pedigree and bulk population methods may be subjected to a "soft selection" or no selection at all, and seeds are harvested from each individual that will contribute to the F_3 (Fig. 14.1). The number of seeds per F_2 plant varies according to the dictates of the program. Because the bulk population method is applied instead of pedigree as a means to reduce costs, population size is an important consideration. Therefore the success of a bulk population program rests on the size of populations, the number of distinct populations, and the manner in which successive populations are sampled.

The F_3 population may be constituted in a myriad of ways. At one extreme, the F_3 may consist of a large number of seeds from a few selected F_2 individuals and at the other extreme, the F_3 may be constituted of a few seeds from each of many individuals. The following equation relationship will be invoked to help sort out the appropriate strategy of how many selections and the number of progeny per selection to include in the succeeding generation bulk:

$$P\left(F_3 \text{ beneficial genotypes}\right) = s\left(F_2 \text{ parent}\right) + t\left(F_3 \text{ family}\right)$$

and

$st = N_3$
$P = $ probability
$N_3 = $ maximum number of individuals that can be accommodated

The plant breeder wishes to maximize $P(F_3$ beneficial genotypes) while minimizing N_3 since resource requirements are directly proportional to the total number of individuals. What is a "beneficial genotype"? It is an individual that contains at least one desirable allele at as many loci as possible that control the desirable phenotype. The equation is not intended to be a mathematical theorem, presented purely to aid the plant breeder in development of breeding strategy. The equation is presented to help the plant breeder to achieve a workable compromise on lines versus populations while also considering resource limitations. For example, a reasonable compromise may be s = 50 and $t = 50$, so $N_3 = 2500$ (in this example). If s = 100, $t = 25$, if s = 25, $t = 100$, etc. Either way, 2500 seeds are harvested from the F_2 and mixed into the same lot without regard to genetic lineage.

Bernardo (2003) showed that selection response in an inbreeding program such as bulk population was largest when the maximum number of different breeding populations (i.e., families) was used along with moderate population sizes. The effect of the number of breeding populations was minor, however, when selection was practiced among the parents or when heritability was <1.0. Large selection responses could be obtained with a wide range of combinations between number and size of breeding populations.

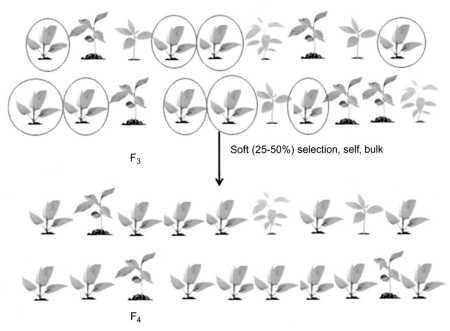

Soft (25-50%) selection, self, bulk

F_3

F_4

FIG. 14.2 Advancement of F_3 selections to the F_4 under the bulk population method.

In the example described in Chapter 13, wherein fruit firmness is to be introduced into a soft-fruited but otherwise desirable tomato, 100 F_2 individuals would be selected for fruit firmness and horticultural quality at a selection intensity of 50%, or 50 individuals chosen to contribute progeny to the F_3. Fruits are harvested from these 50 F_2 selections and seed extracted, and 50 from each fruit are bulked, or mixed, in one large population of 2500 F_3 individuals. Using a planting density typical for staked tomatoes, this would equate to approximately 0.32 acre (=0.129 ha).

The F_3 undergoes selection and is selfed to generate the F_4 (Fig. 14.2). The selection pressure should continue to be low (25–50%) since V_D and linkage disequilibria are usually still substantial within and among F_3 families. Consider the following equation relationship, similar to the one above for the F_3:

$$P\left(F_4 \text{ beneficial genotypes}\right) = u\left(F_3 \text{ parent}\right) + v\left(F_4 \text{ family}\right)$$

and

$$uv = N_4$$

The plant breeder once again strives to maximize P(F_4 beneficial genotypes) while minimizing N_4. As filial families are driven progressively toward fixation, the size of the family necessary to retrieve the total range of variability from the ancestral generation is progressively less. If s and t = 50, the equation may be maximized if u = 1000 and v = 25, representing a selection pressure of 40%. Therefore, N_4 = 25,000 in this particular example. Since the planting density for tomatoes is about 8000 plants per acre (19,768 per ha), the resulting F_4 nursery would consist of over 3 acres (1.214 ha), prohibitively expensive to establish and manage because market tomatoes are usually established from transplants and require intensive management inputs such as precise fertilizer applications, thrive under drip irrigation, need prudent pest and disease control, and staking is necessary to maximize net yield.

Consequently, if N_4 for this example is limited by the plant breeder at 1000 (1/8 acre) due to resource limitations, then hard choices must be made pertaining to u and v. For tomatoes, the main labor requirement for obtaining new generations lies in the extraction of seeds, a much more involved process than it is with small grains. Part of the reason for this is that seeds must be removed from each tomato fruit and the proteinaceous gel that encapsulates each seed must be eliminated. It would be easier to collect more seed from fewer fruits (i.e., low u, high v), but the F_3 is still highly heterozygous (25% per locus). Therefore, at least 50 fruit from different F_3 selections should be sampled (u ≥ 50). If u = 50, then v will be 20 for this example.

Breeding methods for self-pollinated crop species were developed using small grains (wheat, barley, oats, etc.) as the model for applications. It is much easier and cheaper to produce and handle small grain populations than it is for horticultural crop species such as tomato. Obtaining F_4 family seed for wheat or barley is as easy as walking

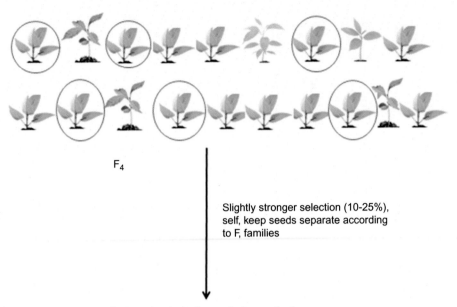

F_4

Slightly stronger selection (10-25%),
self, keep seeds separate according
to F, families

FIG. 14.3   Advancement of F_4 selections to the F_5 under the bulk population method.

down a field row and putting whole intact seed heads into a bag. Typical field densities are also much greater, up to 100,000 per acre, and virtually no management inputs are necessary after the field is prepared. Therefore, N_4 may be much larger for small grains than for horticultural crop species, perhaps 25,000 (1/4 acre), thus making it possible to accommodate larger numbers of F_3 selections (u) and F_4 family sizes (v). Most horticultural crop species require more resources for handling and propagation per unit than are small grains.

Continuing with the bulk population method, seeds from F_4 families (from F_3 selections) are again bulked, or mixed, together (Fig. 14.3). Each F_4 individual is 87.5% homozygous at each polymorphic locus, approaching fixation but still 12.5% heterozygous at each polymorphic locus.

The following equation relationship will again apply for the formation of the next, F_5 bulk population:

$$P\left(F_5 \text{ beneficial genotypes}\right) = w\left(F_4 \text{ parent}\right) + x\left(F_5 \text{ family}\right)$$

and

$$wx = N_5$$

If we, again, choose $N_5 = 1000$ for tomatoes and $N_5 = 25,000$ for small grains, the plant breeder is faced with deciding where to focus limited resources. F_5 individuals within families are 93.75% homozygous at each polymorphic locus, approaching fixation and also more similar to each other since the F_4 parent more monomorphic per locus than is the F_3 parent. Therefore, selection emphasis shifts progressively from individuals to families and the intensity increases in accordance with fixation (~5%). In our examples, hypothetical w and x for tomato may be 100 and 10, respectively, and w and x for barley may be 1000 and 25, equating to selection pressure of 10% in tomatoes for firm fruit and horticultural quality and 4% in barley for stiff straw and agronomic quality. The seeds F_5 selections are once again mixed to form the next bulk population, the F_6.

The following equation relationship pertains to maximizing breeding effectiveness in the F_6 (see Figs. 14.4–14.6):

$$P\left(F_6 \text{ beneficial genotypes}\right) = y\left(F_5 \text{ parent}\right) + z\left(F_6 \text{ family}\right)$$

and

$$yz = N_6$$

The degree of homozygosity has increased to nearly 97% per polymorphic locus among individuals and families in the F_6. Therefore, F_6 progeny within families are less and less distinct from F_5 parents, more and more similar genetically and phenotypically within F_5 families. Progressively less emphasis is placed on the size of families since constituent individuals within families are nearly identical to each other. Selection pressure is very strong, 1–2%, in the F_6. In our examples, assuming that N_i remains constant at 1000 for tomatoes and 25,000 for barley, the respective parameters should

F_5 Strong (5-10%) selection among families → F_6

FIG. 14.4 Advancement of F_5 selections to the F_6 under the bulk population method.

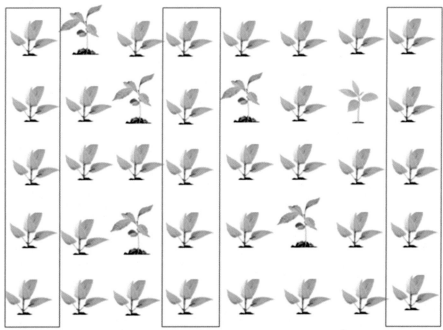

F_6 Strong (2-5%) selection among families → F_7

FIG. 14.5 Advancement of F_5 selections to the F_6 under the bulk population method.

shift even more toward the representation of F_5 parents. For the hypothetical example in tomatoes, let us suppose that $y = 250$ and $z = 4$; for barley $w = 2500$ and $z = 10$.

The selected seeds are again bulked and one to two additional cycles may be accomplished to the F_8 generation. At this juncture, allele fixation is assumed to have been reached, and families are subjected to a very intense cycle of selection to obtain among families final candidate populations to submit for rigorous testing. In the case of tomatoes, perhaps 10–25 F_7 families are screened for fruit firmness in combination with desirable horticultural type, and the best zero to five populations are chosen to test.

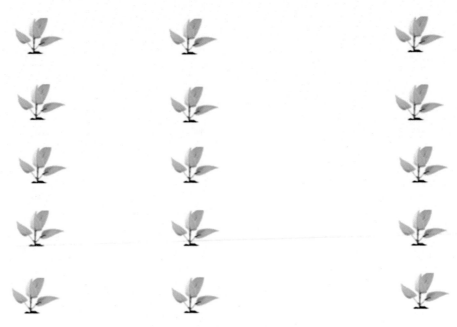

F_8 families advanced to performance trials

FIG. 14.6   Advancement of F_7 selections to the F_8 under the bulk population method.

In summary, the bulk population method proceeds by the generation of a F_1 hybrid, the polymorphisms of which are driven to fixation by cycles of self-pollination. In this regard, the bulk population method is identical to the pedigree method (Chapter 13). The bulk population method diverges from the pedigree method beginning with the F_3 generation when seeds of F_3 families are mixed instead of being kept separate to enable the tracing of F_3, F_4, F_5, etc. individuals and families all the way back to a single F_2. Such records are not typically kept when employing the bulk population method. Instead the constitution of the mixed, or bulked, populations change with progressing generations and level of fixation to maximize the probability of beneficial genotypes that will contribute genetically improved populations by altering the proportion of selected progenitor selections and their progeny. As successive cycles of self-pollination drive heterozygous loci to homozygosity, the emphasis shifts to greater numbers of families and fewer individuals within families, similar to the pedigree method.

The bulk population method is utilized in situations where resource limitations render it impossible to support labor and record-keeping requirements of the pedigree method. Limited resources are an important challenge to the plant breeder. Any plant breeder will be successful if provided with unlimited resources: labor, land, machines, time, and all are equivalent to money. What separates the average from the exceptional plant breeder is the ability to marshal limited resources into the greatest possible genetic impacts. The bulk population method represents an effective compromise in instances where resources are inadequate to support the pedigree method. The bulk population method saves substantial labor and, usually, space, but does not necessarily save time. Evidence shows that care must be exercised in applying selection pressures that are restrictive (i.e., excessively "hard") during the early generations of a bulk population program (Keim et al., 1994). Linkage disequilibria have been shown to be at least partially responsible for this need to relax selection pressure (Ibrahim et al., 1996). Since resource limitations usually lead to the adoption of the bulk population method, it is unlikely that greenhouses or nurseries in distant lands will be incorporated to double-up generations per year to reduce the time necessary to complete a program cycle.

The bulk population method is very applicable, however, in situations where the selection pressure consists of phenotypic response (e.g., tolerance or resistance) to broadly applied environmental factors such as temperature extremes, moisture extremes, soil solute extremes, pathogens or insects, etc. The environmental challenge may be efficiently applied to bulk populations during vegetative growth, then selections made efficiently by choosing vigorous, tolerant, or resistant individuals for the subsequent bulk population. In particular, allowing prolonged and successive exposure of segregating populations of a marginal-value horticultural crop species to a targeted geo-ecological environment has been shown to be an effective way to achieve cultivar adaptation (Nichols et al., 2009).

In species wherein the outcrossing rate is significant, for example, 10% or greater, the bulk population method may be preferable over the pedigree method since the assumption of self-pollination is substantially violated (see Chapter 13 for a discussion of the effects of outcrossing on the drive to fixation). The methods by which populations are managed also enable further random recombination of alleles to occur. The rigid organization of populations in the pedigree system negates this possibility.

The probability of accomplishing breeding objectives is most generally reduced in the bulk population method as compared with the pedigree method. The extent of compromise depends on the situation (species, traits under consideration), population sizes, and the skill of the plant breeder in making selections and correctly constituting bulked populations. Differences in the probability of success of pedigree vs. bulk population also diminish greatly as basic assumptions (100% autogamy, moderate to high phenotype heritabilities, independent gene assortment) are violated This reduction in the probability of a single program must be balanced by the fact that the bulk population method also may allow the plant breeder to invest conserved resources in a broader range of projects. Plant breeders often combine the pedigree and bulk population methods into single programs, using bulked populations during early generations, F_3 through F_5 then switching to the pedigree method to economize on resources.

Since current MAS protocols generally employ genomic derivatives from individuals and not populations or bulks, opportunities for integration of MAS into bulk population scaffolds are limited. The use of QTL as surrogate selection criteria for breeding self-pollinated species is particularly compelling in early generations to overcome the mitigating effects on advances from selection on phenotypic variability due to dominance and linkage disequilibrium. Research reports have appeared that support the benefits of integrating MAS into early-generation bulk population programs. For example, by applying very strong selection for genomic regions from the adapted parents of wide (upland x lowland) rice crosses indicates that, in non-marker-assisted breeding, where genetically distant parents have been used in a bulk population program, genomic regions from an upland type were strongly selected in the upland environments and regions from the lowland type in lowland environments (Steele et al., 2004).

SINGLE SEED DESCENT

The *single seed descent* method, abbreviated SSD, is predicated on the genotypic array of the initial cycle of gene and allele recombination and segregation, the F_2 generation (Tigchelaar and Casali, 1976). As was posited in Chapter 13, F_2 individuals are the portal to all subsequent generations and theoretically contain all of the alleles that will reside in downstream generations that will ultimately become genetically fixed. The number of potential genotypes increases exponentially as the breeding program proceeds, but with correspondingly less genetic divergence from parent to progeny (Schnell et al., 1980).

The SSD method begins with the array of genetic variation in the F_2 generation and then propels specific descendants to a state of fixation as rapidly as possible as selection is gradually introduced by selecting single seeds that descend from a F_2 individual from one generation to the next ($\rightarrow F_3 \rightarrow F_4 \rightarrow F_5 \rightarrow F_6 \rightarrow F_7 \rightarrow$) until fixation (usually F_7 or F_8) is reached. No selection is practiced during the early cycles of the fixation process (F_3, F_4, F_5), and selection pressure gradually increases with each generation and is very strong at the end of the program (F_6, F_7, F_8; Fig. 14.7). Adaptations to the SSD method are applied to accommodate biological attributes of different crop species. For example, legume breeders often use a multiple-seed procedure in which a single pod rather than a single seed is harvested from each plant and bulked (Macchiavelli and Beaver, 2001).

The size of the F_2 population and number of lines carried to fixation are the fundamental parameters that will determine the success of any given SSD program. The effect of population size on resource requirements is obvious, so the goal is to define a number that will result in maximum genotype recovery with the lowest possible resource inputs depending on the species under study and the number and genetic complexity of traits being targeted for improvement. It is much easier to produce and manage small grain populations than it is for tomato, as per our earlier discussions. Likewise, it is much easier to produce and manage tomato populations than it is for a larger-mass woody perennial species such as peach.

The size of the theoretically perfect F_2 population will increase exponentially with the number of traits under selection, and also as the numbers of genes controlling the different traits grows larger. In the peach fruit drop example, let us assume that three major genes control fruit drop, and that horticultural type consists of five main traits, each controlled by five distinct genes, only two of which are polymorphic in the parents. A total of 13 primary polymorphisms will be segregating, therefore, in the F_2. The theoretical perfect F_2 size will be 4^{13} plus an imputed statistical interval (based on V_p) to ensure a 95% or more probability of the recovery of desired genotypes. The project budget dictates that we can manage a population with a maximum size of 100, considering space, labor, and supply requirements. The SSD method, therefore, may necessitate a major compromise from theoretical population size requirements.

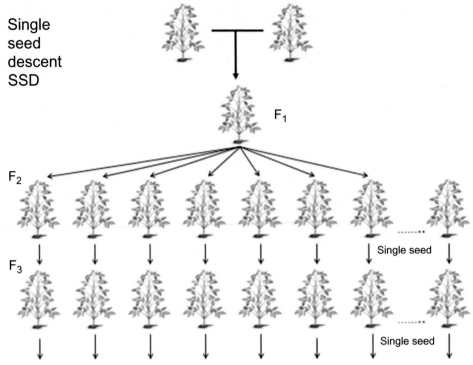

FIG. 14.7   Single seed descent breeding program during early generations ($F_2 \rightarrow F_5$).

In the tomato example, let us assume that fruit firmness is controlled by five major genes and horticultural type by ten component traits with an average of four major genes per trait (or 40 major genes total), only one locus per trait of which is polymorphic in this particular cross. Therefore, a total of 15 genes are polymorphic between the two parents relative to the characters under selection. Thus, the theoretical perfect population size of 4^{15} + 95% statistical confidence interval is extremely large, but the budget will limit the number of F_2 individuals to 100, a much greater compromise than for the peach fruit drop example above. From these examples, that are oversimplifications of real situations, it follows that the size of the F_2 will almost always be a compromise between theoretical needs and resource limitations. Therefore, the SSD method is considered to be more effective in situations involving fewer genes and high heritability as compared to many genes and complex quantitative inheritance.

The SSD method is invoked to save both time and money as compared to the pedigree method. Since no selection is used to truncate the early generations under advancement, plants may be grown under non-adapted environmental conditions or seasons. The challenge is to conserve time resources by advancing as many generations in as little time as possible, using controlled environments and off-shore nurseries where possible. If two generations can be obtained per calendar year for an annual crop species, a SSD program should take less than five years to proceed from the original parental cross to the F_8. With three generations per year, the time reduces to 3.33 years. Cost reductions as compared to pedigree are achieved by virtue of smaller population sizes, shorter time frames, and dispensation of selection pressures during early generation cycles.

The SSD method begins the same way as does the pedigree and bulk population methods, with a cross of individuals chosen for genetic complementation (Fig. 14.7). The same hypothetical examples, the firm tomato, and low fruit drop peach will suffice to illustrate SSD. In the case of low fruit drop peach, the standard variety that tends to exhibit fruit drop is crossed with an individual selected from a breeding population that features strong peduncles and a generally acceptable horticultural phenotype. The aim will be to generate a F_2 population consisting of 100 individuals. If the average fecundity rate for peach is assumed to be 50 progeny per plant, we will need two to three F_1 progeny from the cross to produce the desired F_2. In the tomato example, where the targeted F_2 population size is 1000 and fecundity is 250 per plant, only four F_1 progeny are necessary to produce the targeted number of F_2 individuals.

In both the tomato and peach examples, the F_2 population is grown through reproductive maturity and viable seed maturation. Since no selection will be practiced, it does not necessarily matter where the population is grown. Selective pressures are inadvertently introduced in populations that are manifested by changes in allelic frequencies. For example, selection for early maturation on populations may result in decreased reproductive yield. In a SSD program, a single seed is extracted from all individuals within populations, thus obviating any possibility of selection.

A single seed is harvested from each F_2 individual, and the rest of the seeds are discarded or kept as a backup for prospective F_3 plant mortality. Many potentially desirable genotypes are certainly sacrificed in the SSD method, but that is another inherent compromise. The single seeds, each a F_3 individual, are bulked into a single population, 1000 for tomato and 100 for peach. This pattern is repeated for the F_4 and F_5 and possibly the F_6. Starting with the F_6 or F_7, the method reverts to the same actions as the pedigreemethod, with breeding nurseries grown in the area of adaptation and strong selection practiced among and within families. F_8 plants are then subjected to very intense phenotypic selection, from 1000 to less than five for tomato and from 100 to one to three for peach. Alternatively, the selection process may be extended to two or more cycles, but since the method is chosen partly to achieve economies of time, this is not usually done. The selected entities are then submitted to rigorous multi-year, multi-location testing.

The SSD method is often "cut and pasted" into ongoing breeding programs, particularly when resources are temporarily curtailed or suspended. For example, a plant breeder may embark on a pedigree program and proceed to the F_3 when labor and/or land become unavailable for a protracted time period. The program may be shifted at the F_3 generation from pedigree to SSD since it requires fewer resources, and little or no technical expertise since selection is not practiced enabling the program to continue without complete sacrifice of the potentially valuable genotypes.

The method is most often used in situations where a serious problem develops quickly within a crop species that can be best approached by developing new cultivars through an inbreeding plant breeding program. Grower, processor, or distributor/retailer clientele are more interested in a fast solution to the problem than in a plethora of information and secondary breeding populations pertaining to the trait that is causing their problem. Sponsors are also able to support the program at a more modest level than for the pedigree method. The SSD method is attractive in this situation because it can quickly lead to the development of a genetically improved population at a reasonable cost and with no technological uncertainties. However, both genetic precision and information capture are sacrificed.

MAS can be easily integrated into a SSD breeding program to accomplish early generation selections and allow F_2, F_3, F_4, etc. individuals of higher genetic potential to be targeted (Delannay and Staub, 2010). For example, large populations of individuals may be produced as transplants that are submitted to QTL genotyping. Only F_2 individuals that are selected for the greatest potential are included in the population of mature plants for the harvest of single seeds for the next generation. If early generation selections are possible for traits of high heritability and low interference from V_D and linkage disequilibrium, the SSD breeding program is rendered to be much more efficient (Vilela et al., 2009).

THE DOUBLED HAPLOID METHOD (VIA MICROSPORE CULTURE)

In Chapters 3 and 8 the consequences of different ploidy levels were described and discussed. *Haploidy* is defined as either the gametophytic chromosome (and gene) number or the basic chromosome number of a taxon (usually family), also referred to as the monoploid number or "x" depending on the systematic context. The complement of alleles in a gamete is termed a haplotype. The cytogenetic genome constitution of a diploid sporophyte individual or population consists of two haplotypes and is often described cytogenetically as "2n = rx = s" where *r* is the multiple of basic genome ploidy and *s* is the number of chromosomes. This nomenclature is used to help untangle the basic evolutionary and gametic chromosome number: *Dihaploid* is a gamete with a chromosome number equal to two times the basic number for the taxon. The doubled haploid breeding method, haploidy refers to the gametophytic complement, or haplotype.

Haploids in vascular plants are usually observed only in gametophytes. The sporophyte undergoes meiosis to form haploid gametophytes (male and female) that fuse to form a new diploid sporophyte (Chapters 3 and 10). In gymnosperms and angiosperms, the haploid gametophyte is dramatically reduced in both stature and longevity as compared with the gametophyte generation in "lower" plant phylogenetic taxa (ferns, mosses, etc.), existing only to channel genes into the next generation of sporophytes. While the use of selection on gametophytes for the breeding of corresponding sporophytes has been demonstrated, the sporophyte is the overwhelmingly conspicuous manifestation of the species (Richards, 1986).

Cross- and self-pollination are two mating strategies that are harnessed by plants to achieve divergent evolutionary goals. Cross-pollination fosters long-term preservation of genetic variability, whereas self-pollination favors near-term exploitation, and long-term extinction (Richards, 1986). As has been stressed earlier, cross-pollination promotes heterozygosity and genotype polymorphism, while self-pollination leads to homozygosity, allelic fixation, and monomorphism. The most aggressive method of inbreeding by controlled mating is self-pollination (Fig. 14.8). Other mating strategies that culminate in genetic fixation are versions of sib-mating (full, half, quarter, etc.). Self-pollination

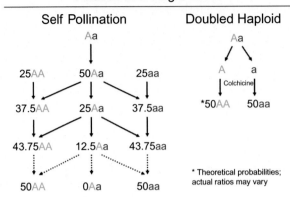

FIG. 14.8   Comparison of the characteristics of inbreeding under self pollination vs. a hypothetical doubled haploid protocol.

assumes that gametes from the same individual will fuse to form the next generation of sporophytes, but if the parent is heterozygous, gametes may be genotypically different resulting in persistent heterozygosity.

What if gametes of the same haplotype could be fused to form a 2n zygote? The result is a sporophyte that is 100% homozygous (Fig. 14.8). If such gamete fusions were possible, genetic fixation from a multiply heterozygous individual could be achieved in a single step. Where can gametes of the same haplotype be found and made to fuse with each other?

In rare instances, egg nuclei that are unfused with sperm nuclei have been observed to continue developing to form an embryo, or haploid sporophyte. Such events can be seen where a tiny, weak plant is germinating among a mass of larger, vigorous seedlings. In many plant species, the phenomenon of the "twin seedling" is observed wherein one of the pairs of seedlings emerging from a single seed is of normal vigor, and the other is weak (Dweikat and Lyrene, 1990; Uno et al., 2002). Cytological observations often confirm that the weakling is indeed a haploid, derived from an unfertilized female gametophyte. If allowed to continue growing, such haploid sporophytes will sometimes flourish to maturity and even attempt reproduction, but haploid plants are incapable of undergoing a normal meiosis to produce viable gametes. If the haploid seedling is treated with colchicine, the natural alkaloid that interferes with the segregation of chromosomes to opposite poles of a dividing cell (Chapter 3), the chromosome number may be doubled to the diploid level, and the plant may regain sexual fertility. The haploid plant will, sometimes, undergo chromosome doubling spontaneously. The doubled haploid plant is also homozygous at all genetic loci throughout the genome (Fig. 14.9).

Inducing a haploid cell to become diploid sporophyte is the essence of the *doubled haploid* (DH) breeding method. The replication of DNA within chromosomes of a haploid sporophyte to form identical daughter chromosomes, followed by a thwarting of mitotic cell division and nuclear restitution in a diploid state, creates a panhomozygote based on the haplotype (Fig. 14.9). The genotype of the gamete was derived from the array of recombinants and segregants realized during the meiotic process by which it was produced. If a plant has the genotype AaBbCc, there are eight possible haplotypes (assuming independent assortment of the loci): ABC, aBC, AbC, ABc, abC, aBc, Abc, and abc. If the chromosomes are doubled, there are eight corresponding diploid genotypes: AABBCC, aaBBCC, AAbbCC, AABBcc, aabbCC, aaBBcc, AAbbcc, and aabbcc. All of the DH plants are completely homozygous but have different genotypes.

The ramifications of the ability to proceed from a multiply heterozygous F_1 directly to an array of totally homozygous segregants are twofold. First, and most importantly, the haploid gamete → dihaploid plant developmental transformation may be inserted into a breeding program for a self-pollinated species with tremendous time savings, and possible conservation of other resources as well. If it takes 0.5 years to proceed from the F_1 plant to the population of doubled haploid individuals, that are genetically fixed, nearly three years have been shaved off the SSD program with three generations per year, and nearly five years with two generations per year. Each step of progressive fixation via filial generations requires space, labor, and supplies, that are also obviated.

The second ramification pertains to linkages in repulsion, or desirable alleles linked to undesirable ones (Chapter 3). As the number of traits included in the breeding program and genes underlying these traits increase, linkage involving the genes under selection is inevitable. In the doubled haploid phenomenon, only a single meiotic cycle that includes heterozygous genotypes occurs in a doubled haploid breeding program. This is probably inadequate for the

Doubled Haploids: "Instant" Pure Lines (assuming no recombination)

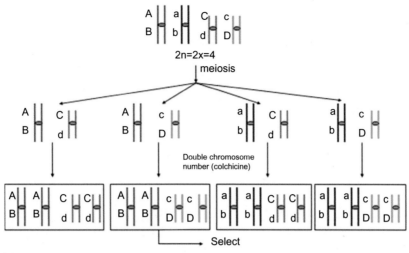

FIG. 14.9 Doubled haploid plants are homozygous at the chromosome, gene, and nucleotide level. This figure illustrates the direct segregation of a heterozygote into sets of doubled haploid homozygotes.

conversion of all repulsion into coupling linkages. Traditional breeding methods, such as pedigree and bulk population, provide opportunities for repulsion linkages to be converted to coupling in several successive early generation filial cycles (e.g., F_2, F_3, F_4, etc.). If the linkages are "tight" (loci linked physically close together), the population of doubled haploids required for a high probability of conversion of repulsion to coupling is inordinately high. Therefore, the single meiosis in the doubled haploid method is both desirable with regard to time savings and undesirable with regard to genetic recombination. This also applies to the independent assortment of alleles at many loci in instances where several traits of complex inheritance are under selection and if the ability to obtain dihaploid plants is hindered.

The DH breeding method is conceptually simple (Fig. 14.10). The two parents are hybridized to produce a F_1, from which gametes are obtained and induced to double their chromosome number. Then the newly diploid gametes must be tricked into switching from being gametophytes into being sporophytes. After this transformation is accomplished, the doubled haploid plants are self-pollinated to produce a population of genetically identical seed-derived individuals, since no direct selection is possible on plants regenerated from tissue cultures. The selfed progeny of the doubled haploids are subjected to strong selection, similar in intensity to that imposed at the end of the SSD program. A small number of selections with the best combination of attributes are then advanced to the rigorous testing phase.

If it were that simple, the DH method would be the predominant strategy used for self-pollinated crop plant species due to truncated time frame and low resource requirements as compared to other breeding methods for self-pollinated crop species. The reader has probably asked themselves during the previous paragraph "how are gametophytes or other haploid cells tricked into becoming sporophytes?" Also "how are haploids/gametes induced to double chromosome number?" Isn't the normal developmental progression for a sporophyte to give rise to a gametophyte via megasporogenesis or microsporogenesis, and a new sporophyte from the fusion of sperm and egg? Gametophytes do not typically become sporophytes in nature in the absence of fusion to form a 2n zygote. The forced, or artificial, transformation of haploids or gametophytes into diploid sporophytes has proven to be a difficult and perplexing challenge. Like many developmental processes, natural or otherwise, this phenomenon is not yet well understood. In some cases, reproducible protocols for the induction of haploid plants have been elucidated; in most other cases such protocols remain elusive (Baenziger, 1996; Maluszynski, 2003). Progress is evident in the elucidation of gene expression patterns during the microspore→haploid plant developmental transformation that will illuminate strategies to increase yields of haploid plants and to broaden the range of plant species in which haploids are accessible for plant breeding and basic plant research (Vrinten et al., 1999; Ferrie and Caswell, 2011).

One approach to the challenge of inducing gametophytes to behave developmentally like sporophytes is to subject them to tissue culture. It is well established that growth regulators or natural phytohormones in combination with other compounds and conditions can incite cells and tissues to undergo developmental transformations. What happens when male and female gametophytes are cultured and challenged with media containing altered phytohormone regimes? In most plant species that have been studied, cultured gametophytes are not transformed into sporophytes,

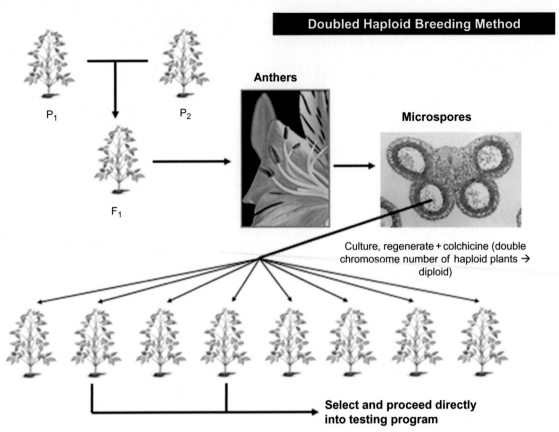

FIG. 14.10   The doubled haploid breeding method, using microspore culture as the example method to access haploidy.

although they may be developmentally changed into something else, usually a tumorous tissue mass. The tumor may subsequently be regenerated into one or more whole plants, but the culturing phase may have introduced genetic variability into the regenerates (i.e., somaclonal variation; see Chapter 8) an undesirable outcome in this context.

In a heretofore limited number of plant species, however, the male gametophyte (developing microspore or mature pollen) has been observed to emulate sporophyte-like growth similar to a sexual embryo but with distinct differences (Fig. 14.11). Such "somatic" embryos (as distinguished from sexual embryos derived from true zygotes) are multicellular entities that exhibit polarity and meristems that appear analogous to the radical and shoot. When somatic embryos

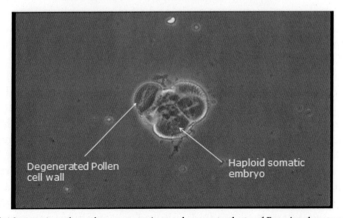

FIG. 14.11   The growth of a haploid somatic embryo from a maturing male gametophyte of *Brassica oleracea* var. *italica* under culture conditions.

are placed on culture medium devoid of growth regulators, they often develop into whole plants. The induction of somatic embryos from male gametophytes of certain plant species is reproducible, but the mechanisms responsible for induction are not well understood. As more is understood about plant development at the molecular level we will be able to devise techniques to induce male gametophytes from a broader spectrum of plant species to form somatic embryos (Henry et al., 1993; Maraschin et al., 2005; Pulido et al., 2006; Hosp et al., 2007; Sharma et al., 2010; Ferrie and Caswell, 2011; Seifert et al., 2016). It is likely that manipulation of a set or sets of transcription factors, such as the MADS-box genomic family presented in Chapter 10, will ultimately lead to the widespread ability to generate haploids at will (Immink et al., 2002).

Through empirical experimentation, a short and slowly growing list of plant species has been developed wherein sporophytes can be obtained from cultured gametophytes. Species within the families Gramineae (*Oryza, Triticum, Hordeum, Avena, xTriticosecale, Secale*), Brassicaceae (*Brassica, Raphanus*), and Solanaceae (*Solanum, Capsicum, Petunia, Nicotiana*) have been discovered to be particularly amenable to this induced developmental transformation. The induction of somatic embryos from microspores has been observed, however, in many economically significant crop species in other plant families. As the utility of this capability lies primarily in plant breeding, most of the research has been done on important food crop species.

During the 1970s through the 1990s, virtually every economic plant species was studied for the ability to grow as, manipulate as, and regenerate plants from cell and tissue cultures, including microspores and/or anthers. The difficulty in expanding the list of species amenable to gametophyte-derived haploid tissue cultures suggests that the developmental pathway from gametophyte to sporophyte is generally difficult to incite. The fact that the list is steadily growing shows it will eventually be possible to produce haploid microspore-derived tissue cultures in all or most plant species. Plant species once thought to be non-regenerable from tissue and cell cultures are now easily manipulated following sustained empirical research. Reports of successes with the regeneration of haploid or dihaploid plants from cultured microspores or microspores in anthers from 1990 to the present include the following horticultural species: coffee *Coffea arabica* (Neuenschwander and Baumann, 1995), cassava *Manihot esculenta* (Perera et al., 2012), oak *Quercus* sp. (Bueno et al., 2003), strawberry *Fragaria X ananassa* (Svensson and Johansson, 1994), phlox *Phlox drummondii* (Razdan et al., 2008), service tree *Sorbus domestica* (Arrillaga et al., 1995), carrot *Daucus carota* (Matsubara et al., 1995), dill *Anethum graveolens* (Ferrie et al., 2011), onion *Allium cepa* (Jakse et al., 2010), *Callerya speciosa* (Huang et al., 2017), sunflower *Helianthus annuus* (Thengane et al., 1994), linseed *Linum usitatissimum* (Nichterlein and Friedt, 1993), primrose *Primula forbesii* (Jia et al., 2014), apple *Malus pumila* and pear *Pyrus communis* (Kadota et al., 2002; Hofer, 2004), camellia *Camellia japonica* (Pedroso and Pais, 1994), Asiatic hybrid lily *Lillium longiflorum* (Han et al., 1996), muskmelon *Cucumis melo* (Lim and Earle, 2009), winter squash *Cucurbita maxima* and pumpkin *Cucurbita moschata* (Dong et al., 2016; Kurtar et al., 2016), gentian *Gentiana triflora* (Doi et al., 2010), apricot *Prunus armeniaca* (Germanã et al., 2011), almond *Prunus dulcis* (Cimã et al., 2017), cyclamen *Cyclamen persicum* (Takamura et al., 2011), citrus *Citrus sinensis* and *C. clementine* (Cardoso et al., 2014), asparagus *Asparagus officinalis* (Feng and Wolyn, 1994), narcissus *Narcissus tazetta* (Chen et al., 2005), African violet *Saintpaulia* spp. (Uno et al., 2016), and poplar *Populus trichocarpa* (Baldursson et al., 1993). The sheer heterogeneity of this list imparts confidence that the range of amenable species will continue to expand, and that applications of the doubled haploid method will be more widespread in the future.

The finding of regenerated plants from cultured anthers and microspores that are diploid and polyploid has prompted suspicions about the origin of the regenerates. The advent of genetic markers has made it possible to demonstrate unequivocally the gametophyte origin of most of these regenerates (Eimert et al., 2003; Vanwynsberghe et al., 2005; Rivas-Sendra et al., 2015). The tissue culture process is often associated with changes in ploidy (Ferrie and Mãllers, 2011; Gu et al., 2014). Diploid regenerates of gametophyte origin are potentially useful, but polyploid regenerates are not directly applicable to the doubled haploid breeding method.

Complicating the picture pertaining to dihaploid plants even further, the ability to undergo this developmental change is also affected by the genotype of the plant contributing the gametophyte, and also by the genotype of the gametophyte (Dwivedi et al., 2015). Thus, certain elements of a gametic array may be more amenable to the formation of somatic embryos than are others. In other words, the doubled haploid plants may not represent a random sample of gametes from the individual on which they were produced (Orton and Browers, 1985; Chen et al., 2001; Cistue et al., 2005).

HAPLOIDS FROM INTERSPECIFIC HYBRIDS

A peculiar observation was made while studying the progeny from certain interspecific and intergeneric hybridizations. In some instances, tissue sectors or even whole plants have been obtained from the hybrids that were not phenotypically intermediate between the species parents but, instead, bore an uncanny resemblance to only one of the parents. Cytogenetic characterizations of these sectors or plants often revealed that they contained only the genome of the resembled parent, with little or no cytological evidence of any genetic contribution from the other parent. The phenomenon, known as *chromosome or genome elimination*, was observed with both reciprocal origins, so the sector or plant must have been derived from a hybrid that contained a composite genome followed by the elimination of the genome of the other parent (Barclay, 1975; Ho and Kasha, 1975; Orton, 1980; Houben et al., 2011).

Recent findings in *Arabidopsis thaliana* have implicated alterations in the centromeric histone protein CENH3 in the chromosome elimination phenomenon (Seymour et al., 2012). The centromere is a defining functional and structural unit of the eukaryotic chromosome. The centromeric kinetochore complex assembles during mitotic and meiotic metaphase and facilitates chromosome segregation. Centromeres contain unique repetitive sequences and are enriched with transposons and retrotransposons. Although how centromere structure and function is determined is still not clearly understood, the binding of CENH3 to centromeric repetitive DNA sequences has been found to play a critical role (Watts et al., 2016).

Chromosome elimination has been reported frequently among interspecific crosses within Gramineae (Barclay, 1975; Ho and Kasha, 1975; O'Donoughue and Bennett, 1994; Riera-Lizarazu et al., 1996; Dwivedi et al., 2015) and, sporadically, in other families (Oberwalder et al., 1998; Chen et al., 2007; Tu et al., 2010). The most widely studied case developed in the laboratory of Dr. Ken Kasha at the University of Guelph (see biography of Dr. Kasha in Chapter 1) was with cultivated barley (*Hordeum vulgare*) x *H. bulbosum*, both 2n = 2x = 14 (Ho and Kasha, 1975; Kasha et al., 1995). If this cross is made multiple times using the wild species as female parent, it is possible to recover high numbers of haploid *H. vulgare* plants germinating from the resulting seeds. The fact that *H. vulgare* plants are derived from presumed hybrids where *H. bulbosum* was the female parent is direct evidence that hybridization had been successfully effected. The haploid plants descended directly from the male gametophyte that fused with the *H. bulbosum* egg nucleus, and the *H. bulbosum* genome was subsequently lost during the early mitotic cell divisions of the developing hybrid embryo.

The doubled haploid breeding protocol presented earlier using microspore culture as the source of haploids (Fig. 14.10) may be modified to accommodate haploids from interspecific hybrids followed by genome elimination as follows: An F_1 hybrid of the "assertive" or targeted parental species is constructed as for the pedigree or other breeding methods for self-pollinated crop species. The F_1 is then hybridized with the other "non-assertive" species the genome of which will be eliminated during interspecific hybrid embryo development. If *H. vulgare* combining lodging resistance with good agronomic type was crossed with *H. bulbosum*, a population of haploid *H. vulgare* plants that embodies the gametic array of polymorphisms between these two parents will result. The haploid plants are then converted to the doubled haploids with colchicine and self-pollinated to produce more seeds. The subsequent homozygous population is subjected to strong selection in an identical manner as Fig. 14.10.

The discovery of haploids and doubled haploids from microspore culture and chromosome elimination has spurred tremendous interest in the use of haploids in plant breeding. Clues to the underlying mechanisms that govern the developmental shift from gametophyte to sporophyte, or that precipitate the segregation of whole genomes from wide hybrids, have been elucidated prompting optimism that applications of haploids will continue to expand. No surefire methods for haploid plant production have yet been developed, however, and the practitioner is often reduced to empirical research on different crosses; chemical or environmental factors, levels, timings; environmental stress treatments; or combinations thereof.

The treatment of haploids in this chapter has focused on applications in self-pollinated crop species. As of this writing, most of the demonstrated applications of haploid technologies lie within this realm. It is also a viable technique for the rapid development of inbred plants and populations of plants for combining ability studies in a hybrid breeding program (Chapter 15), a strategy used mostly in outcrossing plant species. The same biological limitations (i.e., difficulty in obtaining haploid plants) apply to this application as well.

HETEROSIS AND HYBRID CULTIVARS IN SELF-POLLINATED CROP SPECIES

The phenomenon of heterosis will be described in more detail in Chapters 15 and 16. Briefly, heterosis is broadly defined as the value of the F_1 hybrid exceeding the midparent value. Commercially, heterosis is more compelling if the F_1 hybrid exceeds both parents. Consequently, the notion of "mid-parent" (MPH) and "high-parent" (HPH) heterosis has been advanced, although the distinction may be moot at the molecular level. Generally, the observation of inbreeding depression, or the tendency for vigor and reproductive yield to decline with progressive allele fixation, is associated with heterosis.

Heterosis has been studied and applied extensively in cross-pollinated crop species, for example, maize, sorghum, carrot, onion, cucumber, squashes, muskmelon, asparagus, and cabbage-related vegetables (Chapters 15 and 16). Many hybrid horticultural crop species are also propagated and marketed as asexual clones (Chapter 17). The strategy of marketing hybrid crop cultivars instead of pure line and open-pollinated populations to avert seed savings and genotype piracy is a strong business incentive (Chapter 12). By far the biggest roadblock to the successful development of hybrid cultivars of predominantly self-pollinated crop species is the ability to produce large quantities of genetically pure F_1 hybrid seeds. The natural biological attributes of autogamy include low pollen quantities and absence of pollen vector systems (Chapter 10) that are essential for large-scale hybrid seed production.

Intuitively, inbreeding depression is inconsistent with self-pollination. Charlesworth and Charlesworth (1999) concluded that deleterious mutations probably play a major role in causing inbreeding depression. They surmised that it is difficult to account for the very large effects of inbreeding on fitness in outcrossing plants without a significant contribution from variability maintained by selection. Overdominance effects of alleles on fitness components seem not to be important for most phenotypes. Recessive or partially recessive deleterious effects of alleles, some maintained by mutation pressure and some by balancing selection, thus seem to be the most important source of inbreeding depression. Classical genetic studies and modern molecular evolutionary approaches now suggest that inbreeding depression and heterosis are predominantly caused by the presence and dominant masking of recessive deleterious mutations in populations (Charlesworth and Willis, 2009).

Inbreeding should progressively purge deleterious recessive alleles from gene pools, but self-pollinating species retain polymorphisms for dominant and additive alleles that are neutral or beneficial for agricultural applications. Therefore, it is theoretically possible to observe heterosis in self-pollinating crop species. Buti et al. (2013) concluded that heterosis in sunflower (*Helianthus annuus*) resulted from mutations in the cis-regulatory elements of genes, largely related to retrotransposon insertions and/or removals over relatively short evolutionary time frames. Perhaps a similar mechanism is present in other autogamous crop species?

In the inbreeding small grain rice (*Oryza sativa*), heterosis has been described and exploited in hybrids of *indica* x *japonica* types (Cheng et al., 2007) in the form of "super hybrid" rice that features heterotic genomic combinations from *indica* and *japonica* (Singh et al., 2015). Studies of heterosis and hybrid breeding are also gaining prominence in wheat (*Triticum aestivum*). Longin et al. (2013) found that hybrids significantly outyielded the best commercial inbred line variety underlining the potential of hybrid wheat breeding. One crucial limitation in wheat is the lack of divergent heterotic groups that is being addressed by introgressing heterotic genomic segments from related species such as spelt (Akel et al., 2018). Male sterility systems and other methods have been developed to foster cost-effective production of hybrid seeds in both rice (Cheng et al., 2007) and wheat (Whitford et al., 2013).

The business strategic attraction of hybrid cultivars has culminated in studies of heterosis and conveyance to cultivars and cms-based hybrid seed production protocols in other self-pollinated agronomic crop species such as soybean (*Glycine max*; Palmer et al., 2001; Nie et al., 2017), cotton (*Gossypium hirsutum*; Tyagi et al., 2014; Thombre and Mehetre, 1979), sunflower (*H. annuus*; Cheres et al., 1994, 2000).

The allure of heterosis and higher profits associated with hybrid cultivars has led to the development of hybrids in predominantly self-pollinated horticultural crop species, for example, tomato (*Solanum lycopersicum*; Tigchelaar, 1990), pepper (*Capsicum annuum*; Geleta and Labuschagne, 2004), and pea (*Pisum sativum*; Espósito et al., 2014). In processing tomato, Conti et al. (1990) found that hybrids were superior for both marketable and soluble solid yields (+11.0% and +8.7%, respectively, as an average over environments). Heterosis estimates tended to be higher in less favorable environments.

No high-parent heterosis (HPH) was observed for any trait in this study. In fresh market tomato, however, HPH was reported for number of branches per plant, early yield, total yield and fruit firmness (Shalaby, 2013). Several studies have demonstrated that heterosis is both rare and difficult to explain in cultivated tomato (Atanassova et al., 2002; Semel et al., 2006). Inbreds that outyielded the original hybrid were obtained "relatively easily" by Christakis and Fasoulas (2001). They concluded that selection for homozygote superiority on the basis of genetic components of crop yield and quality led to the development of inbreds that outperformed corresponding hybrids.

It is very clear, however, that F_1 hybrid cultivars offer an excellent pathway to stack, or pyramid, multiple dominant disease resistance alleles (for different disease pathogens or races of a specified pathogen) in a single, genetically monolithic population (Erb and Rowe, 1992; Scott, 2005; Hanson et al., 2016). Inbreds with multiple dominant disease resistance alleles (see "vertical resistance" in Chapter 19) are developed by recurrent backcross (Chapter 18) then inbreds that carry complementary alleles are hybridized to produce a F_1 hybrid cultivar that carries all the resistance alleles present in both inbreds (Fig. 14.12). MAS is a powerful tool to facilitate the development of stacked multiple disease resistance in tomato (Foolad and Panthee, 2012).

Hybrid Varieties for Multiple Disease Resistance

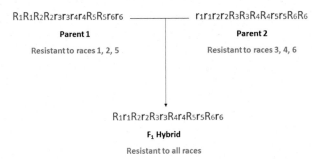

FIG. 14.12 Two complementary inbreds are developed by recurrent backcrossing that have different dominant disease resistant alleles for different pathogens or pathogen races. The corresponding F_1 hybrid is heterozygous at all polymorphic loci, exhibiting resistance to all diseases or races.

Both nuclear/genic and cytoplasmic male sterility have been reported in *S. lycopersicum* (Melchers et al., 1992; Atanassova, 2007). More recently, a conditional nuclear male sterile allele has been discovered that has great potential as a tool for hybrid seed production (Pucci et al., 2017). The existence of effective male sterility systems and the genetic and cultural ability to promote heterogamous floral attributes in tomato (Scott and George Jr, 1980; Tikoo and Anand, 1980), in conjuction with potential pollinating insect vectors, would appear to set the stage for more efficient hybrid tomato seed production. As of the publication of this volume, however, the overwhelming proportion of commercial F_1 hybrid tomato seeds are produced by hand emasculation and pollination. Mating system is, obviously, a very complex biological process that is difficult to subvert.

GENOME SELECTION IN SELF-POLLINATED CROP SPECIES

Genome selection (GS) schemes are proving to be both powerful and cost-effective adjuvants to all breeding strategies for self-pollinated crop species (Marulanda et al., 2016). The breeding strategy "GSrapid" with moderate nursery selection followed by one stage GS and one final stage with phenotypic selection on grain yield had the highest annual selection gain across all strategies, budgets, costs, and variance components. Owing to the very high number of test candidates entering breeding strategies with GS, the costs for the doubled haploid method were much higher per candidate than for the other breeding methods.

SUMMARY OF BREEDING METHODS FOR SELF-POLLINATED CROP SPECIES

Table 14.1 summarizes and contrasts characteristics of the four primary breeding strategies used for self pollinated crop species relative to criteria most germane to the plant breeder.

TABLE 14.1 Comparison of The Breeding Methods for Self-Pollinated Crop Species with Regard to Resource Requirements and Collateral Outcomes

Strategy	Time	Space	Labor	Technical	Genetic Precision
PD	10+ Y	High	High	Med	High
BK[a]	≤10+ Y	Med	Low	Low	Low
SSD[a]	≤10+ Y	Low	Med	Low	Low
DH	≤5 Y	Low	Med	High	Med

[a]*May be able to reduce time requirement with >1 generation per year when selection is not applied.*

References

Akel, W., Longin, C.F.H., Weissman, E.A., Liu, G., Thorwarth, P., Würschum, T., Mirdita, V., 2018. Can spelt wheat be used as heterotic group for hybrid wheat breeding? Theor. Appl. Genet. 131 (4), 973–984.

Allard, R.W., 1960. Principles of Plant Breeding. John Wiley & Sons, New York, NY (USA). 485 pp.

Arrillaga, I., Lerma, V., Perez-Bermudez, P., Segura, J., 1995. Callus and somatic embryogenesis from cultured anthers of service tree (*Sorbus domestica* L.). HortScience 30 (5), 1078–1079.

Atanassova, B., 2007. Genic male sterility and its application in tomato (*Lycopersicon esculentum* mill.) hybrid breeding and hybrid seed production. Acta Hortic. 729, 45–51.

Atanassova, B., Shtereva, L., Balatcheva, E., 2002. Estimation of heterosis for productivity and early yield in F_1 hybrids of tomato (*Lycopersicon esculentum* mill.) mutants differing in their vitality. Acta Hortic. 579, 567–572.

Baenziger, P.S., 1996. Reflections on doubled haploids in plant breeding. In: Jain, S.M., Sopory, S.K., Veilleux, R. (Eds.), In Vitro Haploid Production in Higher Plants. Vol. 1. Kluwer Academic Publishers, Dordrecht (Netherlands), pp. 35–48.

Baldursson, S., Krogstrup, P., Norgaard, J.V., Andersen, S.B., 1993. Microspore embryogenesis in anther culture of three species of *Populus* and regeneration of dihaploid plants of *Populus trichocarpa*. Can. J. For. Res. 23 (9), 1821–1825.

Barclay, I.R., 1975. High frequencies of haploid production in wheat (*Triticum aestivum*) by chromosome elimination. Nature 256 (5516), 410–411.

Bernardo, R., 2003. Parental selection, number of breeding populations, and size of each population in inbred development. Theor. Appl. Genet. 107 (7), 1252–1256.

Borojevic, S., 1990. Principles and Methods of Plant Breeding. Elsevier, Amsterdam (Netherlands). 368 pp.

Bueno, M.A., Gomez, A., Sepulveda, F., Segui, J.M., Testillano, P.S., Manzanera, J.A., Risueno, M.C., 2003. Microspore-derived embryos from *Quercus suber* anthers mimic zygotic embryos and maintain haploidy in long-term anther culture. J. Plant Physiol. 160 (8), 953–960.

Buti, M., Giordani, T., Vukich, M., Pugliesi, C., Natali, L., Cavallini, A., 2013. Retrotransposon-related genetic distance and hybrid performance in sunflower (*Helianthus annuus* L.). Euphytica 192 (2), 289–303.

Cardoso, J.C., Martinelli, A.P., Germanã, M., Antonieta, L., Rodrigo, R., 2014. In vitro anther culture of sweet orange (*Citrus sinensis* L. Osbeck) genotypes and of a *C. clementina* × *C. sinensis* 'Hamlin' hybrid. Plant Cell Tissue Organ Cult. 117 (3), 455–464.

Charlesworth, B., Charlesworth, D., 1999. The genetic basis of inbreeding depression. Genet. Res. 74 (3), 329–340.

Charlesworth, D., Willis, J.H., 2009. The genetics of inbreeding depression. Nat. Rev. Genet. 10, 783–796.

Chen, Y., Kenaschuk, E., Dribnenki, P., 2001. Inheritance of rust resistance genes and molecular markers in microspore-derived populations of flax. Plant Breed. 120 (1), 82–84.

Chen, L.J., Zhu, X.Y., Gu, L., Wu, J., 2005. Efficient callus induction and plant regeneration from anther of Chinese narcissus (*Narcissus tazetta* L. var. *chinensis* Roem). Plant Cell Rep. 24 (7), 401–407.

Chen, L., Qunfeng, L., Zhuang, Y., Chen, J., Zhang, X., Wolukau, J.N., 2007. Cytological diploidization and rapid genome changes of the newly synthesized allotetraploids *Cucumis* × *hytivus*. Planta 225 (3), 603–614.

Cheng, S.-H., Zhuang, J.-Y., Fan, Y.-Y., Du, J.-H., Cao, L.-Y., 2007. Progress in research and development on hybrid rice: a super-domesticate in China. Ann. Bot. 100 (5), 959–966.

Cheres, L.H., Van Fossen, C., Arias, D., Carter, R.L., 1994. Cytoplasmic male sterility in sunflower: origin, inheritance, and frequency in natural populations. J. Hered. 85 (3), 233–238.

Cheres, M.T., Miller, J.F., Crane, J.M., Knapp, S.J., 2000. Genetic distance as a predictor of heterosis and hybrid performance within and between heterotic groups in sunflower. Theor. Appl. Genet. 100 (6), 889–894.

Christakis, P.A., Fasoulas, A.C., 2001. The recovery of recombinant inbreds outyielding the hybrid in tomato. J. Agric. Sci. 137 (pt. 2), 179–183.

Cimã, G., Marchese, A., Germanã, M.A., 2017. Microspore embryogenesis induced through in vitro anther culture of almond (*Prunus dulcis* mill.). Plant Cell Tissue Organ Cult. 128 (1), 85–95.

Cistue, L., Echavarri, B., Batlle, F., Soriano, M., Castillo, A., Valles, M.P., Romagosa, I., 2005. Segregation distortion for agronomic traits in doubled haploid lines of barley. Plant Breed. 124 (6), 546–550.

Conti, S., Sanguineti, M.C., Roncarati, R., 1990. Hybrid performance as compared to parents in processing tomato. Adv. Hort. Sci. 4 (3), 151–154.

Delannay, I.Y., Staub, J.E., 2010. Use of molecular markers aids in the development of diverse inbred backcross lines in Beit alpha cucumber (*Cucumis sativus* L.). Euphytica 175 (1), 65–78.

Doi, H., Takahashi, R., Hikage, T., Takahata, Y., 2010. Embryogenesis and doubled haploid production from anther culture in gentian (*Gentiana triflora*). Plant Cell Tissue Organ Cult. 102 (1), 27–33.

Dong, Y., Wei-Xing, Z., Xiao-Hui, L., Liu, X., Ning-Ning, G., Jin-Hua, H., Wen-Ying, W., Xiao-Li, X., Zhen-Hai, T., 2016. Androgenesis, gynogenesis, and parthenogenesis haploids in cucurbit species. Plant Cell Rep. 35 (10), 1991–2019.

Dweikat, I.M., Lyrene, P.M., 1990. Twin seedlings and haploids in blueberry (*Vaccinium* spp.). J. Hered. 81 (3), 198–200.

Dwivedi, S.L., Britt, A.B., Tripathi, L., Sharma, S., Upadhyaya, H.D., Ortiz, R., 2015. Haploids: constraints and opportunities in plant breeding. Biotechnol. Adv. 33, 812–829.

Eimert, K., Reutter, G., Strolka, B., 2003. Fast and reliable detection of doubled-haploids in *Asparagus officinalis* by stringent RAPD-PCR. J. Agric. Sci. 141, 73–78.

Erb, W.A., Rowe, R.C., 1992. Screening tomato seedlings for multiple disease resistance. J. Am. Soc. Hort. Sci. 117 (4), 622–627.

Espósito, M.A., Bermejo, C., Gatti, I., Guindón, M.F., Cravero, V., Cointry, E.L., 2014. Prediction of heterotic crosses for yield in *Pisum sativum* L. Sci. Hortic. 177, 53–62.

Feng, X.R., Wolyn, D.J., 1994. Recovery of haploid plants from asparagus microspore culture. Can. J. Bot. 72 (3), 296–300.

Ferrie, A.M.R., Caswell, K.L., 2011. Isolated microspore culture techniques and recent progress for haploid and doubled haploid plant production. Plant Cell Tissue Organ Cult. 104 (3), 301–309.

Ferrie, A.M.R., Mãllers, C., 2011. Haploids and doubled haploids in *Brassica* spp. for genetic and genomic research. Plant Cell Tissue Organ Cult. 104 (3), 375–386.

Ferrie, A.M.R., Bethune, T.D., Arganosa, G.C., Waterer, D., 2011. Field evaluation of doubled haploid plants in the Apiaceae: dill (*Anethum graveolens* L.), caraway (*Carum carvi* L.), and fennel (*Foeniculum vulgare* mill.). Plant Cell Tissue Organ Cult. 104 (3), 407–413.

Foolad, M.R., Panthee, D.R., 2012. Marker-assisted selection in tomato breeding. Crit. Rev. Plant Sci. 31 (2), 93–123.

Geleta, L.F., Labuschagne, M.T., 2004. Hybrid performance for yield and other characteristics in peppers (*Capsicum annuum* L.). J. Agric. Sci. 142, 411–419.

Germanã, M.A., Chiancone, B., Padoan, D., Bárány, I., María-Carmen, R., Testillano, P.S., 2011. First stages of microspore reprogramming to embryogenesis through anther culture in *Prunus armeniaca* L. Environ. Exp. Bot. 71 (2), 152–157.

Gu, H., Zhao, Z., Sheng, X., Yu, H., Wang, J., 2014. Efficient doubled haploid production in microspore culture of loose-curd cauliflower (*Brassica oleracea* var. *botrytis*). Euphytica 195 (3), 467–475.

Han, D.S., Niimi, Y., Nakano, M., 1996. Regeneration of haploid plants from anther cultures of the Asiatic hybrid lily 'Connecticut King'. Plant Cell Tissue Organ Cult. 47 (2), 153–158.

Hanson, P., Chee-Wee, T., Ledesma, D., Wang, J.-F., Kwee, L.T., Kenyon, L., Yang, R.-Y., Schafleitner, R., Lu, S.-F., Chen, W., Wang, Y.-Y., Yun-Che, H., 2016. Conventional and molecular marker-assisted selection and pyramiding of genes for multiple disease resistance in tomato. Sci. Hortic. 201, 346–354.

Henry, Y., Bernard, S., Bernard, M., Gay, G., Marcotte, J.L., de Buyser, J., 1993. Nuclear gametophytic genes from chromosome arm 1RS improve regeneration of wheat microspore-derived embryos. Genome 36 (5), 808–814.

Ho, K.M., Kasha, K.J., 1975. Genetic control of chromosome elimination during haploid formation in barley. Genetics 81 (2), 263–275.

Hofer, M., 2004. In vitro androgenesis in apple—improvement of the induction phase. Plant Cell Rep. 22 (6), 365–370.

Hosp, J., de Maraschin, S.F., Touraev, A., Boutilier, K., 2007. Functional genomics of microspore embryogenesis. Euphytica 158 (3), 275–285.

Houben, A., Sanei, M., Pickering, R., 2011. Barley doubled-haploid production by uniparental chromosome elimination. Plant Cell Tissue Organ Cult. 104 (3), 321–327.

Huang, B., Xu, L., Li, K., Fu, Y., Li, Z., 2017. Embryo induction and plant regeneration of *Callerya speciosa* (Fabaceae) through anther culture. Aust. J. Bot. 65 (1), 80–84.

Ibrahim, K.M., Hayter, J.B.R., Barrett, J.A., 1996. Frequency changes in storage protein genes in a hybrid bulk population of barley. Heredity 77, 231–239. pt.3.

Immink, R.G.H., Gadella Jr., T.W.J., Ferrario, S., Busscher, M., Angenent, G.C., 2002. Analysis of MADS box protein-protein interactions in living plant cells. Proc. Natl. Acad. Sci. U. S. A. 99 (4), 2416–2421.

Jakse, M., Hirschegger, P., Bohanec, B., Havey, M.J., 2010. Evaluation of gynogenic responsiveness and pollen viability of selfed doubled haploid onion lines and chromosome doubling via somatic regeneration. J. Amer. Soc. Hort. Sci. 135 (1), 67–73.

Jia, Y., Qi-Xiang, Z., Hui-Tang, P., Shi-Qin, W., Qing-Lin, L., Ling-Xia, S., 2014. Callus induction and haploid plant regeneration from baby primrose (*Primula forbesii* Franch.) anther culture. Sci. Hortic. 176, 273–281.

Kadota, M., Han, D.S., Niimi, Y., 2002. Plant regeneration from anther-derived embryos of apple and pear. HortScience 37 (6), 962–965.

Kasha, K.J., Yao, Q., Simion, E., Hu, T., Oro, R., 1995. Production and application of doubled haploids in crops. In: Sigurbjornsson, B., Machi, S., Dargie, J.D., Gustafson, J.P. (Eds.), Induced Mutations and Molecular Techniques for Crop Improvement. IAEA, Vienna (Australia), pp. 23–38.

Keim, P., Beavis, W.D., Schupp, J.M., Baltazar, B.M., Mansur, L., Freestone, R.E., Vahedian, M., Webb, D.M., 1994. RFLP analysis of soybean breeding populations. I. Genetic structure differences due to inbreeding methods. Crop Sci. 34 (1), 55–61.

Kurtar, E.S., Balkaya, A., Kandemir, D., 2016. Evaluation of haploidization efficiency in winter squash (*Cucurbita maxima* Duch.) and pumpkin (*Cucurbita moschata* Duch.) through anther culture. Plant Cell Tissue Organ Cult. 127 (2), 497–511.

Lim, W., Earle, E.D., 2009. Enhanced recovery of doubled haploid lines from parthenogenetic plants of melon (*Cucumis melo* L.). Plant Cell Tissue Organ Cult. 98 (3), 351–356.

Longin, C.F.H., Gowda, M., Mühleisen, J., Ebmeyer, E., Kazman, E., Schachschneider, R., Schacht, J., Kirchhoff, M., Zhao, Y., Reif, J.C., 2013. Hybrid wheat: quantitative genetic parameters and consequences for the design of breeding programs. Theor. Appl. Genet. 126 (11), 2791–2801.

Macchiavelli, R., Beaver, J.S., 2001. Effect of number of seed bulked and population size on genetic variability when using the multiple-seed procedure of SSD. Crop Sci. 41 (5), 1513–1516.

Maluszynski, M. (Ed.), 2003. Doubled Haploid Production in Crop Plants: A Manual. Kluwer Academic Publishers, Dordrecht (Netherlands). 428 pp.

Maraschin, S.F., de Priester, W., Spaink, H.P., Wang, M., 2005. Androgenic switch: an example of plant embryogenesis from the male gametophyte perspective. J. Exp. Bot. 56 (417), 1711–1726.

Marulanda, J.J., Mi, X., Melchinger, A.E., Jian-Long, X., Würschum, T., Longin, C., Friedrich, H., 2016. Optimum breeding strategies using genomic selection for hybrid breeding in wheat, maize, rye, barley, rice and triticale. Theor. Appl. Genet. 129 (10), 1901–1913.

Matsubara, S., Dohya, N., Murakami, K., 1995. Callus formation and regeneration of adventitious embryos from carrot, fennel and mitsuba microspores by anther and isolated microspore cultures. Acta Hortic. 392, 129–137.

Melchers, G., Mohri, Y., Watanabe, K., Wakabayashi, S., Harada, K., 1992. One-step generation of cytoplasmic male sterility by fusion of mitochondrial-inactivated tomato protoplasts with nuclear-inactivated *Solanum* protoplasts. Proc. Natl. Acad. Sci. U. S. A. 89 (15), 6832–6836.

Neuenschwander, B., Baumann, T.W., 1995. Increased frequency of dividing microspores and improved maintenance of multicellular microspores of *Coffea arabica* in medium with coconut milk. Plant Cell Tissue Organ Cult. 40 (1), 49–54.

Nichols, P.G.H., Cocks, P.S., Francis, C.M., 2009. Evolution over 16 years in a bulk-hybrid population of subterranean clover (*Trifolium subterraneum* L.) at two contrasting sites in South-Western Australia. Euphytica 169 (1), 31–48.

Nichterlein, K., Friedt, W., 1993. Plant regeneration from isolated microspores of linseed (*Linum usitatissimum* L.). Plant Cell Rep. 12 (7/8), 426–430.

Nie, Z., Zhao, T., Yang, S., Gai, J., 2017. Development of a cytoplasmic male-sterile line NJCMS4A for hybrid soybean production. Plant Breed. 136 (4), 516–525.

Oberwalder, B., Schilde-Rentschler, L., Ruoss, B., Wittemann, S., Ninnemann, H., 1998. Asymmetric protoplast fusions between wild species and breeding lines of potato—effect of recipients and genome stability. Theor. Appl. Genet. 97 (8), 1347–1354.

O'Donoughue, L.S., Bennett, M.D., 1994. Comparative responses of tetraploid wheats pollinated with *Zea mays* L. and *Hordeum bulbosum* L. Theor. Appl. Genet. 87 (6), 673–680.

Orton, T.J., 1980. Haploid barley regenerated from callus cultures of *Hordeum vulgare x H. jubatum*. J. Hered. 71, 280–282.

Orton, T.J., Browers, M.A., 1985. Segregation of genetic markers among haploid plants regenerated from anthers of broccoli (*Brassica oleracea* var. *italica*). Theor. Appl. Genet. 69, 637–643.

Palmer, R.G., Gai, J., Sun, H., Burton, J.W., 2001. Production and evaluation of hybrid soybean. Plant Breeding Rev. 21, 263–307.

Pedroso, M.C., Pais, M.S., 1994. Induction of microspore embryogenesis in *Camellia japonica* cv. Elegans. Plant Cell Tissue Organ Cult. 37 (2), 129–136.

Perera, P.I.P., Dedicova, B., Ordonez, C., Kularatne, J.D.J.S., Quintero, M., Ceballos, H., 2012. Recent advances in androgenesis induction of cassava (*Manihot esculenta* Crantz). Acta Hortic. 961, 319–325.

Pickering, R.A., Devaux, P., 1992. Haploid production: approaches and use in plant breeding. Biotech. Agric. 5, 519–547.

Pucci, A., Mazzucato, A., Picarella, M.E., 2017. Phenotypic, genetic and molecular characterization of 7B-1, a conditional male-sterile mutant in tomato. Theor. Appl. Genet. 130 (11), 2361–2374.

Pulido, A., Hernando, A., Bakos, F., Méndez, E., Devic, M., Barnabals, B., Olmedilla, A., 2006. Hordeins are expressed in microspore-derived embryos and also during male gametophyte and very early stages of seed development. J. Exp. Bot. 57 (11), 2837–2846.

Razdan, A., Razdan, M.K., Rajam, M.V., Raina, S.N., 2008. Efficient protocol for in vitro production of androgenic haploids of *Phlox drummondii*. Plant Cell Tissue Organ Cult. 95 (2), 245–250.

Richards, A.J., 1986. Plant Breeding Systems. George Allen & Unwin, London (UK). 529 pp.

Riera-Lizarazu, O., Rines, H.W., Phillips, R.L., 1996. Cytological and molecular characterization of oat x maize partial hybrids. Theor. Appl. Genet. 93 (1/2), 123–135.

Rivas-Sendra, A., Corral-Martínez, P., Camacho-Fernández, C., Seguí-Simarro, J.M., 2015. Improved regeneration of eggplant doubled haploids from microspore-derived calli through organogenesis. Plant Cell Tissue Organ Cult. 122 (3), 759–765.

Schnell, R.J.I.I., Wernsman, E.A., Burk, L.G., 1980. Efficiency of single-seed-descent vs. anther-derived dihaploid breeding methods in tobacco. Crop Sci. 20 (5), 619–622.

Scott, J.W., 2005. Perspectives on tomato disease resistance breeding: past, present, and future. Acta Hortic. 695, 217–224.

Scott, J.W., George Jr., W.L., 1980. Influence of environment and flower maturity on hybrid seed production of exserted stigma tomatoes crossed without emasculation. J. Am. Soc. Hort. Sci. 105 (3), 420–423.

Seifert, F., Bāssow, S., Kumlehn, J., Gnad, H., Scholten, S., 2016. Analysis of wheat microspore embryogenesis induction by transcriptome and small RNA sequencing using the highly responsive cultivar "Svilena". BMC Plant Biol. 16 (1), 97.

Semel, Y., Nissenbaum, J., Menda, N., Zinder, M., Krieger, U., Issman, N., Pleban, T., Lippman, Z., Gur, A., Zamir, D., 2006. Overdominant quantitative trait loci for yield and fitness in tomato. Proc. Natl. Acad. Sci. U. S. A. 103 (35), 12981–12986.

Seymour, D.K., Filiault, D.L., Henry, I.M., Monson-Miller, J., Maruthachalam, R., Pang, A., Comai, L., Chan, S.W.L., Maloof, J.N., 2012. Rapid creation of *Arabidopsis* doubled haploid lines for quantitative trait locus mapping. Proc. Natl. Acad. Sci. U. S. A. 109 (11), 4227–4232.

Shalaby, T.A., 2013. Mode of gene action, heterosis and inbreeding depression for yield and its components in tomato (*Solanum lycopersicum* L.). Sci. Hortic. 164, 540–543.

Sharma, S., Sarkar, D., Pandey, S.K., 2010. Phenotypic characterization and nuclear microsatellite analysis reveal genomic changes and rearrangements underlying androgenesis in tetraploid potatoes (*Solanum tuberosum* L.). Euphytica 171 (3), 313–326.

Singh, S.K., Bhati, P.K., Sharma, A., Sahu, V., 2015. Super hybrid rice in China and India: current status and future prospects. Int. J. Agric. Biol. 17 (2), 1560–1630.

Steele, K.A., Edwards, G., Zhu, J., Witcombe, J.R., 2004. Marker-evaluated selection in rice: shifts in allele frequency among bulks selected in contrasting agricultural environments identify genomic regions of importance to rice adaptation and breeding. Theor. Appl. Genet. 109 (6), 1247–1260.

Svensson, M., Johansson, L.B., 1994. Anther culture of *Fragaria X ananassa*: environmental factors and medium components affecting microspore divisions and callus production. J. Hortic. Sci. 69 (3), 417–426.

Takamura, T., Sakamoto, K., Horikawa, M., 2011. Effect of carbon source on in vitro plant regeneration in anther culture of cyclamen (*Cyclamen persicum* mill.). Acta Hortic. 923, 129–134.

Thengane, S.R., Joshi, M.S., Khuspe, S.S., Mascarenhas, A.F., 1994. Anther culture in *Helianthus annuus* L.: influence of genotype and culture conditions on embryo induction and plant regeneration. Plant Cell Rep. 13 (3/4), 222–226.

Thombre, M.V., Mehetre, S.S., 1979. Cytoplasmic genic male-sterility in American cotton (*Gossypium hirsutum* L.). Curr. Sci. 48 (4), 172.

Tigchelaar, E.C., 1990. Tomatoes for processing in the 90s: genetics and breeding. Acta Hortic. 277, 31–38.

Tigchelaar, E.C., Casali, V.W.D., 1976. Single seed descent: applications and merits in breeding self pollinated crops. Acta Hortic. 63, 85–90.

Tikoo, S.K., Anand, N., 1980. Development of tomato genotypes with exserted stigma and a seedling marker for use as female parents to exploit heterosis. Curr. Sci. 49 (8), 326–327.

Tu, Y.Q., Sun, J., Ge, X.H., Li, Z.Y., 2010. Production and genetic analysis of partial hybrids from intertribal sexual crosses between *Brassica napus* and *Isatis indigotica* and progenies. Genome 53 (2), 146–156.

Tyagi, P., Bowman, D.T., Bourland, F.M., Edmisten, K., Campbell, B.T., Fraser, D.E., Wallace, T., Kuraparthy, V., 2014. Components of hybrid vigor in upland cotton (*Gossypium hirsutum* L.) and their relationship with environment. Euphytica 195 (1), 117–127.

Uno, Y., Ii, Y., Kanechi, M., Inagaki, N., 2002. Haploid production from polyembryonic seeds of *Asparagus officinalis* L. Acta Hortic. 589, 217–224.

Uno, Y., Koda-Katayama, H., Kobayashi, H., 2016. Application of anther culture for efficient haploid production in the genus *Saintpaulia*. Plant Cell Tissue Organ Cult. 125 (2), 241–248.

Vanwynsberghe, L., de Witte, K., Coart, E., Keulemans, J., 2005. Limited application of homozygous genotypes in apple breeding. Plant Breed. 124 (4), 399–403.

Vilela, F.O., do Amaral Júnior, A.T., de Paiva Freitas Júnior, S., Viana, A.P., Pereira, M.G., de Morais Silva, M.G., 2009. Selection of snap bean recombined inbred lines by using EGT and SSD. Euphytica 165 (1), 21–26.

Vrinten, P.L., Nakamura, T., Kasha, K.J., 1999. Characterization of cDNAs expressed in the early stages of microspore embryogenesis in barley (*Hordeum vulgare* L.). Plant Mol. Biol. 41 (4), 455–463.

Watts, A., Kumar, V., Bhat, S.R., 2016. Centromeric histone H3 protein: from basic study to plant breeding applications. J. Plant Biochem. Biotechnol. 25 (4), 339–348.

Whitford, R., Fleury, D., Reif, J.C., Garcia, M., Okada, T., Korzun, V., Langridge, P., 2013. Hybrid breeding in wheat: technologies to improve hybrid wheat seed production. J. Exp. Bot. 64 (18), 5411–5428.

II. BREEDING METHODS

15

Breeding Methods for Outcrossing Plant Species: I. History of Corn Breeding and Open Pollinated Populations

INTRODUCTION

Most plant breeding textbooks address methods pertaining to self-pollinated species first because they are easier to explain and to grasp by readers than are those pertaining to cross-pollinated species. Breeding methods applied for cross-pollinated crop species are more equivocal and mutable than those for self-pollinated crop species, but there are enough common threads to allow for a level of generalization. Population structures that are developed and maintained by outcrossing are more challenging to capture in discrete cultivar entities than they are for self-pollinated populations. We have learned that self-pollinated crop species will be genetically fixed at the conclusion of the breeding program (Chapters 13 and 14). If gametes from different individuals within the population are continually mixed and reshuffled, as they usually are with outcrossing species, the basic parameters that define the plant cultivar are more difficult to attain due to the constant recombination and segregation of alleles resulting in the following phenomena:

- lack of uniformity within populations
- obfuscation of consistent and reproducible differences among populations
- lack of phenotypic consistency from generation to generation

The genetic structure of the population for the maximum relative performance of predominantly outcrossing crop species is one that captures maximum genetic variability or allelic polymorphism. Recall that the progenitor populations in the wild utilize cross-pollination as a strategy for long-term survival of the gene pool (Chapters 2, 4, and 10). Outcrossing allows for the constant unmasking of new alleles and genomic combinations that are initially masked in heterozygous configurations then gradually exposed to the forces of natural selection over evolutionary time. Genomes of wild outcrossing populations typically carry a load of deleterious alleles that are unmasked and result in individual lethality or decreased vitality or fecundity under forced inbreeding. Hence, breeding strategies that have emerged over time tend to minimize strong selection and assortative matings that truncate genetic variability. Simultaneously, strategies for cross-pollinated crop species incorporate disassortative mating and managed selection schemes that allow genetic variability to persist in breeding lines and finished cultivars.

275

Corn, or maize (*Zea mays* L.), is a plant breeding paradigm, the "white mouse" of the discipline, though now rapidly being overtaken in importance and impact by *Arabidopsis thaliana*. Corn has been genetically studied longer and in greater detail than any other plant crop species (*A. thaliana* is not considered to be a crop species). Sustained intensive breeding of corn has demonstrated that selection and controlled mating can eventually eliminate the load of deleterious alleles and intergenic combinations. In some cases, outcrossing crop species may eventually be transformed into self-pollinating crop species and become amenable to the breeding methods described in Chapters 13 and 14. No other crop species rivals maize, however, with regard to the magnitude of applied genetics and breeding research that has been invested during the period 1800 to present. It is instructive, therefore, to review the history of corn breeding as a roadmap for other cross-pollinated species. While most of the knowledge base that has emanated from the body of research on maize genetics and breeding is directly applicable to agronomic crop species, a working knowledge of the history of corn breeding is essential for the student of the plant breeding discipline. The tenets that have been forged from the maize example also provide a solid framework for adaptation in horticulture.

BRIEF HISTORY OF CORN BREEDING

Zea mays resides systematically in Tribe Tripsacaceae of the family Gramineae that includes all other major monocot grain crop species (wheat, rice, oats, barley, rye, sorghum, millet). Corn populations that are now grown commercially descended from a long line of domesticated forms that were derived from wild populations more than 8000 years ago (Galinat, 1988). The cultivated forms bear little resemblance to any wild species. Mangelsdorf and Reeves (1931) conducted an exhaustive study of prospective progenitors of corn and concluded that the initial domestication began in north-central South America. They speculated that, based on clusters of ancient landraces, secondary centers of domestication occurred in the Andes Mountains, Central America, and Mexico. Further, the original ancestor was concluded to be andromonoecious wild pod corn (husks surrounding each seed) that later became extinct (Goodman, 1988).

These findings were subsequently challenged by Noble Laureate George Beadle (see Chapter 1), who championed a hypothesis wherein domesticated corn must have proceeded through a functional intermediate such as teosinte, a related species that still exists in the wild in southern Mexico and Central America. Beadle reasoned that teosinte could be converted directly into palatable food derivatives, such as flour and popped kernels, unlike other wild relatives within the tribe Tripsacaceae. Mangelsdorf and Beadle waged a heated and entertaining debate on the domestication of maize during the 1950s and 60s. Recent evidence from genome and transcriptome sequencing supports Beadle's hypotheses more than Mangelsdorf's, although a complete picture of maize domestication based on molecular evidence is still not fully resolved (Fedoroff, 2003; Wright et al., 2005; Tenaillon and Charcosset, 2011; Liu et al., 2015).

Corn was disseminated throughout the world and discovered to be extremely adaptable within the confines of "temperate" and "tropical" germplasm pools (Liu et al., 2015). Within a relatively short period after its original domestication(s), corn populations were traded to native humans within the Western Hemisphere from Tierra Del Fuego to the northern reaches of Canada and further adapted by progressive mass selection (Wallace and Brown, 1988). By the time the Europeans arrived en masse during the 17th century, corn was a major food crop along the eastern seaboard of North America, used mainly for the production of flour from which flatbreads could be fashioned and cooked over flames (Galinat, 1988). The Europeans took seeds of corn, along with other food crops from the New World (potato, sunflower, tomato, pepper, bean) back to Europe and Asia, where they were adopted and traded (Tenaillon and Charcosset, 2011). Corn is currently grown on every continent and is a major source of sustenance for most geopolitical regions.

The gene pool from which corn was developed was exceedingly heterogeneous, exemplary of an outcrossing species, but molecular studies show that the gene pool was quickly truncated during domestication (Technow et al., 2014). From the tiny central floral rachis that shattered to disperse hard-coated seeds in the wild, the domesticated types that feature prominent staminate and pistillate flowers and a large, sturdy rachis, now called a "cob" were bred by mass selection (Wallace and Brown, 1988). Many primary types or classes of domesticated corn were ultimately developed (Mangelsdorf, 1974):

- *Pod types:* retained the characteristic of both seeds and the whole pistillate flower is enclosed in husks. The genetic control of husk development is relatively simple with very few genes involved.

- *Flint types:* The endosperm is sectored such that soft starches (amylopectins) are in the center and overlaid by harder starches (amyloses). Crosses with teosinte results in a hybrid that bears an uncanny resemblance to modern day popcorn. Flint types have been used in the development of inbreds that contribute to major corn belt hybrids
- *Popcorns:* The endosperm contains only a small proportion of amylopectin, consisting mainly of amylose.
- *Dent types:* The endosperm is sectored such that amylopectin extends from the core of the seed to the crown and amylose encircles the sides of the seed. As the seed dries, the crown shrinks and sinks into the seed, creating a dimple, or "dent". Dent inbreds combine well with flint types to comprise hybrid varieties that are used in wet milling.
- *Sweet types:* The enzymes that combine in the synthesis of starch from disaccharides are low in cellular concentration or are missing. Several single gene mutations have been discovered that impart specific properties to sweet corn: *Su, Se, Sh, Br.* Each mutation encodes a specific enzyme or regulatory pathway in starch biosynthetic pathways.
- *Flour types:* The endosperm consists entirely of amylopectin. Genetic studies have determined that flour types were derived from flint or dent types with specific mutations that inactivated amylose biosynthetic enzymes. Flour types are used mainly for silage. Despite the name, they are not good for flour production or wet milling.
- *Waxy types:* The endosperm consists mainly of amylopectin and the seeds have a relatively high oil content. Waxy types were derived from corn germplasm that was brought to China and subsequently selected for this particular trait.
- *Ornamental or "Indian" types:* The pigmentation of the aleurone layer is under the control of transposable elements that reside within the genomes of ornamental types. The DNA insertion sequences transpose during aleurone development, activating and deactivating anthocyanin pigment biosynthetic genes, resulting in a myriad of seed colors and patterns within and among plants. Ornamental cultivars are very popular for autumn decor.

On a phenotypic dimension separate from seed composition lies the range of plant adaptation. The magnitude of adaptation for corn is extreme as compared with other domesticated plant species as is evident from the importance of corn in nearly all regions of the planet. Corn was first domesticated in low-lying areas close to the equator, characterized as tropical. Tropical corn populations usually exhibit a short day flowering habit characterized by a tendency to flower only following continuous exposure to 12:12 h light-dark cycles. If a population adapted to the tropics is grown during the summer in temperate regions, the result is often plants 20 ft (6.8 m) in height that either flower late in the season, when the day length drops to less than 12 h, or do not flower at all before the first killing frost (Mangelsdorf, 1974).

Populations that are adapted to temperate zones are day length insensitive, or tend towards a long-day flowering habit. Corn cultivars have been developed that encompass the range of latitude daylight characteristics from 0 to 60° North and South latitudes, and from 0 to 12,000 ft (3800 M) in altitude. Genetically specialized populations have been developed that perform well under a broad range of growing conditions (e.g., arid to humid; temperature ranges; soil types from pure sand to pure clay).

Thousands of loci have been characterized and mapped to the 10 linkage groups, including a multitude of genes of agronomic importance: starch composition, floral morphology, plant growth habit, pigmentation, and disease resistance (Jiao et al., 2017). Most of the alleles found in modern-day corn cultivars were present within the wild progenitor populations from which the present domesticates were derived, a testament to the magnitude of latent genetic variability that accumulated as a consequence of outcrossing during millions of years of evolution (Peterson, 1998; Liu et al., 2015).

Beyond the value of corn as a fundamental food source for humans, *Z. mays* is utilized in other ways as well. The wet milling process drives the economics of corn in most developed areas of the world. The market for grain bifurcated into human and non-human uses during the 20th century, before GMO cultivars entered the scene. GMO cultivars are often excluded from human uses due to adverse consumer attitudes towards the application of transgenic technology in food, drugs, and cosmetics. Among the non-human uses of corn are animal feed rations, high fructose syrups, and, more recently, fermentation for the production of ethanol for use as fuel in internal combustion engines. As the planet is depleted of fossil energy reserves this source of solar energy, converted to chemical energy in carbohydrates and lipids by corn and other species, will become pivotal to the sustenance of energy-dependent lifestyles. Conversion from fossil fuels to a plant-based energy system is slowly occurring as ethanol and biodiesel derived from plant agriculture is becoming more prevalent, and as crude oil reserves are progressively depleted (Schwietzke et al., 2009).

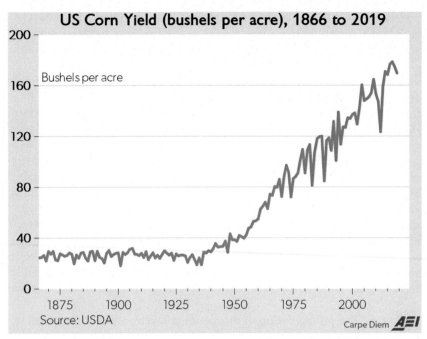

FIG. 15.1   Estimated per bushel corn yields in the U.S. during the period 1866 to 2011. Dramatic yield increases resulted during the transition from open-pollinated to hybrid varieties. Research has shown that most of the gains were attributable to genetic factors. *Used with permission from Mark J. Perry.*

High oil corn types have been developed from the waxy genetic backgrounds for use when the food vegetable oil market overpowers that for wet milling (Goldman, 2004). Certain cultivars are used exclusively for silage because corn stalks store a large quantity of carbohydrates during grain fill. Popcorn types are important contributors to the salty snack food market sector, and sweet corn is consumed by humans as fresh, frozen, canned, and dehydrated food. Finally, the ornamental value of corn in developed countries, particularly in North America, is substantial. In addition to ornamental corn cultivars, many corn growers derive more economic benefit from the harvested stalks and corn mazes for fall ornamentation than from the actual food crop harvest.

Corn is an archetype of the breeding of cross-pollinated crop species. Mass selection by early human cultures for progressively more useful and specialized populations gave way to more intensive, focused breeding during the 20th century. Tremendous gains in economic output were realized as a consequence of these efforts (Fig. 15.1). Most of the advances in corn yield were a direct consequence of genetic and not cultural factors (Pollak, 2003). Biotechnology is now having a great impact on the cultivated corn, particularly resistance to lepidopterous larval insect pests such as corn ear worm (Hurley et al., 2004). GMO crops that are resistant to pests and herbicides present inherent environmental or ecological risks, such as the potential for negative impacts on beneficial or desirable species. A highly-documented example is the negative impacts of Bt on the monarch butterfly (Tschenn et al., 2001). Beyond plant breeding activities, human cultures and communication also played important roles in the overall domestication processes. As humans traveled and progressed, so did the corn populations they had been given by ancestors. By the time that Europeans arrived on the North American continent, corn was regarded so highly by native civilizations that their theology was based largely on the cycle of corn planting, growing, harvesting, and milling (Peterson, 1998).

HYBRID CORN CULTIVARS

Before 1900 all genetic advances by humans in the corn gene pool were by mass selection (Chapter 5), although controlled mating had been shown to be an important factor for sugar beet improvement by 1850. The first publication that described a new breeding method from mass selection was Hopkins (1899) (cited by Hallauer et al., 2010) who is credited with the introduction of the ear-to-row method, adapted from the techniques developed for sugar beet by Vilmorin (see Chapter 1 for Vilmorin's biography). The ear-to-row method was the first form of progeny test, consisting of individuals selected for the next breeding cycle based on the performance of all sibling individuals from a single ear. The method was improved incrementally with detasseling to promote outcrossing, and saving half ears while the other half was tested.

The successes of the ear-to-row method in forging yield increases precipitated more intensive breeding efforts (Hallauer et al., 2010). The negative consequences of inbreeding quickly became apparent, and measures were developed to foster heterozygosity and vigor. Within 20 years, it was apparent that the method was useful to develop populations to a certain level, but that gains were eventually diminished. The notion that hybridity is associated with crop plant performance, or vigor, was not new in 1900. The first published reference to hybrid vigor is attributed to Kölreuter and Knight in the 1760s. Many others subsequently commented on the phenomenon, but the most impact was achieved by Charles Darwin (1895) who published the seminal treatise entitled *The Effects of Cross and Self Fertilization in the Vegetable Kingdom* that documented hybrid vigor among certain of his crosses. Contemporaneously, American William J. Beal reported in 1878 that F_1 hybrid crosses of commercial corn cultivars exhibited promising performance, and suggested the utilization of hybridization as a breeding strategy (Smith and Betrán, 2004).

Edward East of the Connecticut Agricultural Experiment Station and George Shull of Cold Spring Harbor Laboratory (see Chapter 1 for biographies) began studies reported in 1905 on the effects of inbreeding and hybridization on corn vigor (Hallauer et al., 2010). The populations originating from East's pioneering experiments are still being carried forward in the present day. They coined the terms inbreeding depression and hybrid vigor, or heterosis, all still in wide usage. As was discussed earlier and elsewhere in this text, inbreeding depression is the progressive loss of general vigor as individuals within a population are forcefully self pollinated.

Hybrid vigor, or *heterosis*, is the tendency for progeny to exceed the average of parents, or midparent, with regard to a trait of interest. The aggregate result of many heterotic loci is a composite performance level that exceeds both parents, sometimes also called heterosis, but more accurately termed *heterobeltiosis*. Bruce (1910; cited in Paterson et al., 1991) and Jones (1917) offered the first explanations of heterosis based on emerging knowledge of Mendelian genetics. They speculated that the phenomenon could be traced to the cumulative effects of desirable dominant or semi-dominant alleles at many independent loci, fostered by linkages in coupling. East (1936) later hypothesized that the cumulative effects of overdominance were largely responsible for heterosis. He provided an eloquent mechanistic explanation of how gene products from a heterozygous genotype would imbue a more vigorous phenotype than would either homozygote. The debate over the molecular and developmental mechanisms of heterosis raged and continues to the present. Even in the age of accelerating molecular capabilities, research articles on the phenomenon of heterosis have not fully unraveled the underlying mechanisms. The heterosis phenomenon probably includes elements of both the "dominance" and "overdominance" theories.

Following the exciting results reported by East, Shull, and Donald Jones, other scientists conducted important early work with corn hybridization that was crucial to the rapid advances that were to come in the following decades. Most notably, G.N. Collins and later F.D. Richey of the USDA began and sustained work on corn inbreeding that ultimately paved the way to hybrid varieties adapted to the U.S. corn belt. Henry A. Wallace took a different route, envisioning the commercial opportunities that hybrid corn varieties presented. He began a private enterprise in 1913 based on the development of proprietary corn inbreds that ultimately became the Pioneer Hi-Bred Corn Company, currently the largest seed company in the world and a major player in the world of biotechnology. Wallace was a highly charismatic and commanding personality who later rose to political heights as Secretary of Agriculture under President Franklin Roosevelt and Vice President of the U.S. from 1940 to 1944 (see Chapter 1 for Wallace's biography).

The Purnell Act of 1925 effectively federalized U.S. corn breeding efforts, citing the work as being of strategic importance to the future of the food supply. The Act placed all public corn breeding programs under USDA control, in cooperation with the state experiment stations in the North Central Region. This alignment of independent programs enabled for the timely exchange of both information and germplasm that greatly accelerated the progression of corn hybrids onto the commercial agriculture scene. To this day, the Purnell Act is one of the most enduring examples of successful cooperation between the U.S. USDA/ARS and state agricultural experiment stations and extension organizations (Smith, 2009).

Between 1910 and 1920, hybrid corn breeding efforts were focused on F_1 hybrid varieties, according to the hypothesis articulated by Shull (1909). By 1920, however, it was apparent that inbreeding depression would be a fundamental deterrent to the commercial success of hybrid corn varieties (Wallace and Brown, 1988). The major advancement of the late 1920s and 1930s came with the development of an intermediate strategy to overcome inbreeding depression and still reap the benefits of hybrid vigor in populations planted by farmers: higher female vigor through increased genetic variability (Jones, 1922). The strategies advanced by Jones were top cross, 3-way, double cross, and synthetic hybrid populations (Fig. 15.2). All of these types are comprised of seed that is harvested from mildly heterotic female parent populations, thus rendering seed costs more affordable to farmers still most familiar with open-pollinated populations.

A high volume of research, much of it under the auspices of the Purnell Act, was invested into the ability to predict the performance of multi-inbred hybrid combinations such as 3-way and double crosses (Richey and Sprague, 1931;

Hybrid cultivars

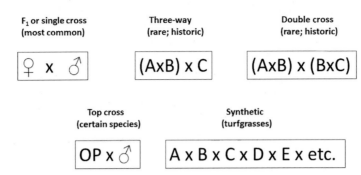

FIG. 15.2 The comparative structures of different forms of hybrid cultivars.

Jenkins, 1934). Much insight into corn genetics and quantitative inheritance in general was gained from these studies. As a consequence of the introduction of double cross and 3-way hybrid strategies, the proportion of U.S. corn acreage attributable to hybrids as compared to open-pollinated cultivars grew from ~1% to 81% during the 16-year period from 1935 to 1951. By 1965, virtually 100% of the U.S. corn crop was produced from hybrid varieties, a substantial proportion still comprised of double-cross and 3-way hybrids (Peterson, 1998).

Plant breeders continued to chart incremental gains of inbred performance in maize after 1965 (Troyer and Wellin, 2009). While the selection criteria were applied mainly to traits contributing to combining ability, the pool of alleles gradually shifted to those that were predominantly desirable types, and undesirable forms contributing to the former genetic load were eliminated. The phenomenon of inbreeding depression was slowly eclipsed to the point that F_1 (also known as single cross) hybrids were economically feasible. By 1995, nearly 100% of the corn crop was comprised of F_1 hybrids (Pollak, 2003). Plant breeders boasted that the performance of inbreds was beginning to rival the best hybrids. The value of hybridity to the industry was overwhelming, however, and no serious consideration has been given to the prospect of pure line corn varieties. Protection from piracy offered by hybrid varieties is a major business consideration for seed companies (Smith, 2009).

The overall method for breeding hybrid varieties has been separated into distinct phases that will be described and analyzed in more depth below. Briefly, the first phase consists of the accumulation of desirable alleles into population gene reservoirs. The second phase pertains to the extraction of inbred lines from gene reservoirs and selection based on general combining ability. Finally, fixed or inbred lines are tested for specific combining ability, the top combinations being advanced to the rigorous testing phase.

The concept of population improvement was solidified after inbreds had been extracted from most accessible sources of open-pollinated populations and tested in hybrid combinations. After all such combinations had been exhausted the only viable means to attain enhanced performance was to produce better inbreds. That reasoning led, ultimately, to the realization that the collection of alleles in the source population was of critical importance to the inbreds extracted from it. The strategies that had been developed for self-pollinated crops were employed for this purpose, most notably the pedigree method (Chapter 13).

Jenkins (1940) and, later in more detail, Hull (1945, 1952) introduced the strategy known as recurrent selection to incorporate the outcrossing mating process into a program aimed at the accumulation of desirable alleles in single populations. Sprague and Brimhall (1950) and a series of studies by Lonnquist (1949, 1951) verified the legitimacy of recurrent selection. Comstock et al. (1949) took the strategy to the next logical step. Since the end product was to be a hybrid variety, and two or more contrasting gene pools were known to maximize heterosis, populations subjected to improvement by recurrent selection should bear a correspondence to one another. Ultimately, they should be selected for the presence of alleles that tend to combine well with each other. This simple concept was refined into what is now known as "reciprocal recurrent selection" (Hallauer et al., 2010).

Breeding and genetic work on corn have continued to pave the way for other outcrossing plant species. The pioneering experiments reported by McClintock (1939, 1956) illuminated the phenomenon of "controlling elements" that later came to be known as insertion sequences and transposons that were later found in all eukaryotic organisms in which they were studied. The maize linkage map eventually became saturated with conventional, isozyme, RFLP, AFLP, RAPD, SSR, SNPs, and many other molecular marker systems that are currently in wide use as surrogates for

"QTL" (Chapter 9). Early problems encountered in the transformation and tissue culture of corn were overcome, and GMO corn varieties exhibiting pest and herbicide resistance are now commonplace. The corn genome has been 100% sequenced in 2009 with regular updates since then (Jiao et al., 2017).

COMBINING ABILITY AND ESTIMATION METHODS

In most modern cultivars of outcrossing plant species, each individual within a population is derived from the fusion of gametes from genetically disparate parents. While mass selection involves the individual within a population, selection for combining ability goes back one generation and is practiced on the parents of the individual. The broad definition of combining ability (CA) is the breeding value of an individual (Sprague and Tatum, 1942). More precisely, the combining ability of the genotype of an individual is the predicted relative performance of progeny from that individual. An individual's phenotype is one component of combining ability but depending on heritability does not constitute a reliable predictive feature. Two inferior individuals may produce gametes that fuse to form a progeny that has superior performance. If the parents are inbreds drawn directly from an outcrossing population this will often be evident due to inbreeding depression.

Earlier in this chapter, during the discussion of the history of corn breeding, the results of Beal (1878) were cited as the first documented report of heterosis. What Beal measured was one component of combining ability. The individual plant is a package wherein a particular combination of genes is parked in time and manifested to the plant breeder or farmer in the form of a phenotype. One way to characterize a given individual is in terms of combining ability. How does the individual, or genotype of that individual, tend to perform when gametes are fused with those from another individual? The performance of progeny from the hybridization of two individuals is termed *specific combining ability* (SCA) since it is not necessarily predictive of progeny resulting from crosses with other individuals (Sprague and Tatum, 1942; Comstock et al., 1949). The mean performance of progeny from a given individual is determined from the pairwise crosses of a set of prospective parent populations called a *diallel* or *half-diallel*, and termed the *average combining ability* (ACA), the impacts of which will be discussed below (Fig. 15.3).

The daunting task of performing a half-diallel of crosses among hundreds or even thousands of individuals was confronted by plant breeders during the early 20th century (Baker, 1978). Was there a simpler way to sift through a population of individuals and determine which will tend to combine well with other parents and which will not? Initially, corn breeders measured the performance of progeny of hybrids between inbreds and open-pollinated standard cultivars, also known as "top-crosses". Populations later known as "testers" were empirically developed

Half-diallel for determination of ACA and SCA

Performance of progeny in inbred x inbred crosses: Avg = ACA; Cells = SCA

	A	B	C	D	E	F	Avg
A		3	[2]	4	3	1	**2.6**
B			[5]	4	3	2	**3.4**
C				[2]	[3]	[1]	**2.6**
D					6	3	**3.8**
E						2	**3.4**
F							**1.8**

FIG. 15.3   A hypothetical half diallel for determining SCA for a cross of two inbreds and estimating ACA for an inbred. The SCA of AxB is 3. The ACA of B is 2.6.

that appeared to constitute excellent predictors of combining ability with any other parent (Richey, 1924; Davis, 1927; Jenkins and Brunson, 1932). The term *general combining ability* (GCA) was established to denote the performance of the progeny of a given individual in crosses with one or more widely acknowledged tester populations (Sprague, 1946). GCA is an estimator of ACA and V_A.

Depending on the genetic constitution of populations used to determine SCA, ACA and GCA may vary (Sprague and Tatum, 1942). As the number of individuals in the program increases, GCA and ACA tend to become more consistent. GCA is much easier to predict than is ACA since it requires only the evaluation of x_i x tester progeny, not the evaluation of all x_i x $x_{j\rightarrow k}$ progeny. While GCA is considered to be an estimate of the additive genetic variability for the traits under selection, SCA effects are also attributable to non-additive variance components. For this reason, the convention is that SCA is often expressed as a deviation from GCA. In a typical breeding program for outcrossing species that employs progeny tests as a basis for selection, GCA will be used during early stages of the breeding program to select for V_A and selection for SCA is applied later in the program for V_D, V_I, V_{GxE}, etc. after the number of candidate genotypes has been reduced to a manageable level.

POPULATION IMPROVEMENT FOR OUTCROSSING SPECIES

In contrast to pure lines for self-pollinated crop species, the genetic structure of finished cultivars of cross-pollinated crop species does not target panhomozygosity. Crop performance is maximized by polyheterozygosity that is manifested phenotypically as heterosis. In the most simplistic case, heterozygosity is captured and sustained in a hypothetically panmictic population that has been bred for enhanced frequencies of desirable alleles and reduced frequencies of undesirable (or sublethal) alleles, known as an open-pollinated population (Fig. 15.4). The hybrid population requires a more targeted and intensive breeding effort that produces two or more parental populations that will be hybridized to constitute the finished product (see Chapter 16). Alternatively, a desirable genotype may be fixed asexually at any point in the breeding program then propagated clonally to constitute commercial quantities of propagules for planting (Fig. 15.4).

There are four primary strategies used to increase the frequencies of desirable alleles and gene combinations within populations of cross-pollinated crop species; leading to one of these alternative cultivar population structures (Fig. 15.4):

- Mass selection
- Recurrent selection for phenotype
- Recurrent selection for combining ability
- Reciprocal recurrent selection for combining ability

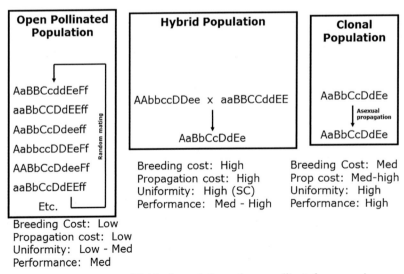

FIG. 15.4   Alternative population genetic structures of finished populations of cross-pollinated crop species.

Mass selection has already been described in Chapter 6. It is the most primitive and intuitive of the alternatives. The plant breeder examines the individuals within a population and chooses the best ones from which to harvest seed for the next generation or planting cycle. There is no control over mating, so only one-half of the gametes in any given selection is controlled. The source of the pollen that fused with the egg on the selected plant, or fruit, is presumed to be a random sample of all compatible and available staminate flowers. If the pollen came from a genotype replete with undesirable alleles desirable alleles selected from the female parent may not result in meaningful progress (Allard, 1999).

Recurrent selection is simply mass selection with control exercised over mating (Fig. 15.5). Plant breeders debate whether there should be a self-pollination between each intermating and selection cycle, but we will consider all permutations under the same banner. In the simplest form, recurrent selection consists of selecting desired individuals within a population prior to mating followed by the intermating of selected individuals. Consequently, both female and male gametes have undergone selection, and genetic progress should be at least 100% faster than for mass selection.

If the trait subjected to selection is associated with reproduction or progeny dispersal, such as fruit or seed, controlled mating becomes problematic due to confounding effects. Since corn falls into this category, it is not surprising that substantial research has been invested in the development of recurrent selection methodologies that account for this limitation. The step of self-pollination between intermating cycles was introduced for this purpose and to unmask recessive alleles to subsequent selection. Each individual within the RC_n population is self-pollinated and seeds are collected. It is immediately apparent that generating and managing all of these populations will invoke resource issues. A portion of the seeds of the selfed progeny of each RC_n individual is grown and resulting the population is evaluated based on phenotype. The remaining seeds are then planted and allowed to intermate.

Recurrent selection for combining ability is, as the name implies, based on the use of a progeny test as the basis for selection of individuals within a population (Fig. 15.6). During early generations, selfed individuals are crossed with a tester line to determine GCA. Individuals with the highest GCA are selected and intermated. This cycle is repeated, and if Vp is sufficiently low, subsequent cycles may be promulgated wherein individual x individual crosses are conducted to determine SCA. The pairwise combinations that exhibit the highest SCA are then intermated.

A synthetic population is, by definition, one that is constituted by the progenies of three or more inbreds (Hallauer et al., 1988). In contrast, a composite population is an admixture of fixed genotypes of a predominantly self-pollinated crop species (Allard, 1999). The conceptual breeding approach is to separate the genetic components of a diverse population then perform selection and controlled mating among those components. At some future time, the selected components are mixed back together to reconstitute a population that contains genetic elements derived from the original population, but with enhanced allelic frequencies and linkages. The composite approach provides for any genetic elements to be added into the mix. A popular strategy is to inbreed with selection for GCA then later for SCA (Lonnquist, 1961).

Recurrent Selection for Breeding OP and Hybrid Varieties

A breeding method used to genetically improve populations: based on selection and controlled mating to increase the frequency of desired alleles

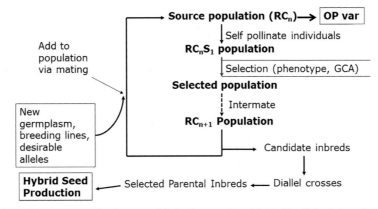

FIG. 15.5 The basic cyclic scheme for recurrent selection to enrich the frequencies of desirable alleles in breeding populations.

Selection Based on Progeny Tests

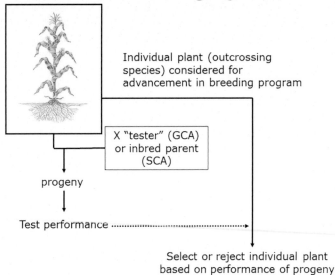

FIG. 15.6　Selection based on progeny tests, also known as combining ability (GCA, SCA).

OPEN POLLINATED POPULATIONS

The process of domestication was associated with the lineage of open pollinated (OP) populations from the wild form to the domesticated derivative in tiny per generation increments (Galinat, 1988). Mass selection was performed on phenotypes of agricultural populations resulting in the gradual change of underlying allelic frequencies over time. Ultimately, all of the populations of domesticated plant species were handed down to recent generations in the form of OP populations (Briggs and Walters, 1997). It was not until the laws of genetics were well established that the powerful implications of altered mating were employed to drive rapid genetic changes, starting in the early 20th century (Duvick, 1996). Yet OP populations persist as viable commercial products. The gardening seed catalogs are replete with OP cultivars and they are favored by organic farmers who shun large commercial seed companies and the hybrid and GMO cultivars they offer. OP cultivars also fit the paradigm for sustainable farming because they may be used to reproduce themselves. In other words, the farmer may save their own seed for the next year's crop.

The "OP population" designation is generally used whenever no control is exercised over mating or where only supplemental pollen vectors may be added to enhance seed yields. The term "OP" is used interchangeably for both outcrossing and inbreeding species. When "OP" is used in conjunction with outcrossing species the implication is that methods will be employed to promote random mating within the population utilized for seed production. The OP mating process is almost always accomplished in an open field particularly since it is a relatively primitive method associated with marginal varietal performance. The use of controlled environments, such as greenhouses, in conjunction with OP populations is usually considered to be too expensive. The field is chosen based on suitability to the species and uniformity of critical adaphic factors such as elevation, soil type, moisture, and wind. The established plants are nurtured to ensure robust flowering and vigorous gametes.

Genetic studies of populations of outcrossing plant species usually assume that mating is completely random. In reality, mating is rarely truly random. There are many reasons for the lack of randomness, primarily the interaction of the underlying genetic and environmental factors that control mating in a complex manner (Richards, 1986). Plants in natural habitats do not align themselves in regular geometric patterns. Rather, they tend to occur where the seed dispersal process from the previous mating cycle happened to deposit the next generation. Agents that promote seed dispersals, such as integumental appendages, wind, water, and animals, are not manifested randomly. In some cases, the genotype of the plant bearing the seed, or of the seed itself, actually affects the seed dispersal process (a GxE interaction).

If the aim is to use open pollination as a method to reproduce a population genetically the choice of individuals from which the next generation will be harvested becomes critical. Population size must be adequate to allow for the accurate representation of allelic frequencies and linkage relationships that are responsible for the most important components of overall economic performance. If the population is inadequate, genetic drift may contribute to altered allelic frequencies, particularly if $P \geq 0.8$ or ≤ 0.2 for loci of key importance. Selection, inadvertent or otherwise, may introduce undesirable genetic changes to the population. For example, if the economic attributes are non-reproductive, care must be exercised to ensure that reproductive characters are not enriched, such as early flowering. At the other extreme, selection may be practiced to eliminate early flowering types, thus correcting for the tendency for early flowering types to be overrepresented in the progeny pool.

Random mating implies that the probabilities of all potential matings within the population are equal. Developmental differences, physical distances between plants, and the spatial behavior of pollen vectors cause the probabilities of matings to be non-random. Relative probabilities are higher for matings of developmentally equivalent individuals that are nearby and lower for matings between individuals that are developmentally asynchronous and physically separated. It is obvious that, for wind-pollinated species, the downwind individuals in a prevailing wind pattern will encounter more pollen sources than will upwind individuals. If the plant species is insect pollinated the nesting and feeding behavior of the vector can greatly affect the patterns of pollen flow. Depending on the targeted species, endemic pollen vectors may be adequate, or may be added artificially. Large powered fans and bee hives are two common examples.

References

Allard, R.W., 1999. Principles of Plant Breeding, second ed. Wiley & Sons, Inc., New York, NY. 254 pp.

Baker, R.J., 1978. Issues in diallel analysis. Crop Sci. 18 (4), 533–536.

Beal, W.J., 1878. Report of the Michigan State Board of Agriculture. vol. 17. Michigan State College, East Lansing, MI, 445–457.

Briggs, D., Walters, S.M., 1997. Plant Variation and Evolution. vol. 512 Cambridge Univ. Press, Cambridge, UK.

Comstock, R.E., Robinson, H.F., Harvey, P.H., 1949. A breeding procedure designed to make use of both general and specific combining ability. Agron. J. 41, 360–367.

Darwin, C., 1895. The Effects of Cross and Self Fertilisation in the Vegetable Kingdom. D. Appleton, Cambridge, UK. 482 pp.

Davis, R. L. 1927. Report of the plant breeder. Rep. Puerto Rico Agric. Exp. Sta., PR (USA), pp. 14–15.

Duvick, D.N., 1996. Plant breeding, an evolutionary concept. Crop Sci. 36 (3), 539–548.

East, E.M., 1936. Heterosis. Genetics 21, 375–397.

Fedoroff, N.V., 2003. Prehistoric GM corn. Science 302 (5648), 1158–1159.

Galinat, W.C., 1988. The origin of corn. Agronomy 18, 1–31.

Goldman, I.L., 2004. The intellectual legacy of the Illinois long-term selection experiment. Plant Breed. Rev. 24, 61–78.

Goodman, M.M., 1988. The history and evolution of maize. Crit. Rev. Plant Sci. 7, 197–220.

Hallauer, A.R., Russell, W.A., Lamkey, K.R., 1988. Corn breeding. In: Sprague, G.F., Dudley, J.W. (Eds.), Corn and Corn Improvement. Amer. Soc. Agron., Madison, WI, pp. 463–564.

Hallauer, A.R., Carena, M.J., Miranda Filho, J.B., 2010. Quantitative Genetics in Maize Breeding. Springer Publ., New York, NY, p. 312. 664 pp.

Hopkins, C.G., 1899. Illinois Agricultural Experiment Station Bulletin 55. University of Illinois, Champaign-Urbana, IL.

Hull, H.F., 1945. Recurrent selection for specific combining ability in corn. J. Am. Soc. Agron. 37, 134–145.

Hull, H.F., 1952. Recurrent selection and overdominance. In: Gowen, J.W. (Ed.), Heterosis. Iowa State Univ. Press, Ames, IA, pp. 451–473.

Hurley, T.M., Mitchell, P.D., Rice, M.E., 2004. Risk and the value of Bt corn. Am. J. Agric. Econ. 86 (2), 345–358.

Jenkins, M.T., 1934. Methods of estimating the performance of double-crosses in corn. Agron. J. 26, 199–204.

Jenkins, M.T., 1940. The segregation of genes affecting yield of grain in maize. J. Am. Soc. Agron. 32, 55–63.

Jenkins, M.T., Brunson, A.M., 1932. Methods of testing inbred lines of maize in crossbred combinations. J. Am. Soc. Agron. 24, 523–530.

Jiao, Y., Peluso, P., Shi, J., Liang, T., Stitzer, M.P., Wang, B., Campbell, M.S., Stein, J.C., Wie, X., Chin, C.S., Guill, K., Regulski, M., Kumari, S., Olson, A., Gent, J., Schneider, K.L., Wolfgruber, T.K., May, M.R., Springer, N.M., Antoniou, E., McCrombie, W.R., Presting, G.G., McMullen, M., Ross-Ibarra, J., Dawe, R.K., Hastie, A., Rank, D.R., Ware, D., 2017. Improved maize reference genome with single-molecule technologies. Nature 546, 524–527.

Jones, D.F., 1917. Dominance of linked factors as a means of accounting for heterosis. Proc. Natl. Acad. Sci. U. S. A. 11, 310–312.

Jones, D.F., 1922. The productiveness of single and double first generation corn hybrids. J. Am. Soc. Agron. 14, 242–252.

Liu, H., Wang, X., Warburton, M.L., Wen, W., Jin, M., Deng, M., Liu, J., Tong, H., Pan, Q., Yang, X., Yan, J., 2015. Genomic, transcriptomic, and phenomic variation reveals the complex adaptation of modern maize breeding. Mol. Plant 8, 871–884.

Lonnquist, J.H., 1949. The development and performance of synthetic varieties of corn. Agron. J. 41, 153–156.

Lonnquist, J.H., 1951. Recurrent selection as a means of modifying combining ability of corn. Agron. J. 43, 311–315.

Lonnquist, J.H., 1961. Progress from recurrent selection procedures for the improvement of corn populations. In: Research Bulletin No. 197. Agricultural Experiment Station of Nebraska, Lincoln, NE. 32 pp.

Mangelsdorf, P.C., 1974. Corn, Its Origin, Evolution and Improvement. The Belknap Press of Harvard University Press, Cambridge, MA. 273 pp.

Mangelsdorf, P.C., Reeves, R.G., 1931. Hybridization of maize, Tripsacum, and Euchlaena. J. Hered. 22, 328–343.

McClintock, B., 1939. The behavior in successive nuclear divisions of a chromosome broken at meiosis. Proc. Natl. Acad. Sci. U. S. A. 25, 405–416.

McClintock, B., 1956. Controlling elements and the gene. Cold Spring Harb. Symp. Quant. Biol. 21, 197–216.

Paterson, A.H., Tanksley, S.D., Sorrels, M.E., 1991. DNA markers in plant improvement. In: Sparks, D.L. (Ed.), Advances in Agronomy. vol. 46. Academic Press, New York, NY, pp. 40–90.

Peterson, P.A., 1998. Development of maize genetics and breeding. In: Peterson, P.A., Bianchi, A. (Eds.), Maize Genetics and Breeding in the 20th Century. World Scientific Press, Singapore, pp. 1–107.

Pollak, L.M., 2003. The history and success of the public-private project on germplasm enhancement of maize (GEM). Adv. Agron. 78, 45–87.

Richards, A.J., 1986. Plant Breeding Systems. George Allen & Unwin, London (UK). 529 pp.

Richey, F.D., 1924. Effects of selection on the yield of a cross between varieties of corn. Tech. Bull. U.S. Dept. Agric. 1209, 1–19.

Richey, F.D., Sprague, G.F., 1931. Experiments on hybrid vigor and convergent improvement in corn. Tech. Bull. U.S. Dept. Agric. 267, 22.

Schwietzke, S., Kim, Y., Ximenes, E., Mosier, N., Ladisch, M., 2009. Ethanol production from maize. In: Kriz, A.L., Larkins, B.A. (Eds.), Molecular Genetic Approaches to Maize Improvement. Springer Publ., New York, NY, pp. 347–364.

Shull, G.H., 1909. A pure-line method of corn breeding. Rep. Am. Breed. Assoc. 5, 51–59.

Smith, J.S., 2009. The Garden of Invention: Luther Burbank and the Business of Breeding Plants. Penguin Press, New York, NY. 354 pp.

Smith, C.W., Betrán, J., 2004. Corn: Origin, History, Technology, and Production. John Wiley & Sons, New York, NY. 949 pp.

Sprague, G.F., 1946. Early testing of inbred lines. J. Am. Soc. Agron. 38, 108–117.

Sprague, G.F., Brimhall, B., 1950. Relative effectiveness of two systems of selection for oil content of the corn kernel. Agron. J. 42, 83–88.

Sprague, G.F., Tatum, L.A., 1942. General vs. specific combining ability in single crosses of corn. Agron. J. 34, 923–932.

Technow, F., Schrag, T.A., Schipprack, W., Melchinger, A.E., 2014. Identification of key ancestors of modern germplasm in a breeding program of maize. Theor. Appl. Genet. 127 (12), 2545–2553.

Tenaillon, M.I.n., Charcosset, A., 2011. A European perspective on maize history. C. R. Biol. 334 (3), 221–228.

Troyer, A.F., Wellin, E.J., 2009. Heterosis decreasing in hybrids: yield test inbreds. Crop Sci. 49 (6), 1969–1976.

Tschenn, J., Losey, J.E., Jesse, L.H., Obrycki, J.J., Hufbauer, R., 2001. Effects of corn plants and corn pollen on monarch butterfly (Lepidoptera: Danaidae) oviposition behavior. Environ. Entomol. 30 (3), 495–500.

Wallace, H.A., Brown, W.L., 1988. Corn and Its Early Fathers. Iowa State University Press, Ames, IA. 141 pp.

Wright, S.I., Bi, I.V., Schroeder, S.G., Yamasaki, M., Doebley, J.F., McMullen, M.D., Gaut, B.S., 2005. The effects of artificial selection on the maize genome. Science 308 (5726), 1310–1314.

Further Reading

Jones, D.F., 1918. The effects of inbreeding and cross-breeding upon development. Conn. Agric. Exp. Station Bull. 207, 5–100.

Breeding Methods for Outcrossing Plant Species: II. Hybrid Cultivars

INTRODUCTION

A hybrid cultivar is constituted of the progeny of the cross of two or more genetically distinct populations. As was related in Chapter 15 (history of corn breeding), hybrid varieties were highly successful and eventually came to supplant open pollinated (OP) cultivars and dominate the market following the first introduction in the 20th century CE. This progression (OP → hybrid) has recurred in many other crop species as well, including certain self-pollinated crop species. Hybrids are attractive to plant breeders and private seed companies not only because they offer a convenient way to capture heterosis in a uniform genetic background, but also because they are not easy to propagate or appropriate by competitors or customers.

Hybrid cultivars are not exclusive to outcrossing plant species. Commercial hybrid varieties are also common in certain predominantly inbreeding species such as tomato. The debate persists as to the degree of heterosis for harvest yield and quality that is exhibited by the major autogamous grain crops such as wheat, rice, oat, barley, cotton, and soybean. If heterosis can be demonstrated, the next question is whether the economic advantage it may provide is adequate to offset the higher cost of hybrid breeding and seed production. Plant breeders are striving to develop biological tools to facilitate the large-scale production of hybrid seeds in most autogamous plant species, for example male sterility. In predominantly self-pollinating crop species such as tomato, pepper, and eggplant, seeds of hybrid cultivars are produced by hand. Most hybrid seed production from hand emasculation and pollination takes place in Asia to mitigate high labor costs.

After the economic benefits of captured heterosis and unit seed production costs are determined, the question is posed whether the additional development and production costs associated with hybrid cultivars are ameliorated by the benefits imparted by heterosis. A relational formula to consider in deciding whether to pursue a hybrid breeding strategy in a targeted crop species or population thereof is as follows (Allard, 1999):

$$V_{hs} - C_{hs} > V_{op} - C_{op}$$

where V_{hs} = the value of hybrid seed in terms of the total value of the corresponding harvest of a unit land area (e.g., 1.0 Ha); C_{hs} = the cost of hybrid seed per unit planted land area (e.g., 1.0 Ha); V_{op} = the value of OP seed per unit land area (e.g., 1.0 Ha); C_{op} = the cost of OP seed per unit land area (e.g., 1.0 Ha).

Since hybrid seed production is inherently more risky than is the production of OP seed, the estimated simple cost term may not be adequate to estimate the validity of the business opportunity. For example, the hybrid cultivar mandates maintenance of stock seeds of two or more parental populations instead of only one, rendering the consequences of a crop loss greater for the hybrid than for the OP cultivar. Costs should be averaged over several years with the assumption that environmental and/or geopolitical (if the seeds are produced in distant countries) factors may culminate in reduced yields more than might be anticipated. Alternatively, the seed contractor may wish to balance the above equation to account for risks that are not readily accounted for. Ultimately the degree of difference between the net estimated equation values (e.g., $V_{hs} - C_{hs}$ and $V_{op} - C_{op}$).

The original studies on the phenomenon of hybrid vigor reported by Beal (1878) involved crosses of corn cultivars that were not genetically or phenotypically uniform. East (1908) and Shull (1908) applied the principles of pure line development published by Johannsen (1903) to produce genetically uniform corn inbreds. They quantified the effects of inbreeding depression on corn seed yields and extolled the benefits of using inbred parents. During the time frame extending from the 1920s through the 1930s, top cross hybrid corn cultivars (Chapter 15) were common on the commercial agricultural scene in the U.S. The top cross is a hybrid between an inbred line and a non-uniform genetic population such as an OP cultivar, the OP used as female parent to offset lower seed yields experienced with inbreds. Interest in the physiological mechanism of hybrid vigor also propelled a wave of academic activities in the 20th century. Expanding interest in the inheritance of quantitative traits in plants was apparent after 1946 primarily due to observations of heterosis in maize hybrids (Hallauer et al., 2010).

Incremental plant improvements were applied to the hybrid breeding concept from the 1920s to the present and nearly all commercially successful contemporary hybrid corn varieties are F_1 types constituted from highly inbred parents. Genetic fixation allows the plant breeder to obtain the maximum concentration of desirable alleles and linkage relationships. Population fixation provides a convenient way to reduce V_P by reducing V_G to 0. Occasionally the term "F_2 hybrid" is used to describe a population but this is a misleading attribution since the population is not hybrid at all in the sense of a controlled cross of inbred parents. Any sexual propagation of a population of F_1 individuals will lead to sacrifices of both heterosis and phenotypic uniformity.

A clear pathway from germplasm to finished cultivar is essential for success in hybrid breeding programs. The seeds sold to farmers in the case of F_1 hybrid cultivars will be constituted of a mass cross of two inbred lines; one contributing female gametes, the other male gametes (Fig. 16.1). The "direction" of the cross (i.e., $♀P_1 × ♂P_2$ vs. $♀P_2 × ♂P_1$) is often crucial to the success of the finished product. It is very important that the cross is cost-effective to perform on a mass scale. Considerable attention will be devoted to this step later in the chapter, as the cost of hybrid seed production must be addressed from the start of the breeding program.

The two inbred lines must combine well together, or exhibit positive SCA, in addition to having mutually high GCA. The breeding method employed must provide for the generation of genotypes that have exceptional levels of combining ability, and also allows for the selection of the best inbred lines for seed production. Finally, the source

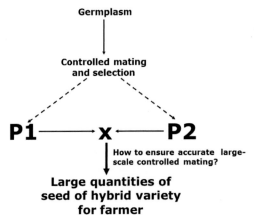

FIG. 16.1　The endpoint of a hybrid breeding program: the production of large quantities of genetically pure F_1 seeds for distribution to farmers for field planting.

population from which inbreds are to be extracted must be appropriate to the breeding objectives at hand. The desired alleles must be present in the highest possible frequencies, and beneficial linkage relationships must be maximized. As the inbreeding process begins, alleles and linkages will be locked into lineages, and the maximum performance will be limited by what was present in the original individuals extracted from the germplasm source(s).

Combining ability is most commonly employed as a criterion for selection as it relates to metric or nondiscrete quantitative traits. Since heterosis is manifested as overall fitness and vigor, the specific traits that are usually targeted include yield or yield components. One component of yield that is often positively impacted by heterosis is stress tolerance. Another that might have a negative impact on yield is stature, leading to yield losses due to plant lodging. As complex traits are dissected further into components, manifestations of heterosis tend to cancel out. For simply inherited traits that the plant breeder plans to incorporate into the hybrid population, the corresponding alleles are incorporated into one or both parents.

INBREEDING DEPRESSION

Inbreeding is a process by which the state of genetic fixation, or homozygosity, is produced. Populations of cross-pollinated crop species usually carry a genetic load of deleterious alleles that are usually masked in heterozygous dominant gene relationships (East and Jones, 1919). If panmictic populations of cross-pollinated crop species are forced to inbreed by self-matings or other matings among related parents, the probability that these deleterious alleles occur in a progeny as a homozygous recessive genotype is elevated. The collective "unmasking" of deleterious alleles during the inbreeding process culminates in the phenomenon of inbreeding depression (ID; East, 1908).

An important requirement of populations that are commercial varieties is that they are relatively uniform, except in rare instances where diversity is an economic attribute. As we have seen under numerous contexts, the plant breeder reduces V_G to as close to 0 as possible, thus minimizing V_P if V_E is well-controlled. One benefit of F_1 hybrid varieties produced from two inbred parents is that all individuals in the finished cultivar are theoretically identical. If the two inbreds are developed such that the representation of desirable alleles and combining ability are maximized, the variety should exhibit both high performance and uniformity.

Inbreeding, however, is not the only process by which genetic fixation may be attained. In Chapter 14, the possibility was raised that haploid gametes could be transformed to the diploid level, then manipulated to become functioning sporophytes. The resulting sporophyte is 100% homozygous, or fixed. This also occurs in nature, albeit rarely.

The most common inbreeding strategy is self-fertilization/pollination, wherein both gametes are derived from the same parent. The gametes from an individual will tend to resemble each other genetically since they were both derived from the same parent. In Chapter 3, the "drive to fixation" that is characteristic of successive cycles of self-pollination was described and presented as the essential force behind the breeding methods used for self-pollinating plant species, culminating in genetically fixed, or pure, lines. The rate of fixation is, theoretically, 50% per locus per generation under recurrent self-pollination. For three unlinked polymorphic loci the proportions of heterozygotes and homozygotes following cycles of inbreeding is as follows (Table 16.1):

TABLE 16.1 Drive to Fixation at 1, 2 and 3 Loci Following a Cross to Produce the F_1 Genotype AaBbCc Followed by Recurrent Cycles of Self-pollination

Generation	% Heterozygous 1 locus	% Homozygous 1 locus	% Homozygous 2 loci	% Homozygous 3 loci
F_1	100.00000	0.00000	0.00000	0.00000
F_2	50.00000	50.00000	25.00000	12.50000
F_3	25.00000	75.00000	56.25000	42.18750
F_4	12.50000	87.50000	76.56250	66.99219
F_5	6.25000	93.75000	87.89063	82.39746
F_6	3.12500	96.87500	93.84766	90.91492
F_7	1.56250	98.43750	96.89941	95.38536
F_8	0.78125	99.28175	98.56866	97.86069

The progression from polyheterozygosity to virtually complete fixation takes approximately ten generations. Linkage is a common phenomenon. Depending on the structure of the genome for a plant species under consideration the probability that two important genes are linked by 20 map units or less is $2.75 \pm 2.25\%$. If the linkage is weaker, 40 map units or less, the probability is doubled. The probabilities that specific pairs of important genes are linked within any given individual plant are additive. Therefore, only 20–200 gene pairs would be needed to reach certainty of finding such a linkage. It is common in plant breeding programs to combine genetically disparate individuals in an attempt to pyramid desirable traits into single genotypes, rendering the probability of repulsion linkages to be relatively high.

The purpose of the inbreeding process in plant breeding schemes is to assort alleles and to select individuals that have the highest proportion of desirable alleles. Linkages in repulsion that remain intact during the inbreeding process are problematic. For the example cited above, the plant breeder would have a relatively short period during which both parental haplotypes exist within a genome, thus enabling a desirable recombination event. Recurrent selection is a method employed by plant breeders to prolong the inbreeding process and to allow more opportunities for desirable recombination events to occur. Successive cycles of self-pollination, selection, and inter-mating result in a slow and prolonged progression to allele fixation, the rate depending on the selection intensity.

The sib-mating process was first advanced as a way to slow down the inbreeding process to allow undesirable repulsion linkages to recombine (Jones, 1916). Siblings are two individuals related by virtue of gametes from the same set of parents. Humans are, of course, incapable of self-mating. Human mating between related individuals is strongly discouraged by society and religious mores; likely tracing from high incidences of mortality and developmental and physiological anomalies among progeny of familial matings.

Any given human individual may carry hundreds of masked recessive deleterious alleles, a few of which may impart future benefits in the form of increased relative fitness (Barrett and Charlesworth, 1991). These recessive alleles remain masked for many generations due to the low relative frequency in the mating population. The carrier individual inevitably passes the alleles to a portion of offspring. The offspring from a sib mating would produce a high (25%) proportion that would carry the deleterious recessive allele in homozygous form. If the parent carried more than one deleterious allele, the chances that sib-mating would generate individuals carrying homozygous deleterious alleles increases proportionately.

Sib-mating within the plant kingdom, however, is not necessarily an undesirable aspect of the natural or artificial mating system. While sib-mating is characterized by a relatively rapid progression to fixation it is incrementally slower than self-pollination. As compared to self-pollination, sib-mating results in homozygosity at half the rate per locus per generation, or 25%. Thus, the same number of generations under sib mating of an individual AaBbCc would generate the following genotype proportions (Table 16.2).

A comparison of the values in Table 16.2 with those of Table 16.1 shows that the drive to fixation is still strong for sib-mating but much slower than self-pollination over succeeding generations. In half sib-matings (matings between individuals that share only one parent) the drive to fixation is proportionately slower. The degree of parental relatedness may be further reduced (e.g., grandparents in common, great-grandparents in common, etc.) and as genetic distance between mated individuals increases, the probability that they will share rare deleterious recessive alleles

TABLE 16.2 Drive to Fixation at 1, 2 And 3 Loci Following a Cross to Produce The F_1 Genotype Aabbcc Followed by Recurrent Cycles of Sib Mating

	% Heterozygous	% Homozygous	% Homozygous	% Homozygous
Generation	1 locus	1 locus	2 loci	3 loci
F_1	100.00000	0.00000	0.00000	0.00000
Sib_2	50.00000	50.00000	25.00000	12.50000
Sib_3	37.50000	62.50000	39.06250	24.41406
Sib_4	28.12500	71.87500	51.66016	37.13074
Sib_5	21.09375	78.90625	62.26196	49.12858
Sib_6	15.82031	84.17969	70.86220	59.65158
Sib_7	11.86523	88.13477	77.67738	68.46078
Sib_8	8.89892	91.10108	82.99407	75.60849

continues to decrease proportionately. Accordingly, social and religious mores against human cousin-matings are more equivocal than are sib-matings, and the genetic consequences are consistent with these societal conventions. Sib- and cousin-matings necessitate more time and resources to achieve fixation than self-pollination, but matings more distantly related than self-pollination also provide more opportunities for the conversion of repulsion into coupling linkages. In practice, however, plant breeders rarely incorporate inbreeding schemes with genetic distances greater than sib- or half-sib matings due to the large number of generations necessary to achieve allele fixation.

It is likely that the phenomenon of ID was common knowledge among farmers by the 19th century; manifested as the progressive decline of overall vigor with successive forced cycles of inbreeding in a cross-pollinated crop species such as maize (Fig. 16.2). Darwin (1876) was among the first to publish a comprehensive study of ID. Thirty-two years later, scientists in the U.S. independently documented ID in maize (East, 1908; Shull, 1908). To reconcile the characteristics of ID with the newly embraced Mendelian genetics platform, two competing theories on the genetic causes of ID were advanced. Shull (1909) articulated the "overdominance" theory that heterosis was a consequence of the collective additive effects of overdominance at individual loci. ID was explained as the pervasive absence of overdominance. Davenport (1908) proposed the "dominance" theory that implicated the notion of genetic load in the heterosis phenomenon. Dominant alleles collectively masked the deleterious effects of recessive alleles in this theory. ID was, therefore, due to the collective unmasking of these deleterious alleles. While there is strong circumstantial evidence for the transition from heterozygosity to homozygosity as the cause of ID, recent molecular studies have demonstrated that there are more factors than collective allele fixation involved with the ID phenomenon (Ritland, 1996; Andorf et al., 2010; Hedrick et al., 2016; Shen et al., 2017).

Charles Darwin played an important role in our understanding of ID. Not only did he publish seminal 19th century papers on the phenomenon, but it has also been asserted that Darwin's family history inadvertently provided evidence that supported the simplistic interpretation of homozygosity and inbreeding depression (Álvarez et al., 2015). Darwin documented the deleterious consequences of self-fertilization on progeny in numerous plant species, and this research led him to suspect that the health problems of his ten children, who were very often ill, might have been a consequence of his marriage to a first cousin.

ID has been documented in a broad spectrum of phenotypes in plants. In some species, such as *Mimulus guttatus*, ID is associated with a wide range of fitness-related traits including seed germination rate and vigor, survival to flowering, and flower, fruit and seed production (Willis, 1993). Not surprisingly, ID is manifested as a reduction in harvest yield, for example, autotetraploid potato (*Solanum tuberosum*; Golmirzaie et al., 1998) and pineapple (Sanewski, 2009). In a broad spectrum of cross-pollinated species, ID is manifested

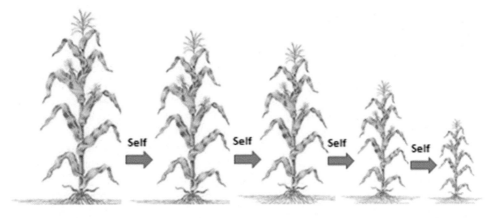

Inbreeding Depression in *Zea mays*

General traits affected: biomass per plant; vegetative growth rate; somatic organ size; reproductive organ size; unit yield of seeds, fruits, storage carbohydrates and lipids; tolerance or resistance to pests, pathogens, and herbivores

FIG. 16.2 The macro-biological manifestation of inbreeding depression: progressive decline of plant size, vigor, and economic yield during successive forced cycles of inbreeding (self-pollination).

as the inability of plants to survive to reproduce, for example maize (Benson and Hallauer, 1994), *Narcissus triandrus* (Hodgins and Barrett, 2006), *Saintpaulia ionantha* (Kolehmainen et al., 2010), and *Cucurbita foetidissima* (Kohn and Biardi, 1995). In almond (*Prunus dulcis*) ID was manifested by reduced pollen germination, decelerated growth of pollen tubes through the style, delayed ovule development at the prezygotic phase, and high fruit drop (Martínez-García et al., 2012). Self-pollination significantly affected germination and survival rates, yield, and to a lesser extent seed weight of *Phaseolus coccineus* (González et al., 2014). In the allotetraploid *Brassica napus*, plant biomass and seed weight were the most conspicuous traits affected by ID (Damgaard and Loeschcke, 1994). Another common trait affected by ID is insect herbivore tolerance, for example, herbivory of *Abrostola asclepiadis* on *Vincetoxicum hirundinaria* (Muola et al., 2011). Hirao (2010) quantified ID at different seedling developmental stages in *Rhododendron brachycarpum* and reported measures of inbreeding depression of 0.891 at seed maturation, 0.122 (but not significant) at seed germination and 0.506 at seedling survival.

As more tools to dissect, sequence, and alter the genome are developed, tests of the dominance and overdominance theories of ID have been applied. The scope of inferences has widened and now includes such facets as the interactions between genes, the relative abundance of major versus minor genes, life cycle stage expression, and mutation rates (Ritland, 1996). For example, deleterious recessive alleles were found to be the main factors responsible for ID in two closely related annual plants, the primarily selfing *Mimulus micranthus* and the mixed-mating *M. guttatus* (Dudash and Carr, 1998). In *Eucalyptus grandis*, observed patterns of genome sequences were consistent with at least several detrimental loci with large effects on ID associated with both parental chromosomes of the 11 pairs. It is likely that 100 or more genes, many with substantial effects on viability, contributed to ID (Hedrick et al., 2016). Busch (2005) demonstrated that self-pollination in natural populations of *Leavenworthia alabamica* was associated with the selective removal of partially recessive deleterious alleles that caused ID. Likewise, a negative association was found between genetic load and self-pollination rate in *Arenaria uniflora*, suggesting that deleterious recessive alleles are the primary source of ID (Fishman, 2001). Benson and Hallauer (1994) concluded from comparisons of ID in selected and unselected populations of maize that epistasis was not an important factor. Results in *Phaseolus coccineus* showed that different deleterious loci are acting at different stages. Inbreeding tended to purge individuals of deleterious recessive alleles to reduce ID (González et al., 2014). A simple genetic model with two types of unlinked loci, "underdominant" and partially dominant, with multiplicative effects on fitness, were found to create an "optimal outcrossing distance" of parents in a hybrid breeding program under a wide range of parameter values (Schierup and Christiansen, 1996).

In a study of the effects of inbreeding and selection intensity on ID, the rate of self-pollination was directly proportional to the level of ID, and the homozygote viability that resulted in maximum ID tended towards one-half the heterozygote viability (Ziehe and Roberds, 1989). Experimental results on strawberry suggested that pedigree inbreeding coefficients were poor predictors of changes in allele fixation when populations are developed through cycles of breeding and selection. Further, ID is of minor importance for strawberry breeding populations managed with adequate population sizes and strong directional selection (Shaw, 1995).

Many investigators have reported on evidence for sex-related effects on ID. In maize, maintaining alleles in the homozygous state over several generations produced a progressive decrease of paternally-imprinted expression that was reversed by allele heterozygosity. These researchers concluded that metastable epigenetic effects were not associated with ID (Auger et al., 2004). A study on ID in *Arabis fecunda* (Brassicaceae) identified significant maternal-parent-by-pollination-treatment interactions for mean seed weight, and dry weight, that were consistent with ID caused by deleterious recessives and varying past maternal inbreeding (Hamilton and Mitchell-Olds, 1994). ID was found to be stronger in pistillate flower function than in staminate flower function in summer squash (*Cucurbita pepo*) (Hayes et al., 2005). In *Brassica napus*, however, both female and male fitness characters showed significant ID following two cycles of self-pollination (Damgaard and Loeschcke, 1994).

ID is not a static descriptor or genetic parameter of a crop species. ID is also not monolithic within a species or under all circumstances. Perhaps most significantly, ID can be changed by controlled mating and selection over time. For example, progressive inbreeding of cross-pollinated crop species such as maize under selection for vigor and combining ability have effectively reduced ID. 22 cycles of recurrent selection reduced ID for 13 of 16 traits in selected maize populations (Benson and Hallauer, 1994). Several studies have shown, collectively, that inbred yields have increased 1.9–3.5 times faster than heterosis yields over a period of 50 years (Troyer and Wellin, 2009). It is clear that ID in maize inbreds is decreasing with progressive selection for inbred plant vigor (Fig. 16.3).

Both the vigor and yield of corn inbreds has increased dramatically since the early 20th century under recurrent selection

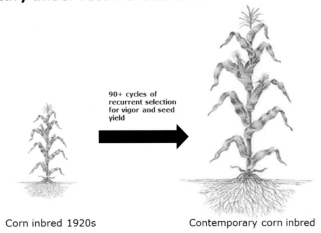

Corn inbred 1920s   Contemporary corn inbred

FIG. 16.3 Recurrent selection for phenotypic vigor and seed yield over the 100-year period from 1920 to 2020 has reduced ID substantially in maize.

HETEROSIS

What is hybrid vigor, or heterosis, and how can it be manipulated beneficially by plant breeders? At a single locus, alleles may interact in a range of patterns (Fig. 16.4). If the value of the heterozygote is at the midpoint between the parents [i.e. $(m + n)/2$], the interaction is classified as "additive" (Chapter 3). As the heterozygote value is greater than $(m + n)/2$ but less than n, the interaction is considered "codominant". If the heterozygote is equal to n (the greater of two parents), the interaction is "complete dominance". Finally, if the value of the heterozygote is greater than n, the allelic interaction is classified as "overdominance".

To expand the scenario from a single locus to a large number of interacting loci of a quantitatively inherited trait, the phenotypic value of a hybrid individual or population may also be defined relative to the parents (Fig. 16.5). If the hybrid value x is equal to $(P_1 + P_2)/2$, the net interaction of all genetic factors is said to be additive. If the hybrid value is in the range $(P_1 + P_2)/2 < x \leq P_2$ the net interaction is said to be heterotic relative to the midparent $[=(P_1 + P_2)/2]$. Finally,

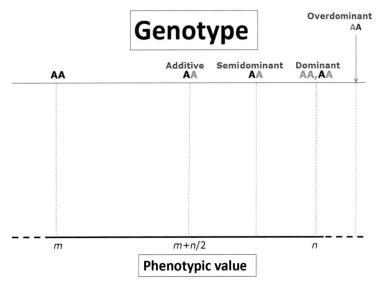

FIG. 16.4 Classification of allelic interactions at a single locus.

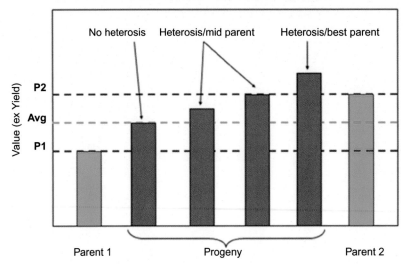

FIG. 16.5   The value of the hybrid relative to parents determines whether the targeted phenotype is additive $[x = (P_1 + P_2)/2]$, heterotic relative to the midparent $[(P_1 + P_2)/2 < x \leq P_2]$, or heterotic relative to the best parent $(x > P_2)$.

if the hybrid value is $x > P_2$ the net interaction is defined as heterotic relative to both parents, a phenomenon termed *heterobeltiosis*. With regard to plant breeding applications, heterobeltiosis is by far the most commercially important form of heterosis. Under some circumstances, the terms "heterosis" and "heterobeltiosis" and considered to be synonymous.

Like ID, heterosis is not a static parameter of plants, populations, or circumstances. Heterosis is highly dependent on many interacting factors, for example the GxE interactions that affect performance and yield. Griffing (1990) showed that heterosis for fruit yield in tomato is highly dependent on environmental factors, particularly nutrient limitations. A similar finding of environment-dependent heterosis was found in wheat, also a self-pollinated crop species (Kindred and Gooding, 2005).

Molecular biology is helping us to better understand what heterosis is and what it is not. We also have a much clearer understanding of the circumstances that are associated with manifestations of heterosis. When the overdominance and dominance (see above) theories of ID and heterosis were first proposed in the early 20th century, ID and heterosis were presumed to be inextricably linked at the molecular and mechanistic levels. In other words, ID was considered as the effective opposite of heterosis. There are enough clear departures from this paradigm in collective research results that it is fair to state that the situation is more complicated than East, Shull, Davenport, and Jones first surmised (Hallauer, 2007).

Guo et al. (2006) found in maize that the proportion of additively expressed genes was positively associated with heterosis while the proportion of expressed genes with a bias towards the paternal parent was negatively correlated with heterosis. This study also concluded that there was no correlation between the over- or under-expression of specific genes in maize hybrids with heterosis. Wei et al. (2015) found that heterosis in maize was controlled by different genetic mechanisms and that over-dominance effects were the main contributors to heterosis for plant-related traits at the single-locus level. In a separate project, non-additive expression patterns suggested that a trans-regulatory mechanism acted early after fertilization in hybrid embryo and endosperm, although the majority of genes showed additive expression levels in the embryo and dosage-dependent expression levels in the endosperm (Jahnke et al., 2010). On the basis of average dominance level in rice, 28.6% of QTL affecting yield-related agronomic traits showed overdominance, 35.7% exhibited partial dominance, and 30% were additive (Wang et al., 2012). Results produced by Meyer et al. (2015) suggested that a combination of dominance, overdominance, and epistasis was involved in biomass heterosis in an *Arabidopsis thaliana* cross. In another study in *A. thaliana* seedlings, heterosis was shown to be associated with non-additive gene expression that resulted from earlier changes in gene expression in the hybrids relative to the parents. Non-additively expressed genes were involved in metabolic pathways critical for plant growth, such as photosynthesis (Zhu et al., 2016). It is quite clear that, depending on the context, heterosis may be associated or correlated with additive, dominance, or overdominance genetic effects.

The picture has become even more complicated as more reports of the bases of heterosis have appeared (Liu et al., 2015). A study of heterosis by Paschold et al. (2010) concluded that nonadditive expression of specific genes in the phenylpropanoid pathway and superoxide dismutase 2 might contribute to manifestations of heterosis in maize roots. Another report showed that the overexpression of a transcription factor LaAP2L1 in *Arabidopsis thaliana* led to markedly enlarged organs and heterosis-like traits. The enlarged organs and heterosis-like traits displayed by plants transformed with 35S::LaAP2L1 were mainly due to enhanced cell proliferation and prolonged growth duration (Li et al., 2013). Transcriptome analysis in F_1 hybrids as compared to parental lines of *Brassica napus* revealed that various phytohormone (auxin and salicylic acid) response genes were significantly altered (Shen et al., 2017). Highly expressed genes in F_1 hybrids of *B. oleracea* were mostly related to yield-contributing characteristics (Jeong et al., 2017). They speculated that the identified genes might be associated with the mechanism of heterosis in *B. oleracea* and provide a foundation to reveal the general complexity of regulatory gene networks associated with the genetic mechanism of heterosis.

Reports of epigenetic genome modifications associated with heterosis have also appeared in the literature. For example, a specific type of cytosine-methylation was found to be positively correlated with grain yield heterosis in maize (Qi et al., 2010). In another study, both *Arabidopsis thaliana* and *Landsberg erecta* hybrids displayed increased DNA methylation across their entire genomes, especially in transposable elements. Growth and vigor of F_1 hybrids were compromised by treatment with an agent that demethylated DNA and by abolishing production of functional small RNAs due to mutations in *Arabidopsis* RNA methyltransferase (Shen et al., 2012). In heterotic hybrids of *B. napus*, the majority of the small interfering RNA (siRNA) clusters had a higher expression level than in the parents, and there was also an increase in genome-wide DNA methylation in the F_1 hybrid (Shen et al., 2017).

Large increases in biomass and yield in high-parent heterosis hybrids have suggested that alterations in bioenergetic processes may contribute to heterosis. Expression of specific alleles and/or post-translational modification of specific proteins correlated with higher levels of heterosis in maize (Dahal et al., 2012). The collective body of knowledge demonstrates that the genome, transcriptome, proteome, and metabolome and interactions of these are implicated in manifestations of heterosis, leading Andorf et al. (2010) and Goff (2011) to propose a systems approach to the study of hybrid vigor.

One important difference between ID and heterosis is that the latter may have distinct parental effects. For example, paternal genome excess F_1 triploids (i.e., $A_fA_f \times A_mA_mA_mA_m \rightarrow A_fA_mA_m$) displayed positive heterosis whereas maternal genome excess F_1 hybrids (i.e., $A_fA_fA_fA_f \times A_mA_m \rightarrow A_fA_fA_m$) displayed "negative" [$x < (P_1 + P_2)/2$] heterosis effects in *A. thaliana* (Fort et al., 2016). The proportion of genes in maize hybrids that exhibit a bias towards the expression level of the paternal parent was negatively correlated with hybrid yield and heterosis (Guo et al., 2006).

One hypothesis advanced in the early 20th century was that heterosis should be correlated with the genetic distance (i.e., unrelatedness) of parents. Overwhelming evidence shows that this is not necessarily the case, for example, yield components of broccoli (Hale et al., 2007); yield of alfalfa (Riday et al., 2003); yield of oilseed rape (Yu et al., 2005; Tian et al., 2017); quality and yield of Chinese cabbage (Kawamura et al., 2016); and pepper fruit quality and yield (Geleta et al., 2004). Conversely, correlations between genetic distance and measures of heterosis were reported in muskmelon (José et al., 2005).

Heterosis has been documented and captured in cultivars of many horticultural crop species. It is not surprising that heterosis has been documented in sweet corn, manifested mainly as earlier flowering, larger plants, and increased yield (Dickert and Tracy, 2002). In broccoli (*B. oleracea* var. *italica*), About half of 36 hybrids examined exhibited high-parent heterosis for head weight and stem diameter, and almost all hybrids manifested high-parent heterosis for plant height and breadth. Heterosis was also found for harvest maturity in the form of earliness (Hale et al., 2007). In *Capsicum annuum* (bell and chili pepper), heterosis has been reported for dry fruit yield per plant, number of fruits per plant, and days to maturity (Marame et al., 2009). Also in *C. annuum*, Singh et al. (2014) reported on the finding of heterosis for earliness, total yield, number of fruits, fruit length, plant height and breadth. In bulb onion (*Allium cepa*), heterosis was documented for bulb yield and weight and disease resistance (Abubakar and Ado, 2008). José et al. (2005) observed heterosis for fruit shape and fruit length in muskmelon (*Cucumis melo*). Heterosis for plant height, number of primary branches per plant, pericarp thickness, fruit firmness, total soluble solids, ascorbic acid content, number of flowers per cluster, number of fruits per plant and total fruits yield per plant were reported in tomato (*Solanum lycopersicum*) (Solieman et al., 2013). Among many traits studied, heterosis was greatest for seed yield in hybrids of *Cuphea lanceolata* (Ali and Knapp, 1996). In chrysanthemum (*Chrysanthemum morifolium*), SCA effects were positively correlated with heterosis for waterlogging tolerance (Su et al., 2017). This account is not exhaustive but exemplifies the broad spectrum of horticultural crop species and traits that are amenable to a breeding strategy aimed at the capture of heterosis.

Studies showing that the magnitude of heterosis has progressively decreased over the past 100 years and that magnitude of ID has also decreased dramatically over this period (Troyer and Wellin, 2009). This realization led to the proposition that replacing preliminary testcross trials with finished-inbred yield trials would greatly increase efficiency without sacrifice of effectiveness (Troyer and Wellin, 2009). Independent studies have confirmed that yield increases of cross-pollinated maize have plateaued over the past 50 years and are now comparable to those of self-pollinated soybean (Egli, 2008). Perhaps hybrids should be replaced entirely by highly selected pure/inbred populations? Since hybrid cultivars impart a measure of proprietary protection as compared to pure lines, this scenario is not likely in the near future.

APPLICATIONS OF MAS FOR HETEROSIS

Dubreuil and Charcosset (1999) suggested that associations among inbreds and populations further proved to be consistent with pedigree data of the inbreds, and provided new information on the genetic basis of heterotic groups. QTL for heterotic phenotypes were identified in allotetraploid *B. napus* (Basunanda et al., 2010). This group speculated that QTL hotspots might harbor genes involved in regulation of heterosis (and also for fixed heterosis in the tetraploid state) for different traits throughout the plant life cycle, including a significant overall influence on heterosis for seed yield. The maize genome sequencing project was completed in 2009 and was recently updated to include more information on polymorphic base loci (Jiao et al., 2017). Many other angiosperm genomes (e.g., rice, wheat, tomato, soybean, *A. thaliana*, etc.) have been completely sequenced, allowing for generalizations about the consensus functions of heretofore obtuse genomic regions. This information will allow plant breeders to pinpoint genomic regions responsible for heterosis to find markers to assist breeders in selecting for these genomic regions, and to engineer heterotic genome regions into hybrid cultivars of the future.

BREEDING STRATEGIES FOR HYBRID CULTIVARS

The genetic structure of the F_1 hybrid cultivar was described in Chapter 15. The hybrid population consists of a monolith of individuals that are enriched for homo- and heterozygous loci that maximize performance relative to targeted phenotypes (Fig. 16.6). These phenotypes most typically involve components of economic benefit, or "yield". Biomass and seed or fruit mass are universal targets for genetic improvement in breeding programs for the development of hybrid cultivars. The capture of genetic synergies into the bioenergetics processes that are translated into enhanced yield in the form of heterosis is the predominant aim of hybrid breeding programs.

Other goals that do not necessarily implicate the phenomenon of heterosis may also drive hybrid breeding programs. Most notably the F_1 hybrid strategy is an efficient and effective strategy to combine desirable dominant alleles in single a population as polyheterozygotes. One example of this breeding objective is the "pyramiding", or stacking, of dominant alleles for resistance to different races of disease pathogens. This approach will be covered in more detail in Chapter 19. Many horticultural quality traits such as pigmentation, habit, maturity, architecture, absence of thorns or rigid trichomes, fruit ripening, fruit abscission, fruit flavor and texture, fruit firmness, etc. are often controlled by one or a few genes the phenotypic expression of which is not accentuated by heterosis. Breeders of horticultural plant species cultivars employ hybrid breeding strategies to construct inbreds that combine to impart desirable additive, codominant, and dominant combinations of alleles controlling quality traits with or without the capture of heterosis.

The process from germplasm to parental inbreds in the most simplistic case involves adaptations of methods described in Chapters 13 and 14 for self-pollinated crop species. The breeder will generally have some prior knowledge of how genotypes that impart desirable phenotype are inherited. An inbred is a pure line genetically. For breeding programs that target simply-inherited or non-heterotic traits, selection criteria are mainly driven by direct

Genetic Structure of F₁ Hybrid Varieties

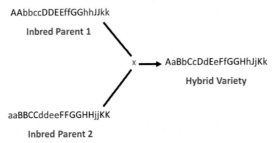

AAbbccDDEEffGGhhJJkk

Inbred Parent 1

X ⟶ AaBbCcDdEeFfGGHhJjKk

Hybrid Variety

aaBBCCddeeFFGGHHjjKK

Inbred Parent 2

FIG. 16.6 The synthesis of the F_1 hybrid archetype from genetically fixed parental populations that were developed by a program of controlled mating and selection to concentrate and configure desirable alleles.

Breeding Hybrid Varieties: General Scheme 1   Breeding Hybrid Varieties: General Scheme 2

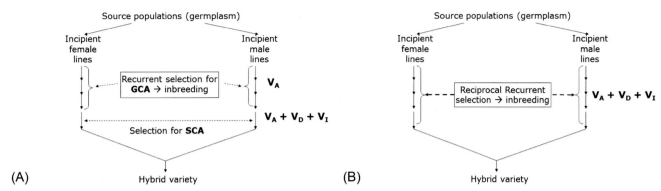

FIG. 16.7 Generalized strategy for the breeding of a F_1 hybrid cultivar; featuring recurrent selection (Chapter 15) for (A) GCA during early generations followed by selection for SCA in later generations or the final stage of inbred parent development; or (B) Reciprocal recurrent selection in both early and late generations.

phenotypes. Consequently, the resulting inbred parents will be expected to combine well and to give rise to a hybrid cultivar that exhibits desirable phenotypes. For many horticultural crop species involving traits of known inheritance and relatively high heritabilities, this simplistic approach to the breeding of parental inbreds is highly effective.

In situations where breeding objectives target phenotypes with complex or quantitative inheritance, a strategy that incorporates selection for combining ability is recommended. Definitions of ACA, GCA, and SCA and the general methods for their estimation were provided in Chapter 15. A strategy for the development of inbred parental lines, or populations, based on combining ability usually begins with recurrent selection for GCA during the early cycles of the program (Fig. 16.7A). The reasons for this are two-fold. GCA estimates V_A that will be the bedrock of the performance of the finished cultivar, although contributions to the forces of heterosis by V_A have been found to be minimal (Falconer and MacKay, 1996). The main reasons that GCA is the preferred early selection criterion are (i) manifestations of V_D, V_I, and V_{GxE} are confounded by early generation allele segregation and linkage disequilibria; and (ii) the number of potential genotypes is extremely high. Since GCA estimates involve the least physical work (number of crosses and line evaluations), it is logical and effective to integrate this strategy to winnow germplasm during early cycles of the breeding program to the most promising set of breeding lines.

After the number of candidate genotypes has been effectively reduced to a manageable and affordable level, the strategy shifts to selection for SCA. Since SCA is estimated by half diallel of candidate parents, the number of crosses and evaluations that are needed is relatively high $\{[n \times (n-1)/2]\}$. At this point in the breeding program, populations will be inbred or nearly inbred and results should be acceptably consistent, especially if evaluations are conducted in multiple locations and years.

Alternatively, the plant breeder may opt to employ reciprocal recurrent selection (RRS) for most or all of the inbred development program. RRS is a strategy by which two populations are selected simultaneously for high GCA and mutually high SCA values (Sleper and Poehlman, 2006). While this necessitates more controlled hybridizations during early generations than recurrent selection for GCA, the higher rate of gain in combining ability often offsets higher costs due to labor and materials.

SOURCES OF BREEDING POPULATIONS

For cross-pollinated species, the plant breeder maintains working germplasm consisting of a set of populations that contain desirable alleles and linkages (Forsberg and Smith, 1980). These populations serve as the raw materials to constitute inbred lines that will impart desirable phenotypes to hybrid cultivars such as general vigor, growth rate, yield partition, maturity window, fruit or foliage quality aspects, disease resistance, and geographical ranges of cultivar adaptation. Since the aim of a hybrid breeding program is to maximize heterosis with regard to targeted phenotypes, and heterosis is promoted in general by heterozygosity of specific underlying genomic loci, the plant breeder may maintain a broad range of populations that represent distinct genetic forms of the crop species. If the capture of heterosis is a fundamental goal of the breeding program, the acquisition and utilization of disparate or genetically disassortative germplasm is not necessarily a fruitful strategy since genetic distance is not strongly and broadly correlated with heterosis (see discussion above).

The business of germplasm acquisition and enhancement (Chapters 7 and 8) is, for the plant breeder, a never-ending enterprise, always a work in progress. New sources of genetic variability are perpetually being added to the mix, and undesirable types removed from source populations by recurrent selection. During the process of population improvement and maintenance, old and new alleles ruminate and recombine in breeding populations, creating new V_G on which selection may yield phenotypic gains.

The breeder of a self-pollinated crop species usually approaches the design of a hybrid breeding program much like the development of a pure line, integrating the additional selection criterion of combining ability. The breeder of a self-pollinated crop species that establishes hybrid cultivars as a breeding objective must always be cognizant of the need to produce large quantities of hybrid seeds. Is the value of the hybrid cultivar sufficient to support the production of hybrid seeds by hand pollination, as with tomato and pepper, or with the assistance of a floral-altering mutation such as male sterility? If so, the presence of these genes in the pool of starting germplasm is of great importance. Introducing these genes into inbred populations that are already developed is not recommended.

RECURRENT SELECTION SCHEMES

The final goal of the hybrid breeding program for a seed-propagated plant species is to develop two or more inbred lines that are hybridized to produce seeds of the heterotic commercial cultivar. The inbreds must possess certain fundamental characteristics to facilitate this objective; mainly that they are sufficiently vigorous that the seed production field may be easily established, and levels of pollen and unit costs of seed production are within acceptable ranges. Recurrent selection schemes have culminated in progressive improvements of inbred performance in corn, as was discussed above and in Chapter 15. For other outcrossing species, ID may be manifested in reduced overall vigor, including seedling establishment and fecundity. In such cases, strong selection for inbred performance during the inbreeding process is essential. In some species with genomes that include a high relative genetic load and ID, the selection of productive inbreds may require many generations, such as with maize (Fig. 16.3).

The situation for the breeder of a hybrid cultivar of a cross-pollinated crop species is similar to that faced by the breeder of self-pollinated crop species, who faces daunting odds (see Table 13.1) and must constantly make trade-offs and compromises during the period from initial cross to the final product (Chapters 13 and 14). The successful plant breeder balances all elements such as population sizes, number of crosses, number of test populations, and number of breeding generations to achieve the best possible result while minimizing the consumption of monetary, time, land, labor, etc. resources. Since the result of the inbred development process should be two or more entities that combine well with each other, a modification of methods used to develop pure line populations (Chapters 13 and 14), known as recurrent selection, is widely applied.

The general cyclic ("recurrent") stepwise process of recurrent selection was described in Chapter 15, Fig. 15.5. Recurrent selection for phenotype is analogous to the pedigree method except with an intermating step. Intermating selected individuals provides more opportunities for desirable linkages in coupling to be captured within the context of the program. The selection criterion is only individual phenotype, not combining ability. The most powerful breeding strategy for cross-pollinated crop species is the heterotic F_1 hybrid. Selection for phenotype is the cheapest strategy but will not directly target alleles and allelic combinations that maximize heterosis.

There are three variants of recurrent selection for combining ability: GCA, SCA, and reciprocal (Welsh, 1981). All of these variants are more expensive to execute than selection for phenotype due to time, labor, materials, and space considerations. A comparison of selection for phenotype vs. selection for combining ability that includes the extra step of the progeny test is presented in Fig. 16.8. The difference between the three variants is manifested by the identity of the individuals that candidates for selection are crossed with to produce the progeny test cross populations.

GCA estimates the additive portion of V_P among the set of parents. GCA of an individual may be determined by calculating the means of all pairwise crosses within a given diallel of n individuals $[n(n-1)]$, or half diallel $[n(n-1)]/2$ (Fig. 16.9). An alternative method is to cross a candidate individual to one or more proven tester lines that consistently serve to predict GCA. A "tester" is a population (usually an established heterozygous OP or segregating breeding line; Fehr, 1987) that when crossed with an individual the progeny exhibit values that are similar to ACA from a diallel or half diallel. Yes, that is a circular definition, but it is accurate.

It is possible to conduct effective selection for high GCA (V_A) regardless of the degree of heterozygosity (V_D) and segregation of interacting loci (V_I). Therefore, during the early stages of inbred parent development while many loci are still heterozygous and segregating into the constituent homozygotes, selection for GCA will still be effective to maximize desirable V_A. Since potentially only a single tester population may be necessary to estimate GCA the total number of crosses that will be needed is equal to the number of incipient inbreds (n), not $n(n-1)$ or $n(n-1)/2$.

Selection Based On Phenotype Vs. Combining Ability

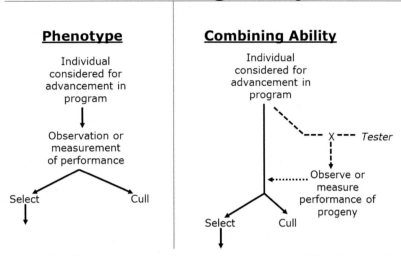

FIG. 16.8 Comparison of selection based on phenotype vs. selection based on combining ability. The genetic identity of the combining ability (CA) predictor differentiates selection for GCA, SCA, and reciprocal.

Half diallel for yield (q/ha) in maize

	IB-1	IB-2	IB-3	IB-4	IB-5	IB-6	IB-7	IB-8	IB-9	IB-10	IB-11	ACA	GCA
IB-1	12.6	42.6	46.3	45.1	45.4	45.9	54.5	45.2	43.1	42.9	42.9	45.5	45.71
IB-2		7.7	41.6	45.4	46.4	48.2	51.5	43.9	41.5	45.7	41.1	44.8	44.82
IB-3			9.1	45.2	44.2	45.7	50.4	45.4	41.7	43.4	36.5	44.1	44.18
IB-4				8.1	42.5	46.8	47.4	44.4	44.3	41.7	42.6	44.5	44.68
IB-5					4.6	47.5	49.8	41.4	47.2	29.9	38.8	43.3	43.18
IB-6						8.3	55.5	46.7	45.5	44.8	42.6	47.1	47.22
IB-7							11.2	50.5	51.9	50.6	43.5	50.6	51.19
IB-8								9.2	44.6	42.5	42.2	44.7	44.58
IB-9									7.4	42.4	41.2	44.4	42.08
IB-10										3.8	40.1	42.4	42.08
IB-11											10.6	41.2	40.78

Adapted from data published by Gama et al. (1995)

FIG. 16.9 An example of a half diallel for yield (q/ha) in maize. Values in the IB×IB cells are estimates of SCA. The mean SCA value for an IB is ACA. GCA is estimated by the performance of progeny of IB×tester crosses. Note the similarity of ACA and GCA values in this example.

During the early stages of inbred development, therefore when the number of potential candidate lines is high and lines are segregating, selection for GCA will reduce the number of crosses to a manageable level. The selection intensity should not be so high that potential inbreds, exhibiting moderate GCA but high SCA, are excluded during the early generations.

Selection for GCA, therefore, involves the performance of progeny from pairwise crosses of candidate individuals with a tester population, then advancing the individuals associated with the best progeny performance. Selection for SCA starts with an intercross of all candidate individuals followed by evaluation of progeny and selection of pairs of candidates that exhibit high mutual SCA. Reciprocal recurrent selection is applied when two populations are under development are targeted as prospective parents in a hybrid or synthetic. Candidates from one population are used as testers for the other, and vice versa. Theory and practice have demonstrated that this is an effective method, albeit resource consumptive due to the larger number of mandated crosses, for the development of two or more populations with high mutual combining ability (Fehr, 1987).

The estimation of SCA requires all pairwise crosses to be done, consisting of a half diallel if reciprocals are deemed to be equivalent. For 20 lines, this mandates 190 crosses, and for 50 the number climbs to 1225. The number of mandated crosses for determining GCA may be lower than for SCA, but the risk is higher that potentially valuable SCA inbred combinations will be sacrificed. The number of mandated crosses increases exponentially as n increases to the point that the program becomes untenable.

RRS also mandates full diallel crossing matrices at each cycle to assess the mutual CA of both incipient parent populations (Fig. 16.10). The starting point is two incipient parental populations, the individuals of which are both selfed and crossed with individuals of the other parental population. Based on the results of progeny tests of the parent $1\times$parent 2 diallel, individuals are selected for advancement to the next cycle.

The archetype for recurrent selection (Chapter 15, Fig. 15.5) includes a self-pollination step to achieve inbreeding and expose recessive alleles to selection. Alternative methods of inbreeding such as sib- or half-sib mating may be substituted for selfing. Depending on the method of inbreeding, incipient inbred lines begin to approach fixation by the S_7, Sib_9, or $HalfSib_{11}$. The strategy at this juncture is to manage the program such that a range of 20–50 lines are identified for the next step in the breeding program, the estimation of SCA. The pair or pairs that deliver the highest performance, representing the epitome of both GCA and SCA for that particular program, will be utilized as female and male parent in the corresponding hybrid variety. Alternatively, a group of selected inbreds with the highest GCA and SCA will be intercrossed and the progeny bulked to constitute a synthetic cultivar.

Recurrent selection has been applied effectively for the development of commercial cultivars of many horticultural crop species. Examples include apple (Kumar et al., 2010); potato (Bradshaw et al., 2009); oil palm (Wong

Reciprocal Recurrent Selection

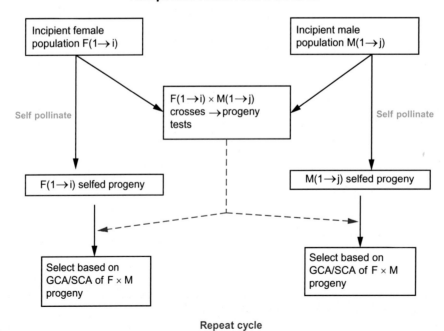

FIG. 16.10   One cycle of reciprocal recurrent selection for combining ability.

and Bernardo, 2008); cucumber (Cramer and Wehner, 2000); red beet (Eagen and Goldman, 1996); Japanese quince (Rumpunen and Kviklys, 2003); passion fruit (Viana et al., 2017); strawberry (Masny et al., 2016); blueberry (Lyrene, 2005); and peach (Lyrene, 2005).

CELL AND MOLECULAR BIOLOGY TOOLS IN RECURRENT SELECTION

Advances in cell and molecular biology have provided a bounty of tools to improve the recurrent selection and inbreeding process. If the plant breeder has access to a protocol for the production of doubled haploids, it is possible to dispense with inbreeding and to proceed directly from the source populations to the candidate inbreds for SCA selection. As compared to the reciprocal recurrent selection, DH can be easily applied to maize in the absence of large populations (Gallais, 2009). If the DH protocol is robust and a large number of incipient parents are produced, it may be necessary to conduct selections in a stepwise fashion, with GCA as an early criterion followed by SCA. Gordillo and Geiger (2008a) showed that by intercrossing a reduced number of selected DH-derived lines for starting a new recurrent selection cycle, the short-term response to selection might be increased, but the population size and, thus, the selection limits in the long run, are diminished. Gordillo and Geiger (2008b) further proposed an algorithm involving the intercrossing lines from different subpopulations to alleviate DH limitations due to small population sizes.

Molecular markers have proven to be excellent independent criteria for measuring the effectiveness of recurrent selection schemes. For example, markers were used during a recurrent selection program in *B. napus* to identify base populations with broad genetic variation for selection of breeding lines with better performance (Yuan et al., 2004). A MAS breeding strategy was developed that involved selecting plants at an early generation with a fixed, favorable genetic background at specific loci followed by a single large-scale marker-assisted selection (SLS-MAS), and maintaining as much allelic segregation in the rest of the genome as possible (Ribaut and Betran, 1999).

Another general strategy, known as marker-aided recurrent selection (MARS), has been developed to describe selection for neutral molecular markers to improve the effectiveness or efficiency of recurrent selection. Xie and Xu (1998) showed that if a large fraction of V_A could be explained by marker loci, the efficiencies of MARS relative to recurrent selection for phenotype was great. This advantage of MARS was most pronounced when family size was small. The availability of known QTL in MARS is most beneficial for traits controlled by a moderately large number of QTL (e.g., 40; Bernardo and Charcosset, 2006).

This advantage for small populations is particularly attractive for species with large perennial individuals. Cros et al. (2014) showed that genomic selection was valuable for reciprocal recurrent selection in oil palm (*Elaeis guineensis*) as it could account for family effects and Mendelian sampling terms despite small populations and low marker density. MARS can also be very effective for species with reduced sexual fertility, for example, potato (Slater et al., 2013). Conversely, independent studies have concluded for maize (Bernardo and Yu, 2007) and oil palm (Wong and Bernardo, 2008) that for a realistic yet relatively small population size of $N = 50$, genome-wide selection is superior to MARS or phenotypic selection with respect to gain per unit breeding program cost and duration.

PRODUCTION OF HYBRID SEED

General considerations for seed production during the breeding process were covered in Chapter 11. The following passages cover special seed production challenges that are limited to hybrid cultivars. The novice plant breeder often finds the accomplishment of a single cross between two distinct individuals to be a daunting task (Wright, 1980). Production of quantities of hybrid seeds in the range of 1,000 to 10,000 is usually possible by hand pollination. The commercial farmer, however, will need many more than that. In the case of corn, where ~7 lbs./acre is the norm for standard planting densities, a farmer producing 500 acres will need 3500 lbs., or 1.75 tons, of seeds. Imagine that every single seed must originate from the fusion of gametes from the two inbred parents developed in the breeding program. This is a truly monumental task!

The monoecious mating system of domesticated corn presents natural features that render this job to be relatively simple. The process of controlled hybridization by emasculation and the detasseling of corn for this purpose were presented in Chapter 10, the same procedure by which commercial seeds of contemporary hybrid corn cultivars are produced. One of the two constituent inbred parent lines is selected as the female parent, usually the higher yielding of the two. The chosen female and male inbreds are planted in a characteristic field configuration that will facilitate mass emasculation and the individual management of the separate lines in the same field (Fig. 16.11).

Typical Hybrid Corn Seed Production field

FIG. 16.11 F_1 hybrid corn production field featuring a 4♀:2♂ configuration. It is extremely critical to understand and control the physical location of potential sources of contaminating pollen. If a field designated for hybrid seed production were planted next to a regular crop production field, the pollen would be driven by wind from one field to the other, and some unwanted pollen would certainly fall onto the pistillate flowers of the female inbred. Therefore, seed production fields are always established with a minimum isolation distance to another potentially contaminating pollen source. The distance varies with the crop species and geographical location. Minimum isolation distances were established empirically, by planting genetic marker lines at varying distances until the frequency of hybrids decreased to below an acceptable level, 0.1% or less. *From http://departmentofagriculture.blogspot.com/2013/03/agricultural-production.html. National Agricultural Library, Agricultural Research Service, U.S. Department of Agriculture.*

The female inbred is usually overrepresented in the seed production field as compared with the male parent since females they will ultimately contribute the seeds of the new cultivar. The male inbred will only contribute pollen for the cross. For a hybrid seed production field, a typical planting configuration is four or six females to one male, but this ratio varies according to the reproductive vigor of the parents and other parameters. The location of the two inbreds in the field is clearly delineated to avoid mixing up the females and males. Advances in GPS technology and remote sensing have made it possible to store coordinates on a computer, from which all subsequent field operations, including the seed harvest are controlled automatically by satellite and tractor-mounted sensors.

The characteristics of the inbred parents are well understood before the seeds are planted into the field. The grower will know exactly when the optimum time presents itself for emasculation, or detasseling, to be accomplished. The dynamics of corn growth based on soil and air temperature and incident light are well established. At the optimum developmental stage, a large cutting machine detassels all of the female inbreds in the field. The pollen from the male inbreds is subsequently shed and falls on all pistillate flowers within the isolation distance, including the male rows. Following pollination, the male inbred rows are removed entirely before potentially contaminating self-pollinated (or male×male) seeds can mature.

As straightforward as this operation would appear, hybrid seed production in corn has historically presented a major challenge. During the early 20th century, ID was prevalent in parental populations making it necessary to employ double cross and three-way hybrids to alleviate ID. Later, as studies of inheritance identified a myriad of mutant forms that had possible utility, male sterile mutations were discovered. There are three main genetic sources of male sterility in corn: S, C, and T, based on cytotype and the interaction with nuclear restorer genes (Weider et al., 2009). All of these cytotypes were used to some extent as tools for the commercial production of hybrid varietal seed, obviating the need to mechanically detassel the female inbreds in a production field and, thus, reducing labor costs. The best source of male sterility for this purpose was T cytoplasm and, by the late 1960s, most of the grain corn acreage in the U.S. was occupied by hybrids produced using T cytoplasm (Levings, 1993). The reader should already know from earlier passages in this text what happened next, and why we have reverted to the use of mechanical detasseling (see Chapter 7 for a discussion of genetic vulnerability and the southern corn leaf blight epidemic of 1970).

Most crop plant species do not exhibit a mating system similar to corn. Floral structures are usually hermaphroditic, and autogamy or partial autogamy are common (Chapter 10) resulting in self-pollinated progeny even within predominantly cross-pollinated species. The most obvious approach to the mass production of hybrid seed for species with hermaphroditic floral structure arrangements is hand-emasculation and pollination (described in detail in Chapter 10). Individual hand-pollinations are simply repeated ad infinitum between two parental inbred lines until adequate amounts of hybrid seed are produced from the resulting fusions of gametes.

How can the large-scale production of pure F_1 hybrid seeds be planned and executed such that the task is cost effective? Alternatively, is it possible to realize hybrid seed values that are sufficiently high such that the cost of hand-pollination is justified? In agronomic food crops such as corn, sunflowers, sorghum, etc. the answer to the second question is "no". For many horticultural crop species, and especially certain ornamental species where the hybrid product has extremely high value, however, the answer to the second question is "yes". In the case of horticultural food crops, hand-pollinated F_1 hybrids are only feasible where the corresponding farm product elicits high consumer demand.

The best example of such a circumstance in horticultural food crop species is the family Solanaceae where hybrid varieties of the tomato, pepper, and eggplant are produced by hand-emasculation and pollination. The mating system for the cultivated tomato is autogamy, or self-pollination. Studies of wild progenitors of the cultivated tomato reveal that outcrossing is prevalent, suggesting that autogamy was a secondary consequence of domestication. The tomato is not only tolerant of inbreeding, but heterosis, and heterobeltiosis, have been demonstrated (Semel et al., 2006).

The value of the tomato crop relative to production and harvesting costs is high especially for certain uses as a perishable commodity. Most of the tomato harvest produced for processing (sauce/paste) is harvested by machines rendering the relative net value to be high as well. Most readers are familiar with tomato fruits. They are full of seeds, up to several hundred per fruit depending on type and cultivar. Each fruit is the result of a single fertilization event. Hand pollination of tomatoes is simple to accomplish: the antheridial cone (fused anthers) is removed in one quick operation, and pollen introduced at the same time. No impervious covering is necessary over the fertilized stigma because tomatoes are generally strongly autogamous. Insect pollinators, however, will venture into tomato flowers, and measures to exclude them should be exercised in open-field hybrid seed production.

While tomato emasculation and pollination operations are relatively simple to accomplish, dexterity and consistency are essential to ensure that all fruits in a block designated for hybrid seed production are, indeed, hybrids and not self-pollinated female parent contaminants. The workers who perform the crosses and the managers who oversee the plots must be well trained and predisposed to a regimented procedure. They also must be attuned to a tedious task for an exceedingly low wage. It can take more than 100 person-hours to produce 1 kg of hybrid tomato seeds. The hybrid tomato seed industry has determined that the enterprise is profitable if the crosses are conducted in a developing area of the world, where labor costs are low and the social structure is supportive of the product goals of accuracy and consistency.

Hand-pollination is not presently feasible for most crop species where heterosis has been demonstrated, due mainly to high production costs associated with low species fecundity, the need for exclusion of contaminating pollen from female flowers, and seed extraction from fruits or capsules. In such cases, the subversion of naturally-occurring or artificially-induced mutations that alter flower morphology or function may be employed. Another way to consider this possibility is the conversion of hermaphrodity to structural or functional monoecy or dioecy. Alterations in floral biology are common within plant species, and may reflect the ongoing strategy used to balance phenotypic replication with the long-term fitness of the gene pool.

Mutations for floral adaptations that are used to produce hybrid seeds include cytoplasmic male sterility, gynoecy, and self-incompatibility (Basra, 2000). As was established in Chapter 10, male sterile floral variants are commonly differentiated from hermaphrodites by a single recessive allele at a *Ms* locus, and there are sometimes many different genetic male sterility systems within a given species. The single recessive nuclear factor has often been found to interact with genetic determinants in the cytoplasm that can overcome the developmental anomaly that leads to male sterility, thus restoring the individual to a state where both sexes are functional. Such systems have been characterized in corn, sorghum, wheat, sunflower, radish, onion, and carrot (Chapter 10).

Economic plant species of the family Cucurbitaceae have been found to possess a broad spectrum of floral structural modifications, ranging from dioecy to hermaphrodity. While the same homologous series is evident in many genera of Curcurbitaceae, the most thoroughly characterized is cucumber (*Cucumis sativus*). Floral expression in cucumber has been found to be controlled by three major genes, *M*, *F*, and *A*. No cytoplasmic components have yet been discovered that interact with these nuclear genes. The heritability of sex expression in cucumber, however, is not 100%. Environmental factors, especially endogenous substances, can greatly affect the phenotype. Silver nitrate, gibberellic acid (GA), and other phytohormones have been demonstrated to induce male or female development in an otherwise hermaphroditic background (Chapter 10).

The possible uses of these systems for mass production of hybrid seed are manifest. By planting a male sterile individual, or functional female, in juxtaposition with a hermaphrodite, and with pollen vectors in place, all seed set on the male sterile plants will be of hybrid origin. Gynoecious lines may be used in the same way. If such systems are used as a tool for efficient seed production care must be exercised that the resulting genotypes will not detract from the economic yield of the resulting hybrid variety (Basra, 2000). If fruit or seed set is a strict requirement of crop performance, sex expression can have a profound impact on the magnitude of yield.

Since the genotype of the inbred parent lines relative to sex expression is paramount to the seed production operation, the plant breeder must perform breeding steps to ensure that the proper genes are in place. Male sterility provides an excellent example of the use of plant breeding to engineer inbred parent lines for hybrid seed production. The underlying genes must be introduced and simultaneously guided into the finished inbreds with the genotypes that will maximize varietal performance. Entire nomenclature systems have been devised to assist breeders and seed production personnel in sorting out potential entanglements of the different lineages that must be managed independently.

The descendants of the "A" or sterile line will be used as the female parent in the hybrid seed production step (Fig. 16.12). The genotype cannot be brought to genetic fixation by self-pollination since it is male sterile. Therefore, the A or S line must be inbred and propagated by sib-pollination. The sib that is used in such matings must have the genotype depicted in the B or maintainer line. This cross will yield progeny that all possess sterile cytoplasm, and segregate 1:1 for the dominant nuclear restorer gene. Thus, the A/sterile and B/maintainer lines are usually distinct genetic entities at the start of the breeding program but that will become genetically identical populations that will differ at the end of the program only with respect to genotype at the *Ms* locus. The C or restorer line will ultimately serve as the male parent in the final cross to produce hybrid seed. It is genetically distinct from the A/sterile and B/maintainer lines, bred to possess combining ability with them. The genotype at the *Ms* locus is critical to harvest economics only if the agricultural product is based on successful sexual reproduction.

In the Cucurbitaceae, where gynoecy is used for the production of hybrid seed, the method used is usually a combination of genotype and chemical induction. Gynoecious, monoecious, or hermaphroditic genotypes may be used in combination with agents that promote the conversion to femaleness (GA). As a rule, less attention is paid to the genotype of the male parent in the hybrid cross, even though cucurbit crops require sexual fertilization for economic yield to be manifested. If the hybrid lacks strong pollen production, a source of pollen called a "pollenizer" may be added to the production field to increase seed yields. The pollenizer merely initiates fruit development, the characteristics of which are entirely determined by the maternal genotype, in this case, the hybrid cultivar.

Other methods have been developed that make use of reproductive genetic systems for the efficient production of hybrid seed from mass controlled matings. One example is the use of sporophytic self-incompatibility. The evolution of self-incompatibility systems in plants for the promotion of outcrossing and long-term fitness was discussed in Chapter 10. The observation that all progeny harvested from two isolated plants that are sexually compatible is of hybrid origin was exploited for the breeding of hybrid varieties of cabbage, broccoli, cauliflower, Brussels sprouts,

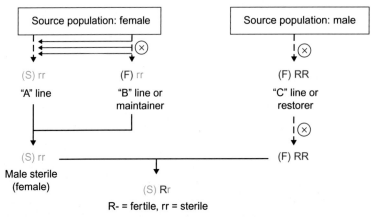

FIG. 16.12 A framework for the incorporation of cytoplasmic male sterility (cms) into a hybrid breeding program to allow for the later use of cms for large-scale production of genetically pure seeds.

and kale (*Brassica oleracea*). The breeding program necessitates the identification of specific SI alleles present in lines, to assure that the resulting hybrid will be of high purity. While certain SI alleles in *B. oleracea* are highly effective for excluding isogenic pollen others are "leaky". Recently, however, a male sterility system has supplanted the use of SI for hybrid seed production of these crops (Singh et al., 2019).

Another example of the use of genetic systems for hybrid seed production is the balanced tertiary trisomic method developed in *Hordeum* by Ramage (1965). This method makes use of a homeologous alien chromosome carrying a recessive male sterility gene. By carefully shepherding the euploid and aneuploid segregants that emerge during each inbreeding cycle, it is possible to engineer a seed production step similar to that described for the cytoplasmic-nuclear system above. Ramage (1965) used balanced tertiary trisomics to demonstrate that cross-fertilization, and resulting heterozygosity, confers a selective advantage over self-pollinating counterparts. The economics of hybrid seed production has never come close to satisfying the fundamental hybrid equation, however, so the details of this system are a curiosity that will be left to the reader to ponder.

References

Abubakar, L., Ado, S.G., 2008. Heterosis of purple blotch (*Alternaria porri* (Ellis) Cif.) resistance, yield and earliness in tropical onions (*Allium cepa* L.). Euphytica 164 (1), 63–74.

Allard, R.W., 1999. Principles of Plant Breeding, second ed. Wiley & Sons, Inc., New York, NY. 254 pp.

Ali, M.S., Knapp, S.J., 1996. Heterosis of *Cuphea lanceolata* single-cross hybrids. Crop Sci. 36 (2), 278–284.

Álvarez, G., Ceballos, F.C., Berra, T.M., 2015. Darwin was right: inbreeding depression on male fertility in the Darwin family. Biol. J. Linn. Soc. 114 (2), 474–483.

Andorf, S., Selbig, J., Altmann, T., Poos, K., Witucka-Wall, H., Repsilber, D., 2010. Enriched partial correlations in genome-wide gene expression profiles of hybrids (*A. thaliana*): a systems biological approach towards the molecular basis of heterosis. Theor. Appl. Genet. 120 (2), 249–259.

Auger, D.L., Ream, T.S., Birchler, J.A., 2004. A test for a metastable epigenetic component of heterosis using haploid induction in maize. Theor. Appl. Genet. 108 (6), 1017–1023.

Barrett, S.C.H., Charlesworth, D., 1991. Effects of a change in the level of inbreeding on the genetic load. Nature 352, 522–524.

Basra, A.S., 2000. In: Basra, A.S. (Ed.), Hybrid Seed Production in Vegetables: Rationale and Methods in Selected Crops. Food Products Press, New York, NY. 135 pp.

Basunanda, P., Radoev, M., Ecke, W., Friedt, W., Becker, H.C., Snowdon, R.J., 2010. Comparative mapping of quantitative trait loci involved in heterosis for seedling and yield traits in oilseed rape (*Brassica napus* L.). Theor. Appl. Genet. 120 (2), 271–281.

Beal, W.J., 1878. Report of the Michigan State Board of Agriculture. vol. 17. Michigan State College, East Lansing, MI. 445–457.

Benson, D.L., Hallauer, A.R., 1994. Inbreeding depression rates in maize populations before and after recurrent selection. J. Hered. 85 (2), 122–128.

Bernardo, R., Charcosset, A., 2006. Usefulness of gene information in marker-assisted recurrent selection: a simulation appraisal. Crop Sci. 46 (2), 614–621.

Bernardo, R., Yu, J., 2007. Prospects for genomewide selection for quantitative traits in maize. Crop Sci. 47 (3), 1082–1090.

Bradshaw, J.E., Dale, M.F.B., Mackay, G.R., 2009. Improving the yield, processing quality and disease and pest resistance of potatoes by genotypic recurrent selection. Euphytica 170 (1–2), 215–227.

Busch, J.W., 2005. Inbreeding depression in self-incompatible and self-compatible populations of *Leavenworthia alabamica*. Heredity 94 (2), 159–165.

Cramer, C.S., Wehner, T.C., 2000. Path analysis of the correlation between fruit number and plant traits of cucumber populations. HortScience 35 (4), 708–711.

Cros, D., Denis, M., Sánchez, L., Cochard, B., Flori, A., Durand-Gasselin, T., Nouy, B., Omoré, A., Virginie, P., Riou, V., Edyana, S., Bouvet, J.-M., 2014. Genomic selection prediction accuracy in a perennial crop: case study of oil palm (*Elaeis guineensis* Jacq.). Theor. Appl. Genet. 128 (3), 397–410.

Dahal, D., Mooney, B.P., Newton, K.J., 2012. Specific changes in total and mitochondrial proteomes are associated with higher levels of heterosis in maize hybrids. Plant J. 72 (1), 70–83.

Damgaard, C., Loeschcke, V., 1994. Inbreeding depression and dominance-suppression competition after inbreeding in rapeseed (*Brassica napus*). Theor. Appl. Genet. 88 (3/4), 321–323.

Darwin, C., 1876. The Effects of Cross and Self-fertilisation in the Vegetable Kingdom. John Murray, London.

Davenport, C.B., 1908. Degeneration, albinism and inbreeding. Science 28, 454–455.

Dickert, T.E., Tracy, W.F., 2002. Heterosis for flowering time and agronomic traits among early open-pollinated sweet corn cultivars. J. Am. Soc. Hort. Sci. 127 (5), 793–797.

Dubreuil, P., Charcosset, A., 1999. Relationships among maize inbred lines and populations from European and North-American origins as estimated using RFLP markers. Theor. Appl. Genet. 99 (3–4), 473–480.

Dudash, M.R., Carr, D.E., 1998. Genetics underlying inbreeding depression in *Mimulus* with contrasting mating systems. Nature 393 (6686), 682–684.

Eagen, K.A., Goldman, I.L., 1996. Assessment of RAPD marker frequencies over cycles of recurrent selection for pigment concentration and percent solids in red beet (*Beta vulgaris* L.). Mol. Breed. 2 (2), 107–115.

East, E.M., 1908. Inbreeding in corn. Rep. Conn. Agric. Exp. Station 1907, 419–429.

East, E.M., Jones, D.F., 1919. Inbreeding and Outbreeding: Their Genetic and Sociological Significance. Lippincott, Philadelphia, PA. 147 pp.

Egli, D.B., 2008. Comparison of corn and soybean yields in the United States: historical trends and future prospects. Agron. J. 100 (3), S79–S88.

Falconer, D.S., MacKay, T.F.C., 1996. Introduction to Quantitative Genetics. vol. 4 Longman Green, Harlow, Essex. 480 pp.

Fehr, W.R., 1987. Principles of Cultivar Development Vol. 1 Theory and Technique. MacMillan Publ. Co., New York, NY. 536 pp.

Fishman, L., 2001. Inbreeding depression in two populations of *Arenaria uniflora* (Caryophyllaceae) with contrasting mating systems. Heredity 86 (Pt 2), 184–194.

Forsberg, R.A., Smith, R.K., 1980. Sources, maintenance, and utilization of parental material. In: Fehr, W.A., Hadley, H.H. (Eds.), Hybridization of Crop Plants. ASA/CSSA Press, Ames, IA, pp. 65–81.

Fort, A., Ryder, P., McKeown, P.C., Wijnen, C., Aarts, M.G., Sulpice, R., Spillane, C., 2016. Disaggregating polyploidy, parental genome dosage and hybridity contributions to heterosis in *Arabidopsis thaliana*. New Phytol. 209 (2), 590–599.

Gallais, A., 2009. Full-sib reciprocal recurrent selection with the use of doubled haploids. Crop Sci. 49 (1), 150–152.

Geleta, L.F., Labuschagne, M.T., Viljoen, C.D., 2004. Relationship between heterosis and genetic distance based on morphological traits and AFLP markers in pepper. Plant Breed. 123 (5), 467–473.

Goff, S.A., 2011. A unifying theory for general multigenic heterosis: energy efficiency, protein metabolism, and implications for molecular breeding. New Phytol. 189 (4), 923–937.

Golmirzaie, A.M., Bretschneider, K., Ortiz, R., 1998. Inbreeding and true seed in tetrasomic potato. II. Selfing and sib-mating in heterogeneous hybrid populations of *Solanum tuberosum*. Theor. Appl. Genet. 97 (7), 1129–1132.

González, A.M., De Ron, A.M., Lores, M., Santalla, M., 2014. Effect of the inbreeding depression in progeny fitness of runner bean (*Phaseolus coccineus* L.) and it is implications for breeding. Euphytica 200 (3), 413–428.

Gordillo, G.A., Geiger, H.H., 2008a. Optimization of DH-line based recurrent selection procedures in maize under a restricted annual loss of genetic variance. Euphytica 161 (1–2), 141–154.

Gordillo, G.A., Geiger, H.H., 2008b. Alternative recurrent selection strategies using doubled haploid lines in hybrid maize breeding. Crop Sci. 48 (3), 911–922.

Griffing, B., 1990. Use of a controlled-nutrient experiment to test heterosis hypotheses. Genetics 126 (3), 753–767.

Guo, M., Rupe, M.A., Yang, X., Crasta, O., Zinselmeier, C., Smith, O.S., Bowen, B., 2006. Genome-wide transcript analysis of maize hybrids: allelic additive gene expression and yield heterosis. Theor. Appl. Genet. 113 (5), 831–845.

Hale, A.L., Farnham, M.W., Nzaramba, M.N., Kimbeng, C.A., 2007. Heterosis for horticultural traits in broccoli. Theor. Appl. Genet. 115 (3), 351–360.

Hallauer, A.R., 2007. History, contribution, and future of quantitative genetics in plant breeding: lessons from maize. Crop Sci. 47 (3), S4–S19.

Hallauer, A.R., Marcelo, J., Carena, J.B., 2010. Quantitative Genetics in Maize Breeding. Springer Science & Business Media, New York, NY. 664 pp.

Hamilton, M.B., Mitchell-Olds, T., 1994. The mating system and relative performance of selfed and outcrossed progeny in *Arabis fecunda* (Brassicaceae). Am. J. Bot. 81 (10), 1252–1256.

Hayes, C.N., Winsor, J.A., Stephenson, A.G., 2005. A comparison of male and female responses to inbreeding in *Cucurbita pepo* subsp. *texana* (Cucurbitaceae). Am. J. Bot. 92 (1), 107–115.

Hedrick, P.W., Hellsten, U., Grattapaglia, D., 2016. Examining the cause of high inbreeding depression: analysis of whole-genome sequence data in 28 selfed progeny of *Eucalyptus grandis*. New Phytol. 209 (2), 600–611.

Hirao, A.S., 2010. Kinship between parents reduces offspring fitness in a natural population of *Rhododendron brachycarpum*. Ann. Bot. 105 (4), 637–646.

Hodgins, K.A., Barrett, S.C.H., 2006. Mating patterns and demography in the tristylous daffodil *Narcissus triandrus*. Heredity 96 (3), 262–270.

Jahnke, S., Sarholz, B., Thiemann, A., Kühr, V., Gutiérrez-Marcos, J.F., Geiger, H.H., Piepho, H.P., Scholten, S., 2010. Heterosis in early seed development: a comparative study of F$_1$ embryo and endosperm tissues 6 days after fertilization. Theor. Appl. Genet. 120 (2), 389–400.

Jeong, S.-Y., Nasar, U.A., Hee-Jeong, J., Hoy-Taek, K., Park, J.-I., Ill-Sup, N., 2017. Discovery of candidate genes for heterosis breeding in *Brassica oleracea* L. Acta Physiol. Plant. 39 (8), 180.

Jiao, Y., Peluso, P., Shi, J., Liang, T., Stitzer, M.P., Wang, B., Campbell, M.S., Stein, J.C., Wie, X., Chin, C.S., Guill, K., Regulski, M., Kumari, S., Olson, A., Gent, J., Schneider, K.L., Wolfgruber, T.K., May, M.R., Springer, N.M., Antoniou, E., McCrombie, W.R., Presting, G.G., McMullen, M., Ross-Ibarra, J., Dawe, R.K., Hastie, A., Rank, D.R., Ware, D., 2017. Improved maize reference genome with single-molecule technologies. Nature 546, 524–527.

Johannsen, W.L., 1903. Ueber erblichkeit in populationen und in reinen linien. Gustav Fischer Verlag, Jena, Germany.

Jones, D.F., 1916. Natural cross-pollination in the tomato. Science 43, 509–510.

José, M.A., Iban, E., Abad, S., Arús, P., 2005. Inheritance mode of fruit traits in melon: heterosis for fruit shape and its correlation with genetic distance. Euphytica 144 (1–2), 31–38.

Kawamura, K., Nagano, A.J., Dennis, E.S., Okazaki, K., Osabe, K., Kaji, M., Shimizu, M., Saeki, N., Fujimoto, R., Kawanabe, G., 2016. Genetic distance of inbred lines of Chinese cabbage and its relationship to heterosis. Plant Gene 5, 1–7.

Kindred, D.R., Gooding, M.J., 2005. Heterosis for yield and its physiological determinants in wheat. Euphytica 142 (1–2), 149–159.

Kohn, J.R., Biardi, J.E., 1995. Outcrossing rates and inferred levels of inbreeding depression in gynodioecious *Cucurbita foetidissima* (Cucurbitaceae). Heredity 75, 77–83. Pt 1.

Kolehmainen, J., Korpelainen, H., Mutikainen, P., 2010. Inbreeding and inbreeding depression in a threatened endemic plant, the African violet (*Saintpaulia ionantha* ssp. *grotei*), of the East Usambara Mountains, Tanzania. Afr. J. Ecol. 48 (3), 576–587.

Kumar, S., Volz, R.K., Alspach, P.A., Bus, V.G.M., 2010. Development of a recurrent apple breeding programme in New Zealand: a synthesis of results, and a proposed revised breeding strategy. Euphytica 173 (2), 207–222.

Levings III, C.S., 1993. Thoughts on cytoplasmic male sterility in cms-T maize. Plant Cell 5 (10), 1285–1290.

Li, A., Zhou, Y., Jin, C., Song, W., Chen, C., Wang, C., 2013. LaAP2L1, a heterosis-associated AP2/EREBP transcription factor of larix, increases organ size and final biomass by affecting cell proliferation in *Arabidopsis*. Plant Cell Physiol. 54 (11), 1822–1836.

Liu, H., Wang, X., Warburton, M.L., Wen, W., Jin, M., Deng, M., Liu, J., Tong, H., Pan, Q., Yang, X., Yan, J., 2015. Genomic, transcriptomic, and phenomic variation reveals the complex adaptation of modern maize breeding. Mol. Plant 8, 871–884.

Lyrene, P.M., 2005. Breeding low-chill blueberries and peaches for subtropical areas. HortScience 40 (7), 1947–1949.

Marame, F., Dessalegne, L., Fininsa, C., Sigvald, R., 2009. Heterosis and heritability in crosses among Asian and Ethiopian parents of hot pepper genotypes. Euphytica 168 (2), 235–247.

Martínez-García, P.J., Dicenta, F., Ortega, E., 2012. Anomalous embryo sac development and fruit abortion caused by inbreeding depression in almond (*Prunus dulcis*). Sci. Hort. 133, 23–30.

Masny, A., Masny, S., Edward, Å., Pruski, K., Dry, W.M.Ä., 2016. Suitability of certain strawberry genotypes for breeding of new cultivars tolerant to leaf diseases based on their combining ability. Euphytica 210 (3), 341–366.

Meyer, R.C., Kusterer, B., Lisec, J., Steinfath, M., Becher, M., Scharr, H., Melchinger, A.E., Selbig, J., Schurr, U., Willmitzer, L., Altmann, T., 2015. QTL analysis of early stage heterosis for biomass in *Arabidopsis*. Theor. Appl. Genet. 120 (2), 227–237.

Muola, A., Mutikainen, P., Laukkanen, L., Lilley, M., Roosa, L., 2011. The role of inbreeding and outbreeding in herbivore resistance and tolerance in *Vincetoxicum hirundinaria*. Ann. Bot. 108 (3), 547–555.

Paschold, A., Marcon, C., Hoecker, N., Hochholdinger, F., 2010. Molecular dissection of heterosis manifestation during early maize root development. Theor. Appl. Genet. 120 (2), 383–388.

Qi, X., Li, Z.H., Jiang, L.L., Yu, X.M., Ngezahayo, F., Liu, B., 2010. Grain-yield heterosis in *Zea mays* L. shows positive correlation with parental difference in CHG methylation. Crop Sci. 50 (6), 2338–2346.

Ramage, R.T., 1965. Balanced teriary trisomics for use in hybrid seed production. Crop Sci. 5, 177–178.

Ribaut, J.M., Betran, J., 1999. Single large-scale marker-assisted selection (SLS-MAS). Mol. Breed. 5 (6), 531–541.

Riday, H., Brummer, E.C., Campbell, T.A., Luth, D., Cazcarro, P.M., 2003. Comparisons of genetic and morphological distance with heterosis between *Medicago sativa* subsp. *sativa* and subsp. *falcate*. Euphytica 131 (1), 37–45.

Ritland, K., 1996. Inferring the genetic basis of inbreeding depression in plants. Genome 39 (1), 1–8.

Rumpunen, K., Kviklys, D., 2003. Combining ability and patterns of inheritance for plant and fruit traits in Japanese quince (*Chaenomeles japonica*). Euphytica 132 (2), 139–149.

Sanewski, G.M., 2009. The effect of different levels of inbreeding on self-incompatibility and inbreeding depression in pineapple. Acta Hort. 822, 63–70.

Schierup, M.H., Christiansen, F.B., 1996. Inbreeding depression and outbreeding depression in plants. Heredity 77, 461–468. pt.5.

Semel, Y., Nissenbaum, J., Menda, N., Zinder, M., Krieger, U., Issman, N., Pleban, T., Lippman, Z., Gur, A., Zamir, D., 2006. Overdominant quantitative trait loci for yield and fitness in tomato. Proc. Natl. Acad. Sci. U. S. A. 103 (35), 12981–12986.

Shaw, D.V., 1995. Comparison of ancestral and current-generation inbreeding in an experimental strawberry breeding population. Theor. Appl. Genet. 90 (2), 237–241.

Shen, H., He, H., Li, J., Chen, W., Wang, X., Guo, L., Peng, Z., He, G., Zhong, S., Qi, Y., Terzhagi, W., Xing, W.D., 2012. Genome-wide analysis of DNA methylation and gene expression changes in two *Arabidopsis* ecotypes and their reciprocal hybrids. Plant Cell 24 (3), 875–892.

Shen, Y., Sun, S., Hua, S., Shen, E., Chu-Yu, Y., Cai, D., Timko, M.P., Qian-Hao, Z., Fan, L., 2017. Analysis of transcriptional and epigenetic changes in hybrid vigor of allopolyploid *Brassica napus* uncovers key roles for small RNAs. Plant J. 91 (5), 874–893.

Shull, G.H., 1908. The composition of field maize. Amer. Breeders' Assoc. Rep. 4, 296–391.

Shull, G.H., 1909. A pure line method of corn breeding. Amer. Breeders' Assoc. Rep. 5, 51–59.

Singh, P., Cheema, D.S., Dhaliwal, M.S., Garg, N., 2014. Heterosis and combining ability for earliness, plant growth, yield and fruit attributes in hot pepper (*Capsicum annuum* L.) involving genetic and cytoplasmic-genetic male sterile lines. Sci. Hort. 168, 175–188.

Singh, S., Dey, S.S., Bhatia, R., Kumar, R., Behera, T.K., 2019. Current understanding of male sterility systems in vegetable *Brassicas* and their exploitation in hybrid breeding. Plant Reprod https://doi.org/10.1007/s00497-019-00371-y.

Slater, A.T., Cogan, N.O.I., Forster, J.W., 2013. Cost analysis of the application of marker-assisted selection in potato breeding. Mol. Breed. 32 (2), 299–310.

Sleper, D.A., Poehlman, J.M., 2006. Breeding Field Crops, fifth ed. Wiley-Blackwell Publ., New York, NY. 432 pp.

Solieman, T.H.I., El-Gabry, M.A.H., Abido, A.I., 2013. Heterosis, potence ratio and correlation of some important characters in tomato (*Solanum lycopersicum* L.). Sci. Hort. 150, 25–30.

Su, J., Zhang, F., Yang, X., Feng, Y., Yang, X., Wu, Y., Guan, Z., Fang, W., Chen, F., 2017. Combining ability, heterosis, genetic distance and their intercorrelations for waterlogging tolerance traits in chrysanthemum. Euphytica 213 (2), 42.

Tian, H.Y., Siraj, A.C., Sheng, W.H., 2017. Relationships between genetic distance, combining ability and heterosis in rapeseed (*Brassica napus* L.). Euphytica 213 (1), 1–8.

Troyer, A.F., Wellin, E.J., 2009. Heterosis decreasing in hybrids: yield test inbreds. Crop Sci. 49 (6), 1969–1976.

Viana, A.P., de Lima e Silva, F.H., Leonardo-Siqueira, G., Rodrigo-Moreira, R., Krause, W., Marcela-Santana, B.B., 2017. Implementing genomic selection in sour passion fruit population. Euphytica 213 (10), 228.

Wang, Z., Yu, C., Liu, X., Liu, S., Yin, C., Liu, L., Lei, J., Jiang, L., Yang, C., Chao, C., Langming, C., Zhai, H., Wan, J., 2012. Identification of Indica rice chromosome segments for the improvement of Japonica inbreds and hybrids. Theor. Appl. Genet. 124 (7), 1351–1364.

Wei, X., Wang, B., Peng, Q., Feng, W., Mao, K., Zhang, X., Sun, P., Liu, Z., Tang, J., 2015. Heterotic loci for various morphological traits of maize detected using a single segment substitution lines test-cross population. Mol. Breed. 35 (3), 287.

Weider, C., Stamp, P., Christov, N., Hüsken, A., Foueillassar, X., Kamp, K.-H., Munsch, M., 2009. Stability of cytoplasmic male sterility in maize under different environmental conditions. Crop Sci. 49 (1), 77–84.

Welsh, J.R., 1981. Fundamentals of Plant Genetics and Breeding. John Wiley & Sons, New York, NY. 290 pp.

Willis, J.H., 1993. Partial self-fertilization and inbreeding depression in two populations of *Mimulus guttatus*. Heredity 71 (2), 145–154.

Wong, C.K., Bernardo, R., 2008. Genomewide selection in oil palm: increasing selection gain per unit time and cost with small populations. Theor. Appl. Genet. 116 (6), 815–824.

Wright, A., 1980. Commercial hybrid seed production. In: Fehr, W.A., Hadley, H.H. (Eds.), Hybridization of Crop Plants. ASA/CSSA Press, Ames, IA, pp. 161–176.

Xie, C., Xu, S., 1998. Strategies of marker-aided recurrent selection. Crop Sci. 38 (6), 1526–1535.

Yu, C.Y., Hu, S.W., Zhao, H.X., Guo, A.G., Sun, G.L., 2005. Genetic distances revealed by morphological characters, isozymes, proteins and RAPD markers and their relationships with hybrid performance in oilseed rape (*Brassica napus* L.). Theor. Appl. Genet. 110 (3), 511–518.

Yuan, M., Zhou, Y., Liu, D., 2004. Genetic diversity among populations and breeding lines from recurrent selection in *Brassica napus* as revealed by RAPD markers. Plant Breed. 123 (1), 9–12.

Zhu, A., Greaves, I.K., Liu, P.-C., Wu, L., Dennis, E.S., Peacock, W.J., 2016. Early changes of gene activity in developing seedlings of *Arabidopsis* hybrids relative to parents may contribute to hybrid vigour. Plant J. 88 (4), 597–607.

Ziehe, M., Roberds, J.H., 1989. Inbreeding depression due to overdominance in partially self-fertilizing plant populations. Genetics 121 (4), 861–868.

Further Reading

Gama, E.E.G., Hallauer, A.R., Ferrão, R.G., Barbosa, D.M., 1995. Heterosis in maize single crosses derived from a yellow Tuxpeño variety in Brazil. Rev. Bras. Genet. 18, 81–85.

Breeding Methods for Outcrossing Plant Species: III. Asexual Propagation

INTRODUCTION

In natural ecosystems, asexual reproduction is a strategy for the short-term exploitation of a habitat by a single, specific, highly fit genotype, or related genotypes (see Chapters 4 and 10). Asexual plant propagation was adapted by early humans as a means to perpetuate desirable phenotypes of certain crop species, primarily woody perennials. Therefore, asexual reproduction, or cloning, was an early tool for crop domestication, and is still widely used by plant breeders. Many variations on asexual propagation are evident from the myriad of plant organs or tissues that are involved with the cloning process in nature and by plant breeders and propagators: shoots, leaves, roots, stems, and apomictic seeds. Runners are modified shoots, tubers are modified roots, bulbs are modified leaves, and corms and rhizomes are modified stems (Richards, 1986). A broad range of modifications is seen in seeds where sexual embryos may be replaced by asexual counterparts of maternal sporophytic origin (apospory; Chapter 10). Many plants are also capable of regenerating roots and shoots from wound surfaces, a phenomenon that has been used for the asexual propagation of woody perennial plant species. Other specific examples of asexual reproduction will be described later in this chapter.

It is not a coincidence that most plant species that exhibit asexual reproduction are derived from outcrossing ancestors. Many plant species are both sexual outcrossers and also reproduce asexually, but very few self-pollinating species also engage in asexual reproduction. Anthropomorphically, the successful plant genotype has two alternative strategies for rapid self-perpetuation: cloning or self-pollination. With self-pollination, prospects for long-term survival of the gene pool are compromised due to allelic fixation. Asexual self-propagation provides an evolutionary safety hatch: reversion to sexual reproduction and a cache of polymorphic loci if things go awry (Cook, 1983).

One may imagine that primitive humans learned quickly that the fleshy, starchy organs such as tubers of certain food crops spontaneously sprouted new foliage and roots and that next year's crop could be established directly from them.

TABLE 17.1   The Top 11 Clonally/Asexually Propagated Crop Species and Mode of Propagation Based on FAO 2006 Statistics (Grüneberg et al., 2009)

Crop species	Mode of asexual propagation	World harvest MT (FAO 2006)
White potato (*Solanum tuberosum*)	Tuber	315 million
Cassava (*Manihot esculenta*)	Hardwood cutting	226 million
Sweet potato (*Ipomoea batatas*)	Sprout cutting	124 million
Sugar cane (*Saccharum officinarum*)	Cane stalk	194 million
Banana/plantain (*Musa × paradisiaca*)	Corm	105 million
Citrus fruits (*Citrus* spp.)	Bud stick graft	89 million
Grape (*Vitis vinifera*)	Hardwood cutting	69 million
Apple (*Malus pumila*)	Bud stick graft	64 million
Yam (*Dioscoea* spp.)	Tuber	51 million
Taro (*Colocasia esculenta*)	Corm	12 million
Strawberry (*Fragaria xananassa*)	Adv. Shoots	4 million

They also observed that the offspring from sporophyte organs were more similar to the plants from which they were derived than were plants from seeds. Any genetic variability that existed within wild populations propagated asexually would have been rapidly depleted following a few cycles of selection.

The list of the most valuable crop species that are propagated asexually prominently includes woody perennial species, both food and ornamental (Table 17.1). These woody perennial species have been mostly adapted to asexual reproduction since no naturally occurring asexual propagules (e.g., tubers, corms, rhizomes, etc.) are evident during natural life cycles (McKey et al., 2010). The ancient technologies of rooting and grafting were developed to enable early breeders to attain uniform stands of the highest performance within a reasonable time period. The minimum generation time for many woody perennial tree species is in the range of 5–10 years and the life span of early humans was not much longer than this.

Under circumstances where consumer demand for agricultural product quality and consistency are very high and coincide with high value of individuals within populations, clonal propagation becomes economically feasible and strategically desirable as a method to capture and protect proprietary genotypes. This situation is particularly common with ornamental crop species such as cut flowers, potted house plants, and many landscape plants produced by the nursery industry. The Plant Patent Act of 1930 (Chapter 12) was the first step toward creating property rights for biological innovation of clonally-propagated crop species. The Act provided for patenting rights of asexually-propagated, and not sexually-propagated, plants. Large commercial nurseries that began to build mass hybridization programs in the 1940s accounted for most of these patents, suggesting that the new intellectual property rights may have helped to encourage the development of a commercial rose and other high-value ornamental woody perennial breeding industries (Moser and Rhode, 2011).

We expect that V_G should be 0 among cloned populations. The genotypes of parent and progeny plants are connected by a continuous succession of mitotic cell divisions with no intervening steps during which genetic variability may be introduced, at least theoretically. The widespread observation of genetic variability from cultured somatic cells and uses of this phenomenon to enhance gene pools were described in Chapter 8. It is possible, therefore, that $V_G > 0$ in clonal populations. Examples of genetic variability within populations of cloned plants will be presented later in this chapter.

In many instances, plant species that engage in the process of asexual reproduction have entirely lost the capacity to undergo sexual reproduction (Richards, 1986). In sugarcane, for example, flowers and pollen are observed, but seeds are rarely set on the panicles (Ortiz et al., 2008). Sexual sterility is also found in many clonal selections of autotetraploid potato (Simmonds, 1997). From a breeding standpoint, sexual sterility presents a severe limitation on the ability to affect further improvements within the clonal line since sex is the fundamental means by which plant breeders forge ongoing genetic improvements (Grüneberg et al., 2009).

Asexual reproduction is one of the fundamental engineered population genetic structures described in Chapter 4. The capability to faithfully replicate genotypes is a boon to plant breeding because it provides the capability to lock in a desirable genetic constitution and to perpetuate it ad infinitum. If a genotype occurs that confers a superior phenotype, there is no need to determine the basis of the connection of genotype to phenotype as is often essential for sexually propagated crop species (Simmonds, 1979). As long as the phenotype is faithfully reproduced concomitantly with clonal propagation, it matters little what the inheritance, gene action, epistasis, fixation or heterozygosity, or epigenetics of the genotype may be. It simply is what it is. On the other hand, the plant breeder often strives to understand the factors that contributed to the development of such a superior genotype to make generalizations about how to produce parallel excellence in the future, or to improve the genotype even more in the future.

Many successful cultivars of clonally propagated crop species were simply selected directly from within earlier commercial varieties and propagated directly. One example is the Russet Burbank potato. The original clone was discovered and selected by famed horticulturist Luther Burbank in 1850 and has been maintained clonally ever since, still grown substantially worldwide (Smith, 2009). The original "Burbank" selection was smooth-skinned. A russeted (brown-skinned) variant of "Burbank" was released in 1902 by L. L. May & Company and as the cultivar "Netted Gem," later changed to "Russet Burbank" (Bethke et al., 2014). No sexual cycle has intervened during this entire period. The Russet Burbank clone is highly sterile, sexually.

The collective experiences in the breeding of clonally propagated crops, therefore, has been quite varied. In certain crop species, for example, potato, an extensive body of knowledge has accumulated on the application of genetic principles to varietal improvement (Bradshaw, 2017). Many potato breeders have championed the transition from clonal to sexual propagation, thus providing them with breeding strategy alternatives that have been used successfully in sexually-propagated crop species. For many other asexually propagated species, the breeding process has been hindered by sterility and long generation times. Woody perennial species are included among the latter class. While woody perennial plant species reproduce sexually in nature, humans have propagated woody perennial species asexually for millennia (Kumar, 2006).

The most simplistic approach to the development of improved varieties of asexually propagated crop species is clonal selection (Acquaah, 2012). Intensive screening of commercial populations often reveals variant individuals or sectoral chimeras on individual plants (see below). It is presumed that such variants or sectors arise from mutations that occurred during the succession of mitotic divisions, the likes of which is propagated until an entire mutant organism or sector thereof is formed. The variant or sector may be submitted into a clonal propagation scheme directly and evaluated for economic performance at the whole population level. This strategy is surprisingly common and successful, particularly for clonally-propagated crop species not addressed by large numbers of trained plant breeders armed with hefty budgets (Grüneberg et al., 2009).

The general approach to breeding more advanced clonally propagated crops is to develop populations of diverse genotypes that feature the highest levels of V_G attainable within a genetic background that is generally adapted to the region and climate of interest. Clonally-propagated species tend to be associated with mating systems that promote outcrossing, as was stated earlier. Attempts to inbreed these species have mostly met with failure due to inbreeding depression (Kumar, 2006; Acquaah, 2012). Intercrossing schemes such as recurrent selection to achieve combinations of many genes, or modified backcross strategies (Chapter 18), are commonly employed.

Since most asexually/clonally propagated crop species are derived from cross-pollinated progenitors, it is appropriate that modified strategies that are effective for cross pollinated species have been found to be effective. A general scheme for the breeding of asexually propagated crop species is depicted in Fig. 17.1. The red dotted line in the diagram is significant, indicating that, unlike sexually propagated crop species, selections may be made for desirable genotypes/phenotypes at any point in the program and advanced into a clonal selection program for commercial introduction of a new cultivar.

Since any selection may be perpetuated directly by clonal propagation, and the particular actions and interactions of genes therein, V_D, V_I, and all variance interaction terms such as V_{GxE} are fully accessible to the plant breeder for genetic improvement. In the case of woody perennial crop species, the primary factors that limit success are land area (due to large plant size) and time (due to perennial habit; Kumar, 2006).

BREEDING OF SELECTED CLONALLY-PROPAGATED CROP SPECIES

A large body of research has been conducted on the veracity of methods to improve the efficiency and effectiveness of breeding schemes for asexually-propagated crop species since these tend to be long-lived and physically space consuming. The most substantial body of results may be found in white potato (*Solanum tuberosum*). While much of base of knowledge and experience may be applied beyond potato, there are also factors that limit the range of application.

Clonal propagation of potato offers important agronomic and genetic advantages to the plant breeder, such as vigorous early growth, higher yields and the quick fixation of hybrid vigor. Bradshaw et al. (1998) confirmed that

Breeding asexually propagated crop species

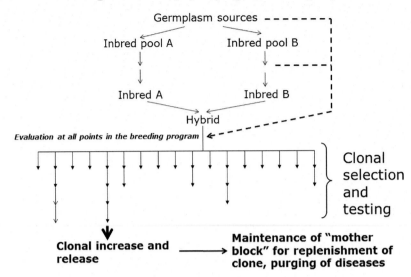

FIG. 17.1   A general strategy for breeding new cultivars of asexually propagated crop species.

selection for visual preference within crosses in the seedling and first clonal generations of potato is very ineffective, but that worthwhile progress can be made from selection in the second clonal generation. Results obtained by Kumar and Gopal (2006) confirmed this finding, that the parameters for tuber yield, average tuber weight and tuber number could only be reliably estimated from second clonal generation and thereafter. Gopal (1998) showed that potato hybrid progeny means could be predicted by the mid-self-values for plant vigor in clonal generations. Obtaining sexual hybrids in potato is challenging. The success rate achieved with potato pollinations in one study was only slightly higher than 30% (Bradshaw et al., 2009).

Inbreeding depression (ID) was found to be very high in potato, in sharp contrast with other Solanaceae crop species. ID in potato is associated with the clonal reproductive system that preserves a high load of deleterious alleles. The paucity of genetic improvement in potato despite intensive breeding efforts reflects the difficulty in eliminating deleterious alleles from potato germplasm due to mating system and autopolyploidy. Further, most deleterious alleles in potato are also locked into repulsion phase linkages and must be converted into coupling to forge needed genetic improvements (Fasoula, 2003).

An investigation was conducted to assess the potential to incorporate wild *Vitis* species in a breeding project for development of the disease-resistant seedless table grape (*V. vinifera*) cultivars. Hybridization was conducted using *V. vinifera* as female parents and the wild Chinese *Vitis* spp. as male parents. In-ovulo embryo rescue was used to develop hybrid plants from the seedless females (Tian et al., 2008).

Seedlessness is one of the most important characteristics for mandarin oranges for the fresh-fruit market. Seedless triploid citrus plants can be recovered by 4x×2x hybridizations using non-apomictic genotypes as female parents. The majority of the plants recovered by Aleza et al. (2012) were triploid (98.3%), but a few diploid, tetraploid and pentaploid plants were also obtained, and their genetic origin was analyzed by simple sequence repeat markers.

In another example, first generation hybrids between California-adapted apricot (*P. armeniaca*) cultivars and Central Asian accessions were generally more productive than the Central Asian parents, but fruits were still too small to be commercially viable. Second generation hybrids, obtained through intercrossing elite F_1 hybrids or by backcrosses to California-adapted cultivars, were very diverse in both fruit and tree characteristics. Fruit sizes adequate for fresh marketing are obtainable in the second generation, and large-fruited clones having significantly elevated °brix levels were also observed (Ledbetter, 2009).

Large macadamia (*Macadamia integrifolia* and *M. tetraphylla*) tree size and slow maturity of the crop pose particular problems for breeding. A tandem selection strategy for kernel recovery in a seedling trial and nut-in-shell yield in a clonal trial produced the highest gain to cost ratio, but was limited in the range of amenable genotypes (Topp et al., 2012). Key macadamia selection traits are cumulative nut yield, tree height, canopy width, kernel recovery, and nut quality. Industry participation has involved the use of grower properties for progeny field trials, review of outcomes by an industry steering group and consultation with industry on the important traits in new cultivars. Since macadamia breeding takes a long time and consumes a large quantity of field space, the cooperation of the industry is of paramount importance to success (Oraguzie et al., 2017).

The plant we think of as the "kiwifruit" is a single clone, "Hayward," of the species *Actinidia deliciosa*. "Hayward" is grown because consumers prefer the fruit type, but improvements in a number of fruiting and horticultural characters are needed. Programs underway include F_1 selection, recurrent selection, and clonal improvement by selection of natural or induced mutations. Approximately 100 *Actinidia* species have been described, but different ploidy levels within the genus restrict interspecific hybridization and gene flow. Clonal *Actinidia* rootstocks are also under evaluation (Ferguson et al., 1990).

Progress in the breeding of plantain and banana has been restricted by the complex genetic structure and behavior of cultivated polyploid *Musa × paradisiaca*. Molecular marker assisted breeding has the potential to dramatically enhance the pace and efficiency of genetic improvement in *Musa* (Crouch et al., 1999).

Coconut foliar decay (CFD) is a disease of coconut (*Cocos nucifera*) associated with infection by coconut foliar decay virus (CFDV) that is endemic on the South Pacific island of Vanuatu, a major source of commercial coconut products worldwide. An improved "Vanuatu Tall" × "Rennell Island Tall" hybrid was identified with a high degree of tolerance to CFD that was subsequently propagated asexually (Labouisse et al., 2011).

Morphological traits and several yield components as well as genotypic and phenotypic variability for vigor were investigated in *Cola nitida* groups A and B. Group A had low vigor and high yield while group B had high general plant vigor and low yield. High parent heterosis for yield from group A×group B was 350%. The best candidates from these diallel crosses were asexually propagated and commercialized (Sié et al., 2009).

Cassava (*Manihot esculenta*) breeding at the earlier stages so far has been mainly based on a mass phenotypic recurrent selection, and very little data are obtained to document breeding progress. However, replicating and blocking of clones in a legitimate statistical field design improved the accuracy of the results and validity of the inferences (Ojulong et al., 2008).

Breeding taro (*Colocasia esculenta*) for improved corm quality is complex, and phenotypic recurrent selection is impaired by the long growth cycle and low vegetative propagation ratio. The protocol developed in a study by Lebot and Lawac (2017) is rapid, cost-efficient, environmentally-friendly. This technique can be used in taro breeding programs for the early detection of undesirable hybrids with high ratios of reducing sugars to sucrose followed by clonal propagation of the best selections.

The French (INRA) breeding program has successively introduced double, clonal, three-way all-male, and F_1 (mixed or all-male) asparagus (*Asparagus officinalis*) hybrids. Evaluation indexes showed that (i) all-male F_1 and three-way hybrids tended to have high total yield and earliness; (ii) F_1 mixed hybrids were likely to have relatively large spear diameter; (iii) double, clonal, and foreign hybrids produced the most attractive spears; (iv) the general superiority of all-male hybrids was confirmed; and (v) the four characters that were the major breeding objectives—early yield, high total yield, large spear size, and attractive spear appearance—were difficult, but not impossible, to combine into one hybrid (Corriols and Dore, 1989).

THE SPECIAL CIRCUMSTANCES OF LONG-LIVED WOODY PERENNIALS

The scepter of very long generation times, up to 10 years in some cases, presents acute challenges to plant breeders. For example, the breeding of fruit crops is one of the most important but also neglected areas of horticultural crop breeding. All but a fraction of the world's apple production is based on cultivars developed before 1910, many grape cultivars date to before 1800 CE, and newer woody perennial crop species such as avocado and macadamia were based on the clonal progeny of single tree selections. Moreover, few important genes have been identified in most woody perennial species due to long generation times and paucity of genetic studies (Mullins, 2006).

There have been substantial improvements, however, in both scions and rootstocks in most major tree fruit crops, but selection within and among existing genotypes has been the main basis of improvement rather than the creation of new genotypes by hybridization (Hough, 1979). Clonal selection and imposed phytosanitary procedures have had a major impact on crop productivity (Mullins, 2006). New technical capabilities such as molecular transformation, genome editing, MAS, and GS will have particularly substantial impacts on the efficiency and effectiveness of breeding protocols for perennial crop species (Mehlenbacher, 1995; Janick, 1998; Dosba, 2003; Arús, 2007; Myles, 2013; van Nocker and Gardiner, 2014).

Many excellent volumes have been published on the breeding of woody perennial crop species, and the reader is referred to those for a more detailed and focused treatment of this challenging area of plant breeding (Janick and Moore, 1996; Adam-Blondon et al., 2011; Kole and Abbott, 2012; Badenes and Byrne, 2012; Al-Khayri et al., 2018; Kibet, 2018).

INTERSPECIFIC AND INTERGENERIC HYBRIDS

Interspecific and intergeneric plant hybrids are often plagued by sexual sterility or genetic instability during meiosis (see Chapter 8). These hybrids, and corresponding derivatives, often exhibit qualities that may be of immediate commercial value. The ability to clonally propagate interspecific and intergeneric hybrids or derivatives offers a pathway to tap this potential. This section will describe examples of the use of clonal propagation to breed or commercialize interspecific and intergeneric hybrids.

In potato, the breeding value of tetraploid F$_1$ hybrids between tetrasomic tetraploid *S. tuberosum* and the disomic tetraploid wild species *S. acaule* was examined. F$_1$ hybrids showed a tuber yield and appearance comparable to those of their cultivated parent, indicating a potential as acceptable breeding stocks despite the 50% contribution to their pedigree from wild *S. acaule* (Watanabe et al., 1994). In another project in potato, a breeding scheme based on the production of progenies with odd ploidy was followed to introduce useful genes from the wild *S. commersonii* (cmm) into *S. tuberosum* (tbr) genome. Hybrids from $5x \times 4x$ crosses were characterized for traits of interest, and the selection was assisted by amplified fragment length polymorphism (AFLP) analysis. Even though aneuploidy has often been associated with a reduction in male and female fertility, most of the hybrids were fertile following crosses with tbr, making it possible to produce viable offspring. Selection of hybrids was based on a two-stage scheme that consisted of conventional phenotypic selection followed by an estimation of the wild genome content still present in order to identify hybrids combining noteworthy traits with a low wild genome content (Iovene et al., 2004). All tuber-bearing selections were amenable to clonal propagation for population increases.

Wide crosses have been explored extensively in *Citrus* spp. as a way to combine desirable phenotypes into single clones. For example, 35 interspecific parental somatic hybrid combinations were accomplished with "acid" (lemons and limes) in 2000, 2001, and 2002 to combine seedlessness, cold-tolerance, and disease resistance. More triploid hybrids were generated from lemon seed progenitors compared to the other acid citrus fruit progenitors (Viloria and Grosser, 2005). Sweet orange is an interspecific hybrid rather than a true species. Cultivars are mutations selected over generations (possibly millennia) of clonally propagating the original hybrid. A new generation of sweet-orange-like hybrids is under evaluation, all with "Ambersweet" as a parent (Stover et al., 2016).

Wide crosses followed by clonal propagation were explored in *Vitis* as a way to expand the gene pool. An investigation was conducted to assess the potential to incorporate these species in a breeding project for development of the disease-resistant seedless cultivars. Hybridization was conducted using *V. vinifera* as female parents and the wild Chinese *Vitis* spp. as male parents. In-ovulo embryo rescue was used to develop hybrid plants from the seedless females (Tian et al., 2008).

In another study, attempts were made to transfer resistance to the Pierce's disease pathogen from wild species into cultivated grape. The inheritance of resistance to *Xylella fastidiosa* (Xf), the bacterium which causes Pierce's disease (PD) in grapevines, was evaluated within a factorial mating design consisting of 16 full-sib families with resistance derived from *V. arizonica* × *V. rupestris* interspecific hybrids. Resulting data indicated that resistance should be relatively easy to pass on from parents to progeny in a breeding program for the development of clonally-propagated PD-resistant grape cultivars, particularly when selection is based on cane maturation scores or stem Xf populations (Krivanek et al., 2005).

Interspecific crosses followed by clonal propagation were attempted in *Rosa* as a strategy to enhance bloom quality and aroma. *Rosa damascena* is the most important scented rose species cultivated for rose oil production. *R. bourboniana* (Edward rose), a related species, is popular due to longer blooming period and ease of propagation. Results indicated that RAPD and SSR markers are useful for hybrid identification of clonally-propagated scented roses (Kaul et al., 2009).

In papaya, morphological, molecular and cytological analyses were performed to assess the hybridity of 120 putative interspecific hybrids of *Carica papaya* × *C. cauliflora*. The number of main leaf veins was intermediate between the two parents while the hermaphrodite flower sex form and the low vigor were distinctive features of these hybrids. Petiole length, stem diameter, leaf length, leaf width and flower color were similar to *C. papaya*, whereas leaf shape, type, serration, venation, petiole hairiness and flower shape were similar to *C. cauliflora*. Cytological analyses revealed that 7–48% of the cells in the interspecific hybrids were aneuploid, suggesting that chromosome elimination had occurred. Selections derived from these interspecific papaya hybrids were to be commercialized by clonal propagation (Magdalita et al., 1997).

The genetic causes of heterosis in aspen tree growth were investigated by comparative genetic analysis of intra- and inter-specific crosses derived from *Populus tremuloides* and *P. tremula*. A new analytical method was developed to estimate the effective number of loci affecting a quantitative trait and the magnitudes of their additive and dominant effects across loci. During the first 3 years of growth, interspecific hybrids displayed strong heterosis in stem growth, especially volume index, over intraspecific hybrids. Heterozygotes newly formed through species combination showed much greater growth than the heterozygotes from intraspecific crosses at a reference locus. Heterosis in aspen growth appeared to be under multigenic control, with a slightly larger number of loci for stem diameter and volume than for height (Li and Wu, 1996). Heterotic tree selections must be propagated asexually for subsequent commercialization.

Substantial experimental efforts have been invested in the development of clonal interspecific and intergeneric *Hydrangia* hybrids as a way to expand the gene pool. One such study assessed the compatibility of interspecific crosses between *H. macrophylla* and *H. angustipetala* as a source of genetic diversity. The interspecific hybrids produced in this study were attractive, fertile plants that are being used in further breeding to develop new cultivars (Kardos et al., 2009). In another study, *H. macrophylla* × *H. paniculata* plants resembled *H. paniculata* in leaf shape and pubescence, and appeared to be less susceptible than *H. macrophylla* to powdery mildew (Reed et al., 2001). An intergeneric hybrid between *Dichroa febrifuga* and *H. macrophylla* was found to be intermediate in appearance between parents, but variability in leaf, inflorescence, and flower size and flower color was observed among the hybrids (Reed et al., 2008).

In clonally-propagated blueberry, diploid *Vaccinium darrowii* was used in breeding tetraploid southern highbush blueberry (*V. corymbosum*) as a source of reduced chilling requirement, adaptation to hot, wet summers, and resistance to leaf diseases. It was found that *V. darrowii* in Florida is quite variable, but that the use in breeding of a wider range of *V. darrowii* accessions could provide beneficial diversity in the gene pool of blueberry (Chavez and Lyrene, 2009).

Cultivated strawberry has a long breeding history of interspecific hybridization followed by polyploidization and clonal propagation to capture desirable genetic variability. For example, two groups of *Fragaria decaploid* (2n = 10x = 70) breeding populations were studied by Ahmadi and Bringhurst (1992). One of these groups was derived from pentaploid (2n = 5x = 35) and hexaploid (2n = 6x = 42) natural or synthetic interspecific hybrids between octoploid (2n = 8x = 56) *F. chiloensis* or *F. virginiana*, both from California, and various *Fragaria* diploids (2n = 2x = 14). The chromosome number of interspecific hybrids was doubled with colchicine or through the naturally generated unreduced gametes. Resulting decaploids combined the genomes of the best octoploid cultivars with those of the above diploid species, facilitating the combination of desirable genes for high yield, day neutrality, and excellent fruit quality into a single genome then asexually-propagated for commercial introduction.

An intergeneric hybrid between *Dendranthema morifolium* (chrysanthemum) variety "Zhongshanjingui" and *Artemisia vulgaris* (mugwort) "Variegata" was attempted to combine ornamental qualities with resistance to aphids. The resulting hybrid was clonally propagated to produce large populations for testing. Aphid resistance in the intergeneric hybrid was found to be a consequence of altered leaf micromorphology and bioactive essential oil content (Deng et al., 2010).

Interspecific hybridization followed by clonal propagation is one of the most important strategies to achieve desirable genetic combinations in *Lilium* sp. where it is necessary to introduce new traits, such as flower color, petal shape, stem size and strength, bloom longevity, and disease resistance to remain competitive in the marketplace. *Longiflorum* × Asiatic, or LA, hybrids and Oriental × Trumpet, or OT, hybrids have become dominant combinations in breeding programs because of their superior performance over Asiatic and Oriental hybrids (Arens et al., 2014).

Attempts have been made in *Allium* sp. to combine desirable traits of garlic and leek. An interspecific hybrid between leek (*A. ampeloprasum*) and garlic (*A. sativum*) was produced by in vitro fertilization using a fertile garlic clone as a pollen donor and an *A. ampeloprasum* ovary culture. The odor compounds of garlic, lacking in leek, were detected in volatiles of the interspecific hybrid. The interspecific hybrid could be propagated asexually by planting bulb cloves (Yanagino et al., 2003). In another study of prospective wild *Allium* bridge species, 14 interspecific hybrids in sexual diploid *A. senescens* var. minor × apomictic tetraploid *A. nutans* crosses, and eight interspecific hybrids in sexual diploid *A. senescens* var. minor × apomictic hexaploid *A. senescens* crosses were produced. Triploid and tetraploid interspecific hybrids exhibited intermediate parental morphological characteristics. Parthenogenesis ranged from 26.0% to 86.0% in five tetraploid interspecific hybrids. Non-parthenogenesis to parthenogenesis segregated in a 3:5 ratio in *A. senescens* var. *minor* × *A. senescens* crosses (Kim et al., 1999).

Interspecific hybrids and clonal propagation were explored in sweet potato as a means of expanding the gene pool. Interspecific hybridization in the genus *Ipomoea* is very difficult due to natural reproductive barriers. Two novel interspecific F_1 hybrids were obtained between *I. batatas* (2n = 6x = 90) and two wild species, *I. grandifolia* (2n = 2x = 30) and *I. purpurea* (2n = 2x = 30). It was found that the clonally-propagated hybrids were quite distinctive in leaf color and morphology, and yielded intermediate storage roots with respect to size and quality as compared to their respective parents (Cao et al., 2009).

APPLICATIONS OF CELL AND TISSUE CULTURE IN BREEDING ASEXUALLY PROPAGATED CROP SPECIES

After new advances in cell and tissue culture were forged during the 1960s–80s, excitement surfaced regarding the prospects that micropropagation could be applied for the clonal propagation of annual and seed-propagated crops. Examples included pea (Griga et al., 1986), *Gerbera* (Murashige et al., 1974), and western white pine (Percy et al., 2000). Despite this initial excitement, successes with micropropagation have not been widespread in sexually-propagated horticultural crop species, although the technique is broadly applicable for purging plants and plant tissues of viral pathogens (Ebi et al., 2000; Bryan et al., 2003; O'Herlihy et al., 2003).

Many potato breeders have successfully bridged the gene pools of sexually incompatible species by protoplast fusion to create new hybrids. These are often sexually sterile and must be be propagated clonally. For example, one project was conducted to evaluate the resistance to *P. infestans* in somatic hybrids between *S. nigrum* and the diploid *S. tuberosum* clone ZEL-1136. Nine *S. nigrum* (+) ZEL-1136 hybrids showed a resistance that was significantly higher than that of *S. nigrum*, while six clones expressed a level of resistance to *P. infestans* similar to that of *S. nigrum*. Results confirmed the effective transfer of late blight resistance of *S. nigrum* into somatic hybrids with cultivated potato (Zimnoch-Guzowska et al., 2003). In another study, *S. cardiophyllum* (2n = 2x = 24) is a wild potato species found to be highly resistant to late blight but that is sexual incompatibility with *S. tuberosum*. Out of 26 regenerates from somatic *S. tuberosum* + *S. cardiophyllum* protoplast fusions, only four were confirmed as true somatic hybrids containing both parental genomes based on molecular markers and phenotypes (Chandel et al., 2015).

In other reports of somatic hybridization in potato, interspecific somatic hybrids between commercial cultivars of potato *S. tuberosum* "Agave" and "Delikat" and the wild diploid species *S. cardiophyllum* were produced by protoplast electrofusion. Somatic hybrids were tested for their resistance to Colorado potato beetle (CPB), to Potato virus Y (PVY), and foliage blight. Results confirmed that protoplast electrofusion could be used to transfer the CPB, PVY and late blight resistance of *S. cardiophyllum* into somatic hybrids with the cultivated potato genome (Thieme et al., 2010).

Scab resistance is one of the most important goals of apple breeding, typically achieved by time-consuming and expensive conventional selection techniques. The genetic modification of a recipient species with genes from a sexually compatible species, a process known as *cisgenesis*, is a promising tool for plant breeding to rapidly develop disease resistant apple cultivars. A cisgenic, scab-resistant population of the apple variety "Gala" expressing the wild apple scab resistance gene *Rvi6* under control of "Gala" regulatory sequences was successfully developed (Jänsch et al., 2014).

Protoplast fusion has been used in *Citrus* sp. to produce novel allotetraploid hybrids for use as parents in crosses with diploids to produce easy-peel, seedless, triploid cultivars (Wu et al., 2005). Forty-three plants were produced from fusions between protoplasts of "Encore" mandarin and "Valencia" sweet orange cultivars, and of "Encore" and "Caffin" clementine mandarin cultivars in which protoplasts were isolated from embryogenic callus of the donors. DNA fingerprinting confirmed that both parents contributed DNA to the hybrids and which were, therefore, allotetraploid (Wu et al., 2005).

Bud sporting, the consequence of sudden variations in genotype or gene expression of somatic cells, leads to the occurrence of phenotypically altered shoots in many asexually propagated plant species. In ornamentals, such as azalea (*Rhododendron simsii*), flower color bud sports are a valuable source of variation. Regeneration of tetraploid petal marginal tissue was performed and led to the production of the first induced tetraploid Belgian azalea. Flower color sports were frequently observed after plant regeneration (De Schepper et al., 2004).

In *Rosa*, apical buds of lateral branches were asexually propagated by cutting and treated with chemical mutagens. The growth and differentiation or morphological changes of the mutagen-treated buds were traced in developing flowers. Embryogenic calli were obtained via adventitious roots induced from the petals and successfully regenerated into intact plants. Results suggested an effective method for easily and rapidly inducing phenotypic variation in flowers of rose and for in vitro multiplication of regenerated plants (Nonomura et al., 2001).

The production of new cultivars and higher quality health-promoting products from *Echinacea* spp. necessitates an understanding of the regulation of plant growth and the production of specific phytometabolites. Studies were designed to generate elite varieties of *E. purpurea* based on regeneration efficiency and desirable chemical profile. Clonal propagation of seedling-derived regenerates and screening for antioxidant potential and concentrations of caftaric acid, chlorogenic acid, cichoric acid, cynarin, and echinacoside led to the identification of 58 unique breeding lines. Results demonstrated the potential for selective breeding of elite, highly regenerative, chemically superior, clonally-propagated cultivars from the naturally occurring gene pool in sexually-propagated populations of *E. purpurea* (Murch et al., 2006).

New genetic variation arising in cell and tissue cultures is undesirable when the goal is genetic fidelity (Orton, 1984). In potato, Intraspecific somatic hybrids have been produced by protoplast fusion in eight combinations involving ten dihaploids (2n = 2x = 24) in an attempt to provide new germplasm for potato breeding. Cytological analysis revealed extensive variation in chromosome number, such as aneuploid, aneusomatic, and mixoploid hybrids. Some hybrids exhibited structurally rearranged chromosomes and had a high frequency of aberrant anaphases (Gavrilenko et al., 1999). In oil palm (*Elaeis guineensis*), random amplified polymorphic DNA (RAPD) analysis using arbitrary 10-mer oligonucleotide primers was employed to investigate the genetic fidelity of somatic embryogenesis-derived regenerates. Of the 387 primers, 73 (19%) primers enabled the identification of polymorphisms between clones. It was concluded by Rival et al. (1998) that the regeneration protocol based on somatic embryogenesis for oil palm clonal propagation did not induce any gross genetic changes. Pinker et al. (2012) used genetic instability in strawberry tissue cultures beneficially. For generative segregation, a time span of 30 months was needed to produce autotetraploid strains. More tetraploid plants were observed from callus culture than from generative segregation. The autotetraploid mutants did not differ in fertility and were used in test crosses with *F.* × *ananassa* to produce hexaploid hybrids.

MARKER-ASSISTED SELECTION IN CLONALLY-PROPAGATED CROP SPECIES

The methods described for developing and utilizing molecular markers as surrogates or coding sequences for targeted traits in Chapter 9 are especially germane to asexually propagated crop species. Many of these species are cross-pollinated derivatives that feature large individual mass, thus requiring enormous field plot areas for testing, and often are characterized by relatively long generation times. Since the source of V_p is often irrelevant in asexually propagated crops (provided that expression is not transient), direct selection on desirable alleles in early generation breeding populations can greatly shorten the total time needed to proceed from germplasm to cultivar, and may also reduce costs by enriching smaller populations with greater proportions of exceptional genotypes.

For example in Rosaceae, the complete genome sequences of apple, peach, and diploid strawberry—one member of each of the three main fruit-producing branches of the Rosaceae family tree—were available by 2010, under the auspices of an international program known as "RosBREED". By 2014, genomic sequence information was routinely used in many US apple, peach, and cherry breeding programs. The application of genome sequences has significantly reduced wasted effort to eliminate poor families and has also reduced the costs to grow and evaluate thousands of seedlings genetically destined to have unacceptable fruit quality or maturity date. Limitations were evident during this period due to lack of knowledge, such as an understanding of genotype by environment (G×E) interactions and loci associated with variation for other valuable attributes (Iezzoni et al., 2016). Objectives for RosBREED in the future are to: (i) develop donor parents with multiple alleles for disease resistance, (ii) enrich breeding families with alleles for disease resistance and superior horticultural quality, (iii) advance selections with alleles for superior fruit quality with improved confidence, (iv) increase the routine adoption of DNA-informed breeding for rosaceous crops, and (v) engage industry stakeholders in project outcomes, evaluation, and adjustments (Iezzoni et al., 2017).

Simple sequence repeat (SSR) markers were developed in *Rubus* spp. from genomic and expressed sequence tag (EST) libraries in red raspberry (*R. idaeus*, subgenus *Idaeobatus*) and also in blackberry (*Rubus* subgenus *Rubus*). The objective of the study was to determine the suitability of SSR markers developed in other *Rubus* species for use in black raspberry. Twenty-seven primer pairs appeared to generate polymorphic markers that will be useful as prospective QTL in *Rubus* breeding programs (Dossett et al., 2010).

Genetic improvement in banana (*Musa*×*paradisiaca*) has been hindered by resource limitations, especially field space. Banana breeders require large tracts of land for growth and maintenance of plant populations. Additionally, long generation time, low levels of fruit set, and low seed viability are characteristic of many genotypes of cultivated banana. A study was conducted to compare different PCR-based marker systems (RAPD, VNTR, and AFLP (see Chapter 9)) for the analysis of breeding populations generated from two diverse *Musa* breeding schemes. In general, there was a poor correlation between the estimates of genetic similarity based on different types of markers (Crouch et al., 1999). Marker systems have improved vastly since 1999, and this limitation is likely due to the paucity of markers available at that time.

The cost-effective use of marker technology is dependent on the nature and timing of the use of such markers. Slater et al. (2013) conducted a systematic study of the cost-effectiveness of integrating MAS into breeding programs for clonally propagated crop species using potato as an archetype. MAS provides the advantage of being applicable at the seedling or an early generation stage. Results indicated that MAS could be applied cost-effectively in the second clonal generation for all models currently employed in potato breeding.

APOMIXIS

Apomixis is a form of asexual reproduction wherein the seed carries an embryo that is a clone of the 2n maternal parent plant instead of a sexual hybrid (2n maternal × paternal) embryo resulting from a fusion of gametes. This form of asexual mating system was discussed in detail in Chapter 10. The phenomenon of apomixis was discovered when segregation ratios for certain crosses were found to be significantly different from predictions based on assumptions of sexual reproduction (Barcaccia and Albertini, 2013).

The progeny of a given cross should reflect the alleles contributed by both the pollen and pistillate parents. In certain species, and under some circumstances, the maternal parental types predominate, often to the complete exclusion of true hybrids. Further examination of the reproductive biology of such plants reveals that cells from the maternal sporophyte supplant the egg and proceed to divide to form an embryo. The source of the sporophytic cell depends on the apomictic species being studied (Hörandl and Hojsgaard, 2012). In some cases, the nucellus is involved and, in others, the suspensor (Koltunow and Grossniklaus, 2003).

Techniques for the manipulation of apomixis in plant breeding are most advanced in Kentucky bluegrass (*Poa pratensis*; Curley and Jung, 2004). Many of the techniques applicable to *P. pratensis* are directly transferable to other apomictic crop species. Apomixis is the most convenient form of clonal propagation because the clone is packaged in a seed. Seeds are easily stored and germinated *en masse* and at will, and much agricultural infrastructure has been established to produce, process, store, package, distribute, and sow seeds, such as mechanical sowing equipment and coating/pelleting methods. Apomictic seeds make it possible to store large populations for relatively long periods and to sow and culture cloned individuals easily and in developmental synchrony.

Apomixis has been observed to be either obligate or facultative among the plant species in which it occurs (Koltunow, 2012). In obligate apomicts, the consequences to plant breeding are the same as if the species were reproduced asexually via vegetative organ, for example, tuber or rhizome. Facultative apomicts, however, are characterized by the observation of both sexual and asexual embryos within mature seeds (Garcia et al., 1999). In Kentucky bluegrass, the incidences of apomixis and true hybridity are conditioned by both genotype (5 loci) and environment (Matzk et al., 2005). Certain genotypes are much more predisposed to apomixis than are others (Bonos et al., 2000). Crosses of sexual × apomictic types usually give rise to progeny that exhibit a range of mating behavior from one parental extreme to the other. Plant breeders have capitalized on this finding by devising a breeding method comprised of the following elements (Fig. 17.2). Several successful cultivars of Kentucky bluegrass have been developed in this manner.

Reports have appeared on research aimed at understanding the genetic and developmental control of facultative apomixis (Koltunow and Grossniklaus, 2003; Matzk et al., 2005; Liu et al., 2012). If facultative apomixis can be manifested in other plant species, the tremendous power and efficiency of this approach may be replicated across a broader spectrum of crop species breeding efforts.

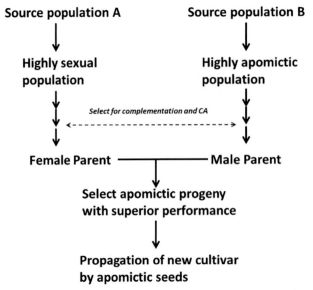

FIG. 17.2   The development of an apomictic cultivar of Kentucky bluegrass by heterotic breeding for combining ability between sexual and apomictic germplasm pools.

THE PLANT CHIMERA

The *chimera* is a mythical creature that consists of parts of two animals fused, for example, a centaur (half human, half horse). This term has been used to describe a similar situation in plants that is not mythical at all: the coexistence of two or more genotypes within a single organism. Chimeras are the basis of the visual phenomenon known as *sporting* or the appearance of a sector on an individual plant that bears a discernible phenotypic difference to the background surrounding it. An example of a plant chimera is a branch bearing red flowers on a plant that consists otherwise of white flowers (Fig. 17.3). While formal genetic verifications of the origin of such events are rarely undertaken, it is presumed that a mutation occurred during the ontogeny of the plant that converted the genotype of the resident cells, tissues, and organs from white to red. Evidence for intra- and inter-cellular exchange of genetic material has been published, but does not appear to contribute substantively to the formation of genetic variability in somatic tissues (Charlesworth and Charlesworth, 2012).

The developmental bases and types of plant chimeras were described in Chapter 8 (section on Induced mutations). Chimeras have significance to plant breeders for two main reasons. The most obvious is that they can provide a potential source of genetic variation upon which selection may be applied for purposes of achieving improvements (Chapter 8; Fig. 17.4). Perhaps red flowers have never been observed for our hypothetical ornamental plant and

FIG. 17.3   An example of a chimera in azalea; a sector of red flowers in a plant bearing otherwise white flowers.

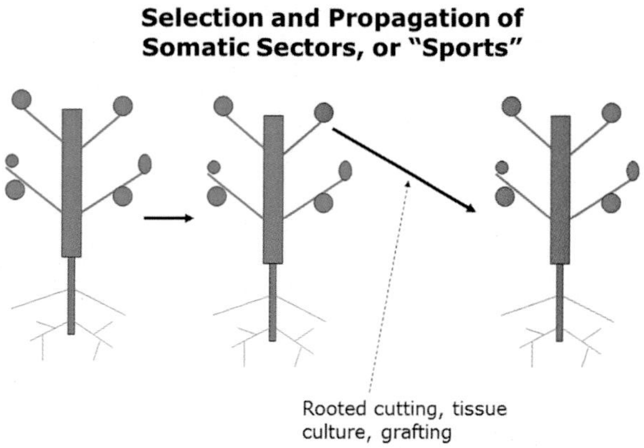

**Selection and Propagation of
Somatic Sectors, or "Sports"**

Rooted cutting, tissue
culture, grafting

FIG. 17.4   The selection and clonal propagation of a chimera, or sport, for the development of a prospective new cultivar.

would elecit incremental market demand if such a product were available. The other reason is that chimeras often persist for long periods of time during perpetual clonal propagation. The phenotype of the individual that consists of a chimeral combination may be conditioned by the chimeral configuration. Accordingly, care must be exercised to plan and execute the propagation procedure such that the developmental of the chimeral tissues is not disrupted.

As was described in Chapter 8, plant (and animal) development proceeds in a manner such that the origin of tissues and organs may be traced back to relatively few primordial cells in the developing embryo (Esau, 1965). Depending on when and where the mutational even occurs during development, the resulting chimera will be manifested in different ways. During the transformation from zygote to the globular embryo, the cell divisions are asymmetrical, leading to different daughter cell types. This ultimately culminates in embryonic cell layers that give rise to entire longitudinal structures in the adult plant.

Chimeras are classified by the location and relative proportion of mutated to non-mutated cells in the apical meristem. The *periclinal chimera* is the most significant and economically important category since they are relatively stable and can be asexually propagated. A mutation in a cell positioned near the apical dome so that the cells produced by subsequent divisions form an entire layer consisting of the mutated type produces a periclinal chimera. The resulting meristem contains one entire layer which is genetically different from the other two layers of the meristem. If, for example, the mutation occurs in layer I (LI), then the epidermal layer of the shoot which is produced after the mutation is of the new genetic type (Esau, 1965).

The thornless blackberry is an example of a LI periclinal chimera. The mutated cells of the epidermal layer of this type produce no "thorns" (more correctly called "prickles"). The thornless epidermis covers a stem whose LII and LIII and entire root meristem cells contain the thorny genotype. This can be demonstrated by regenerating plants from root cuttings of thornless blackberries that revert to the thorny genotype. (Darrow, 1928).

Periclinal chimeras are common in grape and result mainly from somatic mutations that occur naturally during plant growth (Pelsy, 2010). Various types of mutations were shown to be responsible for genetic diversity among clones: point mutations, large deletions, illegitimate recombination, or a variable number of repeats in microsatellite sequences. It was speculated that through these somatic molecular and cellular mechanisms, genotypes of clones drift and slowly evolve. Franks et al. (2002) were able to separate the phenotypes of a periclinal chimera of the grape cultivar Pinot Meunier by tissue culture. In blueberry (*V. darrowii*) that had been treated with the mitotic spindle inhibitor colchicine, several types of plants were identified: (i) plants with a polyploid LI (epidermal tissues) and LII (internal tissues) plant layers, (ii) periclinal chimeras with a polyploid LI layer and a normal diploid LII, and (iii) periclinal chimeras with a polyploid LII layer and a diploid LI (Chavez and Lyrene, 2009).

If the derivatives of the mutated cell do not entirely cover the apical dome, a *mericlinal chimera* is produced. Mericlinal chimeras are generally restricted to one cell layer, as was also the case for the periclinal chimera. Some shoots or leaves which develop from a mericlinal chimera may exhibit the mutant phenotype whereas others do not, tracing back to the cells in the apical dome from which the structures were derived. Certain mericlinal chimeras involve such a small number of cells that only a tiny portion of one leaf may be affected (McMahon et al., 2010).

A *sectorial chimera* results from mutations which affect entire sections of the apical meristem, the altered genotype extending through all the cell layers. Sectorial chimeras are unstable and may give rise to shoots and leaves that are normal. Both normal types and mutated plant tissues and organs can be produced, depending upon the point on the apical dome that are mutated and from which the shoots differentiate (McMahon et al., 2010).

Mericlinal and sectorial chimeras are, by their nature, unstable and the likelihood of propagating plants with the same morphological pattern from these types is low. Therefore, such chimeral types are not used in pure form commercially, converted instead to more stable types beforehand. Periclinal chimeras are very stable and, in some circumstances, chimeral plants are the predominant forms in commerce, for example, variegated flowers.

Any plant propagation technique that does not disrupt the arrangement of embryonic cell layers will result in phenotypic fidelity for periclinal chimeras. Examples of acceptable methods include rooted cuttings, budding, grafting, divisions, and tip layering. Any propagules derived from differentiated adventitious shoots may not result in the desired phenotype since the tissue layers are reformatted. Conversely, adventitious shoot/root formation and plant tissue culture techniques may be used to "dissect" chimeral plants into their genotypic components (Kleyn et al., 2013).

Phenotypes that resemble genetic chimeras but that are, in fact, caused by transient and non-heritable environmental or epigenetic mechanisms are common (Marcotrigiano, 1997). Another source of apparent new phenotypes such as leaf and flower variants may be infection with pathogenic viruses. When using cell culture, grafting, or direct shoot/root regeneration procedures to isolate putative mutant sectors, the question of the underlying cause of the new phenotype should always be addressed by conducting inheritance studies of the phenotype in homogeneous whole plants.

GRAFTING

Grafting is an ancient technology that amounts to an admixture of different genotypes to form an artificial chimera (Fig. 17.5). The method is applied primarily for purposes of the clonal propagation and enhanced performance of woody perennial species. The roots and shoots are handled and bred separately, referred to respectively as rootstocks and scions. In propagation applications, grafting is used to capitalize on the proliferation of shoots from the scion, or grafted or economic entity. Many herbaceous species, even annuals, have also been shown to adapt well to established grafting protocols. Since it is usually easy to propagate genetically acceptable populations of annuals by seed, the time and expense of grafting are a major impediment.

The range of adaptation has been a fundamental barrier to the broad dissemination of potentially superior genotypes of woody perennial plant species. Horticultural research demonstrated a possible way to circumvent this barrier, the grafting of genetically superior scions onto broadly adapted rootstocks or vice versa (i.e., the grafting of broadly adapted scions onto narrowly adapted rootstocks). Most readers are familiar with the apple hydra, a single tree onto which numerous scions of different cultivars have been grafted. Most commercial orchards of fruit trees and vineyards of grapes in North America and Europe consist of trees and vines that are chimeras, superior scions grafted onto selected rootstocks. This capability has greatly expanded the range of scion adaptation dramatically, and in some cases has conferred resistance to diseases and insect pests that attack the root system.

The technology has been successful to the extent that breeding programs in grapes and fruit trees are often bifurcated into the scion and rootstock (Webster, 2003). The scion is bred primarily for specific consumer and market traits, such as fruit and bloom quality, while rootstocks are selected for characteristics that are more attuned to the cost of production. Rootstocks are also selected for the broadest possible range of compatibility with potential scions. The rootstock often produces shoots that may supplant the scion and result in decreased harvest value, but such contaminant trees or vines are usually recognized and eliminated. Rarely, new genetic forms appear at the graft union, and it is speculated that some form of somatic recombination event has occurred or new chimera formed from the juxtaposition of different genotypes (Pelsy, 2010).

The simplistic view of grafting is that the resulting artificial chimera will function according to the additive interaction of traits contributed by the scion and rootstock. That is, the scion will manifest a phenotype as it did on its natural root system, only with the added benefits conferred by the rootstock, such as broader adaptation and biotic resistance. Scientific studies of scion-rootstock interaction have demonstrated, however, that the performance of the chimera is not always predicted from the sum of its parts. The most dramatic example is dwarfing rootstocks that have been developed for fruit trees. Scions grow on standard rootstocks exhibit a characteristic internode length. When they are grafted onto a dwarfing rootstock, however, the internode length is shortened substantially, thus reducing the stature of the mature tree and rendering it easier to manage and harvest. This result is not surprising since it is well known that both root and shoot meristems produce and respond to growth-altering root- and shoot-derived phytohormones.

Dwarfing rootstocks have been discovered in a broad spectrum of woody perennial crop species, including pome fruits (apple, pear, quince), stone fruits (peach, plum, cherry), *Citrus* spp., persimmon, and several nut species (Janick and Moore, 1996). In apple, dwarfing vs. non-dwarfing rootstocks were reported to differ in the number of nodes,

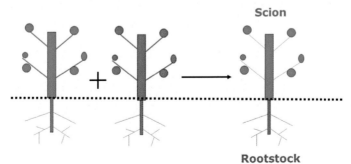

Grafting: An Ancient Technology

FIG. 17.5 The formation of an artificial chimera by grafting the scion (shoot) of one genotype to the rootstock of a different genotype. In this manner, the benefits imparted by both genotypes may be combined into a single individual.

but not other measures of dwarfness (Seleznyova et al., 2003). In another study, a dwarfing apple rootstock promoted flowering, accelerated the transition from juvenility to maturity, and regulated cycles of seasonal growth and termination (Foster et al., 2014). Fazio et al. (2017) found significant differences in the mineral nutrition of plant grown on dwarfing apple rootstocks exhibiting the genes *Dw1* and *Dw2*. In the dwarfing apple rootstock M9, Feng et al. (2017) found low expression of a gene associated with cytokinin biosynthesis caused by methylations in the promoter region, leading to low levels of trans-zeatin biosynthesis in roots. Zheng et al. (2018) reported a novel dwarfing mechanism in perennial woody plants that involves a gene (*MdWRKY9*) controlling brassinosteroid production that resulted in scion dwarfing. It appears that the tools of molecular biology will soon unlock the secrets of this fascinating interaction of two genotypes in a scion-rootstock chimera to produce a novel whole-plant phenotype.

TISSUE AND CELL CULTURE AND ARTIFICIAL SEEDS

During the period from 1970 until approximately the end of the 20th century, optimism was prevalent regarding the prospect that tissue and cell culture technologies could be developed to the point of broad practical application for plant propagation. The rapid emergence of breakthroughs in the successful culturing and regeneration of plants lent credence to this prospect. It was even speculated that plant breeding would be entirely supplanted by a combination of cell culture and molecular biological methodologies. While work continues in the applications of cell and molecular biology in plant improvement, no such wholesale replacement of traditional plant breeding by cell and molecular methods is yet in sight.

The term *micropropagation* was coined to denote the use of cells and small tissue aggregates in the asexual propagation process. While traditional clonal propagation methods rely on developmental modifications that were naturally-occurring, such as tubers, stolons, and bulbils, micropropagation purported to have unlimited applications. The process called for the extraction of cells, proliferation, reorganization into a developmental context, then channeling of resulting whole plants back into a viable production context (Kleyn et al., 2013).

Certain plant species perform such developmental transformations more readily than do others. Early studies demonstrated that tobacco (and relatives) was the organogenic archetype, while carrot (and relatives) was the embryogenic archetype (Litz et al., 1985). The embryogenic model was taken to the level of the ultimate conceptual demonstration, the "artificial seed" (Fig. 17.6). This is an entity formed by the union of a somatic embryo with a semi-solid support matrix such as agar or alginate within which it may be stored and disseminated, and within which the embryo develops further into a viable plant. The artificial seed shares theoretical attributes with apomictic seeds (see above).

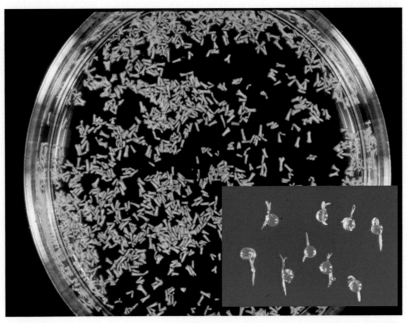

FIG. 17.6   This composite photo depicts a large population of carrot somatic embryos with artificial alginate seeds containing somatic embryos in the inset photo.

Optimism for these scenarios has faded, at least for the foreseeable future. The primary reason is that the methods are challenging, involving an aseptic phase wherein sterile technique is paramount. This requirement necessitates facilities, equipment, and training that mandates costs far exceeding the range of projected benefits. Secondly, the protocols are not consistently effective. Sometimes they work, sometimes they do not, and sometimes the results are in between. Often, no explanation can be found for the lack of consistency (Kleyn et al., 2013).

One factor is certain: the genotype of the entity designated for micropropagation has a substantial impact on success. The chain of events through which the cells pass en route to becoming somatic embryos also appears to be critical for success, and it is challenging to replicate these with accuracy. Finally, genetic conservation cannot be assured. The phenomenon of somaclonal variation is prevalent and unpredictable during the cell and tissue proliferation phase. Studies have demonstrated, however, that regeneration results in an attenuation of genetic variability as compared to the culture from which regenerates were derived, and many studies have implicated somaclonal variation as a culprit in the loss of regeneration capacity (Orton, 1980; Orton, 1984).

While the direct applications of cell and tissue culture in plant propagation in production agriculture are on the wane, the technology plays an important role as a critical step in many breeding programs. Genetic transformation and genome editing are dependent on the ability to regenerate cells and tissues that have been transformed by *Agrobacterium* vectors or biolistics. Also, anther and microspore culture are used extensively by plant breeders to obtain genetically homozygous doubled haploids.

Cell and tissue culture also plays a role in the purging of pathogens from clones. Traditional clonal propagation schemes, especially those that involve large multicellular plant organs such as tubers, are notoriously susceptible to contamination by pathogens. It is relatively easy to purge clonal propagates from infections involving bacterial and fungal pathogens, but virus' and other subcellular entities (e.g., prions) are difficult or impossible to eradicate from whole plants. This is where cell and tissue culture continue to play an important role in the propagation process. It has been demonstrated that the non-retro viruses are completely absent from meristem cells in infected plants. By excising and culturing meristems, or cells thereof, then regenerating plants under sterile conditions, the clonal line may be purged of the viral contaminants (Bryan et al., 2003; O'Herlihy et al., 2003).

References

Acquaah, G., 2012. Principles of Plant Genetics and Breeding, second ed. Wiley/Blackwell, New York, NY. 756 pp.

Adam-Blondon, A.-F., Martinez-Zapater, J.-M., Kole, C. (Eds.), 2011. Genetics, Genomics, and Breeding of Grapes. CRC Press, Boca Raton, FL. 396 pp.

Ahmadi, H., Bringhurst, R.S., 1992. Breeding strawberries at the decaploid level. J. Am. Soc. Hort. Sci. 117 (5), 856–862.

Aleza, P., Juárez, J., Hernández, M., Ollitrault, P., Navarro, L., 2012. Implementation of extensive citrus triploid breeding programs based on 4x × 2x sexual hybridisations. Tree Genet. Genomes 8 (6), 1293–1306.

Al-Khayri, J., Jain, S.M., Johnson, D.V. (Eds.), 2018. Advances in Plant Breeding Strategies: Fruits. Springer, Berlin (GER). 990 pp.

Arens, P., Shahin, A., van Tuyl, J.M., 2014. (Molecular) breeding of *Lilium*. Acta Hortic. 1027, 113–127.

Arús, P., 2007. Integrating genomics into Rosaceae fruit breeding. Acta Hortic. 738, 29–35.

Badenes, M.L., Byrne, D.H. (Eds.), 2012. Fruit Breeding (Handbook of Plant Breeding). Springer, Berlin (GER). 875 pp.

Barcaccia, G., Albertini, E., 2013. Apomixis in plant reproduction: a novel perspective on an old dilemma. Plant Reprod. 26 (3), 159–179.

Bethke, P.C., Nassar, A.M.K., Kubow, S., Leclerc, Y.N., Li, X.-Q., Haroon, M., Molen, T., Bamberg, J., Martin, M., Donnelly, D.J., 2014. History and origin of Russet Burbank (Netted Gem), a sport of Burbank. Am. J. Potato Res. 91 (6), 594–609.

Bonos, S.A., Meyer, W.A., Murphy, J.A., 2000. Classification of Kentucky bluegrass genotypes grown as spaced-plants. HortScience 35 (5), 910–913.

Bradshaw, J.E., 2017. Plant breeding: past, present and future. Euphytica 213 (3), 60.

Bradshaw, J.E., Dale, M.F.B., Swan, G.E.L., Todd, D., Wilson, R.N., 1998. Early-generation selection between and within pair crosses in a potato (*Solanum tuberosum* subsp. *tuberosum*) breeding programme. Theor. Appl. Genet. 97 (8), 1331–1339.

Bradshaw, J.E., Dale, M.F.B., Mackay, G.R., 2009. Improving the yield, processing quality and disease and pest resistance of potatoes by genotypic recurrent selection. Euphytica 170 (1–2), 215–227.

Bryan, A.D., Pesic-VanEsbroeck, Z., Schultheis, J.R., Pecota, K.V., Swallow, W.H., Yencho, G.C., 2003. Cultivar decline in sweetpotato: I. Impact of micropropagation on yield, storage root quality, and virus incidence in 'Beauregard'. J. Am. Soc. Hort. Sci. 128 (6), 846–855.

Cao, Q., Zhang, A., Ma, D., Li, H., Li, Q., Li, P., 2009. Novel interspecific hybridization between sweetpotato (*Ipomoea batatas* (L.) Lam.) and its two diploid wild relatives. Euphytica 169 (3), 345–352.

Chandel, P., Tiwari, J.K., Ali, N., Devi, S., Sharma, S., Luthra, S.K., Singh, B.P., 2015. Interspecific potato somatic hybrids between *Solanum tuberosum* and *S. cardiophyllum*, potential sources of late blight resistance breeding. Plant Cell Tiss. Org. Cult. 123 (3), 579–589.

Charlesworth, B., Charlesworth, D., 2012. Elements of Evolutionary Genetics. Roberts & Co. Publ., Greenwood Village, CO. 734 pp.

Chavez, D.J., Lyrene, P.M., 2009. Production and identification of colchicine-derived tetraploid *Vaccinium darrowii* and its use in breeding. J. Am. Soc. Hort. Sci. 134 (3), 356–363.

Cook, R.E., 1983. Clonal plant populations. Am. Sci. 71 (3), 244–253.

Corriols, L., Dore, C., 1989. Use of rank indexing for comparative evaluation of all-male other hybrid types in asparagus. J. Am. Soc. Hort. Sci. 114 (2), 328–332.

Crouch, J.H., Crouch, H.K., Constandt, H., van Gysel, A., Breyne, P., van Montague, M., Jarret, R.L., Ortiz, R., 1999. Comparison of PCR-based molecular marker analyses of *Musa* breeding populations. Mol. Breed. 5 (3), 233–244.

Curley, J., Jung, G., 2004. RAPD-based genetic relationships in Kentucky bluegrass: comparison of cultivars, interspecific hybrids, and plant introductions. Crop. Sci. 44 (4), 1299–1306.

Darrow, G.M., 1928. Notes on thornless blackberries. J. Hered. 19, 139–142.

De Schepper, S., Leus, L., Eeckhaut, T., Van Bockstaele, E., Debergh, P., De Loose, M., 2004. Somatic polyploid petals: regeneration offers new roads for breeding Belgian pot azaleas. Plant Cell Tiss. Org. Cult. 76 (2), 183–188.

Deng, Y., Chen, S., Lu, A., Chen, F., Tang, F., Guan, Z., Teng, N., 2010. Production and characterisation of the intergeneric hybrids between *Dendranthema morifolium* and *Artemisia vulgaris* exhibiting enhanced resistance to chrysanthemum aphid (*Macrosiphoniella sanbourni*). Planta 231 (3), 693–703.

Dosba, F., 2003. Progress and prospects in stone fruit breeding. Acta Hortic. 622, 35–43.

Dossett, M., Bassil, N.V., Finn, C.E., 2010. Transferability of *Rubus* microsatellite markers to black raspberry. Acta Hortic. 859, 103–110.

Ebi, M., Kasai, N., Masuda, K., 2000. Small inflorescence bulbils are best for micropropagation and virus elimination in garlic. HortScience 35 (4), 735–737.

Esau, K., 1965. Plant Anatomy, second ed. John Wiley & Sons, Inc., New York, NY, pp. 89–112.

Fasoula, D.A., 2003. The effects of clonal propagation on the genetic improvement of potato. Acta Hortic. 579, 67–72.

Fazio, G., Grusak, M., Robinson, T.L., 2017. Apple rootstocks' dwarfing loci relationships with mineral nutrient concentration in scion leaves and fruit. Acta Hortic. 1177, 93–102.

Feng, Y., Zhang, X., Wu, T., Xu, X., Han, Z., Wang, Y., 2017. Methylation effect on IPT5b gene expression determines cytokinin biosynthesis in apple rootstock. Biochem. Biophys. Res. Commun. 482, 604–609.

Ferguson, A.R., Seal, A.G., Davison, R.M., 1990. Cultivar improvement, genetics and breeding of kiwifruit. Acta Hortic. 282, 335–347.

Foster, T.M., Watson, A.E., van Hooijdonk, B.M., Schaffer, R.J., 2014. Key flowering genes including FT-like genes are upregulated in the vasculature of apple dwarfing rootstocks. Tree Genet. Genomes 10 (1), 189–202.

Franks, T., Botta, R., Thomas, M.R., 2002. Chimerism in grapevines: implications for cultivar identity, ancestry and genetic improvement. Theor. Appl. Genet. 104 (2/3), 192–199.

Garcia, R., Asins, M.J., Forner, J., Carbonell, E.A., 1999. Genetic analysis of apomixis in *Citrus* and *Poncitrus* by molecular markers. Theor. Appl. Genet. 99 (3–4), 511–518.

Gavrilenko, T., Thieme, R., Tiemann, H., 1999. Assessment of genetic and phenotypic variation among intraspecific somatic hybrids of potato, *Solanum tuberosum* L. Plant Breed. 118 (3), 205–213.

Gopal, J., 1998. Identification of superior parents and crosses in potato breeding programmes. Theor. Appl. Genet. 96 (2), 287–293.

Griga, M., Tejklova, E., Novak, F.J., Kubalakova, M., 1986. In vitro clonal propagation of *Pisum sativum* L. Plant Cell Tiss. Org. Cult. 6 (1), 95–104.

Grüneberg, W.J., Mwanga, R.O.M., Andrade, M.I., Espinoza, J., 2009. Breeding clonally propagated crops. In: Ceccarelli, S., Guimarães, E.P., Weltzien, E. (Eds.), Plant Breeding and Farmer Participation. FAO, Vienna, pp. 275–322.

Hörandl, E., Hojsgaard, D., 2012. The evolution of apomixis in angiosperms: a reappraisal. Plant Biosyst. 146 (3), 681–693.

Hough, L.F., 1979. Fruit breeding—history, progress, and perspective. HortScience 14 (3), 329–332.

Iezzoni, A., Weebadde, C., Peace, C., Main, D., Bassil, N.V., Coe, M., Fazio, G., Gallardo, K., Gasic, K., Luby, J., McFerson, J., van de Weg, E., Yue, C., 2016. Where are we now as we merge genomics into plant breeding and what are our limitations? Experiences from RosBREED. Acta Hortic. 1117, 1–6.

Iezzoni, A., Weebadde, C., Peace, C., Main, D., Bassil, N.V., Coe, M., Fazio, G., Gallardo, K., Gasic, K., Luby, J., McFerson, J., van de Weg, E., Yue, C., 2017. RosBREED2: progress and future plans to enable DNA-informed breeding in the Rosaceae. Acta Hortic. 1172, 115–118.

Iovene, M., Barone, A., Frusciante, L., Monti, L., Carputo, D., 2004. Selection for aneuploid potato hybrids combining a low wild genome content and resistance traits from *Solanum commersonii*. Theor. Appl. Genet. 109 (6), 1139–1146.

Janick, J., 1998. Fruit breeding in the 21st century. Acta Hortic. 490, 39–45.

Janick, J., Moore, J.N., 1996. Fruit Breeding Volume 1: Tree and Tropical Fruits. Wiley, New York, NY. 632 pp.

Jänsch, M., Paris, R., Amoako-Andoh, F., Keulemans, W., Davey, M.W., Pagliarani, G., Tartarini, S., Patocchi, A., 2014. A phenotypic, molecular and biochemical characterization of the first cisgenic scab-resistant apple variety 'Gala'. Plant Mol. Biol. Report. 32 (3), 679–690.

Kardos, J.H., Robacker, C.D., Dirr, M.A., Rinehart, T.A., 2009. Production and verification of *Hydrangea macrophylla* x *H. angustipetala* hybrids. HortScience 44 (6), 1534–1537.

Kaul, K., Karthigeyan, S., Dhyani, D., Kaur, N., Sharma, R.K., Ahuja, P.S., 2009. Morphological and molecular analyses of *Rosa damascena* x *R. bourboniana* interspecific hybrids. Sci. Hortic. 122 (2), 258–263.

Kibet, T., 2018. Forest Genetics and Tree Breeding. Astral Press, New Delhi. 298 pp.

Kim, B.J., Kwon, Y.C., Kwack, Y.H., Lim, M.S., Park, E.H., 1999. Interspecific hybridization in sexual diploid *Allium senescens* var. minor x apomictic tetraploid *A. nutans* and sexual diploid *A. senescens* var. minor x apomictic hexaploid *A. senescens*. Plant Breed. 118 (5), 439–442.

Kleyn, J., Bridgen, M., Scoggins, H., 2013. Plants from Test Tubes: An Introduction to Micropropagation. Timber Press, Inc., Portland, OR. 269 pp.

Kole, C., Abbott, A.G. (Eds.), 2012. Genetics, Genomics and Breeding of Stone Fruits. CRC Press, Boca Raton, FL. 418 pp.

Koltunow, A., 2012. Apomixis. John Wiley & Sons, Ltd., New York, NY. https://doi.org/10.1002/9780470015902.a0002035.pub2.

Koltunow, A.M., Grossniklaus, U., 2003. Apomixis: a developmental perspective. Annu. Rev. Plant Biol. 54, 547–574.

Krivanek, A.F., Famula, T.R., Tenscher, A., Walker, M.A., 2005. Inheritance of resistance to *Xylella fastidiosa* within a *Vitis rupestris* x *Vitis arizonica* hybrid population. Theor. Appl. Genet. 111 (1), 110–119.

Kumar, N., 2006. Breeding of Horticultural Crops: Principles and Practices. New India Publ. Agency, New Delhi. 207 pp.

Kumar, R., Gopal, J., 2006. Repeatability of progeny mean, combining ability, heterosis and heterobeltiosis in early generations of a potato breeding programme. Potato Res. 49 (2), 131–141.

Labouisse, J.-P., Sileye, T., Bonnot, F., Baudouin, L., 2011. Achievements in breeding coconut hybrids for tolerance to coconut foliar decay disease in Vanuatu, South Pacific. Euphytica 177 (1), 1–13.

Lebot, V., Lawac, F., 2017. Quantitative comparison of individual sugars in cultivars and hybrids of taro [*Colocasia esculenta* (L.) Schott]: implications for breeding programs. Euphytica 213 (7), 147.

Ledbetter, C.A., 2009. Using Central Asian germplasm to improve fruit quality and enhance diversity in California adapted apricots. Acta Hortic. 814, 77–80.

II. BREEDING METHODS

Li, B., Wu, R., 1996. Genetic causes of heterosis in juvenile aspen: a quantitative comparison across intra- and inter-specific hybrids. Theor. Appl. Genet. 93 (3), 380–391.

Litz, R.E., Moore, G.A., Srinivasan, C., 1985. In vitro systems for propagation and improvement of tropical fruits and palms. In: Janick, J. (Ed.), Horticultural Reviews. vol. 7. Avi Publ. Co., New York, NY.

Liu, D., Dong, Q., Sun, C., Wang, Q., You, C., Yao, Y., Hao, Y., 2012. Functional characterization of an apple apomixis-related MhFIE gene in reproduction development. Plant Sci. 185–186, 105–111.

Magdalita, P.M., Drew, R.A., Adkins, S.W., Godwin, I.D., 1997. Morphological, molecular and cytological analyses of Carica papaya X C. cauliflora interspecific hybrids. Theor. Appl. Genet. 95 (1/2), 224–229.

Marcotrigiano, M., 1997. Chimeras and variegation: patterns of deceit. HortScience 32 (5), 773–884.

Matzk, F., Prodanovic, S., Bäumlein, H., Schubert, I., 2005. The inheritance of apomixis in Poa pratensis confirms a five locus model with differences in gene expressivity and penetrance. Plant Cell 17 (1), 13–24.

McKey, D., Elias, M., Pujol, B., Duputié, A., 2010. The evolutionary ecology of clonally propagated domesticated plants. New Phytol. 186 (2), 318–332.

McMahon, M.E., Kofranek, A.M., Rubatzky, V.E., 2010. Plant Science: Growth, Development, and Utilization of Cultivated Plants, fifth ed. Pearson, London. 688 pp.

Mehlenbacher, S.A., 1995. Classical and molecular approaches to breeding fruit and nut crops for disease resistance. HortScience 30 (3), 466–477.

Moser, P., Rhode, P.W., 2011. Did Plant Patents Create the American Rose? Working Paper 16983, National Bureau of Economic Research, Cambridge, MA. http://www.nber.org/papers/w16983.

Mullins, M.G., 2006. Plant Improvement in Horticulture: The Case for Fruit Breeding. The Regional Institute Online Publishing. Archived at: http://www.regional.org.au/au//asa/1980/invited/genetic-exploitation/mullins.htm.

Murashige, T., Serpa, M., Jones, J.B., 1974. Clonal multiplication of Gerbera through tissue culture. HortScience 9 (38 (Sect. 1)), 175–180.

Murch, S.J., Peiris, S.E., Shi, W.L., Zobayed, S.M.A., Saxena, P.K., 2006. Genetic diversity in seed populations of Echinacea purpurea controls the capacity for regeneration, route of morphogenesis and phytochemical composition. Plant Cell Rep. 25 (6), 522–532.

Myles, S., 2013. Improving fruit and wine: what does genomics have to offer? Trends Genet. 29 (4), 190–196.

Nonomura, T., Ikegami, Y., Morikawa, Y., Matsuda, Y., Toyoda, H., 2001. Induction of morphologically changed petals from mutagen-treated apical buds of rose and plant regeneration from varied petal-derived calli. Plant Biotechnol. 18 (3), 233–236.

O'Herlihy, E.A., Croke, J.T., Cassells, A.C., 2003. Influence of in vitro factors on titre and elimination of model fruit tree viruses. Plant Cell Tiss. Org. Cult. 72 (1), 33–42.

Ojulong, H., Labuschangne, M.T., Fregene, M., Herselman, L., 2008. A cassava clonal evaluation trial based on a new cassava breeding scheme. Euphytica 160 (1), 119–129.

Oraguzie, N.C., Watkins, C.S., Chavoshi, M.S., Peace, C., McConchie, C., O'Hare, P., 2017. Overview of the Australian macadamia industry breeding program. Acta Hortic. 1161, 73–78.

Ortiz, R., Dochez, C., Asiedu, R., Moonan, F., 2008. Breeding vegetatively propagated crops. In: Lamkey, K.R., Lee, M. (Eds.), Plant Breeding: The Arnel R. Hallauer International Symposium. Wiley Online Library. https://doi.org/10.1002/9780470752708.ch18.

Orton, T.J., 1980. Chromosomal variability in tissue cultures and regenerated plants of Hordeum. Theor. Appl. Genet. 56, 101–112.

Orton, T.J., 1984. Somaclonal variation: theoretical and practical considerations. In: Gustafson, J.P. (Ed.), Gene Manipulation in Plant Improvement. Plenum, New York, NY, pp. 427–468.

Pelsy, F., 2010. Molecular and cellular mechanisms of diversity within grapevine varieties. Heredity 104 (4), 331–340.

Percy, R.E., Klimaszewska, K., Cyr, D.R., 2000. Evaluation of somatic embryogenesis for clonal propagation of western white pine. Can. J. For. Res. 30 (12), 1867–1876.

Pinker, I., Olbricht, K., Pohlheim, F., 2012. Potentials of callus culture as a breeding tool for polyploidisation of Fragaria vesca L. Acta Hortic. 961, 351–358.

Reed, S.M., Riedel, G.L., Pooler, M.R., 2001. Verification and establishment of Hydrangea macrophylla 'Kardinal' x H. paniculata 'Brussels Lace' interspecific hybrids. J. Environ. Hortic. 19 (2), 85–88.

Reed, S.M., Jones, K.D., Rinehart, T.A., 2008. Production and characterization of intergeneric hybrids between Dichroa febrifuga and Hydrangea macrophylla. J. Am. Soc. Hort. Sci. 133 (1), 84–91.

Richards, A.J., 1986. Plant Breeding Systems. George Allen & Unwin, London. 529 pp.

Rival, A., Bertrand, L., Beulé, T., Combes, M.C., Trouslot, P., Lashermes, P., 1998. Suitability of RAPD analysis for the detection of somaclonal variants in oil palm (Elaeis guineensis Jacq.). Plant Breed. 117 (1), 73–76.

Seleznyova, A.N., Thorp, T.G., White, M., Tustin, S., Costes, E., 2003. Application of architectural analysis and AMAPmod methodology to study dwarfing phenomenon: the branch structure of 'Royal Gala' apple grafted on dwarfing and non-dwarfing rootstock/interstock combinations. Ann. Bot. 91 (6), 665–672.

Sié, R.S., N'Goran, J.A.K., Montagnon, C., Akaffou, D.S., Cilas, C., Dagou, S., Mondeil, F., Charles, G., Branchard, M., 2009. Characterization and evaluation of two genetic groups and value of intergroup hybrids of Cola nitida (Vent.) Schott and Endlicher. Euphytica 167 (1), 107–112.

Simmonds, N.W., 1979. Principles of Crop Improvement. Longman Group, Ltd., London. 408 pp.

Simmonds, N.W., 1997. A review of potato propagation by means of seed, as distinct from clonal propagation by tubers. Potato Res. 40 (2), 191–214.

Slater, A.T., Cogan, N.O.I., Forster, J.W., 2013. Cost analysis of the application of marker-assisted selection in potato breeding. Mol. Breed. 32 (2), 299–310.

Smith, J.S., 2009. The Garden of Invention: Luther Burbank and the Business of Breeding Plants. Penguin Press, New York, NY. 354 pp.

Stover, E., Driggers, R., Hearn, C.J., Bai, J., Baldwin, E., McCollum, T.G., Hall, D.G., 2016. Breeding "sweet oranges" at the USDA U.S. Horticultural Research Laboratory. Acta Hortic. 1127, 41–44.

Thieme, R., Rakosy-Tican, E., Nachtigall, M., Schubert, J., Hammann, T., Antonova, O., Gavrilenko, T., Heimbach, U., Thieme, T., 2010. Characterization of the multiple resistance traits of somatic hybrids between Solanum cardiophyllum Lindl. and two commercial potato cultivars. Plant Cell Rep. 29 (10), 1187–1201.

Tian, L., Wang, Y., Niu, L., Tang, D., 2008. Breeding of disease-resistant seedless grapes using Chinese wild Vitis spp. Sci. Hortic. 117 (2), 136–141.

Topp, B.L., Hardner, C.M., Kelly, A.M., 2012. Strategies for breeding macadamias in Australia. Acta Hortic. 935, 67–72.

van Nocker, S., Gardiner, S.E., 2014. Breeding better cultivars, faster: applications of new technologies for the rapid deployment of superior horticultural tree crops. Hortic. Res. 1, 14022. https://doi.org/10.1038/hortres.2014.22.

Viloria, Z., Grosser, J.W., 2005. Acid citrus fruit improvement via interploid hybridization using allotetraploid somatic hybrid and autotetraploid breeding parents. J. Am. Soc. Hort. Sci. 130 (3), 392–402.

Watanabe, K.N., Orrillo, M., Vega, S., Masuelli, R., Ishiki, K., 1994. Potato germplasm enhancement with disomic tetraploid *Solanum acaule*. II. Assessment of breeding value of tetraploid F_1 hybrids between tetrasomic tetraploid *S. tuberosum* and *S. acaule*. Theor. Appl. Genet. 88 (2), 135–140.

Webster, A.D., 2003. Breeding and selection of apple and pear rootstocks. Acta Hortic. 622, 499–512.

Wu, J.H., Ferguson, A.R., Mooney, P.A., 2005. Allotetraploid hybrids produced by protoplast fusion for seedless triploid *Citrus* breeding. Euphytica 141 (3), 229–235.

Yanagino, T., Sugawara, E., Watanabe, M., Takahata, Y., 2003. Production and characterization of an interspecific hybrid between leek and garlic. Theor. Appl. Genet. 107 (1), 1–5.

Zheng, X., Zhao, Y., Dongqian, S., Shi, K., Wang, L., Li, Q., Wang, N., Zhou, J., Yao, J., Xue, Y., Fang, S., Chu, J., Guo, Y., Kong, J., 2018. MdWRKY9 overexpression confers intensive dwarfing in the M26 rootstock of apple by directly inhibiting brassinosteroid synthetase MdDWF4 expression. New Phytol. 217 (3), 1086–1098.

Zimnoch-Guzowska, E., Lebecka, R., Kryszczuk, A., Maciejewska, U., Szczerbakowa, A., Weilgat, B., 2003. Resistance to *Phytophthora infestans* in somatic hybrids of *Solanum nigrum* L. and diploid potato. Theor. Appl. Genet. 107 (1), 43–48.

18

The Backcross Method

INTRODUCTION

As plant breeding programs predicated on inbreeding or outcrossing within a narrow range of germplasm progressed over time, gains inevitably begin to plateau as frequencies of desirable alleles were maximized. One example of this is the diminishing gains phase in mass selection programs described in Chapter 5. The plateau affirms that the available genetic variability contributing to superior performance has been successfully tapped. Most crop species under domestication and subsequent population selection for over 10,000 years had experienced such a plateau by the late 19th century. Following the discoveries of Mendel and geneticists of the early 20th century, truly remarkable progress has been evident in the output and quality of crop species. Application of Mendel's principles has been the most significant factor responsible for this amazing increase in the rate of genetic improvements.

As gains in the performance of the best cultivars in a given crop species reach a plateau, the number of traits under consideration for improvement is progressively reduced. The situation ultimately arises wherein a cultivar is superior in most respects but falls short with regard to one or few discernible traits. Examples of trait deficiencies include disease resistance, mating behavior, pigmentation, nutritional quality, maturity date, and others. If the genes for improvement exist within the species gene pool, or closely related species, and the trait is simply inherited (≤ 3 genes), the backcross method is an excellent clear choice of breeding method for producing a new cultivar that replicates the flawed superior one with the targeted trait upgrade. Until about 1990, backcross was the exclusive choice for the introduction of genes for targeted traits but new cellular and molecular technologies since then have provided plant breeders with exciting alternatives. The two broad technology-based trait conversion options are genetic transformation or gene editing (e.g., CRISPR-cas9; Chapter 8).

The backcross method is addressed in a separate chapter, exclusive of breeding methods adapted to the mating system, because the strategy is independent of the genetic structure of the starting and targeted populations. The backcross method will always result in the population structure of the recurrent parent regardless of mating system and genome structure as long as adequate population sizes to discourage genetic drift are utilized. The backcross method works equally well, therefore, for self- and cross-pollinating species, although certain modifications (e.g., population size) may be needed to accommodate the mating system. With the progressive improvement of cultivated crop cultivars, the backcross method is more popular than ever, perhaps the most widely practiced of all breeding approaches at this time. One feature that drives popularity is the relatively short time period required to reach project culmination, usually four to five generations.

The backcross method may be used to improve a diverse range of population types from pure lines and inbred parents to open-pollinated populations and clones in crop species that feature a sexual cycle. Even hybrid varieties may be adapted to the backcross method by clearly understanding the inheritance of the trait being transferred. If the desired form of the trait is completely dominant, only one parental inbred must carry the newly transferred allele. If the targeted form of the trait is recessive, both parents must be fixed for the transferred allele.

While the backcross method is conceptually simplistic and intuitive, it was not distilled into a defined process until first proposed by Harlan and Pope (1922). The premise of the backcross method is to identify a source of a desirable gene followed by transfer of the gene to an established cultivar or breeding line by progressively crossing to the established parent, while selecting intensively for both the trait conferred by the gene(s) from the source variety and the phenotype of the established cultivar. Through this process, the natural Mendelian genetic mechanisms of recombination and segregation are employed to effectively cut and splice the gene from the source genome to that of the established cultivar (Figs. 18.1 and 18.2). Because the established cultivar has already been bred intensively to maximize performance, the end population is a clear incremental improvement: the established cultivar with an additional trait added on (or subtracted, if the transferred gene has a suppressing effect).

The backcross method is not as precise as are those of molecular transformation or genome editing, however. Many cycles of recurrent crossing to the established cultivar, or backcrosses, are essential to get rid of the genes from the trait donor population and replace them with those from the established recurrent parent population. The backcross method must always be adapted to account for the heritability and mode of inheritance of the trait being transferred and the genome into which it is being transferred (Bernardo, 2003) in addition to circumstantial practical factors (Mumm, 2007). Linkages in coupling and repulsion in parental haplotypes usually also play a role in the design of a program, although beneficial or detrimental linkage relationships are often only discovered after a program has already been initiated.

Backcross Method for Dominant Traits

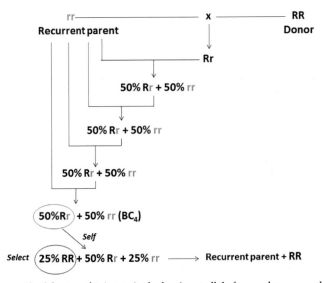

FIG. 18.1   The backcross breeding method for transferring a single dominant allele from a donor population to a recurrent parent population.

Backcross Method: Cytogenetic Perspective (for dominant trait)

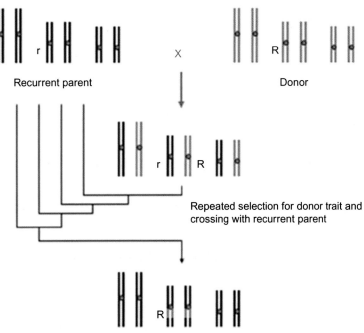

Recurrent parent

X

Donor

Repeated selection for donor trait and
crossing with recurrent parent

FIG. 18.2 Genomic or cytogenetic depiction of the genetic processes from parental to finished populations following a recurrent backcross breeding program. Theoretically, the genome of the end population will possess most of the genome (chromosomes) of the recurrent parent with a small genomic segment containing the transferred gene from the donor parent.

HISTORICAL PERSPECTIVE

The first reference to the method was made by Harlan and Pope (1922) who proposed the strategy and speculated that it would be of great value in transferring genes between populations. Shortly thereafter, Richey (1927) published many of the theoretical underpinnings for the backcross method, such as the expected gene and genotype frequencies over successive generations. He demonstrated that, if the established cultivar and donor gene(s) source population differed in alleles at 100 loci, over 90% would be recovered in the homozygous state following ten backcross generations.

The landmark experiment conducted by Briggs (1930, 1935, 1938) attracted the attention of all plant breeders. The study was begun in 1922 and involved the transfer of a single dominant disease resistance allele from unadapted to adapted populations of wheat. The work included a convincing demonstration that the general background of the established cultivar ("Martin") had been completely restored along with a newly transferred gene conferring bunt resistance. The resulting improved cultivar proved to be a tremendous economic success in the western U.S. Briggs and co-workers continued to employ the backcross method successfully for the transfer of other simply-inherited traits to other established wheat cultivars, thus demonstrating that the strategy was not limited in scope to bunt resistance in wheat.

Suneson (1947) published the results of a study that demonstrated the persistence of traits of the source population following many backcross generations in wheat. Affected traits were diverse, including yield, quality, maturity date, stature, stem stiffness, and disease resistance. It was concluded that the program could be accelerated by the application of negative selection among segregating populations for source traits other than that under transfer.

The backcross method was broadly adopted during the 1930s and 1940s but did not constitute a primary thrust in the overall plant breeding picture until decades later. Thomas (1952) reviewed the progress that had been realized over that period in non-grain crop species. Briggs and Allard (1953) iterated the three fundamental requirements for a successful backcross breeding program: (i) a satisfactory recurrent parent (established cultivar) must exist, (ii) it must be possible to retain a worthwhile intensity of the character under transfer (from the source population) through several backcrosses, and (iii) The genotype of the recurrent parent must be reconstituted by reasonable number of backcrosses executed with populations of manageable size. While those requirements might seem obvious to us now, we must appreciate the tremendous amount of experience that has accumulated since this benchmark publication appeared.

Warschefsky et al. (2014) and many other scientists and government policymakers have expressed alarm over the expanding genetic vulnerability of agriculture. They advocated the expansion of the gene pools and ranges of genetic variability in cultivated crops by backcrossing genes for resistance to pests and pathogens, tolerance to abiotic extremes, and reduced dependence on inputs from wild progenitors and phylogenetic relatives.

THEORETICAL CONSIDERATIONS

The backcross method is based fundamentally on the Mendelian test cross to study the mode of inheritance of a given trait. The test cross is defined as the crossing of a F_1 hybrid to the recessive parent followed by determination of the frequency patterns of trait inheritance among progenies. If the parents are homozygous for polymorphic alleles at a single locus, test cross genotypic segregation ratios will be 1:1 instead of 1:2:1, necessitating a smaller minimum size of the segregating population since, for a single polymorphic gene, 50% are of the desired type instead of 25%. If more than one polymorphic locus is involved, the benefits of using a test cross instead of a F_2 population for testing phenotype segregation ratios are incrementally, and exponentially, greater.

There are two fundamental versions of the backcross method (Figs. 18.1 and 18.3). The source population, from which the desired trait is drawn, is usually referred to as the donor. The established variety, or population to which the desired trait is being transferred, is usually referred to as the recurrent parent. If the trait being transferred is primarily dominant over the allele(s) present in the recurrent parent, the method is a simple algorithm: select for the trait under transfer, cross to the recurrent parent, then repeat the first two steps until the finished population is obtained. If the trait under transfer is recessive, however, it is necessary to insert regular inbreeding steps, usually self-pollination, to produce recessive homozygotes that may be selected for further backcrosses (Fig. 18.3). The frequency of intervening self-pollinations is balanced against the size of the populations that must be carried to ensure that the proportion of desired segregants is not overwhelmed by recurrent parental genotypes that lack the desired allele.

The terminology that is most frequently used to denote breeding populations in a backcross breeding program is BC(x) or BC_x, where x denotes the number of total backcrosses to the recurrent parent that have been affected. If intervening self-pollinations are performed, an "S" term is sometimes added to signify this operation. For example, BC(3) S(1) or BC_3S_1 refers to a population for which three backcrosses have been accomplished then the resulting population was self pollinated once. The notations do not usually spell out if intervening self-pollinations were performed following earlier backcrosses, mandating accurate records to reconstruct the progression and to fully reconcile results with genetic predictions.

The progress of a backcross breeding program is usually measured quantitatively in terms of the proportion of the overall genome of the recurrent that has been recovered. In a standard program involving the transfer of a single dominant allele, the progression is as follows (Table 18.1).

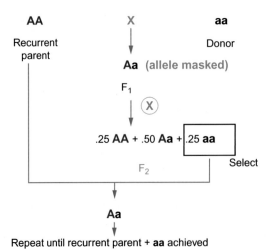

The Backcross Method with Recessive Traits

FIG. 18.3   The backcross method to transfer a desirable recessive allele from a donor population to a recurrent parent. Intermittent self-pollination, or inbreeding, is essential to "unmask" the recessive allele for purposes of selection and elimination of individuals devoid of the allele.

TABLE 18.1 Proportion of the Genome of the Recurrent Parent that is Recovered During the Progression of Generations in a Backcross Breeding Program from F_1 to BC_6 Assuming Random Recombination and Segregation of Genomic Units

Generation	% of genome of recurrent parent
F_1	50.00000
BC_1	75.00000
BC_2	87.50000
BC_3	93.75000
BC_4	96.87500
BC_5	98.43750
BC_6	99.21875

The progression will be the same for the transfer of a recessive trait, except that it will take much longer due to the intervening self-pollinations (Richey, 1927):

$$\text{Proportion homozygous for recurrent parent} = \left[\left(2^r - 1 \right) / 2^r \right]^n$$

where n = number of heterozygous loci in the F_1 and r = number of backcross generations

This progression is not distributed evenly throughout the genome. The portions of the genome that are unlinked to the trait under transfer will be brought back to the type of the recurrent parent more quickly than will any portions that are linked due to persistent linkage disequilibrium. If the trait is controlled by more than one gene, the degree of linkage disequilibrium is compounded. Linkage disequilibrium during backcrossing was first identified by Suneson (1947) where certain characteristics of the recurrent parent were recovered more quickly than were others. Thus, a backcross program may culminate in a population that consists of the genome of the recurrent parent but with a large DNA segment originating from the donor. This segment encompasses the locus bearing the gene with the transferred allele. Such a large chromosomal segment may contain many alleles at important genes from the donor parent and could have a substantially adverse effect on the overall performance of the population.

Since progressively tighter linkage results in a progressively lower frequency of recombinant gametes, the size of the population carried through the backcross process becomes an important consideration. The total number of meiotic events that lead to gametes must be maximized within each backcross population to ensure the highest probability that linkages in coupling are recovered. To obtain recombinant gametes that effectively consist only of a small donor segment, perhaps the locus of interest and less than 5.0 flanking cM of genomic DNA, a rare double cross-over is necessary (Fig. 18.4). The probability of such an event for 10.0cM of genomic DNA during meiosis is 0.25%, or 25 in 10,000. The actual probability is much lower since the degree of interference that distorts double cross-overs increases as the recombinational events are physically closer (Hospital, 2001). In some cases, the degree of distortion may even affect the physical order of markers on linkage groups (Lorieux et al., 1995). Thus, it is more likely that the net result will proceed in increments, with one cross-over occurring during one meiotic cycle and the other in a future generation. It is of critical importance, therefore, that the numbers of individuals in backcross populations be large enough to provide for rare desirable linked recombinants. One way to assure this is to use the backcross (and not recurrent parent) individuals as male parents.

The standard backcross protocol specifies that selection is applied only for the trait under transfer. It is presumed that the natural mechanisms of recombination and segregation will result in the progressive recovery of the alleles and gene organizational features of the established cultivar, or recurrent parent. In reality, selection is usually practiced for both the character under transfer and the desired attributes of the recurrent parent, greatly accelerating the rate at which progress is realized towards a progressive return to the general background of the recurrent parent as the program proceeds, particularly for phenotypes with high narrow sense heritability ($h^2 \geq 0.75$). For traits that exhibit low heritability or are determined by primarily non-additive gene action and interactions, selection will have little effect.

For example, consider a hypothetical example where the recurrent parent is differentiated from a donor population with respect to alleles at ten loci, all unlinked to the trait under transfer. If the h^2 of the phenotypes at all ten loci is 0.80,

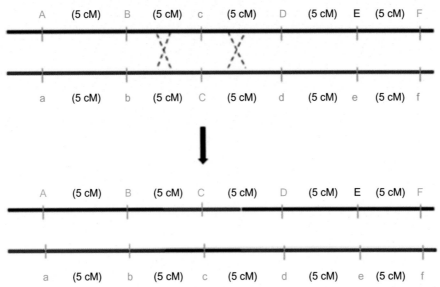

FIG. 18.4 A depiction of the events of a double cross-over leading from a linkage in repulsion (A-B-c-D-E-F) to a linkage in coupling (A-B-C-D-E-F) relative to the gene under transfer.

then it can be said that the h^2 of the recurrent parental phenotype is also 0.80. Applying a selection pressure of 10% per backcross generation, the progress towards the recovery of the genome of the recurrent parent is estimated as:

$$R = ih^2_{ns} \sqrt{V_P}$$

where R is the response to selection, and i is the selection intensity measured in standard deviations from the mean. By calculating the number of genotypes in each generation that are the same as the recurrent parent, the following progression results (Table 18.2; "with selection").

This example is a very dramatic demonstration of the power of selection practiced on phenotypic characteristics of the recurrent parent. In most instances, lower heritability and antagonistic interactions will hinder advances, although progress will nearly always be faster than if no selection is applied.

Many modifications of the backcross strategy under specific circumstances have been proposed to improve the effectiveness or reduce resource requirements. For example, Lewis and Kernodle (2009) used tobacco (*Nicotiana tabacum* L.) as a model system to propose and examine aspects of a modified backcross procedure where transgenic *Arabidopsis thaliana* gene *FT* (*Flowering Locus T*) gene overexpression reduced generation time and accelerate gene transfer during backcrossing. It was demonstrated that constitutive *FT* overexpression dramatically reduces days-to-flower in diverse tobacco genetic backgrounds. The breeder selected for a *FT* transgene insertion and for the trait(s) of interest at each

TABLE 18.2 A Comparison of the Proportion of the Genome of the Recurrent Parent Recovered without and with (Parameters of Selection Described in Text) Selection During Succeeding Generations of a $F_1 \rightarrow BC_6$ Backcross Breeding Program

	% of genome of recurrent parent	
Generation	No selection	With selection
F_1	50.00000	50.00000
BC_1	75.00000	75.00000
BC_2	87.50000	90.75000
BC_3	93.75000	99.99999
BC_4	96.87500	100.00000
BC_5	98.43750	
BC_6	99.21875	

backcross generation, with the exception of the final cycle. Selection in the final generation was conducted for the trait(s) of interest and also against *FT* expression to generate the desired backcross-derived trait conversion.

Under circumstances where the plant breeder is attempting to transfer recessive alleles via backcross, methods featuring two or three stages per cycle may be employed. The two-stage method utilizes alternate backcrossing and self-pollination to identify BC_xS_y [or BC(x)S(y)] plants with the recessive trait while the three-stage method uses two sequential crosses followed by self-pollination. A method to compare the cost-effectiveness of alternative values was presented for cases where the relative costs of cross- and self-pollination and evaluation of S_y progeny are known (Isleib, 1997). Only in cases where the recessive trait is controlled by a single locus can it be more cost-effective to make two sequential crosses to the recurrent parent before self-pollinating rather than to cross- and self-pollinate. Another limitation is that the cost of evaluating BC_xS_y plants must be high relative to the cost of producing BC_xS_0 plants.

The backcross strategy is one of the most widely adopted and adapted of all breeding methods among horticultural crop species. In *Phaseolus vulgaris* (dry and snap bean), diseases cause severe losses worldwide (20–100%) to yield and quality (Singh and Schwartz, 2010). Breeding for resistance using backcross, pedigree, and bulk-pedigree methods to one or two diseases simultaneously is widely and vigorously pursued. For example, common bacterial blight (CBB) caused by *Xanthomonas campestris* pv. *phaseoli* dramatically reduces *P. vulgaris* yield and quality worldwide (Mutlu et al., 2005). Genetic resistance to CBB does not exist in the pinto bean gene pool, the most important dry bean market class in North America. CBB resistance genes from a non-pinto *P. vulgaris* donor parent were introgressed into the recurrent pinto bean parent "Chase" using classical backcross breeding and "intermittent MAS" (see below).

Petunia hybrida is one of the major bedding plants grown worldwide, and the most important character driving consumer demand is seasonal floral display. Floral longevity, or bloom shelf life, has received little direct attention from petunia breeders. Increased floral longevity would enhance the value of many floral ornamental crop species. Four parental genotypes (two with short flower life, two with long flower life) were crossed in a partial diallel mating design to create six F_1 families. F_1 individuals were then self-pollinated and backcrossed to the corresponding recurrent parents to create F_2 and BC families. Results showed the presence of significant additive gene effects for floral longevity indicating that the trait is amenable to transfer by backcross (Krahl and Randle, 1999).

In peanut (*Arachis hypogaea*), high oleic acid level in the seeds has been hypothesized to have a positive impact on roasted sensory quality. A series of lines derived by backcrossing the high-oleic trait into several existing cultivars were compared with the respective recurrent parents (Pattee et al., 2002). Seed oleic acid level had a positive effect on roasted peanut flavor intensity, with high oleic levels increasing flavor by 0.3 flavor intensity units when averaged across all seven background recurrent parent genotypes. The magnitude of flavor improvement varied, however, across different parental background genotypes.

TRANSFER OF GENES ACROSS SPECIES BARRIERS

Phenotypic contrasts of cultivated populations of crop species with corresponding wild progenitors generally show immense differences. Pervasive phenotypic differences between cultivars and wild relatives of a crop species are daunting to plant breeders, but the possibility that the underlying genes may be excised from wild population genomes and moved into cultivars is compelling. If the wild population is sexually compatible with the cultivar, a standard backcross breeding program may be implemented to achieve such a result. Success depends on the degree of phenotypic gulf separating wild from cultivated (necessitating many backcross generations to select for cultivated and against wild traits), the heritable basis of the trait of interest, and how its expression may change in different genetic backgrounds.

Most cultivated crop species are associated with undomesticated populations of the same species or closely related species. The process of speciation is, by definition, associated with reproductive isolation of the gene pools (Chapter 2). The isolation mechanisms may be due to developmental or genetic factors or both. Although certain species may be demonstrated to be closely related by methods such as genomic DNA sequencing, they may still be reproductively isolated. Many examples have been reported, however, of successful hybridization of cultivated species with other related taxa, usually species within the same genus. Hybridization beyond the level of genus is exceedingly rare. In many cases, genomics has resulted in the systematic reorganization of species relationships and what was thought to be an intergeneric hybrid is actually interspecific.

The backcross method has been used successfully for the transfer of desirable traits from different but related species to the economic counterpart. The notion is intuitive: a wild relative population or individual is found to exhibit a phenotypic component of overall crop performance that is deemed to be better than the corresponding cultivated species. Examples of valuable genes discovered in wild relatives of crop species abound: general plant growth habit, cold, heat, drought, salt, anaerobiosis tolerance, disease and insect resistance, day length flowering behavior, nutritional attributes, and aesthetic traits such as color and flavor.

The requirements are the same as for the use of the backcross method for transfer of phenotypic traits between populations within species, with some important caveats. The donor and recurrent parent must, of course, be reproductively compatible at least to the extent that hybrid and backcross progeny are feasible. Additional steps may be necessary such as the use of tissue culture to rescue embryos from aborting seeds. Zygote or early embryo abortion is generally most evident in hybrids from the initial interspecific cross, and manifestations of reproductive isolation usually diminish with progressive backcross generations to the cultivated recurrent parent.

Meiotic cell divisions in the ensuing F_1 and BC_x populations must result in the exchange of DNA between the chromosomes of the donor and recurrent parent for the backcross method to be effective for the transfer genes from wild relatives into crop species. The viability of interspecific hybrids is not adequate evidence that the constituent genomes are pairing and undergoing recombination during prophase I. Chromosome pairing and recombination are absolutely essential for the transfer of the desired gene or genes from the genome of the wild species into the recurrent parent (Fig. 18.4). Any whole chromosomes, or segments thereof, will carry undesirable genes that adversely affect performance of the end product. Gametogenesis and functional meiosis in the original interspecific hybrid and the early backcrosses, therefore, are the most critical factors for the success of the program. An alien addition or substitution line may also be isolated that may be backcrossed to different backgrounds in an attempt to stimulate homologous chromosome pairing and recombination.

Applications of the backcross method in the introgression of wild genes into the gene pools of cultivated crops are wrought with challenges. Among the most daunting of challenges is the distortion of predicted Mendelian segregation ratios among progeny of interspecific hybrids due to genomic incongruities and ploidy inequities. For example, Ochanda et al. (2009) examined the effects of exerting selection on interspecific hybrids before backcrossing. F_1 hybrid plants are generally used to backcross to the adapted lines or populations. An alternative approach is to backcross segregating F_2 individuals selected for agronomic acceptability. Based on undesirable results from these studies, selection before backcross in the process of introgression of exotic germplasm was not recommended.

Many examples of the successful transfer of desirable traits from wild relative species to horticultural crop species via backcross reside in the scientific literature. Charles M. Rick (see Chapter 1 for a biography) was an early proponent of this strategy and conducted important work in Solanaceae to underpin the power of using wild species relatives for gene pool expansion (Rick, 1972). In a recent example with cultivated tomato, ten selected inbred backcross lines (IBCL), from a *Solanum lycopersicum* × *S. pennellii* inbred population with resistance to beet armyworm (BAW; *Spodoptera exigua*), higher fruit mass and fruit yield, were crossed with eight elite cultivated *S. lycopersicum* inbred lines in a Design II mating design (see Chapter 10). BAW resistance in the ten selected IBCL and ICBL-derived F_1 progeny was associated with two undesirable traits: later maturity and larger vine size. Selection of inbred lines was more effective at identifying positive GCA for fruit mass and fruit yield than GCA for BAW resistance (Hartman and St Clair, 1999).

Dominant alleles for resistance to disease pathogens are common targets for backcross-mediated introgression from the gene pools of wild species relatives into the genomes crop species. For example in Brassicaceae, BC_1 and F_2 progenies from triploid F_1 and tetraploid F_1 hybrids between *B. napus* and 2x and 4x *B. oleracea* ssp. *capitata* (cabbage) were studied for their general morphology, resistance to race 2 of *Plasmodiophora brassicae* (the pathogen causing the clubroot disease), chromosome number, and meiotic chromosome behavior. The presence of homoeologous pairing between *B. napus* and *B. oleracea* chromosomes was observed in all the plants and considered advantageous for selecting recombinant progeny in later generations. Resistance to *P. brassicae* race 2 was successfully introgressed into *B. oleracea* using this strategy. The potential use of microspore culture to extract gametic progenies from resistant BC_1 and F_2 plants was suggested (Chiang et al., 1979).

A dramatic example of the power of the application of the backcross method for introgression of genes across species barriers is the development of American chestnut clones with resistance to chestnut blight. Morphological features of leaves and twigs of American chestnut (*Castanea dentate*), Chinese chestnut (*C. mollissima*), their F_1 hybrid, and three successive generations of backcrosses between hybrid populations and American chestnut were examined to determine rate of recovery of the American chestnut morphology after hybridization to capture Chinese chestnut genes for blight resistance (Diskin et al., 2006). 96% of trees in the BC_3 resembled American chestnut and were distinctly different from Chinese chestnut. Backcross breeding appears to be a workable strategy for restoring this species as an important component of eastern U.S. forest ecosystems.

Prohens et al. (2012) demonstrated that many differences exist for plant and fruit morphology among cultivated eggplant (*S. melongena*), *S. aethiopicum* and the corresponding interspecific hybrid. Backcross (to *S. melongena*) progeny exhibited morphological variation with intermediate heritability values for the attributes evaluated. The interspecific hybrid exhibited fruit phenolic content and quality similar to *S. aethiopicum* and was also heterotic for fruit flesh browning. Results demonstrated that interspecific hybridization could be a powerful tool to expand the *S. melongena* eggplant gene pool. In another interspecific backcross study on eggplant, Mennella et al. (2010) produced new interspecific eggplant genotypes bearing useful traits derived from the wild species parents (i.e., resistance/tolerance to plant pathogen fungi) together with nutraceutical and antioxidant properties typical of the cultivated species.

Backcross breeding in zinnias has been successful even with the challenges of variable ploidy levels and gametic chromosome numbers. True-breeding lines of *Zinnia marylandica* (that is a 2n = 4x = 46 allotetraploid of *Z. angustifolia × Z. violacea*) were reciprocally backcrossed with diploid and autotetraploid forms of *Z. angustifolia* (2n = 2x = 22 or 2n = 4x = 44) and *Z. violacea* (2n = 2x = 24 or 2n = 4x = 48; Boyle, 1996). Backcrosses were more successful with *Z. angustifolia* and *Z. violacea* as autotetraploids than as diploids. BC_1 hybrids of *Z. marylandica* and *Z. violacea* have direct commercial potential as seed-propagated bedding plants.

Cytogenetic incongruencies are common when attempting to produce interspecific hybrids (Pasutti et al., 1977). For example, backcrossing the diploid (2n = 2x = 24) F_1 interspecific hybrid between Longiflorum × Asiatic lilies (LA) to Asiatic parents (LAA) resulted in the production of 104 BC_1 progeny plants (Khan et al., 2009). Among these progeny were 27 diploids, 73 triploids (2n = 2x = 36) and 4 aneuploids (2x − 1, 2x + 2 or 2x + 3). Genomic in situ hybridization (GISH) revealed extensive intergenomic recombination among the chromosomes in LA hybrids. A large number of Longiflorum chromosomes were transmitted to the BC_1 progenies from LA hybrids. However, very few Longiflorum chromosomes were transmitted from the BC_1 triploid (LAA) plants to the BC_2 progenies.

An interspecific hybrid combination may not prove to be of any direct value as a prospective product of plant breeding, but such hybrids are often found to be useful as "bridge" intermediates to allow the flow of genes among isolated gene pools. For example, interspecific hybrids of *Diplotaxis siettiana × Brassica rapa* hybrids were completely pollen sterile, and backcrosses with *B. rapa* as male recurrent parent did not yield any seeds (Nanda Kumar and Shivanna, 1993; the synonymous historical species name *B. campestris* is used for *B. rapa* in this paper). This mostly sterile hybrid was then used as a bridge cross to transfer the cytoplasm of *D. siettiana* to two other incompatible cultivars of *Brassica*—*B. juncea* and *B. napus*. Pollinations of the amphidiploid (*D. siettiana × B. rapa*, 2n = 36) with pollen of *B. juncea/B. napus* readily produced seeds without embryo rescue.

Interspecific hybridization followed by recurrent backcrossing is a method used to develop cytogenetic chromosome addition and substitution lines (see Chapter 3). In an example of this strategy, intergeneric crosses were made between *Brassica oleracea* (2n = 18) and *Moricandia arvensis* (2n = 28) utilizing embryo rescue (Bang et al., 2007). In the backcross with *B. oleracea* some of these hybrids produced developed BC_1 plants with 2n = 32 chromosomes. Observations suggest that monosomic addition lines of *B. oleracea* carrying a single chromosome of *M. arvensis* were produced that could offer potential for future genetic and breeding research, together with other novel hybrid progeny developed in this intergeneric hybridization. Mithila and Hall (2013) demonstrated the transfer of auxinic herbicide resistance from *B. kaber* to *B. juncea* and *B. rapa* by traditional backcrossing with in vitro embryo rescue of the interspecific hybrids.

CHANGE OF CYTOTYPE

The backcross method has been used extensively to achieve the desired combination of nuclear and cytoplasmic genotypes, capitalizing on the maternal inheritance of cytoplasmic genomes. By conducting recurrent backcrosses such that the recurrent parent is always the paternal, or male, parent the cytotype will be from the wild relative while the nuclear genome will gradually be constituted of genes contributed by the cultivated recurrent parent.

This technique is used predominantly in circumstances where cytoplasmic male sterility (cms) is used as a tool for the mass production of hybrid seed. If two inbred lines exhibit high SCA, but are both male fertile, it will probably be prohibitively expensive to produce hybrid seed from that particular combination. To develop a cms parent that can serve as female in a large-scale seed production, the higher-yielding or more fecund parent is entered into a backcross breeding program with a donor population that is enveloped by sterile cytoplasm. Following ~6 backcrosses with the high SCA inbred as male parent, the inbred is effectively converted from fertile to sterile. It is presumed that the donor and recurrent population are devoid of nuclear restorer genes. At the conclusion of the program, the original population of the recurrent parent, in fertile cytoplasm, may be used as the maintainer (B-line) for female inbred maintenance (Fig. 18.5).

Scotti et al. (2003) found that nuclear-cytoplasmic interactions can influence fertility and agronomic performance of interspecific hybrids in potato as well as other species. Backcross progeny were produced by crossing a somatic hybrid between *Solanum tuberosum* (tbr) and the wild species *S. commersonii* (cmm) with various potato clones. Genotypes with cytoplasms sensitive to nuclear genes derived from *S. commersonii* and inducing male sterility showed identical mtDNA composition, as based on mtDNA analyses with various PCR-based and RFLP markers.

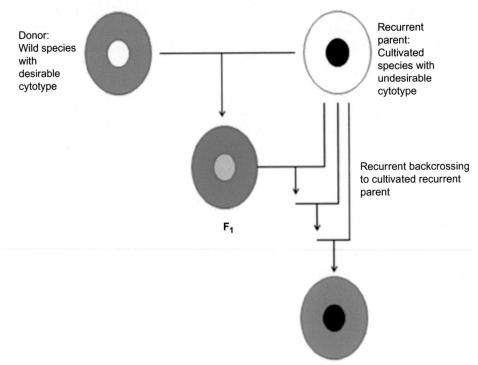

FIG. 18.5 Application of the backcross method to transfer cytotype from a sexually compatible wild species to the somatic cells of a cultivar.

Genotypes with cytoplasms not inducing male sterility in the presence of the cmm nuclear genes showed a different mtDNA organization. Analysis of cpDNA confirmed similarity of cytoplasmic composition in CMS-inducing genotypes and clear differences with the others.

INCREASING THE NUMBER OF RECURRENT PARENTS OR TRAITS UNDER TRANSFER

Two or more parent populations are used in the production of hybrid, synthetic, and composite cultivars. The backcross method may still be used effectively in situations where multiple parentages are involved (Borlaug, 1957). If the trait under transfer is dominant, it may only be necessary to utilize one parent or few in the backcross program for hybrid and synthetic cultivars. If the trait is semi-dominant, additive, or recessive, or if the population is to be perpetuated by open-pollination, it will be necessary to conduct two or more parallel backcross programs to transfer the gene into all parental populations.

If the donor population possesses more than one trait that could benefit a targeted recurrent parent, it is possible to perform breeding operations to facilitate the simultaneous transfer of both or all. The dynamics discussed above are compounded, making it necessary to increase population sizes accordingly. If 50 plants of each BC generation are grown for the transfer of a single trait, 250–500 may be necessary for the simultaneous transfer of two. The plant breeder must take care that the backcross breeding program parameters do not adversely affect recombination and segregation.

If two or more desirable traits are identified in distinct populations, there are three alternative courses that could be taken, all incorporating the power of the backcross method: (i) pyramid the desired traits into one donor population, then embark on a backcross program, (ii) conduct two or more independent, parallel backcross programs, then combine the desired traits by recurrent selection or production of F_1 hybrid populations, and (iii) conduct two or more independent backcross programs in parallel series. The results of all three strategies are similar, but each exhibit distinct advantages and disadvantages. The first method has the lowest time requirement, but mandates larger population sizes to achieve reasonable probabilities that desirable recombination and segregation will be recovered. The third method will take the longest time, but has the least ongoing management costs. The second method lies is intermediate between the two other methods in terms of time and resource requirements.

MODIFICATIONS TO THE BACKCROSS METHOD

Under specific circumstances, the backcross method is surmised to be the most appropriate strategy to reach the intended breeding objective. In some of these cases, modifications to the backcross method can be applied to achieve an acceptable outcome. An example of where modifications of the backcross method may be an effective alternative is when the genotype of the recurrent parent is difficult to define in precise terms. Certain populations reproduced by open-pollination or by asexual reproduction fall into this category.

In OP populations, overall population performance is conditioned by a subtle balance of allele fixation and polymorphisms. Allelic frequencies are critical to the realization of population performance. If frequencies are too high or low, inbreeding depression may result in a depression of overall population performance. If frequencies are not high or low enough, phenotypic variability increases to the point that cultivar integrity is threatened.

The most effective way to ensure that the genetic integrity of cultivars is not undermined by altered allelic frequencies is to perform backcross matings *en masse*, most readily accomplished under circumstances where the trait under transfer is selectable and of high h^2. The best example of such a scenario is a dominant disease resistance allele. Pollen from the backcross populations is used to fertilize a large number (>1000) of females from the recurrent parent OP population followed by screening of the progeny for disease resistance. This procedure may also be performed if the trait is easily distinguished from the corresponding condition in the recurrent parent, but requires more labor.

Alternatively, it may be possible to reconstitute the genetic structure of the recurrent parent population by intercrossing n inbred derivatives. As $n \to \infty$ the identity of the original and reconstituted population is absolute. As $n \to 0$ the reconstituted population diverges progressively from the ancestral source. The challenge to the plant breeder is to identify a value for n that maximizes the intersection of maximum genetic identity and minimum resource requirements. The backcross breeding program must proceed with n parallel populations, each component necessitating labor, time, and space. The practical range of maximization for n in horticultural crop species is ~10–100.

The challenge in adapting backcross to be an effective and efficient breeding strategy is quite different for clonal populations. This population structure is characterized by an infinite number of individuals of the identical genotypic constitution, the structure of which usually defies simple definition. Any interjection of sexual reproductive cycles compromises the integrity of the prized genotype. If the plant breeder wishes to transfer a given trait from a distinct source to a clonal population, it is possible to employ the backcross method, but the result will rarely be consistent with the archetype for the method: the recurrent parent augmented by the transferred trait. This dilemma is especially true for woody perennial species where resource requirements to maintain individuals are relatively high. Since the number of individuals in the breeding program must be reduced due to resource limitations, inbreeding depression in the finished populations may also be an issue of concern. Due to the unique requirements of clonally-propagated cultivars, biotechnological strategies such as molecular transformation and genome editing may be more attractive for trait management than backcross

Paul Hansche of the University of California, Davis first developed the "modified backcross method" for fruit tree breeding to achieve a partial alleviation of the problems of unwanted segregation and inbreeding depression (attributed by Moore and Janick, 1983). The main feature of this strategy is the employment of similar but genetically distinct recurrent parent populations. Rather than to continually mate the nascent backcross population with a single recurrent parent, the idea is to switch to two to three different recurrent parents during the program, with a final ideotype in mind that embodies the desired attributes of the similar cultivars combined with the new desired trait (Figs. 18.6 and 18.7).

APPLICATIONS OF MOLECULAR MARKERS AND MAS IN THE BACKCROSS METHOD

The concept of the use of linked surrogate genes to select for genomic segments carrying desirable genes predates the use of molecular markers. For example, Burton and Werner (1991) evaluated the use of nonlethal genetic markers to locate heterotic chromosome blocks (HCBs) and the development of methods to use the markers to transfer the selected HCBs to a popular inbred, and to increase the yield of a hybrid based on that inbred.

Tanksley et al. (1981) demonstrated the use of linked molecular markers as tools to greatly enhance the efficiency and accuracy of the backcross method. This group used isozyme markers that were limited in number and genome distribution. Function-neutral DNA markers have been much more powerful when inserted into this theoretical framework. Not long after this important proposition and demonstration, Young and Tanksley (1989) showed that RFLP markers could be used in the backcross method to greatly facilitate the speed and precision of introgression of a targeted tomato fruit quality trait.

The backcross method is replete with points at which molecular markers may be inserted and used to enhance the efficiency of a given program. Such markers are extremely powerful because they are employed basically as tags for

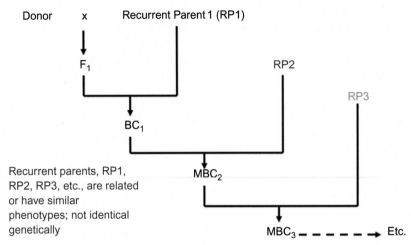

Modified Backcross: Used in cases of intolerance to inbreeding or to enhance genetic variability in breeding populations

Donor   x   Recurrent Parent 1 (RP1)

F_1 RP2

RP3

BC_1

Recurrent parents, RP1, RP2, RP3, etc., are related or have similar phenotypes; not identical genetically MBC_2

MBC_3 – – – – – → Etc.

FIG. 18.6 Modifications to the backcross method to mitigate losses of plant vigor due to small population sizes and resulting inbreeding depression; known as the "modified backcross method".

Concurrent backcross for the development of multiline cultivars

Example: Durable resistance to multiple races of a pathogen

	Donor genes	Number of backcrosses	Recurrent cultivar	Finished Isolines	Multiline
Donor					
Donor 1	R_1R_1	5	A(rr)	$A(R_1R_1)$	
Donor 2	R_2R_2	5	A(rr)	$A(R_2R_2)$	**Isolines composited to constitute a multiline cultivar**
Donor 3	R_3R_3	5	A(rr)	$A(R_3R_3)$	
Donor 4	R_4R_4	5	A(rr)	$A(R_4R_4)$	
Donor 5	R_5R_5	5	A(rr)	$A(R_5R_5)$	

FIG. 18.7 The use of concurrent backcross programs for introgression of resistance genes to different races of a pathogen into the genome of one recurrent parent. The resulting "isolines" are then mixed, or bulked (also known as compositing), to constitute the multiline cultivar.

chromosome segments, and not for any particular function per se. With the advent of RAPDs, AFLPs, SSRs, SNPs, and other robust DNA marker systems (see Chapter 9), it is relatively simple to find polymorphisms that distinguish even closely related populations, in this case the donor and recurrent parent within a mostly monomorphic species.

Studies of the inheritance of molecular polymorphisms coupled with genome sequence information will reveal which ones are useful in distinguishing chromosome segments of the corresponding populations. Ideally, cosegregation studies that involve the trait that will be transferred in the backcross program will identify useful molecular marker loci to select for the trait and other desirable genomic regions. Any linkages that are discovered will be particularly useful for improving the efficiency and effectiveness of the program. Such results will also assist the practitioner in choosing markers that may be employed as tags for the broadest possible representation of the entire genome as possible.

The most simplistic strategy is to select for molecular marker phenotypes associated with the recurrent parent while simultaneously selecting for the presence of the trait under transfer. Since molecular markers have a heritability of 1.00 and are discernible at the seedling stage of development, large populations of seedlings may be ascertained for molecular phenotype, then screened for those having the highest proportion of markers associated with the recurrent parent. The selected individuals are then subjected to phenotypic selection at the whole plant level.

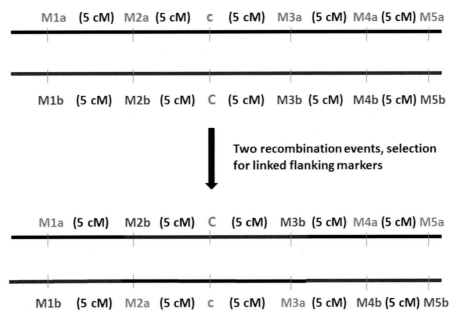

FIG. 18.8 Selecting markers flanking a gene under introgression for the genomic segment carrying that gene while selecting for all other markers of the recurrent parent (see text).

Linkages to the trait under transfer will provide the most powerful tools for subsequent backcrossing. The most favorable scenario consists of a cassette wherein the transferred locus is flanked by four polymorphic molecular loci, two on each side (Fig. 18.8). On each side, a proximal marker is situated as close as possible to the transferred locus, and a distal marker located as close as possible to the proximal locus. For each succeeding backcross generation, seedling selections are conducted wherein selection is applied for the presence of the distal markers of the recurrent parent simultaneously with the proximal markers of the donor population.

A great deal of experimental work on applications of MAS for enhancing the power of backcross breeding has appeared in the scientific literature since 1990. "Advanced backcross QTL analysis" was proposed as a method of combining QTL analysis with cultivar development by Tanksley and Nelson (1996). This protocol was tailored for the discovery and transfer of valuable QTL allele from unadapted donor lines (e.g., landraces, wild populations, wild species, etc.) into established elite inbred lines. Simulations suggested that advanced backcross QTL analysis is effective in detecting additive, dominant, partially dominant, or overdominant QTL. The study concluded that advanced backcross QTL analysis could open the door to exploiting unadapted and exotic germplasm for the quantitative trait improvement of a broad spectrum of crop species.

An excellent example of the power of QTL in a backcross program is selection for altered or enhanced root system structure. Plant breeders are accustomed to performing selection on phenotypes of foliage and fruits, but root systems in the soil are impossible to assess directly without using forensics or inflicting tremendous damage to plants. Steele et al. (2013) demonstrated the use of MAS with backcross to introgress four QTL for root traits into an upland rice cultivar. The introgressed lines and the recurrent parent were grown for six years by resource-poor farmers in upland sites in Eastern India and yields recorded and significantly increased yield under relatively favorable field conditions. Root studies under controlled conditions showed that introgressed populations had longer roots throughout shoot tillering than the recurrent parent (14 cm longer 2 weeks after sowing). They concluded that both improved roots and increased yield could be attributed to the introgression of QTL.

Another class of phenotypes that is particularly amenable to MAS is seed and fruit quality. The determination of these phenotypes may be expensive, technically challenging, or necessitate the sacrifice of seeds. For example, quantitative measurements of amino acid content in seeds are destructive. In a study to avert seed destruction, two elite normal maize inbreds that produced a heterotic experimental hybrid with 45% yield advantage over the standard check were used for conversion to *opaque-2* donors using MAS backcross (Tripathy et al., 2017). The most promising BC_3F_3 introgression lines from cross combinations were subsequently selected based on higher tryptophan and lysine content. The newly developed introgression lines were equivalent in field performance to the recurrent commercial parents with nearly double the lysine and tryptophan content.

"Pyramiding" refers to the introgression of desirable genes into a single individual or population recurrent parent from many different donors. It is an immense challenge when employing the traditional backcross method due to compounding population size requirements and difficulty in selecting for diverse phenotypes that may have confounding interactions. By enabling the plant breeder to select desirable recombinants with precision, MAS provides an effective way to pyramid genes concurrently instead of sequentially.

The simplest form of pyramiding is the introgression of two independent genes from separate donors into a single recurrent parent. Frisch and Melchinger (2001) showed that in such backcross breeding programs with three backcross generations, the least marker data points were required when (i) applying selection strategies consisting of three or four selection steps on the basis of presence of the target genes and selection indices calculated from the marker genotype, (ii) increasing the population size from early to advanced generations, and (iii) merging the target genes in an early generation.

In another study that adapted MAS for QTL to pyramid desirable genes via backcross, Singh et al. (2015) combined genes for resistance to two important rice diseases blast (*Magnaporthe oryzae*) and sheath blight (*ShB*; *Rhizoctonia solani*). Pusa 6B, the Basmati quality cms maintainer line of a popular superfine grain aromatic rice hybrid, was highly susceptible to both the diseases. MAS backcross produced advanced selections carrying both blast and *ShB* resistant genes in the Pusa 6B background that resistant to both pathogens without compromising grain and cooking quality.

What is the most advantageous order in which genomes may be combined with incremental bilateral steps to achieve the desired gene pyramiding result? Ishii and Yonezawa (2007) conducted experiments to answer this question and to exploit MAS for the construction of gene-pyramided lines from multiple donors. They concluded that backcross programs should be performed separately for each gene donor before performing crossing between the donors. When four such introgressed individuals (A, B, C, and D) have been produced, for example, they should be crossed in a schedule like (A×B)×(C×D) in which the number of target markers of A plus B should be as similar as possible to that of C plus D. The genotypes of donor populations should be modified with regard to linked or redundant markers and to minimize the occurrence of repulsion linkages.

In a study on the development of effective methods to pyramid introgressed genes with backcross, the relative effectiveness of two typical marker-based schemes for constructing high-degree gene-pyramided lines, AF (gene Assemblage First) and BF (Backcross First), were contrasted (Ishii et al., 2008). In AF, target genes of all donor parents are assembled onto the genome of a plant first, followed by backcross generations for the recovery of recipient parent genome. In BF, conversely, the backcross is performed first separately for each donor, followed by generations of crossing for the assemblage of target genes. Stochastic calculations reported by these researchers showed that BF was superior to AF when molecular selection is used for both target genes and background markers; with the same number of generations (time) and cost of genotyping, BF produced a much higher recovery of recurrent parent genome than AF.

MAS backcross has been applied with tremendous success in horticultural crop species. For example, acyl-sugars exuded from type IV trichomes mediate multiple pest resistance found in the wild tomato species, *Solanum pennellii*. A MAS breeding program was used to attempt the transfer of the ability to accumulate acyl-sugars to cultivated tomato (Lawson et al., 1997).

In another example, multiple lateral branching (MLB) is a quantitatively inherited trait associated with yield in cucumber (*Cucumis sativus* L.). Quantitative trait loci (QTL) were identified for MLB, and QTL-marker associations were subsequently verified by marker-assisted selection (Robbins et al., 2008). To test the effects of pyramiding QTL for MLB, molecular genotyping was utilized to create two sets (standard- and little-leaf types) of inbred backcross lines possessing various numbers of QTL that promote branching. Although pyramiding QTL for MLB did not uniformly increase the number of lateral branches, pyramiding QTL in inbred lines allowed further characterization of individual QTL involved in MLB.

In another example with tomato, sources of resistance to early blight (EB), caused by *Alternaria solani*, were identified within wild species (e.g., *Solanum hirsutum*) related to and cross-compatible with cultivated tomato (Foolad et al., 2002). A MAS backcross program was conducted to identify and validate QTL for EB resistance in backcross populations of a cross between a susceptible tomato (*S. lycopersicum*) breeding line (maternal and recurrent parent) and a resistant *S. hirsutum* breeding line.

In pepper (*Capsicum annuum* L.), Phytophthora wilt caused by *Phytophthora capsici* is a disease that inflicts major economic losses in regions prone to excess rainfall and poor soil drainage. Thabuis et al. (2004) conducted a study to transfer resistance to *P. capsici* using MAS backcross with QTL linked to four resistance factors from a small-fruited pepper into a large-fruited bell pepper recurrent parent. The MAS backcross program was initiated from a doubled-haploid line issued from the mapping population and involved three cycles. The additive and epistatic effects of the 4 resistance factors were recapitulated and validated in these populations indicating that introgression of four QTL in this MAS backcross program was successful.

In a more general sense, MAS may be used to develop populations that embody an enhanced range of genetic variability to facilitate future breeding efforts. For example, to broaden the genetic base of Beit Alpha cucumber (*Cucumis sativus* L.) for plant improvement, diverse accessions were compared employing a previously defined standard marker array to choose wide-based parental lines for use in backcross introgression (Delannay and Staub, 2010).

COMPARISONS OF BACKCROSS TO MOLECULAR TRANSFORMATION

The proponents of biotechnology and GMOs have cited three fundamental advantages that the methods offer: (i) expansion of the range of traits that may be addressed, since no barriers exist for the insertion or editing of DNA sequences, (ii) for existing traits that may be transferred by traditional methods such as backcross, molecular transformation or genome editing are much faster and more precise, and (iii) the integrity of the recurrent parent is not altered by meiotic cell divisions. A comparison of the traditional backcross strategy and the molecular transformation (or genome editing) strategy for trait conversion is depicted in Fig. 18.9.

There is no question that all of these assertions hold merit but the 3rd assertion is the most defensible. As was discussed in the "Modifications to the backcross method" section above, certain situations are encountered wherein it is difficult or nearly impossible to reconstitute the population genetic structure of the recurrent parent via backcross to the recurrent parent. In such instances, molecular transformation or genome editing are desirable alternatives. Parenthetically, most woody perennial plant species are somewhat recalcitrant to tissue and cell culture. Since current available methods for transformation or genome editing necessitate the culture and regeneration of plant cells, these technologies are not universally available. Fortunately, new protocols to expand the range of amenable crop species for application of cell and molecular technologies are constantly being developed.

The backcross method is subject to biological limits imposed by the tolerance of genome constitution. The method is intended primarily for applications within ranges of sexually compatible plants, usually species. Under certain circumstances the method may be adapted for use with distinct gene pools that are not completely isolated reproductively (see Transfer of genes across species barriers above). The primary differences between backcross and transformation are gene copy number and targeted loci. The traditional backcross method and genome editing are both predicated on the substitution or conversion of one allele for another at a specific, defined locus within the genome. Molecular transformation, however, results in the addition of a functioning gene to the genome, a discrete and somewhat unpredictable DNA insertion into a chromosome. The interaction of inserted gene with the new genome context may, indeed, deliver an entirely unexpected result. Such was the case with the "trans-switch" gene

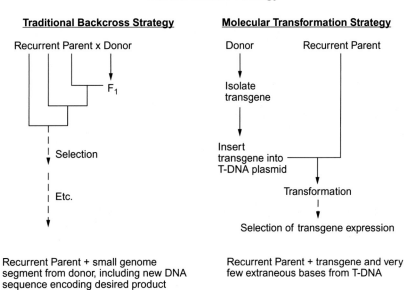

FIG. 18.9 A comparison of the traditional backcross method with molecular transformation.

inactivation system marketed in the early 1990s. The mechanism of inactivation was presumed to be translational interference, but further examination revealed an entirely unknown pathway by which potentially harmful invading DNA sequences are silenced. This discovery led to the identification and uses of miRNAs, a phenomenally powerful tool for gene therapy (Nogoy et al., 2018).

The techniques for plant transformation presently available do not provide for precise placement within targeted genomes. Biolistics often results in multiple copies of the transforming sequences located more or less randomly throughout the genome. The distribution of *Agrobacterium*-mediated T-DNA transformation insertion sites has been demonstrated to be highly non-random in the rice genome (Zhang et al., 2007). They found that T-DNA insertions were biased towards large chromosomes in both absolute number and relative density of insertions. Within chromosomes the insertions occurred more densely in the distal ends and less densely in the centromeric regions. The distribution of the T-DNA insertions was also found to be highly correlated with that of full-length cDNAs but the correlations were highly heterogeneous among the chromosomes. In this study, T-DNA insertions were not found within transposon-related sequences, but were found in sequences with a strong bias toward the 5′ upstream and 3′ downstream regions of the genes. Finally, T-DNA insertions occurred preferentially among the various classes of functional genes such that the numbers of insertions were in excess in certain functional categories but were deficient in other categories.

Kim et al. (2007) found a relatively high frequency of T-DNA insertions in heterochromatic regions, including centromeres, telomeres and rDNA repeats in *Arabidopsis thaliana*. The frequency with which T-DNA insertions mapped to exon, intron, 5′ upstream and 3′ downstream regions closely resembled their respective proportions in the *Arabidopsis* genome. This group also found that T-DNA integration occurs without regard to DNA methylation. They concluded that T-DNA integration may occur more randomly than previously indicated, and that selection pressure might shift the recovery of T-DNA insertions into gene-rich or transcriptionally active regions of chromatin.

For biolistics-mediated transformation, three integration patterns were observed from experiments using fluorescence in situ hybridization (FISH) on extended DNA fibers (fiber-FISH) to visualize three distinct classes of integration sites in the wheat genome: large tandemly repeated integration; large tandem integrations interspersed with unknown DNA; and small insertions, possibly interspersed with unknown DNA. Metaphase FISH showed that the integration of transgenes was located in both hetero- and euchromatic regions of the genome, as well as proximal, interstitial and distal, regions of the chromosomes (Jackson et al., 2001). It appears, therefore, that the integration of a transgene following biolistic cellular introduction is mostly random. Partier et al. (2017) demonstrated that a relatively large (53 kbp) intact segment of DNA could be inserted into the wheat genome via biolistics, far exceeding the transgene size limits of T-DNA-mediated transformation.

The inheritance of the newly inserted transgene may differ from typical Mendelian phenotypic and genotypic ratios. The primary transformant (T_1) is usually hemizygous for the inserted sequence and embedded gene(s). If the inserted sequence is successfully transmitted to gametes, a homozygote for the sequence may be obtained by assortative mating, usually self-pollination. If the inserted gene sequence and function are similar to those of the genomic counterpart, the homozygous transformant may behave as a chromosomal duplication. If the transgene is dissimilar or imparts a new phenotype, the resulting transformant will more likely behave genetically more like a segmental alloploid.

Time requirements of transformation, in contrast with the backcross method and genome editing, are highly dependent on the context of application. At the extremes of short generation time, often less than 30 days in the case of *Arabidopsis thaliana*, a backcross program to BC_4 may be completed in less than 6 months. Transformation and genome editing may require a year or more to the point that genetically altered plants are identified. Additional steps are always needed to attain the desired population structure based on the transformation or genome editing event, perhaps two to three years in total. Consequently, molecular transformation and genome editing do not necessarily impart the time savings that the plant breeder may seek by invoking these strategies.

Technical and economic feasibility remain as daunting obstacles in the widespread adoption of molecular transformation and genome editing as viable alternatives to backcross. It took over 20 years to establish a protocol for transformation of maize despite an enormous magnitude of experimental work. Margins may be too thin for most species to support the technical capabilities necessary to support costs of these technologies. In many species, the minute details necessary to maximize protocol efficiencies are lacking and more investment is warranted. Therefore, investment capital will likely continue to promote further advancements of molecular technologies into plant breeding, especially given the uncertain environment surrounding the consumer acceptance of genetically engineered products.

References

Bang, S.W., Sugihara, K., Jeung, B.H., Kaneko, R., Satake, E., Kaneko, Y., Matsuzawa, Y., 2007. Production and characterization of intergeneric hybrids between *Brassica oleracea* and a wild relative *Moricandia arvensis*. Plant Breed. 126 (1), 101–103.

Bernardo, R., 2003. Parental selection, number of breeding populations, and size of each population in inbred development. Theor. Appl. Genet. 107 (7), 1252–1256.

Borlaug, N.E., 1957. The development and use of composite varieties based upon the mechanical mixing of phenotypically similar lines developed through backcrossing. In: Report of the 3rd International Wheat Conference, pp. 12–18.

Boyle, T.H., 1996. Backcross hybrids of *Zinnia angustifolia* and *Z. violacea*: embryology, morphology, and fertility. J. Am. Soc. Hort. Sci. 121 (1), 27–32.

Briggs, F.N., 1930. Breeding wheats resistant to bunt by the backcross method. J. Am. Soc. Agron. 22, 239–244.

Briggs, F.N., 1935. The backcross method in planting breeding. Agron. J. 27 (12), 971–973.

Briggs, F.N., 1938. The use of the backcross in crop improvement. Am. Nat. 72, 285–292.

Briggs, F.N., Allard, R.W., 1953. The current status of the backcross method of plant breeding. Agron. J. 45, 131–138.

Burton, G.W., Werner, B.K., 1991. Genetic markers to locate and transfer heterotic chromosome blocks for increased pearl millet yields. Crop. Sci. 31 (3), 576–579.

Chiang, M.S., Chiang, B.Y., Grant, W.F., 1979. Transfer of resistance to race 2 of *Plasmodiophora brassicae* from *Brassica napus* to cabbage (*Brassica oleracea* ssp. *capitata*). III. First backcross and F$_2$ progenies from interspecific hybrids between *Brassica napus* and *Brassica oleracea* ssp. *capitata*. Euphytica 28 (2), 257–266.

Delannay, I.Y., Staub, J.E., 2010. Use of molecular markers aids in the development of diverse inbred backcross lines in Beit Alpha cucumber (*Cucumis sativus* L.). Euphytica 175 (1), 65–78.

Diskin, M., Steiner, K.C., Hebard, F.V., 2006. Recovery of American chestnut characteristics following hybridization and backcross breeding to restore blight-ravaged *Castanea dentate*. For. Ecol. Manage. 223 (1–3), 439–447.

Foolad, M.R., Zhang, L.P., Khan, A.A., Nino-Liu, D., Liln, G.Y., 2002. Identification of QTLs for early blight (*Alternaria solani*) resistance in tomato using backcross populations of a *Lycopersicon esculentum* x *L. hirsutum* cross. Theor. Appl. Genet. 104 (6/7), 945–958.

Frisch, M., Melchinger, A.E., 2001. Marker-assisted backcrossing for simultaneous introgression of two genes. Crop. Sci. 41 (6), 1716–1725.

Harlan, H.V., Pope, M.N., 1922. The use and value of backcrosses in small grain breeding. J. Hered. 13, 319–322.

Hartman, J.B., St Clair, D.A., 1999. Combining ability for beet armyworm, *Spodoptera exigua*, resistance and horticultural traits of selected *Lycopersicon pennellii*-derived inbred backcross lines of tomato. Plant Breed. 118 (6), 523–530.

Hospital, F., 2001. Size of donor chromosome segments around introgressed loci and reduction of linkage drag in marker-assisted backcross programs. Genetics 158 (3), 1363–1379.

Ishii, T., Yonezawa, K., 2007. Optimization of the marker-based procedures for pyramiding genes from multiple donor lines: I. Schedule of crossing between the donor lines. Crop. Sci. 47 (2), 537–546.

Ishii, T., Hayashi, T., Yonezawa, K., 2008. Optimization of the marker-based procedures for pyramiding genes from multiple donor lines. III. Multiple-gene assemblage using background marker selection. Crop. Sci. 48 (6), 2123–2131.

Isleib, T.G., 1997. Cost-effective transfer of recessive traits via the backcross procedure. Crop. Sci. 37 (1), 139–144.

Jackson, S.A., Zhang, P., Chen, W.P., Phillips, R.L., Friebe, B., Muthukrishnan, S., Gill, B.S., 2001. High-resolution structural analysis of biolistic transgene integration into the genome of wheat. Theor. Appl. Genet. 103 (1), 56–62.

Khan, N., Zhou, S., Ramanna, M.S., Arens, P., Herrera, J., Visser, R.G.F., van Tuyl, J.M., 2009. Potential for analytic breeding in allopolyploids: an illustration from Longiflorum x Asiatic hybrid lilies (*Lilium*). Euphytica 166 (3), 399–409.

Kim, S.-I., Veena, Gelvin, S.B., 2007. Genome-wide analysis of *Agrobacterium* T-DNA integration sites in the *Arabidopsis* genome generated under non-selective conditions. Plant J. 51 (5), 779–791.

Krahl, K.H., Randle, W.M., 1999. Genetics of floral longevity in petunia. HortScience 34 (2), 339–340.

Lawson, D.M., Lunde, C.F., Mutschler, M.A., 1997. Marker-assisted transfer of acylsugar-mediated pest resistance from the wild tomato, *Lycopersicon pennellii*, to the cultivated tomato, *Lycopersicon esculentum*. Mol. Breed. 3 (4), 307–317.

Lewis, R.S., Kernodle, S.P., 2009. A method for accelerated trait conversion in plant breeding. Theor. Appl. Genet. 118 (8), 1499–1508.

Lorieux, M., Goffinet, B., Perrier, X., de Leon, D.G., Lanaud, C., 1995. Maximum-likelihood models for mapping genetic markers showing segregation distortion. 1. Backcross populations. Theor. Appl. Genet. 90 (1), 73–80.

Mennella, G., Rotino, G.L., Fibiani, M., D'Alessandro, A., Francese, G., Toppino, L., Cavallanti, F., Acciarri, N., Lo Scalzo, R., 2010. Characterization of health-related compounds in eggplant (*Solanum melongena* L.) lines derived from introgression of allied species. J. Agric. Food Chem. 58 (13), 7597–7603.

Mithila, J., Hall, J.C., 2013. Transfer of auxinic herbicide resistance from *Brassica kaber* to *Brassica juncea* and *Brassica rapa* through embryo rescue. In Vitro Cell. Dev. Biol. 49 (4), 461–467.

Moore, J.N., Janick, J., 1983. Methods in Fruit Breeding. Purdue University Press, West Lafayette, IN. 464 pp.

Mumm, R.H., 2007. Backcross versus forward breeding in the development of transgenic maize hybrids: theory and practice. Crop. Sci. 47 (3), S164–S171.

Mutlu, N., Miklas, P., Reiser, J., Coyne, D., 2005. Backcross breeding for improved resistance to common bacterial blight in pinto bean (*Phaseolus vulgaris* L.). Plant Breed. 124 (3), 282–287.

Nanda Kumar, P.B.A., Shivanna, K.R., 1993. Intergeneric hybridization between *Diplotaxis siettiana* and crop brassicas for the production of alloplasmic lines. Theor. Appl. Genet. 85 (6/7), 770–776.

Nogoy, F.M., Nou, I., Song, J.Y., Kang, K.K., Niño, M.C., Yong-Gu, C., Jung, Y.J., 2018. Plant microRNAs in molecular breeding. Plant Biotechnol. Rep. 12 (1), 15–25.

Ochanda, N., Yu, J., Bramel, P.J., Menkir, A., Tuinstra, M.R., Witt, M.D., 2009. Selection before backcross during exotic germplasm introgression. Field Crop Res 112 (1), 37–42.

Partier, A., Gay, G., Tassy, C., Beckert, M., Feuillet, C., Barret, P., 2017. Molecular and FISH analyses of a 53-kbp intact DNA fragment inserted by biolistics in wheat (*Triticum aestivum* L.) genome. Plant Cell Rep. 36 (10), 1547–1559.

Pasutti, D.W., Weigle, J.L., Beck, A.R., 1977. Cytology and breeding behavior of some Impatiens hybrids and the backcross progeny. Can. J. Bot. 55 (3), 296–300.

Pattee, H.E., Isleib, T.G., Gorbet, D.W., Moore, K.M., Lopez, Y., Baring, M.R., Simpson, C.E., 2002. Effect of the high-oleic trait on roasted peanut flavor in backcross-derived breeding lines. J. Agric. Food Chem. 50 (25), 7362–7365.

Prohens, J., Plazas, M., Raigon, M.D., Segui-Simarro, J.M., Stommel, J.R., Vilanova, S., 2012. Characterization of interspecific hybrids and backcross generations from crosses between two cultivated eggplants (Solanum melongena and S. aethiopicum Kumba group) and implications for eggplant breeding. Euphytica 186 (2), 517–538.

Richey, F.D., 1927. The convergent improvement of selfed lines of corn. Am. Nat. 61, 430–449.

Rick, C.M., 1972. Further studies on segregation and recombination in backcross derivatives of a tomato species hybrid. Biol. Zentralbl. 91 (2), 209–220.

Robbins, M.D., Casler, M.D., Staub, J.E., 2008. Pyramiding QTL for multiple lateral branching in cucumber using inbred backcross lines. Mol. Breed. 22 (1), 131–139.

Scotti, N., Monti, L., Cardi, T., 2003. Organelle DNA variation in parental *Solanum* spp. genotypes and nuclear-cytoplasmic interactions in *Solanum tuberosum* (+) *S. commersonii* somatic hybrid-backcross progeny. Theor. Appl. Genet. 108 (1), 87–94.

Singh, S.P., Schwartz, H.F., 2010. Breeding common bean for resistance to diseases: a review. Crop. Sci. 50 (6), 2199–2223.

Singh, A.K., Singh, V.K., Singh, A., Ellur, R.K., Pandian, R.T.P., Krishnan, S.G., Singh, U.D., Nagarajan, M., Vinod, K.K., Prabhu, K.V., 2015. Introgression of multiple disease resistance into a maintainer of Basmati rice CMS line by marker assisted backcross breeding. Euphytica 203 (1), 97–107.

Steele, K.A., Price, A.H., Witcombe, J.R., Shrestha, R., Singh, B.N., Gibbons, J.M., Virk, D.S., 2013. QTLs associated with root traits increase yield in upland rice when transferred through marker-assisted selection. Theor. Appl. Genet. 126 (1), 101–108.

Suneson, C.A., 1947. An evaluation of nine backcross-derived wheats. Hilgardia 17, 501–510.

Tanksley, S.D., Nelson, J.C., 1996. Advanced backcross QTL analysis: a method for the simultaneous discovery and transfer of valuable QTLs from unadapted germplasm into elite breeding lines. Theor. Appl. Genet. 92 (2), 191–203.

Tanksley, S.D., Medina-Filho, H., Rick, C.M., 1981. The effect of isozyme selection of metric characters in an interspecific backcross of tomato—basis of an early screening procedure. Theor. Appl. Genet. 60 (5), 291–296.

Thabuis, A., Palloix, A., Servin, B., Daubeze, A.M., Signoret, P., Hospital, F., Lefebvre, B., 2004. Marker-assisted introgression of 4 *Phytophthora capsici* resistance QTL alleles into a bell pepper line: validation of additive and epistatic effects. Mol. Breed. 14 (1), 9–20.

Thomas, M., 1952. Backcrossing: The Theory and Practice of the Backcross Method in the Breeding of Some Non-Cereal Crops. Commonwealth Agr. Bur., Cambridge. No. 16. 136 pp.

Tripathy, S.K., Devraj, L., Dinesh, M.I., Maharana, M., 2017. Development of opaque-2 introgression line in maize using marker assisted backcross breeding. Plant Gene 10, 26–30.

Warschefsky, E., Varma, P.R., Cook, D.R., von Wettberg, E.J.B., 2014. Back to the wilds: tapping evolutionary adaptations for resilient crops through systematic hybridization with crop wild relatives. Am. J. Bot. 101 (10), 1791–1800.

Young, N.D., Tanksley, S.D., 1989. RFLP analysis of the size of chromosomal segments retained around the *Tm-2* locus of tomato during backcross breeding. Theor. Appl. Genet. 77 (3), 353–359.

Zhang, J., Guo, D., Chang, Y., You, C., Li, X., Dai, X., Weng, Q., Zhang, J., Chen, G., Li, X., Liu, H., Han, B., Zhang, Q., Wu, C., 2007. Non-random distribution of T-DNA insertions at various levels of the genome hierarchy as revealed by analyzing 13 804 T-DNA flanking sequences from an enhancer-trap mutant library. Plant J. 49 (5), 947–959.

CHAPTER

19

Breeding for Disease and Insect Resistance

INTRODUCTION

Where life exists in the universe, it is likely that there will be diversity in proportion to the habitats and ecological niches that are available. No static life form is sustainable in the face of constant change. Opportunities for the capture of radiant or chemical energy appear and disappear during the course of evolution. One of these opportunities is to become a plant parasite or herbivore, that sustain themselves by consuming the organisms that have captured the sun's energy and converted it to carbohydrate chemical energy.

Energy is a fundamental need of all organisms. Life on Earth is driven by energy from the sun that is converted into chemical energy (carbohydrates) by photosynthesizing plants and microbes. This energy flows through ecosystems in many forms, as chemical, thermal, and light, as it sustains the balance of life. All animals, arthropods, fungi, most bacteria, viruses, and even a few plants obtain chemical energy by taking it from other organisms: as herbivores, carnivores, omnivores, and saprophytes. Embedded within this continuum are the pathogens and parasites. Fungi, bacteria, viruses, and viroids are regarded as pathogens if they attack living plants. Insects and other herbivorous arthropods are known as "pests" if they feed on cultivated crop species.

The human experience is instructive. As a species, we are neither herbivores nor carnivores, but omnivores. The structure of the human alimentary tract attests to this dietary mode. Where do we fit in the "food chain"? In the repast, less than 1 million years henceforth, humans were situated near the "top" ecological niches, occupied by fewer and larger organisms. Our natural defenses against carnivores are meager: we have no armor, foul excretions, sharp teeth and claws, exterior camoflage, climbing ability, or extreme speed with which to protect ourselves. Our survival is dependent on cunning behavior, manifested in shelter and food acquisition strategies. As our reasoning ability evolved to be more powerful, our ability to elude predators improved progressively. Even in our current stage of bio-societal development, we still fear the predator, though instances of predation upon humans are remarkably rare.

Likewise, the collective intellect of humans has led to effective treatments to avoid or ameliorate many pathogenic diseases. Immune systems prevent pathogenic diseases in humans and other vertebrates. Plants and many invertebrates possess different systems to avert being consumed by pathogens and parasites, including an impenetrable barrier to potentially harmful microbes and active physiological systems that prevent or slow infesby biotic invaders. Active and passive mechanisms to avert disease and herbivory vanish after death, and saprophytic organisms rapidly invade tissues and cells. Forensic entomology has been adopted as a powerful tool to establish the time and circumstance of human death in the justice system. Based on the quantities and species of arthropods present within a corpse, the time of death can be accurately predicted (Benecke, 2001).

Terrestrial plants are presented with greater challenges in averting pathogens and parasites than are motile animals. Plants are firmly rooted in one location per generation and must overcome a myriad of incidental biotic and abiotic stress factors. Many physical and chemical measures are deployed by plants to make themselves less attractive to potential pathogens or herbivores e.g., thorns, leaf trichome secretions, impervious coverings, synthesis of offensive compounds that are toxic or non-palatable, etc. Plants also have an additional arsenal of weaponry to ward off pathogens that will be covered later but do not possess a multi-tiered antigen-antibody-based immune system similar to that of vertebrates. It is incorrect, therefore, to state that any plant possesses "immunity" to a pathogen or pest, although such characterizations appear frequently in scientific literature.

Pathogens and parasites are organisms that derive chemical energy from other organisms during a prolonged state of function. The "perfect" parasite is like the taxman, always taking a portion of what the host is making but never putting it out of business. A dead host cannot pay taxes. The more productive the host is, in fact, the more chemical energy that is available for the parasite. Evidence strongly suggests that hosts and parasites have coevolved for long periods of geological history to achieve the delicate balance between the two organisms (Shepherd and Mayo, 1972). Where this balance is effective, manifestations of plant damage or dysfunction due to herbivory and disease are low. Changes in the climatic and environmental conditions under which crops are grown have also resulted in the appearance of new diseases (Boyd et al., 2013).

When organisms are removed from their native ecosystems, taken thousands of miles away to a new geo-ecological location, and grown in monocultures, herbivory and disease tend to become relatively more severe (Day, 1974). Modern techniques of agriculture have greatly accentuated the potential impacts of disease and herbivory by the implementation of monocultures (see Chapter 2). By setting the stage with a population of identical hosts growing on a homogeneous environment, all that is needed is the corresponding pathogen to complete the disease equation (see Disease Triangle below). The monoculture also enhances the disease dynamic by providing a means for inoculum amplification.

Modern agriculture, with its vast monocultures of lush fertilized crops, provides an ideal environment for pests, weeds, and diseases. Even with existing crop protection measures, yield losses of approximately 33% due to disease and herbivory occur globally. Traditional crop breeding programs are limited by the time taken to move resistance traits into elite crop genetic backgrounds and the limited gene pools in which to search for novel resistance. Resistance based on single genes does not protect against the full spectrum of pests, weeds, and diseases, and is more likely to break down as pests evolve counter-resistance. Resistance genes can be stacked, however, to make it harder for pests to evolve counter-resistance and to provide multiple resistances to different attackers (Bruce, 2012).

One approach to the reduction of undesirable impacts due to insect herbivory and pathogenesis is the breeding of hosts that are genetically conditioned to avoid the injury and loss of economic yields. Examples of physical and chemical factors that have been exploited as bases of resistance are provided later in this chapter. The desired phenotype, lack of damage in the presence of pathogens or herbivores, is the result of a complicated cascade of interactions between two or more discrete organisms and the environment. In many cases, herbivorous arthropods are vectors for diseases as well. Organisms are dynamic, prone to change to adapt to changing environments and ecosystems. Invertebrate herbivores and fungal, bacterial, and viral pathogens exhibit reproductive and population attributes that enable them to mutate and adapt to changes rapidly. Wholesale extermination or exclusion of arthropod herbivores or pathogens foments the selection and rapid expansion of populations of individuals able to overcome the agents of extermination or exclusion. Consequently "resistance management" has become an important strategy for both horticulturists and plant breeders.

Breeding for disease resistance is, inevitably, an important component of most programs, and such efforts date to the dawn of the modern plant breeding era (Biffen, 1905, 1912; Orton, 1909). There is no predominance of disease losses in inbreeding vs. outcrossing species, so the heritable bases for avoidance of disease and herbivory must be incorporated into the all of breeding strategies described in Chapters 13–18. Ancestors of crop species and their wild relatives were exposed to a broad range of pathogens, parasites, and herbivores during the course of evolutionary history, and genes for resistance are often present within extant gene pools. This chapter addresses the specific challenges and strategies in the development of new cultivars that feature disease and pest resistance.

The bodies of literature in the fields of plant pathology and economic entomology are large, and the reader is advised to consult further with more detailed accounts (Agrios, 2005; Pedigo and Rice, 2009; Sambamurty, 2009; Schmann and D'Arcy, 2009; David, 2015). Likewise, applications of plant breeding methods for the development of plant cultivars that exhibit resistance or tolerance are substantially documented (Nelson, 1973; Russell, 1978; Maxwell and Jennings, 1980; Dhan, 1986; Niks et al., 2011). This textbook will provide a cursory review of these disciplines for purposes of illuminating classical and contemporary plant breeding concepts.

THE DISEASE CONCEPT

A pathogen is a type of parasite that is associated with a specific phenotype, known commonly as a disease. The specific disease will always depend on the interactions of three factors: *host*, *pathogen*, and *environment*. Disease is a phenotype, as contrasted with "healthy," associated with an individual (host). The disease phenotype is a manifestation of a biological battle between a host and pathogen for the host's energy reserves. The host generally puts up a fight that may halt, retard, or contain the proportion of host tissues consumed by the pathogen. The disease phenotype is dependent on specific interactions with a pathogen and environment that favors the interaction of host and pathogen such that disease results, known as the "disease triangle" (Stevens, 1960; Fig. 19.1):

Each of the three factors must be present within a certain range for the disease phenotype to be manifested. If a compatible host and pathogen are present, but the environmental requirements for disease are absent, the disease phenotype will not occur. Likewise, if the host and pathogen are not compatible but the environment is conducive, disease will not proceed. The disease triangle is, of course, a simplistic representation of very complex biological and abiological interactions. In some cases, disease results from more than one pathogen. Organisms that are not even considered to be pathogenic may invade living plant tissues under certain specific environmental conditions.

Concept of the Disease Triangle

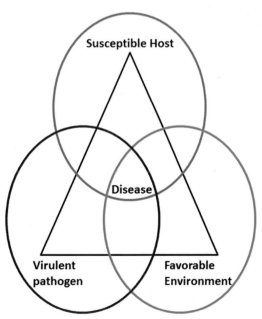

FIG. 19.1 A depiction of the disease triangle concept first described in this format by Stevens (1960). Disease only occurs with the combination of virulent pathogen and susceptible host genotypes and conducive environmental parameters. Outside of this area of intersection, no disease is observed.

FIG. 19.2 Plant disease symptoms on horticultural crop species, illustrating the broad range of different phenotypes. *(A) From https://en.wikipedia.org/wiki/Black_rot_(grape_disease)#/media/File:Guignardia_bidwellii_04.jpg. (B) From https://commons.wikimedia.org/wiki/File:UncinulaTulasneiLeaf.jpg. (C) From https://commons.wikimedia.org/wiki/File:Turnip_yellow_mosaic_virus_2.jpg. (D) From https://www.staff.ncl.ac.uk/ethan.hack/BIO2003-disease-pictures.html. (E) From https://en.wikipedia.org/wiki/Alternaria_solani#/media/File:Alternaria_solani_-_leaf_lesions.jpg. (F) From https://www.flickr.com/photos/zaqography/3649144255/. (G) From https://en.wikipedia.org/wiki/Apple_scab#/media/File:Holding_Apple_with_scab.jpg. (H) From https://www.flickr.com/photos/fsegarra/3146366612/. Used with permission from Frank R. Segarra. (I) From https://www.flickr.com/photos/scotnelson/32925478015.*

While plant diseases may be classified according to the affected organs or tissues or the features of the lesions, this is a truly diverse and obtuse range phenotypes to contemplate (Fig. 19.2). Disease is not a common or expected outcome of the interaction of a host with a fungal, bacterial, viral, or sub-viral organism. Host plants and animals coexist with thousands of different species of microbes that may be present in very high populations, but the overwhelming majority of these microbes never incite any recognized disease symptoms. Certain microbes establish relationships with plants and animals that are mutually beneficial such as exophytes, endophytes, mycorrhizae, and symbionts. A subclass of such beneficial associations is those that diminish the incidence and severity of diseases, a scenario that is potentially problematic for farmers using pesticides and fungicides that destroy beneficial organisms, allowing pathogens to thrive and diseases to proceed.

The farmer has little control over the weather, but may be able to alter the environment in subtle ways to discourage diseases. Examples include enhanced soil drainage, windbreaks, impervious coverings, and the application of biological or abiotic factors to discourage disease or arthropod herbivory. He/she likewise cannot usually control the presence and population sizes of pathogens, although some influence may be exercised by the use of pesticides and/or fungicides. Multi-cropping and inter-cropping can reduce pathogen populations by rendering the crop ecosystem to be less prone to compatible host-parasite combinations. The farmer can also choose to plant cultivars with known resistance to pathogens.

Disease may also be considered an altered course of host development due to the interaction with a specific biotic or abiotic factor that is both characteristic and predictable. Plant pathology literature is replete with examples of disease descriptions for domesticated crop species. While diseases are also prevalent in wild plant populations, little attention is paid to them. The source of pathogen inoculum is, however, often from or involving wild plant populations closely juxtaposed with crops.

Among the millions of organisms that interact with the potential host, what characteristics distinguish pathogens from non-pathogens? Plant pathogens possess structural and physiological modifications that enable them to gain access to the vascular elements and inter- and intra-cellular fluids that contain the host's energy and nutrient reserves. Fungal pathogens often exhibit specialized structures, such as *haustoria* or *infection pegs* that enter plant tissues through leaf stomates or surface wounds (Fig. 19.3). These specialized invading structures are modified in specific ways to overcome the defenses of the targeted host, or to evade physical barriers.

Once inside the plant host, pathogens can move or grow and, ultimately, locate and consume chemical energy reserves. As the pathogen navigates and feeds within the host, it follows a characteristic path that is manifested by the host as a set of progressive symptoms. A given disease is associated with the predictable progression (or *etiology*) of these symptoms. Pathogens often undergo reproduction within or upon the host and asexual or sexual fruiting bodies, and spores are visible in or on disease lesions. Characteristic spores or fruiting bodies are useful, therefore, to diagnose the underlying causes of diseases.

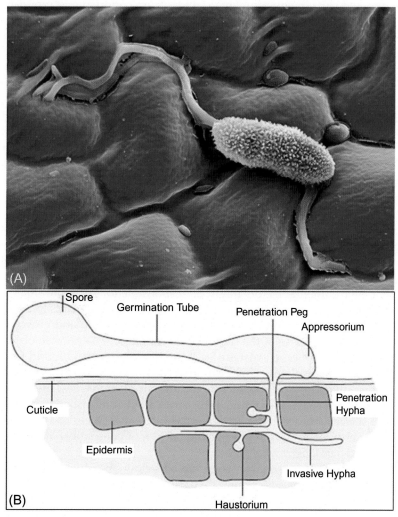

FIG. 19.3 (A) Hyphae germinating from a fungal spore on a leaf surface; and (B) penetration of the leaf surface and internal cells by the infection peg, appressorium and haustoria. *(B) From Meng, S., et al., 2009. Common processes in pathogenesis by fungal and oomycete plant pathogens, described with Gene Ontology terms. BMC Microbiol. 9, Article number: S7. Originally adapted from Schumann, G.L., 1991. Plant Diseases: Their Biology and Social Impact. American Phytopathological Society, St. Paul, MN.*

Plant pathologists have developed a unique nomenclature to describe plant disease phenotypes. Rots and blights can occur on either aerial or submerged plant organs. Necrotic spots and changes in pigmentation, such as yellowing, are primarily foliar. Mildews have a sheet-like appearance, and rusts are foliar or surface lesions that are overtly orange-colored. Wilts are usually pervasive, involving loss of turgor in sectors of or entire plants. The common name of a disease is based on visual symptoms that are diagnostic. There is a tendency for similar pathogens to incite corresponding symptoms on hosts, and to attack plants using similar strategies (Elliott, 1958). Thus, many of the pathogens responsible for powdery mildew diseases on a wide range of hosts are related, and the same generalization applies for other diseases (Agrios, 2005).

The dynamics of infectivity, or passage among hosts, are often diagnostic for a given disease. Each pathogen proceeds through an infective phase while completing the life cycle. In some cases, the pathogen sporophyte and gametophyte may be entirely distinct, one being an obligate pathogen the other a saprophyte. In others, the gametophyte nuclei may coexist in a fungal mycelium as heterokaryons for extended periods, not fusing until the completion of a life cycle. In still others, the gametophyte phase may be absent or greatly reduced. Curiously, many of the most prominent pathogens fall into the latter category, although one would presume that sexual reproduction is more conducive to the generation of virulence via mechanisms of recombination and segregation during successive generations (Agrios, 2005).

Diseases ultimately inflict damage to individual hosts within the crop population resulting in a negative impact on harvest yield or quality. In severe instances, the crop monoculture is destroyed by the disease, resulting in absolute loss. In other cases, however, a host population may exhibit severe symptoms, and later resume normal growth, presumed to be mostly due to a shift in environmental factors from conducive to non-conducive to the disease. In such cases, damage may appear severe, but economic losses may be minimal. Horticultural crops are particularly affected by diseases since the value of derived products is usually defined by a narrow set of qualitative features.

A cursory examination of the host population that has endured a disease epidemic, or unchecked progression of a pathogen, often reveals a range of effects. The range of disease severity, like everything in the realm of host-pathogen interactions, depends on the situation. Where apparently normal, healthy plants are observed within a background of mostly dead or dying compatriots the question arises: why did a few plants thrive in the presence of the pathogen and environmental conditions conducive to disease while the majority of others perished under the same circumstances?

The plant breeder hypothesizes that the healthy plants contain genes for resistance to the pathogen. The plant breeder hybridizes a dying, diseased plant with a vigorous, healthy one followed by the screening of segregating (F_2 or BC) populations that will lead to an understanding of the heritable basis of hypothetical disease resistance. Segregation patterns for resistance and susceptibility vary with different host-pathogen systems. Segregating populations very often exhibit clear, discernible disease phenotypic classes, 3:1 healthy:diseased within the F_2, and 1:1 within the $F_1 \times$ healthy test cross (or backcross, BC). Infrequently the inheritance of disease resistance is the converse of the above, with the disease susceptible phenotype appearing to be dominant. A continuum of disease symptom severities is sometimes observed in F_2 and BC populations, ranging from diseased to healthy. In many instances, the segregating populations exhibit no discernible patterns at all, suggesting that the original resistant and susceptible phenotypes had no genetic basis. What environmental factors might be responsible for such a dramatic difference in the disease phenotype within a small physical distance between resistant and susceptible plants in a field?

Plant pathologists are experts at isolating pathogens from the environment and growing and studying them under controlled conditions. Different isolates of the same species of pathogen are often observed to be identical in all respects except for host pathogenicity. Some isolates of the pathogen are usually found to be pathogenic on specific host species, or cultivars, while other isolates exhibit different disease reactions on a set of genetically-diverse hosts. If pathogen isolates are differentiated by pathogenicity on different plant species, they are usually regarded as distinct taxonomic groups called *forme specialis*. If pathogen isolates are distinguished based on pathogenicity to different cultivars within a plant species, however, they are referred to as *races*. Races of a pathogen occur commonly, and are distinguished based on a characteristic and predictable pattern of compatible and incompatible interactions on different host genotypes, called a *host differential* (Table 19.1).

Race-specific host resistance will be covered in more detail below. As was stated earlier, this *vertical resistance* is often characterized by host disease manifestations that are "all or none" depending on the compatibility of the host and pathogen (host resistance genotype; pathogen race) as compared to *horizontal resistance* that is manifested in varying levels of disease manifestations on different host genotypes or cultivars. For horizontal resistance, the extremes of disease severity are observed infrequently as compared to vertical resistance (Fig. 19.4). It is also well established that these general patterns of resistance fall into two categories with regard to inheritance. Vertical resistance is usually controlled by dominant/recessive alleles at single or a very few loci with large individual effects. In

TABLE 19.1   Hypothetical Example of a Host Differential Used to Identify Races of a Pathogen

Host population	Race A	Race B	Race C	Race D	Race E
McKinley	D	ND	ND	ND	ND
Whitney	ND	D	ND	ND	ND
Washington	ND	ND	D	ND	ND
Ranier	ND	ND	ND	D	ND
Teton	ND	ND	ND	ND	D

D, disease; *ND*, no disease.

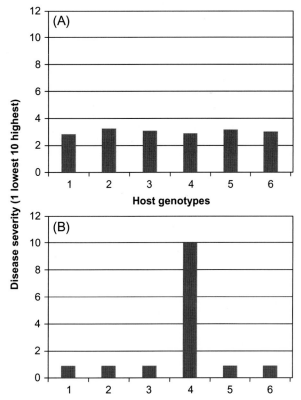

FIG. 19.4   Patterns of pathogenicity of a single pathogen isolate on different host genotypes. (A) Horizontal resistance with little host-pathogen specificity; (B) vertical resistance with a high level of host differentiation of pathogen genotypes.

contrast, horizontal resistance is quantitatively inherited, controlled by the products of many mostly genes that have small individual additive effects.

What, exactly, are the underlying factors responsible for compatible and incompatible host-pathogen interactions? In cases where the difference is found to be simply inherited, or consisting of discrete explainable classes, the fundamental feature is recognition. The pathogen is enveloped in a cell membrane and wall, and specific substances, usually proteins or glycoproteins, are embedded therein. The host possesses passive exclusion barriers, including lignin and cork, and thick waxes that coat the cuticular layers that is adequate to exclude most common saprophytic microbes. Plants also express internal passive barriers to the spread of foreign invaders, such as xylem gums and tyloses (Nelson, 1973).

Pathogens have developed the capability to overcome such passive barriers, but must still prevail over more specific host defense systems. In incompatible (no disease) interactions, the host has developed the ability to recognize one or more substances and linked this recognition event to the synthesis and localized deposition of antagonistic substances or, in the extreme cases, localized cell death (*hypersensitivity*, or *apoptosis*) to prevent the potential pathogen from gaining entry to the living host. Examples of plant antimicrobial responses include pathogenesis-related (PR) proteins such as chitinases and peroxidases, antimicrobial substances known as phytoalexins, and multifactorial cascade systems such as systemic acquired resistance (SAR), or induced systemic resistance (ISR) involving salicylic acid or jasmonic acid (Schmann and D'Arcy, 2009).

A compatible interaction occurs, manifested later by disease, when the host fails to recognize any substances on the pathogen surface. Races are usually distinguishable based on the presence or absence of compounds in the pathogen cell walls that are recognized by the host. The pathogen-recognition compounds are usually glycoproteins, a class of compounds that also elicit allergenic reactions in vertebrates with immune systems.

The appearance of a new pathogen race is thought to represent a mutation that either alters or removes a specific glycoprotein from the pathogen cell wall. Consequently, a set of hosts that could identify the pathogen race loses this ability, and disease is observed on plant genotypes or cultivars that were previously established to be disease-free. The host will remain susceptible to this new race until a corresponding genetic change occurs (e.g., recognition of another glycoprotein in the pathogen cell wall) that enables the host to identify the pathogen race, and to link recognition with destruction or exclusion [e.g., toxin production or localized cell death (apoptosis)].

Returning to the scenario where the diseased×healthy cross gave rise to segregating populations wherein a continuum of disease symptom severities was observed, is there a similar genetic and physiological scenario at play here? If so, how is the existence of a continuum of host disease severity explained? For example, are there manifestations of passive barriers to pathogen entry or spread that may be variable among individuals within a population, leading to a continuous range of disease response? The recognition of pathogen races can play a role in the overall range of non-discrete disease responses but is not the only factor responsible for the differences in disease severity. It is presumed that more than one gene is involved and that two or more physical or physiological mechanisms are interacting, and also with environmental fluxes, to create the continuum. Where disease phenotypic variance has been fully partitioned for populations exhibiting non-discrete disease severities, V_D and V_I have been found to be relatively high. Hence, while broad sense h^2 for the disease resistance phenotype may be greater than 0.75, narrow sense h^2 is often quite low.

While the subsequent discussion deals with the two extremes of heritability patterns, simple/discrete vs. complex/non-discrete, there are intermediate patterns of segregation that may aptly be described as possessing elements of either or both. The plant breeder must always investigate an unknown host-parasite system anew, without the presumption of the general application of lessons learned from previous studies or so-called model systems.

THE GENE-FOR-GENE THEORY

The discrete patterns of disease, denoted by host genotypes, and corresponding discrete patterns of host-pathogen compatibility conforming to prescribed genotypes of both organisms ultimately led Harold H. Flor (see Chapter 1 for a biography) to develop the gene-for-gene theory of pathogenicity and virulence in the late 1940s (Flor, 1956). He postulated the existence of resistance (R/r) genes in the host and corresponding avirulence (A/a) genes in the pathogen, working on flax (*Linum usitatissimum*) rust (pathogen *Melampsora lini*). The R genes conditioned apoptosis, or hypersensitivity, while the A genes in the pathogen conditioned evasion of the host's recognition capabilities. Host resistance R was hypothesized to be dominant over susceptibility r, and pathogen avirulence A was dominant over virulence a.

Later, de Wit (1981) extended Flor's hypotheses to the study of tomato (*S. lycopersicum*) leaf mold pathogen (*Cladosporium fulvum*). There are many races of the leaf mold pathogen, differentiated by a panel of tomato host genotypes (Joosten and de Wit, 1999). In race 9, a glycoprotein was isolated (*Avr9*) that was present in diseased tissues. The isolated glycoprotein alone could cause a host resistant to race 9 to exhibit the same HR response as the pathogen. Resistance to race 9 was found to be inherited as a single dominant allele at the *Cf9* locus, an R gene. It was hypothesized that the Cf9 gene product facilitated the recognition of the Avr9 glycoprotein, and linked recognition to the HR response and host resistance.

Avr4, a similar glycoprotein to Avr9 causes avirulence in tomato with the *cf4* gene. When a specific gene for R is bred into a host, the gene enables the host to recognize the gene product for virulence in the pathogen (Stergiopoulos et al., 2010). That product is considered the "avirulence" gene (*avrA*) of the pathogen that corresponds to the plant resistance gene (R). More than one resistance gene may exist in a host, often called "pyramiding" or stacking more than one resistance gene in a particular plant. For example, a single tomato plant may carry both *cf4* and *cf9* resistance genes.

Flor (1956) and Van der Plank (1963) amalgamated these observations into one unified theory, widely known as "Gene-For-Gene" (GFG). The elements of the GFG Theory are as follows:

- For each dominant gene for resistance in the host there is a corresponding dominant gene for avirulence in the pathogen, or

- For each *gene that confers virulence in the pathogen*, there is a corresponding *gene that confers resistance in the host*.
- Host resistance genes for resistance are generally dominant (R) and genes for susceptibility recessive (r).
- Pathogen avirulence genes (*Avr*; inability to incite disease) are generally dominant (Avr) and genes for virulence are recessive (avr) (note that the pathogen is often virulent during the haploid phase of the life cycle, so dominant allelic interactions are irrelevant).
- Disease is a consequence of the specific interaction between an *Avr* effector gene product of the pathogen with a *R* gene product of the host that detects the effector and elicits a resistance response, such as hypersensitivity (Chisholm et al., 2006).
- Only when the pathogen goes undetected will the disease phenotype proceed.
- If the host recognizes pathogen effectors, defense responses are activated, and disease is prevented.
- Different race-specific pathogen *Avr*-host *R* may act independently; many may be present in the same individual (e.g., resistance *stacking*).

The interaction of host and pathogen may be regarded as the analogy of a cat and mouse where the mouse is trying to eat the cat instead of the other way around. The initial genetic change from incompatible to compatible disease response resulted in the change of surface compounds such that the host (cat) could no longer detect (see) the pathogen (mouse). The host then undergoes a genetic change in the recognition apparatus that enables it to detect the pathogen. Recognition may be for the same altered surface compound, that it now recognizes, or a different compound entirely. The mechanism of recognition is envisioned as a "lock-key" form of interaction, similar in concept to the correspondence of an epitope and the antibody that recognizes it in vertebrate organisms (Niks et al., 2011).

The range of disease responses with changing host and pathogen genotypes is as follows (Table 19.2).

Resistance, or host-pathogen incompatibility (ND), occurs when the host has a specific gene for resistance (*R*) the gene product of which recognizes the gene product of the corresponding pathogen gene for avirulence (*Avr*). Disease, or host-pathogen compatibility (D), can occur in a number of ways: (i) the pathogen *Avr* gene codes for an effector, but the host (rr) lacks the R gene to detect it; (ii) the host possesses an R allele, but the pathogen lacks the corresponding gene that encodes the elicitor or inducer (a); or (iii) the pathogen lacks the gene for the elicitor or inducer (a) and the host lacks the ability (r) to detect the same elicitor or inducer (Fig. 19.5).

If there are multiple races of the pathogen present, it is possible that single host individuals may be resistant to all of them because pathogen avirulence genes and host resistance genes are not usually allelic, but entirely separate loci. The joint disease expectation of two independent host resistance-pathogen avirulence pairs is as follows (Table 19.3).

Host-pathogen resistance loci, where mapped in the genome, have been found to exist at multiple sites on linkage groups. In some cases, however, they are closely linked and may share DNA sequence homology, suggesting that they may share a common ancestral gene. If a new race of a pathogen appears, and a host population is identified that carries a corresponding resistance gene, the backcross method (Chapter 18) will usually be employed to move the gene into the desired host population. In cases where more than one resistance gene is introduced into a targeted host, sequential or multilateral backcross programs may be undertaken, referred to as resistance gene *pyramiding or stacking*.

The GFG Theory has had a profound and lasting effect on how scientists, and especially plant breeders, have devised strategic approaches to the incorporation of disease resistance into commercial varieties. The discrete race-specific nature of resistance that is conditioned by single host genes has come to be known as vertical disease resistance, while the non-race specific, continuous (non-discrete) and multi-genic forms are referred to as horizontal disease resistance.

TABLE 19.2 Summary of predicted disease responses in the presence of host *R* and pathogen *Avr* (A/a) genotypes (i.e. the gene-for-gene theory of pathogenesis).

Pathogen	Host resistance	
Avirulence	Phenotype[a]	
Genotype	R_1–	r_1r_1
A_1	ND	D
a_1	D	D

[a] R_1– = R_1R_1 or R_1r_1.
ND, no disease; D, disease.

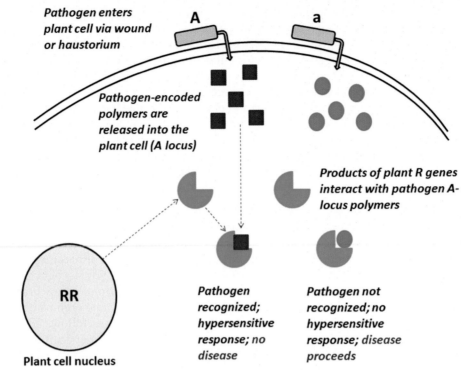

FIG. 19.5   This illustration embodies a conceptual description of the steps involved with race-specific resistance based on localized cellular apoptosis, or hypersensitivity; an example of a gene-for-gene host-pathogen interaction.

TABLE 19.3   Joint inheritance of disease phenotype of two corresponding "gene-for-gene" host R (resistance) and pathogen Avr (A/a) loci.

Pathogen Avirulence Genotype	Host resistance phenotype[a]			
	R_1–R_2–	R_1–r_2r_2	$r_1r_1R_2$–	$r_1r_1r_2r_2$
A_1A_2	ND	ND	ND	D
A_1a_2	ND	ND	D	D
a_1A_2	ND	D	ND	D
a_1a_2	D	D	D	D

[a] R_1– = R_1R_1 or R_1r_1; R_2– = R_2R_2 or R_2r_2.
ND, no disease; D, disease.

HORIZONTAL AND VERTICAL RESISTANCE

Van der Plank (1963) first advanced the current framework under which vertical and horizontal resistance were defined and contrasted. The concept of horizontal resistance invariably raises the consciousness of the distinction between "resistance" and "tolerance" (Parlevliet, 1977; Parlevliet and Zadoks, 1977). If a very slight disease response is observed within a population that includes both completely diseased and completely healthy individuals, is it referred to as levels of resistance or tolerance? In contrast, if a plant exhibits a disease response, but very little pathogen is discovered, what can be said of the host resistance to or tolerance of the pathogen?

Vertical resistance (VR) provides a population genetic framework that inevitably leads to the continuous appearance of new pathogen virulences and pathogenicities. Three problem areas exist concerning diseases caused by microbial pathogens: (i) the emergence of new diseases, (ii) the loss, or breakdown, of disease resistance bred into plants, and (iii) the development of pathogen resistance to chemical control substances. Evidence points to both host plant resistance and chemical susceptibility being overcome through point mutations in the pathogen. Because the population sizes of diseases such as virus, bacteria, rusts and powdery and downy mildews are so large and generation times are short, a broad spectrum of point mutations is constantly appearing. Experiences in managing plant disease control reveal the remarkable speed and the practical impact of adaptation in wild microorganism populations to changes in their environment, and the difficulty of modulating adaptation (Hollomon and Brent, 2009).

Horizontal resistance can be explained by a polygenic system where the individual genes are vertical and operating on a gene-for-gene basis with virulence genes in the pathogen (Parlevliet and Zadoks, 1977). The frequencies of the resistance and virulence genes are such that the effective frequencies of resistance genes tend to be negatively related to the magnitude of the gene effect, explaining why major genes often occur at low frequencies, while minor genes appear to be frequent. It is in this way that the host and the pathogen, both as extremely variable and vigorous populations, can co-exist. Horizontal resistance (HR) and VR as meant by Van der Plank (1963), therefore, do not represent different kinds of resistance; they represent, rather, polygenic and oligogenic resistance that is based on the same underlying physiological mechanisms.

Disease and pathogen behavior are not necessarily the same. Pathogenic disease is a phenotype that results from the interaction of a host and a pathogen (Day, 1974). The pathogen may or may not be present at the point in the host where disease lesions are observed. It may be located at some distant point from where the disease is observed. In other cases, the host may be loaded with pathogens, but exhibit little or no disease. The grower does not care, concerned only if the disease phenotype adversely affects the economic outcome.

We have established a clear definition of what vertical resistance is and basic ways it differs from horizontal resistance (Fig. 19.4). A more comprehensive description of the nature of horizontal resistance is as follows (Lindhout, 2002; Michelmore, 2003; St Clair, 2010):

- HR is controlled by several to many genes.
- HR is more or less equally effective against all "races" of a pathogen (conversely, no host genetic resistance system distinguishes races of a pathogen).
- HR is more "durable" than vertical resistance.
- HR is also referred to as: non-specific, polygenic, durable resistance, and quantitative or partial resistance.

What is meant by "durable"? Conceptually, it is like comparing a wall of a single stone to one comprised of many bricks. If the single large stone can be overcome, the wall may be easily breached. While the individual brick is smaller in incremental effects to prevent entry, many more of them must be overcome. Vertical resistance (VR) may be more absolute than HR but the host possessing VR genes is only one pathogen gene mutation away from a compatible, disease response. In contrast, the breakdown of HR would require the loss of many independent genes simultaneously, or multiple mutations of the pathogen to overcome host barriers. The brick wall analogy extends to the building process. It is much easier to craft vertical than horizontal resistance due to the number of genes and heritability differences.

Another factor at play in the durability of HR relates to population dynamics and genetics. The absolute resistance conditioned by single-gene vertical systems presents a large population of pathogens with enormous selection pressure. Any mutation from avirulence to virulence within the population will be subject to tremendous positive selective forces as the newly virulent pathogen successfully enters the host and begins to consume energy resources and reproduce. In contrast, the selective pressures exerted by HR are much less severe. Experimental evidence has borne out the prediction that HR may last indefinitely, while VR breaks down relatively more frequently.

The inheritance of HR is difficult to study due to the involvement of a relatively large number of genes with small incremental effects on the disease phenotype. Understanding the inheritance patterns of genes underlying HR is essential for developing breeding strategies involving MAS. In experiments to elucidate the heritable basis of rice blast (caused by the fungal pathogen *Magnaporthe grisea*), rice (*Oryza sativa*) genotypes, along with their F_2 segregants derived from a complete diallel mating design were exposed to pathogen inoculum to determine combining ability and gene actions for the number and size of sporulating lesions developed on the plants, and area under the disease progress curve. Results showed that both additive and non-additive gene actions were involved in HR (Mulbah et al., 2015).

On the other hand, VR is much easier to study at the genetic and molecular level than HR because of the involvement of fewer genes with larger effects. Zhao et al. (2009) conducted experiments to elucidate the molecular basis of race specificity of bacterial blight of rice caused by the bacterial pathogen *Xanthomonas oryzae* pv. *oryzae*, a clear example of VR. Leucine-rich repeat (LRR)-kinase plasma membrane protein encoding sequences were found to mediate race-specific resistance to this disease. Plants carrying different LRR domains have different resistance spectra and the functions of the R genes are regulated by developmental stage. Experimental results suggest that the gradually increased expression of LRR sequences plays an important role in progressively enhanced *X. oryzae* pv. *oryzae* resistance during rice development.

Both VR and HR systems against late blight caused by the fungal pathogen *Phytophthora infestans* have been characterized in white potato (*S. tuberosum* L.). VR genes for late blight have been reported to be readily and rapidly overcome, and potato cultivars with quantitative and durable resistance to a broad spectrum of *P. infestans* races are not commonly available. The potato cultivar "Sarpo Mira" has been shown to exhibit durable resistance to *P. infestans*. Rietman et al. (2012) studied the resistance of "Sarpo Mira" in a segregating population by matching the responses to *P. infestans* effectors with race-specific resistance to differential strains. Resistance to late blight in "Sarpo Mira" was found to result from the combination of four pyramided qualitative VR genes and a novel quantitative HR gene. Effector-based resistance breeding was proposed as a strategy to facilitate selection and recombination of VR and HR genes that may culminate in more durable resistance to late blight.

Pushpa et al. (2014) advanced the study of concomitant VR and HR in potato further. Nontargeted metabolic profiling of resistant and susceptible potato cultivars, using high-resolution liquid chromatography and mass spectrometry, was applied to elucidate the quantitative resistance mechanisms against *P. infestans* (the US-8 pathotype). Sequencing of the coding genes of enzymes found to be involved in the resistance phenotype revealed single-nucleotide polymorphisms between resistant and susceptible genotypes, and the amino acid changes caused missense mutations altering protein functions. Hydroxycinnamic acid amides deposited in cell walls of resistant hosts inhibited pathogen colonization thus reducing lesion expansion. It was speculated that resistant and susceptible alleles of these genes could be either used as markers for MAS breeding programs or stacked into elite cultivars through cisgenic approaches.

More recently, a non-race specific rice (*Oryza sativa* L.) disease resistance gene was described in great detail. Through a genome-wide association study, Li et al. (2017) reported the identification of a natural allele of a transcription factor that confers non-race-specific resistance to blast. A survey of 3000 sequenced rice genomes revealed that this allele exists in 10% of rice populations. The allele caused a single nucleotide change in the promoter of the *bsr-d1* gene, which results in reduced expression of the gene through the binding of the repressive MYB transcription factor and, consequently, an inhibition of H_2O_2 degradation and enhanced disease resistance.

There are numerous examples of vertical genes that have been effectively deployed for decades of intensive crop culture without breaking down. Scott et al. (2011) proposed the existence of a third category of disease resistance to describe such resistance genes that defy the definitions of VR and HR: *oblique resistance*. For example, oblique resistance was observed in tomato (*S. lycopersicum*) in response to bacterial spot (*Xanthomonas* spp.) race T1. Hypersensitive resistance conferred by two or three genes was found, whereas VR is defined as being monogenic.

PLOIDY AND DISEASE RESISTANCE

Does genome ploidy level have an effect on plant resistance to disease pathogens and herbivorous pests? Polyploids have been espoused as a means to achieve higher levels of resistance or tolerance to biotic and abiotic stresses (Chen et al., 2017; Hias et al., 2018). While polyploids contain and express more copies of R genes, it is not intuitive why they would be more resistant than diploids since all genes are present in the same proportions. Innes et al. (2008) sequenced an approximately 1 million-bp region in soybean (*Glycine max*) centered on the *Rpg1-b* disease resistance gene and compared this region with another duplicated 10–14 million years ago. They found that, in contrast to low-copy genes, nucleotide-binding-leucine-rich repeat disease resistance gene clusters have undergone dramatic species/homoeologue-specific duplications and losses. This may help to explain why evolutionary forces following polyploidization may enhance biotic resistance, but does little to illuminate the effects on resistance in artificially-induced polyploids vs. diploids.

Ercolano et al. (2004) assessed the resistance to *Erwinia carotovora* subsp. *carotovora* and potato virus X (PVX) among haploids ($2n = 2x = 24$) extracted from tetraploid ($2n = 4x = 48$) *Solanum tuberosum*. They reported that R genes were expressed in haploids that were not expressed in the corresponding source tetraploids. This demonstrates that gene dosage and expression dynamics may vary with ploidy level.

In apple, autotetraploid trees (*Malus × domestica*) have been reported to exhibit better overall performance in the field than do diploids. Fungal diseases seriously affect the apple industry and, due to the long growth period of autopolyploid woody plants, little information is available on breeding disease-resistant artificially induced autotetraploids. Two autotetraploid apple cultivars had relatively lower disease symptom severity following infection by *Alternaria alternata* and *Colletotrichum gloeosporioides* than in the corresponding diploids. Real-time quantitative PCR analysis revealed that disease resistance-related genes were remarkably up-regulated in these autotetraploid trees (Chen et al., 2017). In another study of cv. "Royal Gala" apple transformed with R genes, three of seven diploid transformants were significantly more resistant to *Erwinia amylovora* than the non-transformed control and, in one case, a tetraploid transgenic line was significantly more resistant than the diploid shoot from which it was derived (Liu et al., 2001).

Hias et al. (2018) determined the influence of artificial genome doubling on the response of three *Malus × domestica* genotypes that had different levels of tolerance to apple scab (*Venturia inaequalis*). Based on visual symptom evaluation and real-time PCR quantification of *V. inaequalis* DNA in apple leaves, increased resistance was observed in the neotetraploid form of the monogenic resistant genotype compared to its diploid progenitor. Results suggested that polyploidization may be a viable strategy in apple scab resistance breeding programs.

Another study on the effects of host resistance at different genomic ploidy levels was conducted with musk-melon (*Cucumis melo*). The major cause of powdery mildew in muskmelons is the fungus *Sphaerotheca fuliginea*. There are several cultivar- and season-specific races of this pathogen. Experiments were conducted to determine whether powdery mildew resistance could be manifested at the haploid level from two disease-resistant melon lines. The responses of haploid and diploid plants to powdery mildew were observed to be identical (Kuzuya et al., 2003).

Generalizations about the effect of polyploidization on disease resistance are not yet possible. Rather, the plant breeder should be aware that this strategy is one of many potential alternatives to consider. Each host-pathogen system must be investigated to determine the effect of polyploidy on resistance to confirm the validity of this approach.

BREEDING FOR DISEASE RESISTANCE

Populations of plant pathogens and the biotic vectors that often spread them are increasing worldwide as they overcome the pesticides, agricultural practices, and biocontrols that once held them in check (Seifi et al., 2013). Concomitantly, many pesticides are being banned due to environmental concerns. Expanding global trade and travel are helping to spread viral, bacterial, and fungal plant pathogens into new areas, while global warming is allowing insect vectors to expand their ranges (Moffat, 2001). All of these trends point to the need for effective and rational plant breeding programs to develop new generations of cultivars that feature durable, cost-effective, and environmentally friendly mechanisms to avoid or limit damage due to disease and herbivory (Li et al., 2013).

"Disease resistance" is a misnomer since disease is the host's response to the pathogen and environment in combination with damage and sporulation resulting from the pathogen. Genetic resistance is ultimately of the host to the pathogen, but the term "disease resistance" is nevertheless in wide usage. The disease phenotype is, of course, a result of the interaction of three fundamental factors described earlier: host, pathogen, and environment. Disease is, therefore, unlike most other phenotypes addressed by plant breeders in that not one but two or more organisms are involved. Moreover, the pathogen is highly dynamic, adapted to rapid genetic changes to keep up with the hosts on which they depend for survival.

Plant breeders usually work closely with plant pathologists in the development and management of programs aimed at disease resistance. Training and experience with pathogens and disease are invaluable for the formulation of methods to screen host plants for genetic resistance and susceptibility. The pathogen must be properly isolated, identified, and propagated. If any of these steps are performed incorrectly, the entire program may be unsuccessful. The correct pathogen may be acquired, but the essential pathogenic properties may be lost if the pathogen is not maintained or propagated properly. Many pathogens alternate between saprophytic and pathogenic forms during their life cycle, and prolonged culture in the absence of the host can habituate saprophytic behavior. In some cases, the sporophyte and gametophyte exhibit entirely different modes of energy procurement, including pathogen, endo- or ectophyte, or saprophyte. It is a common practice to culture the pathogen directly on the host periodically to sustain pathogenicity since prolonged time periods under saprophyte conditions results in pathogen habituation and non-pathogenicity.

New students of plant pathology are familiarized with Koch's Postulates, named after the scientist who first articulated the concepts in the form of a guiding principle (Agrios, 2005). The main feature of Koch's postulates is that if an isolated organism is purported to be the causative agent of a disease, it must be demonstrated that the organism is capable of inciting the disease once again. Among the questions that plant pathologists always strive to answer is "what is the mode of pathogen entry?" Are there certain developmental times or inoculum types that are more effective than others for inciting disease?

This principle of Koch's Postulates is important within the realm of plant breeding as well. A given isolate must be demonstrated to incite the disease in the same manner over an extended time period. If the breeding program continues for many years, the host resistance and susceptibility observed in year one must be the same as in year five or ten.

The plant breeder, alone or in concert with a collaborator in the field of plant pathology, must be able to combine a host and pathogen under prescribed environmental conditions and incite disease symptoms consistently and reproducibly. When disease is incited on a genetically heterogeneous host population, it must be possible to distinguish individuals that embody resistant as opposed to susceptible genotypes. If that is accomplished, the resistance genes may be effectively incorporated into new cultivars using one of the breeding methodologies described in Chapters 13–18.

In a MAS resistance gene stacking study involving eight QTL in lettuce (*L. sativa*) that were associated with resistance to downy mildew (caused by the obligate fungal pathogen *Bremia lactucae*), only three out of ten double combinations showed an increased resistance effect under field conditions. This group targeted complete race-nonspecific resistance to lettuce downy mildew, as was observed for the nonhost wild lettuce species *L. saligna*. Genetic dissection of race-nonspecific resistance in *L. saligna* revealed several QTL for resistance with field infection reductions of 30–50%. Seven of ten homozygous breeding lines with stacked introgression segments showed a similar level of infection as the most resistant parent, revealing epistatic interactions with "less-than-additive" effects (den Boer et al., 2014).

SCREENING METHODS

The plant breeder often embarks on the development of new techniques to distinguish among genetically resistant and susceptible individuals within a population, accomplished for one or both of two primary reasons: (i) to reduce resource requirements (e.g., time, space, and labor), and (ii) to increase the h^2 of the disease phenotype. The standard screening technique against which the effectiveness of other alternatives is measured is the field epiphytotic. This technique calls for the establishment of a plant population, usually analogous to commercial parameters, and the introduction of a pathogen at the appropriate developmental time, and in the most effective form of inoculum. Ideally, the field epiphytotic closely emulates actual conditions under which plants under commercial production may contract the same disease, ensuring the relevance of results.

The field epiphytotic suffers from two potential drawbacks: (i) uncontrolled environmental parameters quite often adversely affect disease severity and uniformity, and (ii) depending on the host and disease, inordinate time, space, and labor may be necessary to promulgate an effective genetic resistance screening operation. Due to the second factor, costs may constitute a major drain on program resources. For these reasons, alternative methods to screen plants for disease resistance are frequently pursued.

Improved screening methods incorporate common elements including the use of enclosures and devices to precisely control environmental parameters such as temperature, humidity, and light. Individual plant inoculations of the pathogen may be accomplished in a manner that is more consistent than is the field epiphytotic. Further, the disease phenotype may be dissected into smaller units, each examined for correlative value to the whole phenotype. For example, a hypersensitive response on a newly emerging seedling may be adequate to signal the presence of underlying genes for vertical resistance to the pathogen and can lead to the development of a *seedling assay* that is both effective and much less expensive than the field epiphytotic.

Ideally, it is possible to use seedlings to screen for resistance genotypes to many different disease pathogens simultaneously. In practice, however, disease phenotypes interact extensively, and it is usually difficult or impossible to sort out the individual effects attributable to one pathogen or another. If two distinct diseases both culminate in host death, only one responsible pathogen can be convicted of the murder.

Intensive research has illuminated the underlying factors responsible for many disease phenotypes at the molecular level. Armed with this knowledge, it is sometimes possible to devise effective screening strategies based on reduced levels of organismal complexity. At the extreme of this spectrum, a toxin or enzyme excreted by the pathogen during pathogenesis is exposed to isolated cells from the host. A response by the host cell to the pathogenic factor, such as dead or reduced growth, may correlate well with resistance genotype. If molecular toxins are involved, the possibility exists that resistance may be both generated and screened in vitro at the cellular level. In practice, however, very few examples have been elucidated wherein the disease phenotype is the consequence purely of defined molecular/cellular interactions.

In an example of a successful simplified disease screening assay, pod rot of cocoa (*Theobroma cacao*), caused by several species belonging to the fungal genus *Phytophthora*, is the main cause of cocoa harvest losses worldwide. A leaf disc test was developed that was well correlated at the genetic level to the more expensive and time-consuming test previously used (Nyasse et al., 2007). Other examples of devised disease resistance screening protocols devised for specific crop species and diseases are described later in this chapter.

MARKER-ASSISTED BREEDING (MAS) FOR DISEASE RESISTANCE BREEDING

The disease phenotype is usually difficult, time-consuming, and resource intensive to produce on demand. Disease is also usually manifested in a broad spectrum of characteristics and degrees of expression, resulting in depressed realized heritability due to high V_D, V_I, V_{GxE}, and V_E variance components. This situation is especially cogent for HR, or quantitative, resistance. MAS is providing a way to select more effectively for disease resistance alleles by dramatically increasing h^2. Although quantitative resistance loci provide partial and durable resistance to a range of pathogen species in different crops, the molecular mechanism of quantitative disease resistance has remained largely unknown. Recent advances in the characterization of the genes contributing to quantitative disease resistance and plant-pathogen interactions at the molecular level provide clues to the molecular bases of broad-spectrum resistance and durable resistance. Knowledge is also turning quantitative resistance genes with minor effects into a productive resource for crop protection via biotechnological approaches (Kou and Wang, 2010).

Selecting superior coffee (*Coffea arabica*) genotypes is facilitated by MAS, and this technique is particularly suitable for transferring disease resistance alleles because it nullifies environmental effects and allows selection of resistant individuals in the absence of the pathogen. Molecular markers linked to two major genes for resistance to coffee rust and coffee berry disease (CBD) were identified and validated. Eleven true-breeding coffee rust resistant were identified by MAS. MAS also allowed the identification of sources of CBD resistance for use in preventive breeding for resistance to this serious disease (Alkimim et al., 2017).

Many common bean (*Phaseolus vulgaris*) improvement programs use MAS to facilitate cultivar development. Several recent germplasm releases have used molecular markers to introgress and or pyramid major genes and QTL for disease resistance. As the integration of genomics in plant breeding advances, the challenge will be to develop molecular tools that also benefit breeding programs in developing countries. Genomic examination of complex traits such as quantitative disease resistance should help bean breeders devise more effective selection strategies. Transgenic breeding methods for bean improvement are not well defined, nor efficient, as beans are recalcitrant to regeneration from cell cultures (Beaver and Osorno, 2009). MAS has been shown to hold promise as a method to facilitate pyramiding common bacterial blight resistance between *P. vulgaris* gene pools. The cost of conventional disease resistance breeding was estimated at US$1.55 per plant compared to $2.03 for MAS-assisted breeding (Duncan et al., 2012).

Also in *P. vulgaris*, combining QTL is a preferred strategy for improving bacterial blight resistance, but interactions among different QTL for the same resistance gene were unknown. Segregation for resistance among $BC_6:S_2$ plants derived from $BC_6:S_1$ plants that were heterozygous for both QTL did not deviate significantly from expected ratios of 9 resistant:3 moderately resistant:4 susceptible. These results indicated that breeders would realize greatest gains in resistance to common bacterial blight of *P. vulgaris* by selecting breeding materials that are fixed for both QTL (Vandemark et al., 2008).

Beet curly top virus (CTV) is an important virus disease of *P. vulgaris* in the semiarid regions of the United States, Canada, and Mexico and the only effective control strategy is genetic resistance. Larsen et al. (2010) set out to determine if a *P. vulgaris* landrace contained novel genes for resistance to CTV. Genetic analyses revealed random amplified polymorphism DNA (RAPD) markers associated with a major-effect quantitative trait loci (QTL) from this landrace that exhibited stable expression for 3 years. A RAPD marker was converted to a sequence-characterized amplified region (SCAR) and used to locate the QTL on linkage group 6 of the *Phaseolus* core map. This landrace was found to possess novel resistance to CTV conditioned by at least two genes, one with major the other minor effect.

A precursor to developing efficient breeding programs for polygenic resistance to pathogens should be a greater understanding of genetic diversity and stability of resistance QTL in plants. Partial resistance genes are considered to be more durable than monogenic resistances. The diversity and stability of partial resistance to *Aphanomyces euteiches* in pea (*Pisum sativum* L.) towards pathogen variability were determined. A total of 135 additive-effect QTL corresponding to 23 genomic regions and 13 significant epistatic interactions associated with partial resistance to *A. euteiches* in pea were identified. Results confirmed the complexity of inheritance of partial resistance to *A. euteiches* in pea and provided a solid foundation for the choice of consistent QTL for use in MAS schemes to increase current levels of resistance to *A. euteiches* in pea breeding programs (Hamon et al., 2011).

MAS has been used effectively to breed potato (*Solanum tuberosum*) for resistance to late blight, caused by *Phytophthora infestans*. Breeding for late blight resistance has been a challenge because the race-specific resistance genes introgressed from wild potato *S. demissum* have been short lived and breeding for "horizontal" or durable resistance has achieved only moderate successes. A high-level of late blight resistance was identified in a wild potato relative, *S. bulbocastanum* subsp. *bulbocastanum*, controlled mainly by a single resistance allele *RB*. A QTL was developed the *RB* allele and was used to select for resistance in breeding populations derived from the potato x *S. bulbocastanum* somatic hybrids. Results demonstrated that MAS was an effective strategy for the transfer of the *RB* allele into potato using the backcross breeding method (Colton et al., 2006).

Most popular commercial potato cultivars are susceptible to *Verticillium dahliae*, a fungal pathogen causing Verticillium wilt disease, though some cultivars with relatively high resistance also exist. One hundred thirty-nine potato cultivars and breeding selections were analyzed for resistance to *V. dahliae* and for the presence of the microsatellite marker allele STM1051-193 that is closely linked to the resistance quantitative trait locus located on the short arm of chromosome 9. Early and very early potatoes are usually more susceptible to Verticillium wilt regardless of disease resistance genotype, though the pattern of the allele effect is always the same. Results showed that the STM1051-193 allele could be used for MAS, but potato maturity class also needs to be considered when making the final decision about the plant resistance level (Simko et al., 2004).

A locus with a strong effect on *P. infestans* (the late blight pathogen) resistance was mapped to the end of potato chromosome XI in the vicinity of the *R3* locus. Marker 45/XI exhibited the strongest linkage to the resistance locus and accounted for between 55.8% and 67.9% of the variance in the mean resistance scores noted in the detached leaflet assays. Following MAS backcross breeding, ten breeding lines containing a late blight resistance locus from *S. tuberosum* and *S. phureja* donor cultivars were obtained (Tomczyåska et al., 2014). In a separate study, there was strong evidence that the two indistinguishable QTL for foliage maturity type and late blight resistance on chromosome 5 may actually be a single gene with pleiotropic effects on both traits. Two QTL for resistance to late blight showed a significant epistatic interaction suggesting that QTL for late blight resistance affect each other's expression (Visker et al., 2003).

A study was conducted to determine the efficiency of conventional phenotype selection (CS) vs. MAS in breeding for maize streak virus (MSV) resistance in Uganda. Both breeding approaches were effective in generating resistant genotypes, but disease incidence was higher in populations under CS (79%) than MAS (62%). This difference was likely a consequence of the lower h^2 of the disease phenotype in CS as compared to the QTL phenotypes in MAS. An equal number of lines generated by MAS and CS displayed high yield potential and MSV resistance in testcrosses. If molecular laboratory facilities are accessible, MAS would be recommended in breeding over CS for MSV resistance breeding (Abalo et al., 2009).

Wilt disease caused by *Phytophthora capsici* in pepper is among the most damaging factors to *Capsicum* spp. crops worldwide. A major gene for resistance to *P. capsici* was discovered and a tightly linked QTL was identified (Thabuis et al., 2004). A modified recurrent backcross breeding strategy was initiated to transfer the resistance factors from the donor population to superior cultivars. The resistance phenotype and allelic frequencies strongly depended on backcross population and screening severity. A loss of resistant QTL alleles was observed in the BC_1, particularly for the low-effect QTL, whereas better conservation of the resistant QTL alleles was observed in subsequent backcross generations. Changes in the allelic frequencies of loci not linked to resistance QTL and for horticultural traits across the breeding process indicated that the recovery of the recipient parent genome was not significantly affected by the selection for resistance.

MAS is a particularly powerful tool for unraveling the immensely complex layers of genetic and genetic × environment interactions in a highly polyploidy crop species like strawberry (*Fragaria × ananassa*). The *Fragaria* genus, including 21 wild and cultivated species, contains genetic sources of diseases resistance that are quite rich but not fully exploited in breeding for resistance. Usefulness of different molecular techniques and high throughput technologies for the dissection of genetic resistance mechanisms and the explanation of plant diversity in relation to pathogens at the DNA level will allow this genetic variability to be effectively manipulated. A model of a comprehensive exploration of the strawberry genome, including the generation of resistance markers and identification of genes involved with induction or regulation of plant response to pathogen attack, is potentially very powerful for polyploidy strawberry (Korbin, 2011).

In some instances, MAS has been found to be inappropriate as a strategy to accelerate or otherwise improved the efficiency or precision of breeding programs. One such example is in *Rosa* spp. where the most important fungal diseases are black spot, powdery mildew, botrytis, and downy mildew. Rose rosette, a lethal viral pathogen, is also emerging as a devastating disease in North America. QTL have been identified for resistance to black spot and powdery mildew using the technique of genotyping by sequencing (GBS, see Chapter 9) to generate thousands of markers. GBS will provide plant breeders with the ability to more readily identify useful linked markers. Although there is much potential for QTL, most rose breeders are not currently using MAS, primarily because a good set of marker/trait associations that illuminate a path to stable disease resistance is not yet available (Debener and Byrne, 2014).

MOLECULAR BASES AND APPROACHES TO THE BREEDING OF PLANT DISEASE RESISTANCE

A thorough understanding of the etiology (progression) of plant diseases may provide suggestions as to how plants may be engineered to ward off the responsible pathogens. For example, in cases where simple diffusible toxins are excreted by the pathogen to weaken the host while acquiring energy reserves, the exclusion or chemical alteration of the toxin may lead to resistance. Another approach is to capitalize on structural components of the pathogen that could lead to vulnerability. A diagnostic attribute of fungi is that the cell walls that envelop mycelia contain the carbohydrate compound chitin. Chitin is not found in plants, but is also a structural component of insect exoskeletons. Chitinases have been discovered in saprophytic bacteria that have been advanced as possible sources of resistance to fungal pathogens (Day, 1992).

Bacteria are characterized by a myriad of distinctions in cellular structure and function as compared to eukaryotes. Most antibiotics work by targeting the biochemical machinery, such as unique pathways and DNA transcription and translation, to selectively kill prokaryotic bacteria within infected eukaryotic tissues. Antibiotics also work well for the eradication of bacterial pathogens from plants. With the possibility that transformation may be used to alter host plants, the engineering of genomes to encode products that disrupt bacterial functioning is quite possible. Care must be taken, however, that eukaryotic plant plastids and mitochondria are sheltered from such systems since they function similarly to prokaryotes (Broekaert, 1996).

The best demonstration of the power of molecular approaches for breeding disease resistance lies with viral pathogens. Most economic plant species are attacked by a spectrum of viral pathogens, the infective/lytic cycle being similar to those of bacteria and mammals. The viral disease phenotypes are distinct: twisting, gnarled growth patterns, stunting, and alterations in pigmentation.

The genomes of most plant pathogenic viruses consist of single-stranded RNA. The infective RNA is injected into a host cell, leaving the proteinaceous coat behind. During the ensuing lytic cycle, new virus particles are synthesized from the ssRNA, the RNA viroid and protein coats produced separately then packaged together. The coat protein gene is encoded on the viroid, not subverted from the plant machinery, presenting a clear "choke point" in the lytic cycle upon which a resistance strategy may be based.

The tomato disease resistance allele *Pto* confers resistance to strains of the bacterial pathogen *Pseudomonas syringae* pv tomato expressing the avirulence gene *avrPto*. Transformation of *Nicotiana benthamiana* with *Pto* results in specific resistance to *P. syringae* pv *tabaci* strains carrying *avrPto*. The resistant phenotype is manifested by strong inhibition of bacterial growth and the ability to exhibit a hypersensitive response (Rommens et al., 1995).

The basis of this resistance was an expressed antisense sequence of corresponding viral coat proteins transformed into the host genome by *Agrobacterium* vectors (Ramu et al., 2011). The resistance was observed to be nearly absolute, with no disease symptoms observed under a wide range of growing conditions. The introduced coat protein antisense sequences may now be treated as a dominant gene, as for vertical resistance. Since the mechanism of resistance is exclusive of host and pathogen recognition and involves the pervasive disruption of the lytic cycle, it is difficult to imagine how the pathogen could undergo simple mutations to overcome the resistance.

Ralstonia solanacearum (Rso) is a causal agent of bacterial wilt disease in a wide range of horticultural crops. Rso strains are heterogeneous in nature and are therefore problematic regarding both classification and development of disease resistance. Rso pathogen-associated molecular patterns (PAMPs) and effector proteins are secreted into plant cells where they respectively activate and suppress plant immunity, thereby affecting Rso virulence. The introduction into plants of known pattern recognition receptors (PRRs) that recognize Rso PAMPs was suggested as a possible strategy to confer resistance to a large number of strains. Conserved "core" effectors from Rso pathotypes could be used to identify and deploy nucleotide-binding leucine-rich repeat (NLR) resistance genes in a targeted crop cultivar. Stacking multiple NLRs that recognize Rso effectors would provide durable disease resistance by minimizing the chance for Rso to evade the implemented resistance (Jayaraman et al., 2016).

Potato (*S. tuberosum*) leaves infected with the late blight pathogen (*P. infestans*) produced a serine protease inhibitor (PLPKI) with specificity for microbial proteases. PLPKI inhibited the activity of extracellular proteases produced by two pathogens of potato, *P. infestans* and *Rhizoctonia solani*, but was inactive against proteases secreted into the culture media by the binucleate *Rhizoctonia* N2 that is non-pathogenic on potato. Western blot analyses showed a positive correlation between the levels of PLPKI and the degree of HR, showing its highest accumulation in a highly resistant clone (Feldman et al., 2014).

Verticillium wilt (*Verticillium dahliae*) is an economically important disease of many high-value horticultural crops that is difficult to manage due to the long viability of pathogen resting structures, wide host range, and the inability of fungicides to reach the pathogen once it is in the plant vascular system. In chili pepper types of *C. annuum*, breeding for resistance to Verticillium wilt is especially challenging due to the limited resistance sources. Homologs of the tomato *Ve1* resistance gene have been characterized in diverse plant species, and interfamily transfer of *Ve1* within Solanaceae confers race-specific resistance. Queries in the chili pepper WGS database in NCBI with *Ve1* and *Ve2* sequences identified one open reading frame (ORF) with homology to the tomato *Ve* genes. A homozygous haplotype was identified for the susceptible accessions and resistant accessions. A cleaved amplified polymorphic sequence (CAPS) molecular marker was developed within the coding region of *C. annuum Ve* and screened diverse germplasm that has been previously reported as being resistant to Verticillium wilt in other regions. Based on phenotyping using a New Mexico *V. dahliae* isolate, the marker could select resistance accessions with 48% accuracy (Barchenger et al., 2017).

DETAILED EXAMPLES OF BREEDING HORTICULTURAL CROPS FOR DISEASE RESISTANCE

A good way to appreciate the diversity of strategies that plant breeders have utilized to develop cultivars with disease resistance is to review examples of specific breeding programs for host resistance to particular pathogens. The sections that follow provide a closer look at three such scenarios in a diverse range of horticultural crop species: (i) the downy mildew pathogen of sweet basil, (ii) eastern filbert blight pathogen of hazelnut, and (iii) fusarium yellows pathogen of celery.

Example 1: Breeding for Resistance to the Downy Mildew Pathogen of Sweet Basil

Downy mildew, caused by *Peronospora belbahrii*, is a relatively new disease of basil (*Ocimum* spp.) in the United States (Roberts et al., 2009). The disease renders the crop unmarketable due to a "sooty" sporulation bloom and discoloration on leaves (Fig. 19.6). Efforts to identify sources of resistance to this pathogen were undertaken as a first step towards breeding new resistant sweet basil cultivars. Thirty *Ocimum* sp. cultivars and advanced breeding lines were evaluated for susceptibility to the basil downy mildew pathogen (*P. belbahrii*) in field trials. Popular commercial sweet basil (*O. basilicum*) cultivars were found to be among the most susceptible to this disease pathogen (Fig. 19.7). Disease symptoms and sporulation of *P. belbahrii* on *O. × citriodorum* and *O. americanum* "spice" cultivars were present but far less predominant than on most *O. basilicum* cultivars evaluated. Certain spice cultivars exhibited no visible disease symptoms (Wyenandt et al., 2010; Wyenandt et al., 2015).

FIG. 19.6 Disease symptoms of downy mildew caused by *Peronospora belbahrii* on leaves of sweet basil (*O. basilicum*). *Courtesy of Dr. C. Andrew Wyenandt, Rutgers University.*

FIG. 19.7   A comparison of downy mildew susceptible sweet vs. resistant spice basils. *Courtesy of Dr. C. Andrew Wyenandt, Rutgers University.*

Barriers to hybridization among *Ocimum* species are variable. Some species may be hybridized relatively easily while others are more distantly-related phylogenetically and gene flow is more difficult to attain. Bridging these species is but one challenge; sorting out the genes that control foliar volatile compound levels and chilling resistance presents further daunting complications. Dudai et al. (2010) hybridized two cultivars of spice basil that differed in their aromatic volatile profiles: One parent was a methyl chavicol type and the other a eugenol type. In the F_2 generation approximately 35% eugenol and 15% methyl chavicol types were recovered. Thus, when hybridizing *O. basilicum* with *O. × citriodorum* or *O. americanum*, care must be exercised to target resistance in combination with the "sweet Italian" leaf morphology and volatile content for selection (Dudai et al., 2010).

Next, screening methods for visualizing and selecting the downy mildew disease phenotype were developed. Pyne et al. (2014) developed a rapid approach to screen and evaluated downy mildew response at the cotyledon and true leaf growth stages under controlled environmental conditions (Fig. 19.8). Four accessions exhibited little or no sporulation at either growth stage, three of which showed other symptoms including chlorosis and necrosis.

Independently, Ben-Naim et al. (2015) screened 113 populations of *Ocimum* spp. (83 germplasm accessions and 30 commercial cultivars) for resistance to *P. belbahrii* at the seedling stage in growth chambers and during three seasons in the field. Most *O. basilicum* entries were highly susceptible, whereas most entries belonging to *O. americanum*, *O. kilimanadascharicum*, *O. gratissimum*, *O. campechianum*, or *O. tenuiflorum* were highly resistant at both the seedling stage and in the field. F_1 plants of two crosses were highly resistant, F_1 plants of 24 crosses were moderately resistant, and F_1 plants of one cross were susceptible suggesting full, partial, or no dominance of the resistance gene(s), respectively.

After a suitable and reproducible disease screening protocol was developed, effective studies of the heritable bases of resistance were possible. The commercial spice basil cultivar was identified as resistant and hybridized with a susceptible sweet basil inbred line to generate a full-sibling family. All siblings in the F_1 and BC_1 (to resistant parent) populations were resistant providing, strong evidence that inheritance of resistance was conferred by dominant alleles. Segregation ratios in the F_2 and backcross to the susceptible parent populations exhibited chi-square goodness of fit to the two-gene complementary and recessive epistatic models (Pyne et al., 2015).

Ben-Naim et al. (2018) reported on the transfer of a resistance gene from the highly resistant tetraploid wild basil *O. americanum* var. *americanum* to the susceptible *O. basilicum* "Sweet basil." F_1 interspecific hybrid plants were resistant indicating that the gene controlling resistance is dominant, but the F_1 was self-sterile due to the substantial genetic distance between the parents. F_1 plants were pollinated with the susceptible parent and 115 BC_1S_1 embryos were rescued using tissue culture methods. BC_1S_1 plants segregated 5:1 resistant/susceptible suggesting that resistance in F_1 was controlled by two dominant genes.

Expressed sequence tag simple sequence repeat (EST-SSR) and single nucleotide polymorphism (SNP) markers were developed and used to map the DM resistant × susceptible F_2 population. Disomic segregation was observed

Develop simple greenhouse screening method

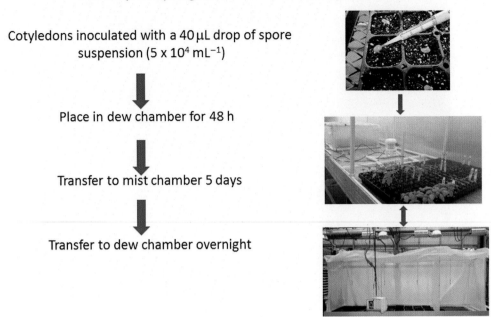

Cotyledons inoculated with a 40 μL drop of spore
suspension (5 x 10⁴ mL⁻¹)

Place in dew chamber for 48 h

Transfer to mist chamber 5 days

Transfer to dew chamber overnight

FIG. 19.8 A rapid, reproducible, and relevant screening protocol for downy mildew of sweet basil based on inoculation of cotyledons and scoring of pathogen sporulation. *Courtesy of Dr. C. Andrew Wyenandt, Rutgers University.*

in both SNP and EST-SSR markers providing evidence that the *O. basilicum* genome structure is allotetraploid, thus allowing for subsequent analysis of the mapping population as a diploid intercross. A single major QTL explained 21–28% of the phenotypic variance and demonstrated dominant gene action. Evidence was also found for an additive effect between the two minor QTL and the major QTL associated with downy mildew susceptibility (Pyne et al., 2017).

Since the inheritance of DM resistance was discovered to be qualitative, the adaptation of a traditional backcross strategy to transfer resistance alleles from the spice to the sweet basil cultivar was deemed to be feasible. Due to sterility barriers, difficulty in recovering legitimate recombinants, and polyploidy, the backcross process was more difficult to accomplish than originally thought. Because basil had not been extensively studied genetically breeding efforts and the development of sweet basil with resistance to *P. belbahrii* were severely hindered. Pairwise cluster analyses of polymorphic loci revealed three major and seven subpopulation clusters in *Ocimum*. The constituent "k3" cluster is a rich source of DM resistance, but introgression of resistance into commercially important "k1" populations is impeded by reproductive barriers as demonstrated by multiple sterile F_1 hybrids (Pyne et al., 2018).

New DM resistant sweet basil cultivars ("Rutgers Obsession DMR," "Rutgers Devotion DMR," "Rutgers Passion DMR," and "Rutgers Thunderstruck DMR") were released for sale and distribution during 2018–19, demonstrating the success of plant breeding to solve this challenging problem (Branson, 2018).

Example 2: Breeding for Resistance to the Eastern Filbert Blight Pathogen of Hazelnut

Eastern filbert blight (EFB) is an economically significant disease of European hazelnut (*Corylus avellana*) in the U.S. Currently, EFB only occurs in North America, but with imminent geographical translocation of inoculum, the disease will eventually appear in Europe and Asia where *C. avellana* is cultivated extensively. Since chemical controls have thus far been ineffective for EFB management, genetic resistance is the only viable disease control strategy to this fungal disease caused by *Anisogramma anomala* (Peck) E. Muller (Mehlenbacher et al., 1991). This pathogen is an obligate biotroph of *Corylus* spp. and infects only living tissues, producing stromata on the surface of the stem (Figs. 19.9–19.11). Recent evidence shows that *A. anomala* is a more genetically diverse pathogen than was originally thought, complicating the strategy of breeding resistant cultivars (T. J. Molnar, personal communication).

Greenhouse and field screening of *Corylus* spp. germplasm was undertaken to study the inheritance of known EFB resistance and to identify new sources for a prospective breeding program (Coyne et al., 1998). It was found

FIG. 19.9 Overt disease manifestations of Eastern Filbert Blight; spore stromata on a woody stem of European hazelnut (*Corylus avellana*). *Courtesy of Dr. Thomas J. Molnar, Rutgers University.*

FIG. 19.10 Closer view of *A. anomala* spore-bearing stromata on a woody stem of European hazelnut (*Corylus avellana*).

that the *C. avellana* cultivar "Gasaway" expressed a high level of resistance to this disease that was conferred by a single dominant gene. An independent source of resistance in *C. avellana* was the selection "Zimmerman" in which the resistance gene was found to be attached to two independent centromeres producing a 3:1 resistant:susceptible segregation ratio (Lunde et al., 2006). The "Gasaway" resistance gene has been incorporated into many genetic backgrounds via backcross under the assumption that this source of resistance would be highly resilient over time and geographical locations. Recent results have shown, however, that the "Gasaway" R-gene is either breaking down or is not universally effective against all A. anomala isolates (Muehlbauer et al., 2018).

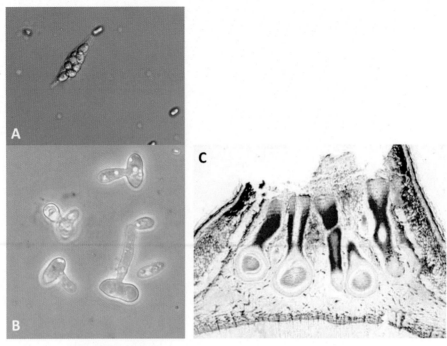

FIG. 19.11 (A) *A. anomala* only reproduces by ascospores; there are no asexual conidial spores; (B) germinating ascospores of *A. anomala* in culture; (C) histological cross-section of a stroma showing asci from which ascospores are ejected.

Six *Corylus* species relatives of cultivated hazelnut were screened to identify new resistance genes. *C. cornuta* Marshall var. *cornuta*, *C. cornuta* var. *californica*, *C. heterophylla*, and *C. blume* were highly resistant to EFB, as were most *C. americana* genotypes and one *C. colurna* clone tested, but *C. jacquemontii* was highly susceptible. In several cases, hybrids of these species with susceptible *C. avellana* were also resistant indicating that resistance was dominant. Further searches for EFB resistance among *Corylus* spp. were conducted by Chen et al. (2007). *C. avellana* accessions from Spain and Finland, 5 *C. americana* × *C. avellana* hybrids, 4 *C. colurna* × *C. avellana* hybrids, and one *C. heterophylla* var. *lutchuensis* × *C. avellana* hybrid exhibited complete resistance to EFB.

New *Corylus* species germplasm collections and acquisitions from western and central Asia were screened for resistance by Molnar et al. (2007) and Leadbetter et al. (2016). Specifically, 605 *C. avellana* seedlings from germplasm collected from the Russian Federation and the Crimean peninsula of Ukraine were inoculated with spores of *A. anomala* and disease responses were evaluated. Eight accessions showed no signs of the pathogen or symptoms of the disease. RAPD markers tightly linked to the single dominant resistance gene from "Gasaway" (Chen et al., 2005) were not present in all resistant seedlings suggesting that they represent novel sources of genetic resistance to EFB.

A linkage map for *C. avellana* was constructed using RAPD and SSR markers and the 2-way pseudo-testcross approach (Mehlenbacher et al., 2006). Eleven linkage groups were identified corresponding to the haploid chromosome number of hazelnut (n = x = 11). The maps were relatively dense with an average of 2.6 cM between adjacent markers. The "Gasaway" locus for resistance to EFB was mapped to chromosome 6R for which two additional markers tightly linked to the dominant allele were identified and sequenced. Additional sources of EFB resistance and associated genetic results were documented recently by Molnar et al. (2018). Based on co-segregation with SSR markers an independent source of EFB resistance from the cultivar "Ratoli" was assigned to linkage group 7 (Sathuvalli et al., 2011). Recent results showed that a single QTL region associated with a new source of EFB resistance from a southern Russian accession was located on hazelnut linkage group (LG) 2 (Honig et al., 2019).

EFB symptoms require a relatively long time to appear following inoculation of plants or under natural conditions. Efforts were undertaken, therefore, to develop an effective disease resistance screening method that was both quicker and relevant to resistance observed under field conditions. A new protocol was developed that incorporated elements of host plant pre-conditioning and controlled environments that reduced the time required for distinguishing resistant vs. susceptible *C. avellana* seedlings from about 15 to six to seven months (Molnar et al., 2005). Later, DNA probes were developed to improve the sensitivity of the protocol and to shorten the time to distinguish resistant from susceptible genotypes (Molnar et al., 2013).

It is clear that all the elements needed to develop resistance to EFB in European hazelnut have been successfully developed. Therefore, using linked markers and, eventually, R gene sequences to select for qualitative dominant resistance, is an effective way to ensure that this crop species will continue to present a viable crop opportunity.

Example 3: Breeding for Resistance to the Fusarium Yellows Pathogen of Celery

Fusarium yellows was first found in a celery field near Kalamazoo, MI (USA) in 1914 (Lacy et al., 1996a). The disease was shown to be caused by the fungal pathogen *Fusarium oxysporum* f.sp. *apii* (R. Nels. & Sherb.) and is associated with foliar yellowing, growth stunting, withering of petioles and leaves, and a red to brown discoloration of water-conducting tissues of the roots, crown, and petioles (Fig. 19.12). The disease spread and grew more serious over time and by 1931 was widespread in North America. Fusarium yellows resistant cultivars began to appear in 1939 and a purportedly "immune" (a misuse of the animal-based term) cultivar ("Tall Utah 52-70") was introduced in 1952. This cultivar originated from a single-plant selection in resistance screening field trials. Within a few years after the introduction of this new cultivar and later selections most of the celery acreage in North America was converted to new resistant cultivars, and the disease was presumed to have been vanquished.

A report appeared in 1978, however, that the Fusarium yellows disease had recurred in California on "Tall Utah 52-70," starting in 1959 (Hart and Endo, 1978). The original strain that was pathogenic on the pre-1952 "blanched" or self-blanching celery cultivars was given the name "race 1" and the new strain that was also pathogenic on the "green" "Tall Utah 52-70" was named "race 2." A new strain, that was pathogenic on "Tall Utah 52-70" but not on self-blanching cultivars, was later found to be unrelated to race 2 and was named "race 3" (Puhalla, 1984). Race 3 is not as aggressive as race 2, and it has not been a serious challenge to celery growers.

By 1987, race 2 had infested most celery production areas in North America including British Columbia, California, Michigan, New York, and Texas. In 2016, race 2 was discovered in celery production fields in Argentina (Lori et al., 2016). Interestingly, Fusarium yellows never appeared in celery production areas of central Florida. It was speculated that summer flooding of fields in Florida results in anaerobic fermentation at high temperatures that keeps populations of *F. oxysporum* f. sp. *apii* in check (Lacy et al., 1996a).

Spores of the pathogen can be spread readily by the movement of infested soil or infected transplant seedlings. The pathogen survives in soil for long periods of time as overwintering chlamydospores that persist even in the absence of the host, rendering short-term crop rotation ineffective as a control or avoidance method for Fusarium yellows. Attempts to mitigate pathogen inoculum in soil with fungicides have been unsuccessful (Lacy et al., 1996a).

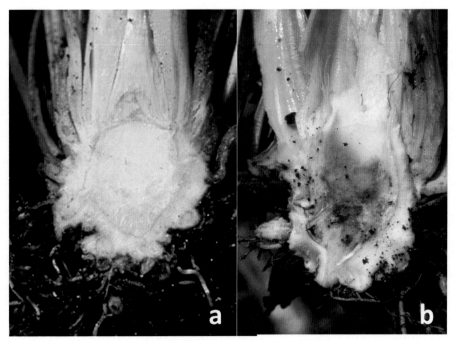

FIG. 19.12   (A) Transverse sectional view of the crown of a healthy celery plant, and (B) view of the crown of a susceptible celery genotype infect by *F. oxysporum* f. sp. *apii* showing vascular discoloration that is diagnostic of the Fusarium yellows disease.

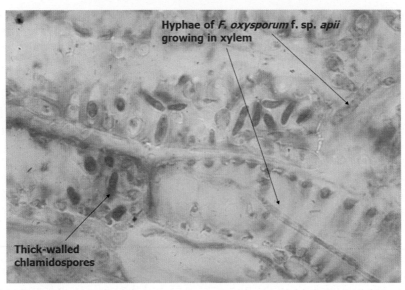

FIG. 19.13   Photomicrograph of hyphae and chlamydospores of *F. oxysporum* f. sp. *apii* race 2 observed in histological preparations of xylem tissues of petioles of susceptible celery cultivar ("Tall Utah 52-70R").

Studies were conducted that shed light on the early stages of the Fusarium yellows disease and responses by the host to pathogen invasion. Root apices from *Apium graveolens* L. cultivars resistant and susceptible to *F. oxysporum* f.sp. *apii* race 2 were studied at various times after inoculation, using light and electron microscopy to determine structural response(s) of the hosts during penetration and colonization by the pathogen (Jordan et al., 1988). Penetration was intercellular and intracellular and involved mechanical and enzymatic mechanisms. Hyphae of the pathogen and chlamydospores were observed in xylem tissues of susceptible hosts (Fig. 19.13). Callose deposits, that formed in vascular tissue as the fungus colonized it, were two and three times greater in the epidermis and four and nine times greater in the cortex of the resistant than in two susceptible hosts, respectively. Hyphal counts in the cortex of the resistant host were 50% fewer than in the susceptible hosts.

Celery has a biennial life cycle resulting in a relatively long generate on time, approximately 16 months from seed to seed. The plant is also slow-growing under cultivation, and Fusarium yellows disease symptoms take a long time to appear following inoculation. Many different strategies were tested in attempts to develop a disease screening protocol that was both efficient and accurate, including cell and molecular-based strategies (Orton, 1982). The only method that proved to be reliable and reproducible was developed and described by Schneider (1984). This protocol involves the stepwise culture of the pathogen on a complex medium (potato dextrose) followed by a period of saprophytic growth on enriched barley straw. The pathogen/straw mixture is dried and mixed with potting soil at a fixed ratio then celery seedlings are transplanted into the inoculated soil in pots and grown under greenhouse or controlled environmental conditions. Disease symptoms typically appear eight to ten weeks after transplanting. Crowns of plants are then split and scored for vascular discoloration using a numerical scale.

A comprehensive collection of *A. graveolens* germplasm was developed, and accessions of var. *dulce*, var. *rapaceum* (celeriac), and var. *secalinum* (smallage) were screened for resistance to *F. oxysporum* f. sp. *apii* races 1 and 2 (Orton et al., 1984a). Several accessions of var. *rapaceum* exhibited an excellent level of resistance, but PI 169001 was deemed to be the best resistance gene donor due to overall plant phenotype. Crosses of PI 169001 with a selection from "Tall Utah 52-70R" (OXN40) that exhibited heritable partial resistance gave rise to F_2 and BC segregation ratios that were consistent with two independent unlinked resistance genes, one with dominant and the other additive allelic interactions (Orton et al., 1984a).

The PI 169001 and OXN40 resistance genes were introgressed into a green commercial celery genetic background by recurrent backcross. After two backcross cycles, a population was selected that exhibited a high level of resistance to race 2 and also a relatively acceptable *dulce* celery phenotype. This population was released to the seed industry as "UC-1" (Orton et al., 1984b). Subsequently, seed companies and researchers used UC-1 as a donor in further backcross breeding efforts and race 2-resistant populations combined with commercially acceptable green celery type were released for cultivation (Quiros et al., 1993).

Another approach to the development of new cultivars with resistance to *F. oxysporum* f. sp. *apii* race 2 has been to select among regenerated somaclones of commercial celery cultivars. This strategy of enhancing genetic variability with somaclonal variation was described in Chapter 8. The advantage is that resistance may be incorporated directly into an existing commercial cultivar without adulterating any of the valuable phenotypic features that define its market value. In one such effort, plants regenerated from suspension cultures of the race 2 susceptible cultivar "Tall Utah 52-70R" were screened for resistance and several were selected (Heath-Pagliuso et al., 1988). One designated "UC-T3" was deemed to have the most promise and was studied further. First generation (S_1) progeny, second generation (S_2) progeny, and backcross (BC) progeny of UC-T3 were evaluated for resistance to race 2. The lowest ranking S_2 family in both the lightly infested and heavily infested fields was significantly more resistant to race 2 than was "Tall-Utah 52-70R." The resistance was concluded to be heritable and controlled by at least two dominant genes (Heath-Pagliuso and Rappaport, 1990). Independently, Lacy et al. (1996b) reported the release of another race 2 resistant population ("MSU-SHK5") selected from plants regenerated from tissue cultures of a slightly resistant cultivar "Tall Utah 52-70HK." While it does not appear that either "UC-T3" or "MSU-SHK5" have been grown commercially, it is likely that these genotypes have been used by seed companies in their ongoing celery breeding efforts.

The pathogen, *F. oxysporum* f. sp. *apii*, has been classified as "fungi imperfecti" since no sexual reproductive cycle has ever been observed (Agrios, 2005). Despite the lack of sexual reproduction for the generation of new mutational recombinants, new races of the pathogen continue to appear. It is likely that *F. oxysporum* f. sp. *apii* does not exemplify a typical gene-for-gene system wherein pathogen races are closely related by recent derivation. Support for this notion was first presented by Puhalla (1984) based on classical criteria for discerning relationships in *Fusarium* such as colony morphology and heterokaryon formation. More recently, molecular evidence for the independent appearance of certain races of *F. oxysporum* f. sp. *apii* have been published (Epstein et al., 2017).

F. oxysporum species complex (FOSC) isolates were obtained from celery with symptoms of Fusarium yellows between 1993 and 2013 primarily in California. In 2013, new highly virulent clonal isolates, designated race 4, were discovered in production fields in Ventura County, California. Analyses of a 10-gene dataset comprising 38 kb showed that *F. oxysporum* f. sp. *apii* is polyphyletic. Race 2 is nested within clade 3, whereas the evolutionary origins of races 1, 3, and 4 are within clade 2. Based on 6898 single nucleotide polymorphisms from the core FOSC genome, race 3 and the new highly virulent race 4 are highly similar (Nei's Da = 0.0019) suggesting that *F. oxysporum* f. sp. *apii* race 4 evolved from race 3 (Epstein et al., 2017). An updated *F. oxysporum* f. sp. *apii* race differential is depicted in Table 19.4. It is clear that race 4 constitutes a significant threat to celery growers worldwide prompting plant breeders to start searching for sources of resistance that will serve as donors in a new cycle of backcrosses.

TABLE 19.4 An updated host differential to distinguish races of *F. oxysporum* f. sp. *apii*.

	Pathogenicity of isolate on host				
	Apium graveolens				
FOA isolate	**UC-1**	**Tall Utah 5270R**	**Fordhook**	*Tithonia rotundifolia*	**Race Designation**
ATCC 18142	ND	ND	D	HD	1
ATCC 15636	ND	ND	D	D	1
France 3	ND	ND	HD	D	1
C2444 Santa Maria, CA	ND	ND	HD	ND	1a
XM10 Santa Maria, CA	ND	ND	HD	ND	1a
3ER2 Salinas, CA	ND	HD	HD	ND	2
212P1 Oceano, CA	ND	HD	HD	ND	2
6PIA Oxnard, CA	ND	HD	HD	ND	2
Puhalla 1 Salinas, CA	ND	HD	ND	??	3
Puhalla 2 Oxnard, CA	ND	HD	ND	??	3
Epstein Camarillo, CA	HD	HD	HD	??	4

ND, no disease; *D*, disease; *HD*, high disease intensity; *FOA*, *Fusarium oxysporum* f. sp. *apii*.

EXAMPLES OF AND EXPERIENCES WITH DISEASE RESISTANCE IN OTHER HORTICULTURAL CROP SPECIES

Viral, bacterial, and fungal diseases of ornamental plants cause major economic losses due to diminished productivity and quality. Chemical methods are available for control of fungal diseases and, to a lesser extent, for bacterial diseases, but there are no economically effective chemical controls for viral diseases except to control biotic vector species. Genetic transformation allows for the introduction of genes for specific, or in some instances broad spectrum, disease resistance into plant genotypes that have been selected for desirable horticultural characters. In contrast, the introduction of natural resistance by traditional breeding may take many cycles of breeding to combine disease resistance with desirable ornamental quality (Hammond et al., 2006).

An important need in long-lived forest trees is for disease resistance that withstands new and dangerous pathogens, mutations, and genetic shifts in pathogens. Experience from agriculture has allowed modeling of pathogen-host systems and the genetic variations within hosts and pathogens that permit coexistence. Overall, the diversity of behavioral of models, of the nature of resistance and virulence genes, and of the biology of both hosts and pathogens precludes any unique formula for forest ecosystem stability. Genetic diversity offers risk buffering for susceptibility to a new and serious pathogen or pathogen genotype (Burdon, 2001).

Scab caused by the pathogen *Venturia inaequalis* is considered the most important fungal disease of cultivated apple (*Malus × domestica*). 16 monogenic resistances against scab have been found in different *Malus* spp. and some of them are currently used in apple breeding for scab-resistant cultivars. To overcome long generation times and high breeding nursery maintenance costs, cloning of disease resistance genes and the use of the cloned genes for the transformation of high-quality apple cultivars is a strategic alternative. Toward this end, a bacterial artificial chromosome (BAC) contig spanning the apple scab resistance locus was constructed. The next step will be to identify and incorporate the scab resistance gene into a transformation vector for incorporation into the apple genome (Galli et al., 2010).

Pierce's disease (PD) is among the most important factors for predicting the success of grape production worldwide. The inheritance of resistance to *Xylella fastidiosa* (Xf), the bacterium which causes PD in *Vitis* sp., was evaluated in a factorial mating design consisting of 16 full-sib families with resistance derived from *V. rupestris × V. arizonica* interspecific hybrids. Direct estimation of bacterial populations on the host yielded the highest broad-sense heritability for resistance indicating that this measure of resistance was the least affected by V_E. Narrow-sense heritability of PD resistance was moderately high (0.37–0.52). Complex segregation analysis using the computer program "Statistical Analysis for Genetic Epidemiology" (SAGE™) strongly affirmed the existence of a major gene for PD resistance, accounting for 91% of the total genetic variance for PD resistance (Krivanek et al., 2005).

The cranberry (*Vaccinium macrocarpon*) fruit rot complex can cause severe crop loss and requires multiple fungicide applications each year to minimize damage within economic thresholds. 70% of nearly 600 *V. macrocarpon* genotype observational plots exhibited severe rot while approximately 6% showed some level of resistance. Families from resistant parents had a higher frequency of resistant progeny indicating the presence of additive genetic effects in these resistant trial entries and the potential for improving resistance through breeding. A few resistant progenies originated from susceptible parents suggesting non-additive variance for field fruit rot resistance also exists. DNA fingerprinting of resistant accessions identified several distinct types, offering potentially different sources of genetic resistance (Johnson-Cicalese et al., 2009).

Leaf rust caused by the fungus *Hemileia vastatrix* is the most devastating disease of arabica coffee (*Coffea arabica*). Experiments were conducted to gain insight into the mechanism of introgression into *C. arabica* of a leaf rust resistance gene from *C. liberica* and to identify linked molecular markers. Amplified fragment length polymorphism (AFLP; see Chapter 9) analysis of a population subset using 80 different primer combinations revealed that at least half of the total polymorphisms observed in the population were associated with introgression of *C. liberica* chromosome fragments. Linkage analysis revealed only three distinct introgressed fragments following a traditional backcross program corresponding to a total genomic length of 52.8 cM (Prakash et al., 2004).

Verticillium wilt, a vascular disease caused by the soilborne fungus *Verticillium dahliae* Kleb., currently represents the major cultivation constraint in many olive (*Olea europaea*) growing areas. Only a few traditional cultivars have exhibited high levels of disease resistance to *V. dahliae*. An olive breeding program was initiated aiming at obtaining new cultivars displaying both high levels of disease resistance and good horticultural characteristics (Arias-Calderón et al., 2015).

White potato (*S. tuberosum*) is plagued by a large number of fungal, bacterial, and viral disease pathogens. Development of potato varieties resistant to soft rot and early blight has been hindered by the scarcity of resistant germplasm. A diploid wild species, *S. brevidens*, exhibits significant resistance to both diseases. Using both molecular and cytogenetic approaches, Tek et al. (2004) demonstrated that a single copy of chromosome 8 from *S. brevidens* replaced a *S. tuberosum* chromosome 8 in a resistant selection.

Fusarium tuber rot, caused by *Fusarium solani*, is a major source of losses of tuber quality and quantity in caladium (*Caladium×hortulanum*) during storage and production. The effect of temperature on radial mycelial growth of nine *F. solani* isolates in vitro was determined, and all responded similarly to temperature variables, with optimal growth predicted to be at 30.5°C. The relationship of these temperatures to disease development was then determined for the most aggressively pathogenic *F. solani* isolate, and it was found that disease development in inoculated tubers was most extensive at low temperatures. The interaction between *F. solani* isolates and caladium cultivars was statistically highly significant indicating that cultivars were not equally susceptible to different pathogenic isolates of *F. solani* (Goktepe et al., 2007).

The fungal pathogen *Sclerotinia sclerotiorum* that causes "white mold" diseases is known to infect more than 400 plant species. It is a widespread problem in common bean (*P. vulgaris*) in the United States, causing more than 30% average yield losses. It was discovered that certain accessions of *P. coccineus* (commonly known as scarlet runner bean) possess a relatively higher level of resistance to *S. sclerotiorum* and can be used to introgress resistance into *P. vulgaris* (Schwartz et al., 2006).

Tomato chlorosis virus (ToCV) (genus Crinivirus, family Closteroviridae) is an emerging threat to tomato crops worldwide. Symptoms on fruits are not obvious but yield losses occur through decreased fruit size and number. Control of ToCV epidemics is difficult because the virus is transmitted by several whitefly vector species and wide host ranges allow the vectors to survive in wild habitats. Two sources of resistance to ToCV were identified, each derived from interspecific hybrids of *Solanum lycopersicum×S. peruvianum* and *S. chmielewskii*. Resistance was manifested by the impairment of virus accumulation and disease symptom expression, both under natural infection and after challenging with ToCV in controlled inoculations. Genetic control of resistance to ToCV infection was conferred by a major locus with mainly additive effects but also partial dominance for higher susceptibility. Also, an additive×dominance epistatic interaction with at least one additional gene was evident (García-Cano et al., 2010).

Also in tomato, tomato yellow leaf curl disease (TYLCV) is a devastating disease caused by a complex of begomoviruses. Almost all breeding for TYLCV resistance has been based on the introgression of the *Ty-1* resistance allele derived from *S. chilense*. Fluorescence in situ hybridization (FISH; see Chapter 3) analysis revealed two chromosomal rearrangements between *S. lycopersicum* and *S. chilense* in the genomic region of the *Ty-1* introgression. All recombination events were located on the long arm beyond the inversions, showing that recombination in the inverted region was absent (Verlaan et al., 2011).

A breeding program was developed from a *S. lycopersicum×S. pimpinellifolium* cross followed by several generations of self-pollination with applied selection for resistance to tomato yellow leaf curl virus (TYLCV) and tomato yellow leaf curl Sardinia virus (TYLCSV). Response to TYLCV infection of P_1, P_2, F_1, F_2, BC_1, and BC_2 generations fitted, for this line, led researchers to conclude that the resistance from *S. pimpinellifolium* was under monogenic control with partial recessive allelic interaction and incomplete penetrance (Pérez de Castro et al., 2007).

Resistance to anthracnose in chili pepper, caused by *Colletotrichum capsici* and *Co. acutatum*, was investigated in *C. baccatum* and *C. chinense*. Frequency distributions of disease scores in F_2 and BC_1 populations suggested a single recessive gene responsible for the resistance at mature green fruit stage and a single dominant gene for the resistance at ripe fruit stage. Based on phenotypic data, the two newly identified genes, *co4* and *co5*, from *C. baccatum* appeared to be different loci from the *co1* and *co2* previously identified from *C. chinense* and will be valuable sources of resistance to anthracnose in *Capsicum* spp. breeding programs (Mahasuk et al., 2009).

A muskmelon (*Cucumis melo*) breeding line and six plant introductions exhibited partial resistance to cucurbit leaf crumple virus (CuLCrV) in naturally infected field tests and controlled inoculation greenhouse tests. One accession was completely resistant in two greenhouse tests. Genetic resistance to CuLCrV in muskmelon was found to be recessive. Resistance in an accession appeared to be allelic with resistance in the other six cultivars based on F_1 data (McCreight et al., 2008).

Both lettuce mosaic virus (LMV) and corky root disease caused by *Sphingomonas suberifaciens* are diseases of *Lactuca* sp. that inflict serious yield and quality losses. Resistant×susceptible lettuce F_1 hybrids were backcrossed to the susceptible parent once, and the BC_1 and BC_1S_1 generations were screened for resistance to LMV. Seven lines showing resistance to LMV and corky root disease caused by *S. suberifaciens* were selected. From field observations, these breeding lines also had moderate resistance to the downy mildew pathogen (*Bremia lactucae*; Mou et al., 2007).

"Big vein" is also an important disease of lettuce. This disease is incited by Mirafiori lettuce big vein virus and vectored by the soil-borne fungus *Olpidium brassicae* (Woronin) P.A. Dang. The lettuce wild relative species *L. virosa* exhibits resistance to this pathogen. Experiments were conducted to determine the inheritance of resistance and the possibility of introgressing the gene into *L. sativa*. Following the successful production of *L. sativa×L. virosa* hybrids, plants were selected from resistant BC families were used as parents to create BC_2 progeny from crosses with high partial-resistant cultivars, intermediate partial-resistant cultivars, and susceptible cultivars to test for the presence of transgressive segregants. Complete resistance to big vein was not recovered in segregating populations, possibly a consequence of insufficient sampling of BCS$_2$ progeny or tight linkages. Variation for partial resistance was observed in all BC generations, and transgressive segregants were identified among BC$_2$ families from crosses using partially resistant and susceptible parents (Hayes and Ryder, 2007).

Both white rust and downy mildew are extremely important foliar diseases of spinach (*Spinacia oleracea*). Resistance of spinach to white rust (*Albugo occidentalis*) and races 3 and 4 of downy mildew (*Peronospora farinosa* f. sp. *spinaciae*) was quantified on several cultivars and breeding lines in separate field inoculation experiments. Cultivars and breeding lines that were selected for white rust resistance had significantly higher levels of field resistance to both white rust and races 3 and 4 of downy mildew relative to the experimental controls (Brandenberger et al., 1994).

Alternaria black spot of cruciferous vegetables, incited by different species of *Alternaria*, is a serious threat to Brassicaceae crop species throughout the world. The black spot pathogens, *A. brassicae* (Berk.) Sacc. and *A. brassicicola* (Schw.) Wiltsh., have a wide spectrum of *Brassica oleracea*, *B. rapa*, and *B. napus* hosts, such as head cabbage, Pak choi, Chinese cabbage, cauliflower, broccoli, kale, and other cultivated and wild crucifers. Infected seeds with spores on the seed coat or mycelium under the seed coat are the main avenue of distribution for these pathogens. The most economically feasible method of disease control is the development of resistant cultivars of Brassicaceae crop species, since transgenic strategies have not been successful. Black spot-resistant genotypes have not been reported in cultivated *Brassica* species, although cultivars differ in level of disease severity (Nowicki et al., 2012).

INSECT RESISTANCE AND TOLERANCE

Entire volumes have been devoted to the phenomenon of plant resistance or tolerance to herbivorous insects (Russell, 1978; Maxwell and Jennings, 1980; Dhan, 1986). Three fundamental strategies for breeding insect pest resistance have emerged from experimental studies: (i) introducing resistance transgenes into the host genome, (ii) exploiting natural variation in resistance already present in the crop gene pool, and (iii) introgressing resistance from sexually compatible wild relative species (Hervé, 2018). Option ii has been plagued by the lack of public support for GMO products, defined as plants that retain transformed interspecific DNA sequences. Methods to excise these extraneous DNA sequences are being developed to overcome this perception (Cotsaftis et al., 2002)

The phenomena of insect herbivory and nematode parasitism are consistent with many of the same tenets of the "disease triangle" (see above). Host and pathogen or herbivore genotype, environment, and plant developmental status play important roles in the degree of host/pest/parasite interaction for both disease and herbivory. For example, plant physiological growth stage and growing environment significantly interact with celery (*A. graveolens*) genotypes to influence the expression of genetic resistance of celery to the beet armyworm, *Spodoptera exigua* (Hubner). Celery genotypes were more resistant to beet armyworm during the warm season and as plants matured. Both plant age and growing environment proved interact significantly with plant genotype to magnify or suppress expression of genetic plant resistance to pests (Diawara et al., 1994).

The best example of a success story in insect pest resistance breeding strategy (i) is GMO varieties of agronomic crop species (corn, cotton) transformed with sequence variants of the *cry* gene from *Bacillus thuringiensis* (Mohan Babu et al., 2003). The crystal protein is produced in adequate quantities within transformed plant tissues for toxicity, and consumed by herbivores of the Lepidoptora and Coleoptora families where alkaline digestive antagonism results in disruption of feeding damage and, eventually, death of the herbivore. The toxin has no discernible effect on vertebrates, resistant to the toxin because they possess an acidic gut. The success of this story is dampened, however, by concerns that the insect pest populations will develop resistance to the crystal protein. Manifestations of Bt resistance in arthropod pest populations are already beginning to appear.

Before Bt, attempts to incorporate insect resistance or tolerance into plant breeding programs had met with mostly unremarkable results. Why has breeding for resistance to disease been so much more successful than for insect pest resistance? The answer may lie in a comparison of the mode of damage inflicted by the responsible biotic agents. Pathogens must gain entry to the interior of the plant by devious means, usually molecular trickery. Arthropods, in contrast to pathogens, possess specialized mouthparts to break and tear plant tissues or subsume plant fluids directly from the vascular system. The arthropod digestive system then extracts energy and nutrients from tissues or fluid similar to vertebrate herbivores. Since arthropod and vertebrate digestion are so similar, plants that are resistant to arthropod pests are often also problematic for vertebrate herbivores. The molecular interplay and subtle mechanisms that provide clues for disease resistance mechanisms are not as obvious for invertebrate herbivore pests. Insects have the ability, however, to detect volatile plant compounds in nanogram quantities, and have evolved physical and molecular mechanisms to evade or overcome plant molecular defenses. These characteristics have been successfully exploited for pest management and host pest resistance.

An excellent example of the application of engineered *Cry* genes for pest resistance is lepidopterous pests of broccoli (*B. oleracea* var. *italica*). Metz et al. (1995) used *A. tumefaciens* to transform flowering stalk explants of five broccoli genotypes with a construct containing the neomycin phosphotransferase gene and a *B. thuringiensis* (Bt) gene. Selected plants that gave 100% mortality of susceptible larvae allowed survival of a strain of diamondback moth (*Plutella xylostella*) that had evolved resistance to Bt in the field. F_1 hybrids between resistant and susceptible insects did not survive. Analysis of progeny from 26 resistant transgenic lines showed 16 that gave segregation ratios consistent with a single T-DNA integration. Later, Liu et al. (2012) showed that ovipositing *P. xylostella* adults could not discriminate between Bt and non-Bt or spinosad-treated and untreated hosts.

Insect-resistant crops is a very effective way to control insect pests in agriculture, and the development of such crops can be greatly enhanced by knowledge on plant resistance mechanisms and the genes involved. Plants have evolved diverse ways to cope with insect attack that has resulted in natural variation for resistance. Scientific studies of the molecular genetics and transcriptional background of this variation have facilitated the identification of resistance genes and processes that lead to resistance against insects. Insect-resistance mechanisms are still unclear at the molecular level, and exploiting natural variation with novel technologies (e.g. QTL) will contribute greatly to the development of insect-resistant crop varieties (Broekgaarden et al., 2011).

Differences do exist within many plant species with regard to insect resistance/tolerance. Some populations or individuals are less affected by a given insect pest than are others. The genetic basis for these differences, however, is usually very complicated. Resistance is often traced to the expression of a toxic compound that discourages feeding. Unfortunately, the toxin is also quite often effective on vertebrates as well.

A broad spectrum of factors has been found to be responsible for degrees of pest tolerance. For example, aromatic compounds released into the atmosphere in minute quantities, parts per trillion, are used by insects to find species on which they are adapted to feed. Minor differences in the structure or levels of these compounds can affect incident pest populations.

To illustrate this possibility, *Prunus davidiana*, a wild stonefruit species with poor fruit quality related to cultivated peach (*P. persica*), has been used as a source of resistance to pests and diseases in breeding programs. Two genotypes of *P. davidiana* were studied for fruit biochemical composition and compared to three genotypes of *P. persica* and two *P. persica* × *P. davidiana* hybrids. Correlations of compounds and levels with resistances were established. Fruit of *P. davidiana* clones had higher malic acid, neochlorogenic and crypto-chlorogenic acids and lower sucrose concentrations than the fruit of all *P. persica* genotypes. *P. persica* × *P. davidiana* hybrids had intermediate values between their parents for neochlorogenic acid concentration (Moing et al., 2003).

Other plant structures such as cuticular hairs or trichomes, and chemical excretions thereof, have been found to play a role in arthropod host preference (Steffens and Walters, 1991). Gross morphological alterations may occur, such as conversion of fibrous to tap roots and shedding of foliage. Damaged plant surfaces often undergo suberization and/or accumulate polyphenols that are unpalatable to insects and other invertebrate herbivores. Tolerant plants may exhibit an altered reaction to developmental signals, such as the failure to form galls to protect eggs and pupae.

Secondary plant metabolites are potentially of great value for providing robust resistance in plants against insect pests. Such metabolites often comprise small lipophilic molecules (SLMs), and can be similar also in terms of activity to currently used insecticides, for example, the pyrethroids, neonicotinoids and butenolides that provide more effective pest management than the resistance traits exploited by breeding. Crop plants mostly lack the SLMs that provide their wild ancestors with resistance to pests. Advances in genetic engineering of secondary metabolite pathways that produce insecticidal compounds and, more recently, SLMs involved in plant colonization and development, for example, insect pheromones, offer specific new approaches, but which will require more sophisticated insertion strategies than any that have been developed to date (Birkett and Pickett, 2014).

Antibiosis mechanisms are not always equally effective against different developmental stages of arthropod pest species, having a great impact on the breeding of cultivars with pest resistance with no adverse impacts on acceptance by the farmer or consumer. For example, both nymphal and adult spittlebugs [*Aeneolamia varia* (F.), *Aeneolamia reducta* (Lallemand), and *Zulia carbonaria* (Lallemand)] cause serious economic damage to susceptible brachiaria grass pastures in tropical climates. Both life stages are xylem feeders: nymphs feed primarily on roots and stems, whereas the adults feed mainly on foliage. Experimental studies revealed major inconsistencies between reaction to nymphs and reaction to adults on the same host genotype. Correlations between nymphal and adult damage scores were low, suggesting that resistance to the different life stages is mostly independent (Lófez et al., 2009; Cardona et al., 2010).

Arthropod pests and parasitic nematodes are attracted to hosts by intricate volatile chemical mechanisms that can be manipulated for breeding pest resistance. For example, sedentary plant endoparasitic nematodes can cause detrimental yield losses in crop plants making the study of detailed cellular, molecular, and whole plant responses to them a subject of importance. Plant susceptibility/resistance is mainly determined by the coordination of different signaling pathways including specific plant resistance genes or proteins, plant hormone synthesis and signaling pathways, and reactive oxygen signals that are generated in response to nematode attack. Crosstalk between various nematode resistance-related elements is an integrated signaling network regulated by transcription factors and small RNAs at the transcriptional, posttranscriptional, and/or translational levels (Li et al., 2015).

Despite that insects are much larger than microbial pathogens, and are mostly diploid, the population genetic dynamics of pathogenicity and herbivory are remarkably similar among the two disparate phylogenetic groups. Populations of habituated insect pests are notoriously mutable. Any barrier that is placed between them and their food source inevitably serves as a selective agent for mutations that result in the eventual ineffectiveness of the original barrier. As a consequence, genetic insect resistance, where initially successful, has almost always been overcome by mutant populations of pests.

MAS AND BREEDING FOR INSECT PEST RESISTANCE

The implementation and adoption of MAS in breeding for disease resistance is advanced compared to the implementation of MAS for insect and abiotic stress resistance. Examples of breeding in common bean using molecular markers reveal the role and success of MAS in gene pyramiding, rapidly deploying resistance genes via marker-assisted backcrossing, enabling simpler detection and selection of resistance genes in absence of the pathogen, and contributing to simplified breeding of complex traits by detection and indirect selection of quantitative trait loci (QTL) with major effects.

Cumulative mapping of disease resistance traits has revealed new resistance gene clusters while adding to others, and has reinforced the co-location of QTL conditioning resistance with specific resistance genes and defense-related genes. MAS breeding for resistance to insect pests has been accomplished for bean pod weevil, bruchid seed weevil, leafhopper, thrip, bean fly, and whitefly, including the use of arcelin proteins as selectable markers for resistance to bruchid seed weevil (Miklas et al., 2006).

A study was conducted to determine the specific chromosome(s) of resistant radish (*Raphanus sativus*) carrying the gene(s) for nematode resistance as a prerequisite to convert *Brassica napus* from a host into a trap crop for beet cyst nematode (BCN) *Heterodera schachtii*. The number of radish chromosomes in *Raphanobrassica* segregants was determined by fluorescence in situ hybridization, using a *Raphanus*-specific DNA probe and species source type was confirmed with polymorphic RAPD markers. Five distinct *B. napus-R. sativus* chromosome addition lines (comprising the whole set of nine radish chromosomes) were selected and crossed to *B. napus*. Chromosome d had a major resistance effect, whereas the presence/absence of the other radish chromosomes had nearly no influence on cyst number. BCN resistance was independent of the glucosinolate content in roots (Peterka et al., 2004). In subsequent experiments, a dominant major QTL explaining 46.4% of the phenotypic variability was detected in a proximal position of chromosome d (Budahn et al., 2009).

EXAMPLES OF AND EXPERIENCES WITH NEMATODE AND INSECT HERBIVORE RESISTANCE IN HORTICULTURAL CROP SPECIES

Undesirable apple fruit quality traits are frequently associated with pest- and disease-resistant cultivars and may be related to physiological resource allocation mechanisms. A study was conducted to evaluate the association between insect resistance and fruit quality in apple. There was a positive correlation between codling moth (*Cydia pomonella*) fruit infestation and fruit firmness. Additionally, a positive correlation was identified between shoot infestation by green apple aphid (*Aphis pomi*), fruit number as well as sugar content. The positive relationship of increased infestation by some pest insects and quality-determining fruit characteristics such as firmness or sugar content points to a possibly increased necessity for plant protection measures in apple cultivars producing high-quality fruits. One possible explanation of higher pest infestation in cultivars producing fruits with high quality is a tradeoff between resource allocation to defensive secondary metabolites or fruit quality (Stoeckli et al., 2011).

Resistance to the dagger nematode *Xiphinema index* is an important objective in grape rootstock breeding programs. This nematode not only causes severe feeding damage to the root system, but it also vectors grapevine fanleaf virus, the causal agent of fanleaf degeneration and one of the most severe viral diseases of grape. The dynamics of nematode numbers, gall formation, and root weight loss were investigated using a range of genotypes, soil mixes, and pot sizes over a 52-week period. Quantitative trait loci (QTL) for *X. index* resistance were found in *V. vinifera* among 255 marker loci. Results revealed that *X. index* resistance is controlled by a major QTL near marker VMC5a10 on chromosome 19 (Xu et al., 2008).

Brown et al. (2009) introgressed resistance to *Meloidogyne chitwoodi* from *Solanum bulbocastanum* into the gene pool of cultivated potato (*S. tuberosum*) using traditional backcrossing. A single dominant gene was found to be responsible for resistance to race 1 of the parasitic nematode. An additional form of resistance was discovered in certain advanced backcross clones; this form of resistance was inherited as a single dominant gene and mapped to chromosome 11 where other resistance factors were located. Stacking the two genotypes was speculated to be an effective strategy for durable *M. chitwoodi* resistance in *S. tuberosum*.

The aphid species *Macrosiphum euphorbiae* (Thomas) and *Myzus persicae* (Sulzer) (Hemiptera: Aphididae) are responsible for substantial yield reductions in potato (*S. tuberosum*) production due to direct phloem-feeding and by vectoring pathogenic viruses. Breeding aphid resistance from *S. chomatophilum* into the germplasm pool of cultivated potato presents an alternative means to control infestations and viral diseases. Aphid resistance from *S. chomatophilum* plant parts was assessed among accessions through evaluations of aphid performance and by assessing the impact of resistance on different aphid developmental stages. Accession and plant physiological age, but not aphid developmental stage, influenced all life-history parameters (Pompon et al., 2010).

The onion thrip, *Thrips tabaci* Lindeman (Thysanoptera: Thripidae), a worldwide pest of onion, *Allium cepa* and relatives, can reduce yield by more than 50%, and is even more problematic when it transmits Iris yellow spot virus (family Bunyaviridae, genus Tospovirus, IYSV). Because *T. tabaci* is difficult to control with insecticides and other strategies, field studies on onion resistance to *T. tabaci* and IYSV were conducted. Eleven of the 49 cultivars tested had very little leaf damage and were considered resistant to *T. tabaci*. The visual assessment indicated that all resistant cultivars had yellow-green colored foliage, whereas the other 38 had blue-green colored foliage. Two onion populations had the lowest infestations of *T. tabaci* suggesting the presence of strong antibiosis and/ or antixenosis. The other nine cultivars had variable infestations of *T. tabaci* indicating a possible combination of categories of resistance. Results indicated that potential exists for developing onion resistance to *T. tabaci* as part of an overall integrated pest management strategy, but suggest difficulties in identifying resistance to IYSV (Diaz-Montano et al., 2010).

The red spider mite *Tetranychus evansi* can cause up to 90% yield losses in cultivated tomato (*S. lycopersicum*) production fields. The wild tomato relative *S. hirsutum* is very resistant to arthropod herbivory. Studies on prospective sources of pest resistance in *S. hirsutum* have often found that leaf trichomes and their secretions were implicated. To better understand relationships among resistance, repellency, and 2,3-dihydrofarnesoic acid, a trichome-borne sesquiterpenoid spider mite repellent, two tomato (*S. lycopersicum*) cultivars were interbred with a highly resistant/ spider mite repellent accession of *S. hirsutum*. Backcross and F_2 generations were produced with each tomato cultivar (Snyder et al., 2005).

Further attempts were made to develop screening methods for tomato resistance to *T. evansi*. It is known that trichome morphology and chemical compounds present in tomato leaves are important to successful infestation by *T. evansi*. Resistance to *T. evansi* was evaluated in 84 arbitrarily chosen tomato accessions, and a significant difference in the number of *T. evansi* adults/leaf disk was found among accessions. The resistance mechanism of the tested tomato accessions was determined to be antixenosis (Fernandes et al., 2015).

Root maggot resistance from canola (*Brassica napus*) that originated from the weedy crucifer, *Sinapis alba*, was transferred to rutabaga (*B. napus* var. *napobrassica*) by traditional backcrossing. A population of doubled haploids was developed from *B. napus* × *S. alba* F_1 plants and screened in a field with a high population of root maggots. Resistant and susceptible isolines were identified from different crossing groups, and these isoline pairs were used to develop a biochemical selection protocol based on HPLC profiles where glucosinolates can be present as an aid to resistance breeding. Olfactory signals that attract root maggot are the isothiocyanates that are volatile breakdown products of glucosinolates (Malchev et al., 2010).

Nymphs and alates of the lettuce (*L. sativa*) aphid *Nasonovia ribisnigri* (Mosley) (Homoptera: Aphididae) were tested on ten lettuce cultivars with and 18 cultivars without the *N. ribisnigri* resistance gene *Nr*. Resistant and susceptible plants were identified after 3 days using whole plant bioassays. Longer-term no-choice tests using single leaves or whole plants resulted in no survival of *N. ribisnigri* on resistant plants, indicating great promise of the *Nr* gene for management of *N. ribisnigri*. Leaf disc bioassays were found to be ineffective for *N. ribisnigri* resistance screening on *L. sativa* (Liu and McCreight, 2006). Subsequently, McCreight (2008) found two new and potentially unique sources of resistance to *N. ribisnigri* in the wild lettuce relatives *L. serriola* and *L. virosa*.

II. BREEDING METHODS

Leafminer *Liriomyza sativae* (Diptera: Agromyzidae) damage is one of the main crop production issues for musk-melon (*C. melo*) production worldwide. Leafminer resistance is manifested by the death of larvae soon after they begin feeding on the leaf mesophyll and the result is leaf mines that are small and insignificant in terms of yield reduction. Populations with contrasting levels of resistance were obtained from the progenies of crosses of susceptible and resistant selections followed by successive self-pollinations used by the pedigree breeding method. One gene with complete dominance conditions resistance, and the mechanism of resistance was shown to be antixenosis (Celin et al., 2017).

Host-plant resistance could be a useful tool for managing weevil species *Cylas puncticollis* and *C. brunneus* that are major insect pests of sweet potato (*Ipomoea batatas*) in Africa. *Cylas* spp. resistance was evaluated in 134 sweet potato cultivars and landraces over two seasons in two agroecologically diverse locations. Several sweet potato cultivars expressed resistance to *Cylas* spp. and resistance characteristics were demonstrated to be quantifiable and thus potentially useful in plant-breeding (Muyinza et al., 2012).

METHODS TO MINIMIZE NEW RESISTANT GENOTYPES OF PATHOGENS AND PESTS AND ALSO AVOID DAMAGE TO BENEFICIAL SPECIES

Biological control is an important ecosystem service delivered by natural enemies of arthropod and nematode pests that can also reduce selection pressure for new resistant pest pest resistance genotypes. Together with breeding for plant defense, biological control constitutes one of the most promising alternatives to pesticides for controlling herbivores and parasites in sustainable crop production. Induced plant defenses may be promising targets in plant breeding for resistance against arthropod pests because they are activated upon herbivore damage and costs to the crop plant are only incurred when the defense is needed. By focusing on inducible resistance traits that are compatible with the natural arthropod enemies of pests and, specifically, traits that foster large sustainable communities of natural enemies, plant breeders can engineer more durable pest resistance systems (Pappas et al., 2017).

Kennedy et al. (1987) identified critical agricultural and ecological factors for assessing different host resistance modalities (antibiosis, antixenosis, tolerance) with respect to levels of insect resistance within the context of pest management requirements of different crops and cropping systems. They focused on the problem of maximizing the durability of insect resistance by minimizing selection for new and more virulent biotypes. Depending on context, the use of a particular modality and level of resistance may simplify pest management and reduce crop losses. Knowledge of the genetic variability of the target pest with regard to plant resistance and an understanding of the direct biological effects of the resistance on the insect is also essential. Selection pressure for virulent insect biotypes exerted by resistant crop cultivars was shown to be dependent on the modality of resistance as well as the agricultural and ecological context in which it is deployed.

Allegations of purported effects on the monarch butterfly, *Danaus plexippus* L., following continuous exposure of larvae to the pollen of plants transformed with Bt/*Cry* have appeared in popular media outlets. Dively et al. (2004) conducted enlightening experiments on the effects of natural deposits of *Bacillus thuringiensis* (Bt) and non-Bt milkweed pollen. This study was more pertinent to the question of collateral effects of Bt/*Cry* on the monarch butterfly since milkweed is the preferred food source for this insect, not *Z. mays*. Experimental exposure levels were similar to within-field levels that monarch butterfly populations might experience in typical field host populations (unlike some of the studies quoted in the popular media). 23.7% fewer larvae exposed to these levels of Bt/*Cry* pollen during anthesis reached the adult stage as compared to untreated controls. Exposure also prolonged the developmental time of larvae by 1.8 days and reduced the weights of both pupae and adults by 5.5%. The sex ratio and wing length of adults were unaffected by Bt/*Cry*. When considered over the entire range of the U.S. corn belt that represents only 50% of the breeding population of monarch butterflies, the risk to larvae associated with long-term exposure to Bt corn pollen is 0.6% additional mortality.

Dively (2005) extended this experimental strategy to explore the effects of Bt/*Cry* on a broader spectrum of arthropod species. A field experiment was conducted over three years to assess the effects of transgenic field corn expressing stacked lepidopteran-active *B. thuringiensis* (Bt)-derived VIP3A and Cry1Ab proteins on non-target arthropods. More than 500,000 arthropods were examined, representing 203 taxonomic groups in 112 families and 13 orders; 70% were saprovores, 13% were herbivorous insects, 14% were predators, and 3% were parasitoids. Biodiversity and community-level responses were not significantly affected by expression of the stacked VIP3A and Cry1Ab proteins. Significant changes in certain taxa did occur in the Bt plots, which were indirectly related to plant-mediated factors, prey density responses, and the absence of plant injury. Arthropod communities in the insecticide-treated plots displayed both negative and positive changes in the abundances of individual taxa. Changes in non-target communities in plots previously exposed to insecticides and the Bt hybrid did not carry over to the following growing season.

Genetic transformation of plants for insect pest control involves the insertion of genes that code for toxins that may be characterized as endogenous biopesticides. Some of these toxins, for example, Cry proteins of *Bacillus thuringiensis*, have a range of biological activity that extends beyond the targeted pests. Natural enemies of herbivores have received increasing attention because predatory arthropods are an important component of insect pest control. Natural enemies of herbivorous insects have largely been ignored in plant breeding programs although many examples show that plant breeding impacts the efficacy of biological control. Sustainable pest management will only be possible when negative effects on non-target, beneficial arthropods are minimized. The toxins produced in Bt plants retain their toxicity when bound to the soil, so the accumulation of these toxins is likely to occur. Earthworms may function as intermediaries through which the BT toxins pass on to other trophic levels within the soil ecosystem (Groot and Dicke, 2002). The biological control function provided by natural enemies is regarded as a protection goal that should not be harmed by the application of any new pest management tool.

Plants producing Cry proteins from the bacterium, *B. thuringiensis* (Bt), have become a major tactic for controlling pest Lepidoptera on cotton and maize, and risk assessment studies are needed to ensure they do not harm important natural enemies. However, using Cry protein-susceptible hosts as prey often compromises such studies. To avoid this problem, Tian et al. (2013) utilized pest Lepidoptera, cabbage looper (*Trichoplusia ni*), and fall armyworm (*Spodoptera frugiperda*) that were resistant to Cry1Ac produced in broccoli, Cry1Ac/Cry2Ab produced in cotton, and Cry1F produced in maize. There were no differences in any of the fitness parameters regardless if green lacewing (*Chrysoperla rufilabris*) predators consumed prey that had consumed Bt or non-Bt plants.

Maize "plus-hybrids" are cultivars consisting of a population mixture of cytoplasmic male-sterile (CMS) hybrids and unrelated male-fertile hybrids engineered to ensure adequate pollination of the whole stand. Experiments showed that plus-hybrids could make a large contribution to the coexistence of transgenic and conventional maize by biocontainment, that is, eliminating or reducing the release of transgenic pollen in *B. thuringiensis* or herbicide-tolerant populations (Munsch et al., 2010). It is likely that this strategy can be adapted to other crop species and growing systems.

The consensus from many independent experimental studies on non-targeted effects of insect resistance, and especially of transformed Bt genotypes, on crop and natural ecosystems is that impacts are minimal. Interactions between biological control agents (insect predators, parasitoids, and pathogens) and traditional pest-resistant or GMO crops exceed simple toxicological relationships, a priority for assessing the risks of GMO crop cultivars to non-targeted species (Lundgren et al., 2009).

References

Abalo, G., Tongoona, P., Derera, J., Edema, R., 2009. A comparative analysis of conventional and marker-assisted selection methods in breeding maize streak virus resistance in maize. Crop Sci. 49 (2), 509–520.

Agrios, G.N., 2005. Plant Pathology, fifth ed. Academic Press, New York, NY. 952 pp.

Alkimim, E.R., Caixeta, E.T., Sousa, T.V., Pereira, A.A., de Oliveira, A.C.B., Zambolim, L., Sakiyama, M.S., 2017. Marker-assisted selection provides arabica coffee with genes from other *Coffea* species targeting on multiple resistance to rust and coffee berry disease. Mol. Breed. 37 (1), 6.

Arias-Calderón, R., León, L., Bejarano-Alcázar, J., Belaj, A., de la Rosa, R., Rodríguez-Jurado, D., 2015. Resistance to *Verticillium* wilt in olive progenies from open-pollination. Sci. Hort. 185, 34–42.

Barchenger, D.W., Rodriguez, K., Jiang, L., Bosland, P.W., Hanson, S.F., 2017. Allele-specific CAPS marker in a *Ve1* homolog of *Capsicum annuum* for improved selection of *Verticillium dahliae* resistance. Mol. Breed. 37 (11), 134.

Beaver, J.S., Osorno, J.M., 2009. Achievements and limitations of contemporary common bean breeding using conventional and molecular approaches. Euphytica 168 (2), 145–175.

Benecke, M., 2001. A brief history of forensic entomology. Forensic Sci. Int. 120 (1–2), 2–14.

Ben-Naim, Y., Falach, L., Cohen, Y., 2015. Resistance against basil downy mildew in *Ocimum* species. Phytopathology 105 (6), 778–785.

Ben-Naim, Y., Falach, L., Cohen, Y., 2018. Transfer of downy mildew resistance from wild basil (*Ocimum americanum*) to sweet basil (*O. basilicum*). Phytopathology 108 (1), 114–123.

Biffen, R.H., 1905. Mendel's law of inheritance and wheat breeding. J. Agric. Sci. 1, 4–48.

Biffen, R.H., 1912. Studies in inheritance of disease resistance II. J. Agric. Sci. 4, 421–429.

Birkett, M.A., Pickett, J.A., 2014. Prospects of genetic engineering for robust insect resistance. Curr. Opin. Plant Biol. 19, 59–67.

Boyd, L.A., Ridout, C., O'Sullivan, D.M., Leach, J.E., Leung, H., 2013. Plant-pathogen interactions: disease resistance in modern agriculture. Trends Genet. 29 (4), 233–240.

Brandenberger, L.P., Correll, J.C., Morelock, T.E., McNew, R.W., 1994. Characterization of resistance of spinach to white rust (*Albugo occidentalis*) and downy mildew (*Peronospora farinosa* f.sp. *spinaciae*). Phytopathology 84 (4), 431–437.

Branson, K., 2018. Rutgers scientists develop new varieties of sweet basil. Rutgers Today, 2018. June 17, https://news.rutgers.edu/rutgers-scientists-develop-new-varieties-sweet-basil/20180601#.W2OwXsIpC1v.

Broekaert, W.F., 1996. Antifungal proteins and their application in the molecular breeding of disease-resistance plants. Acta Hort. 355, 209–211.

Broekgaarden, C., Tjeerd, A., Snoeren, L., Dicke, M., Vosman, B., 2011. Exploiting natural variation to identify insect-resistance genes. Plant Biotech. J. 9 (8), 819–825.

Brown, C.R., Mojtahedi, H., Zhang, L.H., Riga, E., 2009. Independent resistant reactions expressed in root and tuber of potato breeding lines with introgressed resistance to *Meloidogyne chitwoodi*. Phytopathology 99 (9), 1085–1089.

Bruce, T.J.A., 2012. GM as a route for delivery of sustainable crop protection. J. Exp. Bot. 63 (2), 37–541.

Budahn, H., Peterka, H., Mousa, M.A.A., Ding, Y., Zhang, S., Li, J., 2009. Molecular mapping in oil radish (*Raphanus sativus* L.) and QTL analysis of resistance against beet cyst nematode (*Heterodera schachtii*). Theor. Appl. Genet. 118 (4), 775–782.

Burdon, R.D., 2001. Genetic diversity and disease resistance: some considerations for research, breeding, and deployment. Can. J. For. Res. 31 (4), 596–606.

Cardona, C., Miles, J.W., Zuñiga, E., Sotelo, G., 2010. Independence of resistance in *Brachiaria* spp. to nymphs or to adult spittlebugs (Hemiptera: Cercopidae): implications for breeding for resistance. J. Econ. Entomol. 103 (5), 1860–1865.

Celin, E.F., da Silva, F.D., de Oliveira, N.R.X., de Cássia Souza, D.R., de Aragão, F.A.S., 2017. Simple genetic inheritance conditions resistance to *Liriomyza sativae* in melon. Euphytica 213 (5), 101.

Chen, H., Mehlenbacher, S.A., Smith, D.C., 2005. AFLP markers linked to eastern filbert blight resistance from OSU 408.040 hazelnut. J. Am. Soc. Hort. Sci. 130 (3), 412–417.

Chen, H., Mehlenbacher, S.A., Smith, D.C., 2007. Hazelnut accessions provide new sources of resistance to eastern filbert blight. HortScience 42 (3), 466–469.

Chen, M., Wang, F., Zhang, Z., Fu, J., Ma, Y., 2017. Characterization of fungi resistance in two autotetraploid apple cultivars. Sci. Hort. 220, 27–35.

Chisholm, S.T., Coaker, G., Day, B., Staskawicz, B.J., 2006. Host-microbe interactions: shaping the evolution of the plant immunity response. Cell 124, 803–814.

Colton, L.M., Groza, H.I., Wielgus, S.M., Jiang, J., 2006. Marker-assisted selection for the broad-spectrum potato late blight resistance conferred by gene RB derived from a wild potato species. Crop Sci. 46 (2), 589–594.

Cotsaftis, O., Sallaud, C., Breitler, J.C., Meynard, D., Greco, R., Pereira, A., Guiderdoni, E., 2002. Transposon-mediated generation of T-DNA- and marker-free rice plants expressing a Bt endotoxin gene. Mol. Breed. 10 (3), 165–180.

Coyne, C.J., Mehlenbacher, S.A., Smith, D.C., 1998. Sources of resistance to eastern filbert blight in hazelnut. J. Am. Soc. Hort. Sci. 123 (2), 253–257.

David, B.V., 2015. Elements of Economic Entomology. Brillion Publishing, New Delhi. 398 pp.

Day, P.R., 1974. The Genetics of Host-Pathogen Interaction. W. H. Freeman, San Francisco.

Day, P.R., 1992. Plant pathology and biotechnology: choosing your weapons. Annu. Rev. Phytopathol. 30, 1–13.

de Wit, P.J.G.M., 1981. Physiological Studies on Cultivar-Specific Resistance of Tomato Plants to *Cladosporium fulvum*. Landbouwuniversiteit te Wageningen, Wageningen. 128 pp.

Debener, T., Byrne, D.H., 2014. Disease resistance breeding in rose: current status and potential of biotechnological tools. Plant Sci. 228, 107–117.

den Boer, E., Pelgrom, K.T.B., Zhang, N.W., Visser, R.G.F., Niks, R.E., Jeuken, M.J.W., 2014. Effects of stacked quantitative resistances to downy mildew in lettuce do not simply add up. Theor. Appl. Genet. 127 (8), 1805–1816.

Dhan, P.S., 1986. Breeding for Resistance to Diseases and Insect Pests. Springer-Verlag, New York, NY. 222 pp.

Diawara, M.M., Trumble, J.T., Quiros, C.F., White, K.K., Adams, C., 1994. Plant age and seasonal variations in genotypic resistance of celery to beet armyworm (Lepidoptera: Noctuidae). J. Econ. Entomol. 87 (2), 514–522.

Diaz-Montano, J., Fuchs, M., Nault, B.A., Shelton, A.M., 2010. Evaluation of onion cultivars for resistance to onion thrips (Thysanoptera: Thripidae) and iris yellow spot virus. J. Econ. Entomol. 103 (3), 925–937.

Dively, G.P., 2005. Impact of transgenic VIP3A x Cry1Ab lepidopteran-resistant field corn on the nontarget arthropod community. Environ. Entomol. 34 (5), 1267–1291.

Dively, G.P., Rose, R., Sears, M.K., Hellmich, R.L., Stanley-Horn, D.E., Calvin, D.D., Russo, J.M., Anderson, P.L., 2004. Effects on monarch butterfly larvae (Lepidoptera: Danaidae) after continuous exposure to Cry1Ab-expressing corn during anthesis. Environ. Entomol. 33 (4), 1116–1125.

Dudai, N., Chaimovitsh, D., Fischer, R., Belanger, F., 2010. Aroma as a factor in the breeding process of basil. Acta Hort. (860), 167–171.

Duncan, R.W., Gilbertson, R.L., Singh, S.P., 2012. Direct and marker-assisted selection for resistance to common bacterial blight in common bean. Crop Sci. 52 (4), 1511–1521.

Elliott, F.C., 1958. Plant Breeding and Cytogenetics. McGraw-Hill, New York.

Epstein, L., Sukhwinder, K., Chang, P.L., Carrasquilla-Garcia, N., Guiyun, L., Cook, D.R., Subbarao, K., O'Donnell, K., 2017. Races of the celery pathogen *Fusarium oxysporum* f. sp. *apii* are polyphyletic. Phytopathology 107 (4), 463–473.

Ercolano, M.R., Carputo, D., Li, J., Monti, L., Barone, A., Frusciante, L., 2004. Assessment of genetic variability of haploids extracted from tetraploid (2n = 4x = 48) *Solanum tuberosum*. Genome 47 (4), 633–638.

Feldman, M.L., Andreu, A.B., Korgan, S., Lobato, M.C., Huarte, M., Walling, L.L., Daleo, G.R., Wehling, P., 2014. PLPKI: a novel serine protease inhibitor as a potential biochemical marker involved in horizontal resistance to *Phytophthora infestans*. Plant Breed. 133 (2), 275–280.

Fernandes, M.E.S., Fernandes, F.L., Silva, D.J.H., Picanço, M.C., Jham, G.N., Alves, F.M., 2015. Resistance of tomato accessions from the horticulture germplasm bank to red spider mites: exploring mite preference patterns and antixenosis mechanism. Int. J. Pest Manag. 61 (4), 284–291.

Flor, H.H., 1956. The complementary genic systems in flax and flax rust. Adv. Genet. 8, 29–54.

Galli, P., Patocchi, A., Broggini, G.A.L., Gessler, C., 2010. The Rvi15 (Vr2) apple scab resistance locus contains three TIR-NBS-LRR genes. Mol. Plant-Microbe Interact. 23 (5), 608–617.

García-Cano, E., Navas-Castillo, J., Moriones, E., Fernández-Muñoz, R., 2010. Resistance to tomato chlorosis virus in wild tomato species that impair virus accumulation and disease symptom expression. Phytopathology 100 (6), 582–592.

Goktepe, F., Seijo, T., Deng, Z., Harbaugh, B.K., Peres, N.A., McGovern, R.J., 2007. Toward breeding for resistance to *Fusarium* tuber rot in *Caladium*: inoculation technique and sources of resistance. HortScience 42 (5), 1135–1139.

Groot, A.T., Dicke, M., 2002. Insect-resistant transgenic plants in a multitrophic context. Plant J. 31 (4), 387–406.

Hammond, J., Hsu, H.T., Huang, Q., Jordan, R., Kamo, K., Pooler, M., 2006. Transgenic approaches to disease resistance in ornamental crops. J. Crop Improv. 17 (1–2), 155–210.

Hamon, C., Baranger, A., Coyne, C.J., McGee, R.J., Le Goff, I., L'Anthoene, V., Esnault, R., Riviere, J.-P., Klein, A., Mangin, P., McPhee, K.E., Roux-Duparque, M., Porter, L., Miteul, H., Lesne, A., 2011. New consistent QTL in pea associated with partial resistance to *Aphanomyces euteiches* in multiple French and American environments. Theor. Appl. Genet. 123 (2), 261–281.

Hart, L.P., Endo, R.M., 1978. The reappearance of *Fusarium* yellows of celery in California. Plant Dis. Rep. 62, 138–142.

Hayes, R.J., Ryder, E.J., 2007. Introgression of novel alleles for partial resistance to big vein disease from *Lactuca virosa* into cultivated lettuce. HortScience 42 (1), 35–39.

Heath-Pagliuso, S., Rappaport, L., 1990. Somaclonal variant UC-T3: the expression of *Fusarium* wilt. Theor. Appl. Genet. 80 (3), 390–394.

Heath-Pagliuso, S., Pullman, J., Rappaport, L., 1988. Somaclonal variation in celery: screening for resistance to *Fusarium oxysporum* f. sp. *apii*. Theor. Appl. Genet. 75 (3), 446–451.

Hervé, M.R., 2018. Breeding for insect resistance in oilseed rape: challenges, current knowledge and perspectives. Plant Breed. 137 (1), 27–34.

Hias, N., Svara, A., Keulemans, J.W., 2018. Effect of polyploidisation on the response of apple (*Malus* × *domestica* Borkh.) to *Venturia inaequalis* infection. Eur. J. Plant Pathol. 151 (2), 515–526.

Hollomon, D.W., Brent, K.J., 2009. Combating plant diseases—the Darwin connection. Pest Manag. Sci. 65 (11), 1156–1163.

Honig, J.A., Muehlbauer, M.F., Capik, J.M., Kubik, C., Vaiciunas, J.N., Mehlenbacher, S.A., Molnar, T.J., 2019. Identification and mapping of eastern filbert blight resistance quantitative trait loci in European hazelnut using double digestion restriction site associated DNA sequencing. J. Am. Soc. Hort. Sci. (in press).

Innes, R.W., Ameline-Torregrosa, C., Ashfield, T., Cannon, E., Cannon, S.B., Chacko, B., Chen, N.W.G., Couloux, A., Dalwani, A., Denny, R., Deshpande, S., Egan, A.N., Glover, N., Hans, C.S., Howell, S., Ilut, D., Jackson, S., Lai, H., Mammadov, J., del Campo, S.M., Metcalf, M., Nguyen, A., O'Bleness, M., Pfeil, B.E., Podicheti, R., Ratnaparkhe, M.B., Samain, S., Sanders, I., Ségurens, B., Sévignac, M., Sherman-Broyles, S., Thareau, V., Tucker, D.M., Walling, J., Wawrzynski, A., Yi, J., Doyle, J.J., Geffroy, V., Roe, B.A., Maroof, M.A.S., Young, N.D., 2008. Differential accumulation of retroelements and diversification of NB-LRR disease resistance genes in duplicated regions following polyploidy in the ancestor of soybean. Plant Physiol. 148 (4), 1740–1759.

Jayaraman, J., Segonzac, C., Cho, H., Jung, G., Sohn, K.H., 2016. Effector-assisted breeding for bacterial wilt resistance in horticultural crops. Hort. Environ. Biotech. 57 (5), 415–423.

Johnson-Cicalese, J., Vorsa, N., Polashock, J., 2009. Breeding for fruit rot resistance in *Vaccinium macrocarpon*. Acta Hort. Iss. 810 (1), 191–198.

Joosten, M.H.A.J., de Wit, P.J.G.M., 1999. The tomato—*Cladosporium fulvum* interaction: a versatile experimental system to study plant-pathogen interactions. Ann. Rev. Phytopathol. 37, 335–367.

Jordan, C.M., Endo, R.M., Jordan, L.S., 1988. Penetration and colonization of resistant and susceptible *Apium graveolens* by *Fusarium oxysporum* f.sp. *apii* race 2: callose as a structural response. Can. J. Bot. 66 (12), 2385–2391.

Kennedy, G.G., Gould, F., Deponti, O.M.B., Stinner, R.E., 1987. Ecological, agricultural, genetic, and commercial considerations in the deployment of insect-resistant germplasm. Environ. Entomol. 16 (2), 327–338.

Korbin, M., 2011. Molecular approaches to disease resistance in *Fragaria* spp. J. Plant Protect. Res. 51 (1), 60–65.

Kou, Y., Wang, S., 2010. Broad-spectrum and durability: understanding of quantitative disease resistance. Curr. Opin. Plant Biol. 13 (2), 181–185.

Krivanek, A.F., Famula, T.R., Tenscher, A., Walker, M.A., 2005. Inheritance of resistance to *Xylella fastidiosa* within a *Vitis rupestris* x *Vitis arizonica* hybrid population. Theor. Appl. Genet. 111 (1), 110–119.

Kuzuya, M., Hosoya, K., Yashiro, K., Tomita, K., Ezura, H., 2003. Powdery mildew (*Sphaerotheca fuliginea*) resistance in melon is selectable at the haploid level. J. Exp. Bot. 54 (384), 1069–1074.

Lacy, M.L., Berger, R.D., Gilbertson, R.L., Little, E.L., 1996a. Current challenges in controlling diseases of celery. Plant Dis. 80 (10), 1084–1091.

Lacy, M.L., Grumet, R., Toth, K.F., Krebs, S.L., Cortright, B.D., Hudgins, E., 1996b. MSU-SHK5: a somaclonally derived *Fusarium* yellows-resistant celery line. HortSci. 31 (2), 289–290.

Larsen, R.C., Kurowski, C.J., Miklas, P.N., 2010. Two independent quantitative trait loci are responsible for novel resistance to beet curly top virus in common bean landrace G122. Phytopathology 100 (10), 972–978.

Leadbetter, C.W., Capik, J.M., Mehlenbacher, S.A., Molnar, T.J., 2016. Hazelnut accessions from Russia and Crimea transmit resistance to eastern filbert blight. J. Am. Pomol. Soc. 70 (2), 92–109.

Li, Y., Huang, F., Lu, Y., Shi, Y., Zhang, M., Fan, J., Wang, W., 2013. Mechanism of plant-microbe interaction and its utilization in disease-resistance breeding for modern agriculture. Physiol. Mol. Plant Pathol. 83, 51–58.

Li, R., Rashotte, A.M., Singh, N.K., Weaver, D.B., Lawrence, K.S., Locy, R.D., 2015. Integrated signaling networks in plant responses to sedentary endoparasitic nematodes: a perspective. Plant Cell Rep. 34 (1), 5–22.

Li, W., Zhu, Z., Chern, M., Yin, J., Yang, C., Li, R., Cheng, M., He, M., Wang, K., Wang, J., Zhou, X., Zhu, X., Chen, Z., Wang, J., Zhao, W., Ma, B., Peng, Q., Chen, W., Wang, Y., Liu, J., Wang, W., Wu, X., Li, P., Wang, J., Zhu, L., Li, S., Chen, X., 2017. A natural allele of a transcription factor in rice confers broad-spectrum blast resistance. Cell 170, 114–126.

Lindhout, P., 2002. The perspectives of polygenic resistance in breeding for durable disease resistance. Euphytica 124 (2), 217–226.

Liu, Y.B., McCreight, J.D., 2006. Responses of *Nasonovia ribisnigri* (Homoptera: Aphididae) to susceptible and resistant lettuce. J. Econ. Entomol. 99 (3), 972–978.

Liu, Q., Ingersoll, J., Owens, L., Salih, S., Meng, R., Hammerschlag, F., 2001. Response of transgenic Royal Gala apple (*Malus* x *domestica* Borkh.) shoots carrying a modified cecropin MB39 gene, to *Erwinia amylovora*. Plant Cell Rep. 20 (4), 306–312.

Liu, X., Chen, M., Onstad, D., Roush, R., Collins, H.L., Earle, E.D., Shelton, A.M., 2012. Effect of Bt broccoli or plants treated with insecticides on ovipositional preference and larval survival of *Plutella xylostella* (Lepidoptera: Plutellidae). Environ. Entomol. 41 (4), 880–886.

Lófez, F., Cardona, C., Miles, J.W., Sotelo, G., Montoya, J., 2009. Screening for resistance to adult spittlebugs (Hemiptera: Cercopidae) in *Brachiaria* spp.: methods and categories of resistance. J. Econ. Entomol. 102 (3), 1309–1316.

Lori, G.A., Malbran, I., Mourelos, C.A., Wolcan, S.M., 2016. First report of *Fusarium oxysporum* f. sp. *apii* Race 2 causing Fusarium yellows on celery in Argentina. Plant Dis. 100 (5), 1020.

Lunde, C.F., Mehlenbacher, S.A., Smith, D.C., 2006. Segregation for resistance to eastern filbert blight in progeny of 'Zimmerman' hazelnut. J. Am. Soc. Hort. Sci. 131 (6), 731–737.

Lundgren, J.G., Gassmann, A.J., Bernal, J., Duan, J.J., Ruberson, J., 2009. Ecological compatibility of GM crops and biological control. Crop Prot. 28 (12), 1017–1030.

Mahasuk, P., Taylor, P.W.J., Mongkolporn, O., 2009. Identification of two new genes conferring resistance to *Colletotrichum acutatum* in *Capsicum baccatum*. Phytopathology 99 (9), 1100–1104.

II. BREEDING METHODS

Malchev, I., Fletcher, R., Kott, L., 2010. Breeding of rutabaga (*Brassica napus* var. *napobrassica* L. Reichenb.) based on biomarker selection for root maggot resistance (*Delia radicum* L.). Euphytica 175 (2), 191–205.

Maxwell, F.G., Jennings, P.R. (Eds.), 1980. Breeding Plants Resistant to Insects. John Wiley & Sons, New York, NY. 683 pp.

McCreight, J.D., 2008. Potential sources of genetic resistance in *Lactuca* spp. to the lettuce aphid, *Nasanovia ribisnigri* (Mosely) (Homoptera: Aphididae). HortScience 43 (5), 1355–1358.

McCreight, J.D., Liu, H.Y., Turini, T.A., 2008. Genetic resistance to cucurbit leaf crumple virus in melon. HortScience 43 (1), 122–126.

Mehlenbacher, S.A., Thompson, M.M., Cameron, H.R., 1991. Occurrence and inheritance of resistance to eastern filbert blight in 'Gasaway' hazelnut. HortScience 26 (4), 410–411.

Mehlenbacher, S.A., Brown, R.N., Nouhra, E.R., Gokirmak, T., Bassil, N.V., Kubisiak, T.L., 2006. A genetic linkage map for hazelnut (*Corylus avellana* L.) based on RAPD and SSR markers. Genome 49 (2), 122–133.

Metz, T.D., Roush, R.T., Tang, J.D., Shelton, A.M., Earle, E.D., 1995. Transgenic broccoli expressing a *Bacillus thuringiensis* insecticidal crystal protein: implications for pest resistance management strategies. Mol. Breed. 1 (4), 309–317.

Michelmore, R.W., 2003. The impact zone: genomics and breeding for durable disease resistance. Curr. Opin. Plant Biol. 6 (4), 397–404.

Miklas, P.N., Kelly, J.D., Beebe, S.E., Blair, M.W., 2006. Common bean breeding for resistance against biotic and abiotic stresses: from classical to MAS breeding. Euphytica 147 (1–2), 105–131.

Moffat, A.S., 2001. Finding new ways to fight plant diseases. Science 292 (5525), 2270–2273.

Mohan Babu, R., Sajeena, A., Seetharaman, K., Reddy, M.S., 2003. Advances in genetically engineered (transgenic) plants in pest management an overview. Crop Prot. 22 (9), 1071–1086.

Moing, A., Poessel, J.L., Svanella-Dumas, L., Loonis, M., Kervella, J., 2003. Biochemical basis of low fruit quality of *Prunus davidiana*, a pest and disease resistance donor for peach breeding. J. Am. Soc. Hort. Sci. 128 (1), 55–62.

Molnar, T.J., Baxer, S.N., Goffreda, J.C., 2005. Accelerated screening of hazelnut seedlings for resistance to eastern filbert blight. HortScience 40 (6), 1667–1669.

Molnar, T.J., Lombardoni, J.J., Muehlbauer, M.F., Honig, J.A., Mehlenbacher, S.A., Capik, J.M., 2018. Progress breeding for resistance to eastern filbert blight in the eastern United States. Acta Hortic. 1226, 79–85.

Molnar, T.J., Zaurov, D.E., Goffreda, J.C., Mehlenbacher, S.A., 2007. Survey of hazelnut germplasm from Russia and Crimea for response to eastern filbert blight. HortScience 42 (1), 51–56.

Molnar, T.J., Walsh, E., Capik, J.M., Vidyasagar, S., Mehlenbacher, S.A., Rossman, A.Y., Zhang, N.A., 2013. Real-time PCR assay for early detection of eastern filbert blight. Plant Dis. 97 (6), 813–818.

Mou, B., Hayes, R.J., Ryder, E.J., 2007. Crisphead lettuce breeding lines with resistance to corky root and lettuce mosaic virus. HortScience 42 (3), 701–703.

Muehlbauer, M., Capik, J.M., Molnar, T.J., Mehlenbacher, S.A., 2018. Assessment of the 'Gasaway' source of resistance to eastern filbert blight in New Jersey. Sci. Hort. 235, 367–372.

Mulbah, Q., Shimelis, H.A., Laing, M.D., 2015. Combining ability and gene action of three components of horizontal resistance against rice blast. Euphytica 206 (3), 805–814.

Munsch, M.A., Stamp, P., Christov, N.K., Foueillassar, X.M., Hüsken, A., Camp, K.-H., Weider, C., 2010. Grain yield increase and pollen containment by plus-hybrids could improve acceptance of transgenic maize. Crop Sci. 50 (3), 909–919.

Muyinza, H., Talwana, H.L., Robert, O., Mwanga, M., Stevenson, P.C., 2012. Sweetpotato weevil (*Cylas* spp.) resistance in African sweetpotato germplasm. Int. J. Pest Manag. 58 (1), 73–81.

Nelson, R.R., 1973. Breeding Plants for Disease Resistance: Concepts and Applications. Pennsylvania State Univ. Press, State College, PA. 401 pp.

Niks, R.E., Parlevliet, J.E., Lindhout, P., Bai, Y., 2011. Breeding Crops With Resistance to Diseases and Pests. Wageningen Academic Publishers, Wageningen. 198 pp.

Nowicki, M., Nowakowska, M., Niezgoda, A., Elåbieta, K., 2012. Alternaria black spot of crucifers: symptoms, importance of disease, and perspectives of resistance breeding. Veg. Crops Res. Bull. 76 (1), 5–19.

Nyasse, S., Efombagn, M.I.B., Kebe, B.I., Tahi, M., Despreaux, D., Cilas, C., 2007. Integrated management of *Phytophthora* diseases on cocoa (*Theobroma cacao* L.): impact of plant breeding on pod rot incidence. Crop Prot. 26 (1), 40–45.

Orton, W.A., 1909. The development of farm crops resistant to diseases. In: U.S. Dept. of Agric. Yearbook 1908, pp. 453–464.

Orton, T.J., 1982. Breeding celery disease resistance and improved quality. In: Pusateri, F. (Ed.), California Celery Research Program 1980–1981 Annual Report. California Celery Research Advisory Board Publ., Salinas, CA, pp. 41–62.

Orton, T.J., Durgan, M.E., Hulbert, S.D., 1984a. Studies on the inheritance of resistance to *Fusarium oxysporum* f. sp. *apii* in celery. Plant Dis. 68 (7), 574–578.

Orton, T.J., Hulbert, S.D., Durgan, M.E., Quiros, C.F., 1984b. UC1, *Fusarium* yellows-resistant celery breeding line. HortScience 19 (4), 594.

Pappas, M.L., Broekgaarden, C., Broufas, G.D., Kant, M.R., Messelink, G.J., Steppuhn, A., Wåckers, F., van Dam, N.M., 2017. Induced plant defences in biological control of arthropod pests: a double-edged sword. Pest Manag. Sci. 73 (9), 1780–1788.

Parlevliet, J.E., 1977. Plant pathosystems: an attempt to elucidate horizontal resistance. Euphytica 26 (3), 553–556.

Parlevliet, J.E., Zadoks, J.C., 1977. The integrated concept of disease resistance; a new view including horizontal and vertical resistance in plants. Euphytica 26 (1), 5–21.

Pedigo, L.P., Rice, M.E., 2009. Entomology and Pest Management, sixth ed. Prentice Hall, Upper Saddle River, NJ. 784 pp.

Pérez de Castro, A., Diez, M.J., Nuez, F., 2007. Inheritance of tomato yellow leaf curl virus resistance derived from *Solanum pimpinellifolium* UPV16991. Plant Dis. 91 (7), 879–885.

Peterka, H., Budahn, H., Schrader, O., Ahne, R., Schutze, W., 2004. Transfer of resistance against the beet cyst nematode from radish (*Raphanus sativus*) to rape (*Brassica napus*) by monosomic chromosome addition. Theor. Appl. Genet. 109 (1), 30–41.

Pompon, J., Quiring, D., Giordanengo, P., Pelletier, Y., 2010. Characterization of *Solanum chomatophilum* resistance to 2 aphid potato pests, *Macrosiphum euphorbiae* (Thomas) and *Myzus persicae* (Sulzer). Crop Prot. 29 (8), 891–897.

Prakash, N.S., Marques, D.V., Varzea, V.M.P., Silva, M.C., Combes, M.C., Lashermes, P., 2004. Introgression molecular analysis of a leaf rust resistance gene from *Coffea liberica* into *C. arabica* L. Theor. Appl. Genet. 109 (6), 1311–1317.

Puhalla, J.E., 1984. Races of *Fusarium oxysporum* f. sp. *apii* in California and their genetic interrelationships. Can. J. Bot. 62 (3), 546–550.

II. BREEDING METHODS

Pushpa, D., Yogendra, K.N., Gunnaiah, R., Kushalappa, A.C., Murphy, A., 2014. Identification of late blight resistance-related metabolites and genes in potato through nontargeted metabolomics. Plant Mol. Biol. Rep. 32 (2), 584–595.

Pyne, R.M., Koroch, A.R., Wyenandt, C.A., Simon, J.E., 2014. A rapid screening approach to identify resistance to basil downy mildew (*Peronospora belbahrii*). HortScience 49 (8), 1041–1045.

Pyne, R.M., Koroch, A.R., Wyenandt, C.A., Simon, J.E., 2015. Inheritance of resistance to downy mildew in sweet basil. J. Am. Soc. Hort. Sci. 140 (5), 396–403.

Pyne, R., Honig, J., Vaiciunas, J., Koroch, A., Wyenandt, C., Bonos, S., Simon, J., 2017. A first linkage map and downy mildew resistance QTL discovery for sweet basil (*Ocimum basilicum*) facilitated by double digestion restriction site associated DNA sequencing (ddRADseq). PLoS ONE 12 (9), e0184319.

Pyne, R.M., Wyenandt, C.A., Simon, J.E., Vaiciunas, J., Honig, J.A., 2018. Population structure, genetic diversity and downy mildew resistance among *Ocimum* species germplasm. BMC Plant Biol. 18 (1), 69.

Quiros, C.F., D'Antonio, V., Greathead, A.S., Brendler, R., 1993. UC8-1, UC10-1, and UC26-1: three celery lines resistant to Fusarium yellows. HortScience 28 (4), 351–352.

Ramu, S.V., Sreevathsa, R., Sarangi, S.K., Udayakumar, M., 2011. Plant protection through RNAi: alternative for pesticides and chemical approaches to pest control. In: Grover, L.M. (Ed.), Genetically Engineered Crops: Biotechnology, Biosafety, and Benefits. Nova Sci. Publ, New York, NY, pp. 211–222.

Rietman, H., Bijsterbosch, G., Cano, L.M., Lee, H.-R., Vossen, J.H., Jacobsen, E., Visser, R.G.F., Kamoun, S., Vleeshouwers, V.G.A.A., 2012. Qualitative and quantitative late blight resistance in the potato cultivar 'Sarpo Mira' Is determined by the perception of five distinct RXLR effectors. Mol. Plant-Microbe Interact. 25 (7), 910–919.

Roberts, P.D., Raid, R.N., Harmon, P.F., Jordan, S.A., Palmateer, A.J., 2009. First report of downy mildew caused by a *Peronospora* sp. on basil in Florida and the United States. Plant Dis. 93, 199.

Rommens, C.M.T., Salmeron, J.M., Oldroyd, G.E.D., Staskawicz, B.J., 1995. Intergeneric transfer and functional expression of the tomato disease resistance gene *Pto*. Plant Cell 7 (10), 1537–1544.

Russell, G.E., 1978. Plant Breeding for Pest and Disease Resistance 1st Edition. Elsevier Publ., Amsterdam. 496 pp.

Sambamurty, A.V.S.S., 2009. A Textbook of Plant Pathology. I. K. International Publishing House, Pvt., Ltd., New Delhi. 424 pp.

Sathuvalli, V.R., Chen, H., Mehlenbacher, S.A., Smith, D.C., 2011. DNA markers linked to eastern filbert blight resistance in "Ratoli" hazelnut (*Corylus avellana* L.). Tree Genet. Genomes 7 (2), 337–345.

Schmann, G.L., D'Arcy, C.J., 2009. Essential Plant Pathology, second ed. Amer. Phytopathol. Assoc., Washington, DC. 384 pp.

Schneider, R.W., 1984. Effects of nonpathogenic strains of *Fusarium oxysporum* on celery root infection by *F. oxysporum* f. sp. *apii* and a novel use of the Lineweaver-Burk double reciprocal plot technique. Phytopathology 74 (6), 646–653.

Schwartz, H.F., Otto, K., Teran, H., Lema, M., Singh, S.P., 2006. Inheritance of white mold resistance in *Phaseolus vulgaris* x *P. coccineus* crosses. Plant Dis. 90 (9), 1167–1170.

Scott, J.W., Jones, J.B., Hutton, S.F., 2011. Oblique resistance: the host-pathogen interaction of tomato and the bacterial spot pathogen. Acta Hort. Iss. (914), 441–447.

Seifi, A., Visser, R.G.F., Bai, Y., 2013. How to effectively deploy plant resistances to pests and pathogens in crop breeding. Euphytica 190 (3), 321–334.

Shepherd, K.W., Mayo, G.M.E., 1972. Genes conferring specific plant disease resistance. Science 175, 375–380.

Simko, I., Haynes, K.G., Jones, R.W., 2004. Mining data from potato pedigrees: tracking the origin of susceptibility and resistance to *Verticillium dahliae* in North American cultivars through molecular marker analysis. Theor. Appl. Genet. 108 (2), 225–230.

Snyder, J.C., Thacker, R.R., Zhang, X., 2005. Genetic transfer of a twospotted spider mite (Acari: Tetranychidae) repellent in tomato hybrids. J. Econ. Entomol. 98 (5), 1710–1716.

St Clair, D.A., 2010. Quantitative disease resistance and quantitative resistance loci in breeding. Annu. Rev. Phytopathol. 48, 247–268.

Steffens, J.C., Walters, D.S., 1991. Biochemical aspects of glandular trichome-mediated insect resistance in the Solanaceae. Am. Chem. Soc. Symp. Ser. 449, 136–149.

Stergiopoulos, I., van den Burg, H.A., Ökmen, B., Beenen, H.G., van Liere, S., Kema, G.H.J., de Wit, P.J.G.M., 2010. Tomato Cf resistance proteins mediate recognition of cognate homologous effectors from fungi pathogenic on dicots and monocots. Proc. Natl. Acad. Sci. U. S. A. 107 (16), 7610–7615.

Stevens, R.B., 1960. In: Horsfall, J.G., Dimond, A.E. (Eds.), Plant Pathology, An Advanced Treatise. vol. 3. Academic Press, New York, NY, pp. 357–429.

Stoeckli, S., Mody, K., Dorn, S., Kellerhals, M., 2011. Association between herbivore resistance and fruit quality in apple. HortScience 46 (1), 12–15.

Tek, A.L., Stevenson, W.R., Helgeson, J.P., Jiang, J., 2004. Transfer of tuber soft rot and early blight resistances from *Solanum brevidens* into cultivated potato. Theor. Appl. Genet. 109 (2), 249–254.

Thabuis, A., Lefebvre, V., Bernard, G., Daubeze, A.M., Phaly, T., Pochard, E., Palloix, A., 2004. Phenotypic and molecular evaluation of a recurrent selection program for a polygenic resistance to *Phytophthora capsici* in pepper. Theor. Appl. Genet. 109 (2), 342–351.

Tian, J.-C., Jurat-Fuentes, J.L., Xiang-Ping, W., Li-Ping, L., Romeis, J., Naranjo, S.E., Hellmich, R.L., Wang, P., Earle, E.D., Shelton, A.M., 2013. Bt crops producing Cry1Ac, Cry2Ab and Cry1F do not harm the green lacewing, *Chrysoperla rufilabris*. PLoS ONE 8 (3), 1–6.

Tomczyåska, I., Stefaåczyk, E., Chmielarz, M., Karasiewicz, B., Kamiåski, P., Jones, J.D.G., Lees, A.K., Jadwiga, A., 2014. A locus conferring effective late blight resistance in potato cultivar Sárpo Mira maps to chromosome XI. Theor. Appl. Genet. 127 (3), 647–657.

Van der Plank, J.E., 1963. Plant Diseases: Epidemics and Control. Academic Press, New York.

Vandemark, G.J., Fourie, D., Miklas, P.N., 2008. Genotyping with real-time PCR reveals recessive epistasis between independent QTL conferring resistance to common bacterial blight in dry bean. Theor. Appl. Genet. 117 (4), 513–522.

Verlaan, M.G., Szinay, D., Hutton, S.F., de Jong, H., Kormelink, R., Visser, R.G.F., Scott, J.W., Bai, Y., 2011. Chromosomal rearrangements between tomato and *Solanum chilense* hamper mapping and breeding of the TYLCV resistance gene Ty-1. Plant J. 68 (6), 1093–1103.

Visker, M.H.P.W., Keizer, L.C.P., van Eck, H.J., Jacobsen, E., Colon, L.T., Struik, P.C., 2003. Can the QTL for late blight resistance on potato chromosome 5 be attributed to foliage maturity type? Theor. Appl. Genet. 106 (2), 317–325.

Wyenandt, C.A., Simon, J.E., McGrath, M.T., Ward, D.L., 2010. Susceptibility of basil cultivars and breeding lines to downy mildew (*Peronospora belbahrii*). HortScience 45 (9), 1416–1419.

Wyenandt, C.A., Simon, J.E., Pyne, R.M., Homa, K., McGrath, M.T., Zhang, S., Raid, R.N., Li-Jun, M., Wick, R., Guo, L., Madeiras, A., 2015. Basil downy mildew (*Peronospora belbahrii*): discoveries and challenges relative to its control. Phytopathology 105 (7), 885–894.

Xu, K., Riaz, S., Roncoroni, N.C., Jin, Y., Hu, R., Zhou, R., Walker, M.A., 2008. Genetic and QTL analysis of resistance to *Xiphinema index* in a grapevine cross. Theor. Appl. Genet. 116 (2), 305–311.

Zhao, J., Fu, J., Li, X., Xu, C., Wang, S., 2009. Dissection of the factors affecting development-controlled and race-specific disease resistance conferred by leucine-rich repeat receptor kinase-type R genes in rice. Theor. Appl. Genet. 119 (2), 231–239.

Glossary

Adaptation, broad The tendency for a population to perform consistently across all environments.

Adaptation, specific The tendency for a population to perform differently according to environment.

Addition line Following interspecific hybridization, the hybrid is recurrently backcrossed to one of the parents. One potential outcome of this process is the appearance of an individual with the entire genome of one parent plus one pair of chromosomes from the other parent, know as an addition line.

Additive The interaction of alleles that is characterized by the phenotype of a heterozygote being between the values of the two parents. Complete additivity is also known as *codominance* and is defined by the heterozygote being halfway between the parents.

Agamospermy The formation of a seed without sexual reproduction.

Agrobacterium tumefaciens The bacterial pathogen that causes crown gall disease of woody perennial plants. The disease is incited by the introduction of T-plasmid DNA (T-DNA) into host plant tissues followed by transformation of *A. tumefaciens* genes into the host genome that cause the host to form a tumor on which only the pathogen can feed. The T-plasmid has been modified for use as a tool to achieve transformation in plants.

Allele Different forms of a gene due to differences in DNA sequence that culminate in a detectable difference in phenotype. A single gene may have many different alleles. If alleles interact in a heterozygote in a non-additive manner, dominance is greater than 0.

Allelic frequency Within a population the relative proportion of a specified allele among all potential haplotypes.

Allelomorph Synonymous with allele.

Allopolyploid The genome of an individual, genotype, or species is comprised of multiple sets of haplotypes that are different, usually from closely related species that have marginal sexual compatibility. This condition is expressed in terms of the number of sets of haplotype genomes: Allotriploid, allotetraploid, allopentaploid, etc.

Amphidiploid An amphiploid that behaves sexually like a true diploid (bivalent meiotic chromosome pairing).

Amphimixis The union of the sperm and egg in sexual reproduction.

Amphiploid An individual or species the genome of which contains haplotype genomes from two or more distinct species (see allopolyploid).

Androdioecy Individuals bear either hermaphrodite or staminate flowers.

Andromonoecy Individuals bear both hermaphrodite and staminate flowers.

Aneuploid An individual the genome of which is comprised of partial (not partial) haplotype genomes. Euploids are also characterized by chromosome numbers different from multiples of x.

Angiosperm A plant species that has true flowers and produces seeds enclosed within a carpel. The angiosperms are a large group and include herbaceous plants, shrubs, grasses, and most trees.

Animal and Plant Health Inspection Service (APHIS) A unit of the USDA that is charged with preventing plants, animals, and microbes from entering the U.S. that could adversely affect agriculture and human health. They are the organization that enforces the U.S. Federal Insecticide, Fungicide, and Rodenticide Act (FIFRA).

Anisogamy Individual progeny are not the product of the sexual fusion of gametes; instead, they are derived from maternal somatic cells.

Antibiosis An antagonistic association between two organisms in which one is adversely affected; one general strategy for breeding insect and nematode resistance.

Antisense A strategy of silencing mRNA function by introducing another DNA sequence by transformation the base sequence of which is in the complementary configuration to the genomic "sense" sequence. If the "sense" sequence is ATGCGTA, the antisense sequence is TACGCAT. Experimental evidence has shown that when the antisense sequence is transcribed, the sense mRNA is prevented from translation into a gene product.

Antixenosis An association between two organisms wherein an individual possesses characteristics that cause pest individuals from forming an antagonistic interaction (such as herbivory). Examples in plants repelling insects include leaf trichomes and trichome exudates or volatile allelochemicals.

Apomixis The process of meiosis during gametogenesis is bypassed developmentally such that the sexual zygote is supplanted by an asexual diploid somatic cell from the female sporophyte.

Apoptosis Localized cell death. In plant pathology this phenomenon is associated with race-specific and vertical disease resistance. Also known as hypersensitivity.

Apospory Development of the embryo from sporophyte tissue other than the archesporium (incipient egg), usually the maternal nucellus.

Asexual propagation The production of many individuals from a single individual by the results of successive mitotic cell division cycles without any intervening sexual (meiotic) cell divisions. Also known as vegetative propagation or "cloning."

Autopolyploid The genome of an individual, genotype, or species is comprised of multiple sets of the same haplotype. This condition is expressed in terms of the number of sets of haplotype genomes: autotriploid, autotetraploid, autopentaploid, etc.

Avirulence gene A locus in a pathogen that is part of the gene-for-gene resistance-avirulence system. An allele of this locus causes a substance to be produced, in many instances a glycoprotein, that the host detects eliciting a resistance (hypersensitivity/apoptosis) response. Other alleles at this locus may devoid the glycoprotein or cause changes such that the host no longer detects it, resulting in disease.

Bacillus thuringiensis A soil bacterium that is pathogenic on insect larvae. Different strains of *B. thuringiensis* are pathogenic on different species of insect. Certain *B. thuringiensis* strains produce δ endotoxins that are deposited as crystal proteins; these are the source of insecticidal activity. *Cry* loci encode the Bt endotoxins and have been excised from *B. thuringiensis*, cloned, and introduced into plants to produce Bt cultivars that are insect-resistant.

Backcross Following the hybridization of two genetically distinct individuals to create a F_1 hybrid, the subsequent mating of the F_1 individual to another that possesses one of the parental genotypes.

Bayesian statistics A type of model based on the degree of belief based on prior knowledge about the event or circumstance usually derived from previous experiments. Bayesian models are based on Bayes' Theorem: $P(A \text{ given } B) = [P(B \text{ given } A) \times P(A)]/P(B)$ where $P(B) \neq 0$.

Biodiversity The genetic variability that is unmanaged; the sum total of genetic variability present in Earth's natural and engineered habitats.

Biolistic A method of introducing foreign DNA into a plant cell pursuant to achieving transformation that uses rapid acceleration and DNA encapsulation. In effect, encapsulated DNA is "shot" into plant cells and tissues using a gun. The biolistic device is often referred to as a "gene gun."

Bivalent Two homologous chromosomes form a paired structure during meiotic prophase I and metaphase I that is either ring- or rod-shaped.

Breeders' seed The population that is considered to embody the highest level of genetic fidelity to the cultivar or inbred population that is subject to a seed increase.

Bulk A mixture of seeds or individuals of different origin or genotype.

Bulked segregant analysis (BSA) This approach identifies molecular markers associated with a trait of interest by genotyping DNA extracted from bulked samples of individuals at the trait's phenotypic extremes.

C-value A standardized measure of the total amount of DNA in a specified genome based on chromatin-specific dyes and light absorbance.

Center of Diversity A geographical area where a species, either domesticated or wild, was first domesticated and where genetic variability for crop improvement may be found.

Center of Origin A geographical area where the wild progenitors of a domesticated species first evolved.

Centromere A specialized structure of a chromosome for attachment to the microtubulin fibers of the mitotic and meiotic spindle followed by the separation of chromatids and movement to derivative cells. Recombination in the region of the centromere is often depressed as compared with other regions of the chromosome. Chromosomes are often "constricted" at or near the centromere, and this region is also referred to as a *primary constriction*.

Certified seed The population of individuals produced with parents from foundation seed using methods prescribed by the plant breeder (e.g., open pollination; enclosed cage with pollinators, etc.).

Chiasma A cytogenetic structure observed during prophase I of meiosis that is analogous to the exchange of chromatin between paired homologous chromosomes. The physical observation of a recombination event. Plural: chiasmata.

Chimera An individual that consists of a combination or mixture of cells that two or more distinct genotypes.

Chromatid In a binary chromosome, one of the two constituent rod-like elements.

Chromosome The subcellular organelle in which DNA and genes exist. Chromosomes are comprised of *chromatin*, an organized amalgam of nucleic acids and proteins. Chromosomes are also associated with DNA packaging mechanisms, DNA replication and repair enzyme complexes, transcription factors and molecular complexes, and mitotic and meiotic cell division functions related to the distribution of genes in descendant cells. Chromosomes are the visible manifestation of linkage groups. The number and morphology of chromosomes are used as criteria in systematics. The chromosome may be a single or a binary structure (see *chromatid*).

Chromosome banding The use of dyes to produce consistent patterns of binding and non-binding to chromatin that results in a reproducible pattern that is useful for identifying specific chromosomes or chromosome.

Chromosome elimination Following interspecific hybridization or somatic hybridization to form a cell or embryo that combines the genomes of disparate species, the genome of one parent is systematically excluded or degraded such that only the genome of the other parent remains. The genome loss process usually proceeds in increments of whole chromosomes. In embryos of sexual hybrids, the remaining genome is usually haploid. Therefore, chromosome elimination has been used as a method to access haploidy.

Cisgenesis Organisms that have been engineered using a process in which genes are artificially transferred (transformed) between organisms that could otherwise be conventionally bred.

Cleistogamy The pollination of a hermaphrodite flower while the calyx and/or corolla (lemma/palea in monocots) is/are closed, thus preventing heterogamy.

Colchicine A biological alkaloid derived from extracts of corms of the autumn crocus (*Colchicum autumnale*) that has many pharmaceutical and biological uses including binding to and inhibition of microtubulin, the muscle-like fibers that pull apart chromosomes during mitosis and meiosis. In plant breeding, colchicine is used to disrupt chromosome disjunction to induce polyploids to form. Colchicine is also used in cytogenetics to improve the resolution of metaphase chromosomes.

Combining ability An individual plant's tendency to transmit desirable characters to progeny following hybridization.

Composite cultivar A population comprised of two or more distinct genotypes that are mixed to constitute a single entity; usually by mixing, or bulking, of seeds of these genotypes.

Consultative Group for International Agricultural Research (CGIAR) The umbrella non-profit international organization that assumed control in 1971 of the International Agriculture Research Centers that were established after World War 2 by the Rockefeller Foundation.

Controlled environment agriculture Crop or plant production in structures such as greenhouses where key environmental parameters such as light quantity and quality, temperature, humidity, atmosphere, soil composition, and soil moisture are monitored and controlled.

copyDNA (cDNA) Reverse (RNA→DNA) oligonucleotides of mRNA transcripts.

Cosegregation Genes (usually one related to plant performance and the other a marker) that do not assort independently during meiosis. Therefore, they are linked, or cosegregate.

CRISPR/Cas9 CRISPR = Clustered Regularly Interspaced Short Palindromic Repeats, a type of complex molecular unit in prokaryotes that detects and changes the DNA of viral pathogens rendering them non-infective. The CRISPR/Cas9 is an invention comprised of an engineered CRISPR with a cas9 nuclease that, when combined with synthetic guide RNA can locate and modify genomic DNA sequences. CRISPR/Cas9 is a powerful tool in genome editing, one of several that are currently in widespread usage.

Cultivar A population of a plant species that has been genetically improved through plant breeding to the point that it is phenotypically distinct and successful in commerce or otherwise useful in some context.

Cytogenetics The study of the behavior of chromosomes during mitosis and meiosis as it relates to the science of genetics.

Cytotype The collective plastid DNA and mitochondrial DNA genotypes of an individual.

Deleterious allele or mutation An allele of a gene that imparts undesirable phenotypic effects on individuals and populations as compared to the wild-type or other prevailing alleles.

Diallel All pairwise crosses among a finite set of individuals or populations including self-pollinations and reciprocals. If there are x entities, the diallel would consist of x^2 hybrid progeny sets.

Dichogamy Separation in time or space between pollen maturation and dehiscence and stigma receptivity within a flower such that autogamy may not occur.

Dihaploid A restitution gamete or one that is produced by a tetraploid that contains two haplotype genomes.

Dihybrid cross The hybridization of two individuals that differ with respect to genotypes at two loci, performed to investigate the joint inheritance of these two genes.

Dioecy Individuals only bear pistillate or staminate flowers. A panmictic population usually consists of approximately 50% female and 50% male individuals.

Diploidization The tendency for a newly-formed polyploidy entity to exhibit progressively more bivalent and less multivalent pairing over many generations. Bivalent pairing leads to a higher proportion of viable gametes, or higher fitness, presumed to drive this process. Gene products that enforce homologous over homeologous chromosome pairing are the basis of this phenomenon.

Diplospory Development of the embryo sac from the archesporium, often following irregular meiosis; a form of agamospermy.

Discounted cash flow analysis (DCF) A valuation method used to estimate the attractiveness of an investment opportunity. DCF analyses use future free cash flow projections and discounts them using a required annual rate to arrive at present value estimates. A present value estimate is then used to evaluate the potential for investment. DCF is an excellent way to determine the net value of a plant breeding project because long time frames, monetary fluctuations, and economic risk are considered.

Disease resistance The active exclusion from or severe growth restriction of a pathogen in a plant that results in the absence of the disease phenotype.

Disease tolerance The tendency of an individual or population to exhibit a desirable level of performance or to mitigate the disease phenotype despite the presence of a pathogen.

Disease triangle The disease phenotype only occurs when a susceptible host is infected with a pathogenic agent under environmental conditions that allow the disease process to proceed.

Disease vector A biotic or abiotic agent that carries one or more pathogens and plays a role in the spread of the disease(s) to hosts.

Domestication The process by which wild or natural populations of plants and animals are changed genetically according to the needs of humans. Fossil evidence shows that the first domestication efforts began about 60,000 years ago. In plants, mass selection is presumed to have played a major role in the domestication of species.

Dominance The interaction of two alleles in a diploid individual that is characterized by the parental phenotype of one allele expressed by a heterozygote.

Double cross hybrid A population that is derived from four inbred parents that are hybridized to produce two F_1 hybrids (F_{1a} and F_{1b}) then F_{1a} and F_{1b} are hybridized to produce the double cross hybrid.

Double digest restriction-site associated DNA sequencing technology (ddRAD-seq) This is a reduced representation sequencing technology that samples genome-wide enzyme loci by next-generation sequencing. The ddRAD strategy is economical, time-saving, and requires little technical expertise or investment in laboratory equipment.

Drive to fixation Starting with an individual that is heterozygous at many loci, successive cycles of self-pollination leads to progeny that are progressively more homozygous at a rate of 0.50 per locus per generation.

Emasculation The removal of male gametes from a hermaphrodite flower.

Embryo rescue The phenomenon of embryo abortion is commonly observed following interspecific hybridization. Embryo rescue involves the excision of the embryo prior to abortion followed by culturing on an artificial medium that allows it to continue to grow and develop into a mature hybrid plant.

Epidemic The natural spread of a pathogen that results in the widespread occurrence of a disease in a population at a particular time and place.

Epiphytotic A plant disease that tends to recur sporadically and to affect large numbers of susceptible plants.

Epistasis The effects of interactions of two or more genes that determine or affect a specified phenotype is termed epistasis. If genes interact in a non-additive manner, epistasis is said to be present, in particular, the suppression of the effect of one such gene by another (i.e., $V_I > 0$). If genes interact in an additive fashion, epistasis is said to be absent (i.e., $V_I = 0$).

Etiology Chronological steps in the progression of the disease phenotype.

Euploid An individual or species the genome of which is comprised of entire (not partial) haplotype genomes. Euploids are also characterized by chromosome numbers in multiples of x.

Euchromatin Chromatin that is resistant to binding by dyes thus appearing microscopically translucent. Euchromatic regions of the genome are the sites of actively transcribed DNA sequences.

Fecundity A measurement or assessment of the ability of an individual or species to produce progeny; often measured in the number of progeny produced per individual.

Filial Generation or generations after the parental generation.

FISH (fluorescence in situ hybridization) A cytogenetic technique wherein a DNA oligonucleotides is covalently bonded to a fluorescent dye moiety, then annealed to chromatin or mitotic metaphase chromosomes that have been affixed to glass slides. The fluorescent dye indicates where the genomic DNA sequence complementary to the oligonucleotide is physically located within the genome (i.e., which chromosome and location on that chromosome).

Fitness The relative contribution of an individual as compared to all other individuals in a population to the pool of derived progeny that constitute the new population.

Fixation, genetic The genotype of an individual is 100% homozygous such that natural or forced self-pollination yields progeny that are genetically identical to the parent.

Foundation seed The population of individuals produced with parents from breeders' seed using methods prescribed by the plant breeder (e.g., open pollination; enclosed cage with pollinators, etc.).

Freedom to operate (FTO) A formalized process that generates a formatted report that articulates the ownership and licensing status of critical technologies for a targeted product or service.

Functional genomics The study of how genomes are structured and expressed to facilitate known biological processes in biochemistry and biophysics, physiology, reproduction, development, growth, energy transduction, transpiration, etc.

Gametophyte The haploid phase of the plant life cycle. In angiosperms and gymnosperms, gametophytes are bifurcated developmentally into a female (megagametophyte) and male (microgametophyte) forms.

Gene-for-gene theory of vertical resistance For each dominant gene for resistance in the host, there is a corresponding dominant gene for avirulence in the pathogen or, for each gene that confers virulence in the pathogen, there is a corresponding gene that confers resistance in the host. Host resistance genes for resistance are generally dominant (R) and genes for susceptibility recessive (r).

Gene pool, extant The total range of germplasm available for genetic improvement of a given species by sexual hybridization.

Generation time The elapsed time required for an individual, population, or species to undergo a complete life cycle, measured in days or years. In plant breeding, the term "seed to seed" is often applied to the same definition.

Genetic assortative mating Selection of parents based on similarity of genotypes.

Genetic disassortative mating Selection of parents based on dissimilarity of genotypes.

Genetic drift The tendency for allelic frequencies to change due to random effects associated with small population sizes.

Genetic load The sum of all masked recessive deleterious alleles that exist in a population.

Genetic vulnerability A condition brought about by the cultivation of plant populations that contain insufficient levels of genetic variability to cope with environmental fluxes such as thermal, moisture, salt, or biotic stresses.

Genetically Modified Organism (GMO) An individual that contains a genome with DNA sequences that were derived from a different species through the process of genetic transformation.

Genome The complete set of genes or genetic material present in a cell, tissue, or individual.

Genome saturation The mapping of polymorphic marker loci to the point that the entire genome is linked within ~2 cM to one of the markers. Depending on the magnitude of the genome, 300–500 mapped marker loci may be necessary to achieve saturation. Saturation ensures that all potential genes are linked to one or more markers.

Genome selection A form of marker-assisted selection in which genetic markers covering the entire genome are used so that all quantitative trait loci (QTL) are in linkage disequilibrium with at least one marker.

Genome-wide association studies (GWAS) These studies utilize collections of diverse, unrelated lines that are genotyped and phenotyped for traits of interest, and statistical associations are established between DNA polymorphisms and trait variation to identify genomic regions where genes governing traits of interest are located.

Genomic Breeding Value (GEBV) This is a statistic that is a predictor of how well a plant will perform as a parent for crossing and generation advance in a breeding pipeline based on the similarity of its genomic profile to other plants in a training population.

Genomic imprinting An epigenetic phenomenon in hybrids of two parents that causes genes to be expressed according to the patterns of one of the parents.

Genotype The genomic DNA sequence or set of alleles possessed by a specified sporophyte or derived gametophyte.

Genotype by environment interactions (GxE) The relative effect of genotype on phenotype is conditional on the environment. If $V_{GxE}=0$, a population is said to exhibit broad adaptation. If V_{GxE} is relatively high (as compared to V_p), a population is said to exhibit specific adaptation.

Genotyping-by-sequencing (GBS) A highly multiplexed genotyping system involving DNA digestion with different enzymes and the construction of a reduced representation library, which is sequenced using an NGS platform. It enables the detection of thousands of SNPs in large populations or collections of lines that can be used for mapping, genetic diversity analysis, characterizing polymorphism and orthologs in polyploids, and evolutionary studies.

Germplasm The raw material of plant breeding; any source of genes that may be incorporated into a plant breeding program. Traditionally, germplasm was distributed and maintained as seeds or pollen, but with the ability to transform plants, DNA sequences are also considered to be germplasm.

Germplasm release If a specific crop species is addressed adequately by plant breeders in the private sector, breeders in the public sector may develop and release improved populations that are not sufficiently refined to compete in the market, but are potentially valuable sources of genes and genetic combinations for the private sector. Such a strategy is termed a "germplasm release."

Germplasm repository A facility in or from which seeds, clonal somatic tissues and cells, pollen, and DNA sequences for plant breeding purposes are organized, characterized, stored, replenished, and disseminated.

Global Seed Vault A facility located on Svalbard Island (Norway) in the North Atlantic Ocean wherein seeds of plant species are stored under low temperature and humidity for access in case of global catastrophe.

Gymnosperm A plant species that has seeds unprotected by an ovary or fruit. Gymnosperms include the conifers, cycads, and ginkgos.

Gynodioecy Individuals bear either hermaphrodite or pistillate flowers.

Gynoecous A flower that contains only functional female structures; male structures are either absent or non-functional.

Gynomonoecy Individuals bear both hermaphrodite and pistillate flowers.

Half diallel All pairwise crosses among a finite set of individuals or populations including self-pollinations and half of the reciprocals. If there are x entities, the half diallel would consist of $(x^2+x)/2$ hybrid progeny sets.

Half-sib (sibling) mating Individuals that have one of two parents in common.

Haploid When used in reference to the plant life cycle, haploid refers to the gametic genomic constitution. When used in reference to ploidy, the basic genomic constitution of the indicated taxonomic group.

Haplotype The genotype of a gametophyte.

Hardy-Weinberg equilibrium In the absence of selection, mutation, drift, and introgression, population allelic frequencies tend to remain consistent within populations of progeny.

Hemizygous A diploid individual that possesses only one allele for a specified gene.

Herbivore An animal (mollusk, arthropod, mammal) that consumes plant tissues as the main dietary source of energy and nutrients.

Heritability The proportion of phenotypic variance that is attributable to genetic factors. The term is abbreviated as h^2 and is estimated by calculating V_G/V_p. h^2 may be expressed as a proportion ($0 \leq h^2 \leq 1$) or a percentage ($0 \leq h^2 \leq 100$). h^2 is usually synonymous with "broad sense heritability" (see "narrow sense heritability" above), but different researchers use the term in different ways.

Herkogamy Physical separation of anthers and stigma within a flower that prevents autogamy from occurring.

Hermaphrodite A gamete-bearing structure, or flower, that contains both functional male and female gametes.

Heterobeltiosis Heterosis in which the best parent is exceeded by the F_1 hybrid; also referred to as high parent heterosis (HPH).

Heterochromatin Chromatin that is not resistant to binding by dyes thus appearing microscopically dense. Heterochromatic regions of the genome are the sites DNA sequences that are not actively transcribed.

Heterogamy Individual progeny are derived from fusions of gametes from different individual parents.

Heteromorphy The co-existence of two or more genetically controlled hermaphrodite floral types within a population. The pin and thrum floral types of buckwheat are used to exemplify this phenomenon.

Heterosis The tendency of a hybrid individual to exhibit a phenotype or phenotypes that exceed the average of the parents is termed "heterosis." In cases where heterosis is of a magnitude that the phenotype or phenotypes of both parents are exceeded, the circumstance is known as heterobeltiosis, or high parent heterosis.

Heterozygous A diploid individual that possesses different alleles for a specified gene.

Homeologue Chromosomes that are related phylogenetically but that are distinct by DNA sequence divergence and/or differences in chromosome structure caused by DNA additions, deletions, or inversions during the course of population reproductive isolation and forces of evolution. Under some circumstances, such as the presence or absence of genes and gene products that promote or discourage homologous chromosome pairing, homeologous chromosomes undergo homology-driven pairing during meiosis.

Homogamy Pollen maturation and dehiscence within a flower occur simultaneously to allow autogamy to be affected.

Homologue Chromosomes that are nearly identical with respect to constituent DNA sequence.

Homozygous A diploid individual that possesses two identical alleles for a specified gene.

Horizontal resistance (HR) The disease is present in many degrees among individuals and populations. The degree of host resistance to the pathogen is inherited quantitatively. Resistance in the host does not depend on the genotype (or race) of the pathogen. HR has been demonstrated to be more "durable" than vertical resistance.

Horticulture The science and art of growing plants to produce fruit, nut, spice, specialty phytochemical, vegetable, ornamental, and nursery products. The field of horticulture also includes plant conservation, landscape restoration, soil management, landscape and garden design, construction, and maintenance, and arboriculture.

Host differential A set of host genotypes that are used to discern the pathogenic race or strain of a specific pathogen by observing the pattern of disease vs. non-disease caused by the pathogens on hosts.

Hybrid An individual or population that combines haplotypes from two distinct parents. The hybrid population structure is used extensively as a basis for cultivars in many crop species.

Ideotype A simplified graphic representation of a phenotype.

Inbred An individual or population that has been driven towards panhomozygosity by successive disassortative matings (self-pollination, sib- or half-sib) or by doubled haploidy.

Inbreeding depression The reduced biological fitness of a given population as a result of inbreeding or polyhomozygosity/panhomozygosity.

Intellectual property Any idea or know-how that was developed without public disclosure that may be subject to ownership and protection by prevailing laws.

Interspecific hybridization The fusion of gametes from different species to form a hybrid the genome of which consists of the haplotypes of both species. Isolation barriers present limits on the genetic distance between species that may be sexually hybridized. Somatic hybridization may be used to circumvent sexual isolation barriers. Interspecific hybrids are often unstable cytogenetically both in somatic tissues and during meiosis.

Introgression The transposition of an allele or alleles or genomic DNA sequences from one population to another by recurrent mating. In nature, the new allele or alleles are subjected to natural selection. In plant breeding, the allele or alleles are subjected to artificial or surrogate (e.g., marker assisted) selection.

Isolation distance The minimum distance between two populations to ensure that no mixture of the gametes from the populations occurs. This term is used primarily for replication or increase of a population genetic structure by open pollination. Factors such as vector biology and gametophyte longevity and accessibility are used to determine the recommended isolation distance.

Isozyme Different forms of enzymes that catalyze the same biochemical transformation. Isozymes may be distinguished by amino acid sequencing or, more readily, by observing different electrophoretic migration velocities on a stained gel matrix. Isozyme loci were the first markers used as surrogates in MAS strategies.

Karyotype The number and morphology of the chromosomes that constitute the genome of an individual or species. The karyotype is usually a graphic or photographic representation of metaphase chromosomes ordered from largest to smallest.

Landrace A population of a plant species that is partially domesticated with respect to present-day cultivars of that species. Landraces are considered important sources of valuable genes for plant breeding.

Lethal dose 50 (LD$_{50}$) The dose of a mutagen or toxic substance that produces 50% mortality.

Linkage A departure from an independent assortment of genes in segregating populations (e.g., 9:3:3:1 in F_2 with dominance). Linkage is a consequence of the physical proximity of genes on chromosomes or linkage groups; the distance between genes is negatively associated with the degree of departure from the random assortment. A "tight" linkage is one in which two genes are very close together such that recombination events between loci during meiosis is rare. If two genes are very tightly linked, one may be used as a selection criterion for the other, the theoretical basis of marker-assisted selection.

Linkage in coupling Two desirable alleles are linked together is cis configuration.

Linkage in repulsion Two desirable alleles at linked loci are in the trans configuration. In other words, one desirable and one undesirable allele are linked in cis configuration.

Locus The physical site within the genome of a species or individual at which a specific gene is located.

MADS-box A physical genomic region consisting of genes, the products of which are DNA binding proteins that modulate transcription, or "transcription factors." Different configurations have different functions in development and physiology. One major class of MADS-box genes is specific to floral development. MADS is an acronym for MCM1 from *Saccharomyces cerevisiae*; AGAMOUS from *Arabidopsis thaliana*; DEFICIENS from *Antirrhinum majus*; SRF from *Homo sapiens* indicating that these DNA sequences have been highly conserved during evolutionary time.

Maintainer Applicable during the development of inbreds for hybrid seed production using cytoplasmic male sterility. An individual or population that possesses the nuclear genotype rr (sterile) and the cytoplasmic genotype F (fertile); isogenic otherwise with the "A" line (incipient female parent) except for the cytoplasmic genotype that is S (sterile).

Male sterility A heritable system that causes hermaphrodite flowers to become gynoecious by a variety of mechanisms that eliminate male gametes. Male sterility may be controlled by genes that are nuclear, cytoplasmic, or interaction of nuclear and cytoplasmic. The cytoplasmic genes that affect male sterility have been shown to be mitochondrial.

Map, cytogenetic The representation of the genome of a species by morphological features and in situ hybridizations of chromosome and segments of chromosomes.

Map, genetic The representation of the genome of a species by linkage of genes as determined in crosses of polymorphic parents.

Marker A genomic DNA sequence that may be visualized using electrophoresis, activity dyes for gene products, direct affinity dyes, homology-based hybridization with dye or radioisotope tag, and homology-based PCR amplification. Marker loci may be located anywhere in the genome and are not necessarily associated with gene products or expression.

Marker-aided recurrent selection (MARS) The application of selection for neutral-function molecular markers to improve the effectiveness or efficiency of recurrent selection.

Marker-assisted back-crossing (MABC) In this form of marker-assisted selection, a genomic locus (gene or QTL) associated with a desired trait is introduced into the genetic background of an elite breeding line through several generations of backcrossing.

Marker-assisted selection (MAS) A method in which quantitative trait loci (QTL) are used as highly heritable surrogates to select for desirable alleles at linked loci that control the targeted phenotype.

Mass selection A plant breeding method by which desirable individuals are recurrently selected, and seeds from those individuals are harvested and mixed to constitute the next generation. Only the female parent is controlled in this method. Most crop species were domesticated by humans prior to the 20th century CE by mass selection.

Material Transfer Agreement (MTA) A written document that spells out the terms to which signed parties agree in sharing proprietary information or materials. The enforcement of the MTA is by torte or contract law.

Megagametophyte The mature female gametophyte, analogous to the egg cell.

Megasporogenesis The process by which the egg cell is produced from sporophyte tissues.

Meiosis The cellular cytogenetic process by which a sporophyte produces haploid gametes. The process is characterized by recombination of homologous genomic segments within chromosomes and the reduction in ploidy from 2n to n that results in the unequal distribution of alleles to gametes known as *segregation*.

Mendel's Laws (i) Genes exist in two copies, (ii) one copy of each gene from each parent, and (iii) some genes mask others, or are dominant over others.

Metabolome The total number of metabolites present within an individual, cell, or tissue.

Microgametophyte The mature male gametophyte, analogous to pollen.

microRNA (miRNA) A short (20–25 base) non-coding RNA molecule that functions in mRNA silencing and post-transcriptional regulation of gene expression.

Microsporogenesis The process by which pollen is produced from diploid sporophyte tissues.

Micropropagation Asexual or clonal increases of individuals or populations performed using single somatic cells or small tissue masses as the basic unit of propagation.

Mitosis The cellular cytogenetic process by which one cell gives rise to two genetically identical derivative cells. Mitosis is the basis of cloning. On rare occasions, genetic variability is introduced into individuals during successive mitotic cell cycles.

Mitotype A mitochondrial haplotype

Monoculture The absence of genetic variability and biodiversity in agricultural ecosystems.

Monoecy A mating system wherein separate pistillate and staminate flowers are borne on individuals.

Monoploid A genome consisting of an entire set of genes from the basic haplotype of a taxonomic unit that includes a polyploidy series of evolutionary derivatives.

Monosomic A diploid individual that only has one copy of a specific chromosome instead of two.

Multiline cultivar A population that is an admixture of pure lines that are identical to each other at most genetic loci, but differ with regard to one or a few loci that are central to the issue of vulnerability.

Multi-parent advanced generation inter-cross (MAGIC) A type of multi-parent population developed from four to eight diverse founder lines, generated to increase the precision and resolution of QTL mapping because of the larger number of alleles and recombination events compared to bi-parental mapping populations.

Mutagen A chemical (e.g., ethyl methanesulfonate) or physical agent (e.g., electromagnetic wave lengths such as UV radiation or ionizing radiation [γ or X-rays]) that increases the natural or spontaneous mutation rate.

Mutagenesis The use of a mutagen on an individual or gamete to create new genetic variability.

Mutation A naturally-occurring or induced change to the genome or genomic DNA of an organism that is inherited by its progeny.

Mutation breeding A plant breeding method that incorporates a mutagenesis step followed by selfing and selection to produce new desirable alleles.

Mutation rate The frequency or probability that a specific mutation will occur under a standard circumstance of time and genomic measure. Typically, the mutation rate is expressed in terms of per locus per generation. This parameter is highly variable according to organism and locus, ranging from 10^{-4} to 10^{-8} under natural conditions.

n The gametic chromosome number.

Narrow sense heritability The proportion of phenotypic variance that is attributable to additive genetic factors. The term is abbreviated in this textbook as H^2_{NS} and is estimated by calculating V_A/V_P. H^2_{NS} may be expressed as a proportion $(0 \leq H^2 \leq 1)$ or a percentage $(0 \leq H^2 \leq 100)$.

National Laboratory for Genetic Resources Preservation (NLGRP) An operating unit and facility of NPGS that provides for comprehensive redundancy of all germplasm accessions maintained at Plant Introduction Stations and Clonal Repositories. NLGRP is located in Fort Collins, CO (USA) and operates in cooperation with Colorado State University.

National Plant Germplasm System (NPGS) A unit of the U.S. Department of Agriculture USDA), Agricultural Research Service (ARS) that is concerned with the collection/acquisition, cataloging, long-term storage, and dissemination of seeds and asexual propagules to worthy germplasm users. The internet-based Germplasm Resources Information Network (GRIN) is the main portal by users into this organization.

Nested association mapping (NAM) NAM combines advantages of linkage and association mapping and eliminates disadvantages of both; it takes into consideration recent and historical recombination events, facilitating high resolution mapping.

Net present value (NPV) A statistic used in connection with the discounted cash flow analysis to valuate a business opportunity. NPV = {a[1 − (1 + i)^N]}/i where a = annual net cash flow; i = interest or discount rate; and N = number of years.

Open pollination A population produced by the uncontrolled mating of a set of parent plants. Abbreviated OP.

Ortholog Two DNA sequences that are related phylogenetically, the term usually applied to genes coding functional gene products, the products of which having similar or dissimilar functions.

Outcrossing rate The percentage of matings within a population that originate from two different parents.

Overdominance The interaction of alleles that is characterized by the phenotype of a heterozygote being greater than the values of either of the two parents.

Panheterozygous A diploid genotype that is 100% heterozygous at all constituent loci.

Panhomozygous A diploid genotype that is 100% homozygous at all constituent loci. A doubled haploid individual is an example of a panhomozygote.

Panmictic mating See random mating.

Parasite An organism that lives in or on another organism (the host) and benefits by deriving nutrients at the host's expense.

Parental gamete Following the hybridization of two unlike individuals, the genotype of a specified gamete is the same as one of the parents.

Parthenocarpy A condition during the formation of a fruit in which the presence of seeds is not a necessary condition, leading to seedless fruits. Typically, if seeds fail to develop, fruits are aborted. Parthenocarpy is conditioned by genes that uncouple seed and fruit development.

Passport Within the context of NPGS and germplasm management, the passport is a spate of standardized information on each PI that facilitates the organization of the GRIN database and provides information to prospective germplasm users.

Patent A government authority or license conferring a right or title to an invention or process for a set period, especially the sole right to exclude others from making, using, or selling the invention. In effect, a patent is a limited monopoly for a period of up to 20 years. An invention is patentable if it is novel, useful, and non-obvious.

Pathogen A bacterium, virus, fungus, or other microorganism that can cause disease.

Pathogen race Different isolates of a pathogen exhibit different reactions on a host genotype. Different host genotypes exhibit different compatibility reactions to the same pathogen isolate, explained by the existence of pathogen isolates that differ only with respect to genotype at the *Avirulence* (*Avr*) locus; known as races. The existence of pathogen races is usually associated with vertical disease resistance.

Pedigree A system of familial relationships, or genealogical record.

Pedigree book or database An informal record of genetic and parental relationships and relative performance of breeding lines maintained by the plant breeder.

Phenogenesis The developmental process by which a specific phenotype occurs.

Phenome The set of all phenotypes expressed by a cell, tissue, organ, individual, population, or species.

Phenotype The outward appearance or metric measurement of an individual.

Phenotypic assortative mating Selection of parents based on similarity of phenotypes.

Phenotypic disassortative mating Selection of parents based on dissimilarity of phenotypes.

Phenotypic plasticity The ability of an individual to alter phenogenesis according to the environment to improve fitness. Phenotypic plasticity tends to decrease h^2.

Phenotypic variance The imputed value for a specified population and phenotype is $V_P = 1/n \sum (x - \mu)^2$ where n = sample size, x = value of individual, and μ = population mean. V_P may be partitioned into constituent sources of variation V_G (genetic), V_E (environmental), and V_{GxE} (genotype × environment interactions).

Phytohormone A molecule that is synthesized and excreted by an individual plant that controls or regulates germination, growth, metabolism, reproduction, or other physiological and developmental factors.

Phytosanitary certificate A document issued by a representative of the agricultural authority certifying that a specified lot of biological material is free from contamination by undesirable abiotic and biotic agents. Such a certificate is usually required to move plant material and seeds between jurisdictions.

Pistillate flower A flower that bears only female gametes.

Plant breeding Any human method of plant propagation that results in a permanent, heritable change.

Plant breeding algorithm An iterative process that starts with an objective followed by germplasm acquisition, recurrent cycles of controlled mating and selection, testing of the best selected populations vs. commercial standards, and release of the population(s) into commerce.

Plant introduction (PI) A population that is maintained as germplasm by an individual or organization, usually as seeds or asexual propagules. The PI is generally assigned a name or number and a set of descriptors that are used by plant breeders to determine utility in breeding programs. The NPGS lists all PIs on the GRIN internet-accessible database. PI is also known as *accession*.

Plant Variety Protection (PVP) Plant breeders are granted up to 25 years of exclusive control over new, distinct, uniform, and stable sexually reproduced or tuber propagated plant varieties; under the Plant Variety Protection Act of 1970, Federal Statute.

Plasmid An extrachromosomal DNA element. The term generally applies to prokaryotes and cytoplasmic organelles. Plasmids are usually circular, possessing no 5′ or 3′ ends, but may be linear. In plant breeding, plasmids are primarily used to achieve genetic transformation, for example the T-plasmid of *Agrobacterium tumefaciens* that is used to introduce and integrate into the host genome desirable genes that are absent from the host gene pool.

Pleiotropy A phenomenon wherein a gene appears to exert effects on two or more independent phenotypes. In plant breeding, pleiotropy is usually a concern when a gene exerts positive effects on one phenotype and negative effects on the other(s).

Plus hybrid A population that consists of a genetic blend of cytoplasmic male-sterile (CMS) hybrids and unrelated male-fertile hybrids to ensure the pollination of the entire population. This type of population structure has been used primarily in maize.

Pollen quartet The four products of microsporogenesis are briefly attached in a four-unit structure prior to separating and becoming independent pollen grains. Also known as a "pollen tetrad."

Pollen vector A biotic or physical agent that conveys pollen in 3-dimensional space from the source plant. Insects, birds, mammals, wind, and water are the most important pollen vectors.

Pollenizer A male fertile diploid population that is interplanted with a parthenocarpic triploid population to allow pollen to stimulate fruit set on the triploid plants. This strategy is used primarily in crop species of Cucurbitaceae.

Pollination, cross Progeny that result from the fusion of male and female gametes that originate from different individuals.

Pollination, self Progeny that result from the fusion of male and female gametes from the same individual.

Polyheterozygous A diploid individual, the genotype of which is heterozygous for alleles at many or most genomic loci.

Polyhomozygous A diploid individual, the genotype of which is homozygous for alleles at many or most genomic loci.

Polymerase chain reaction (PCR) A technique used in molecular biology to amplify a single copy or a few copies of a DNA oligonucleotide by several orders of magnitude, generating thousands to millions of copies of the oligonucleotide. The technique combines temperature fluxes with controlled timing intervals on DNA-nuclease-buffer mixtures.

Polymorphism A condition at a genomic locus where differences in DNA sequence are found. Polymorphism is measured to estimate genetic variability with respect to the number of alleles and the frequency of predominant vs. rare alleles.

Polyploid An individual, population, or species that possess more than two haplotypes in the genome of the sporophyte. Multiples of the number of copies of the haplotype genome are indicated by an epithet: Triploid = 3 copies; tetraploid = 4 copies; pentaploid = 5 copies, etc.

Polytene chromosome A type of chromosome that consists of multiple fused chromatids and is found in many animal species but is uncommon in plants. Polytene chromosomes exhibit prominent banding patterns that are useful in identifying chromosome segments, and have been used extensively to relate cytogenetic to genetic genome maps in insects (e.g., *Drosophila melanogaster*).

Prebreeding All activities designed to identify desirable characteristics and/or genes from unadapted materials that cannot be used directly in breeding populations and to transfer these traits to an intermediate set of materials that breeders can use further in producing new cultivars.

Progeny test The breeding value of an individual is assessed by measuring the performance of the corresponding progeny in specified hybrid combinations.

Promoter A relatively short (<100bp) DNA sequence that initiates the transcription of a gene, usually in response to binding of the sequence with a transcription factor. The base sequence of the promoter is the source of the specific and unique manner of gene expression. The promoter is usually located at the 5′ end of the coding sequences of the gene, but sequences at the 3′ end may also modulate gene expression.

Propagule A specialized multicellular structure with the function of overwintering, multiplying, and distributing individuals of a population or species. Propagules may be sexual (true seeds) or asexual (apomictic seeds, bulblets, stolons, etc.) in origin.

Protandry The form of dichogamy where the male gametes in a flower are viable before the stigma is receptive.

Proteome The entire complement of proteins that is or can be translated from mRNA by a cell, tissue, or individual.

Protogyny The form of dichogamy where the stigma is receptive in a flower before the male gametes are viable.

Protoplast The plant cell after the rigid polysaccharide cell wall is removed, bordered only by a lipid membrane. Protoplasts are used in plant breeding to produce somatic hybrids, cybrids, and to introduce macromolecules packaged in liposomes physically.

Pseudogene A genomic DNA segment the sequence of which is similar or identical to a functional gene but has been changed to be non-functional or exhibit a different function than the original gene.

Pyramid In the plant breeding context, pyramid is used to describe a process by which desirable alleles that affect a specific phenotype are introduced into a single individual or population in a stepwise fashion, usually by backcross breeding.

Quadrivalent A structure that may be observed during meiotic prophase I and metaphase I in a polyploidy individual that is comprised of four homologous chromosomes paired together. The structure usually appears cytologically as a ring or a "v" shape. The quadrivalent is unstable at anaphase I and often leads to the unequal distribution of haplotypes.

Qualitative A phenotype that is controlled by few genes (1–3), each with a relatively large incremental effect.

Quantitative A phenotype that is controlled by many genes, each with a relatively small incremental effect. Also known as "multigenic."

Quantitative trait locus/loci (QTL) A locus or loci that are tightly linked to a gene or genes that control a quantitatively-inherited phenotype. QTL are used as surrogates to select for desirable alleles at linked loci in a method known as marker-assisted selection (MAS).

Radioisotope An unstable constitution of an element that emits energy in the form of mass (α and β particles) and electromagnetic radiation (γ rays) in the range of 10^{-10} to 10^{-12} m λ range (also known as "ionizing radiation"). γ rays can penetrate biological tissues and damage DNA, causing mutations. Synthetic radioisotopes that are byproducts of nuclear energy generation such as ^{60}Co and ^{137}Cs are popular as sources of mutagenic ionizing radiation in mutation breeding programs due to accessibility and relatively long half-life (5–30 years).

Random mating A theoretical situation in which all female gametes in a population have an equal probability of fusing with all male gametes.

Randomized complete block design (RCBD) The physical array of experimental entries is organized according to a statistical model. Treatment populations are placed at locations within a contiguous block according to random assignment. Each block contains all the treatments, so is said to be complete. The blocks are replicated a prescribed number of times (usually 3–8) at a specific location and time. RCBD is used extensively in agricultural experimentation, including the testing of populations for overall performance pursuant to the selection of new cultivars.

Recessive The interaction of two alleles in a diploid individual that is characterized by the total absence of a parental phenotype associated with one allele in a heterozygote.

Reciprocals, or reciprocal hybrids A♀ × B♂ vs. B♀ × A♂. Reciprocal hybrids are usually not genetically equivalent due to differences between the parents in genes located in the cytoplasm (e.g., plastids, mitochondria, and siRNA).

Reciprocal recurrent selection (for combining ability) The selection of individuals of two distinct populations based on the performance of their progeny after individuals from the two populations are systematically hybridized.

Recombinant gamete Following the hybridization of two, unlike individuals, the genotype of a specified gamete is different from both of the parents. Recombinant is also referred to as "non-parental."

Recombinant inbred line (RIL) An immortal mapping population consisting of fixed (inbred) lines in which recombination events between chromosomes inherited from two inbred strains are preserved. RILs are generated by crossing two divergent parents followed by several generations of inbreeding to achieve homozygosity.

Recurrent selection A plant breeding method applied to obtain genetically variable populations and derived progenies that are evaluated, selected, and cyclically recombined in an orderly, dynamic, and continuous process with the objective of increasing the frequencies of desirable alleles.

Registered seed The population of individuals produced with parents from certified seed using methods prescribed by the plant breeder (e.g., open pollination, enclosed cage with pollinators, etc.).

Resistance (R) gene A host (plant) locus that may have one too many alleles that is part of the gene-for-gene resistance-avirulence system. An allele of this locus enables the host to detect the glycoprotein from the pathogen *Avr* gene, resulting in no disease (a localized hypersensitive/apoptosis event). Other alleles (r) at this locus may not enable the host to detect the glycoprotein resulting in disease.

Response to selection The change in population mean or standard deviation from one generation to the progeny generation by phenotypic or genotypic selection.

Restitution gamete The result of a mutation in which the meiotic metaphase II plates are not perpendicular but parallel, leading to the fusion of two haploid nuclei leading to a diploid gamete.

Restorer The incipient male inbred in a hybrid breeding strategy that utilizes cytoplasmic male sterility to facilitate seed production. The restorer must have the nuclear genotype RR.

Restriction endonuclease or enzyme A bacterial enzyme that cuts double-stranded DNA at a specific oligonucleotide sequence that is usually 4–7 bases in length. Over 60 different REs have been found based on species source and oligonucleotide specificity. Certain types of restriction enzymes cut DNA such that a single-stranded "tail" exists on the cut ends, allowing DNA to be recombined by the homology of the nucleotide base sequence of the tail. The discovery of REs portended the development of methods to produce recombinant DNA.

Seed production The art and science of using natural plant reproductive systems to amplify genotypes accurately.

Seedling assay A disease screening method that utilizes a uniform population of young plants, a pathogen inoculation protocol, and controlled environment to induce disease symptoms for discerning resistant vs. susceptible host genotypes.

Selection In plant breeding and genetics, selection is the differential fitness of genotypes or individuals within a population under natural or artificial conditions that results in a permanent change in allelic frequencies.

Self-incompatibility, gametophytic The SI phenotype of pollen is determined by its genotype.

Self-incompatibility, sporophytic The SI phenotype of pollen is determined by the genotype of the female that produced the pollen.

Self-incompatible (SI) An individual bearing hermaphrodite flowers that is incapable of self-pollination due to a physiological interaction between self-pollen and the stigma surface and style tissues resulting in a failure of pollen grains to germinate or pollen tubes to grow.

Sequence-based mapping (SbM) An approach requiring deep sequencing (5x to 8x genome coverage) of two DNA pools derived from individuals from the phenotypic extremes of a segregating population, to identify candidate genes associated with a phenotype of interest.

Sib (sibling) mating Individuals that have the same two parents.

Somaclonal variation A new phenotypic variant that is observed in cell cultures or plants regenerated from cell cultures. Some somaclonal variation events have been shown to be transient (epigenetic) while others are heritable, a consequence of point mutations, transposon movement, and genomic reorganization during cell and tissue culture. Somaclonal variation is used as a method to enhance the range of available genetic variability.

Somatic hybrid A cell, tissue, or regenerated individual from the fusion of protoplasts from two genetically distinct donors. Usually, the donors are sporophytes of individuals that are not sexually compatible (i.e., that are separated by isolation barriers). The fusion can result in a hemizygous cell or may culminate in cells, tissues, and plants that have all or most of the genome of one donor and little to none of the other. Unlike a zygote, a somatic hybrid is also a "cybrid," containing cytoplasmic genetic determinants, or organelles, from both protoplast donors.

Sporophyte The diploid phase of the angiosperm and gymnosperm life cycle. The sporophyte transitions to gametophytes via meiosis. Haploid gametophytes or spores then fuse to form new diploid sporophytes.

Sport A sector on an individual plant that bears a discernible phenotypic difference to the background surrounding it.

Staminate flower A flower that bears only male gametes (pollen).

Stenospermocarpy A mutation that causes seeds to abort during the early developmental stages, usually before the seed coat is deposited. Such mutants are used to breed so-called seedless table grape cultivars.

Step trials A strategy for testing breeding lines for relative performance that is predicated on identifying the most desirable candidates for possible commercial release as cultivars. Initially, field trials will have many entries with small plot sizes and no replication. As the program proceeds, the number of entries decreases (as entries are culled), the plot size increases, and replication are added to allow statistical inferences to be imputed and applied.

Sublethal An allele or mutation that imparts the phenotypic effect of depressing plant performance profoundly to the point of near mortality.

Substitution line Following interspecific hybridization, the hybrid is recurrently backcrossed to one of the parents. One potential outcome of this process is the appearance of an individual with all chromosome pairs except one from one parent plus the corresponding homeologous pair of chromosomes from the other parent, known as a substitution line. Substitution lines are useful for contrasting gene expression in homeologous chromosome units.

Supernumerary chromosome A segment of chromatin that may or may not possess all the constituent structural and functional attributes of a chromosome. These entities contain genetic information that is supplemental to the haplotype, and they are replicated and distributed more randomly during mitosis and meiosis than are genomic chromosomes. The significance is largely unknown but probably related to transposons in evolution. Supernumerary chromosomes are also known as "B" chromosomes.

Syntenic The term used to describe two polymers that are identical in sequence; applied most often to nucleic acids (e.g., DNA).

Synthetic A population derived from mixed progeny from the intercross of 3 or more inbreds.

Target Enrichment Sequencing (TES) Target enrichment sequencing (TES) is a powerful method to enrich genomic regions of interest and to identify sequence variations. TES is useful for SNP identification in non-model species where a genome reference not available.

Targeting-induced local lesions in genomes (TILLING) A reverse genetics approach for the rapid discovery and mapping of induced causal mutation responsible for traits of interest.

Telomere The end of a chromosome that is characterized by tandemly repeated oligonucleotide sequences and unique genetic and molecular functions.

Test cross The cross of a heterozygote to a parental homozygote to examine the inheritance of a phenotype among progeny.

Top cross hybrid A population of progeny from the hybridization between individuals of inbred and OP populations.

Totipotency The innate ability of a plant cell or tissue to regenerate into a whole functional plant.

Trade secret Intellectual property that is protected without any procedural formalities for an unlimited time period. To qualify for protection by most worldwide justice systems, the following criteria must be met: (i) the information must be maintained as demonstrably secret, (ii) the information must have commercial value because it is a secret, and (iii) the information must have been subject to reasonable steps by the rightful holder of the information to keep it secret.

Training population (TP) A genotyped and phenotyped reference breeding population used to develop a model to predict genomic-estimated estimate breeding values for genomic selection.

Transcriptome The sum total of all the messenger RNA (mRNA) molecules transcribed from the genes of an individual.

Transgene A segment of DNA that is cloned in bacterial plasmids, purified, then transformed into a cell, tissue, or individual. The transgene DNA may be from any source and of any function.

Transposon A genomic DNA sequence that is capable of excising and changing locations within the genome. First discovered and characterized in the mid 20th century CE by Barbara McClintock (see Chapter 1 for a biography), transposons can insert themselves into functioning genes, usually turning them off. Many different and unrelated transposons have been described that behave differently according to the environment or developmental status. Some transposons contain all the genes necessary for excision and insertion, while other depend on genes outside of the boundaries of the transposon. A large proportion of the genome of eukaryotes is comprised of recent or ancient transposons suggesting that they play an important role in evolution. Also known as transposable elements (TE) or controlling elements.

Triploid An individual or population that possesses three haplotype genomes. Such an individual is sexually sterile due to the absence of bivalent pairing during meiotic prophase I and metaphase I resulting in aberrant chromosome segregation at anaphase I. Triploids are used extensively in plant breeding to produce parthenocarpic seedless cultivars of fruit-bearing species.

Trisomic A diploid individual that has three copies of a specific chromosome instead of two.

Truncating selection The highest fitness values are exhibited by individuals or genotypes at or near the center of the population distribution.

Twin seedling Two plants emerging from a single seed that originates from the true zygote and a fertilized or unfertilized egg nucleus. Depending on their origin, twin seedlings may be diploid + diploid, diploid + haploid or haploid + haploid. In plant breeding, twin seedlings are sought as a potential source of haploid plants, especially if other methods (e.g., anther/microspore culture) are not available.

Ultraviolet (UV) electromagnetic radiation Partially visible light characterized by wavelengths (λ) in the range of 10^{-7} to 10^{-8} nm. UV radiation can penetrate short distances into biological tissues and cause damage to DNA resulting in mutations. UV λ is emitted by our sun and are likely one cause of spontaneous mutations in nature.

Univalent A chromosome that does not pair with another during meiotic prophase I and metaphase I; the univalent fails to attach to the meiotic metaphase spindle and moves randomly to one of the derivative telophase I cells.

Utility patent, plant A patent granted for biological products and processes emanating from a plant possessing a specific genotype, including plant materials and uses. A fundamental feature of utility patents applied to plants is that they may be issued to cover many genotypes that are subject to a similar set of novel uses. Because of potential for abuses, utility patents on plant genotypes carry additional requirements to prove novelty/distinction and non-obviousness.

Vertical resistance An "all or none" response that is characterized by qualitative inheritance. Resistance in the host depends on the genotype (or race) of the pathogen. Vertical resistance has been shown to be less "durable" than horizontal resistance.

Whole genome re-sequencing (WGRS) A strategy to sequence an individual genome where short sequence reads generated by NGS are aligned to a reference genome for the species, providing information on variants, mutations, structural variations, copy number variation, and rearrangements between and among individuals, based on comparison to the reference genome.

x The lowest, or basic, haploid chromosome number of an indicated taxonomic unit.

Index

Note: Page numbers followed by *f* indicate figures and *t* indicate tables.

Printed in the United States
By Bookmasters